Engineering Tools for Environmental Risk Management – 4

T0136196

Engineering Tools for Environmental Risk Management – 4

Risk Reduction Technologies and Case Studies

Editors

Katalin Gruiz

Department of Applied Biotechnology and Food Science, Budapest University of Technology and Economics, Budapest, Hungary

Tamás Meggyes

Berlin, Germany

Éva Fenyvesi

Cyclolab, Budapest, Hungary

CRC Press
Taylor & Francis Group
Boca Raton London New York

CRC Press is an imprint of the
Taylor & Francis Group, an **informa** business

A BALKEMA BOOK

Published by:
CRC Press/Balkema
P.O. Box 447, 2300 AK Leiden, The Netherlands
e-mail: Pub.NL@taylorandfrancis.com
www.crcpress.com – www.taylorandfrancis.com

First issued in paperback 2020

© 2019 by Taylor & Francis Group, LLC
CRC Press/Balkema is an imprint of the Taylor & Francis Group, an informa business

No claim to original U.S. Government works

ISBN 13: 978-0-367-73193-9 (pbk)
ISBN 13: 978-1-138-00157-2 (hbk)

Visit the Taylor & Francis Web site at
http://www.taylorandfrancis.com

and the CRC Press Web site at
http://www.crcpress.com

Typeset by Apex CoVantage, LLC

British Library Cataloging-in-Publication Data
A catalogue record for this book is available from the British Library

Library of Congress Cataloging-in-Publication Data
Names: Gruiz, Katalin, editor.
Title: Risk reduction technologies and case studies / editors, Katalin Gruiz,
    Tamás Meggyes & Éva Fenyvesi.
Description: Leiden, The Netherlands : CRC Press/Balkema, [2019] |
    Series: Engineering tools for environmental risk management ; 4 |
    Includes bibliographical references and index.
Identifiers: LCCN 2018048279 (print) | LCCN 2018050830 (ebook) | ISBN
    9781315778754 (ebook) | ISBN 9781138001572 (hardcover : alk. paper)
Subjects: LCSH: Soil remediation.
Classification: LCC TD878 (ebook) | LCC TD878 .R56 2019 (print) |
    DDC 628.5/5—dc23
LC record available at https://lccn.loc.gov/2018048279

Table of contents

3 Natural attenuation in contaminated soil remediation   95

K. GRUIZ

É. FENYVESI, K. GRUIZ, E. MORILLO & J. VILLAVERDE

Preface

The four volumes of the book series "Engineering Tools for Environmental Risk Management" deal with environmental management, assessment & monitoring tools, environmental toxicology and risk reduction technologies. This last volume focuses on engineering solutions usually needed for industrial contaminated sites, where nature's self-remediation is inefficient or too slow. The success of remediation depends on the selection of an increasing number of conventional and innovative methods. This volume classifies the remedial technologies and describes the reactor approach to understand and manage in situ technologies similarly to reactor-based technologies. Technology types include physicochemical, biological or ecological solutions, where near-natural, sustainable remediation has priority.

A special chapter is devoted to natural attenuation, where natural changes can help achieve clean-up objectives. Natural attenuation and biological and ecological remediation establish a serial range of technologies from monitoring only to fully controlled interventions, using 'just' the natural ecosystem or sophisticated artificial living systems. Passive artificial ecosystems and biodegradation-based remediation – in addition to natural attenuation – demonstrate the use of these 'green' technologies and how engineering intervention should be kept at a minimum to limit damage to the environment and create a harmonious ecosystem.

Remediation of sites contaminated with organic substances is analyzed in detail including biological and physicochemical methods.

Comprehensive management of pollution by inorganic contaminants from the mining industry, leaching and bioleaching and acid mine drainage is studied in general and specifically in the case of an abandoned mine in Hungary where the innovative technology of combined chemical and phytostabilization has been applied.

The series of technologies is completed by electrochemical remediation and nanotechnologies.

Monitoring, verification and sustainability analysis of remediation provide a comprehensive overview of the management aspect of environmental risk reduction by remediation.

Abbreviations

2,4D	2,4-dichlorophenoxyacetic acid
2,4-DCP	2,4-dichlorophenol
A.f.	*Acidithiobacillus ferrooxidans*
A.t.	*Acidithiobacillus thiooxidans*
ABC	alkaline barium calcium (desalination process)
ABS	alkybenzene sulfonates
AC	activated carbon
ALD	anoxic limestone drains
AMD	acid mine drainage treatment
AOPs	advanced oxidation processes
ARD	acid rock drainage
ASAP	accelerated site assessment procedure
ASP	aspartic acid
BATNEEC	best available technology not entailing excessive costs
BCD	beta-cyclodextrin
BNP	bimetallic nanoscale particle
BMPs	best management practices
BOD	biological oxygen demand
BTEX	benzene, toluene, ethylbenzene and xylenes
CAH	chlorinated aliphatic hydrocarbons
CAPEX	capital expenditures
CARD-FISH	catalyzed reporter deposition fluorescence *in situ* hybridization technique
CBA	cost-benefit assessment
CCP	combined chemical and phytostabilization
CD	cyclodextrin
CDEF	cyclodextrin-enhanced soil-flushing technology
CDT	cyclodextrin technology
CEC	cation exchange capacity
CHCs	chlorinated hydrocarbons
CKD	cement kiln dust
CMC	critical micelle concentration
CMBCD	carboxymethyl beta-cyclodextrin
CNT	carbon nanotube
COD	chemical oxygen demand
CP	chlorophenol

CPEO	Center for Public Environmental Oversight
CRD	catalytic reductive dehalogenation
CRM	conceptual risk model
CSIA	compound-specific isotope analysis
CSM	conceptual site model
CWA	chemical warfare agent
DAS	dispersed alkaline substrates
DCE	dichloroethylene
DCR	dispersion by chemical reaction
DDD	dichlorodiphenyldichloroethane
DDE	dichlorodiphenylchloroethane
DDT	dichlorodiphenyltrichloroethane
DGGE	denaturing gradient gel electrophoresis
DNA	deoxyribonucleic acid
DNAPL	dense non-aqueous phase liquid
DNT	dinitrotoluene
DO	dissolved oxygen
DPVE	dual phase vacuum extraction
DTA	direct toxicity assessment
DW	dry mass
EAP	environmental action program
EBPR	enhanced biological phosphorus removal
EBQC	effect-based quality criteria
ECB	erosion control blankets
EDTA	ethylenediaminetetraacetic acid
EEA	European Environmental Agency
EFM	engineered fiber matrix
EKSF	electrokinetic soil flushing
ENA	enhanced natural attenuation
EOM	extractable organic material
EPA	U.S. Environmental Protection Agency's
EPH	extractable petroleum hydrocarbon
EPS	extracellular polymeric substances
EQC	environmental quality criteria
ERD	enhanced reductive dechlorination
ERH	electrical resistance heating
ESEM	environmental scanning electron microscope
ETV	environmental technology verification
EZVI	emulsified zero-valent iron
FA	fly ash
FCR	food chain reactor
FFTs	fiber filtration tubes
FGM	flexible growth medium
FISH	fluorescence *in situ* hybridization technique
FRTR	Remediation Technologies Screening Matrix and Reference database
FTIR	Fourier transform infrared spectroscopy
GC	gas chromatography

GC EPH	GC-measured extractable petroleum hydrocarbon
GCW	groundwater circulation wells
GEM	genetically engineered microorganisms
GSR	green and sustainable remediation
GW	groundwater
HCH	hexachlorocyclohexane
HOC	hydrophobic organic compounds
HPBCD	hydroxypropyl beta-cyclodextrin
HPTRMs	high performance turf reinforcement mats
HQC	Hungarian quality criteria
HRC	hydrogen release compound
ICP-AES	inductively coupled plasma atomic emission spectroscopy
ICP-MS	inductively coupled plasma mass spectrometry
ICP-OES	inductively coupled plasma optical emission spectrometry
IPT	integral pumping test
ISAS	*in situ* (in-well) air stripping
ISCO	*in situ* chemical oxidation
ISCR	*in situ* chemical reduction
ISRM	*in situ* redox manipulation
ISTD	*in situ* thermal desorption
K_D	soil–water partition coefficient for inorganic constituents
K_{ow}	octanol–water partition coefficient for organic constituents
K_p	soil–water partition coefficient for organic constituents
LCA	life cycle assessment
LCM	life cycle management
LDS	low density sludge
LEV	local exhaust ventilation
L.f.	*Leptospirillum ferrooxidans*
LKD	lime kiln dust
LM-NIP	lactate-modified nanosized zero-valent iron particles
LNAPL	light non-aqueous phase liquid
MBBR	moving bed biofilm reactors
MBRs	membrane bioreactors
MCA	multi-criteria analysis
MCPA	4-chloro-2-methylphenoxyacetic acid
MDS	minimum data set
MeBCD	methyl-beta-cyclodextrin
MIMO	multi-input/multi-output (control algorithms)
MNA	monitored natural attenuation
MPN	most probable number
MPVE	multiple phase vacuum extraction
MW	microwave heating
NA	natural attenuation
NAD(P)H	nicotinamide adenine dinucleotide
NAP	net acid-generating potential
NAPL	non-aqueous phase liquid
NIP	nanoscale zero-valent iron particles

NOM	natural organic matter
NPL	National Priorities List
NRP	no-risk product
NRS	Nanomaterial Research Strategy
nZVI	nano-scale zero-valent iron
OCP	organochlorine pesticide
OLD	oxic limestone drains
OM	outer membrane
ORC	oxygen release compound
ORP	oxidation reduction potential
PAA	polyacrylic acid
PAHs	polycyclic aromatic hydrocarbons
PBDEss	polybrominated diphenyl ethers
PCBs	polychlorinated biphenyls
PCDD/F	polychlorinated dibenzo-p-dioxins and -furans
PCFs	polychlorinated dibenzofurans
PCE	perchloroethylene
PCP	pentachlorophenol
PCR	polymerase chain reaction
PCTs	polychlorinated terphenyls
PDS	peroxydisulfate ($S_2O_8^{2-}$)
PEC	predicted environmental concentration
PGM	peptone-glucose-meat-extract
PLB	pulsed limestone bead
PLFA	phospholipid fatty acid
PMS	peroxymonosulfate (HSO_5^-)
PNEC	predicted no effect concentration
POP	persistent organic pollutants
PPT	pressure pulse technology
PRB	permeable reactive barrier
PS	periplasmic space
P&T	pump & treat technology
PVC	polyvinyl chloride
QC	quality critera
qPCR	quantitative polymerase chain reaction
RAMEB	randomly methylated beta-cyclodextrin
RAPS	reducing and alkalinity producing system
RBA	risk – benefit assessment
RBQC	risk-based quality criteria
RCAs	recycled concrete aggregates
RCR	risk characterization ratio
RDX	hexahydro-trinitrotriazine, explosive
REC	risk reduction, environmental merit and costs
REE	rare earth elements
RNA	ribonucleic acid
ROM	run of mine ore
RR	risk reduction
RX	halogenated organic compounds
SA	sustainability assessment

SAC	soil air circulation systems
SAMMS	self-assembled monolayers on mesoporous supports
SAPS	successive RAPS
SBRs	sequencing batch reactors
SDS	sodium dodecyl sulfate
SEA	socioeconomic assessment
SEAS	surfactant-enhanced air sparging
SEF	surfactant-enhanced soil flushing
SEM	scanning electron microscope
SEP&T	surfactant-enhanced pump & treat technology
SER	steam-enhanced remediation
SF	soil function
SIP	stable isotope technique
SMS	spent mushroom substrate
SND	simultaneous nitrification/denitrification
SOD	soil oxidant demand
SOMS	swellable, organically modified silica
SPSH	six-phase soil heating
SQC	soil quality criteria
SQI	soil quality indicator
SRB	sulfate-reducing bacteria
SS	steel shots
S/S	stabilization/solidification
SSI	soil service indicator
SuRF-UK	UK Sustainable Remediation Forum
SVE	soil vapor extraction
SVOC	semivolatile organic compound
SWOT	analysis of strengths, weaknesses, opportunities and threats
TCE	trichloroethylene
TDS	total dissolved solids
TE	total extract
TEM	transmission electron microscopy
TEVES	thermal enhanced vapor extraction system
TIC	total inorganic carbon
TOC	total organic carbon
TNT	2,4,6-trinitrotoluene
TPH	total petroleum hydrocarbons
TRMs	turf reinforcement mats
VC	vinyl chloride
VCH	volatile chlorinated hydrocarbon
VOC	volatile organic compound
WRP	waste rock piles
WS	water soluble
WTP	willingness to pay
WWTP	waste water treatment plant
XPS	X-ray photoelectron spectroscopy
XRD	X-ray diffractometry
XRF	X-ray fluorescence
ZVI	zero-valent iron

About the editors

 **Katalin Gruiz** is Associate Professor at Budapest University of Technology, Budapest, Hungary. She graduated in chemical engineering at Budapest University of Technology and Economics in 1975, received her doctorate in bioengineering and her Ph.D. in environmental engineering. Her main fields of activities are these: teaching, consulting, research and development of engineering tools for risk-based environmental management, development and use of innovative technologies such as special environmental toxicity assays, integrated monitoring methods, biological and ecological remediation technologies for soil and water, both for regulatory and engineering purposes. Prof. Gruiz has published 70 scientific papers, 25 book chapters, 43 conference papers, and edited 9 books and a special journal edition. She has coordinated a number of Hungarian research projects and participated in European ones. Gruiz is a member of the REACH Risk Assessment Committee of the European Chemicals Agency. She is a full-time associate professor at Budapest University of Technology and Economics and heads the research group of Environmental Microbiology and Biotechnology.

 **Tamás Meggyes** is Research Coordinator in Berlin, Germany. He is specializing in research and book projects in environmental engineering. His work focuses on fluid mechanics, hydraulic transport of solids, jet devices, landfill engineering, groundwater remediation, tailings facilities, and risk-based environmental management. He has contributed to and organized several international conferences and national and European integrated research projects in Hungary, Germany, United Kingdom, and the US. Tamás Meggyes was Europe editor of the Land Contamination and Reclamation journal in the UK and a reviewer of several environmental journals. He was invited by the EU as an expert evaluator to assess research applications and by Samarco Mining Company, Brazil, as a tailings management expert. In 2007, he was named Visiting Professor of Built Environment Sustainability at the University of Wolverhampton, UK. He has published 130 papers including 14 books and holds a doctor's title in fluid mechanics and a Ph.D. degree in landfill engineering from Miskolc University, Hungary.

Éva Fenyvesi is senior scientist and founding member of CycloLab Cyclodextrin Research and Development Ltd, Budapest, Hungary. She graduated as a chemist and received her Ph.D. in chemical technology at Eötvös University of Natural Sciences, Budapest. She is experienced in the preparation and application of cyclodextrin polymers, in environmental application of cyclodextrins and in gas chromatography. She participated in several national and international research projects, in the development of various environmental technologies applying cyclodextrins. She is author or co-author of over 70 scientific papers, 10 chapters in monographs, over 50 conference presentations, and 14 patents. She is an editor of the Cyclodextrin News, the monthly periodical on cyclodextrins.

Chapter 1

Contaminated site remediation: Role and classification of technologies

K. Gruiz

Department of Applied Biotechnology and Food Science, Budapest University of Technology and Economics, Budapest, Hungary

ABSTRACT

The first three chapters of this book aim to identify and classify remediation technologies from the point of view of physicochemical or biological technology and the main physical phases where contaminants occur and should be treated. The remediation principle is based on the reactor approach, which suggests that remediation technologies can be characterized by quantifiable parameters such as mass transport; input and output; physicochemical or biological transformation; and mass balance in the same way as in any engineering technology even if the processes are performed *in situ* or *ex situ*.

Planning requires the knowledge of tools, operations, modes of application, expected impacts of the chosen technological parameters and overall efficiencies. All positive and negative impacts, risks and benefits of the technology should be predicted beforehand and confirmed after completion. It is essential to verify the technologies, especially in the case of new technologies, new locations, or unknown geochemical and soil conditions. Environmental, social, and economic risks and benefits should be planned and assessed in order to keep their acceptable balance. Evaluation of the efficiencies and verification of the technology require a multi-skilled team that can select and determine the scope of application and the key parameters and prepare the monitoring plan to acquire the necessary (measured) data for a complex evaluation. The evaluation should cover the direct impacts of processes during their application and the overall impact of the activity on the local environment and its users. In addition to local aspects, the assessment should include watershed-scale and global impacts, and sustainability analysis in the widest context. Potential damage before, during and after remediation (clean-up, rehabilitation) of a small site or a large area may have extensive and long-term impacts on the health of the ecosystem, the human population around the watershed or the global atmosphere, but also on the region's social, economic, and cultural landscape. The first three chapters give a general overview of remediation technologies and some of them will be discussed in detail. The last chapter will discuss technology verification and sustainability assessment of environmental remediation.

1 INTRODUCTION

The term "remediation" means healing the environment, re-establishing a state in which water and soil can fulfill their natural role and provide services for the benefit of mankind. Biogeochemical element and water cycling and provision of a habitat for the ecosystem are essential for accomplishing nature's role. The services that the environment provides for anthropogenic purposes include water resources, agricultural production and forestry, urban and recreational land uses for special economic, social or cultural needs (for more information see Volume 1 of this book series Gruiz *et al.*, 2014).

According to the engineering approach, environmental remediation is similar to other physicochemical, biological, and agrotechnologies used for the treatment of solid or liquid-phase materials. All liquid-, solid-, or slurry-treatment technologies can be implemented either in the environment in the initial location of the (contaminated) material (i.e. *in situ*) or in a separate treatment plant in open, semi-open, or closed systems (see also Chapter 2). The management of soil remediation is responsible for harmonizing remediation with spatial planning, selecting the best-fitting technology to the contaminant(s), to the environment, and to the land use; planning and establishing the treatment plant; implementing the technology; and monitoring and controlling technological performance and environmental efficiency. Remedial technologies are extremely versatile: the same soil treatment technology can be used *in situ*, in a heap or in a built reactor. Whichever is applied, the treated material is reused, typically in the environment, while ensuring the required quality of the products, the remediated soil and water.

One must distinguish between impacts of the technology during and after implementation. During remediation, increased emissions to the atmosphere, groundwater, and surface waters can be expected and emission control technologies should be applied to manage this risk. Post-remediation environmental risks and benefits manifest themselves during the reuse of the treated material or site. Positive and negative impacts of remediation may cover local to watershed or global scales, immediate and short- to long-term impacts. Planning should pay special attention to technological conditions of *in situ* and subsurface technologies such as methods applied, processes, risks, and the desired efficiency. Key parameters are to be monitored during implementation. Short-term, long-term, site-specific, and regional or global impacts of an *in situ* remediation technology may differ depending on the actual environment (sensitivity of land use, urgency of the clean-up) and the interactions between the technology and the environment and contaminants. Well-known, demonstrated and generally verified technologies still do not fully guarantee success because each case is different. The evaluation of the feasibility and sustainability of a technology requires a case-by-case approach and the validation of the predicted values, expected trends and achieved results, i.e. the complete verification of the remediation case.

2 SUSTAINABLE ENVIRONMENTAL REMEDIATION

Contaminated site management and environmental risk reduction favor site remediation for contaminated land. Prevention does not work "retroactively" and only restrictions such as the prohibition of certain land uses or users may reduce the risk but they cannot improve the environment itself. However, the large number of sites (nearly one million contaminated sites awaiting reclamation in Europe alone) justifies the use of other environmental risk reduction (RR) options, which may include temporary interventions such as restrictions in uses and other preventive measures to control emissions, further risk increases, or land downgrading. Economic aspects such as the necessary expenditure (financial and labor), the importance and benefits of future land uses, or aesthetic issues may also play a role in choosing the most appropriate risk reduction option. On the other hand, remediation may have much wider applications than repairing contaminated land. Contamination is only one possible type of soil or land impairment. Organic matter loss, desertification, salinization, sodification, erosion, landslides, compaction, sealing, and other land takes by human "civilization" cause serious degradations globally. The consequences are fertility decline, loss of ecosystem diversity and low-quality agricultural products, i.e. low-quality food and vulnerable or impaired human health. Soil health is essential for life, food production, human and ecosystem health, and for buffering climate changes and other adverse impacts such as droughts and floods. Therefore, it is important that soil remediation maintains soil quality, and continuously compensates for soil degradation, humus loss, and nutrient decline – including not only the lack of macronutrients but also mezo- and microelements and other biologically active molecules.

The 2012 report of the Joint Research Centre "State of the Soil in Europe" (JRC, 2012) mentions nearly 3.5 million potentially contaminated sites in Europe alone, while EEA (EEA, 2015a) cited 1 million identified and 2.5 million estimated contaminated sites in total in 2015. The origins of contamination are waste disposal and treatment (40% of the sites) and industry and commerce (35% of the sites). These sites need detailed investigation and in some cases remediation as well. The number of remedied sites is about 100,000 across the EU, i.e. 4% of the estimated number to be managed. The specific number of potentially contaminated sites per 1000 people is two to four in the EU (EEA, 2015b; ESDAC, 2013a,b,c).

US Superfund (2018) applies a practical approach and maintains a National Priorities List (NPL, 2017) with a manageable number of priority sites (identified as the most urgent among the total). It currently lists 1337 sites: remediation has been carried out at 1189 of them since 1983 and 392 have been deleted from the list (NPL Action, 2017).

Remediation includes the reduction of physical (erosion, landslide), chemical (ignition, explosion, and corrosion), ecological (aquatic and terrestrial ecosystem), and human health risks in all environmental compartments, i.e. air, surface water and sediments, and soil and subsurface water. When dealing with these risks, the aim of remediation is to re-establish a good quality of the environment with healthy biological diversity and activity and safe use for humans. The ethics behind remediation involves restoring natural diversity and the quality of water and soil over the long term, as well as protecting natural resources and ecological services from adverse anthropological impacts. A feasible solution should achieve optimal efficiency and harmony between future land uses and the selected remediation technology. The management team needs exhaustive information on the nature of deterioration, the site, short- and long-term spatial plans, and on the best available technologies.

Compared to the preparatory work and the necessary knowledge, remediation technologies (equipment and processes) are rather simple in most cases and can be adopted from other industries such as mining, agriculture, and from nature itself.

Reducing the risk posed by contaminants in soil includes the treatment of all physical phases: soil air (soil gas and vapor), soil water (moisture, pore water, groundwater, seepage water, leachate), and solid phases (soil, base rock, soil slurry or deposited sediment). Soil phases are linked surface waters and sediments via transport pathways of infiltration, convection, flooding, erosion, etc. Consequently, soil remediation may apply most of the clean-up technologies used for contaminated waters, liquid and solid wastes, agricultural land, and the environment in general.

Sustainable remediation or "green remediation" has two main goals: (i) maximizing environmental benefit in the long term and (ii) minimizing the footprint of clean-up activities throughout the remediation project. To fulfill the second requirement, innovative technologies are needed that are more efficient compared to those in conventional use and that meet site-specific requirements and rely on best management practices and new strategies to fulfill sustainability requirements.

"More efficient" in this context means that the technology works faster and more effectively meets the site-specific requirements, achieves the target without causing further risks or damage, and serves long-term land uses in harmony with the environment and its ecosystem.

"**Cost efficiency**" implies that the specific cost is lower than or equal to that of conventional technologies. This approach only applies when options are compared that are equivalent in terms of environmental benefits. A more generic economic efficiency assessment compares remediation costs to all benefits such as better environmental quality, human health, and future land uses. A socioeconomic assessment that assumes the monetization of environmental, societal, and health issues is required to quantify all these consequences of remediation. It makes a quantitative comparison of the contaminated and remedied state of the environment. "Environmentally efficient" technologies reduce environmental and human-health risk locally (on and near the site) within the shortest possible time and off site and/or globally over the long term.

"**Eco-efficiency**" assumes that the technology does not endanger the environment by causing additional emissions, material and energy utilization and that the equipment and buildings used do not detract from the natural scenery. Eco-efficiency also includes improved quality and quantity of the ecosystem services after remediation compared to the preceding status. Combining all three may result in sustainable remediation and is a specific requirement of remediation programs in most European countries (GRI, 2006; Paganos *et al.*, 2013; EC, 2013; EEA, 2014a; EEA, 2015b; EEB, 2012) and in the US (US EPA, 2008, 2009; Fiksel *et al.*, 2009, 2012; NRC, 2011). WHO (2013) also focuses on contaminated sites and their remediation.

The 7th EAP (2013) provides that "land is managed sustainably in the European Union, soil is adequately protected and the remediation of contaminated sites is well underway" by 2020 and commits the EU and its Member States to "increasing efforts to reduce soil erosion and increase soil organic matter, to remediate contaminated sites and to enhance the integration of land use aspects into coordinated decision making involving all relevant levels of government, supported by the adoption of targets on soil and on land as a resource, and land planning objectives." It also requires that "the Union and its Member States should also reflect as soon as possible on how soil quality issues could be addressed using a targeted and proportionate risk-based approach within a binding legal framework" (EAP 7th, 2013; Endl & Berger, 2014; Van Liedekerke, 2014). The legal tools are supported by information collected by the European Soil Data Centre (ESDAC, 2017) with ETC SIA (2013) having published on soil evaluation instruments or with Payá Pérez *et al.* (2015) having published on European success stories.

2.1   Measuring sustainability

Decision makers need measurable indicators to assess, track, and equitably weigh integrated human health, socioeconomic, environmental, and ecological factors in order to foster sustainability in constructed and natural environments. Several systems of sustainability indicators have been recommended and used for sustainability assessment (Sikdar, 2003; Vyas & Kumaranayake, 2006; Wilson *et al.*, 2007; Mayer, 2008; Singh *et al.*, 2009; Zamagni *et al.*, 2009; US EPA, 2010; WBCSD, 2011):

-   The common classification covered by most of the legal tools (air, water, land, human health, ecology) can also be used as the basis of sustainability evaluation as is done in the US EPA's reports on the environment (e.g., ROE, 2015).

- Another way to categorize indicators is as follows (Fiksel *et al.*, 2012):
  - o Air, climate and energy;
  - o Chemical safety for sustainability;
  - o Homeland security research;
  - o Human health risk assessment;
  - o Sustainable and healthy communities;
  - o Safe and sustainable water resources.

- System-based indicators, which is a third type of classification, include four major categories (Fiksel *et al.*, 2012):
  - o Adverse outcome indicates destruction of value due to adverse impacts on individuals, communities, business enterprises or the natural environment. It may cover exposure, risk or impact, losses, impairments, e.g. health impacts by air or water pollution, life cycle footprint of energy use, etc.
  - o Resource flow indicates pressures associated with the rate of resource consumption, including materials, energy, water, land or biota depletion, e.g. material flow volume, greenhouse gas emissions, water treatment efficacy, recycling rate, and land use.
  - o System condition indicates the state of the systems being considered. Indicators are: health, wealth, satisfaction, growth, dignity, capacity and quality of life. Examples: air, water and land quality, infrastructure, employment, income, etc.
  - o Value creation, including both economic values and the value of well-being through enrichment of individuals, communities, business enterprises or the natural environment. Some of the indicators are profitability, economic output, income, capital investment, and human development. Examples: cost reduction, increased energy efficiency, less transport, etc.

Sustainable remediation requires a well-informed selection of available technologies. The biological and physicochemical basis of remediation must be understood and our knowledge of the "black box" of soil expanded. To properly select and design engineering approaches and tools, information is needed about the properties and environmental impacts of the technological tools, the possible dangers and necessary safety measures. This information enables ecologically efficient technologies to be successfully designed and implemented, including biotechnologies that can utilize the natural capacity of soil microbes or plants. Verified technologies and standard information from their databases are highly supportive. Some countries (US, Canada, Japan, China) do operate such an environmental technology verification (ETV) system and, in addition, databases on the technology and the demonstration or routine application are available (CLU-IN, 2017; CLU-IN, verified, 2017; Canada verified, 2017; EU verified, 2017). The Canadian system accepts the claims with the performance testing prepared by a third-party organization and starts with the assessment based on the data provided by the technology owner. The US model – started in 1995 and first introduced in 2001 – is based on the collected results of tests and made public by an organization responsible for verification. Thus, some experience is available in the US and Canada, but this still is short of a complete and uniform global system.

In Europe, ETV is a new tool aimed at improving the development and wider use of environmental technologies and helping innovative environmental technologies reach the market (EU ETV, 2017a). A current pilot program supports innovative ideas that can benefit the environment and health but have not been accepted by the market for no other reason than that they are new and experience with them is lacking (EU ETV, 2017b; EU, verified 2017).

Summing up, the EU initiative verifies the purpose, characteristics, conditions, and performance of an innovative and demonstrated technology. The Statement of Verification provides the following information:

–   An independent proof of verified performance parameters;
–   A way to validate innovative technological features that satisfy specific user needs;
–   A tool to demonstrate an added value for the environment.

The EU ETV has not included soil remediation technologies yet in spite of some available information from the projects under EURODEMO (2004–2008), EURODEMO+ (2017), and the Hungarian MOKKA (2004–2008). These projects established databases for demonstrated innovative European soil remediation technologies and thus they can be considered preparatory projects for the ETV database of soil remediation technologies. The concept of life cycle assessment used for energy, waste, and water technologies is not included because it is not entirely compatible with soil remediation technologies.

2.2   Remediation technologies – classification, innovation, biotechnologies

The technological databases of the EUGRIS (2017) and ENFO (2017) projects provide further information on innovative technologies in Europe. Practical documents on sustainability issues can be found on the websites of the European NICOLE (2017) and SNOW-MAN (2017) networks or in the US Sustainable Remediation Forum white paper (2009). The "Practical Framework" of ITRC (2011) or the NICOLE Sustainable Remediation (2017) pages also give valuable information.

A holistic approach is necessary to fulfill economic, ecological, and human requirements simultaneously. Rather than resort to drastic solutions that work against nature, the engineer should use soft and gentle technologies that provide benefits to humans and the environment.

Excavation and disposal (33% on average) and other *ex situ* treatments (19% biological, 7% physicochemical, and 7% thermal, 33% in total) are currently still the dominant technologies in Europe (EEA, 2014b). From an average of 7%, the highest rate of *in situ* bioremediation is practiced in The Netherlands (20%), whereas it is not applied at all in some other countries (France and UK). Estonia (not included in the 13 countries shown in Figure 1.1) gives priority to *in situ* bioremediation, applying it almost exclusively. Figure 1.1 shows the average shares of various remediation technologies in 13 European countries.

European statistics and other literature sources indicate that the basis for classification is the *location* of the treatment: i.e. whether it is performed *in situ* or *ex situ*. This approach is secondary in the classification system used in this book because most of the technologies can be applied both *in situ* and *ex situ* depending on site characteristics and case management. Technologies used for soil remediation usually consist of simple operations such as

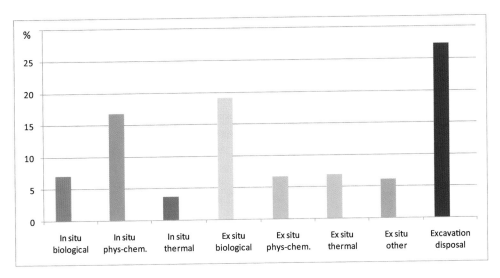

Figure 1.1 Most frequently applied remediation technologies for contaminated soil in 13 European countries.

pumping, injection, heating, extraction, mixing, washing, etc. This chapter will classify soil remediation according to

— The phase of the soil to which the technology is applied;
— The interaction of the technology with the contaminant classified into the main groups of immobilization, mobilization, transformation or degradation;
— The type of the key processes: physical, chemical, biological, ecological, or a combination thereof.

Practitioners do not put enough effort into justifying the selection of the technology or into its classification – most probably due to lack of information about the available technological options. Most professionals are familiar with one or a few technologies, and they offer the well-known, regularly practiced technology for all kinds of problems (e.g. excavation for solid soil and "pump and treat" for groundwater).

Small-scale and pilot experiments prior to technology planning and implementation are very helpful in technology selection and enhance the accuracy of efficiency prediction. Nevertheless, preliminary experiments are often missing from site-remediation tools, although a simple experiment can help decide whether a technology is suitable to a certain site. Even the best-fitting technological parameters can be estimated in advance using small-scale and pilot experiments.

In the literature, the methods used for the treatment of contaminated soil are generally grouped into three categories: "dig and dump" (excavation and containment); *ex situ* (on-site or off-site treatment after removal); and *in situ* remediation (without removal). The reactor approach recommended by the author of Chapter 2 does not consider these groups as main categories (in the case of *in situ* treatments, the soil volume itself is considered the reactor), but priority is given to the characteristics of the contaminants and the changes that they undergo during remediation. The *in situ/ex situ* differentiation is not clear in many cases because the water may be treated *ex situ* while the solid phase is treated *in situ* or vice versa. The decision on the best combination of *ex situ* and *in situ* treatments is a management issue and takes into consideration urgency, sensitivity of the surrounding environment, price of excavation and transport, time and labor requirement, etc.

Most of the remedial technologies (with a few exceptions) can be carried out both *in situ* and *ex situ*. Generally, operations that require extensive homogenization and/or which have to be carried out in a slurry phase are difficult to perform *in situ*. Extremely high temperatures are also difficult to apply *in situ*; however, there are examples that prove this wrong (e.g. *in situ* vitrification at significant depths or subsurfaces inaccessible for excavation). Chemical reagents and/or additives that could cause great risk due to the openness of the reactor may give rise to special concern (see Chapter 2).

Biotechnologies are classified according to the extent of engineering actions required since bioremediation is usually based on natural processes that are monitored or supervised – also called monitored natural attenuation – or can be enhanced after identifying the bottlenecks (see Chapter 3). The extent and type of action used to enhance natural remediation show great variety. Aeration, pH adjustment, nutrient supplementation, or the exploitation of the natural humidification process all need minor technological interventions with moderate effects. However, changing the temperature or the redox potential and applying mobility/availability-enhancing additives or microbial inoculants can have a stronger effect, though additives or physicochemical parameters must always be homogeneously distributed in the solid soil. Depending on the extent of the action required, soil remediation technologies based on natural processes include the following (Gruiz, 2009, Chapter 3):

- Natural attenuation (NA);
- Monitored natural attenuation (MNA);
- Enhanced natural attenuation (ENA);
- *In situ* bioremediation;
- *Ex situ* bioremediation.

3 CLASSIFYING REMEDIATION TECHNOLOGIES

As mentioned previously, the *in situ/ex situ* classification is worth redefining because every soil phase can be treated *in situ* or *ex situ* by itself, regardless of the other phases. However, if one of the soil phases is treated either by an *in situ* or *ex situ* method, the choice of treatment method for the other two soil phases (or a fourth, i.e. the contaminant phase, which occurs in some cases) is limited.

It is important to note that physicochemical and biological processes occur in the soil even if the practitioner does not want them to occur or even if the remediation technology is not based on them. Even if we pump out the groundwater or extract the soil gas and treat groundwater or soil gas *ex situ*, the "treatment" of the remaining soil gas and groundwater (including soil moisture) will continue *in situ* in the soil. If soil gas is extracted in order to remove the volatile polluting substances, redox potential is increased in the vented soil and this will have chemical consequences (e.g. oxidation state of the contaminants) and biological consequences (e.g. activation of the aerobic soil microbiota and enhancement of biological processes). Pumping out water generates increased subsurface water flows, lowers the water table and, as a consequence, causes higher redox potential in the drained, now three-phase, soil layer.

3.1 Technology classification defined by the contaminants

Soil remediation technologies can also be classified according to the characteristics of the contaminant. Physical, chemical, and biological characteristics of the contaminants determine basically the choice of technology, which depends on whether the contaminant tends to be strongly sorbed; is volatile, water-soluble or can be made water-soluble; or can be degraded, mobilized, or immobilized either physically, chemically, or biologically. The dynamic nature and tendency of changes (partitioning among physical phases, changes in mobility or accumulation) and the impact of the technology will determine technological efficiency; thus it is advisable to assess it in preliminary experiments. It is also important to know whether the contaminating substance consists of a single component or more; whether these components are similar to each other (e.g. petroleum hydrocarbons, chlorinated biphenyls) or completely different (e.g. PAHs and metals); or whether they make up a sequence according to their physical – chemical – biological characteristics. Even in the case of a single contaminating chemical substance, it may be necessary to apply different technologies for different soil phases. If there are several contaminants in the soil, it is likely that some of the soil phases need combined remediation.

The choice of technology will be influenced by the type and strength of interactions between the contaminant and the soil, i.e. the partition of the contaminant between the soil phases. This can be estimated from the physicochemical characteristics of the contaminant and the characteristics of the soil. If the uncertainty of the estimated interaction is high, laboratory experiments must be used to determine partition and the partition's dependence on the planned technological operations and parameters (pumping, aeration, heating, washing, additives, etc.). These data are almost always needed for planning and optimizing – both for physicochemical and biological treatments.

The forecast on the fate and behavior of the contaminants in soil is based to a large extent on their (i) degradability (photodegradability, hydrolytical instability and biodegradability), (ii) partitioning among physical phases (proportionate to Henry's constant for gas-liquid and to K_{ow} for liquid-solid partitioning) and their (iii) adverse effects/toxicities or other biological impacts (see also Gruiz *et al.*, 2015; Hajdu & Gruiz, 2015). When designing the technology, the behavior in the actual soil should be estimated from known substance characteristics. Substance-based estimates (Henry, K_{ow}, biodegradability, hazards) may vary significantly depending on soil type, current soil status, microbial activity, the ecosystem's sensitivity, and on climatic and seasonal conditions. Some of the behavioral characteristics of the contaminant in real soil can be estimated using models, but experiments are needed to answer most of the questions. Since uncertainty is generally high, large-scale and pilot plant experiments should be used to simulate real situations and to acquire refined information to aid in the selection of the technology and planning of the parameters of operation.

3.2 Technologies based on contaminant mobilization or immobilization

In line with the reactor approach (see Chapter 2), classification according to the fate of the contaminant and the changes during performance of the technology takes priority. The risk posed by the contaminant can be reduced by removal or various eliminations after converting it into a more mobile, more volatile, water-soluble, desorbable, and more bioavailable form than the original. The opposite option is not to remove but to irreversibly immobilize contaminants as much as possible.

Mobilized contaminants can be removed in unchanged (physical, chemical, or biological extraction) or in transformed, or partially or completely degraded forms. Transformation may cover volatilization, hydrolysis, complex formation, solubilization/micelle formation, photodegradation, thermal degradation, chemical decomposition, oxidation/reduction, and biological transformation and degradation.

Remediation based on immobilization involves a change of the contaminant into a stable form in which it cannot be transported and its harmful biological effects cannot be manifested; in other words, it cannot become volatile or water-soluble and it cannot be transported from the solid phase to the aqueous or gaseous phase to plants or animals and food chains. The main processes are thermal stabilization, sorption, precipitation, chemical condensation, polymerization, oxidation, and biosorption.

Tables 1.1 and 1.2 show the classification of technologies according to whether the remediation immobilizes or mobilizes the contaminant.

Table 1.1 Remediation based on the mobilization of soil contaminants.

Physicochemical characteristics of the contaminant	Contaminated/treated medium		
	Solid-phase soil	Groundwater	Soil air
Volatile	Remediation based on *in situ* desorption, volatilization, solubilization, dispersion chemical or biodegradation	Remediation based on *in situ* photocatalytic degradation; abiotic reduction or oxidation; aerobic and anaerobic biodegradation, reductive dechlorination	Photocatalytic degradation; chemical transformation or degradation, biodegradation
	Phytovolatilization, phytodegradation		
	Soil-vapor desorption, exhaustion and treatment of the separated vapor by conventional gas/vapor clean-up technologies	Stripping technology (air stripping) *in situ* or *ex situ*	Extraction of the soil gas and treatment by conventional air/gas clean-up technologies
	Enhanced desorption by high air flow or by thermal technologies, *in situ* or separate gas/vapor treatment	Heat-enhanced evaporation from groundwater	*In situ* or *ex situ* thermal degradation or catalytic combustion
Water-soluble	Remediation based on *in situ* desorption, solubilization, dispersion, chemical or biodegradation	Remediation based on *in situ* solubilization, abiotic reduction or oxidation, aerobic and anaerobic biodegradation	Remediation based on chemical transformation oxidation, reduction and biodegradation
	Phytoremediation: extraction, degradation		
	Remediation based on electrokinetic operations		
	Remediation based on soil washing	"Pump & treat": groundwater extraction and conventional water-treatment	Soil-vapor extraction and separate gas-phase treatment
	Thermal enhancement of desorption and mobilization	Groundwater treatment in permeable reactive barriers or in simple or cascade reactive soil zones	Air stripping
	Thermal degradation	Thermally enhanced solubilization and degradation	Thermally enhanced desorption and degradation

Physicochemical characteristics of the contaminant	Contaminated/treated medium		
	Solid-phase soil	Groundwater	Soil air
Sorbable	Remediation based on physical, chemical or biological transformation or degradation	Remediation based on *in situ* desorption, mobilization, solubilization, abiotic and biotic degradation	Remediation based on *in situ* desorption, abiotic and biotic degradation
	Phytoextraction, phytodegradation		Phytovolatilization
	Electrokinetic operations		
	Remediation based on physical, chemical or bioleaching and following leachate treatment	"Pump & treat": groundwater extraction and conventional treatment	Soil gas/vapor extraction and conventional *ex situ* treatment
	Desorption enhancement by additives: emulsifiers, water-solubility-enhancing agents, biosurfactants, etc.	Mobilization by solubilization, emulsification by physicochemical and biological methods or additives, then groundwater extraction	Soil gas/vapor extraction enhanced by increased air flow
	Thermal-desorption-based remediation technology	Remediation based on mobilization by heat	Soil gas/vapor extraction enhanced by thermal treatment (low, medium, high temperature)
	Chemical extraction by organic or inorganic solvents		
	Grain-size fractionation Soil incineration Soil pyrolysis		
	Soil vitrification (partly stabilization)-based technologies		

This kind of functional classification makes it easier for both professional decision makers and laymen owners and consumers to understand the nature of the technologies, for many misunderstandings still persist: a typical case is confusing the technologies with the operations. "Groundwater pumping," "biodegradation," or "pneumatic disaggregation," etc. are often called technologies although they are conventional operations or processes applied for the special purpose of soil remediation. These operations can be used in entirely different technologies and for various purposes. For example, groundwater pumping is used for lowering the water table, setting groundwater flow direction, recycling groundwater, treating groundwater on the surface in different physicochemical-biological technologies. Soil gas extraction may have the purpose of treatment on the surface, recycling or heating soil gas, *in situ* bioventing, etc.

Table 1.2 Remediation based on the immobilization of contaminants.

Physicochemical characteristics of the contaminant	Contaminated/treated medium		
	Solid phase of the soil	Groundwater	Soil air
Volatile	Physical immobilization by shifting partition (e.g. by sorption) toward the solid phase		
	Chemical immobilization by reactive soil additives (*in situ*) or reactive filling of the reactor (*ex situ*)		
	Biological immobilization: direct or indirect biological contribution		
	Containment		
Water-soluble	Physicochemical immobilization (sorption, precipitation)		
	Biological immobilization directly, e.g. in living cells or in biofilms, or indirectly by producing immobilizing agents or creating environmental conditions such as pH, redox beneficial for immobilization		
	Enhanced sorption-based technologies, e.g. by immobilizing additives, sorption and precipitation enhancing or solubility and mobility reducing physicochemical or biological impacts		
	Phytostabilization	Biofiltration & rhizofiltration	Vapor uptake by roots
	Chemical oxidation/reduction		
	Other chemical reactions such as condensation or polymerization		
	Immobilization/stabilization by *in situ* reactive zones, e.g. low redox zone or limestone for metals in acidic drain water		
		Immobilization on the filling of fix or replaceable permeable reactive barriers	
	Containment, encapsulation of the contaminated soil volume		
Sorbable	Sorption-based technologies and sorption enhancement		
	Physicochemical stabilization, solubility reduction, precipitation		
	Chemical oxidation/reduction		
	Biological immobilization		
	Phytostabilization	Rhizofiltration	
	Block formation with physicochemical stabilization		
	Vitrification, ceramic embedding with thermal stabilization		

When planning remediation and selecting the appropriate technology combinations, one has to consider that not only does the mobility and mobilizability of contaminants differ, but mobility of the soil phases is also greatly dissimilar depending on the hydrogeological characteristics of the site and the interventions which may influence it. Common transport of a solid-bound substance consists of desorption, solubilization in water and transport by diffusion or water flow, and by uptake by plants or animals, followed by food chains or food webs. The technology can modify and utilize these steps.

Before interpreting Table 1.1, an explanation is necessary for why degradation is included in the main group of mobilization. Degradation typically reduces the size of the molecule. This reduced size generally leads to increased mobility. Degradation is often partial; even if a substance can be degraded completely in theory, it never happens fully in the environment. The resulting temporary final products or intermediates may be harmful. Only some of the degradation processes leads to harmless mineral end products, so we consider it reasonable not to create a separate class for the different types of degradations, which would allow the misleading conclusion of ultimate degradation with harmless end products. The classification system presented considers certain contaminant transformations and degradations as special mobilization types where a chemical or biological impact changes the molecule structure, thus making the molecule more mobile and available.

Those processes that ensure ultimate photodegradation, hydrolysis or biodegradation with the possibility of the contaminant's complete elimination without significant amounts of harmful residues are highly favorable and recommendable for the remediation. However, intermediates and degradation products should be thoroughly assessed.

Tables 1.1 and 1.2 show that the prime residence phase of a contaminant is the result of partitioning. This means that its presence in soil air, soil water, or soil solid may be dominant but still partial: some part of the contaminant is present in other interacting phases. The selection of a technology from one specific column of the tables indicates that the equilibrium is overwhelmingly shifted into one direction and the majority of the contaminant is present in one certain physical phase. Partitioning of the contaminants shows great variety; therefore the same technology may appear in two neighboring columns. If a contaminant is distributed to a similar extent between two soil phases, two different technologies may be necessary for remediation of the two physical phases concerned. Dynamic tests and/or fugacity models can be used to calculate the quantitative measure of mobilization tendency (by definition, fugacity is the measure of the tendency of a substance to escape from a heterogeneous system).

Water-soluble contaminants may occur in each of the soil moisture, the capillary fringe, or in the groundwater (GW). In addition to the water dissolved/solubilized forms, the contaminant may appear as a non-aqueous phase liquid (NAPL).

The scheme in Figure 1.2 provides a summary overview on the most common mobilization-based technological options and Figure 1.3 on the most important technological operations that play a role in the contaminant-mobilization-based soil remediation.

Both mobilization and immobilization are based on a shift in partitioning in the desired direction (increased or reduced mobility) by physicochemical or biological modification of the contaminant, the soil itself, or soil conditions (temperature, pH, redox potential, materials present). The tables show very few implications for the implementation type "*in situ*" or "*ex situ*" as both can be realistic options depending only on managerial considerations.

Figure 1.4 provides a summary overview of immobilization-based technological options and Figure 1.5 on the technological operations in contaminant-immobilization/stabilization-based soil remediation.

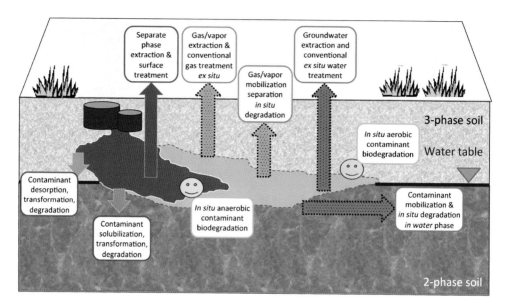

Figure 1.2 Bases of remediation of soils polluted by mobilizable and degradable contaminants.

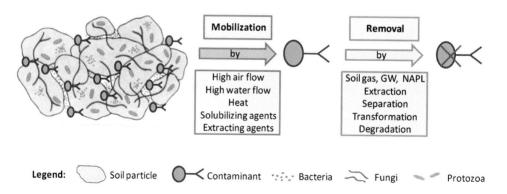

Figure 1.3 Mobilization and removal or degradation of soil contaminants.

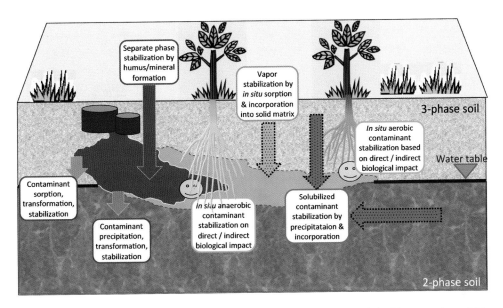

Figure 1.4 Immobilization/stabilization-based remediation for non-mobilizable, non-degradable contaminants.

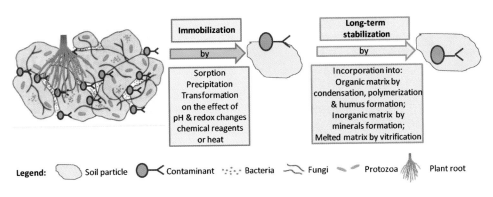

Figure 1.5 Stabilization of non-mobilizable, non-degradable soil contaminants.

3.3 Soil treatment technologies based on physical, chemical, and biological processes

Purely physical, chemical, or biological processes can rarely be used and a combination of several methods is more typical, e.g. biological processes in combination with physical, chemical, or thermal treatment in order to enhance the prime biological process. Combined technology applications are needed to remediate several soil phases with various contaminants. Treatments may target the contaminating molecules, the soil phases, or both. Groundwater extraction is just an operation most often targeting the groundwater together with its content without contaminant-specific considerations. Chemical treatment of any soil phase is typically a reaction between the contaminant and a reagent, e.g. organic substances and peroxide. Physical treatment may change environmental conditions, water flows, particle sizes, electric charge, etc. Living organisms may cause direct physicochemical changes in the contaminant (typically biotransformation, i.e. a chemical change in the contaminant caused by biomolecules produced *in situ*) or may have indirect effects via changing the conditions (e.g. consumption of air, lowering redox potential by oxygen and nitrate consumption, producing biosurfactants or hydrogen sulfide [H_2S] that react with the contaminating metals).

The most common *physical soil treatment* technology is extraction of contaminated soil gas and groundwater for surface treatment. These simple physical operations have a series of chemical and biological consequences in the soil: the space that was occupied by the extracted phases is now filled with new (fresh) contaminated or less-contaminated air or water, but air can also take the place of water. This means that a new equilibrium will be established in the soil, accompanied by sorption–desorption, dissolution–precipitation, evaporation–precipitation, etc. Soil gas extraction not only enhances evaporation and desorption in the pore volume of the three-phase soil, but it provides fresh air for the microorganisms living in soil pores, activating their metabolic functions and activity. If the reduction in water level brought about by groundwater extraction desaturates a layer of the originally saturated soil, this increases redox potential so that formerly anoxic conditions become aerobic. This leads to major changes in the composition and functioning of soil microbiota and may also directly influence the chemical form of contaminants.

Physicochemical reactions are used extensively for the treatment of contaminants, mainly for those that occur in dissolved and sorbed forms. The objective of the chemical transformation can be the following:

– Enhancement of mobility (volatilization, solubilization by enhanced dissolution, emulsification, or dispergation in water, and enhancing accessibility/availability);
– Immobilization (transforming the substance into an insoluble form by precipitation, sorption enhancement, condensation, polymerization, incorporation into stable minerals or humus aggregates); or
– Full or partial chemical degradation, or otherwise changing the components (e.g. by transformation, condensation, polymerization) that are responsible for adverse effects such as toxicity, mutagenicity, etc. (e.g. using dechlorination).

The applicable chemical reactions are photolysis, photodegradation, hydrolysis, oxidation, reduction, substitution, dechlorination, condensation, polymerization, etc.

Thermal processes increase the temperature of the soil by applying low, medium, high, or extremely high temperatures. Soil can be heated both *ex situ* or *in situ* using temperatures as high as 100–120°C for a short period and without damaging the soil and its biology. The 100–800°C temperature range is used to enhance desorption, mainly *ex situ*, in closed reactors, but *in situ* application is becoming more and more widespread. Surprisingly, soil can easily be revitalized after being heated up to 100–350°C over a short time. The high-temperature thermal methods can also be used for deeper soil layers or surface water sediments up to temperatures that melt the silicates (1200°C or higher: this involves vitrification).

Electrokinetic technologies use electric potential difference for the transportation within the soil and collection of charged molecules on the electrodes placed into soil (see Chapter 9).

Biotechnologies can be applied to both soil gas and groundwater and to the entire soil. Among the various biotechnologies, those based on biodegradation play a special role since the biodegrading organisms fully or partially utilize the contaminant in their energy-producing processes, thus removing them from the environment. In addition to energy production, a portion of organic materials and catabolic products is assimilated, i.e. it is incorporated into the protoplasm of soil microbes through biosynthesis. Another mechanism is cometabolic degradation when the xenobiotic substance is accepted by the microorganism's degrading enzyme system (when the xenobiotic's 3D structure is similar to common enzyme substrates). However, energy cannot be produced from this process (the enzymes of the entire electron-transport chain do not accept the xenobiotic as a substrate), so an additional energy source (applied as an additive in the remediation technology) is necessary to continuously maintain the degradation process (see more in Chapter 5). The cometabolic processes may be accompanied by the hazard of cometabolic end products, which can be more toxic than the initial contaminant itself. Therefore, a technology based on cometabolic degradation needs careful design and checks.

Some biotechnologies use specific – separately produced – enzymes of micro- or macroorganisms and apply them in an organism-free form to the soil: these are the so-called enzyme technologies. Most of the contaminants, especially mixtures, require the concerted cooperation of several enzymes, and that is why the use of enzymes for soil remediation needs a thorough selection of commercially available products or a tailor-made enzyme preparation produced by organisms adapted to the target mixture.

In addition to biotransformation- and biodegradation-based soil remediation, the indirect biological processes that change the pH or redox potential of the soil are also significant. Risk reduction of the contaminants can be achieved by the physicochemical consequences of the biologically induced changes in certain soil parameters in the following ways:

– Chemically oxidizing the contaminants;
– Transforming them into an insoluble chemical form;
– Precipitating them;
– Leaching them from the soil by acidification.

These processes can serve as a basis of remediation.

Cometabolic and symbiotic contaminant transformations can be classified within the group of indirect biological effect, where one of the partners makes the other enter into action or finalize biotransformation. Examples for that are the aerobic soil bacteria consuming soil oxygen to give room for facultative and anoxic microorganisms or the rhizosphere community stimulating the growth of remediating plants by nutrient supply. Figure 1.6 provides an overview of indirect biological effects playing a role in physicochemical and biological transformation of soil contaminants.

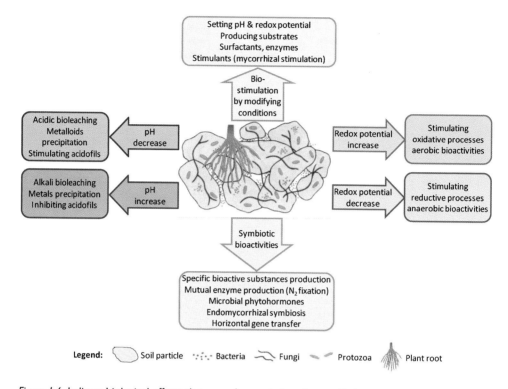

Figure 1.6 Indirect biological effects that may play a role in soil remediation.

Technologies that use plants or plant microorganisms make up a separate group of bio-technologies. The enzyme system of plants can extract and accumulate primarily metal-type contaminants from the soil, evaporate (volatilize), degrade, and transform the volatile ones or immobilize (stabilize) the soil and its contaminants. With the help of plant microorganisms, it is possible to further decrease the risk posed by soil contaminants. The bacteria of the rhizosphere can directly transfer the mineralized organic contaminants next to the plant's roots. The "living machine" type technologies (see more in Chapter 4) are based on this phenomenon. They are used for the remediation of groundwater and the soil's root zones and also for surface waters and sediments. When toxic metals are also present, they may accumulate in the root zone and/or be taken up by the plant material of the artificial ecosystem called "living machine."

A detailed overview of soil bioremediation is presented in Chapter 5.

4   TREATMENT OF INDIVIDUAL SOIL PHASES

As indicated earlier, contaminants may be partially present in each physical soil phase depending on their partitioning between gaseous, liquid, and solid phases (characterized by K_{ow} and Henry's law) and the properties of the soil (sandy, loamy, clayey). This partial presence can theoretically vary between 1 and 99% in each physical soil phase but dominant presence typically occurs in only one or two: water-soluble contaminants are mainly distributed between water and solid, volatile ones between gas and liquid or gas and solid, sorbable ones are chiefly bound to the solid soil phase and only a small fraction to soil water or soil gas. Unfortunately, this "small" part may still pose high risk, given that water and air are mobile phases and their exposure routes (respiration, drinking) result in more intake of mobile contaminants by sensitive target organs than ingestion of soil solid or plant uptake. This fact is mirrored by risk-based soil quality criteria: SQC_{soil} is generally several orders of magnitude greater than $SQC_{groundwater}$.

Since soil is a dynamic system, contaminant partition never reaches equilibrium. Partitioning therefore changes all the time, and the direction of the processes depends on concentration gradients and environmental conditions. The source of pollution, i.e. which soil phase initially hosted a contaminant and how much of it, is the prime determining factor of contaminant distribution. If a sorbable contaminant enters the groundwater, a large part of it is sorbed to the solid phase, which temporarily reduces the concentration in water but creates a risky accumulation. If a large amount of a water-soluble contaminant exceeding sorption capacity enters the three-phase soil, precipitation-infiltration and gravitation will sooner or later transport it to the groundwater.

4.1 Soil gas and vapor

Soil contaminants contained in the gaseous and vapor phase can be treated after extraction together with the contaminated soil gas or groundwater. Desorption and/or volatilization (enhanced by air flow and heat input) can purge contaminants from the solid soil phase and air stripping can do the same from water. Figure 1.7 shows the simplest method for soil air extraction from the vadose zone and its following *ex situ* treatment. Air resupply in this case is ensured by atmospheric air inflow through the soil surface. Instead of a simple extraction well, a combined well can also be used, which contains a treatment unit based on any physicochemical or biological gas/vapor treatment method.

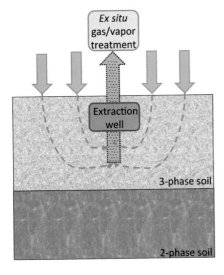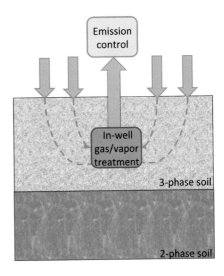

Figure 1.7 Ex situ and in-well soil air treatment.

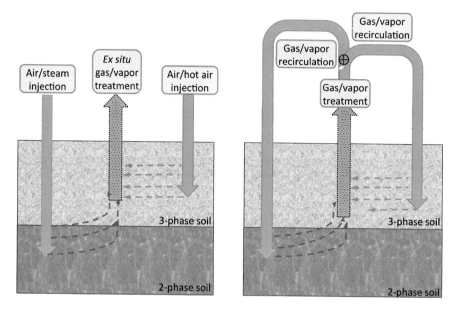

Figure 1.8 Enhancement of soil gas/vapor extraction from the two- and three-phase soil by air and heat injection and recirculation.

Figure 1.8 shows the cases when soil gas flow is enhanced by higher rate flow and/or heat. The flow can be enhanced and controlled by passive air inlet into a certain soil depth or by the use of air injection wells under pressure. Another option for gas flow control and treatment enhancement can be recirculation. Figure 1.8 shows an optional soil gas treatment (with an inserted treatment unit) and recirculation alternatively into the three- or two-phase soil.

Contaminated soil air is treated after extraction in the same way as any contaminated air or gas. Contaminant-specific sorption (on solid filters or sorbers, in wet scrubbers, or in biofilters) can be used to separate, recover, or transform gaseous or vapor-state contaminants. Physicochemical and thermal degradation, e.g. wet oxidation or incineration (with or without a catalyst), photochemical degradation, and destruction by oxidation or by other chemical reactions, as well as biodegradation are feasible options for treating gaseous and vapor-state soil contaminants. The prerequisite is that transport processes, reactions and reactors ensure the proper treatment technology including the correct mass balances, sufficient residential and contact times, and other technological conditions. Operations and technologies as well as their optimal compilation should consider contaminant vapor, soil characteristics, and the size of the area and select *in situ* or *ex situ* physical, chemical, or biological methods accordingly. Polluted soil gas can be treated using filters, sorbers, or reactors in the soil and controlled subsurface air flow instead of using external equipment.

A special case of soil air/gas treatment is the removal of the accumulated products of biological activities – typically CO_2, the inhibitory end product of aerobic biodegradation – and supply oxygen at the same time to optimize aerobic biodegradation.

If the remediation of the contaminant is more efficient in three-phase soil due to the use of gas-phase-based mobilization or biodegradation, groundwater lowering is helpful.

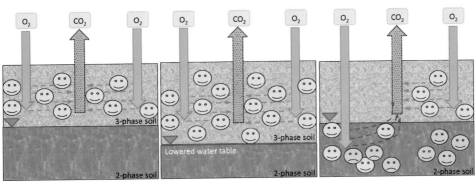

Figure 1.9 Soil aeration to enhance aerobic biodegradation in three- and two-phase soil by raising air flow in the vadose zone, deepening three-phase soil by groundwater lowering or by aerating two-phase soil.

For example the capillary fringe polluted by volatile or semivolatile contaminants can more easily be treated by vapor suction from or aerobic biodegradation in the aerated three-phase soil (see Figure 1.9).

While volatile contaminants can be removed together with the soil gas and the efficiency increased by higher flow rate, the semivolatiles should be desorbed by thermal enhancement at various temperatures (from 30°C to 800°C) both by a subsurface heating system or in desorber equipment combined with contaminated air treatment technologies.

4.2   Soil water and groundwater

Contaminated groundwater can be collected and treated in free surface wells or ponds, in extraction or extraction/injection wells, or first extracted and then treated by any existing batch or continuous technology. Groundwater is often treated *in situ* in the soil without using wells just by *in situ* aeration, heating, pH adjustment, or by directly injecting reactive agents. Groundwater flow can be utilized to create an *in situ* flow-through reactor packed with the soil itself or a reactive filling (permeable reactive barriers = PRB) (Figure 1.10).

Extracted polluted groundwater is treated like any wastewater in simple reactors or complex reactor systems. Most frequently, degradation- or removal-based technologies are applied to eliminate the contaminant. The technology is usually performed on site after water extraction, and may be phase separation, decanting, filtration, physical sorption, chemisorption, precipitation from solution, or other chemical transformations, mainly oxidation or reduction. Other common methods are stripping (removal of gas or vapor from water); thermal treatment to evaporate volatile contaminants; high-temperature thermal treatment for the degradation of organic matter (wet oxidation, thermal degradation); and methods based on electrokinetic and biological processes (mainly biodegradation). In most cases the suitable selection is a combination of the methods mentioned earlier.

Conventionally, the groundwater contaminated by volatile contaminants is extracted and stripped on the surface but water-dissolved volatile contaminants can also be stripped in

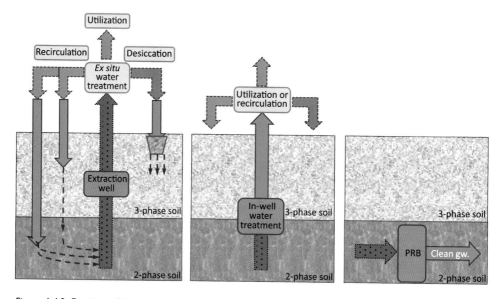

Figure 1.10 Ex situ and in situ groundwater treatment options: after extraction, in well or by PRB.

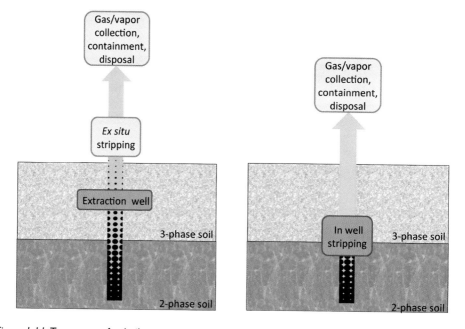

Figure 1.11 Treatment of volatiles-contaminated groundwater by ex situ or in-well stripping.

subsurface wells. In this case, contaminated air from the stripper is extracted to the surface and treated further, contained, or disposed of as shown in Figure 1.11.

In situ groundwater treatment (aeration or additives, etc.) is applied to two-phase soil but it can also be used for a separate aqueous phase collected in wells, for example in wells specially designed for stripping, skimming, and other physicochemical or biological/ biochemical treatments. The residential time of the water in the well should be specified according to the time requirement of the treatment process. Another innovative solution is the application of underground permeable reactive barriers, which are reactive walls (Meggyes *et al.*, 2009) or reactive zones arranged in the path of groundwater flow and filled with reactive material. Reactive barriers are underground trenches filled with contaminant-specific sorbents, filters, biofilters, alkalic or acidic, oxidizing or reducing filling, etc. and their thickness ensures the necessary contact time. Figure 1.12 shows the two main types of arrangement, the continuous permeable barrier and the funnel and gate type assembly. The funnel serves to direct the water flow to the gate and the "gate" may represent a wide variety of water treatment solutions from a simple, filled flow-through system to multi-unit subsurface treatment lines with vessels and reactors to collect, treat, and drain – thus fully controlling the groundwater. The passive systems utilize the energy of the groundwater flow and no additional energy source is implemented for the transport of the water to be treated. An important requirement of these underground reactive structures is that they must not adversely change water pathway and flow rate.

Full or partial recirculation of *ex situ* treated groundwater may be necessary: the treated groundwater can be returned into either the saturated or unsaturated soil zone. To return groundwater into the unsaturated zone, infiltration is applied via the surface with the help of shallow or deep infiltration ditches, trenches, underground injectors, perforated tubes or a network of tubes. The large amounts of water infiltrating through the unsaturated zone can wash out not only contaminants but also nutrients and useful soil components, which means that extensive recirculation can lead to soil damage. The washing out of the water-soluble contaminants from the unsaturated zone to the groundwater is acceptable only under limited conditions and with perfect emission control. Recirculation or guiding the groundwater treated in reactive barriers or reactive zones in the correct direction could also be necessary; in such cases the passive system should be supplied by pumps, taps, and tanks.

To control contaminated groundwater flow, continuous water extraction or circulation wells can be used. A stably maintained depression directs the dissolved contaminant flow toward the extraction well and limits the spread of the contaminant (the scheme on the left on Figure 1.13).

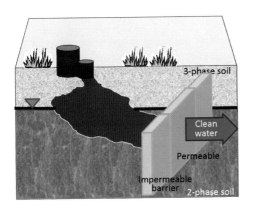 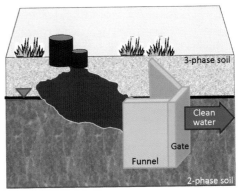

Figure 1.12 In situ groundwater treatment with continuous and funnel and gate-type permeable reactive barriers.

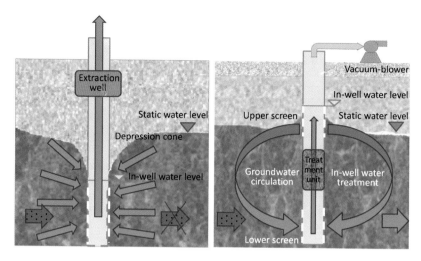

Figure 1.13 Controlling subsurface water flow by depression and circulation/vacuum wells.

The treatment of contaminated groundwater in wells is based on a vacuum created in the well, high enough to lift the water, but not to pump it out to the surface; thus the water level will rise in the well and reach the unsaturated zone. The well is perforated in such a way that the higher column of water infiltrates the unsaturated soil. Therefore, a flow develops between the well bottom and the perforated part at a higher level in the unsaturated zone. The well becomes a flow-through reactor in which water treatment takes place, enabling stripping, heating, and the addition of additives or chemical reagents. Nutrients and other additives can also be introduced into the groundwater or the soil through the well. The well can be considered a mixed reactor or a vertical tube reactor that uses recirculation and the portion of the well between its bottom and the perforation is continuously replaced depending on the groundwater flow rate. In addition, the well can be used to control water cycling in its impact zone, following an arbitrarily created 3D pattern. Wells that already perform this function are called circulation wells (the scheme on the right in Figure 1.13). Additives and reagents injected into the well can be sent into the soil volume reached by the circulating water and the groundwater flowing through the volume reached by the circulation. Reactive units can be built into circulation wells utilizing the created water flow in the well volume, which function as a reactor for in-well air stripping (air injection is necessary), for additives and reagents addition or for containing a permeable reactive packing that the water flows through.

Analogously, the permeable reactive barrier can also be considered a flow-through solid-phase reactor or a filled column. The reactive process may be adsorption, absorption, oxidation, reduction, or any other chemical reaction or biological transformation, including biodegradation. The polluted groundwater flows continuously through the filled reactor, driven by the elevation difference, i.e. the natural groundwater flow. These natural flow circumstances can be influenced by manipulating the flow route and pressures (Meggyes *et al.*, 2009). Hydrogeology of the surrounding soil is an important "technological parameter" in these cases as it (also) determines flow parameters.

Chemical oxidation is one of the most widespread degradation-based methods for organic groundwater pollutants such as chlorinated solvents and petroleum hydrocarbons.

Persulfate is generally used in the form of peroxydisulfate ($S_2O_8^{2-}$). It easily dissolves in water, leaves no harmful by-products and is easier to handle compared to peroxide or ozone. Persulfate itself is reactive but sulfate radicals derived from the persulfate by activation degrade a wider range of contaminants and work faster. The most common activator is iron II (ferrous iron) but zero valent iron (ZVI) and many other agents can be applied as catalysts: UV, heat, high pH, transition metals, and hydrogen peroxide.

Permanganate in the form of potassium ($KMnO_4$) or sodium permanganate ($NaMnO_4$) is applied in solution for chlorinated solvents such as perchloroethylene (PCE), trichloroethylene (TCE) and vinyl chloride (VC). It is not efficient enough for petroleum products.

When hydrogen peroxide is mixed with iron III (ferric iron) (Fenton's reagent), hydroxyl radicals (OH) are formed that efficiently oxidize both chlorinated solvents and petroleum hydrocarbons. This is a very rapid, non-specific (for a wide range of organic compounds), and cheap oxidizing agent for remediation. An application risk to manage is the production of explosive off-gases. Chelated iron can catalyze the reaction at neutral pH.

Biological methods for groundwater remediation are used both *ex situ* and *in situ*.

Aerobic and anaerobic reactor technologies are feasible options for **ex situ biological** groundwater treatment. They include the following:

- Activated sludge or biofilm reactors;
- Pond treatment;
- Rhizospheric wastewater treatment using wetland and aquatic plants combined with microbes in man-made ponds, marshes or wetlands in vertical flow-through or tidal arrangements;
- Ecoengineering technologies, e.g. the combination of pond treatment and rhizospheric wastewater treatment for water and sediment treatment, known as "living machines" (see details in Chapter 4).

In situ biological treatment of the groundwater can be performed in the saturated zone of the soil where the redox potential is usually low or negative. Thus, there are two options to influence the biological process: either the redox conditions are pushed towards more intensive aerobic biodegradation or the original redox potential is maintained and the effectiveness of the existing biological processes is enhanced by redox potential and pH adjustment. Our choice depends on what redox potential is optimal for the microbiota to degrade the contaminants. Other possibilities for intensification of the biological groundwater decontamination are addition of nutrients and energy sources, adjustment of the temperature and the ion/dissolved material concentrations or addition of specific microorganisms. Figure 1.14 summarizes the *in situ* biological groundwater treatment options based on the microbiological and plant-based transformation of the contaminants.

When groundwater contaminants originate from the solid phase, partitioning modifying agents are useful tools aimed at either solubilizing or stabilizing. Solubilizing agents facilitate a shift in partition toward the liquid phase (mobilization by desorption, dissolution, solubilization, micelle formation and emulsification, etc.). Stabilizing agents strengthen sorption of contaminant molecules onto (adsorption) or into (absorption) the solid phase by physicochemical processes (condensation, polymerization, humidification, or other type of incorporation).

Continuous control of the "reaction mixture" is needed to achieve enhanced *in situ* groundwater remediation, i.e. the groundwater being treated and the environmental conditions in the soil should be monitored. The technological parameters can be specified based

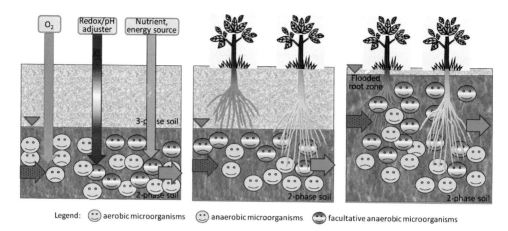

Legend: 😊 aerobic microorganisms   😐 anaerobic microorganisms   😁 facultative anaerobic microorganisms

Figure 1.14 In situ groundwater remediation based on intensified and controlled microbial and plant activities.

on technology monitoring data, primarily by controlling material flow (inputs and outputs) and the concentrations of contaminants, metabolites and additives, the pH, and the redox potential (see also Gruiz & Fenyvesi, 2016).

Air, ozone, or compounds that supply oxygen (hydrogen peroxide, magnesium peroxide, ORC: oxygen release compound, etc.) can be introduced into the saturated zone to make it more aerobic. Since hydrogen peroxide is toxic to the soil microbiota, a finely tuned dosage is needed to prevent overloading the soil with it.

Some peroxide derivatives do not dissolve well. They decompose slowly and are only slightly toxic, and they can therefore provide the aerobic microorganisms with oxygen over a long period of time in the following process: $CaO_2 + 2H_2O = Ca(OH)_2 + 2O_2 + H_2O$.

Calcium peroxide (CaO_2) and magnesium peroxide (MgO_2) serve as long-term sources of oxygen as they are orders of magnitude less water-soluble than sodium percarbonate ($2Na_2CO_3 \times 3H_2O_2$).

Some oxidants produce large amounts of reduced products, which may cause clogging and reduce the permeability of the soil: these are limiting factors for *in situ* application but may be useful in *ex situ* technologies.

Alternative forms of respiration such as nitrate and sulfate respiration (using facultative anaerobic microorganisms) are applied to low-redox-potential microbiological degradation. These processes can be optimized by ensuring constant concentrations of nitrates and sulfates.

Obligate anaerobic microorganisms that work under negative redox potential use carbonate respiration. This type of respiration can be optimized by ensuring negative redox potential (also in order to compensate for the higher value of inflowing groundwater). This can be done by supplying biodegradable substrates for nitrate- and sulfate-respiring facultative anaerobes, which can lower the redox potential by utilizing nitrates and sulfates.

If the soil microbiota cannot directly transform the contaminant into energy but can decompose it through cometabolism, an energy source must be provided.

It is important to take into consideration in planning and maintaining the technology that the *in situ* soil/groundwater volume – the "quasi-reactor" – is in most cases a flow-through system with inputs and outputs characteristic of the geological/pedological context. Thus, compensating for the changes caused by the ongoing decontamination does not suffice to maintain optimal conditions in this reactor volume. It is also necessary to counteract the changes due to uncontrolled inflows causing dilution, concentration, pH, and redox potential changes and sometimes unknown material flows and other physicochemical or biological impacts.

Similar to the *ex situ* groundwater treatment, a multi-stage water treatment process within the soil can be devised creating different conditions and spatial gradients in the designated soil volumes. Figure 1.15 shows the cascade arrangement of a biological water treatment option and Figure 1.16 demonstrates the *ex situ* and the *in situ* placement of this layout.

By manipulating the treatment volumes within the soil, stable conditions can be reached in favor of the desired physicochemical or biological process. Typical *in situ* interventions are these: aeration or injection of oxygen release compounds, pH and redox adjusters, surfactants, chemical reagents, nutrients, energy sources for cometabolic degradation, etc. While the technological stages of the *ex situ* groundwater treatment are established in independent reactors in a cascade layout, the designated volumes of the soil can be considered reactors that have different optima and the natural or artificial groundwater flow directs the water through the specifically designed soil volumes. This way an aerobic stage can be added (via aeration or another type of oxygen supply) after an anaerobic groundwater treatment step (lowering the redox potential by substrate addition) (see Figure 1.16) or an alkalic volume after an acidic one, and so on. The residence time can be determined from the flow rate and the soil volume's dimension in the flow direction. In practice the reaction time will determine the necessary size of this soil volume; this is equal to the pass time of the groundwater through the treatment volume.

Other bio- and ecotechnologies that use vegetation for *in situ* groundwater treatment are man-made ponds and wetlands and *in situ* rhizofiltration (see Chapter 4). The essence of the latter is usually that the root-zone microorganisms degrade the contaminants and transform

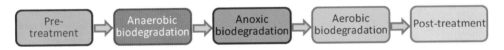

Figure 1.15 Generic scheme of cascade arrangement for biological groundwater treatment. The stages and the sequence are optional.

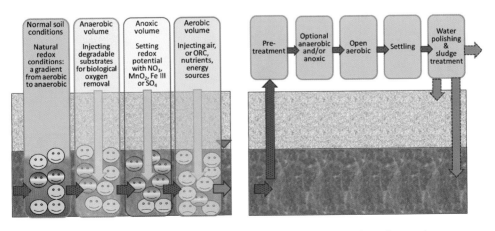

Figure 1.16 A two-stage arrangement for *in situ* (left) and *ex situ* (right) biological groundwater treatment: the anaerobic is followed by an aerobic biodegradation stage.

them into a mineralized form to be utilized by plants. Non-degradable contaminants are immobilized at the same time. Rhizofiltration treatment can be used as follows:

- In surface waters and sediments;
- Under periodically flooded (tidal type) wetland-like arrangements (wetlands, ponds, reactors);
- In wetlands, ponds or in hydroponic reactors with horizontal water flow;
- In water flowing underground; in this case, a good water permeability through the soil must be ensured over the long term;
- In wastewater (using any of the previously listed solutions).

4.3   Treatment of the solid-phase soil

Treatment of the three-phase soil is possible by treating the three physical phases together without separating them or only one or two phases after separation. It depends on both the contaminant(s) (type, number and partitioning between soil phases) and the soil (type, texture, groundwater level and flux). The whole soil (three-phase soil) can be treated either *ex situ* after excavating it from the original location or *in situ* in its original location. *In situ* soil treatment can be carried out as follows:

- In a completely undisturbed soil volume (analogous to a solid-phase batch reactor);
- In an undisturbed solid phase (analogous to a flow-through filled column);
- By extracting and recirculating the contaminated mobile soil phases (analogous to a feedback reactor). The feedback can be either connected to the saturated or the unsaturated soil layers.

The summary Figure of 1.17 differentiates the so called "*in situ*" soil remediation versions, emphasizing that it is important to design which soil phases are treated *in situ* and which are not.

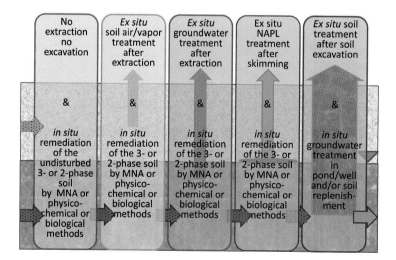

Figure 1.17 Version of *in situ* soil remediation: the solid phase is treated *in situ*, the mobile phases are optionally treated *in situ* or *ex situ*. If the solid phase is removed, soil remediation is called *ex situ*.

4.3.1 *Physicochemical methods*

Of the physical and chemical methods that can be used to treat the whole soil, methods based on *mobilization* of the contaminants are most widespread:

– *Extraction/recirculation of soil gas* in order to remove the volatile components from the soil, thus enhancing desorption and activating the soil microbiota.
– *Groundwater extraction and recirculation* to remove the contaminant from the whole soil, i.e. both from soil water and soil solid. Increased water flow can enhance desorption and dissolution. Mobilizing agents can be added with recycled water. This case can be considered *in situ* soil washing. It is worth differentiating between two-phase and three-phase soil washing. While two-phase soil can be purged both by vertical and horizontal water flows, the three-phase soil can only be by vertical flows, except when temporarily flooded. Spontaneous or directed vertical water flows for leaching the three-phase soil can be combined with the use of additives and the leachate can be collected before reaching the water table or alternatively extracted together with the groundwater. The latter solution is acceptable only when the groundwater is also contaminated and should be extracted anyway.
– *Ex situ soil washing* means the mechanical and/or chemical washing of the three-phase soil or the solid phase only. Its application is recommended for water-soluble contaminants and for those that can be transformed into a water-soluble form. One of the versions of mechanical washing involves the application of high-pressure washers. Here, when applying a strong shear force, not only are the water-soluble contaminants removed from the surface of the soil particles, but also the absorbed contaminants.

These types of treatments are only feasible for *in situ* soil remediation if the dominant part of the contaminant is removable by the mobile and extractable soil phases. If the solid phase functions as an infinite source due to a high solid/liquid partitioning, the removal of the mobile soil phases has low remediation efficiency. *Ex situ* washing combined with particle size fractionation and physicochemical treatments can rescue the solid phase from most of the contaminants and is useful for both soil and sediments.

Nowadays, *ex situ* soil washing is carried out in mobile soil treatment facilities that integrate a series of physical and chemical processes and the individual operations can be purposefully combined (Figure 1.18). Several types of contaminated soil can be treated in these facilities because the technology can be adjusted flexibly to the contaminant and soil type. Treatment operations in slurry-state soils and sediments combine homogenization; grain-size fractionation; precipitation; sedimentation; flotation; washing with water or with additives; extraction; implementation of chemical reactions; aeration; and biological treatment. Sieves, jiggers, centrifuges, cyclones, and filters are used for the separation of the treated phases or those to be treated. Conveyor belts and screws are used to transport the soil to be treated. The same facilities are also suitable to treat dredged sediment slurries as well as wastes that occur in slurry form or as silts.

In situ *soil washing* is a method in which the solid phase is washed *in situ*, and the washing water is treated *ex situ* after extraction. *In situ* water washing can affect the saturated zone of the soil by transporting the pollution there (Figure 1.19). The following

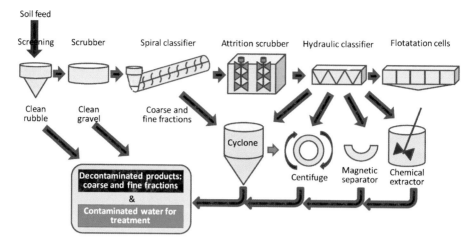

Figure 1.18 Scheme of a versatile soil washing plant.

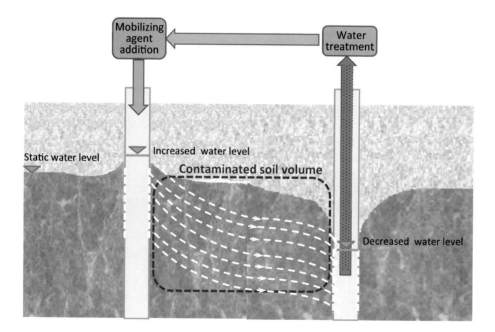

Figure 1.19 In situ washing of the contaminated soil volume between two wells.

options can protect the uncontaminated saturated zone against the negative impacts of the washing water:

– Using a hydraulic barrier (negative pressure in the relevant soil volume);
– Extracting the used water downstream of the groundwater flow and recycling it upstream.

If water washing is extended to the unsaturated zone of the soil, then the washing water has to flow through the contaminated unsaturated soil volume (infiltration from the surface, a ditch, perforated pipe network or injection), and it has to be gathered in the aquifer and extracted from there. This kind of operation increases the environmental risk since it contaminates the groundwater (temporarily); thus, it is advised to only apply it in cases where the groundwater is already contaminated and further pollution can be satisfactorily prevented using a hydraulic barrier or by surrounding the treated site with a cut-off wall or permeable reactive barrier.

Another basic requirement for using *in situ* soil washing is that the contaminant should be water-soluble or easily solubilizable to preclude the necessity of vast amounts of washing water to remove a contaminant from the soil. The proportions are determined not only by water solubility of the contaminant but also by the partition coefficient between the solid phase and the groundwater ($K_p = c_{soil}/c_{water}$), which is also influenced by the characteristics of the soil. The determination of K_p is fundamental in planning every operation based on transport between soil phases. The partition coefficient of single organic contaminants can be calculated from the K_{ow}, but measured data are more reliable as they are based on actual site conditions.

Properly chosen detergents (surfactants, tensides) can increase the effectiveness of soil washing, but they can also be harmful to soil biota; thus its *in situ* application may be accompanied by an increased environmental risk. The use of natural solubilizing or complexing agents is usually preferred (Fenyvesi *et al.*, 2009). A large number of soil microorganisms produce so-called biosurfactants to mobilize their nutrients. In spite of several years of research and development in this field and relative availability of biosurfactants (produced for food and cosmetics industry purposes), biosurfactant-facilitated remediation is uncommon and mainly used for toxic metal mobilization (Hong *et al.*, 2002; Mulligan & Wang, 2006; Wang & Mulligan, 2009).

Soil grain-size fractionation can itself be the basis of decontamination, since a large percentage of the contaminants are bound to colloidal organic matter (humus) or clay fractions that make up only a small portion of the soil (usually less than 5–10%). If the separation of these polluted fractions according to grain size or density can be achieved, then a large percentage of the soil becomes harmless and generally only the humus or clay fractions will require further treatment.

Solvent extraction can be applied using a solvent in place of water. Any kind of solvent that dissolves contaminants and does not cause an unacceptable level of damage to the soil can be used for the extraction. Acidic, alkaline, or organic solvent extractions are mainly used *ex situ* due to operational needs and the risks posed by the emissions but a properly controlled *in situ* "quasi-reactor" can also provide a solution.

4.3.2   Thermal methods

Thermal methods can be efficient for the treatment of the whole soil when contaminated with volatile, volatilizable, thermally mobilizable, or degradable contaminants. In some cases stabilization can also be achieved using thermal techniques. Raising the temperature by a few degrees enhances volatility, solubility, and the extent of desorption, and will thus enhance mobility and/or biological availability and consequently the removal of the contaminant. An increase in temperature will change the distribution of the contaminant and make it more homogeneous. It may activate microorganisms. Soil heating can be applied both *ex situ* and *in situ*. It can be implemented by injection or by otherwise admitting hot air, vapor, or hot water, which can heat the soil and may be removed by continuous flow. Soil can be heated by electrodes and an electric current (electrical resistance heating), by an electrothermal method (electrical heating using fluid movement), electromagnetic heating, or by vibrations at various frequencies (e.g. microwave).

When planning an increase in temperature, one has to consider the following:

– The heat absorption of the actual soil;
– Thermal conductivity and thermal diffusivity of the actual soil;
– The optimum temperature of the remediation process, e.g. evaporation of the contaminant, the temperature optimum of the relevant physicochemical processes or activation of the biota (typically microorganisms that perform the biodegradation).

Many soil-living communities dislike temperatures higher than the usual 12–15°C; thus, there is no point in increasing the biological availability of the contaminant if the microbiota is deactivated at the same time. That is why the *in situ* remediation technologies that use temporary heating have proved more successful: sudden heating allows a part of the contaminant to desorb; microbiota is left to work for a few days; then the mobilizing thermal treatment is repeated. In addition to enhancing microbial activity, soil temperatures also greatly influence seed germination, seedling emergence and growth, and root development. Soil heating in agriculture is a common technology for plant growth enhancement. In the case of soil contaminated with human or plant pathogens, soil heating aims at sterilizing the soil where the lethal temperature is the governing parameter.

Temperatures higher than biological systems can tolerate are mainly applied *ex situ*, except when soil removal/excavation is technically unfeasible (located under buildings or in other inaccessible locations) or the disturbances would increase the risk, e.g. due to increased contaminant discharge into air or water or unacceptable harm to employees working on the site. For *in situ* soil heating, the heating apparatus is placed into the soil and the off-gas treatment on the surface of the soil. For *ex situ* treatment, portable and stable equipment can be applied.

Low- (180–350°C) and high-temperature (400–800°C) desorption of contaminants is carried out in rotating furnaces with gravitational transport or inner conveyor belts or screws, excluding air and applying indirect heating. The contaminants evaporated at high temperatures are transferred from solid to vapor phase and removed from the soil. The exhausted gaseous/vapor phase is treated in multiple stages by cyclones, washers, sorbers, and incinerators. Low-temperature treatment may cause an acceptable damage to the soil: humus will not be damaged permanently and some of the microorganisms can survive and become revitalized. The product can also be utilized as sterilized soil. When the soil is wet, there is a greater need for energy for heating. Some contaminants are more volatile in the presence of water vapor. Contaminant mixtures with compounds of different volatilities may leave significant residual contamination behind.

Incineration, i.e. the complete chemical oxidation through heat, requires a higher temperatures than desorption does. When applying catalysts, a lower temperature can achieve the same perfect oxidation. Incineration destroys the fundamental characteristics of soil: it becomes a dead material that often cannot be revitalized because the soil's organic matter also burns away. If the temperature is not high enough or oxygen is limited, preventing complete oxidation, incomplete combustion produces harmful residues in the soil. The treated soil/solid material can be utilized as neutral filler in geotechnical constructions or revitalized and used for plant cultivation. As an alternative, it can be utilized as sterile soil for biotechnological purposes (e.g. in plant cloning).

Pyrolysis or wet oxidation is the chemical decomposition/degradation of soil contaminants under exclusion of air. Organic materials are transformed into gaseous components and a solid residue containing fixed carbon and ash. Depending on the contaminant, oil/tar residues can be condensed from off-gases by cooling. Pyrolysis typically works under pressure and at operating temperatures of 400–800°C. Pyrolysis gases require further treatment.

Vitrification – i.e. turning soil material into glass – can decontaminate soil combined with the production of useful products (ceramics). Vitrification is performed via the melted form of the material at extremely high temperatures (1600–2000°C). Vitrification comprises several processes ranging from thermal desorption through incineration (where oxygen is present) to pyrolysis and to melting the matrix. The *in situ* vitrified material is left in place, so its result is equivalent to containment or controlled disposal. *Ex situ*, it is reasonable to limit this expensive method to the vitrification of certain fractions of grain-size-fractionated soil: since the colloid-sized fractions (humus and clay) contain the contaminants, treatment of this small portion can render the whole soil harmless. Moreover, utilization of the separated fractions is more advantageous: cleaned gravel and sand can be sold and vitrified clay can be utilized as processed building material.

Off-gas treatment is a key element of thermal soil treatment technologies, both *in situ* and *ex situ*. Reactor technology makes it relatively easy to collect gaseous/vapor phases. However, complete collection of these potential airborne contaminants needs special technology and extra care when *in situ* thermal technologies are used. Treatment technologies applied to the extracted or otherwise collected gas/vapor from soil are the same as those used for common air, vapor, and off-gas treatments (e.g. condensation, washing, absorption, incineration).

4.3.3   Electrokinetic remediation

Electrokinetic solutions are widely used in countries where cheap energy is available. The electrokinetic process is suitable for the removal of certain ionic contaminants in the soil and those transported by groundwater. Electrokinetic solutions can also be combined with surface extraction of groundwater or with biological or thermal treatment. Electrokinetic methods are introduced in Chapter 9.

4.3.4   Biological and ecological methods

Biological and ecological methods should be given priority in every case when the soil should be maintained as part of the terrestrial ecosystem (both natural and agroecosystem).

Bioremediation uses soil microbiota as the core catalyzer of the biotechnological process based on biotransformation, biodegradation, bioleaching, or biostabilization. Phytoremediation uses specific plants or trees for remediation purposes, e.g. removal by volatilization, extraction, or degradation, as well as stabilization by stopping the transport of contaminants or contaminated soil. (There is a detailed overview on soil bioremediation in Chapter 5.)

Ecological solutions combine the use of soil microorganisms and plants, particularly in the root zone (plant roots together with rhizosphere microorganisms) for filtering, sorbing, extracting, transforming, and degrading contaminants from the soil.

Biological and ecological remediations are soft technologies using biologically acceptable conditions (environmental temperatures, pH values, and concentrations). They are more compatible to the ecosystem than technologies using drastic temperatures, pH values, chemical reagents, and additives, and killing soil microbiota and requiring revitalization if the treated material is going to be used as an ecosystem soil and not building material or industrial raw material.

The short summary on the classification of soil remediation technologies in this chapter is the basis for the next, more detailed summary on the reactor approach and the following presentation of individual soil remediation technologies with emphasis on those that are innovative and environmentally efficient. The classification introduced here is based on the contaminant, the soil phases, the basic processes, and the technology itself. Neither operations, nor equipment, nor management aspects have high priority here. Operations and equipment are secondary since they have to fulfill the requirement of a problem-specific technology. Any specific problem can trigger the development of innovative equipment. Management, planning, and sustainability will be discussed in Chapter 11.

REFERENCES

Canada Verified (2017) *Current Verified Technologies*. ETV Canada. Available from: http://etvcanada. ca/home/verify-your-technology/current-verified-technologies. [Accessed 8th October 2017].

CLU-IN (2017) *Providing Information About Innovative Treatment and Site Characterization Technologies*. Available from: https://clu-in.org; https://clu-in.org/remediation; https://clu-in.org/characterization/#90. [Accessed 8th October 2017].

CLU-IN, verified (2017) *Hazardous Waste Clean-Up Information. On-line Characterization and Remediation Databases Fact Sheet*. US EPA Clue-in. Available from: https://clu-in.org/s.focus/c/pub/i/1386. [Accessed 8th October 2017].

EAP 7th (2013) *General Union Environment Action Programme to 2020 'Living Well, Within the Limits of Our Planet'*. Decision No 1386/2013/EU of the European Parliament and of the Council of 20 November 2013. Available from: http://eur-lex.europa.eu/legal-content/EN/TXT/PDF/?uri=CELEX:32013D1386&from=EN. [Accessed 8th October 2017].

EC (2013) In depth report 'Soil Contamination: Impacts on Human Health'. *Science for Environmental Policy*, September 2013, Issue 5. Available from: http://ec.europa.eu/environment/integration/research/newsalert/pdf/IR5_en.pdf. [Accessed 8th October 2017].

EEA (2014a) *Environmental Indicator Report 2014: Environmental Impacts of Production-Consumption Systems in Europe*. Available from: www.eea.europa.eu/publications/environmental-indicator-report-2014. [Accessed 8th October 2017].

EEA (2014b) *Most Frequently Applied Remediation Techniques for Contaminated Soil*. Figure 7 in: Progress in Management of Contaminated Sites. EEA. Available from: www.eea.europa.eu/data-and-maps/indicators/progress-in-management-of-contaminated-sites-3/assessment. [Accessed 8th October 2017].

EEA (2015a) *Progress in Management of Contaminated Sites* (CSI 015). European Environment Agency. Available from: www.eea.europa.eu/data-and-maps/indicators/progress-in-management-of-contaminated-sites-3/assessment. [Accessed 8th October 2017].

EEA (2015b) *SOER 2015 – The European Environment – State and Outlook*. A Comprehensive Assessment of the European Environment's State, Trends and Prospects, in a Global Context. Available from: www.eea.europa.eu/soer. [Accessed 8th October 2017].

EEB (2012) *EEB Position on the 7th Environmental Action Programme "Staying Within Ecological Boundaries"*. European Environmental Bureau. Available from: www.eeb.org/index.cfm/activities/sustainability/7th-environmental-action-programme. [Accessed 8th April 2017].

Endl, A. & Berger, G. (2014) *The 7th Environment Action Programme: Reflections on Sustainable Development and Environmental Policy Integration. European Sustainable Development Network.* Available from: www.sd-network.eu/quarterly%20reports/report%20files/pdf/2014-March-The_7th_Environment_Action_Programme.pdf. [Accessed 8th October 2017].

ENFO (2017) *Environmental Information.* Available from: www.enfo.hu/drupal/en. [Accessed 8th October 2017].

ESDAC (2013a) *Progress in the Management of Contaminated Sites in Europe.* Available from: http://eusoils.jrc.ec.europa.eu/content/progress-management-contaminated-sites-europe-0#tabs-0-description=1. [Accessed 8th April 2017].

ESDAC (2013b) *Soil Threats Data.* Available from: http://eusoils.jrc.ec.europa.eu/resource-type/soil-threats-data. [Accessed 8th April 2017].

ESDAC (2013c) *Potentially Contaminated Sites Per 1000 Person.* ESDAC Datasets. Available from: www.eea.europa.eu/data-and-maps/indicators/progress-in-management-of-contaminated-sites-3/assessment. [Accessed 8th October 2017].

ESDAC (2017) *European Soil Data Centre.* Available from: http://esdac.jrc.ec.europa.eu. [Accessed 8th October 2017].

ETC SIA (2013) *Land Planning and Soil Evaluation Instruments in EEA Member and Cooperating Countries* (with inputs from Eionet NRC Land Use and Spatial Planning). Final Report for EEA from ETC SIA.

EU ETV (2017a) *Environmental Technology Verification in Europe.* Eco-Innovation at the Heart of European Policies. Available from: https://ec.europa.eu/environment/ecoap/etv. [Accessed 8th October 2017].

EU ETV (2017b) *EU Environmental Technology Verification.* EC. Available from: http://ec.europa.eu/environment/etv/ [Accessed 28th May 2017].

EUGRIS (2017) *Portal for Soil and Water Management in Europe.* Available from: www.eugris.info/ [Accessed 8th October 2017].

EURODEMO (2005–2008) *European Coordination Action for Demonstration of Efficient Soil and Groundwater Remediation.* Available from: www.umweltbundesamt.at/eurodemo & www.eugris.info/displayProject.asp?ProjectID=4500&Aw=EURODEMO&Cat=Project. [Accessed 28th May 2016].

EURODEMO+ (2017) EU *Projekt zur Unterstützung innovativer Sanierungstechnologien auf dem Markt.* Available from: www.umweltbundesamt.at/eurodemo. [Accessed 9th October 2017].

EU Verified (2017) *Verified Technologies in Europe.* Available from: https://ec.europa.eu/environment/ecoap/etv/verified-technologies_en. [Accessed 8th October 2017].

Fenyvesi, É., Leitgib, L., Gruiz, K., Balogh, G. & Murányi, A. (2009) Demonstration of soil bioremediation technology enhanced by cyclodextrin. *Land Contamination and Reclamation,* 17(2), 611–618.

Fiksel, J., Eason, T. & Frederickson, H. (2012) *A Framework for Sustainability Indicators at EPA* (Ed. Eason, T.). National Risk Management Research Laboratory, US-EPA, Washington, DC, USA.

Fiksel, J., Graedel, T., Hecht, A.D., Rejeski, D., Sayler, G.S., Senge, P.M., Swackhamer, D.L. & Theis, T.L. (2009) EPA at 40: Bringing environmental protection into the 21st century. *Environmental Science & Technology,* 43(23), 8716–8720.

GRI (2006) *Sustainability Reporting Guidelines: Global Reporting Initiative.* Amsterdam, The Netherlands. Available from: www.globalreporting.org/resourcelibrary/G3-Guidelines-Incl-Technical-Protocol.pdf. [Accessed 8th October 2017].

Gruiz, K. (2009) Soil bioremediation: A bioengineering tool. *Land Contamination & Reclamation,* 17(3–4), 543–552.

Gruiz, K. & Fenyvesi, É. (2016) In-situ and real-time measurements in water monitoring. In: Gruiz, K., Meggyes, T. & Fenyvesi, É. (eds.) (2014) *Engineering Tools for Environmental Risk Management. Volume 3. Site Assessment and Monitoring Tools.* CRC Press, Boca Raton, FL, USA. pp. 181–244.

Gruiz, K., Meggyes, T. & Fenyvesi, É. (eds.) (2014) *Engineering Tools for Environmental Risk Management: Volume 1. Environmental Deterioration and Contamination – Problems and Their Management*. CRC Press, Boca Raton, FL, USA.

Gruiz, K., Molnár, M., Nagy, Z.M. & Hajdu, C. (2015) Fate and behavior of chemical substances in the environment. In: Gruiz, K., Meggyes, T. & Fenyvesi, E. (eds.) *Engineering Tools for Environmental Risk Management: Volume 2. Environmental Toxicology*. CRC Press, Boca Raton, FL, USA. pp. 71–124.

Hajdu, C. & Gruiz, K. (2015) Bioaccessibility and bioavailability in risk assessment. In: Gruiz, K., Meggyes, T. & Fenyvesi, E. (eds.) *Engineering Tools for Environmental Risk Management: Volume 2. Environmental Toxicology*. CRC Press, Boca Raton, FL, USA. pp. 337–400.

Hong, K.J., Tokunaga, S. & Kajiuchi, T. (2002) Evaluation of remediation process with plant-derived biosurfactant for recovery of heavy metals from contaminated soils. *Chemosphere*, 49(4), 379–387.

ITRC (2011) *Green and Sustainable Remediation: A Practical Framework*. Available from: www.itrcweb.org/GuidanceDocuments/GSR-2.pdf. [Accessed 9th October 2017].

JRC (2012) *The State of the Soil in Europe*. Joint Research Centre. Available from: http://eusoils.jrc.ec.europa.eu/ESDB_Archive/eusoils_docs/other/EUR25186.pdf. [Accessed 9th October 2017]. doi:10.2788/77361. [Accessed 8th April 2017].

Mayer, A.L. (2008) Strengths and weaknesses of common sustainability indices for multidimensional systems. *Environment International*, 34(2), 277–291. doi:10.1016/j.envint.2007.09.004.

Meggyes, T., Csővári, M., Roehl, K.E. & Simon, F.G. (2009) Enhancing the efficacy of permeable reactive barriers. *Land Contamination & Reclamation*, 17(2), 635–650.

MOKKA (2004–2008) *MOKKA: Innovative Decision Support Tools for Risk Based Environmental Management*. Available from: http://enfo.hu/mokka/index.php?lang=eng&body=mokka; www.mokkka.hu/index.php?lang=eng&body=mokka. [Accessed 8th October 2017].

Mulligan, C. & Wang, S. (2006) Remediation of a heavy metal contaminated soil by a rhamnolipid foam. *Engineering Geology*, 85(1–2), 75–81. doi:10.1016/j.enggeo.2005.09.029.

NICOLE (2017) *Network for Industrially Contaminated Land*. Available from: www.nicole.org/ [Accessed 8th October 2017].

NICOLE Sustainable Remediation (2017) *Sustainable Remediation Roadmap*. NICOLE. Available from: www.nicole.org/uploadedfiles/2010-wg-sustainable-remediation-roadmap.pdf. [Accessed 8th October 2017].

NPL (2017) *National Priorities List*. Available from: www.epa.gov/superfund/superfund-national-priorities-list-npl. [Accessed 8th October 2017].

NPL Action (2017) *Number of NPL Site Actions and Milestones by Fiscal Year*. Available from: www.epa.gov/superfund/number-npl-site-actions-and-milestones-fiscal-year. [Accessed 8th October 2017].

NRC (2011) *Sustainability and the US EPA (The Green Book)*. The National Academies Press, Washington, DC, USA. ISBN 10: 0-309-21252-9.

Paganos, P., Van Liedekerke, M., Yigini, Y. & Montanarella, L. (2013) Contaminated sites in Europe: Review of the current situation based on data collected through a European Network. *Journal of Environmental and Public Health*, 2013, Article ID 158764. doi:10.1155/2013/158764.

Payá Pérez, A., Peláez Sánchez, S. & Van Liedekerke, M. (eds.) (2015) Remediated sites and brownfields: Success stories in Europe. A report of the European Information and Observation Network's National Reference Centres for Soil (Eionet NRC Soil). doi:10.2788/406096.

ROE (2015) *Report on the Environment, 2015*. US EPA, Office of Research and Development. Available from: www.epa.gov/sites/production/files/2015-09/documents/roe_factsheet_07-17-15_not508c.pdf. [Accessed 8th April 2017].

Sikdar, S.K. (2003) Sustainable development and sustainability metrics. *Journal of American Institute of Chemical Engineers (AIChe)*, 49(8), 1928–1932. doi:10.1002/aic.690490802.

Singh, R., Murty, H., Gupta, S. & Dikshit, A. (2009) An overview of sustainability assessment methodologies. *Ecological Indicators*, 9(2), 189–212.

SNOWMAN (2017) *Knowledge for Sustainable Soils.* Available from: http://snowmannetwork.com/ [Accessed 8th October 2017].

US EPA (2008) Green remediation: Incorporating sustainable environmental practices into remediation of contaminated sites. Office of Solid Waste and Emergency Response, US-EPA.

US EPA (2009) Development and evaluation of sustainability criteria for land revitalization. NSCEP US EPA. Washington, DC, USA, EPA/600/R-09/093.

US EPA (2010) San Luis Basin sustainability metrics project: A methodology for evaluating regional sustainability. EPA Number: EPA/600/R-10/182.

US EPA (2018) *Superfund.* Available from: https://www.epa.gov/superfund [Accessed 1st November 2018].

US Sustainable Remediation Forum White Paper (2009) Sustainable remediation white paper – Integrating sustainable principles, practices, and metrics into remediation projects. *Remediation Journal*, 19(3), 5–114. doi:10.1002/rem.20210.

Van Liedekerke, M. (2014) Progress in the management of contaminated sites in Europe. Technical Report EUR 26376 EN. doi:10.13140/RG.2.1.4213.5444.

Vyas, S. & Kumaranayake, L. (2006) Constructing socio-economic status indices: How to use principal components analysis. *Health Policy and Planning*, 21(6), 459–468. Available from: http://heapol.oxfordjournals.org/content/21/6/459.full. [Accessed 8th October 2017].

Wang, S. & Mulligan, C. (2009) Arsenic mobilization from mine tailings in the presence of a biosurfactant. *Applied Geochemistry*, 24(5), 928–935.

WBCSD (2011) *Guide to Corporate Ecosystem Valuation: A Framework for Improving Corporate Decisionmaking.* World Business Council for Sustainable Development (WBCSD). Available from: www.wbcsd.org/contentwbc/download/573/6341. [Accessed 8th October 2017].

WHO (2013) *Report of Two Workshops on Contaminated Sites and Health.* Syracuse (Italy) 2011 and Catania (Italy) 2012. Available from: www.euro.who.int/__data/assets/pdf_file/0003/186240/e96843e.pdf. [Accessed 8th October 2017].

Wilson, J., Tyedmers, P. & Pelot, R. (2007) Contrasting and comparing sustainable development indicator metrics. *Ecological Indicators*, 7, 299–314.

Zamagni, A., Buttol, P., Buonamici, R., Masoni, P., Guinee, J.B., Huppes, G., Heijungs, R., van der Voet, E., Ekvall, T. & Rydberg, T. (2009) *CALCAS D20 Blue Paper on Life Cycle Sustainability Analysis.* Institute of Environmental Sciences, Lieden University, Leiden, The Netherlands.

Chapter 2

In situ soil remediation: The reactor approach

K. Gruiz

Department of Applied Biotechnology and Food Science, Budapest University of Technology and Economics, Budapest, Hungary

ABSTRACT

Remediation can be classified according to contaminant type, soil phase and the core transformation process, as shown in Chapter 1. The goal of remediation is to reduce adverse effects and the resulting environmental and human risks posed by the contaminants. This goal can be achieved by their elimination, degradation or other transformation to non-hazardous or less hazardous and non-mobile forms. This chapter applies the reactor approach, an interpretation common in chemical and bioengineering, to describe the overall remediation technology. This description is based on a transformation process that (i) is characterized by material balances; (ii) takes place in reactor volumes; (iii) is equipped with machinery to perform certain operations; and (iv) aims to design, operate, and maintain "the plant," i.e. the processes in the reactor.

What this means in the context of soil remediation is that processes take place in real reactors or in quasi-reactors. Quasi-reactors have no walls or other built boundaries, but they are limited by the boundaries of the impact volume of operation, usually by an engineered volume – with *in situ* soil remediation as an example. The entire technology, equipment, processes, material transport, inputs to and outputs from the reactor, and optimal technological parameters and their control, including mass balances, should be planned and implemented as for any other technology. Reactors can be characterized by size, type, and the process taking place in them. They can be closed, semi-closed, or entirely open, but always containing a certain volume of soil or range of operation.

Soil and groundwater remediation covers a sequence of operations, the mass transport route within the reactor or from one reactor to another in a cascade arrangement, and the mass balance of the individual steps and the whole process.

Remediation as a "reactor" differs from other engineering devices in its extreme complexity and heterogeneity and the limited accessibility of the solid material in which the transformation takes place, which hampers predictability and design. Pilot testing, high versatility, and a combination of technologies, suitable monitoring, and control may ensure efficient soil treatment.

Soil remediation is redefined in this chapter based on a generic engineering approach: a controlled transformation produces valuable end products in a reactor, and the processes can be characterized by mass and energy balances.

The author believes that this approach is necessary to clarify confusion regarding terminology and the way of thinking that still prevails in environmental remediation. Today's practice often uses terms of processes or operations instead of those for treatment technologies and confuses technological considerations with managerial aspects. The reactor approach will help *in situ* natural or near-natural bio- and ecotechnologies to obtain recognition and will contribute to their spread and acceptance.

I INTRODUCTION

The reactor approach is based on the fact that natural soil processes obey physical, chemical, and biological/ecological laws, and thus all these processes can be described by common engineering tools (mathematical models) and engineers can intervene in a predictable way. The technologies can be classified according to the degree of engineering interventions, starting with the most careful "soft interventions" such as just monitoring natural remedial processes or modifying *in situ* environmental conditions to boost beneficial processes or suppress harmful and risky ones. Soft interventions can move soil conditions into biologically acceptable ranges of temperature, pH, redox potential, salt, and additive concentrations without destroying soil life. More intrusive technologies strongly change physical, chemical, and biological soil processes and soil characteristics. However, the accompanying degradation in the soil is temporary, with the treated material remaining a "living soil" and continuing to be healthy after remediation. The most drastic type of technological intervention kills soil biota and destroys soil composition and structure. Such extremely degrading remedial technologies can be characterized by the ratio of the reduction of risk posed by the contaminant to the damage caused by soil excavation and treatment.

Conventional classification of remediation into types of *in situ* or *ex situ* is an oversimplification (Gruiz, 2009) that ignores key parameters such as type, extension, concentration, and physical phase of the contaminants and ignores soil type and hydrogeology. The classification system recommended in Chapter 1 of this volume is based on this information, which is crucial from a technological point of view. Another drawback of the *ex situ–in situ* distinction is that it is exclusively based on the solid soil phases, although mobile soil phases (soil gas or groundwater) can be treated both *ex situ* and *in situ* regardless of the treatment of the solid phase. A more precise definition considers the process as a combination of *ex situ* treatment of mobile soil phases and *in situ* treatment of the solid phase in cases where the mobile (gaseous or liquid) soil phases have been transported from their original place to the surface. When soil gas or groundwater is treated *ex situ*, the technology always fails to remove a part of the mobile phase, which is then affected by *in situ* technology. Assuming a more complex situation, the mobile phase controlled by the technology can be treated not only by surface but also by underground operations, optionally in a certain soil volume (a quasi-reactor) or in a real built underground reactor. A more precise definition can be formulated based on the reactor approach: *in situ* soil remediation is a soil treatment procedure where the solid soil phase remains in its original place. The processes and the operations take place and the machinery needed is arranged *in situ* in the soil (subsurface), while soil vapor and/or groundwater may be treated simultaneously also *in situ* or *ex situ* (i.e. on the surface) after being extracted from the ground. Mobile soil phases can be treated in subsurface reactors after removal or *in situ* in their original arrangement and state without being removed.

The phrase "*in situ* groundwater treatment" is interpreted and used in a broad sense. It encompasses a range of treatments from completely passive and monitored natural attenuation through *in situ* manipulation using intensive engineering tools. The latter can involve either injection of substances into the groundwater or treatment of the groundwater in special wells, underground reactors or reactor systems. Subsurface permeable reactive barriers and reactive soil zones are also used increasingly for groundwater remediation: these are reactors of various geometries filled with a reactive medium that enables contaminant-specific physicochemical or biological processes. These subsurface installations also implement *in situ* technologies, special features being that the water is not pumped to the surface and that the installation cannot be seen. Moreover, the equipment is not placed into the soil to be treated but into a separate underground space.

The reactor approach considers the soil volume treated *in situ* as a reactor, and for those cases where the machinery is placed or material is delivered in the soil, it is classified as a quasi-reactor that does not have walls, i.e. it is open on all sides and in direct contact with the surrounding non-contaminated soil or external atmosphere (surface soil layer). On the other hand, the terms "built *in situ* reactor" or "subsurface reactor" are used for a real reactor vessel placed or constructed underground.

The openness of the *in situ* reactor has different meanings from the perspective of soil phases since there are distinct transport processes within the soil gas (without a pressure difference, this mainly involves diffusion and partition), in the groundwater (flow, diffusion and partition), and in the solid phase of the soil (partition and diffusion).

A reactor that is open on all sides has no walls, but this does not mean that it is boundless. Its limits are defined by the boundaries of the pollution, by the treated soil volume and by the range of operation of the technology. The boundaries of the pollution and the technology's range of impact may or may not be identical, and the treated volume may be both smaller and larger than the volume occupied by the pollution.

There can be further divisions within the *in situ* technologies depending on whether or not an undisturbed soil volume is being treated. There is a full spectrum: from systems that are largely undisturbed through systems where the soil gas or groundwater is only circulated and those where the solid phase is extensively mixed, for example, by slurry preparation.

The operations and processes in *in situ* quasi-reactors should also be redefined or given a suitable interpretation. Most of the operations are identical to *ex situ* processes used in tanks or flow-through reactors. However, the effect of the operation is different, first, because the system is a heterogeneous solid-phase system, and, second, because the system is open. This means that the impacts within the quasi-reactor are neither homogeneous nor consistent as in a normal batch (homogeneous or gradient) or flow-through reactor (more or less linear gradient), but it shows various gradients in three dimensions, influenced by inner (treated soil heterogeneity and operational parameters) and outer (surrounding soil conditions and fluxes) parameters and heterogeneities.

Verification of remediation technologies, particularly verification of *in situ* remediation is the key for their use and acceptance. A user-friendly verification tool and uniform practice are necessary to increase trust in innovative, mainly *in situ*, biological and ecological remediation technologies. To establish such a uniform verification methodology for environmental remedial technologies (Gruiz *et al.*, 2008), our way of thinking must be changed with regard to *in situ*, biological, and ecological technologies: they must be considered to be engineering tools and prepared, selected, planned, maintained, controlled and evaluated in the same way as any other technologies in engineering practice (see Chapter 11).

Uncertainties appear to be larger in the case of *in situ* technologies, as engineers lose total control over processes (what is a fiction when working with soil anyway). On the other hand, when using soft and environmentally friendly technologies, i.e. interventions in harmony with natural soil processes, better results can be achieved by supporting natural processes with engineering tools and facilitating their optimal performance.

2 TYPES OF REMEDIATION REACTOR

The design of the reactor can vary considerably, according to whether *ex situ* or *in situ* technology is applied. In the case of *in situ* technology, the "reactor" does not even resemble the usual type of reactor that is normally surrounded by walls. It can be a completely open soil volume in a water-saturated or unsaturated soil zone, the internal space of a water-supply well or the pore volume of a natural or engineered compartment of the soil, which contains the soil gas and groundwater to be treated.

Reactors can be classified according to the following criteria:

– The phase of the soil: gas-, liquid-, slurry- (mud-), or solid-phase reactor;
– The concentration gradient: stirred homogeneous or heterogeneous tank reactor, tube reactor, or a filled column with a gradient;
– The number of technologies applied simultaneously, in accordance with the number of the treated soil phases or contaminants;
– The number of consecutive technologies: single-stage, multi-stage (the latter version can be arranged in series or parallel), or cascade reactor;
– Its connection with the exterior: closed, semi-open, or open;
– Concentration change with time: periodic, continuous, or quasi-continuous;
– Input, output and material flux: output only, input only, or involving recirculation;
– Redox potential: the treated volume is aerobic, anoxic, or anaerobic.

The most common remedial technologies arise from combinations of the aforementioned classification criteria. The same criteria can be applied to both *in situ* and *ex situ* remediation.

2.1   Reactors and operations of *ex situ* remediation technologies

According to the soil phase to be treated, *ex situ* technologies can use the following reactors:

1   ***Reactors for soil gas treatment*** are identical to off-gas or air-treating reactors and work in continuous or periodic mode, with or without recirculation.
2   ***Reactors for groundwater treatment*** are identical to the reactors used for water and wastewater treatment, in which physical, chemical, or biological treatments or a combination thereof are applied to the soil liquid phase. Considering the reactor itself and its operation, it can be a periodic-batch, continuous-batch or filled flow-through reactor (see Figure 2.1). In-well treatment uses the extraction well's inner volume as a reactor for treating the residential groundwater in passive (without pumping) or in active (with pumping) mode (Figure 2.2).

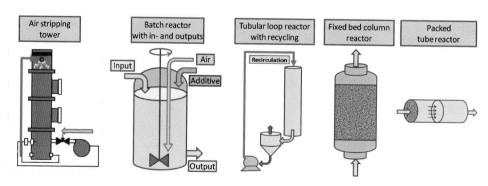

Figure 2.1 Typical reactor types for *ex situ* groundwater treatment.

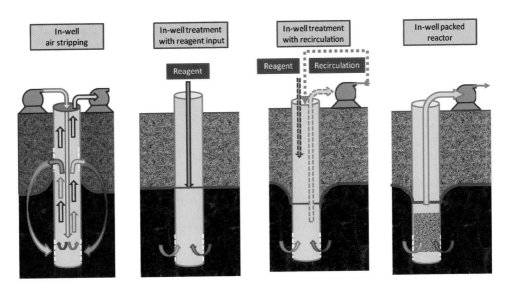

Figure 2.2 Typical in-well groundwater treatment solutions.

3   ***Reactors for treating the solid phase or the whole soil*** can be three-phase or two-phase systems. Slurries of different densities, ranging from dense silts to water-like thin suspensions, can be treated by slurry reactors, washers, sedimentation equipment, phase separators, sieves, cyclones, and flotation cells. The soil slurry/sediment passes through an arbitrary combination of this equipment in an optimal order, tailored to the quality and quantity of contaminants and soil texture. A washing plant usually includes mixers, gravel and sand separators, equipment for organic solvent extraction, and a washer to remove the extractant from the slurry. Post-treatment units for wash waters, extractant and different grain-size solids are also linked.

   Three-phase soil-treatment technologies can be based on physical, chemical, thermal, or biological processes or a combination thereof. There is a wide range of solid-phase reactors and quasi-reactors. Solid materials can be amassed in heaps or prisms or simply layered on a solid surface, and prisms can be equipped with injectors and collector-tube systems. Simple tank reactors are widely used for longer-term treatments with little or no equipment. More complex versions of tank reactors contain stirring, aeration, injection, and water-collection devices. Another reactor type, the tube reactor, is equipped with conveyor belts or pulleys for solid material transfer. Screw reactors ensure solid-phase handling, transport, and stirring.

2.2   *In situ* quasi-reactors

An *in situ* treated soil volume that has no physical walls but is still limited – by the dimensions of the pollutant plume or the range of the operations – can be considered a reactor and it is termed a quasi-reactor. It is characterized by the type of contaminant, the core process on which the remediation is based, the operations and equipment that ensure the process taking place, and the scope of engineering activities.

Applicability and verification of *in situ* remediation is based on the ratio of risk or damage of the intervention to the benefits and the avoidance of damage that would have been caused by the eliminated problem. This should be assessed in light of future land use.

2.3   Physical processes used for *in situ* soil remediation

The most frequently used physical processes in soil remediation are soil aeration and soil gas/vapor extraction, groundwater extraction and injection and soil heating. The processes can be classified according to the intensity of intervention and the type of impact, i.e. whether it is identical to natural processes, in harmony with or opposed to the nature of the soil.

- Soft physical interventions originating from agricultural soil reclamation include mechanical soil loosening, regulation of the soil–water regime by irrigation, drainage, and watering as well as mild heating. The target of these operations is the upper soil layer up to a depth of about one meter.
- Manipulation of the soil-air regime includes the following: soil loosening, aeration by passive wells (atmospheric air input) or by ventilation, soil gas and vapor exhaustion from the vadose zone, and air stripping from the aquifer by a vacuum.
- Soil moisture and the soil–water regime, groundwater flow and level can be manipulated by irrigation, drainage, watering or flooding the soil, water extraction, injection and recirculation using simple and special extraction wells, mechanical and hydrological barriers, or permeable walls.
- Manipulation of soil air and water may target soil pH, redox potential, or temperature changes and the concentrations and partitions of contaminants and additives.
- Manipulation of the solid phase covers loosening and mixing the soil (typically the surface layer), and slurry formation in order to mix additives. *In situ* soil mixing of the deeper layers is one of the most invasive and energy-demanding operations in soil and is mainly used for geotechnical purposes.
- Thermal reclamation and remediation vary from mild heating to *in situ* vitrification (1200°C) to increasing the evaporation and desorption rate, contaminant mobility, or thermal destruction of contaminants and melting soil silicates. Heating can be used for biological control, to activate or inactivate living organisms, microorganisms or seed germination (Section 3.3).
- Additives for correcting soil texture and structure or for soil conditioning are discussed in Section 2.4, "Chemical processes for *in situ* soil remediation."

Figure 2.3 is a summary scheme showing frequent *in situ* operations to control soil air and groundwater flows during soil remediation. Air and water inputs and discharges shown can be applied in an arbitrary combination, depending on the characteristics of the soil and groundwater, the pollutants and the activity of soil microbiota. Several cases are shown in Figure 2.3 to explain the essence of *in situ* soil treatment and the reactor approach. The operations shown are optional. The outline shows a pollution incident that affects both the saturated and unsaturated zones. The arrows in the figure are integrated fluxes of soil mobile phases, contaminants, and additives used for *in situ* treatment.

An *in situ* treatment of the solid phase may be connected with the extraction of soil air with or without treatment on the surface. Atmospheric air can be injected into the soil at an atmospheric or higher pressure, either into the saturated or the unsaturated zone to improve living conditions for the soil microbiota.

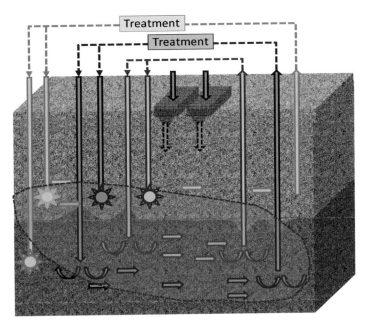

Figure 2.3 Schematic arrangement of *in situ* operations in soil remediation (yellow: air/gas/vapor; blue: water/solution/liquid contaminant; red: reagents/additives).

An *ex situ* water treatment can be combined with *in situ* solid-phase and gas or vapor treatment. Water can be collected by a drainage system, extracted by water extraction wells, or just circulated within the soil using groundwater circulation wells (see Section 2.3). Circulation wells can perform in-well water treatment, and purpose-built installations can facilitate out-of-the-well subsurface treatment of the circulated water. Alternatively, extracted water is treated on the surface, (partly) returned or not returned to the soil. The treated water can be re-admitted from the surface through shallow ditches into the unsaturated zone in order to adjust soil moisture or use for continuous or periodic flooding and washing of the solid phase. Water treated *ex situ* can be directly reinjected into the groundwater, primarily to raise the groundwater level and encourage the groundwater to flow towards the depression wells. Water infiltration systems and injection wells are also used to input liquid, dissolved or suspended additives or reagents. Their transport and distribution are greatly influenced and limited by the sorption (filtering) capacity of the soil and by heterogeneous pore distribution. Gas injection wells can be used for injecting any gas-phase additives (reagents) and hot air or moisture.

2.3.1   Wells and engineered underground reactors for groundwater treatment

Underground water treatment facilities are often called *in situ* reactors, probably because the early reactors were placed into the groundwater flow path, and the water to be treated was flowing through the treatment space more or less passively due to gravity. Later technological developments have managed to fully control groundwater flow by applying specific pressure differences and various flow-directing tools (barriers and gates, circulation cells, etc.) or by using cascade reactor systems and special reactors filled with fixed or removable fillings. The term "*in situ*" is used today for all groundwater treatment technologies that do not pump groundwater to the surface and for all subsurface water treatment reactors.

There are three main types of engineered *in situ* groundwater treatment reactors:

–   Vacuum wells, groundwater circulation wells, and these wells equipped with in-well reactors for (e.g. packed bed reactors) for water treatment (see also Figure 2.2).
–   Permeable reactive barriers (PRB).
–   Engineered *in situ* reactive soil zones, also referred to as "treatment zone," an exemplary quasi-reactor within the soil, with minimal engineering. These soil zones include constructed subsurface structures or the soil itself with or without engineering manipulation (e.g. additives or reagents input).
–   Figure 2.4 shows three technological solutions for groundwater treatment: the permeable reactive barrier and a larger reactive zone filled with contaminant-specific reactive medium. Both are placed in the way of the groundwater. These constructions function as packed bed reactors of various geometry. The third case outlines a soil with intrinsic reactive character, e.g. capable of biodegradation. Figure 2.5 shows reagent placement by injection or through an infiltration trench. The activation a certain soil volume this way creates an *in situ* treatment zone. Activation may involve redox manipulation, or the addition of mobilizing agents or chemical reagents.

2.3.1.1   Permeable reactive barriers and in situ reactive soil zones

Permeable reactive barriers (PRBs) are underground structures intercepting a contaminant plume and channeling the groundwater flow horizontally or vertically through the reactive material in the permeable wall of the reactor, employing a natural or artificial pressure difference. Some types of PRBs use funnels and gates, i.e. the entire cross-section of the barrier is not permeable. The contaminated groundwater interacts with the reactive material of the reactor on a relatively short pathway. The interaction results in reduced risk by immobilizing or transforming/degrading the contaminant.

Engineered reactive soil zones – installed also in the pathway of the groundwater flow, but unlike PRBs, these have larger volumes ensuring a longer pathway for the groundwater

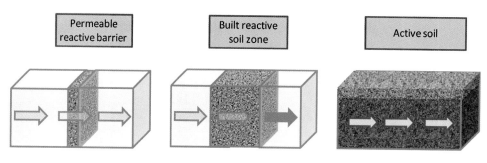

Figure 2.4 Groundwater treatment in the soil using (i) reactive barrier, (ii) reactive soil zone, or just (iii) the soil as an active medium.

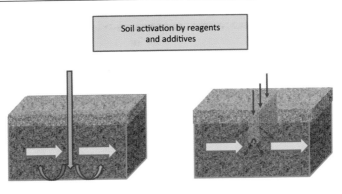

Figure 2.5 Creating an *in situ* treatment zone: activation of the soil by injecting or infiltrating additives and reagents.

flow, consequently longer contact times between the contaminant and the reactive soil volume. With these conditions, reactive zones provide a good opportunity for soft, natural, and ecological processes.

PRBs and reactive soil zones are underground versions of the filled (packed) flow-through reactors. Both of them may work as simple reactors or as cascade reactor systems. These built and *in situ* real reactors have well designed dimensions, represented by built or natural gates and barriers, and an inflow and outflow. Many innovations exist in the field of PRBs (Debreczeni & Meggyes, 1999); their operation can rely on physical, chemical, or biological treatment of the contaminants (Meggyes *et al.*, 2001; Meggyes *et al.*, 2009; Roehl *et al.*, 2005; Simon *et al.*, 2001; Simon *et al.*, 2002; Simon & Meggyes, 2000). The technological solutions vary from the simplest gravitational flow-through reactor to the more sophisticated, strictly controlled and regulated multi-stage reactors with distributor and collector systems (Meggyes & Simon, 2000; Roehl *et al.*, 2007; Meggyes, 2010). Another direction of development favors a suitable combination of PRBs and specialized circulation wells or removable cartridges instead of fixed packing.

Engineered reactive soil zones differ from PRBs in size and degree of openness: they generally have large volumes, enabling a long pathway, i.e. a long residence time for the contaminant to interact with the reactive medium and suitable gradients to develop in the groundwater. Reactive zones do not often have walls. Groundwater flow in reactive zones is typically gravitational. The reactive filling is a cheap, natural or partly artificial material, or the soil itself, and is specialized for pH or redox adjustment, sorption, chemical reaction, typically oxidation, reduction or degradation, as well as for biological transformation. The combination of a PRB and a reactive soil zone can have built walls similar to PRBs and dimensions similar to reactive zones.

When classifying subsurface reactive installations according to the reactor approach, there is a continuous transition from passive *in situ* groundwater treatment to greatly equipped wells, PRBs, including built or otherwise engineered reactive soil zones and subsurface reactor systems (see also US EPA, 1998; Simon *et al.*, 2002; Meggyes, 2009).

Active and passive systems

Both reactive barriers and reactive soil zones can work as passive and active systems. In passive systems, the groundwater is transported through the reactive medium by the natural/gravitational groundwater flow and the reactive bed is established when the system is being built and is not manipulated later on; neither are additives injected into or water or products extracted from the passive reactor. Dissolved groundwater contaminants react with the reactive material and, as a result, are sorbed, precipitated, or chemically and biologically degraded. Active systems use injection and extraction wells, hydraulic barriers, circulation wells, or any operation for directing the groundwater to an *in situ* (subsurface) reactor or one located on the surface where the treatment is carried out. The active system needs regular maintenance and continuous reagent addition and several operations such as pumping, mixing, separating, etc. (see more in Chapter 4.4).

2.3.1.2 Reactive barriers, reactive soil zones, and multi-stage subsurface reactors

A reactive soil volume is a quasi-reactor, where the conversion of contaminants and the reduction of risk, i.e. the remedial process, is occurring. It can be a spontaneously formed and naturally working, a passive (no operations or maintenance) *in situ* quasi-reactor, or an engineered subsurface reactor with high operation and maintenance requirement. Passive or active, both are flow-through, usually packed reactors (open or surrounded by permeable walls), in a single or cascade arrangement.

The physicochemical conditions and/or active microorganisms will modify, degrade, or stabilize the pollutant in such a reactive soil zone when the contaminated water or soil gas flows through the reactor volume. Suitable environmental parameters (pH, redox potential, reagent concentration, etc.) in a passive or active operation, or a microbiota with conversion ability and activity can ensure groundwater or soil gas/vapor decontamination. Biologically, reactive zones often need soft interventions to maintain their activity over the long term such as nutrient or air injection, introduction of alternative electron acceptors, or addition of chemical agents for supplementary treatment.

Underground reactive soil zones use a variety of natural or artificial remedial processes and engineering operations and can be natural, basic, or sophisticated. They can be built for the treatment of groundwater, seepage, landfill leachate, wastewaters, or mine drainage. Long-term permeability, gravitational transportation of the water to be treated, water level gradients and the right composition (organic matter content, alkalinity, and acidity) should be planned and controlled.

The reactive zones created *in situ* can be quasi- and real reactors, depending on whether only the processes or the volume itself are manufactured. Both single- and multi-stage or cascade reactive zones may function as a so-called passive system, i.e. they can work without energy and material input (no additives and no injection, only gravitational water flow). Some of them do not require labor at all, and so they can be controlled remotely (Jarvis & Younger, 2001; Johnson & Younger, 2006).

The well-known root-zone treatment (rhizofiltration) of wastewater is closely related to the built reactive soil zones used in groundwater treatment. These artificially established zones are entirely open reactors, but their properties (pH, redox potential, microbial status, etc.) differ from those of the surrounding soil. They interfere directly with both the groundwater and the solid phase. They can be located close to the surface or at greater depths.

Reactive zone systems

Reactive zones for various water treatment purposes can be connected successively to each other, for example an anaerobic zone can be followed by an aerobic one for the treatment of groundwater contaminated with chlorinated hydrocarbons. This arrangement produces

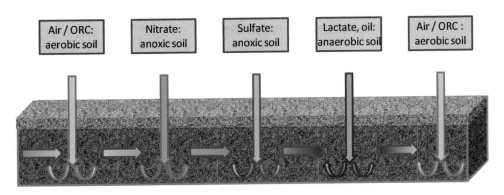

Figure 2.6 In situ treatment zone train created by redox manipulations.

a long flow path, along which the contaminated water meets different conditions and can be treated by consecutive processes. An anaerobic zone is created by injecting adequate substrates (e.g. lactate, special oils, or fats) into the groundwater to provide a substrate for sulfate-reducing microorganisms. These organisms will use the sulfate content of the anoxic groundwater arriving at this soil volume and gradually decrease the redox potential to a value suitable for chloro- or carbonate-respiring microorganisms to achieve reductive dehalogenation. Following reductive dechlorination, the aerobic stage (an oxygen release compound = ORC addition to the saturated soil zone) ensures full biological degradation of chlorinated hydrocarbons and the produced organic compounds such as methane. A treatment zone train is shown in Figure 2.6.

Reactive permeable barriers differ from the reactive zones because they only come into direct contact with the groundwater and not with the solid phase of the soil, and the flow path is relatively short, as the gradients formed on this short path do not play a significant role. Some of them have permeable walls, but in some modern applications only the filling is permeable, not the wall. The filling is placed into a container through which the groundwater flow is channeled.

This brief overview illustrates that soil remediation usually applies the same processes as chemical and biological industries. The main tool is also similar: the method of mass balance or material balance is used to calculate the "reactor" capacity. This is the basic calculation method of chemical engineering: the mass that enters a system must, by the principle of conservation of mass, either leave the system or accumulate within the system. Its feasibility is best demonstrated in the method used for technology verification (Gruiz *et al.*, 2009, Chapter 11 in this volume).

2.3.1.3 Active and passive systems for the treatment of acid mine drainage

In situ passive and active systems are frequently applied to acid mine drainage treatment (AMD) (see also Chapter 7 in this volume), pH neutralization (using carbonate rock), metals precipitation (in hydroxide or sulfide forms), and organic matter providing alkalinity and reducing conditions. Low acidity and low flow rate AMD can be treated in passive systems such as oxic and anoxic limestone drains or limestone beds, limestone diversion wells, aerobic or anaerobic wetlands, systems containing reducing and/or alkali-producing organic materials, permeable reactive barriers, slag leach beds, sulfide passivation beds, microbial reactor systems, etc.

Active treatment systems are needed for a high range of acidity and high flow rates by appropriate operational parameters. These operations incur a high cost, so AMD treatment can be of low economic efficiency. Fixed or small portable plants can be applied for treating pumped AMD, implementing reagent addition, mixing to AMD and separation and disposal of the resulting sludge. Subsurface (*in situ*) active installations may provide pH control and/ or metal precipitation, electrochemical concentration, biologically mediated redox control (e.g. sulfate reduction), ion exchange, sorption, flocculation, filtration or crystallization (Hammarstrom *et al.*, 2003; PIRAMID, 2003; Taylor *et al.*, 2005; Trumm, 2010; Trumm & Watts, 2010; Younger *et al.*, 2002).

2.3.2 Vacuum wells and groundwater circulation

Vacuum wells can be used not only for soil vapor or water extraction and pumping these mobile phases to the surface but also for establishing arbitrary water flows, i.e. *in situ* groundwater circulation and recirculation between any depths of the subsurface soil. These groundwater circulation wells are modified groundwater wells with two or more screens and in various depths. Groundwater can be extracted from the aquifer through the lower screen, for example, and released through the upper screen into the aquifer or the vadose zone according to purpose. Using such installations, contaminated water is not pumped to the surface but circulated within the soil to accelerate contaminant mobilization by water and/or additives, natural attenuation, or any other physical, chemical, or biological processes. The enhanced water exchange established in the well also provides good opportunity for in-well groundwater treatment.

Summing up, circulation wells can be used just for circulating the water within the well's range of impact or for *in situ* treatment – inside or outside the well. Water just circulated or also treated continues to flow downstream, out from the well's range of impact (Figure 2.7).

The direction of groundwater circulation in these wells can be changed from upward to downward and it is possible to build stacked-flow and multiple circulation cells within the soil. Circulation can enhance percolation, soil washing, and mixing, and the addition, delivery, and mixing of air, reagents, microbiological nutrients, and liquid additives into the circulation well, which are then transported further with the circulating groundwater (see more in Alesi, 2008; US EPA Clue in, 2017).

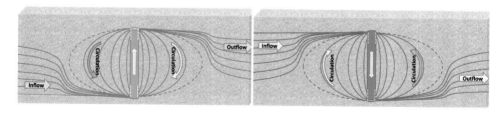

Figure 2.7 Circulation wells with an upper and a lower screen allowing *in situ* vertical groundwater circulation by pumping and recharging the same well. The groundwater circulates several times before flowing downstream.

Flexible water circulation schemes can precisely control groundwater flow and direct it through a reactive soil zone or a reactor built into the well. The reactor is usually arranged between a lower and upper screen and facilitates physicochemical (sorption, oxidation, or reduction) or biological processes (biosorption or transformation/degradation). These reactive treatment wells comprise subsurface tube reactors where the water's flow rate is determined by the applied vacuum and the flow direction/circulation by the position of the screens in the well. The hydrogeological situation, primarily water permeability of the surrounding soil matrix also has a strong influence.

Summing up, groundwater circulation and vacuum wells can be used for the following:

– Vapor extraction from the vadose or capillary zone;
– In-well stripping, i.e. the removal of vapor-state contaminants from groundwater;
– Recovery of non-aqueous phase liquid type contaminants (both light and dense);
– Groundwater recycling to intensively wash, percolate or flush unsaturated and saturated soil zones;
– Circulating water for any purpose, e.g. to pass through a reactor placed into the well between two screens (inflow and outflow);
– Aerating the groundwater, flushing and mixing additives into soil, accelerating bio-degradation by aeration and nutrient addition, and introducing additives and reagents into the aquifer to enhance physical, chemical or biological transformation of the contaminant;
– Hydraulic reactive barriers that are similar to conventional permeable reactive barriers, but instead of permeable walls, vacuum wells direct the water through an in-well reactor or a reactive soil volume.

2.3.2.1   Vacuum wells and circulation wells for volatile-contaminated soil and groundwater

Vacuum vapor extraction wells are used to extract gases or vapors from three-phase soil, from the capillary zone or from two-phase soil. The combination of gas/vapor extraction with water extraction is another solution commonly used.

To extract gas or vapor from three-phase soil, the screen in the well is placed close to the water table in the vadose zone. The extracted volatile substance can be soil air/gas with low-risk components of microbiological origin (such as CO_2, nitrous oxide N_2O, nitric oxide NO, methane, etc.), or contaminant vapors (desorbed from the solid phase or evaporated from groundwater) with various scales of risk. Extracted volatiles are most commonly treated on the surface by well-controlled common gas/air treatment technologies.

The capillary zone can be cleaned from volatile contaminants at best after the water table is depressed to a certain extent. The capillary fringe at a contaminated site generally concentrates water-insoluble contaminants by collecting them and those that are less dense than water, so they float on the surface of the water. Depending on the groundwater fluctuation level, this pollution is smeared in a thinner or thicker layer. The capillary fringe also collects vapors, which spontaneously escape from the two-phase soil.

Special wells are used to extract vapor and water at the same time: vapor from the capillary zone and water from the two-phase soil. Accordingly, one of the screened sections of these wells is located in the three-phase soil and the other under the water table in the two-phase soil. The common two-phase extraction uses a combined well and the extracted gas/vapor and groundwater are removed in two separate ways (extracted by two different pumps or blowers). The dual-phase extraction applies a high vacuum and the contaminant vapor and the contaminated water are removed together through the same conduit (extracted by one single vacuum pump) and separated and treated in a surface installation.

In-well stripping is carried out in circulation wells where the contaminated water is lifted by the vacuum to the upper screen (located in the vadose zone) without pumping water to the surface. This way the well works as a flow-through reactor and is characterized by an inflow (lower screen) and outflow (upper screen), the reactor size (the well volume between the two screens), and the groundwater flow rate. To strip volatile contaminants out from the contaminated groundwater, air is injected into the in-well reactor volume from the surface. Cleaned groundwater is discharged into the vadose zone through the upper screen, and a negative well pressure transports contaminant vapor to the surface. Negative pressure is generally provided by a blower mounted above the ground. The exhausted soil air containing vapor is treated by an air decontamination system comprising traps, filters, activated carbon sorbents, and catalyzed gas combustion (see also Alesi, 2008 and US EPA Clue in, 2017).

2.3.2.2 Vacuum wells for dual-phase and multiphase extraction

Dual-phase or multiphase extraction is implemented by vacuum-enhanced extraction to remove hydrocarbon vapor, contaminated groundwater, or separate-phase petroleum-type contaminants one by one or in combination. A high vacuum is established in the well by a blower on the surface. This well may be screened above and/or below the water table and the position of the extraction tube's end is adjusted to the soil depths from where the contaminants should be extracted. The water table lowered around the well favors vapor extraction from the newly formed three-phase soil and also for trapping the non-aqueous free phase from the surface of the water. In addition to the vacuum system, submersible pumps (for separate water extraction), and slurping/bioslurping (suction of the free phase by a vacuum through a slurp tube) can also be applied. The extracted mixture of vapor and liquid contaminants and contaminated water are treated after separation by surface installations using common air and water treatment technologies.

Dual-phase extraction is mainly used for volatile soil contaminants, fuels, and solvents, which form a separate phase from water. It can be applied in heterogeneous soil formations, but an unknown or inaccessible location and distribution of the contaminant and low permeability of the soil may limit its applicability (see also CPEO, 2017; FRTR, 2016; US EPA, 2016; US EPA, 1999; Figure 2.8).

2.3.2.3 Circulation wells for water-soluble or mobilizable soil contaminants

Groundwater circulation and *in situ* treatment is the main field of application for groundwater circulation wells (GCW). It may concern the aquifer exclusively or both the vadose zone and the aquifer. Placing a GCW in a contaminant plume in the aquifer, the well creates a hydrolytically controlled spherical soil zone from where water is directed into the well. Hydraulically controlled flushing may result in efficient contaminant elimination both from the two- and three-phase soil. Circulation ensures enhanced water flow that extracts mobilizable sorbed contaminants by physical (just high water-solid ratio), chemical (addition

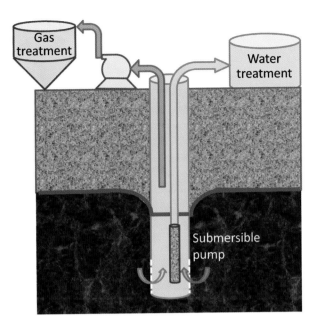

Figure 2.8 Dual-phase extraction.

of surfactants, solvents, reagents) or biological (biosurfactant production, transformation, degradation) methods. The physically or chemically mobilized contaminants may be further treated in in-well flow-through reactors by separation, solid phase extraction, sorption, or chemical transformation. The treated water can be discharged into the aquifer or the vadose zone and the contaminant removed for surface handling. The use of GCW for water-dissolved contaminants is demonstrated by these two examples:

1   *In situ* reduction of chlorinated solvents or toxic metals by using zero-valent iron (ZVI) can be intensified by GCW. ZVI is placed into a permeable reactor vessel and groundwater flow through the ZVI-packed reactor is hydraulically controlled and directed by the GCW. The packed reactor can be located inside the well, outside the well near the entry screen, or arbitrarily placed in the circulation path to ensure that the water flows through. The aim and use of this arrangement is similar to the ZVI-filled permeable reactive barriers. It only differs in the mode of controlling the water flow.

2   Aerobic microbiological degradation can be enhanced by injecting atmospheric air and supplemental nutrients into the circulation well for further distribution in the impact zone. To increase the concentration of the beneficial microorganisms, a sorbent-packed bioreactor can be placed into the well that concentrates the contaminant and supplies the (naturally occurring) microbial cells adhered to the sorbent with air and nutrients. Thus the specific microorganisms will propagate intensively and catalyze biodegradation both in the reactor itself and in the impact zone after leaving the bioreactor. Specialized artificially propagated microorganisms can also be used in such an in-well placed packed bioreactor.

2.3.2.4 GCW for strongly sorbed contaminants

For chemical or biological transformation of poorly mobilizable soil contaminants (typically aged contaminants with large molecular weight that are dominantly sorbed on a solid), GCW-aided delivery can be applied. The additives, nutrients, or reagents (which can enhance the chemical or biological availability and transformation) are injected into the well and spread via the hydrolytically controlled water circulation. This way the additives and reagents get more easily to their destination and in contact with the target molecules (even if they are sorbed on solid surfaces). As a result, the efficiency of the key process can be enhanced. An intensifying effect occurs mainly in the impact area of the well, and to a lesser extent downstream from the well. In such applications the cycled water is the carrier medium and is responsible for additive or reagent delivery to the place of use. For example, *in situ* chemical oxidation (ISCO) can be intensified this way: the dissolved reagent is injected into the well, and the hydrolytically controlled water cycling delivers and uniformly spreads permanganate or peroxide into the contaminated soil.

Beside significant benefits, there are several shortcomings of GCWs in soil and groundwater remediation such as these:

– Its limited use when vertical permeability of the soil is small or strongly variable;
– The design and monitoring requirement can be a disproportionate burden to hydrogeologically heterogeneous sites;
– The intervention results in pH or redox changes, and the chemical reactions, may cause precipitation and clogging.

Several solutions are offered to reduce these shortcomings, for example:

– Denser well placement by using simplified, small-diameter wells and inexpensive construction, e.g. direct push technology (Borden & Cherry, 2000);
– Suitable combinations of permeable reactive barriers and GCWs;
– Multipurpose wells (e.g. the same well for the vadose zone, the capillary fringe and the aquifer) with adaptable screening and flow-direction adjustment (upward or downward).

In addition to the technological possibilities and benefits, there are several managerial advantages of GCW-aided soil remediation such as not requiring pumping or surface treatment, the water drawn can be reintroduced to the aquifer after treatment, environmental impact (treatment, storage, land take by occupying the surface) is reduced, and thus the permission procedure may be easier.

2.3.3 Soil heating

The temperature of the soil can be raised to shift the distribution between physical soil phases, i.e. partition of the sorbed or otherwise bound pollutants towards the more mobile vapor and dissolved forms. Heat can be applied both to saturated and unsaturated soil zones. A wide range of soil-heating technologies are used in practice to intensify mobilization (desorption, dissolution, degradation) of pollutants. Up to a certain scale, heating also has a beneficial effect on biodegradation. Solid-bound water (soil moisture) is significantly affected by heating/mobilization; the result is increased soil dehydration, so moisture replacement may be necessary. Above 100°C, steam is produced from soil moisture, which has an intensive stripping and flushing effect primarily on steam-volatile substances (both contaminants and natural soil material). High temperatures can mobilize higher boiling contaminants (PAHs, PCBs, dioxins) and may cause their direct degradation, i.e. thermal destruction.

Heat-enhanced soil remediation has several advantages such as a relatively short treatment duration (from a few months to a year) and suitability for non-uniform subsurface conditions and contaminant distribution. A disadvantage is the increased desiccation in the vadose zone and potential cooling (due to energy loss) in the aquifer in the case of high groundwater influx. The identification of the contaminant source and focusing the treatment on it significantly increases the overall efficiency of the technology, including cost efficiency. Future land use determines if thermal degradation of the soil matrix and soil microbiota can be accepted or not. A temperature higher than 70°C selectively degrades soil microorganisms; 300°C kills most of them and may also cause humus degradation. Above 500–600°C, thermal treatment completely converts soil into dead material: not only microorganisms, but humus may also be fully destroyed, and toxic residues are left behind in the worst case. Up to a certain scale of degradation, soil can be revitalized depending on the duration and temperature of the treatment. Therefore, thermal treatment requires a thorough study on the impacts and residual risks.

The most widespread heating method for *in situ* thermal desorption of contaminants is electrical resistance heating with closely spaced electrodes for energy input. Chlorinated solvents and petroleum hydrocarbons are typical target contaminants for thermal desorption (see also Johnson *et al.*, 2009; Triplett Kingston, 2008).

2.3.3.1   *Thermal treatment to enhance desorption and vapor recovery*

Thermal treatment intensifies *in situ* thermal desorption (ISTD), resulting in a higher-efficiency remediation and shorter operation time, and is applicable to heterogeneous and low-permeability formations. Volatile and semivolatile compounds require a wide range of temperature. Heat can reach deep soil layers and pollutants under buildings and other structures.

Thermal treatment of soil can be combined with vapor extraction when contaminants are volatile or semivolatile, but the real aim of thermal treatment is to better dissolve contaminants in a washing/flushing liquid, increase availability for chemical reagents, or enhance bioavailability for the soil microbiota. High temperatures can also stabilize certain contaminants.

Thermal treatment is often based on ***indirect heating***, which is performed by injecting hot air or steam through wells or small-diameter pushed injectors. However, ***direct heating***, with the *in situ* conversion of some type of energy to heat up the soil, is more energy efficient. Soil heating is most often used for contaminant desorption both in *ex situ* and *in situ* soil remediation. High-temperature thermal degradation and vitrification is only applied *in situ* for extremely toxic contaminants and inaccessible subsurface contaminants. *Ex situ* thermal technologies such as high temperature desorption, wet oxidization, and vitrification are extensively used for strongly contaminated soil, sediment, and hazardous waste.

2.3.3.2   *Indirect heating*

Indirect soil heating is implemented by injecting hot air, hot water, or steam either *in situ* or *ex situ* into the soil through wells around a smaller or within a larger contaminant source. The impact zone then keeps growing radially around the injection wells. When soil gas and/

or groundwater is simultaneously extracted, the heat transfer medium pushes the mobilized contaminant(s) toward the soil vapor-extraction wells in the vadose zone or in the groundwater wells in the saturated zone.

Hot air is applied to volatile or semivolatile organic contaminants. *Hot water* enhances contaminant expulsion or solubilization from the solid phase both in three- and two-phase soils, displaces organic contaminants (e.g. hydrocarbons) from the pores and reduces the viscosity of liquid contaminants (e.g. NAPL). *Steam* can also be used to heat soil and groundwater and mobilize contaminants, especially those that are steam-volatile. In addition, steam destroys certain contaminants, mainly under reductive or oxidative conditions.

2.3.3.3 Direct soil heating

The three main tools for direct soil heating are electrical resistance, microwave, and radio frequency heating.

Electrical resistance heating (ERH) uses electrical current flowing through the contaminated soil volume between electrodes and produces heat across the electrical resistance of the soil. Volatile and semivolatile contaminants are stripped *in situ* by hot soil air or flushed by hot groundwater or steam. Temperature primarily depends on the contaminant, while electrode spacing and operating time are mainly determined by soil characteristics.

A special type of soil heating using electrical current leads to *in situ* vitrification, i.e. the melting of contaminated soil, including strongly bound stable contaminants. The temperature in this treatment varies between 1,600 and 2,000°C. Mobilizable organic substances (both soil components and contaminants) are removed or destroyed at this high a temperature (thermal degradation). Stable contaminants such as toxic metals, including radionuclides, are incorporated after cooling down into the glass-like amorphous material formed from the melted sand and clay minerals (see more in Beyke & Fleming, 2005; ERH, 2007; Heron *et al.*, 2005).

Microwave heating (MW) has been used from the 1990s for the remediation of soils contaminated with hydrocarbons and more recently for *in situ* groundwater treatment. MW has the potential to significantly reduce treatment times and costs mainly due to a better heat transfer compared to indirect heating (Falciglia *et al.*, 2013; Falciglia *et al.*, 2016; Falciglia *et al.*, 2017). Its application for hazardous wastes was first proposed by Dauerman (1992). Since that time, several applications have been published, e.g. for PCBs by Huang *et al.* (2011) or hexachlorobenzene removal by Lin *et al.* (2010). Microwave absorbers, e.g. MnO_2, can further intensify heat transfer in soil.

Radio frequency heating uses electromagnetic energy to heat soil and enhance soil vapor extraction. Heated soil volume is bounded by two rows of vertical ground electrodes with energy applied to a third row midway between the ground rows. The three rows act as a buried triplet capacitor. The technique can heat soils to over 300°C (Triplett Kingston, 2008; Johnson *et al.*, 2009).

2.3.3.4 Soil heating to intensify biodegradation-based in situ remediation

Heat increases the efficiency of biodegradation because pollutants become more accessible and better distributed. While desorption (mobilization) steadily increases with temperature, the activity of the microbes exhibits a maximum. At the beginning, around the biological optimum temperature, mobilization and microbial activity may support each other, but if mobilization requires a temperature that is too high to be biologically acceptable, it is worth separating the two processes into two phases: a short-term, intensive heating phase followed by a longer-term biodegradation phase. Biodegradation experiments on soil polluted with residual heavy fuel oil (mazout) showed that the microbial community of the soil had better degrading capabilities at 15°C than at 30°C, but temporary short-term heating increased the bioavailability of the contaminant without restricting microbial activity (Molnár *et al.*, 2003).

2.4   Chemical processes for *in situ* soil remediation

Chemical tools in soil remediation are aimed either at modifying physicochemical soil properties or at decreasing the damage caused and the risk posed by the contaminants by transforming or degrading the contaminant itself.

2.4.1   *Chemical reagents and additives*

Soft chemical methods aim at compensating disadvantageous chemical and physical soil properties by adding natural or man-made chemical substances to stabilize soil acidity, alkalinity or salinity to supply microorganisms and plants with nutrients and to control soil organic content and overall structure by adding organic material, humus, or conditioning substances. Soft interventions are applied to natural and agroecosystems, where conservation of soil as a habitat and its ecosystem is an important issue for future land use. In cases where soil-endogenous microbiota will be utilized for bioremediation, similar requirements should be fulfilled in order to keep the active microbiota alive.

Soft chemical interventions are also used as agrotechnologies and called *soil reclamation* or *soil amelioration*. These are also useful solutions in soil remediation, especially as concerns the upper soil layer, and the original quality and use of the deteriorated, degraded, or contaminated soil should be restored.

Not all chemicals used for soil remediation are natural or near-natural, but most of them are synthetic substances applied for modifying soil structure and mobilizing or stabilizing contaminants. Some of them are very reactive and hazardous and are used to transform or degrade contaminants chemically. See train in Figure 2.6.

Chemical reagents (direct oxidants or reductants) can facilitate contaminant conversion by themselves and the optimal technological parameters should support the chemical reactions. They can also encourage certain natural or technological processes (by setting pH or redox potential) or can have an indirect remediation effect (nutrients, sorbents, catalyzers).

The chemical additives used are summarized in the following list (see also Maslov, 2009).

1   *Modifying the pH* of acidic or alkaline soils: the most frequently used chemicals are $CaCO_3$ (lime), $CaSO_4$ (gypsum), $CaCl_2$ (calcium chloride), and H_2SO_4 (sulfuric acid). Treatment options for solonetz and soda-saline-alkali soils are the addition of natural organic matter and mineral fertilizers. Soil liming is one of the most ancient agrotechnologies. Liming materials include the following:

   – *Limestone*, typically calcite and dolomite, burnt and hydrated lime; marl or the shells themselves; industrial by-products such as slags or alkaline ashes;
   – *Waste materials* are increasingly used for soil liming; 2 billion tons per year of diverse alkaline residues are produced and available globally (Gomes *et al.*, 2015;

SOILUTIL, 2018). They include food industry wastes and by-products, e.g. shells, refuse lime from sugar beet processing, alkaline paper mill waste, flue dust from the cement industry, lime from alkaline washers used in air pollution control systems, wastes from water softening plants and several other industrial and mining wastes such as red mud from bauxite processing (aluminum production), steelworks slags, fly ash from coal combustion, concrete crusher fines, rock wool waste, flue gas desulphurization waste, waste from the Solvay process (sodium carbonate production), calcium carbide production, and (chromium, lead) ore processing (see application cases: Chang *et al.*, 2013; Goulding, 2016; Molnár *et al.*, 2016; Pértile *et al.*, 2012; Ujaczki *et al.*, 2016a,b; Zambrano, 2007).

2   *Adjusting redox potential* (Eh) in the soil is a basic tool to control chemical and biological processes. The most advantageous redox potential is determined by favorable chemical or microbial processes such as chemical oxidation or reduction, microbiological transformation, and plant-microbe cooperation. Oxidants (electron acceptors) have significant buffering capacity, so the following additives can set the Eh and maintain the remedial process:

–   *Atmospheric air introduction* by venting or air injection can increase oxygen concentration in soil air. Atmospheric air has lower CO_2 and higher O_2 content compared to soil air and can thus activate aerobic microorganisms and plant roots.
–   *Aquifer oxygenation* – as an alternative to energy-intensive aeration of the groundwater – can be carried out by oxygen release compounds (ORC) such as $Ca(OH)_2$, supplying molecular oxygen for aerobic biodegradation in the aquifer over the long term (e.g. ORC supply, 2017).
–   *Alternative electron acceptors* such as nitrate, manganese (IV), iron (III), sulfate, and CO_2 in favor of anoxic and anaerobic microbial activities.
–   *Strong oxidation* reagents are used to induce direct oxidation reactions with the following contaminants:

   o   Permanganate, in form of potassium permanganate ($KMnO_4$) and sodium permanganate ($NaMnO_4$). Relatively slow oxidation is advantageous, but the resulting MnO_2 causes clogging in the soil.
   o   Peroxide or the mixture of ferrous or ferric iron salts and hydrogen peroxide. Iron plays the role of a catalyst in the production of hydroxyl radicals from peroxide.
   o   Persulfate, or more precisely peroxydisulfate in the form of a sodium salt ($Na_2S_2O_8$). It works with a ferrous iron, Fe(II) catalyst to produce reactive sulfate radicals.
   o   *Ozone (O_3)* is 12 times more soluble than O_2 and, being a gas, minimal residue is left over after oxidation. It is very reactive and, in addition to the target molecules, may oxidize all oxidizable soil components.

–   *Direct chemical reduction* of contaminants in soil with reducing agents to convert contaminants into less toxic or less mobile chemical forms. The chemical agent is injected into or otherwise fills permeable reactive barriers or other surface or subsurface flow-through reactors. Contaminant reduction is used for water-soluble chlorinated toxicants, very toxic chromium VI to be reduced to chromium (III), and for *in situ* elimination of dense non-aqueous contaminant layers, typically chlorinated solvents (e.g. trichloroethylene). The best-known reducing agent is metallic iron (also called zero-valent iron or ZVI) and its variations such as micro or nano ZVI (increased reactive surface) or palladium- or silver-coated ZVI. Polysulfides or sodium dithionite are also frequently used as reducing agents.

–   *Indirect chemical reduction* covers isolation of the treated soil volume from atmospheric air by encapsulation or flooding by water, by establishing oxygen-consuming chemical and biological processes that produce reductive conditions in the soil and groundwater.

3   *Soil structure amendment* by additives includes soft interventions using conditioners, humus or natural and synthetic polymers that upgrade stability and the water regime to strengthen soil functions. Another type of intervention is soil stabilization; in this case natural soil function is secondary, and geotechnical goals such as base (sub-base) stability or landslide management get priority, and the used stabilizers cause drastic changes in soil structure, e.g. solidification or block formation (see next paragraph and Chapter 7).

–   *Conditioners* or structure-forming materials can increase porosity, optimize infiltration properties and water permeability, resulting in increased water and aggregate stability;
–   *Humus* and extracts from humus;
–   *Biochar* is a newly rediscovered natural soil amendment, used both for structural and functional purposes for degraded and contaminated soils. Biochar is produced from organic waste by a thermochemical decomposition technology called pyrolysis. It is a material of high eco-efficiency complying with carbon sequestration, normalizing global element cycling, soil structural amendment and buffering of the soil nutrient regime (see Chapter 5).
–   *Polymers* interacting directly with the soil matrix: binding both to humus and clay minerals (coagulation and formation of strong hydrogen bridges):

    o   Inorganic polymer silicates, e.g. sodium silicate;
    o   Cellulose-based polymers: viscose, methylcellulose, carboxymethyl cellulose;
    o   Synthetic polymers, copolymers, polymer salts (acrylic, metacrylic and maleic acid polymers), polyacrilamide, polyacrylonitrile, non-ionic polyvinyl alcohol or calcium and ammonium lignosulfonates (Maslov, 2009);
    o   Peat glue: a mixture of peat, perlite and a polymer.

–   *Polymer-based emulsions and dispersions*: bitumen, latex, polyvinyl acetate-based emulsions, polyurethane resins, or carbamide-formaldehyde can form a thin film on the surface of the microparticles, resulting in a looser soil structure.
–   *Polymer foams*: inert polymers that do not react directly with the soil, but their foam-like structures modify (loosen) soil structure and increase water absorption after being mixed in.
–   *Swelling polymer hydrogels* are more practical than foams: they can increase their volume to 300x by water uptake. They are ideal amendments for sandy and arid soils and can significantly increase water retention and amend soil–water regimes.
–   *Stabilizing agents* (see also Chapter 8) may increase the partitioning ratio of the contaminant between soil solid and liquid by modifying the chemical form of the contaminant, the soil solid or the soil water or by modifying the interaction

between contaminant and soil-solid particles. Addition of sorbents and stabilizing agents such as clay and lime or the listed polymers can modify soil solid to bind contaminants more strongly. Solidifying agents such as bitumen (not environmentally friendly) or cement incorporate contaminants into the new block-like solid structure.

– *Mobilizing agents* (see also Chapters 5 and 6) may have a significant impact on contaminant availability for the transformation process, whether it is physical (e.g. removal of the contaminant), chemical (e.g. chemical transformation), or biological (e.g. biotransformation). The mobilizing agents' availability-enhancing effect acts against the strong binding of contaminants to soil solid causing a bottleneck in soil remediation. No matter which concept is used for remediation (removal or transformation), the lowering of the solid/water partition, i.e. the ratio of solid-bound to mobile/solubilized contaminant forms, is a step that increases the efficiency of the main conversion process. Mobilizing and solubilizing agents help contaminants with limited mobility/solubility to move from the solid to aqueous phase and enhance *in situ* physicochemical or biological soil treatment, which usually includes remediation based on soil washing, flushing, percolation or chemical- or biodegradation. Several mobilizing agents are available on the market, the most widespread ones being synthetic surfactants, but biosurfactants are also rapidly becoming more available. Synthetic surfactants and biosurfactants are not the only mobilizing and solubilizing agents used for strongly bound, persistent contaminants. Other agents include these: solvents, complex-forming molecules, micelle-forming agents, cyclodextrins, dendrimers, and other nanomaterials. A number of developments and ideas are being implemented in this field (also see Chapters 5 and 6).

4   *Biologically active amendments* such as nutrients, hormone-like stimulants, additives to adjust environmental conditions so that they are beneficial for soil biota:

   – *Additives to adjust pH and redox potential* that are beneficial for the required biota;
   – *Alternative electron acceptors* enhancing anoxic biological degradation;
   – *Supplemental nutrients*, typically N and P sources when the contaminant supplies only C, or an energy-providing cometabolite for cometabolic degradation or micronutrients required by the key species of the soil biota;
   – *Sorbents and soil structural amendments*, especially to ensure better surfaces and a better habitat for soil biota including microorganisms, plants, and soil-dwelling animals.

5   *Nanomaterials* is a multifunctional group of chemicals applicable to mobilizing, delivering, sorbing or otherwise stabilizing, oxidizing, reducing, etc. contaminants, additives or the soil's structural compartments. Thus, their introduction as a separate group is not justifiable based on their use in technologies. Rather their novelty, their common large surface-to-volume ratio and their similar production processes may account for their being grouped together.

The introduction of chemical substances into soil or groundwater has required proper dispersion technologies to be developed. Traditional agrotechnologies such as irrigation or spraying can mix liquid or dissolved additives into the upper soil layer while solid additives are applied using tilling, plowing, harrowing and digging. However, homogeneous application poses a challenge even in this relatively simple task. Deeper soil layers can be reached by injecting liquid chemical agents into the contaminated area or placing solid additives into the path of contaminant plumes in front of subsurface reactors, permeable reactive barriers or engineered reactive zones.

Chemical additives can fulfill their role by themselves (e.g. chemical oxidation), as part of a major technology (surfactant-enhanced biodegradation, or biochar-enhanced bioremediation), or as part of combined technologies such as a combination of chemical stabilization with phytoremediation of soils contaminated with toxic metals. These applications will be discussed later in the book in several chapters reviewing certain technologies (see Chapters 3, 5, and 6).

2.5 Technologies based on biological transformation and its intensification

A number of enhanced biotechnologies have been developed and implemented in the last 20 years. Some developments concentrate on aerobic biodegradation primarily in the vadose zone. Its optimal performance requires aeration and a suitable amount of moisture and supplemental nutrients (in addition to the contaminant) for microbial growth. If the contaminant is mobilized by enhanced desorption or solubilization, bioavailability is also enhanced, which further intensifies aerobic degradation.

Main desorption tools are increased soil airflow and heating, and solubilization can be enhanced by surface-active, solvent-type, or complexing agents. Aerobic degradation in the aquifer needs more engineered intervention to saturate the aquifer with oxygen or supply alternative oxygen sources (peroxides) or electron acceptors (e.g. nitrates) by injection. The addition of emulsifiers or other solubilizing agents is generally necessary to make sorbed contaminants water-soluble. Providing anoxic or anaerobic processes for decontamination in the aquifer may be more feasible compared to creating aerobic conditions. Those biochemical pathways and microbiological processes suitable for decontamination/degradation should be identified first, then the technological parameters should be established to create an optimal environment (temperature, pH, redox potential, nutrient supply, etc.) for those microorganisms that can catalyze the decontamination process. Alternative electron acceptors such as nitrate, $Fe(III)$, $Mn(IV)$, sulfate, or CO_2 are necessary for efficient degradation by microorganisms in the absence of atmospheric oxygen. The product of the electron-transport chain is nitrite or ammonia, $Fe(II)$, $Mn(III)$, sulfide, and acetate or methane accordingly. In cases where the soil does not contain microorganisms in a suitable and sufficiently high concentration for decontamination, the soil can be inoculated by separately grown living microorganisms in the form of a cell suspension injected into either the three- or two-phase soil.

Innovative technological solutions for bioremediation are based on one or more parameters and the additives necessary to reach maximal performance.

2.5.1 Bioventing

Bioventing is the technology that supplies the aerobic soil microorganisms living and working in the vadose zone with a sufficient amount of atmospheric oxygen necessary for maximum-scale aerobic biodegradation. Venting in the context of soil remediation requires the normal air exchange. The soil air in the pore volume should be exchanged by exhausting used air with high CO_2 and letting fresh, atmospheric air with high O_2 into the same space.

The most appropriate remediation in the three-phase soil for biodegradable pollutants of good or poor water solubility is a biodegradation-based technology implemented *in situ*. The soil's aerobic microbiota should be kept in an active state by ensuring optimum conditions for their growth and energy-producing/degrading metabolic activity. In addition to fresh air, this means the provision of suitable soil moisture content and supplementary nutrients – generally N and P, which occur in disproportionately low concentration in a polluted soil compared to the carbon source represented by contaminants.

The groundwater can be extracted parallel to venting in order (i) to treat it on the surface if contaminated or (ii) to lower in this way the water table and increase the thickness of the three-phase soil. Lowering the water table increases the soil volume being treated, including the capillary fringe that may significantly concentrate contaminants.

Figure 2.9 shows a solution that has been frequently applied by the author's research team since 1990 (Gruiz *et al.*, 1996; Gruiz & Kriston, 1995; Fenyvesi *et al.*, 2009; Molnár *et al.*, 2005; Molnár, 2007; Molnár *et al.*, 2007). The treated volume of the contaminated three-phase soil, i.e. the volume under the operations' impact, represents the *in situ* quasi-reactor: it is in fact the space between the two rows of passive air injection wells (C1, C2, C3 and C4, C5, C6) with the air extraction well row (B1, B2, B3) in the middle. In terms of reactor technology, the soil-solid and adherent water (i.e. soil moisture) of this volume represent a fixed bed through which the third phase, i.e. soil air/gas, flows. Due to the pressure

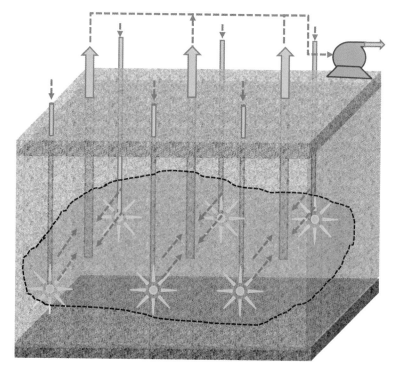

Figure 2.9 In situ bioventing to enhance biodegradation in the vadose soil zone: used air is exhausted by a mild vacuum, fresh air is introduced by passive wells.

difference between the soil-gas extraction wells (B) and the wells for fresh air inlet from the atmosphere (C) into the soil, fresh air starts to flow in and air with a higher oxygen and lower carbon dioxide content travels directly to the vicinity of the biofilms (in the microcapillaries), while, at the same time, the soil air exhaled by microorganisms is removed. This is the mechanism of ventilation. When volatile pollutants are present, the extracted gas must be treated *ex situ* (to prevent emissions into the air). This is useful from the soil's perspective because it decreases pollutant concentration in the soil, and gas (vapor) extraction and biological degradation are additional benefits.

The activity of aerobic microorganisms can be greatly enhanced by manipulating the unsaturated soil zone. However, airflow must not be allowed to desiccate the soil; if there is a real risk of this happening, moisture must be supplemented by irrigation or infiltration. In addition to an *in situ* treatment of the solid phase, groundwater can be treated either *ex situ* or *in situ.*

The system, which seems static at first sight, is greatly responsive to any impacts: airflow can generate concentration gradients in the soil, which has an impact on the air/water, air/solid and water/solid partitioning of the contaminants and other mobilizable soil components (including soil moisture). The process pairs involved are evaporation-precipitation, sorption-desorption, oxidation-reduction, and diffusion. If airflow increases or contaminant partitioning between soil phases changes, microbiological activity also changes immediately.

2.5.2   Intensification of bioremediation

Additives, pH and redox adjustment, heating, bioavailability-enhancing agents, physico-chemical agents that are suitable to help bioremediation with physical or chemical reactions, and, finally, microbial inoculation can increase the efficiency of bioremediation. The technological solutions enhancing bioremediation are discussed in detail in Chapter 5.

The interventions and additives described in Section 2 may represent the core process of a technology or be part of a complex technology involving several processes. This is why they are presented independently of the technology. Another specialty of the flexible use of the interventions and additives presented is the measure of application: the scale of heating, the concentration of the additive, etc. For example, surfactants in high concentration can be used for soil washing or in a biologically acceptable low concentration for bioavailability enhancement. Biologically acceptable quantities from any additive allow the microorganisms to survive, while larger quantities used for physicochemical transformation of the contaminants may irreversibly deteriorate soil microbiota. All these considerations must be implemented in the technology application.

3   COMPARISON OF *IN SITU* AND *EX SITU* REACTORS USED IN SOIL REMEDIATION

This comparison considers technological differences and risks arising from various structures, applications and locations of *ex situ* and *in situ* remediation processes. An analysis of the differences and causes explains that consequences and environmental risks arising from the technologies mainly depend on the pollutant's physicochemical characteristics modified by the design (*ex situ* or *in situ*) plus discharges and emissions. Risk estimation requires detailed information about emissions and reactor types: while *in situ* reactors are typically open, i.e. they have no walls or other closed/sealed boundaries (quasi-reactors), neither do *ex situ* soil treatments usually imply closed reactors. Risks posed by emissions from excavations and *ex situ* technologies are often higher than those from *in situ* applications. Impacts on the individual environmental compartments may be different, e.g. the atmosphere is less affected by *in situ* than *ex situ* technologies, while groundwater can often be better protected when the soil is excavated and treated *ex situ*. Thorough risk assessment and analysis are necessary to calculate the overall risks and compare *in situ* and *ex situ* implementations. In addition to environmental and human risks, economic and spatial planning also play a role in decision making (see also Gruiz *et al.*, 2014).

Let us consider, for example, a moderately volatile toxic pollutant. *Ex situ* biological treatment in layers or prisms creates a large free surface that is in contact with the atmosphere and through which the toxic substance can escape into the air. The very same soil treated by *in situ* bioremediation combined with gas extraction and *ex situ* gas treatment may be fully controlled and risk-free. Treatment in a fully closed reactor on the surface with gas emission control also poses some risk in terms of emissions (failure, leak, maintenance, repair), but the large emission caused by soil excavation and additional machinery use surpasses it many times over. Comparing excavation and reactor construction costs and the problems arising from the limited size of the reactor shows that the *in situ* solution is more advantageous from the perspective of both cost efficiency and environmental efficiency.

In addition to environmental and human risks and direct costs of remediation implementation, spatial planning, current and future land uses, accessibility, etc. should also be taken into consideration when choosing between *in situ* or *ex situ* soil remediation.

In the following section, the relationship with the environment, the internal characteristics, and the processes taking place within real and quasi-reactors will be discussed in detail.

3.1 The boundaries, openness, and interaction of the reactor with its surroundings

3.1.1 The boundaries of the reactor

Most *ex situ* reactors have physical boundaries or walls, and they can be fully or partly surrounded by these. The boundaries can be artificial (e.g. steel, concrete) or natural (a natural watertight layer, etc.). Built *in situ* reactive soil zones arranged inside the soil have no walls or other boundaries and are integral parts of the subsurface system, but their material (man-made filling) and permeability differ from that of the natural surroundings. Permeable reactive barriers (PRBs) are also inserted into the subsurface soil system, but they are only in direct contact with the groundwater phase through their water-permeable walls, which do not allow contact between the solid phases of the soil and the PRB.

In situ quasi-reactors rarely have built boundaries, but in some cases a volume treated *in situ* over the long term is surrounded by an underground barrier just for emission control. It is also possible that the quasi-reactor is surrounded by natural isolating layers, e.g. *in situ* soil treatment in an area between a lower and an upper watertight clay layer. In cases like this, water and air transport occur only horizontally. In the most common case, there is neither an artificial interface nor an isolating layer around the *in situ* treated soil, rather the maximum range of operations and scope of impacts (the physical distance over which the processes exert their effects) and the relevant natural processes determine the boundaries of the quasi-reactor.

Figures 2.10 and 2.11 show the boundaries in a quasi-reactor and in a reactor vessel in homogeneous and in heterogeneous matrices.

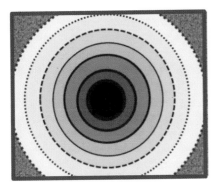
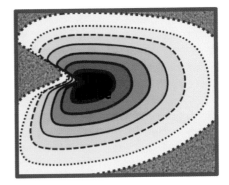

Figure 2.10 Soil treatment in a reactor vessel: the boundaries of the vessel and soil resistance limit the impact range – homogeneous and heterogeneous matrix.

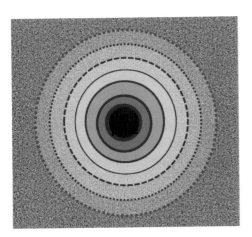
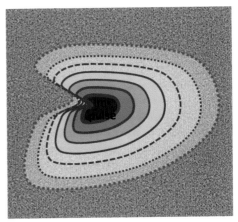

Figure 2.11 Soil treatment in a quasi-reactor, i.e. directly in soil: the boundaries are equal with impact range limited only by soil heterogeneity.

3.1.2 The openness of the reactor

Some of the reactors used for *ex situ* treatment of the excavated soil are closed (e.g. steel tank reactors, isolated soil pools with a surface cover, concrete pools with a surface cover, etc.). However, some of them are partly open (e.g. isolated soil ponds or concrete pools with an open surface, or prisms or thin layers of soil isolated from the working area surface only by a bottom liner) and some are fully open: for example, prisms or soil layers that are not fully isolated. Analyzing the *in situ* quasi-reactors, the same combinations can be found, but, logically, the physically open reactors are more common and the contacting environmental phases are different: an *ex situ* open reactor has a dominant contact with the atmosphere. Their bottom surface may meet the soil below when isolation is lacking, and the leachates can get into the soil and interact with soil waters when they are not collected. The interaction of *ex situ* reactors covers mainly emission and some interaction with atmospheric air (ventilation, gas exhaustion) or water (rainwater or wash-water).

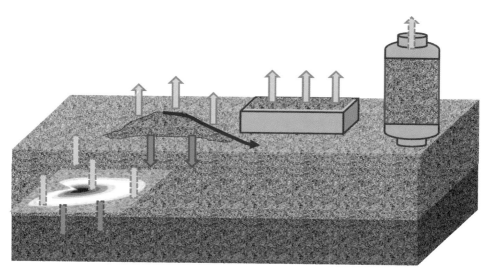

Figure 2.12 Potential emission to air (yellow) and water (blue) from *in situ*, heap, open tank, and closed reactors.

In situ quasi-reactors have extensive contact with soil solid and groundwater: this availability to contact is responsible both for remediation (e.g. while groundwater flows through) and emission (contaminant diffusion into soil air and surrounding soil solid). Their openness (emission) can be limited by natural isolation (i.e. a watertight layer) or by operations such as the depression of the water table in order to establish a hydraulic barrier or by built structures (subsurface barriers).

The built underground reactors interact with their surroundings on a limited scale, and the interactions are mostly controlled (see Figure 2.12).

3.1.3   Contacted and endangered environmental phases

The design and the artificial or natural boundaries of the reactor (or quasi-reactor) limit or enable the interaction of the polluted (treated) soil with certain environmental compartments and phases, and thus limit or enable also material discharge into the environment. Most of the *ex situ* soil reactors are open, and contaminated soil can easily get into contact with the external atmosphere. During the treatment of the soil in a thin layer or in a prism, the polluted soil is separated from the local soil and, with the incorporation of a proper collector system, also from the groundwater. Soil treatment plants (e.g. soil washing plants) may ensure the control of water and solid, but air, volatiles and small-size particulate matter are typically less controlled or not at all. There is no direct contact with the external atmosphere in the *in situ* treated subsurface soils. The treated three-phase soil does not directly get into contact with the groundwater, and contaminant transport can be easily prevented by lowering the water table. The contaminated saturated soil is in direct contact with other subsurface waters. However, the transport of the contaminated water can be restricted by the application of a hydraulic barrier or a PRB.

3.1.4 The treated volume

In real reactors (above or below ground) that are completely closed or those that have boundaries but are open to the atmosphere, the treated volume can be the gas phase, liquid phase, solid phase or all three, as well as a slurry within the physical boundaries. The treated volume is the useful volume of the reactor. The treated volume of an *in situ* quasi-reactor, which is fully open, is determined as the volume under the impact of the technological operations. Safety is paramount during the planning process, and the operations and processes must reach the whole volume that poses an unacceptable risk. Since the risk posed by a pollutant depends on its mobility and transport, a transport model should be used to determine the volume to be treated and its underground position and extension as well as the changes over time.

3.1.5 Emission transport and pathway

The pathways of emission from open, semi-open, and closed reactors are discussed according to the soil phases involved.

Contaminated soil gases and vapors are usually released into the atmosphere or the surrounding soil air through diffusion and sometimes by convection. Artificial venting influences the flux of the gas phase: compared to diffusion it can promote more effective transport processes, and – provided that ventilation is applied – the vented soil air, gases, and vapors can be fully controlled. Other possible ways to control soil gas transport are covering the surface of the soil or working under a tent that provides a controlled air space.

Contaminated groundwater is carried by convection, in accordance with the flow conditions of the groundwater, and it transports the dissolved pollutants. The pollutants may enter the clean groundwater, solid soil, or soil air from the contaminated soil moisture and pore water, via precipitation-infiltration, mixing, diffusion, or partitioning. The most dangerous case is when the contaminated pore water (originating from the surface or from underground sources) moves downwards and enters the groundwater or the deeper subsurface waters.

The pollutants may form a *separate liquid phase* in the soil, called a non-aqueous-phase liquid (NAPL) layer, a separate phase, which can be either volatile or non-volatile, less dense or denser than water. The transport processes and mobility of these liquids are varied – the most typical are those hydrocarbons that float on the surface of the groundwater and only mix with the water to a limited extent, but they move semi-independently by more and more extensive spreading on the surface of the water in the capillary fringe due to water table level fluctuation and by partial sorption onto the adjacent solid phase. The supernatant layer is smeared onto the soil solid and the thickness of this smear varies according to the fluctuation in the level. This leads to the contamination of large amounts of unsaturated soil. As a result, increased biological activity may lead to intensive biosurfactant production, which encourages the pollutants to enter the groundwater. High-density non-aqueous-phase liquids (DNAPL) act differently to the supernatant liquids; they are located at the bottom of the water table in a separate phase layer or lens, directly above the watertight layer. They remain hidden and may serve as a virtually unlimited supply of multiple risks. The watertight layer does not necessarily mean a barrier for certain DNAPLs such as some chlorinated solvents that can cut tunnel-like passageways/gaps in the watertight clay layer, through which they can easily enter the subsurface waters of greater depths.

Contaminated three-phase soil can easily contaminate soil gas and groundwater by partition, diffusion, and convection driven by gravitation or facilitated by infiltrated rainwater. The same transport processes occur during *in situ* soil treatment, so depending on the extent of risk (toxicity of the contaminant and sensitivity of the groundwater, etc.), it must be decided whether to apply *in situ* treatment or the removal of the contaminated soil volume.

Emission and transport from the aquifer require a different consideration as the contaminant is typically transported in a plume from a distant source. Additional transport can move the plume toward sensitive water resources. To protect them, the flow direction of the plume should be changed (by hydraulic barriers) or the plume decontaminated by permeable reactive barriers or by reactors placed into recirculation wells.

3.2 Inside the reactor

3.2.1 Soil phases to be treated

The three main phases of the soil – gas, liquid, and solid – can be treated together or separately, and a wide range of *in situ* and *ex situ* solutions and their combinations are available.

Soil gases and vapors can be treated both *ex situ* and *in situ*. *Ex situ* treatment is worthwhile when most of the (volatile) contaminants occur in the soil gas phase and the pollutant can be removed from the whole soil by soil-gas extraction. Also, sorbed volatile contaminants can be treated by mobilizing and removing soil gases from the soil by means of ventilators or vacuum pumps, which is fairly straightforward. If it is not possible to alter the partition of the volatile substance between the soil solid and gas phases, and it cannot be shifted artificially (e.g. by heating) towards the gas (vapor) phase by enhanced desorption, it is advisable to treat the soil as a whole and not to separate the gas phase. Soil gases can be removed from the liquid phase by stripping, which can be applied both *in situ* without water extraction and *ex situ* after water extraction.

Groundwater can be treated independently of or together with the solid phase. Water and solid phases can even be treated partly together and partly separately. The decision depends on the partition of the contaminant between the phases. If it is largely dissolved or emulsified in the groundwater or it can be transferred into the groundwater by solubilization, emulsification, or other manipulation, it is reasonable to treat the groundwater separately from the solid phase, thus exploiting its easy mobility and extractability from saturated soil. Thus, the partition of the pollutant between the phases is crucial in decision making. The partition between the soil phases ($K_p = C_{soil}/C_{water}$) calculated from the K_{ow} (the octanol–water partition coefficient) is not always authentic since the modified forms of the pollutant and its interactions with the soil matrix and soil microbiota (i.e. spontaneous biosurfactant production) can result in a K_p that differs from the calculated version by several orders of magnitude.

Liquid pollutants that form a *separate phase* should be treated separately. The supernatant can be slurped or skimmed off from the surface of the groundwater. It can also be separated from the groundwater *ex situ* after it is extracted to the surface together or partly together with the groundwater. Phase separation can be facilitated by additives and implemented both *ex situ* and *in situ*, e.g. in wells. DNAPL can rarely be extracted as a separate phase, even if the volume is large and the location is well known, but they are typically dissolved in or mixed with water. DNAPL–water ratio is crucial: if only the dissolved amount can be removed with water, the concentrated DNAPL piled up in permeable sand or gravel lenses will provide a long-term supply of groundwater contamination. Dissolution can be increased by *in situ* enhanced contaminant mobility via solubilization using surfactants, cosolvents or other mobilizing agents such as cyclodextrins. The other way of intensifying removal is the maximal approximation of the location of the lens and increasing DNAPL-water ratio by mixing.

The treatment of **silts and slurries** (dredged sediments) is rarely performed *in situ*, since their excavation (dredging) is mostly justified by river or harbor management. Transportation of the slurry is also feasible, so therefore their treatment in a well-controlled soil/sediment-treatment plant is more efficient than *in situ* treatment. Such complex soil/sediment treatment plants are also called soil-washing plants because of the watery slurry form of the soil/sediment, although they perform generally much more than just washing the soil. They apply grain-size fractionation prior to the remedial treatment that consists of physical, chemical, thermal, and biological steps or an arbitrary combination thereof. Grain-size- and density-based classification (cycloning, flotation, sedimentation) is particularly beneficial because it can provide clean, utilizable course fractions (gravel and sand) as early as in the first stage, leaving a relatively small amount (10–20%) that needs to be further treated. The treatments yield utilizable fine fractions (clay for ceramics) and only a minimal amount of hazardous waste.

Under special conditions, for example *in situ* containment by stabilization or block formation, slurry-phase treatment can be carried out *in situ* by adding a stabilizing agent, e.g. cement, lime or fly ash, etc. Excavation and transportation of soil can be avoided when using this method and, as a result, the hazards from pollutants can be reduced, the physical and static stability of the treated soil volume can be increased, and erosion prior to construction can be prevented.

"Solid soil" usually means **three-phase soil**, the treatment of which can be carried out either *ex situ* or *in situ*. *In situ* remediation methods often combine the *in situ* treatment of the solid phase with the *ex situ* treatment of soil air and groundwater, but "*ex situ*" almost always means treatment of the three phases together.

Two-phase soils are rarely excavated. Instead, following the excavation of three-phase soil (or a part of the saturated soil after lowering the groundwater level), the groundwater is most often treated *in situ* separately in ponds, or in water-treatment reactors *ex situ*.

3.2.2 Homogeneity and heterogeneity

Soil air and groundwater are usually homogeneous after extraction. Slurry-phase soil is normally homogenized and treated in reactors by mixing. Mixed-type tank reactors can achieve high-grade homogeneity.

In situ soil treatment is always carried out in a heterogeneous medium. Gas and liquid phases can be homogenized by continuous recirculation, but not the solid phase – since it will maintain its original heterogeneous structure. Homogeneity in soil cannot be achieved, nor is it a goal, since heterogeneity, concentration gradients, and biological gradients are often key drivers of the technology.

Technologists have to consider soil and pollutant heterogeneity. The heterogeneous distribution and composition of the pollutants are due to the source of pollution, transport processes or the physicochemical changes and partitions occurring in the contaminated area over a long period of time. Pollutants can move in the soil due to gravity, diffusion and capillary forces. They may be partitioned between the physical soil phases: they can be sorbed from the groundwater to the solid phase and desorbed from the solid phase and enter the gas (vapor) or liquid phase. Over time, changes in chemical and biological processes can influence propagation, movement and phase changes of the pollutants. The soil solid phase filters some of the water-dissolved substances. Leaching into the water from the solid also can occur, depending on the dynamic equilibrium and natural soil conditions. Volatile substances accumulate in the gas phase, water-soluble substances gather in the water phase, and sorbable substances accumulate in the solid phase. However, this can lead to major and unpredictable heterogeneity in pollution, depending on conditions, type of soil, and distribution of contaminants in soil phases. Pollutant sources and transport have an important role in creating contaminant heterogeneities. In addition to primary pollutant sources, secondary sources may emerge within the soil due to invisible accumulations. These secondary accumulations and their effects cause not only heterogeneities but also upset the transport models. Further types of heterogeneity occur if the pollutants have multiple components. Depending on their ability to partition between the phases, the pollutants are introduced into the soil phases and become further fractionated during their movement. The example of petroleum derivatives, which can have several hundred components, shows that the components may act differently in the soil, and have different interactions with the soil phases and the microbiota. As a result, selection occurs within the components: some become diluted over time and over greater volumes, while others will accumulate.

Heterogeneity, including microbiological heterogeneity, can play a particularly significant role in biological soil treatment. The heterogeneity in the soil's biological system can have three sources:

- Hydrogeological and geological heterogeneity: three-phase (unsaturated) or two-phase (saturated) soil; water table; the type and layers of the soil; pollutant-binding capacity; air and water content; organic matter (humus) content and clay content of these layers, etc. These heterogeneities play a key role and must be considered by technologists, especially during *in situ* soil treatment.
- Environmental-parameter-related soil heterogeneities usually appear in the form of gradients, for example, redox potential and humus content decreasing as functions of depth. The transition between the external temperature and soil temperature at a certain depth can create a positive or negative gradient depending on the season. Humidity gradients will depend on the extent of irrigation or precipitation. Different gradients can emerge if water comes from the surface or if capillary water is absorbed from the groundwater. These gradients may occur in either (non-mixed) reactors containing the solid phase or in *in situ* quasi-reactors.
- Heterogeneity arising from the partition of the contaminants: the concentration and toxicity of the pollutants are greater in the source and decrease with increasing distance from the source. They are greater along the transport routes and smaller concentrations prevail further away from the routes due to dilution during transport. Rarely, accumulation can happen during transport. Where the utilizable substrate can be found in greater amounts, the number of soil microorganisms will be greater; the amount of carbon dioxide produced will increase; the level of oxygen will decrease; metabolites and end products will appear; and toxicity will usually decrease but may also increase (intermediate or end products of toxic degradation). These heterogeneities usually characterize solid-phase-containing non-mixed reactors. In extreme cases, the heterogeneity of multi-component pollutants can lead to the accumulation of very toxic, persistent

residues causing further disposal problems. Typical gradients can emerge in both *in situ* and *ex situ* treatment prisms depending on the environmental parameters and due to pollutant transport within the soil. These gradients and heterogeneities can either be utilized or eliminated using technological processes. Heating and ventilation can keep the temperature of a static prism or an *in situ* soil volume constant, and ventilation can also influence redox conditions. If the goal is to replace aerobic microorganisms with facultative anaerobic ones, a lower optimum redox potential should be set by reducing the frequency of ventilation in order to provide ample time for the decreased redox potential to establish and for the facultative anaerobic microorganisms to work. The development of such gradients can be encouraged or inhibited, depending on the requirements of the processes responsible for soil remediation.

3.2.3   Reactor types classified according to the concentration gradient

Mixing in homogeneous batch reactors used for groundwater or soil slurry treatment can prevent spatial concentration gradients from being developed. In flow-through reactors, near-linear concentration profiles develop. Flow-through reactors can be filled by packing and operate as fixed or fluidized bed reactors. Two-phase soil can be considered to be a special type of fixed-bed flow-through reactor filled with soil solid. Groundwater is the liquid phase flowing through the pores of the granular packing. Initial heterogeneity of the solid phase and gradients due to time-dependent pollutant transport are superposed in such a continuously working *in situ* quasi-reactor.

An *ex situ*, e.g. tank reactor for water treatment, can work both in batch or flow-through mode. A batch reactor has neither inflow nor outflow during the treatment, but a simple tank reactor can work with or without recirculating the treated water or slurry or by recirculating the mobile soil phase only when the reactor is packed. A spatial or time-related pollutant concentration gradient can develop in continuous reactors built for groundwater or slurry treatment. In a tube reactor, the pollutant exhibits an almost linear concentration gradient due to the core transformation process.

The treatment of contaminated soil and groundwater combined with air and/or water recirculation in a soil volume is equivalent to a continuously operated filled reactor. The spatial pollutant concentration in the circulating air and water is nearly constant and decreases as a function of time. At the same time, the solid phase shows a heterogeneous distribution in space and time. Heterogeneity, as already mentioned, can be caused by hydrogeological and external conditions and by pollutant partition. Heterogeneity in *ex situ* reactors is mainly caused by environmental conditions (precipitation, evaporation, temperature, wind, etc.) or technological parameters (irrigation, injection, venting, etc.) while *in situ* technologies are mainly affected by hydrogeological conditions and by the fate and transport of the pollutant.

Remediation can be based on heterogeneities or gradients within quasi-filled groundwater treatment reactors if a redox gradient develops in a biodegradable contaminant plume. The higher the concentration, the more intensive the biodegradation, so therefore microorganisms may completely consume atmospheric and alternative oxygen sources, which can result in a negative redox potential in the source zone. Redox potential will decrease in lesser scale alongside the plume because the contaminant concentration is lower and biodegrading microorganisms need fewer electron acceptors. Natural or artificial heterogeneities or gradients may result in a quasi-technological sequence of coupled reactor volumes with different technological conditions. Such a type of quasi-reactor sequences with decreasing redox potential can be used for the biodegradation of contaminant mixtures, microbiologically moderated chemical stabilization (immobilization) of toxic metals or for dehalogenation of organic contaminants.

3.2.4 Reactor design: single-stage, multi-stage, and cascade types

Ex situ batch and continuous homogeneous reactors (water and slurry treatment) and *in situ* water treatment may involve multiple stages. The stages of *ex situ* reactors can apply to different technological processes and operations. With regard to *in situ* treatment, this means different conditions within the solid volumes where the groundwater flows through: redox potential (aerobic, anoxic, anaerobic), nutrient supply, temperature, residence time, etc.

3.3 Processes and parameters within the reactor

3.3.1 Aeration

Aerobic biodegradation or other chemical and biological oxidizing processes require oxygen, and this can be derived from the air or from chemical compounds. Aeration methods, such as ventilation, injection of air and sometimes oxygen or ozone can be applied to introduce atmospheric air into the soil. Here too, a distinction must be made between the aeration of the water, slurry, three-phase or two-phase soils, and the aeration of *ex situ* or *in situ* reactors.

When water or slurry is aerated, the aim is to increase the concentration of oxygen dissolved in the water. The efficiency is low because the contact time of air bubbles is short compared to the time required for dissolution. The aeration of saturated soil by injecting air or by recirculating dissolved oxygen-saturated water is an expensive method. Instead, introducing hydrogen peroxide or other peroxide compounds or oxygen-releasing chemical compounds (ORCs) into the groundwater can be used as alternative solutions.

In the three-phase soil, the pressure difference caused by the effect of ventilation generates a current within the pore volume, which provides fresh air for the respiration and biodegradation of the soil microbes. In addition to introducing oxygen, gaseous metabolic products are also removed. Both *ex situ* and *in situ* aerations are performed through perforated wells/tubes or injectors introduced into the three-phase soil for exhaustion or injection, and the pressure difference needed to move the air is generated by a ventilator or vacuum pump or by a compressor.

3.3.2 Redox potential setting

An aerobic, anoxic or anaerobic environment can be created in real or quasi-reactors. Soil exhibits an inherent natural gradient from a high redox potential near the surface through a low, but still positive, redox potential in the upper layers of the saturated soil zone to a negative redox potential in the deeper saturated soil layer.

The technology applied can control both aerobic and anaerobic conditions inside closed *ex situ* reactors and in deeper soil layers. Aerobic technologies are preferred in open reactors or in the upper layer of the *in situ* treated soil, while anaerobic technologies are preferable in the naturally anaerobic deeper soil layers. Depending on the character of the pollutant, it may be necessary to

- Change an initially anaerobic soil to an aerobic one or vice versa;
- Exclude the air from an initially aerobic soil;
- Apply various alternative electron acceptors to adjust a certain redox potential and initiate or enhance the required biological degradation, e.g. petroleum hydrocarbons are most degradable under aerobic conditions while some chlorinated organic substances are most easily broken down under strictly anaerobic conditions;
- Initiate transformation processes, e.g. acidic bioleaching of metals from sulfidic ores under aerobic conditions or the stabilization of mercury or other toxic metals in sulfidic form under anaerobic conditions.

Redox potential can be controlled by aeration or the exclusion of air, by chemical additives or indirectly through microbiological activities. The reduction of the soil's redox potential to lower or negative values can be based on the activity of those microbes that can utilize (consume for energy production) alternative electron donors, e.g. an easy-to-degrade sugar, starch or hydrocarbon without supplying any atmospheric molecular oxygen. After having used (oxidized) these carbon sources, the redox potential drops. To remove nitrates from the soil, for example, the nitrate-reducing (nitrate-respiring) soil microbes can be stimulated by providing high-level sugar, hydrocarbons, BTEX or other suitable substrates. If the redox potential has to be reduced further, the same can be done to the sulfates by supplementing sulfate-reducing bacteria together with a biodegradable substrate. After reducing all available sulfates, the redox environment will become advantageous for microbiological chlorine respiration and methanogenesis.

It must be noted that monitoring redox potential in the soil has special problems: if we try to directly measure redox potential or to analyze redox indicators from a groundwater sample, it must be realized that the water sample taken from a well is a mixture representing the whole vertical length and not the spatial redox distribution. *In situ* measurement of redox potential (by soil-specific redox sensors, e.g. Vorenhout *et al.*, 2004) is required if redox potential adjustment is part of the technology and monitoring is necessary.

3.3.3 *Moisture content*

The moisture content of the soil is an important technological parameter. Certain soil treatments can only be effective at a specific moisture content: it is reasonable to use high-temperature thermal treatment in a dry soil and chemical reactions in homogeneous slurries. The biological processes are strongly dependent on the moisture content. Water demand of microorganisms and plants can differ according to species, and in general the water activity cannot be decreased below the lower limit of biological availability (the ability of microorganisms or plant roots to absorb and use the water in the soil). Biological availability of the pollutant to be degraded/transformed also depends on the moisture content since many processes enhancing availability include hydrolysis/enzymatic hydrolysis. The moisture content of the three-phase soil can be adjusted using conventional watering and trickle watering from the surface or by injecting water (through an injector or a perforated pipe system) both *in situ* and *ex situ*. There are additional possibilities *in situ* that can provide the capillary water by maintaining a constant high groundwater level or temporarily flooding/flushing the unsaturated soil. Another concept is the use of natural or man-made polymeric additives (swelling gels) that can keep moisture in their molecular structure for a long time and release it later than natural soil components and thus supplying moisture for soil biota. Maintaining a constant and homogeneous moisture content is an important technological task in soils since bioremediation efficiency may be limited by moisture content. Homogeneous moistening of the soil *in situ* or in static heaps (prisms) is strongly limited by other gradients and requires special care. In the practice of *ex situ* treatment, the three-phase soil is overwetted and the excess moisture is carefully disposed of, e.g. the leachate is collected. *In situ*, the soil is temporarily flooded, and the excess water enters the groundwater. In such cases, the resultant risk should be assessed and managed (e.g. by hydraulic barriers and extraction). Ensuring moisture by capillary forces from an elevated groundwater level is a good solution from an environmental risk perspective but can help only a certain layer above the water table (the capillary fringe).

3.3.4 *Additives*

Adding and distributing dissolved, solubilized, emulsified, or suspended substances homogeneously into soil is an even more complicated task than just moistening it. It is feasible in water or in a slurry phase, but it is a major issue in solid phases both *ex situ* and *in situ*. In the case of liquid phase addition, first, soil permeability, water/solute conductivity and the soil's filtering effect must be taken into account. The filtering effect is based on a mixture of processes such as ion exchange, ad- and absorption, desorption as well as partitioning between phases (evaporation-precipitation, dissolution-precipitation) resulting in the retention of the additive substance in soil solid. The transport and flow of the injected liquid substance may be limited by its local sorption and its intrusion into the capillaries blocking the transport/distribution and damaging the soil. The distribution of soil heterogeneities and the concentration gradient must also be considered. Water-soluble soil additives can be introduced by flooding the soil volume to be treated. The solution of the additive fills the pore spaces and a fairly homogeneous partitioning can be achieved. If the water-soluble material is sprayed onto the soil surface and infiltrates through the surface or is introduced through a deeper-placed drain or well system or is injected, a filtering effect takes place. This effect means that the solute substance will partition between the solid phase of the soil and soil water. Organic and inorganic additives – unless they are markedly polar – usually attach to the solid phase in a 10^2–10^6 times higher concentration compared to water. This means that an additive dissolved in water and diffusing from the injection source will quickly be adsorbed onto a solid surface and a steep concentration gradient will develop. The partitioning can be shifted towards the water phase by solidifying agents such as solvents of surfactants.

If the additive is not water-soluble but is in a suspension or emulsion, the situation is more difficult and depends on polarity and stability of the suspension/emulsion. Solid, non-soluble substances can only be introduced homogeneously into homogeneous groundwater or slurry reactors. If the reactors/quasi-reactors are filled with solid soil (both *ex situ* or *in situ* types), the suspended additive can be dissolved consistently using high-intensity injection or the three-phase soil can be flooded with water containing the additive. This time-consuming sorption and partition do not proceed instantly during sudden flooding. Additives, solubilizing agents, tensides, or special nanoparticles may aid this process to a certain extent.

A solid additive may be introduced into the soil in solid form by mixing. This can be done in relatively thin soil layers using agrotechnologies. Wetting agents may be necessary for hydrophobic agents. Only the suspension form of solid additives can be introduced into deeper soil layers.

3.3.5   Material output from the soil

Removing or recycling mobile soil phases is easy and similar in the *ex situ* and *in situ* technologies, but the solid phase can only be removed in *ex situ* reactors.

Soil air can be exhausted by a ventilator or a vacuum pump and collected through a vertical, horizontal, or inclined perforated pipe system. *In situ*, air extraction wells are applied, which – in terms of installation – are similar to water extraction wells, but only the disposition of the screen differs. Vented air can be treated in a controlled way, depending on its pollutant content.

Soil moisture can be collected by a drainage system; a perforated, vertically or horizontally installed pipe; or a ditch system from where water is pumped (*in situ*) in accordance with depth and accessibility, or is transported by a pipeline, e.g. gravitationally (*ex situ*) to the treatment site. When *ex situ* soil treatment is applied, seepage water (leachate) is collected in a ditch system, in an area above the isolating layer.

Groundwater is extracted by extraction wells from the saturated soil zone: the wells or well curtains can be free-surface wells, depression wells or vacuum wells equipped with submersible or surface-installed pumps.

Removal of mobile or mobilizable pollutants can take place along with soil gas (volatilizable) or groundwater (water-soluble) extraction. Vapor and gas are extracted through collection pipe systems by ventilators and pumps. If the pollutant is explosive, special explosion-proof equipment must be used. If the pollutant forms a separate phase, it can be extracted from the top or bottom of the groundwater, together with or separated from groundwater using submersible pumps, "scavenger" pumps or pumps connected to slurping tubes. Submersible pumps will pick up the layer according to its location, and when the pollutant phase forms a very thin layer, it will pump a groundwater-pollutant mixture to the surface. This mixture will have to go to phase separation. The scavenger pump is equipped with a hydrophobic filter, which can separate the phases and extract the organic phase only, even if the layer is only 1 mm thick. The supernatant layer can be slurped by a weak vacuum or skimmed from the free surface of wells, ditches or lakes mechanically or manually or by using sorbents.

3.3.6   Recirculation

Soil air/soil gas recirculation: there are four main options available for a three-phase soil:

– Recirculation of a part or the whole of the soil gas without treatment;
– Treatment of the soil gas after extraction, followed by the recirculation of the treated gas;
– The extraction of soil gas and introduction of the fresh (atmospheric) air;
– Exhaustion of soil air and introduction of the additive-enriched air.

Groundwater recirculation: the possible solutions in a two-phase soil are as follows:

– Recirculation of a part or the whole of the groundwater;
– Treatment of the groundwater after extraction, followed by the recirculation of the treated water;

- Extraction of the groundwater and introduction of additive-enriched water into the groundwater;
- Extraction of the groundwater and introduction of additive-enriched water into the unsaturated soil by flooding or temporarily increasing the water level.

In practice, when the slurry phase is treated,

- The whole slurry is recirculated, or
- Water is recirculated only after phase separation (sedimentation, filtering, cycloning, etc.).

3.4 Microorganisms in the reactor

3.4.1 Soil microbiota and its modification

Some 99% of the soil's microbiota lives in biofilms tightly attached to the surface of the solid phase of micro-pores and micro-capillaries. Only a very small percentage of the microorganisms can be found in the groundwater, they either initially lived in the water or originated from detached cells of the biofilm. The biofilm is a mucosal layer that consists of extracellular polymeric substances produced by the microorganisms and serves as a glue to attach microbes to the particle surface and to retain water, solubilize and fix/retain nutrients and other substances in this special habitat of soil microbiota. Biofilms generally represent a highly organized biological system where microorganisms form structured, coordinated, and functional communities. Similar mucous material covers plant root surfaces and the intracellular space and capillary vessels of higher organisms. Mainly aerobic and facultative anaerobic bacteria, fungi, single-cell plants, and animals live in soil pores and capillaries of the three-phase soil. In the two-phase soil, depending on the redox conditions, mainly facultative anaerobic or obligate anaerobic bacteria are present, in biofilms or in lesser extent in free forms.

If the soil conditions change, microbiota will also change, immediately responding to actual conditions. The changes can be followed by species diversity and density as well as community structure and function, more recently via metagenome analysis (Gruiz, 2015, 2016a). Any changes in soil conditions such as the result of natural processes (dilution, acidification, soil compaction, flood, etc.) or technological operations (aeration, irrigation, air and water extraction, pH or redox potential change, homogenization, injecting additives, pollution, etc.) will affect soil microbiota (Gruiz, 2009, Chapters 3 and 5 in this volume).

Bioremediation applies controlled changes that aim at efficiently rendering the pollutants harmless. If the technology is based on the activity of aerobic microorganisms, aeration of the three-phase soil should be chosen, rather than slurry-phase soil treatment where the redox conditions favor the facultative anaerobic microorganisms, and it takes immense efforts to aerate the slurry to keep the aerobic bacteria active and alive. The same must be considered when the groundwater level is increased or reduced during *in situ* remediation: an aerobic community should never be underwater for a long period of time, and vice versa: anaerobes should not be aerated if their initial activity should be maintained.

A balanced natural microbiota is a community with a complex composition and a complicated web of mutual collaborative relationships. If the community has already adapted to the existing environmental conditions and pollutants – and can utilize every available organic non-contaminating and contaminating substance as a substrate – this ability should be used as the basic condition of remediation. When technologists ensure optimal conditions for the activity of existing microorganisms, the method is known as enhanced natural biodegradation. The difficulty is caused by the fact that every contaminated soil has its own path of evolution, and as the remediation proceeds, the optimum requirements may change according to the changing diversity, especially in the case of mixed contaminants. This is why it is important to continuously monitor those indicators that indicate remediation progress and the needs of the microbiota responsible for the core catalysis.

The conditions can be more flexibly changed in *ex situ* technologies to suit the microbiota than in *in situ* ones. Moreover, the microbiota itself can be modified. If the biodegradation of the contaminants requires alternating aerobic and anaerobic conditions, the technological parameters can be changed by saturating the soil with water and supporting the facultative anaerobic microbes by additives, and then the water can be removed so that the aerated volume ensures advantageous conditions for the aerobic ones. Some persistent chlorinated pesticides can be biodegraded by technologies applying strongly controlled alternating redox potentials.

The microbiota can be manipulated by mixing artificially cultivated microorganisms with the native microorganisms. This procedure can have two objectives:

- Increasing population density by returning the same microorganisms after an isolated cultivation;
- Improving the quality by introducing other microorganism(s) that are different from the native ones.

Introducing a microorganism into the solid soil faces the same difficulties as the non-water-soluble solid substances: the filtering effect of the soil does not let the cells spread far from the point of introduction. However, the microorganisms can distribute themselves over the long term if the treated soil provides a suitable habitat.

Foreign microorganisms may favorably supplement the indigenous soil community, but competition may easily develop between native and foreign microorganisms. If the native organisms prevail, then the inoculation was a wasted effort; but if the foreign microorganisms win with a "knockout," the destruction of native ones may become a disadvantage over the long term. A good example of this situation is the addition of fast-action aggressive hydrocarbon utilizers into soils polluted with a mixture of petroleum hydrocarbons. Isolated and artificially cultured species that are accustomed to a laboratory culture medium (usually rich culture media ensuring fast growth) or a growth medium containing only one or a few components of the targeted hydrocarbon mixture, will feed on easily utilizable nutrients and ready-to-degrade pollutants on re-entering the soil and leave only hard-to-degrade, unbalanced residues for the native microorganisms. Thus, they disturb the well-balanced nutritional community (of commensal microorganisms) where the species share everything and consume all the "leftovers." The consequence is that the easy-to-degrade pollutants will degrade faster, but a large amount of residue will remain, and this residue is more difficult to degrade than if the inoculum would not have been applied (Gruiz, 2009, Chapter 3 in this volume). Broad-spectrum bacterial mixtures or the enrichment of the natural microbiota (when it works slowly but otherwise properly) can provide an appropriate solution. Isolated petroleum degrading enzymes of microbial or higher animal origin can also be utilized in some cases as an eco-friendly soft technology.

Foreign microorganisms often fail to survive in the soil, but their genes can endure even after their death, and horizontal gene transfer may transplant them into native microorganisms. However, these processes cannot yet be used in a controlled manner for enhancing the efficacy of bioremediation.

3.4.2 The evolution of the soil

Living soil that is treated *ex situ* or *in situ*, will evolve in its own unique path. The initial stages of evolution are the ones affected by the presence of contaminants. Pollutant-tolerant and pollution-utilizing species gain an advantage, while the more sensitive species decline in numbers or disappear altogether. The privileged species affect the conditions of the whole community but – unless the contaminant is very toxic – those microbial communities prevail that adapt to the new conditions and learn to utilize the contaminant as a new substrate.

The second stage that influences the soil's own evolution is a technological intervention. This intervention changes soil conditions, modifies the distribution of the species and initiates adaptive mechanisms such as mutations, subsequent selections and horizontal gene transfer. The effect of an intervention into the soil ecosystem is generally stronger in *ex situ* technologies, given that the technological parameters may largely differ from biologically acceptable conditions. *Ex situ* treatments do not typically conserve the soil microecosystem because the variety of geotechnological reuses is large. It is most important to conserve the initial biological state when the original ecosystem and the (agricultural or natural) land use is supposed to be maintained or the land use becomes more sensitive (e.g. transition from industrial use to residential use).

3.4.3 Contaminant mobility enhancement

Remediation based on mobilization is often limited by a reduced mobility of the contaminant, both *ex situ* and *in situ*. Physicochemical mobility such as volatility, water solubility, and partition towards more mobile phases can be enhanced as part of the technology by suitably adjusting technological parameters such as temperature, pressure, flow conditions, additives etc. If contaminant mobility cannot be increased in order to establish a feasible and efficient mobilizing technology, the concept has to be changed and a remediation method based on contaminant immobilization or stabilization must be implemented.

Biodegradation and bioavailability have a special role in remediation because biodegradation itself may be an efficient process, as it results in harmless products such as CO_2 or ammonia, the same as during natural organic material degradation. Biological processes basically occur in the aqueous phase biofilms, which develop on micro surfaces that are supplied by nutrients and by oxygen from diffusion. This is the habitat of biodegrading microbes in the soil. Non-polar organic pollutants find it difficult to intrude into these biofilms because they are hydrophobic and cannot be dissolved in water. The two most important steps that facilitate contaminants to get closer to the microorganisms are these:

– A partition of the pollutant between the soil phases toward the aquatic phase, including the biofilm;
– An intensive contact and interaction between solubilized/mobilized contaminants and the microbes.

The equilibrium partitioning of the organic pollutants, initially attracted to the solid phase (humus), has to be shifted towards the biofilm-integrated water phase of the soil. This can be done by using surfactants (surface-active agents such as tensides, detergents, soaps, emulsifiers), polarity-enhancing or micelle-forming substances, complexing molecules, solvents or molecular encapsulating substances. The aforementioned substances can be of synthetic or natural origin and they may be produced by the soil's own microbiota (biosurfactants). If biosurfactant production occurs in the soil, it can be enhanced. If the native microorganisms are not capable of producing biosurfactants in adequate quantities and of the required quality, remediation must enhance their availability.

Soil temperature greatly influences contaminant availability: continuous or temporary heating increases the bioavailability of some organic pollutants by increasing their solubility or volatilization. The vaporized organic compounds change their distribution in the soil: from a continuous liquid film (which covers soil particles in a biologically unavailable, even inaccessible form) to a dispersedly sorbed form. When subjected to a temperature increase of a few degrees Celsius, e.g. volatile hydrocarbon molecules can vaporize, disperse, and condense on more distant colder surfaces. Thus, the original continuous hydrophobic layer transforms into a disperse form, sorbed on a much bigger solid surface, making these tiny oily islands easily accessible to biofilm microorganisms.

Optimized flux of soil gas and groundwater make interaction with biofilms more intensive, and, as a result, biodegradation becomes more efficient. In the aquifer, microbial attack and biodegradation occur only on the plume surface. By increasing the surface of the plume, biodegradation can be activated and its efficiency increased. A contaminant plume can be directed and distributed by adjusting the pressure of the groundwater by simple pumping.

3.4.4 Revitalization

Soil revitalization means the restoration of the soil's vitality and fertility or the establishment of biological activity for future soil uses. Since *ex situ* processes are more aggressive, revitalization of soils treated *ex situ* may be necessary, mainly after applying solvents, strong acids or alkali, oxidizing chemicals such as ozone or peroxides, reducing agents, harmful solubilizing agents, and high-temperature thermal desorption. Damage to the delicate balance of the soil's microflora can be traced back to several causes: the pollutant itself can destroy essential microorganisms directly by its toxicity or indirectly by privileging contaminant-utilizing species at the expense of others. A lack of nutrients can wipe out certain species entirely. If the technology works at higher moisture levels and a lower redox potential than the optimum of the normal microbiota, it can reduce the population of strongly aerobic species. Conversely, drying out the soil by ventilation is disadvantageous for facultative microbes living in the pore water.

As far as *ex situ* treatments are concerned, additives, nutrients, organic supplements, balancing additives, loosening substances or microorganisms needed for the revitalization can be mixed into the soil in the same reactor after treatment. Revitalization of the surface layer of soils treated *in situ* is similar to *ex situ* revitalization. Deeper layers are usually revitalized based on the natural groundwater flow and the necessary additives are injected into groundwater wells upstream of the target area. However, problems are similar to those of material input in general: soil heterogeneity, a filtering effect and limited transport of the revitalizing additives can impair the treatment.

3.5   Comparison of *ex situ* and *in situ* reactors for soil remediation

Following the overview of reactor types and technological solutions, Table 2.1 displays six different reactor types typically used for soil remediation:

1   *Ex situ* closed reactor (a conventional reactor);
2   *Ex situ* open reactor (an open tank or basin, heap or prism);
3   *In situ* surface soil layer (the top 40–50 cm);
4   *In situ* three-phase soil (including the deep unsaturated layers next to the groundwater);
5   *In situ* groundwater and saturated soil: undisturbed natural solid phase;
6   *In situ* groundwater: treatment by PRBs or reactive soil zones.

4   TECHNOLOGY DESIGN

A detailed analysis of differences and similarities between *ex situ* and *in situ* soil remediation allows conclusions to be drawn for their design.

The design of an *ex situ* technology is based on the contaminant type and concentration, the soil type and amount. When this information is available, designing and dimensioning the reactor and determining the amount of reagents, additives and the aeration rate – based on material balances – are relatively easy. The extent of heterogeneity, moisture content, and temperature can also be influenced and optimized if necessary. In spite of a relatively controllable situation, some soils may behave in unexpected ways. Remediation plans have to be supported by laboratory and pilot experiments to minimize the risk posed by uncertainties. The plans will enable the proper selection of the technology itself and the technological parameters, which can ensure optimal conditions for the core process. If the design is not validated by the pilot experiments or the full-size technology applications, longer residence times, higher recirculation rates, more additives, etc. can compensate for any alterations.

The design of *in situ* remediation should be more "dynamic" and much greater uncertainties should be managed in the subsurface: heterogeneity is generally greater than in *ex situ* remediation and the efficiency of processes such as aeration and reagent-, additive-, or nutrient-addition is lower. On the other hand, the unchanged position of the solid soil is not the complete satisfactory condition for an "*in situ*" technology, but the soil's natural structure and internal ecosystem must also be maintained using soft technologies. A healthy natural structure and an active ecosystem may greatly improve the efficiency of remediation by their buffering capacity and resistance to adverse effects and conditions and can help restore soil quality and activity.

The design of an *in situ* technology starts with a dynamic technology-related assessment of the contaminated site. The design of groundwater remediation relies heavily on the hydrogeology of the site.

Biodegradation-based remediation of the unsaturated soil zone requires information about the activity of indigenous microbes, their biodegrading or transforming capacity, the optimum conditions to reach their maximum activity, as well as their response to activating, biodegradation-enhancing interventions. When going deeper towards the saturated zone, the most suitable types of respiration and any bottlenecks of the biological conversion must be identified, and the most effective enhancement must be chosen. The best information can be acquired by *in situ* assessment of the site-specific properties of the soil and the soil

Table 2.1 Technological comparison of ex situ and in situ soil remediation.

Reactor characteristics	Ex situ closed reactor	Ex situ open reactor	In situ or ex situ surface soil layer	In situ three-phase soil	In situ groundwater & saturated soil	PRB & reactive soil zone
Physical limits	Built walls	Partly built walls Partly natural borders No limits from above	No built borders Partly natural borders No limits from above Covered top or tent	No built borders Partly natural borders Underground isolating wall around the treated volume	No built borders Partly natural borders Underground isolating wall around the treated volume	PRB: built walls, placed underground Reactive zone: partly built
Openness	Closed	Semi-open or Open	Open (completely) Semi-open (from above) Artificially closed	Open (completely) Semi-open Artificially closed	Open (completely) Semi-open Artificially closed	PRB: closed Reactive zone: semi-open or open
Treated volume	Within the physical borders	Within the physical borders	Within the range of operation Within artificial borders	Within the range of operation Within artificial borders	Within the range of operation	PRB: within built borders Reactive zone: within zone
Contact with environmental compartments	None	Atmosphere Surrounding soil Groundwater	Atmosphere Groundwater/layer water Surrounding soil	Atmosphere Groundwater/layer water Surrounding soil	Groundwater/layer water Surrounding soil	PRB: only groundwater
Treatable phases	A/GW/2S/3S/ slurry	GW/2S/3S/slurry	A/3S	A/GW/3S	GW/2S	GW/2S
Type of emission	Controlled	Gas/vapor into atmosphere Runoff: into surface water Infiltration: GW/S Solid: erosion/deflation	Gas/vapor into atmosphere and soil air Runoff: into surface water Infiltration: GW/S Solid: erosion/deflation	Gas/vapor into atmosphere and soil air Runoff: into surface water Infiltration: GW/S Solid: erosion/deflation	Gases and vapor into soil air GW: into the surrounding groundwater or surface water	Gases and vapor into soil air GW: into the surrounding groundwater or surface water

(Continued)

Table 2.1 (Continued)

Reactor characteristics	Ex situ closed reactor	Ex situ open reactor	In situ or ex situ surface soil layer	In situ three-phase soil	In situ groundwater & saturated soil	PRB & reactive soil zone
Homogeneity	Homogeneous: A/W/S/slurry Heterogeneous: 2S/3S	Homogeneous: A/W/S/slurry Heterogeneous: 2S/3S	Gradient: A/W Heterogeneous: 3S	Gradient: A/GW Heterogeneous: 3S	Gradient: GW Heterogeneous: GW/2S	Homogeneous GW Gradient: GW
Characteristics and dependence of the whole soil heterogeneity	Homogeneous slurry Homogenized 3S: depending on Environmental parameters The partition of pollutant	Depending on: Technology type environmental parameters The partition of pollutant	Depending on: Technology Hydrogeological parameters Environmental parameters The partition of the pollutant	Depending on: Technology Hydrogeological parameters Environmental parameters The partition of the pollutant	Depending on: Technology Hydrogeological parameters Environmental parameters The partition of the pollutant	Depending on: Hydrogeological parameters Environmental parameters The partition of the pollutant
Design	Single-stage: A/W/S/slurry Multi-stage: W/slurry Cascade: W/slurry	Single-stage: A/W/S/slurry Multi-stage: W/slurry Cascade: W/slurry	Single-stage: 3S	Single-stage: 3S	Single-stage: GW/2S Multi-stage: GW Cascade: GW	Single-stage: GW Multi-stage: GW Cascade: GW
Classification according to concentration gradient	Homogeneous tank reactor in batch mode Homogeneous tank reactor with recirculation Heterogeneous tank reactor with recirculation Continuous homogeneous Cont. heterogeneous/gradient	Homogeneous tank reactor in batch mode Heterogeneous tank reactor in batch mode Homogeneous tank and recirculation Heterogeneous tank and recirculation Continuous homogeneous Continuous heterogeneous	Heterogeneous quasi-tank reactor in batch mode Heterogeneous quasi-tank reactor with recirculation (A/W)	Heterogeneous quasi-tank reactor in batch mode Heterogeneous quasi-tank reactor with recirculation (A/GW)	Heterogeneous quasi-tank reactor in batch mode Heterogeneous quasi-tank with recirculation (A/GW) Heterogeneous continuous quasi-tank reactor with GW flow	Tank reactor with homogeneous filling and GW flow Quasi-tank reactor with homogeneous or heterogeneous filling and GW flow

Aeration/oxygen supply	Ventilation: exhaust/pressure Injection Chemical redox control Anaerobic respiration	Ventilation: exhaust/pressure Injection Chemical redox control Anaerobic respiration	Tilling/mixing/ventilation: Injection Chemical redox control Anaerobic respiration (flood)	Ventilation: exhaust/pressure Injection Chemical redox control Anaerobic respiration	Injection Chemical redox control Anaerobic respiration forms	Injection Chemical redox control Anaerobic respiration forms
Redox potential	Aerobic/anoxic/anaerobic	Aerobic/anoxic	Aerobic/anoxic	Aerobic/anoxic	Aerobic/anoxic/anaerobic	Aerobic/anoxic/anaerobic
Moisture content and its manipulation	3S: irrigation/infiltration/injecting/temporary flooding 2S: saturated with water Slurry: suspension in water	3S: spraying/infiltration/injecting/temporary flooding 2S: saturated with water Slurry: suspension in water	3S: spraying/infiltration/injecting/temporary flooding	3S: spraying/infiltration/injecting/temporary flooding/uptake by capillary forces	2S: saturated Increase in water level Increase in water flow	2S: saturated
Form of additive input/inflow into the reactor	Homogeneous reactor: water-soluble or solid Heterogeneous reactor: dissolved in water/emulsified/suspended	Homogeneous reactor: water-soluble or solid Heterogeneous reactor: dissolved in water/emulsified/suspended	Dissolved in water/emulsified or suspended Mixable solid	Dissolved in water/emulsified or suspended	Dissolved in water/emulsified or suspended	Built into PRB or reactive zone
Output/outflow	Soil air Pore water/soil moisture Groundwater/leachate Slurry Pollutant	Soil air Pore water/soil moisture Groundwater/leachate Slurry Pollutant	Soil air Pore water/soil moisture Groundwater/leachate Pollutant	Soil air Pore water/soil moisture Groundwater/leachate Pollutant	Groundwater Pollutant	Groundwater

(Continued)

Table 2.1 (Continued)

Reactor characteristics	Ex situ closed reactor	Ex situ open reactor	In situ or ex situ surface soil layer	In situ three-phase soil	In situ groundwater & saturated soil	PRB & reactive soil zone
Recirculation	Soil air Groundwater/ leachate Slurry	Soil air Groundwater Drain water/ leachate Slurry	Soil air Runoff Drain water/ leachate	Soil air Runoff Groundwater Drain water/ leachate	Groundwater	Groundwater
Microbiota	Aerobic: W/3S Facultative: W/S/ slurry Anaerobic: W/S/ slurry	Aerobic: W/S Facultative: W/S/ slurry Anaerobic: W/S/ slurry	Aerobic: 3S Facultative: W/2S	Aerobic: 3S Facultative: GW/2S	Aerobic: aerated GW Facultative: GW/2S Anaerobic: GW/2S	Aerobic: aerated PRB Facultative: GW Anaerobic: GW
Modification of the microbiota	By optimizing technological parameters and by inoculation	By optimizing technological parameters and by inoculation	By optimizing technological parameters and by inoculation	By optimizing technological parameters and by inoculation	By optimizing technological parameters and by inoculation	By optimizing technological parameters and by inoculation
Evolution	Modified Induced	Modified Induced	Own natural Modified Induced	Own natural Modified Induced	Own natural Modified Induced	Own natural Modified Induced
Revitalization	Spontaneous Artificial	Spontaneous Artificial	Spontaneous Artificial	Spontaneous Artificial	Spontaneous Artificial	Not relevant
Agent that increases availability	Addition into the reactor	Addition into the reactor	Injecting into the soil Flooding/infiltrating	Injecting into the soil Flooding/infiltrating	Injecting into groundwater	Not relevant
Monitoring	Soil air Groundwater/ pore water Whole soil Slurry	Soil air Groundwater/ pore water Whole soil Surry	Soil air Pore water/ groundwater Whole soil	Soil air Pore water/ groundwater Whole soil	Pore water/ groundwater Whole soil	Groundwater Filling
Regulation	Based on A/W/S/ slurry	Based on A/W/S/ slurry	Based on A/W	Based on A/W	Based on A/GW	Based on GW
Post-monitoring and aftercare	Soil quality control Groundwater quality control	Soil quality control Groundwater quality control	Soil quality control Environmental monitoring	Soil quality control Environmental monitoring	Soil and GW quality control Environmental monitoring	GW quality control Environmental monitoring

A: air; GW: groundwater; S: soil; slurry: slurry phase; 2S: saturated soil; 3S: unsaturated soil; W: water ecosystem. A variety of contaminated site assessment strategies and assessment tools are available today for sampling, analyzing and testing in mobile or remote laboratories. An accelerated site assessment procedure (ASAP) applies direct-push sampling methods combined with an on-site mobile laboratory. If this demanding technology is not available or is not acceptable from a financial point of view, simple and affordable innovative assessment tools such as *in situ* rapid chemical, biological, and toxicological tests, field handheld or other mobilized measuring devices, *in situ* real-time sensors and push-and-pull devices can be applied in the first assessment step (Gruiz, 2016b). The assessment should be combined with immediate evaluation and on-site decision making on the next assessment step or the suitable remediation technologies (Triad approach, 2015).

Push-and-pull technology is a dynamic *in situ* method for studying the response of the soil. It uses controlled air injection or test solutions (reagents, additives, nutrients, oxygen release compounds, etc.) or tracers and measures the response at the same point. Push-and-pull can determine the feasibility of a planned remediation, predict certain technological parameters and collect information for technology design. Push-and-pull has provided useful experience about natural biological processes and their enhancement in the saturated soil zone.

On completing the problem- and site-specific assessment and evaluation, the design of both *in situ* and *ex situ* remediation shall survey all suitable technologies for the relevant problem/site. The usually long list of technologies has to be narrowed down based on priority issues for management purposes such as urgency, available resources, staff, know-how, site accessibility, site management, future land uses, etc. After narrowing down the list to a few technologies, a comparative study is needed that considers technological, environmental, and socioeconomic efficiencies and cultural issues. The monetized efficiencies of the technological options are compared (the reference base is the zero option, meaning that no action is done on the site), and the most efficient remediation method is selected, designed, and implemented (Gruiz *et al.*, 2009, Chapter 11).

5 MONITORING

Monitoring *in situ* remediation can be aimed at two different operation targets:

1 *Technology monitoring*: measuring the technological parameters to follow the progress of the technology, ensure its optimal functioning and control the efficiency of the technology application (based on the material balance).
2 *Environmental monitoring*: observing the emissions or other adverse impacts of the technology application on the surroundings or the wider environment. In addition to the impact of the technology itself, the materials used, the transport and other connected operations should also be monitored.

Long-term monitoring and aftercare of remediated sites are key issues in environmental management. The continuous monitoring of the environment and the soil is equally important in order to observe as early as possible the damage to and deterioration of soil functions. This kind of generic monitoring with the goal of early warning and early intervention is one of environmental management's priority tasks. Long-term, preventive monitoring requires tools that are different from those of a one-off urgent assessment of contaminated sites where damage is obvious, as for the latter, the site is relatively small, assessment cost is not a priority issue and soft and long-term working assessment tools are rarely needed.

Today, sampling and analysis still face problems as they lack robust methods for many soil problems in spite of a large number of innovations implemented. Physicochemical soil analyses still fail to provide a reliable basis for modeling site-specific biological and ecological effects, even if measurements are as closely spaced as one centimeter. Methods to measure and predict probable values for the biological effects of a pollutant, its biological availability, the response of the actual ecosystem and the risk posed on them are unfortunately still in their infancy. Remote sensing provides extensive information, but it cannot be interpreted without the knowledge of the relevant ecosystem. This demonstrates that not only are the methods undeveloped, but so is the interpretation and evaluation of the results. Integrated testing and evaluation of the physical, chemical analytical, biological, and (eco) toxicological methods as recommended by applied sciences and practical developers (Gruiz, 2005, 2016b) can provide a solution to this problem.

5.1 Technology monitoring

Reactors used for *ex situ* soil remediation – containing the more or less homogeneous soil phases (water, slurry) or the heterogeneous whole soil – are easy to access, so sampling is a fairly straightforward issue. A sampling plan, which takes into account heterogeneities and gradients in *ex situ* reactors has maximum impact on the interpretability and quality of the monitoring results. This influences the success of control operations and the overall technology efficiency to a large extent. Technology monitoring is usually focused on:

–   The decrease of the contaminant concentration in the soil;
–   The increase of the transformation/degradation product in the soil or in soil air/water;
–   The activity of the catalyst or biocatalyst;
–   The concentration of the additives;
–   Direct testing of the adverse effect of the pollutants;
–   The beneficial effect of the remediation.

Most of these parameters can be measured in the three-phase soil using physicochemical, biological, or ecotoxicological methods and evaluated in an integrated way.
Sampling the (whole) soil for *in situ* remediation faces the following difficulties:

1   Spatial heterogeneities of soil and pollutants may exceed several times the time-related decrease in the amount of the pollutant.
2   It is often necessary to change and adjust the flow conditions of the soil air and groundwater, and leave the solid phase undisturbed.
3   Core sampling based on drilling changes the flow conditions of soil air and groundwater.
4   Sampling and analysis of soil air, soil moisture and/or groundwater must have priority in aftercare monitoring of *in situ* soil remediation.
5   The physicochemical and microbiological processes in the soil must be modeled based on air and water data. This is only possible if the basic remediation processes are known and the mathematical functions/equations describing the relationships between the measured parameters and the soil conditions are understood.
6   Data about the mobile soil phases represents the average over the treated soil volume and no information about the internal gradients and heterogeneities is available.

A groundwater sample taken from a well or from the extracted water is a mixture of samples from different depths, different redox conditions and this applies to all resulting parameters (redox indicators, chemical and microbial activities and products, etc.).

The essence of the transformation/catalysis facilitated by soil remediation is that a harmless product is produced from the pollutant:

$$\text{Contaminant} \xrightarrow{\text{transformation/catalysis}} \text{NRP}$$

This equation indicates that the pollutant is transformed by the catalyst into a no-risk product (NRP). The transformation can be triggered by a physical agent, chemical reagent or biologically active molecule or organism. The same equation implies the physical destruction, chemical transformation, degradation, biological, or enzymatic transformation of the pollutant, regardless of whether the process means mobilization or immobilization.

Several products may be produced from one contaminant (mainly from the large and difficult-to-degrade substituted molecules), and risky ones (RP) may also occur among them. Soil pollutants are typically complex, consisting of several components and producing an increasing number of products.

$$\text{Contaminant} \xrightarrow{\text{transformation/catalysis}} x\,\text{NRP} + y\,\text{RP}$$

$$n\,\text{Contaminant} \xrightarrow{\text{transformation/catalysis}} nx\,\text{NRP} + ny\,\text{RP}$$

Several options are available for the aftercare of this type of soil remediation (based on the transformation activity): measuring the pollutant decrease, transformation product increase or transformation activity based on mass, concentration and the scale of the physicochemical or biological transformation activity.

The decrease in *pollutant content* in the soil and/or groundwater during soil bioremediation is tantamount to a decrease in the substrate that provides energy for the soil microorganisms, so therefore biological activity will be proportional to the substrate concentration. This does not apply to harmful contaminants, which are best to be monitored by their effects. Their transformation to harmless components decreases their adverse effects (toxicity, mutagenicity, carcinogenicity, reproductive toxicity, etc.), which can be measured by toxicity tests. The environmental risk of the soil can be directly extrapolated from the measured adverse effect (Gruiz *et al.*, 2009; Gruiz, 2016b).

5.1.1 *Product generation*

The intermediate and end products of the transformation process (e.g. chemical oxidation or bioremediation) are harmless products in ideal cases (e.g. NO_2, HCl, NH_4^{2+} etc.). The process can be followed by measuring the concentration of the end products.

If harmful intermediates or side-products are anticipated, it is expedient to base the monitoring on their adverse effects (see more in Gruiz, 2016a,b). Adverse effects can be checked – at least at the end of remediation – by providing evidence about the acceptable quality of the soil.

5.1.2 The activity of the transformer/catalyst

The catalyst can be a physical agent, a chemical or biochemical reagent or the biological activity of living organisms. The activity or the quantity of the catalyst can be monitored; the activity is generally easier to measure in soil than the concentration of a rather small quantity of a not completely homogeneously distributed, reactive compound or a microorganism.

There are a number of options available to monitor biological transformation: cell concentration and its changes (the total number of cells in the soil: aerobic bacteria, fungi, etc.); the specific transformation or degradation capability of the microorganisms (hydrocarbon, PAH- or PCB-degrading, metal mobilizing, etc.); the number of particularly tolerant microorganisms (metal-tolerant); genetic markers (indicator genes); biochemical markers (enzymes responsible for specific biochemical potential), etc.

Those conditions in the soil that directly influence the activity of the transforming physical, chemical or biological agents and the rate and quality of the remedial process are also good indicators in technology monitoring. If the technology is based on oxidation or reduction, the redox potential and redox indicators can be monitored. If the technology is temperature-dependent, the temperature is an easy parameter to measure. In bioremediation, the nitrogen or phosphorus content (the N and P source) of the soil and/or groundwater, redox conditions and their changes are primary parameters (atmospheric O_2, NO_3^-, SO_4^{2-}, Fe^{3+}) directly influencing the activity of the microbiota.

In samples originating from relatively homogeneous, mobile soil phases, the transformation end products or the non-transformed residues can be detected directly by analytical methods. Volatile or water-soluble products are most suitable for monitoring among the detectable ones because they enter the relatively easy-to-sample mobile phases through partitioning. Those pollutants or products that are bound to the solid phase can only be analyzed from solid soil. Sampling and analysis problems may lead to inaccurate concentrations, also because spatial heterogeneities may exceed the short-term changes caused by remediation.

Nutrient and air supply and temperature and humidity are the key technological parameters of bioremediation, which need to be around the optimum of the transforming microbial consortium.

Thus, the technology can be regulated based on the environmental parameters measured, and the conditions can be set at the optimum required for the catalyst, be it chemical or biological.

5.2 Environmental monitoring and post-remediation care

The aim of environmental monitoring during remediation is to prevent the surrounding environment from being polluted and increasing the risk to humans and the ecosystem. Environmental monitoring plays a particularly important role in the case of *in situ* technologies, since the contaminated and treated soil volume is in direct contact with the atmosphere, the groundwater and the surrounding soil. Environmental monitoring of an *in situ* quasi-reactor must be performed within or directly outside the range specified by the pollution and the *in situ* processes.

Environmental remediation monitoring that relies on the mobilization of pollutants includes both the measurement and control of the emissions and the monitoring of the preventive measures. For example, the groundwater cannot leave the treated volume, due to the hydraulic barrier being applied simultaneously with the mobilization of the water-soluble pollutant, and the monitoring should determine the efficacy of the preventive measures.

Post-monitoring takes place after terminating the remediation. Post-remediation monitoring on the site is needed only when the groundwater is still exposed to risk. In an *ex situ* soil remediation, post-monitoring usually ends with a check on the treated soil or at the site of its utilization. If the soil is returned to the environment and is not completely risk-free after treatment, comprehensive monitoring is necessary, similar to that of contaminated sites. An integrated methodology (physicochemical, biological, and environment toxicological testing) is most appropriate in both cases. The future use of the soil must be known for the selection of a suitable analytical and risk assessment methodology so that the actual results can be compared with the soil-use-related acceptable risks.

More stringent requirements must be fulfilled after completion of *in situ* soil treatment because soil-use-specific quality requirements must be met not only by the treated soil volume, but also by the whole site and its surrounding, including the local environmental compartments, e.g. surface and subsurface waters, natural habitats, and human land uses such as residential, agricultural, or recreational, etc. and the exposures of humans via environment. Environmental monitoring is necessary to control each transport route from the treated soil and to prove that the exposure of the habitats is under a certain threshold, and that no adverse effects are posed at the receptor organisms because it is not possible to look inside the black box of the *in situ* treated volume and assess each kilogram of soil to ensure that its quality is unlikely to cause any problem. Thus, only a long-term negative result of the source (the *in situ* treated soil volume), transport pathways and the potentially exposed receptors can determine that the treated soil is harmless and that the previously contaminated site has been cleaned up.

6 VERIFICATION

The reactor approach and related engineering tools such as material balance, time requirement, optimum maintenance, regulation and emission control, etc. make it possible to verify remediation, including *in situ* remediation. Only a verified, technologically, environmentally and socioeconomically efficient technology will be acknowledged, enter the market, become popular and be widely used. The main reason for aversion to and refusal of *in situ* biological or ecological remedial technologies is a lack of insight into these technologies and a lack of opportunities for verification. However, there are other reasons too; one of them is its time requirement. In some cases, such as with major investments, a cost-benefit analysis may suggest that an expensive but fast remediation has to be preferred to an inexpensive but long-lasting one. On the other hand, sustainable soil quality maintenance requires *in situ* biological and ecological remedial methods, which may already exist, and making them better known and enhancing their image by using a prudent verification system so they may contribute to their use in wider circles (Gruiz *et al.*, 2008) (see more in Chapter 11).

7 SUMMARY

In situ soil treatment methods should be considered genuine technologies and proper engineering tools to which the planning, design, monitoring, control, and regulation and verification procedures can be applied as they are to any other technologies. This also holds for bioengineering and ecoengineering tools that utilize their inherent or modified remedial potential based on the soil microbiota or ecosystem. The same is true for other environmental compartments such as surface waters and sediments, runoffs, leachates, and mine outflows.

The reactor approach enables engineers to apply traditional engineering tools to *in situ* remediation, bioremediation, and eco-remediation. This involves designing and planning the technology, calculating the treated volumes, establishing the material balance, monitoring the technological parameters and developing an optimum by controlling and regulating processes, controlling emissions and verifying the remediation, i.e. providing evidence for its efficiency.

While the reactor walls set exact boundaries to a tank reactor, an *in situ* quasi-reactor is only limited by the pollutant plume geometry and the processes' range of action that define boundaries somewhat ambiguously. Gradients and the impact area of operations, rather than the confines of the pollution, determine the treated volume, and a hydraulic barrier is applied instead of a built one, etc. The effect and efficiency of the technological intervention can be controlled and regulated with the help of an adequate technology monitoring based on the material balance. The efficiency of the technology can be quantified based on these monitoring data.

The direct contact of the quasi-reactor with the surrounding environment, due to possible discharge and/or mutual material transport, may pose a higher risk compared to *ex situ* treatment in real reactors. Environmental impacts such as discharge of hazardous substances can be controlled by engineering tools, e.g. hydraulic barriers, soil-gas extraction or PRBs. The results of these preventive measures are validated by environmental monitoring, i.e. measuring the parameters that characterize the environmental risk outside of the quasi-reactor.

The technological efficiency of *in situ* technologies can be increased considerably by using the reactor approach, i.e. by handling the contaminated soil volume as a reactor with material and energy input and output and the transformation process in between. As with an *in situ* technology designed and implemented as any other technology based on mass and energy balances, process control/regulation and technology monitoring can be as efficient from a technological point of view as *ex situ* ones.

The environmental efficiency of *in situ* technologies relevant to energy and water consumption and to using and saving natural resources is much greater than that of *ex situ* technologies, which are always burdened by excavation and transport. Emission control and environmental monitoring can increase the environmental efficiency of *in situ* technologies.

Quantified socioeconomic efficiency of *in situ* technologies is often much greater compared to *ex situ* ones because of smaller expenditures (no excavation and transport), soft, eco-friendly technologies (less emissions and energy consumption), and maintaining the original soil structure, surface, and former land use (or permitting more valuable land uses). Another advantage of *in situ* methods is the conservation of soil quality and the preservation of the soil as a habitat.

However, heterogeneities will continue to cause uncertainties and problems for *in situ* soil remediation. Innovative soil characterization, existing *in situ* testing methods and geophysical mapping of the soil will further reduce the uncertainties and disadvantages of *in situ* remediation in the near future.

REFERENCES

Alesi, E.J. (2008) *General Principles of Groundwater Circulation Well (GCW) Technology for Soil and Groundwater in-situ Remediation*. Available from: www.eco-web.com/edi/00384.html. [Accessed 22nd May 2017].

Beyke, G. & Fleming, D. (2005) *In situ* Thermal Remediation of DNAPL and LNAPL using electrical resistance heating. *Remediation*, 15(3), 5–22.

Borden, R.C. & Cherry, R.S. (2000) *Direct Push Groundwater Circulation Wells for Remediation of BTEX and Volatile Organics*, Idaho National Engineering and Environmental Laboratory, Idaho Falls, ID, USA, 83415.

Chang, Y.T., Hsi, H.C., Hseu, Z.Y. & Jheng, S.-L. (2013) Chemical stabilization of cadmium in acidic soil using alkaline agronomic and industrial by-products. *Journal of Environmental Science and Health Part A Toxic/Hazardous Substances & Environmental Engineering*, 48(13), 1748–1756. doi: 10.1080/10934529.2013.815571.

CPEO (2017) *Dual Phase Extraction*. Center for Public Environmental Oversight. Available from: www.cpeo.org/techtree/ttdescript/dualphex.htm. [Accessed 22nd May 2017].

Dauerman, L., Windgasse, G., He, Y. & Lu, Y. (1992) *Microwave Treatment of Hazardous Wastes: Feasibility Studies*. The Hazardous Management Research Centre, New Jersey Institute of Technology, Newark, New Jersey, USA. Published online in 2016: doi.org/10.1080/08327823.1992.11688167.

Debreczeni, E. & Meggyes, T. (1999) Construction of cut-off walls and reactive barriers using jet technology. In: Christensen, T.H., Cossu, R. & Stegmann, R. (eds.) *Seventh Waste Management and Landfill Symposium*. CISA Environmental Sanitary Engineering Centre, Cagliari, Italy, IV. pp. 533–540.

ERH (2007) *Electrical Resistance Heating: Design and Performance*. Criteria Battelle, presentation from Remediation Innovative Technology Seminar (RITS). Available from: https://clu-in.org/download/contaminantfocus/dnapl/Treatment_Technologies/RITS_2007A_ERH.pdf. [Accessed 28th June 2018].

Falciglia, P.P., Scandura, P. & Vagliasindi, F.G.A. (2016) An overview on microwave heating application for hydrocarbon-contaminated soil and groundwater remediation. *Oil Gas Research*, 2, 110. doi:10.4172/2472-0518.1000110.

Falciglia, P.P., Scandura, P. & Vagliasindi, F.G.A. (2017) Modelling of *in situ* microwave heating of hydrocarbon-polluted soils. *Journal of Geochemical Exploration*, 174, 91–99. doi:10.1016/j.gexplo.2016.06.005.

Falciglia, P.P., Urso, G. & Vagliasindi, F.G.A. (2013) Microwave heating remediation of soils contaminated with diesel fuel. *Journal of Soils and Sediments*, 13, 1396–1407. doi:10.1007/s11368-013-0727-x.

Fenyvesi, É., Leitgib, L., Gruiz, K., Balogh, G. & Murányi, A. (2009) Demonstration of soil bioremediation technology enhanced by cyclodextrin. *Land Contamination and Reclamation*, 17(3–4), 611–618.

FRTR (2016) *Remediation Technologies Screening Matrix and Reference Guide*. Available from: https://frtr.gov/matrix2/section4/4-37.html. [Accessed 8th April 2017].

Gomes, H.I., Mayes, W.M., Rogerson, M., Stewart, D.I. & Burke, I.T. (2015) Alkaline residues and the environment: A review of impacts, management practices and opportunities. *Journal of Cleaner Production*, 112(4), 3571–3582. doi:10.1016/j.jclepro.2015.09.111.

Goulding, K.W.T. (2016) Soil acidification and the importance of liming agricultural soils with particular reference to the United Kingdom. *Soil Use and Management*, 32(3), 390–399. doi:10.1111/sum.12270.

Gruiz, K. (2005) Biological tools for the soil ecotoxicity evaluation: Soil testing triad and the interactive ecotoxicity tests for contaminated soil. In: Fava, F. & Canepa, P. (eds.) *Innovative Approaches to the Bioremediation of Contaminated Sites*. Soil Remediation Series, No. 6, INCA, Venice, Italy. pp. 45–70.

Gruiz, K. (2009) Soil bioremediation: A bioengineering tool. *Land Contamination and Reclamation*, 17(2), 543–552.

Gruiz, K. (2015) Terrestrial toxicology. In: Gruiz, K., Meggyes, T. & Fenyvesi, É. (eds.) *Environmental Toxicology*. CRC Press, Boca Raton, FL, USA. pp. 149–155.

Gruiz, K. (2016a) Monitoring and early warning in environmental management. In: Gruiz, K., Meggyes, T. & Fenyvesi, É. (eds.) *Site Assessment and Monitoring Tools*. CRC Press, Boca Raton, FL, USA. pp. 255–259.

Gruiz, K. (2016b) Integrated and efficient characterization of contaminated soil. In: Gruiz, K., Meggyes, T. & Fenyvesi, É. (eds.) *Site Assessment and Monitoring Tools*. CRC Press, Boca Raton, FL, USA. pp. 1–98.

Gruiz, K., Fenyvesi, É., Kriston, É., Molnár, M. & Horváth, B. (1996) Potential use of cyclodextrins in soil bioremediation. *Journal of Inclusion Phenomena and Macrocyclic Chemistry*, 25, 233–236.

Gruiz, K. & Kriston, É. (1995) In-situ bioremediation of hydrocarbon in soil. *Journal of Soil Contamination*, 4(2), 163–173.

Gruiz, K., Molnár, M. & Feigl, V. (2009) Measuring adverse effects of contaminated soil using interactive and dynamic test methods. *Land Contamination and Reclamation*, 17(3–4), 443–460.

Gruiz, K., Molnár, M. & Fenyvesi, É. (2008) Evaluation and verification of soil remediation. In: Kurladze, G.V. (ed) *Environmental Microbiology Research Trends*, Nova Publishers, Hauppauge, NY, USA. pp. 1–57.

Gruiz, K., Sára, B. & Vaszita, E. (2014) Risk management of chemicals and contaminated land: From planning to verification. In: Gruiz, K., Meggyes, T. & Fenyvesi, É. (eds.) *Environmental Deterioration and Contamination*. CRC Press, Boca Raton, FL, USA. pp. 227–312.

Hammarstrom, J.M., Sibrell, P.L. & Belkina, H.E. (2003) Characterization of limestone reacted with acid-mine drainage in a pulsed limestone bed treatment system at the Friendship Hill National Historical Site, Pennsylvania, USA. *Applied Geochemistry*, 18, 1705–1721.

Heron, G., Carroll, S. & Nielsen, S. (2005) Full-Scale removal of DNAPL constituents using steam-enhanced extraction and electrical resistance heating. *Ground Water Monitoring & Remediation*, 25(4), 97–107.

Huang, G., Zhao, L., Dong, Y. & Zhang, Q. (2011) Remediation of soils contaminated with polychlorinated biphenyls by microwave-irradiated manganese dioxide. *Journal of Hazardous Material*, 186, 128–132.

Jarvis, A.P. & Younger, P.L. (2001) Passive treatment of ferruginous mine waters using high surface area media. *Water Research*, 35(15), 3643–3648.

Johnson, K.L. & Younger, P.L. (2006) The co-treatment of sewage and mine waters in aerobic wetlands. *Engineering Geology*, 85, 53–61.

Johnson, P., Dahlen, P., Triplett Kingston, J., Foote, E. & Williams, S. (2009) State-of-the-practice overview: Critical evaluation of state-of-the-art in situ thermal treatment technologies for dnapl source zone treatment. ESTCP Project ER-0314.

Lin, L., Yuan, S., Chen, J., Wang, L., Wan, J. & Lu, X. (2010) Treatment of chloramphenicol-contaminated soil by microwave radiation. *Chemosphere*, 78, 66–71.

Maslov, B.S. (ed) (2009) *Agricultural Land Improvement: Amelioration and Reclamation*. Volume II. EOLSS Publications, UNESCO eBook. Available from: https://www.eolss.net/outline components/Agricultural-Land-Improvement-Amelioration-Reclamation.aspx. [Accessed 8th October 2018].

Meggyes, T. (2010) Preventing pollution caused by mining activities. In: Sarsby, R.W. & Meggyes, T. (eds.) *Construction for a Sustainable Environment*. CRC Press/Balkema, Taylor & Francis Group, Boca Raton, FL. pp. 23–38.

Meggyes, T., Csővári, M., Roehl, K.E. & Simon, F.-G. (2009) Enhancing the efficacy of permeable reactive barriers. *Land Contamination and Reclamation*, 17(3–4), 635–650.

Meggyes, T. & Simon, F.-G. (2000) Removal of organic and inorganic pollutants from groundwater using permeable reactive barriers. Part 2. Engineering of permeable reactive barriers. *Land Contamination & Reclamation*, 8(3), 175–187.

Meggyes, T., Simon, F.-G. & Debreczeni, E. (2001) New developments in reactive barrier technology. In: Sarsby, R.W. & Meggyes, T. (eds.) *The Exploitation of Natural Resources and the*

Consequences: Proceedings of the 3rd International Symposium on Geotechnics Related to the European Environment, June 21–23, 2000. Thomas Telford, London, UK. pp. 474–483.

Molnár, M. (2007) *Intensified Bioremediation of Contaminated Soils With Cyclodextrin: From the Laboratory to the Field.* PhD Thesis, Budapest University of Technology and Economics, Hungary.

Molnár, M., Fenyvesi, É., Gruiz, K., Leitgib, L., Balogh, G., Murányi, A. & Szejtli, J. (2003) Effects of RAMEB on bioremediation of different soils contaminated with hydrocarbons. *Journal of Inclusion Phenomena and Macrocyclic Chemistry*, 44, 447–452.

Molnár, M., Gruiz, K. & Halász, M. (2007) Integrated methodology to evaluate bioremediation potential of creosote-contaminated soils. *Periodica Polytechnica, Chemical Engineering*, 51(1), 23–32. doi:10.3311/pp.ch.2007-1.05.

Molnár, M., Leitgib, L., Gruiz, K., Fenyvesi, É., Szaniszló, N., Szejtli, J. & Fava, F. (2005) Enhanced biodegradation of transformer oil in soils with cyclodextrin: From the laboratory to the field. *Biodegradation*, 16, 159–168.

Molnár, M., Vaszita, E., Farkas, É., Ujaczki, É., Fekete-Kertész, I., Tolner, M., Klebercz, O., Kirchkeszner, C.S., Gruiz, K., Uzinger, N. & Feigl, V. (2016) Acidic sandy soil improvement with biochar – A microcosm study. *The Science of the Total Environment*, 563–564, 855–865. doi:10.1016/j.scitotenv.2016.01.091.

ORC supply (2017) *Oxygen Release Compounds.* Available from: www.environmental-expert.com/companies/?keyword=oxygen+release+compound. [Accessed 22nd May 2017].

Pértile, P., Albuquerque, J.A., Gatiboni, L.C., da Costa, A. & Warmling, M.I. (2012) Application of alkaline waste from pulp industry to acid soil with pine. *Revista Brasileira de Ciência do Solo*, 36(3), May/June. doi:10.1590/S0100-06832012000300024.

PIRAMID Consortium (2003) Engineering guidelines for the passive remediation of acidic and/or metalliferous mine drainage and similar wastewaters. *European Commission 5th Framework RTD Project no. EVK1-CT-1999-000021 Passive in-situ Remediation of Acidic Mine/Industrial Drainage (PIRAMID).* University of Newcastle Upon Tyne, Newcastle Upon Tyne, UK. 166 pp.

Roehl, K.E., Meggyes, T., Simon, F.-G. & Stewart, D.I. (eds.) (2005) *Long-term Performance of Permeable Reactive Barriers.* Trace Metals and Other Contaminants in the Environment. Volume 7 (Series editor: Nriagu, J.O.). Elsevier, Amsterdam, The Netherlands, Boston, MA, USA, Heidelberg, Germany & London, UK. p. 326. ISBN: 0-444-52536-4.

Roehl, K.E., Simon, F.-G., Meggyes, T. & Czurda, K. (2007) Improvements in long-term performance of permeable reactive barriers. In: Sarsby, R.W. & Felton, A.J. (eds.) *Geotechnical and Environmental Aspects of Waste Disposal Sites.* Taylor & Francis Group, London, UK. pp. 163–172.

Simon, F.-G. & Meggyes, T. (2000) Removal of organic and inorganic pollutants from groundwater using permeable reactive barriers. Part 1. Treatment processes for pollutants. *Land Contamination & Reclamation*, 8(2), 103–116.

Simon, F.-G., Meggyes, T. & McDonald, C. (eds.) (2002) *Advanced Groundwater Remediation: Active and Passive Technologies.* Thomas Telford, London, UK. p. 356. Available from: www.ttbooks.co.uk/advanced-groundwater-remediation. [Accessed 14th March 2015].

Simon, F.-G., Meggyes, T., Tünnermeier, T., Czurda, K. & Roehl, K.E. (2001) Long-term Behaviour of Permeable Reactive Barriers Used for the Remediation of Contaminated Groundwater. *Proceedings 8th International Conference on Radioactive Waste Management and Environmental Remediation, 30 September–4 October 2001.* ASME, Tech. Inst. Royal Flemish Soc. of Engineers, Belgian Nuclear Society, Bruges, Belgium.

SOILUTIL (2018) *Waste Application on Soil Management Concept and Results.* Available from: www.enfo.hu/en/elearning/7369. [Accessed 14th March 2015].

Taylor, J., Pape, S. & Murphy, N. (2005) A summary of passive and active treatment technologies for Acid and Metalliferous Drainage (AMD). Earth Systems. *Fifth Australian Workshop on Acid Drainage, 29–31 August 2005*, Fremantle, Western Australia.

Triad Approach (2015) *A New Paradigm for Environmental Project Management.* Available from: www.clu-in.org/conf/itrc/triad_021005/ [Accessed 14th March 2015].

Triplett Kingston, J.L. (2008) *A Critical Evaluation of in situ Thermal Technologies*. PhD Dissertation, Department of Civil and Environmental Engineering, Arizona State University.

Trumm, D. (2010) Selection of active and passive treatment systems for AMD – Flow charts for New Zealand conditions. *New Zealand Journal of Geology and Geophysics*, 53(2–3). doi:10.1080/002 88306.2010.500715.

Trumm, D. & Watts, M. (2010) Results of small-scale passive systems used to treat acid mine drainage, West Coast Region, South Island, New Zealand. *New Zealand Journal of Geology and Geophysics*, 53, 229–239.

Ujaczki, É., Feigl, V., Farkas, É., Vaszita, E., Gruiz, K. & Molnár, M. (2016a) Red mud as acidic sandy soil ameliorant: A microcosm incubation study. *Journal of Chemical Technology and Biotechnology*, 91(6), 1596–1606. doi:10.1002/jctb.4898.

Ujaczki, É., Feigl, V., Molnár, M., Vaszita, E., Uzinger, N., Erdélyi, A. & Gruiz, K. (2016b) The potential application of red mud and soil mixture as additive to the surface layer of a landfill cover system. *Journal of Environmental Sciences*, 44, 189–196. doi:10.1016/j.jes.2015.12.014.

US EPA (1998) *Permeable Reactive Barrier Technologies for Contaminant Remediation*. EPA/600/R-98/125. Available from: https://clu-in.org/download/rtdf/prb/reactbar.pdf. [Accessed 22nd May 2017].

US EPA (1999) *Multi-Phase Extraction: State-of-the-Practice*. Available from: https://clu-in.org/download/remed/mpe2.pdf. [Accessed 22nd May 2017].

US EPA (2016) *Dual-Phase Extraction*. Chapter XI in: How to Evaluate Alternative Cleanup Technologies for Underground Storage Tank Sites: A Guide for Corrective Action Plan Reviewers (US-EPA). Available from: www.epa.gov/sites/production/files/2014-03/documents/tum_ch11.pdf. [Accessed 8th April 2017].

US EPA Clue in (2017) *Ground-Water Circulating Wells*. Available from: https://clu-in.org/techfocus/default.focus/sec/Ground-Water Circulating Wells/cat/Overview/ [Accessed 22nd May 2017].

Vorenhout, M., van der Geest, H.G., van Marum, D., Wattel, K. & Eijsackers, H. (2004) Automated and continuous redox potential measurements in soil. *Journal of Environmental Quality*, 33(4), 1562–1567. doi:10.2134/jeq2004.1562.

Younger, P.L., Banwart, S.A. & Hedin, R.S. (2002) *Mine Water: Hydrology, Pollution, Remediation*. Kluwer Academic Publishers, Dordrecht, The Netherlands.

Zambrano, M., Parodi, V., Baeza, J. & Vidal, G. (2007) Acids soils pH and nutrient improvement when amended with inorganic solid wastes from kraft mill. *Journal of the Chilean Chemical Society*, 52(2), 1169–1172.

Chapter 3

Natural attenuation in contaminated soil remediation

K. Gruiz

Department of Applied Biotechnology and Food Science, Budapest University of Technology and Economics, Budapest, Hungary

ABSTRACT

Waters and soils have an almost infinite healing capacity, including negative feedback mechanisms responsible for reinstating an optimal equilibrium in element cycling and ensuring an acceptable habitat for the biota. In addition to negative feedback processes, high diversity and flexibility of the biota is required to ensure a dynamic equilibrium and adaptation to actual conditions. The biota is constantly short of nutrients, which leads to maximal exploitation of the available resources, thus even contaminants may serve as nutrient supply for aquatic and soil microorganisms. This way, contaminants become part of the biogeochemical element cycling and, depending on their persistency, may disappear or accumulate in the environment. Water and soil processes have considerable impact on the atmosphere as well, as hydrogeochemical element cycling may result in gas emissions/release with a long residence time in the atmosphere. Biological CO_2 sequestration and high soil organic matter content could mitigate the release of greenhouse carbon dioxide into the air.

If the natural buffering and balancing system of the environment does not function properly, many of the unregulated processes lose their capacity of self-control. In spite of being aware of this, we hardly learn from it, and our environmental management is still not based on it. One of the positive lessons learned is to use natural attenuation in environmental remediation. Unfortunately, this "attenuation" in the environment is often identified/interpreted as a concentration decrease. However, from the point of view of environmental load, the total contaminant amount and, from the perspective of the ecological and human receptors, the specific land use risk posed should be considered. Those natural processes are acceptable in environmental remediation, which can diminish adverse impacts of contaminants and irreversibly reduce the environmental risk posed to the ecosystems and humans via the environment.

A proper attenuation process reduces the contaminant quantity in parallel with its concentration. Decreasing the concentration without removing the contaminant may affect increasingly large areas. In the short term this may not be visible locally, but a generic increase in the contaminants' background concentration poses a high risk to element cycles and ecosystem stability at watershed or even global scale. This phenomenon has been experienced in the last century, and today nobody knows what the result of these changes will be, such as the increases in the background values and the additional environmental loads, as surplus to the already existing levels. We can only hope that the Earth finds a new equilibrium that is still beneficial to the human species.

Natural attenuation and engineered (enhanced) natural attenuation used in environmental remediation are considered real technologies/processes occurring in quasi-reactors (see Chapter 2) operating with or without engineering interventions. Depending on the scale of intervention, management activity is denominated differently such as monitored natural attenuation, engineered or enhanced natural attenuation, soft remediation, close-to-nature bio- or ecotechnologies, *in situ* bioremediation, phytoremediation, artificial wetlands, living machines, etc. Many of these technologies include almost no intervention, while others can be highly controlled and regulated bio- or ecotechnologies.

This chapter discusses the process itself and the monitoring of natural attenuation as well its application as a remedial option and its management as a nature-based technology.

I INTRODUCTION

Chemical contaminants in the environment are subject to several physical, chemical and biological impacts, and while interacting with the environment and with the actual ecosystem members they are transported, diluted or converted by physicochemical and biological (biochemical) processes or incorporated into soil structures or living cells and organisms. In fortunate cases, contaminants are transformed into less hazardous or non-hazardous products or degraded (partially or fully). The contaminant transport and fate processes in contaminated areas are determined both by the properties of the chemical substance and the characteristics of the environment.

The inherent properties of the contaminant such as its volatility, water solubility, partitioning between physical phases (described by K_{ow}), and its availability and degradability serve as the basis to forecast the contaminant's fate and effects in the environment. The properties of the contaminant however are strongly modified by environmental properties and conditions.

The spontaneous physical, chemical, and biological processes involving contaminants in the environment are the same as or very similar to natural processes, which play a considerable role in element cycling and regulate material balances at a global scale. Interactions in the environment can, in most cases, involve the contaminants in the element cycles. In other cases it is not possible to take the contaminant along through the entire biochemical pathway because its continued advancement is blocked at a certain step, the reason being that the natural biochemical apparatus is faced with unfamiliar conditions, an unfamiliar substrate, toxic by-products, etc. without being equipped with alternative biochemical protection mechanisms.

Natural attenuation plays an increasingly important role in contaminated land remediation. It provides generally a low-cost means of remediating contaminated soil and groundwater. The cost items are the long-term monitoring and the soft interventions to enhance natural processes if necessary. Assessment and modeling of contaminant transport and fate is essential in evaluating natural attenuation as a remediation option. The information provided is used for planning, and the evaluation (verification) of the technology and its efficiency in decreasing the risk to humans and the ecosystem.

2 NATURAL ATTENUATION (NA) AND RELATED ECOTECHNOLOGIES FOR SOIL REMEDIATION

When planning and implementing eco-efficient technologies for environmental risk reduction, it is worth exploiting the existing natural risk reduction capability of the soil, taking into account the actual natural attenuation processes in the soil. Topography, hydrogeology, and soil structure are responsible for the transport processes while moisture content, pH, redox potential, and ion-concentrations are key parameters influencing chemical and biological transformations. The *cell factory*, responsible for the biology-based changes in the soil is in the focus of biotechnologies playing a role in environmental risk reduction. The technological parameters are selected/adjusted so that they do not damage the soil ecosystem but rather support the useful (risk-reducing) cell community, and do not increase the hazard and the risk, not even temporarily or outside the treated soil volume. If such a temporary risk increase cannot be avoided, additional protective technologies should be planned and implemented. The optimal technological conditions can be formed spontaneously in natural and artificial ecosystems or created in the engineered system as part of the technology control. Key conditions are the size of the quasi or engineered *in situ* reactor, the distribution and transport of the contaminant and the potential additives, as well as the functionally selected and adjusted technological parameters (flow conditions, residence time, temperature, pH, osmosis pressure, nutrient and oxygen supply, redox potential, special additives, biological and ecological parameters, i.e. species diversity and activity, etc.).

Natural attenuation is an intrinsic remediation process, meaning that the risk of a contaminant being present in the environment is reduced by natural physicochemical and biological-ecological processes. These processes can be destructive, causing disappearance by degradation/breakdown or something non-destructive, such as when the concentration decrease is attributed to dilution, dispersion, or volatilization. The concentration decrease of a certain contaminant alone is not satisfactory. Concentration decrease only due to non-destructive transport processes may cause diffuse pollution and a higher risk than the original concentration. Risk-reducing natural attenuation can be considered a passive, *in situ* remedial process and soil remediation can be based on it. Even if it is considered a passive remedial option, this does not mean doing nothing. In order to manage the risk, the monitoring of NA is necessary.

Monitored natural attenuation (MNA) refers to the deliberate application of natural attenuation processes for contaminated land remediation and intends to ensure the fulfillment of site-specific remedial objectives, i.e. a reduced and acceptable level of environmental risk within an acceptable time frame. MNA provides data for the control and verification of the remediation and to justify the comparable efficiency and sustainability with conventional, highly engineered remediation (see also CIR, 2000a; Alvarez & Illman, 2005; Looney *et al.*, 2006).

Engineered natural attenuation (ENA) covers real technological interventions to stimulate and enhance the natural attenuation processes with the input of additives, substrates, or activators, etc. (see also SNOWMAN, 2018; Clue-in, 2018a,b).

Natural attenuation-based remediation is an innovative remedial approach that relies on natural attenuation processes in contrast to highly engineered, costly remediation with complex machinery and energy demand. Nevertheless, soft interventions may be necessary to enhance, accelerate, and mildly modify natural processes. The scale of intervention covers a wide range of manipulations from slight ventilation or other physicochemical methods, through bioavailability enhancement of nutrient supply for the biota, including soil microorganisms and plants to artificially formed or modified ecosystems.

Engineered natural attenuation and engineered ecotechnologies overlap largely; a distinction is made by professionals because natural attenuation is mainly used in the context of groundwater plume and the ecotechnologies in the much wider context of soil, sediment, or surface water remediation by bio- and ecoengineering measures.

Ecoengineering encompasses methods to force an ecosystem to remediate a contaminated environment. It integrates ecology, biology, biochemistry, genetics, engineering sciences – especially bioengineering and computer engineering. The tools of ecoengineering cover a large range, starting from the exploitation of natural processes without human intervention (which means only the observance and monitoring of the processes), through various levels of engineering and ecological manipulations to the artificial ecosystems (artificial mesocosms, living machines).

The application areas of the natural processes-based ecotechnologies focus on the following:

1   Environmental technologies, waste treatment and waste utilization technologies: i.e. protecting and ensuring an acceptable environmental status of surface waters and soils exposed to permanent contaminants load, treatment of runoff waters prior to entering the receiving water body, development of shore protection strips along surface water bodies, waste water treatment, utilization of wastewater, and wastewater sludge in soil.
2   Managing the self-healing of contaminated areas by bioremediation and phytoremediation, enhancing the efficiency of natural reduction in contaminant concentration or adverse effect, as well as ensuring beneficial environmental conditions and healthy material balance for the natural ecosystem.
3   Prevention of soil degradation and amendment of degraded soils, i.e. addition of nutrients and increasing nutrient binding capacity of the nutrient-deficient soils (macro-, meso- and microelements), organic matter-deficient soils, prevention of acidification, secondary salinization and sodification, erodibility, compaction, etc.
4   Studying prevailing trends within the ecosystem and harmonizing the anthropogenic impacts with the desired healthy ecosystem status. "Engineering" of the ecosystems in the service of sustainable development.
5   Incorporating the ecological aspect into all management and engineering activities, from the selection of technology through planning to implementation.
6   Developing methods and tools to quantitatively measure ecological characteristics and ecological risk.

Thus, natural attenuation is an exceptional tool for long-term maintenance of surface and groundwater, soil and sediment quality and in the sustainable management of diffuse pollution (primarily the risk of water catchments), degraded soils and contaminated land. There are even today many abandoned historically polluted (known as superfund) sites all over the world. Abandoned industrial and mining sites, storage and commercial facilities, and ports are left behind, relying on time to be cured, without any monitoring or intervention, but often with unclear and changing ownership status. "Time" in most cases does its work since nature is able to mobilize astounding forces for survival under changed conditions, even adapting to the presence of contaminants, e.g. by utilizing them as nutrient and setting a new equilibrium in element cycling.

The mass balance calculation helps to estimate if the input rate exceeds the natural attenuation rate, in which case a growing plume is expected (Figure 3.1B); a shrinking plume is expected for the opposite condition, where natural attenuation exceeds the input rate. Engineering interventions intend to decrease the input or increase the attenuation rate. MNA works best where the source of pollution has been removed, as shown in Figure 3.1C. A non-removed active source results in growing plumes in all directions (compared to Figure 3.1A and B). Surface input must be terminated or buried waste removed and disposed of properly. After the source is removed, natural processes may successfully cause the residual amount in the soil and groundwater to be settled and the plume to shrink. The soil and groundwater are monitored regularly to make sure they are cleaned up (Kueper et al., 2014; Clue-in, 2018b).

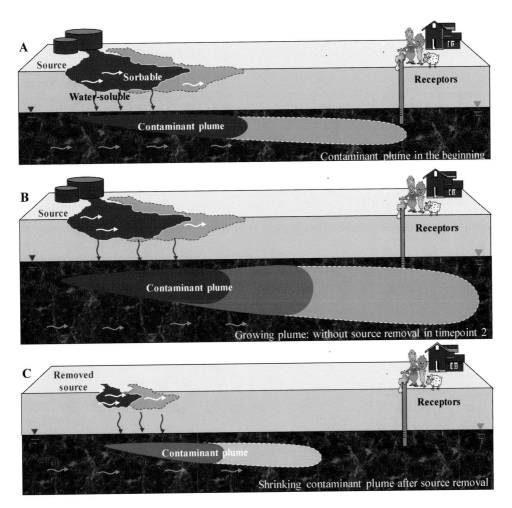

Figure 3.1 Contaminants plume (A) without (B) and after (C) source removal.

Interventions enhancing natural attenuation, even source removal, isolation, or the intensification of the natural attenuation processes, cover a wide range of processes, operations, and complex technologies. These interventions utilize existing natural processes and combine them with directed mass flow and/or enhanced degradation. The scheme in Figure 3.2 illustrates the merely natural processes, the engineered natural processes, and the growing rate of engineering.

Ecoengineering applications of natural attenuation processes in the risk management of abandoned contaminated sites have not yet become widespread because they do not require "serious" machines, "spectacular" equipment, but "only" scientific knowledge on processes occurring out of sight below the surface. Switching from an approach centering on technology to the improvement of environmental knowledge would result in wider acceptance and a breakthrough for the consideration and application of engineered natural attenuation as a technology and ecoengineering tools for corrective or remedial purposes.

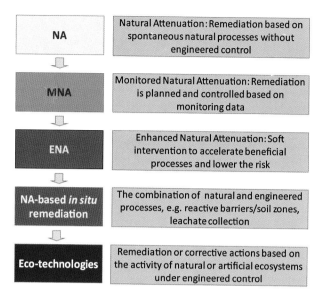

Figure 3.2 Increasing scale of engineering in utilizing natural attenuation in environmental remediation.

More respect to and a realistic attitude towards nature and the environment assumes complex environmental/ecological knowledge. The abandoned contaminated land on one hand and the accidental contamination spills on the other have demonstrated that nature itself is able to fight for pollution mitigation and in fortunate cases remediate itself without human intervention.

For example, in Hungary after the withdrawal of the Soviet army, investigations in the area of the former Soviet army military areas showed high petroleum pollution. The country had not had enough financial resources at that time for immediate remediation/utilization of the assessed sites, thus quite often the remediation had started only after 10–15 years, once the land had been sold or the regional development or spatial planning determined its utilization. At many sites, contaminated with various petroleum hydrocarbons – such as fuels, heating oil, car and lubricant oils and greases – the contamination had attenuated spontaneously due to the degradability of its constituents. Pollution from industrial and mining activities, as well as from the disposal and landfilling of hazardous wastes – such as chlorinated benzenes and phenols or mine wastes containing lead, cadmium, uranium, or other metals – belongs to a different category than petroleum hydrocarbons, given that these are persistent, hardly degradable or non-degradable substances. The subsurface transport and dilution of these persistent and toxic contaminants gradually affect water supplies and produce a continuously growing environmental and human health risk.

To expand the spread of soft technologies (i.e. bio- and ecotechnologies) based on natural processes, the following requirements can be considered:

1   Identification, assessment and modeling of natural attenuation processes. Distinguishing the risk-reducing natural processes from the risk-enhancing processes based on monitoring. Assessing the risk quantitatively.

2 Improvement of the knowledge level of engineers, landowners, environmental policy makers, managers, and other stakeholders.

3 Calling and turning the attention of the relevant contaminated site policy and management representatives toward soft, clean, and inexpensive remediation such as engineered natural attenuation and soft interventions. Beyond the solutions to the acute problems, drawing attention to the necessity of long-term thinking and planning; long-term sustainment of environmental quality and ecosystem services; monitoring of contaminated and deteriorated land and its continuous management. Early interventions using cost- and eco-efficient methods.

4 Introduction of the dynamic system of risk-based land use in order to optimize land uses and harmonize land quality and land use. The less contaminated land should get the most demanding future land use (e.g. residential, recreational) and the future industrial and commercial or other less demanding land uses should be implemented in former "brownfields" after remediation. Another economical efficiency-increasing solution can be the utilization of the land during the remediation. Subsurface natural attenuation and soft *in situ* technologies often leave the surface free and allow for its temporary utilization.

5 We must endeavor to change the demand and availability of new processes in environmental technology market to make low-cost bio- and ecoengineering soil treatments more competitive while simultaneously reducing reliance on high profit but costly physicochemical and unjustified soil replacement technologies.

3 THE FATE AND BEHAVIOR OF CONTAMINANTS IN THE SOIL

The fate of contaminants in the environment is determined by the physicochemical properties of the chemical substance and by the environment's physicochemical-biological-ecological conditions and characteristics. In addition to the environmental compartments – air, surface water and sediments, soil, and groundwater – the biota as a fundamentally important component also has to be considered.

The chemical and biological processes may transform the contaminants into non- or less harmful materials compared to the original compound. However, in some cases the transformation may lead to a more hazardous product and a higher risk than existed beforehand.

The contaminant concentration at a given site may decrease due to diffusion, dilution, and/or partition between phases along the transport pathway. However, concentration decrease does not necessarily mean risk reduction, rather in most cases it may actually increase the risk if, for example, the contaminant from the solid soil phase gets into the groundwater, threatening a more sensitive land use or water supply compared to the previous situation, where the contaminant did not interact with groundwater. A hydrophilic low-risk contaminant bound to the soil when getting into water may reach dangerously high concentrations. The groundwater may transport contaminants to sensitive areas, e.g. from an out-of-use subsurface area to a region of water bases. Relatively low-concentration contaminants can accumulate in soil volumes, e.g. in the rhizosphere of high sorption capacity and biological access. Degradation may yield toxic products/compounds, increasing the risk while the concentration of the original contaminant decreases. Finally, the expansion of the contaminated area that appears to dilute the original pollution source or plume brings an elevated risk to new areas. Even if the risk is acceptable at these newly exposed areas, the background levels will be higher. This tendency continuously increases global background levels and speeds up disadvantageous ecological trends.

Therefore, contaminated soil assessment must take into account not only the contaminant concentration but also the contaminant mass input, i.e. the total amount reaching the environment. Physicochemical and biological pollutant degradation is capable of reducing both concentration and amount, but dilution results in a concentration decrease without mass reduction. Physicochemical-biological immobilization reduces the available quantity without decreasing the total amount of contaminants. These processes may lead to real risk reduction by decreasing the mass of either the contaminant or the available contaminant, provided irreversible immobilization takes place. Irreversible immobilization/stabilization leads to a reduced risk with the mass present in the environment unchanged, but easily reversible processes may create a constant threat, referred to as "chemical time bombs." In the case of toxic hazard, degradation can decrease the contaminants' potential adverse effects, unless toxic intermediaries and/or by-products are formed. The risk due to the contaminants is proportional to the material balance and its changes so that predictions and estimations about NA time requirement and efficiencies should be based on the contaminants' mass balance.

3.1   Processes in the soil

The main physicochemical and biological processes that the contaminant may undergo in the soil are reviewed here.

3.1.1   Physicochemical processes in the soil

−   Physical transport of the contaminant in gas, vapor, liquid as a separate phase, dissolved in water or solid, or sorbed on solid and in slurry form. The driving force may be gravity or pressure difference.
−   Diffusion is a transport process driven by concentration gradients: the molecules move from higher toward lower concentrations;
−   Dilution is the result of the process that decreases the concentration by adding and mixing more water to the solution. The total solution amount will increase, but the dissolved amount remains unchanged.
−   Evaporation–condensation is a process pair in the phase transition of volatile liquids or solids caused by environmental conditions (typically temperature) or concentration differences.
−   Dissolution–precipitation is a process pair covering the phase transition of water-soluble substances on the effect of temperature, concentration differences or cosolvents. Many of the chemical and geochemical reactions in the soil are accompanied by phase transition, so these transformations can be utilized for mobilization or stabilization of contaminants, mainly metals. Some redox potential-dependent transformations of metals, where phase transition results in non-soluble metal species (see also Figure 3.3), are these:

   ○ $Cu^{2+} + 2\ (FeOH) = 2\ (FeO)Cu + 2\ H^+$
   ○ $CrO_4^{2-} + 3\ Fe^{2+} + 4\ OH^- + 4\ H_2O = 3\ Fe(OH)_3 + Cr(OH)_3$
   ○ $Ni^{2+} + Fe^{2+} + HS^- = Fe\ Ni\ S + H^+$
   ○ $Pb^{2+} + HCO_3^- = PbCO_3 + H^+$

−   Sorption–desorption: this process pair covers phase transitions from the point of view of the solid phase of the soil, which can bind substances on its surface both in gas/vapor, or liquid/dissolved (e.g. ionic) forms and release the molecules into the gas or liquid phase

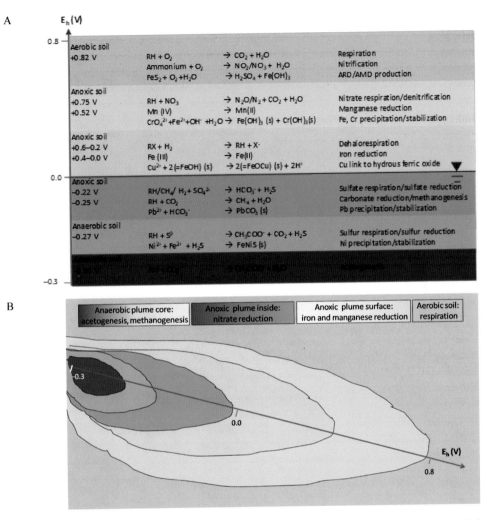

Figure 3.3 Redox potential-dependent processes in the soil and in a contaminant plume: A – vertical redox gradient in soil; B – central redox gradient in a contaminant plume.

of the soil after desorption. The sorbed amount of a hydrophobic organic contaminant can be many thousands or millions of times greater than the dissolved amount in the equilibrium aquatic phase.

− Partition between phases means a condition-dependent distribution of the contaminant between the three physical phases of the soil. Even if a contaminant is mainly split between two phases, e.g. between soil solid and liquid, some (maybe a very small amount) will be found in the third phase in the soil gas.

− Abiotic reactions such as photochemical reactions, hydrolysis, oxidation-reduction-chemical transformation, condensation, etc. contribute greatly to environmental transformation and making chemicals harmless. Photodegradation in water can reach a significant level and may result in complete elimination, but in deeper water layers and

in soil it is negligible. Highly reactive substances may be quickly transformed chemically under soil conditions. The analyses of degradation/transformation products and the assessment of their risk are always necessary.
– Weathering of rocks and soil formation are part of natural soil evolution, involving several physicochemical and biological processes and interactions. Regarding contaminants, it may cause mobilization by fragmentation, dissolution, leaching, runoff, erosion, etc. Natural immobilization/stabilization may take place during soil formation such as clay mineral formation, or microstructure/aggregate formation.
– Humus formation and fossilization result in organic structural building blocks. It is a complex process in which large organic contaminant molecules or small particulate matter can be incorporated into organic soil structure elements.
– Disintegration and leaching of the humus materials and podzolization are soil degradation processes, which may result in poor sorption capacity of the soil and the release of incorporated contaminants.

3.1.2   Biochemical and biological processes in the soil and groundwater

Several spontaneous responses are produced by the biota, which may create abilities to survive in the presence of contaminants and/or to utilize them. Knowing the internal (organism-dependent) and external (environment-dependent) conditions that led to the adaptation of an active transformation by the biota, technologists can also apply as engineering tools the same conditions to enhance biological transformation. Natural processes to be considered are the following:

– Adaptation of the ecosystem and the microorganisms is a typical and complex process in the soil. The microbial communities containing up to 109 cells in 1 cm³ of soil include complex food webs that behave as one organism with a highly versatile metagenome. The metagenome is the total genome of several thousand individual species. This metagenome can adapt swiftly to conditions by changing the distribution of its component genes according to the requirement of the available substrates, the environmental conditions, and the toxic materials. The adaptation mechanisms may be due to one or more of the following: (i) the owner microorganisms of the desired genes will propagate themselves faster, (ii) the required/beneficial genes will be switched on if formerly inactive, and (iii) the best genes will be distributed after propagation by horizontal gene transfer, i.e. by donating the useful gene to the neighboring microorganisms.
– Extracellular enzymes are secreted by microorganisms into the soil, but intracellular and cell-bound enzymes may also survive the producing organisms (Shukla & Varma, 2011). These enzymes are bound to the humus and are responsible for several soil activities, degrading processes, catalysis, etc., generally speaking for the health of the soil. They play an important role in contaminant degradation and can be used as additives in soil remediation after their production by means of various biotechnologies.
– Partitioning of the chemical substance between the environment and the biota is a process that is similar to but less passive than physicochemical partitioning. Living cells and their membranes do not only sorb molecules having affinity to the cell surface but also ensure certain selectivity (not binding to any molecules except for those already known and preferred). They may function as a pump when transferring the molecules

taken up through the membrane into inner organs/tissues, making free again the binding sites (structural templates of the substance) on the membranes.

–   Microbiological transformation, mineralization, and cometabolism are the common biochemical processes taking place in the cells. The cells try – and in many cases succeed – to use the contaminants similarly to natural substrates they commonly use for biosynthesis and energy production by degradation, incorporation, or accumulation. The preferred biochemical pathway is determined by the suitable conditions (e.g. redox potential), the need for energy sources, and the biodegradation products so it is essential to be aware of the mechanisms of biodegradation or transformation to control the natural process and manage the accompanying and residual risks.

The redox potential-dependent biodegradation (see Figure 3.3) and the necessary electron acceptors are demonstrated by the equations elaborated for benzene by Wiedemeier *et al.* (1995a):

–   Aerobic respiration: $C_6H_6 + 7.5\ O_2 \rightarrow 6\ CO_2 + 3\ H_2O$
–   Denitrification: $C_6H_6 + 6\ NO_3^- + 6\ H^+ \rightarrow 6\ CO_2 + 6\ H_2O + 3\ N_2$
–   Iron reduction: $C_6H_6 + 60\ H^+ + 30\ Fe(OH)_3 \rightarrow 6\ CO_2 + 78\ H_2O + 30\ Fe^{2+}$
–   Manganese reduction: $C_6H_6 + 30\ H^+ + 15\ MnO_2 \rightarrow 6\ CO_2 + 18\ H_2O + 15\ Mn^{2+}$
–   Sulfate reduction: $C_6H_6 + 7.5\ H^+ + 3.75\ SO_4^{2-} \rightarrow 6\ CO_2 + 3\ H_2O + 3.75\ H_2S$
–   Methanogenesis: $C_6H_6 + 4.5\ H_2O \rightarrow 2.25\ CO_2 + 3.75\ CH_4$

The vertical redox gradients in soils and the related geobiochemical reactions are similar to the redox gradient that evolved in the plume from the center toward the plume surface. As the plume is moving, it becomes more and more anoxic, finally anaerobic in the middle, since the biodegradation in the surface layer that encounters the new soil consumes soil air, then the nitrate, sulfate, oxidized iron, and manganese until no electron acceptor remains (see Figure 3.3). In addition to the lack of electron acceptors, shortage of water also contributes to the inhibition of biodegradation.

–   Plant transformation processes cover those biochemical processes from volatilization through degradation to bioaccumulation that change the location, form, and concentration of the contaminant.
–   Bioaccumulation results in a higher concentration in a cell, in tissue, in an organ or in an organism (plant, fungal, animal, or human) relative to the environment. Hyperaccumulating organisms such as some special plant can be used for the bioextraction of otherwise hardly mobilizable contaminants, e.g. certain metals.
–   Biomagnification means multiple bioaccumulations along the food chain, resulting in the worst-case deadly concentrations in top predators or predator-eating humans even if the accumulation rate is not very high in the individual food chain stages.

The processes in the soil are actually process pairs and their effects are shifted in one direction or the other depending on the equilibrium conditions, while trying to establish a homeostatic plateau. These processes can be controlled with the environmental parameters determining the equilibrium. In the case of process pairs, the main aim is to set equilibrium that results in an acceptably low risk. In the case of processes not involving balancing pairs or processes shifted in one direction, i.e. contaminant transport and dilution, the risk can be reduced mainly by removal of the source or by a physical or chemical barrier.

Some of the environmental processes trigger negative feedback while others trigger positive feedback loops. The systems controlled by negative feedback can establish long-term steady-state conditions, i.e. the great amount of biomass entering the soil seasonally – providing energy and a growth substrate for the microorganisms – enhances the growth (multiplication) of soil microorganisms. Once the microorganisms have consumed the organic matter, the decrease in nutrient supply reduces their number so their population reaches the usual "resting" level. Moreover, the food chain members relying on these micro-organisms return to their average level with some delay, so the whole of the food chain, even the members of the highest trophic levels, is limited by the nutrient source decrease. Until the sources occur repeatedly, the cycle continues unchanged from season to season. Mean-while, the processes with positive feedback strengthen the ecosystem and divert it from the equilibrium status. According to a typical trend, once a certain limit has been exceeded, the system cannot revert to the original equilibrium with the compensating forces of a negative feedback. In a fortunate situation, new (different from the previous) equilibrium values may be established, but the forces driven by positive feedback may divert the system from this equilibrium status repeatedly or continuously. A typical example of a positive feedback is the continuous decrease of soil organic matter (humus degradation, leaching and no humus supplementation) involving less nutrient retention/bonding, the result of which is decreasing biomass production, i.e. the natural process responsible for soil organic matter replacement. The outcome is complete humus degradation, soil nutrient loss, low nutrient plants, a low water holding capacity of the soil, no plant growth, podzolization and a high risk of soil erosion.

Weathering of sulfidic rocks is also a typical risk-increasing process that cannot be grouped into an antonymic process pair, thus representing a positive feedback. During weathering, chemical and biological oxidation of sulfides (in the presence of oxygen) results in sulfates, contributing to an ever-growing mobilization of toxic metal cations contained in the rock until the sulfides get exhausted. During this process, the solid form (s) pyrite becomes soluble (aq), and the iron hydroxide precipitates (s):

$$4\ FeS_2\ (s) + 15\ O_2 + 14\ H_2O = 16\ H^+\ (aq) + 8\ SO_4^{2-}\ (aq) + 4\ Fe(OH)_3\ (s)$$

The lack of sulfate reduction – which is far slower than oxidation and in addition requires anoxic conditions – leads to extreme acidification and metal leaching. It may occur that acid mine and acid rock drainage represents a pH of zero or very close to it, and the metal content reaches the order of magnitude of 1% in the leachate.

The characteristics of the natural processes differ in the case of degradable (mainly organic and C-, N-, or S-containing inorganic) and non-degradable contaminants (i.e. toxic metals species). It is advisable to distinguish processes occurring in the unsaturated, three-phase and in the saturated two-phase soils and those occurring in the sediments because, due to different redox potentials, the prevailing physicochemical and biological processes are different.

Most of the natural processes are ambivalent in terms of environmental risk; in fortunate cases they diminish but may also increase environmental risk. Apart from intensifying/accel-erating the useful processes by engineering tools, the aim of monitoring and human inter-vention is to reduce the risk-enhancing processes and support natural remedial potential.

Mobilization is one of the natural processes to be highlighted, given that it is the prerequisite of natural attenuation by contaminant removal. Removal may occur with and without degradation. Mobilization and removal without degradation results in the transport of the contaminant into environments not yet contaminated. Isolation of such processes from an unaffected environment can control risk (e.g. by collection and treatment of leachates and runoffs). Removal combined with slow degradation (e.g. aquatic photo- or biodegradation) pose a temporary risk to a limited area so it can be controlled by soft engineering tools, e.g. by hydrostatic barriers. The impact of natural mobilizing processes has to be controlled based on risk assessment. This risk should then be compared to the benefit resulting from mobilization, e.g. shortened treatment time.

The factors influencing NA (pH, redox, chemical form, biological activity, etc.) can be applied as technological parameters to manipulate a naturally evolved situation. For example, to increase the transformation rate and force transformation processes in the desired direction, the biodegradation rates of certain chemicals can be increased using *in situ* biostimulation and/or bioaugmentation (see more in Chapter 6):

– Biostimulation, i.e. the enhancement of the activity of indigenous microorganisms by adding nutrients, surfactants, alternative electron acceptors or cometabolites.
– Bioaugmentation is a manipulation of adding the best possible biological catalyst, e.g. microorganisms having specific degrading capability to a contaminated medium. Instead of microorganisms, any microbial product – typically enzymes – can be applied.

3.2   Mobilization and immobilization

Both mobilization and immobilization cover several physicochemical and biological processes and play a priority role in natural attenuation as well as in soil remediation in general.

3.2.1   Mobilization

In terms of mobilization, the following questions are frequently raised:

– Can the pollutant be mobilized and to what extent?
– Is there any mobilization in the contaminated area?
– Is there any equilibrium between mobilization and immobilization?
– Is the mobilization advantageous or disadvantageous to the ecosystem of the area, and what is the relationship between damage and benefits?
– Is the mobilization detrimental to the humans using the area?
– Could mobilization be the basis of a site-specific remediation technology?
– Does the physicochemical or biological mobilization prevail?
– Would mobilization pose a risk in the case of an *in situ* technology?
– Can the extent of mobilization be measured? Does it show a constant, a decreasing, or an increasing trend?
– What are the measures and additives able to increase or decrease mobilization at the respective site?

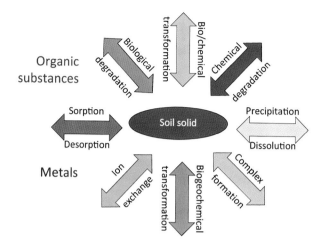

Figure 3.4 Fate of contaminants in soil.

Mobilization is a complex process including contaminant transport and fate (Figure 3.4):

– Transport processes such as diffusion and advection result in:

   o   Surface runoff,
   o   Seepage, and
   o   Infiltration.

– Fate processes resulting in contaminant translocation:

   o   Evaporation,
   o   Dissolution,
   o   Desorption, and
   o   Partition shifted towards the more mobile phase.

– Fate processes resulting in degradation:

   o   Photodegradation,
   o   Hydrolysis, and
   o   Biodegradation.

– Changing of the chemical form of the contaminant.

These processes in turn are all dependent on exterior conditions, primarily temperature, pH and redox potential, polarity, affinity and binding capacity of the medium, as well as on non-target chemicals present in the soil.

Mobilization often occurs spontaneously in a contaminated area, but the opposite may also take place, namely spontaneous immobilization.

Spontaneous mobilization can be as follows:

– Influenced by changing the exterior conditions (temperature, pH, redox potential, sorption);
– Mitigated with physical and hydrogeological barriers (decreasing the groundwater level, construction of reactive barriers); or
– Reduced or stopped by chemical immobilization.

The physical and chemical immobilization may involve biological systems, i.e. modification of redox potential or pH by microorganisms, fungi, and plants (see Figures 3.4–3.6):

– Establishing reductive conditions below ground by oxygen consumption;
– Metal precipitation by means of the sulfides produced by sulfate-reducing microorganisms;
– Metal leaching by acid-producing fungi;
– Modification of the soil's sorption capacity by plants and microbes (i.e. sorption-based filtering effect of the root area).

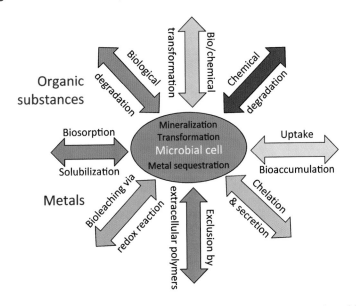

Figure 3.5 Impact of soil microorganisms on organic and inorganic contaminants (metals).

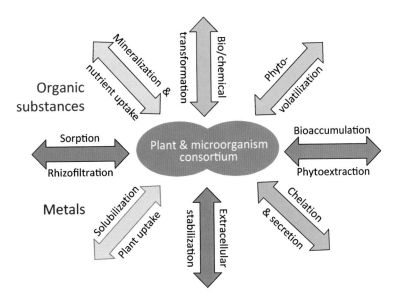

Figure 3.6 The role of plant and microorganisms consortium in contaminants' transport and fate.

In case the mobilization is detrimental, the extent of the risk has to be assessed and mitigated if necessary. Risk mitigation can be affected by hydrogeological barriers, directed material flows, isolation, etc. Should the mobilization be useful then its extent has to be assessed and intensified to become the basis of remediation.

Mobilization is generally simultaneously useful and potentially detrimental. Therefore, adequate measures shall be developed and implemented for its monitoring and manipulation. Technologically, this means separation/confinement in space of the technologically useful process from the process detrimental to the environment. For example, biological leaching of metals from sulfidic rock without any confinement results in leaching of metals (the residual rock will be depleted in metals, plants may spontaneously settle on the metal-depleted material after the pH reverted to normal), but the leached metal load transported by runoff contaminates the surface waters. For this reason, uncontrolled disposal into the environment of mining waste containing toxic metals is very dangerous.

3.2.2   Immobilization and accumulation

Spontaneous immobilization of contaminants in soil, sediment or biological systems may significantly lower the contaminant load in the water phase but represent at the same time, a potential contaminant source, from where contaminant release may occur upon changing environmental conditions. This potential release depends on the scale of the following:

–   The stability of the contaminant characterized by its water solubility and partitioning between the solid and the liquid phases, which is strongly shifted toward the solid in the case of high stability.
–   Irreversibility of the immobilizing process: the potency of the process to revert in changing conditions due to water content and flux, turbulence, pH, redox potential, ion-concentrations, etc.

Immobilization/stabilization may include various processes (see Figures 3.4–3.6):

–   Filtration of particulate matter (by constraint).
–   Sorption and filtration of dissolved substances: inorganic contaminants mainly sorb on clay minerals, organic contaminants, on humus materials.
–   Biosorption can be a passive physicochemical process when biomass passively binds and concentrates the contaminants on its cell surface or in the cell structure; but it can include an active process too, when the contaminant taken up is also transformed into a more stable form for longer-term storage or degraded.
–   Physical isolation based on natural processes.
–   Chemical immobilization based on chemical transformation and precipitation, due to pH (e.g., metal ions by limestone or other calcareous rocks) or redox (metal ions by reduction under water) conditions.

The contaminants accumulated and concentrated by spontaneous immobilization can reduce the risk in the long term, but this establishes a "chemical time bomb" in the soil with a lower or higher likelihood to explode. Usually, a combination of the aforementioned processes occurs.

For example, a contaminant in the sediment under anaerobic conditions may be immobile (many of the metals are in a non-soluble sulfide form), but if flooding relocates it onto the surface of the flood plain soil, it becomes mobile after oxidation due to formation of ionic metal forms. Another example is the contaminated runoff or wastewater entering a wetland, where intensive filtration, sorption, biosorption, chemical transformation, and precipitation may occur. These processes may retain the contaminants and result in a good quality outflow. The wetland soil becomes increasingly contaminated, the non-degradable compounds get concentrated and the degradable ones are degraded. When utilizing or draining such a wetland, these risks must be considered.

Soils and sediments, depending on their particle size distribution, will filter and bind contaminants at different concentrations. The fine fractions and the soil layers in the aquifer or certain sediment layers or deposits in rivers consisting of fine particles can be extremely contaminated, while sand or gravel is in most cases not contaminated.

In summary, a thorough contaminant and site assessment is needed before making deciding about a remediation based on immobilization or stabilization. Especially the risk posed by sites (soils and sediments) highly polluted with immobilized contaminants should be primarily assessed.

4   METHODOLOGICAL BACKGROUND OF NATURAL PROCESS MONITORING

Natural processes and the way they may attenuate or increase the risk of contaminants in the environment are reviewed in this chapter. The methodology required to assess natural processes will be introduced first, followed by details on data collection, measurement and test methods, processing and evaluation of measured data and interpretation of the results.

To explore and determine properly the environmental processes and the concept of monitoring (location, time interval, sample amount, type of measurements and evaluation), the area concerned should be thoroughly tested for potential damage likely to be caused by the contaminants present and transported in the environment and their adverse effects. This risk can be determined based on the *conceptual model* of the site that combines the contaminant's transport model (from the source via transport pathways through the receptors) and the exposure model (ecological and human receptors using the land and the exposure pathways typical for the land uses and the land users) (see Figure 3.7).

Measurements and tests employed for the assessment of contaminated sites and for natural attenuation monitoring represent a complementary integrated methodology (Leitgib *et al.*, 2005; Gruiz, 2016a; Gruiz *et al.*, 2016a). "Integrated" means that the traditionally used physicochemical results are completed with biological and ecotoxicity results to get a full picture of the black box of the "contaminated land." Soil is a complex system within which the contaminant, or very often a multi-component contaminant blend, interacts with the soil matrix with the physical phases and chemical constituents as well as with the soil ecosystem. A soil ecosystem is a continuously changing, flexible, and adaptive community able to convert the contaminant into energy or stabilize/neutralize it while striving for survival. Consequently, any of the indicators of interaction, reactions, and responses of the system components can be included in the integrated methodology.

Several methodological innovations enable NA assessment and monitoring such as the *in situ* and real-time monitoring tools (Gruiz *et al.*, 2016c), the direct push methods (Nemeskéri *et al.*, 2016), statistical methods (Gruiz *et al.*, 2015b), 2D and 3D analytical or numerical modeling tools, developed especially for NA evaluation and prediction such as NAS (Widdowson *et al.*, 2005; NAS, 2018) for the NA time requirement estimation, the BIOPLUME (1997) and BIOSCREEN (1997) tools for biodegradation-based NA, and the BIOCHLOR (2002) or RT3D (2018) for chlorinated solvents (see also Chapelle, 2004; US EPA, 2011, US EPA Models, 2018).

Mass balance (i.e. the equal input and output mass) should be the basis of all calculations, including biobalance. The mass balance can be determined most properly from flux (mass per time) data and expressed as the ratio of or the difference between loading and attenuation. From mass balance data, the nature of the soil pollution or the subsurface plume (shrinking, stable, or growing) as well as the attenuation time can be derived.

Mass and flux data of the contaminants cannot be determined only from the groundwater flux because contaminants, even the dissolved ones, do not move together with the water, due to their different scale of retention by the soil matrix compared to the water. Alternatively, several specific assessment and monitoring methods, as well as transport and fate modeling (both for forecasting and long-term evaluation), are used to quantify contaminant flow, dispersion, and elimination from the environment.

– Aquifer mapping using geophysical methods, e.g. tomography (Hoffmann & Dietrich, 2004) and the use of tracers for flux measurement (Gödeke *et al.*, 2004);
– Special pumping tests, e.g. multiple wells and mathematical models such as the integral pumping test (IPT) for quantitative characterization of contaminant plumes by consideration of linear, instantaneous sorption/degradation (Bayer-Raich *et al.*, 2004, 2006);
– Direct push methods for mass transfer (Mass transfer, 2015);
– Assessing soil matrix capacity to immobilize contaminants, e.g. reactive multi-tracer test, and modeling contaminant retention based on these results (Wachter *et al.*, 2000);
– Characterization of biodegradation by compound-specific isotope analysis (CSIA), an analytical method measuring the ratios of naturally occurring stable isotopes in environmental samples. Its use for the detection of degradation and identification of contaminant sources is based on the ratio of ^{12}C to ^{13}C, which changes in a systematic way in the course of degradation (both biotic and abiotic) (Hunkeler *et al.*, 2008; US EPA, 2008; CSIA, 2015);
– Characterization of microbiological populations and their adaptation/biodegradation by molecular methods, e.g. identifying indicator genes of enzymes or other active proteins in the metagenome responsible for interactions with the contaminant. These proteins may protect the organism from a toxicant or be responsible for the utilization of a substance as a nutrient (Haack and Bekins, 2000; Haack *et al.*, 2004; MNA-DNA, 2011; USGS DNA, 2018);
– Characterizing biodegradation by the stable isotope technique (DNA/RNA-SIP) to reveal the consumption of contaminants as energy substrates. The contaminant is enriched with a heavier stable isotope – typically ^{13}C – that is incorporated into organisms utilizing the contaminant as a substrate. 16S rRNA is the most informative biomarker because more than 120,000 sequences are available from databases. After extracting the RNA from the soil, it is separated to normal and heavy fractions by isopycnic centrifugation. Heavy RNA reveals carbon assimilation, i.e. the presence and activity of contaminant degraders, while sequencing heavy RNA identifies which organisms can consume the contaminant. With the increasing availability of DNA techniques, the method has become popular as shown by a large number of publications (Morasch *et al.*, 2001, 2002; Vieth *et al.*, 2002, 2003; Richnow *et al.*, 2003; Mancini *et al.*, 2003; Meckenstock *et al.*, 2004; Griebler *et al.*, 2004; Kopinke *et al.*, 2005) and is still in widespread use today (Sims, 2013; Kuder *et al.*, 2014);

- Measuring toxicity of the plume directly gives a toxicity value (lethality, inhibition of biological activity, mutagenicity, genotoxicity, immunotoxicity, endocrine disruption, etc.) directly associated with environmental or human health risk (Gruiz *et al.*, 2015a, 2016b; Gruiz, 2016a);
- Applying an integrated approach that combines more methods and evaluates the result in an integrated way (Gruiz, 2016a,b);
- Long-term monitoring and evaluation of all results versus time based on a risk-time curve, denominated in this chapter as "risk profile."

The *risk profile* curve (see Section 4.4 and Figures 3.10–3.13) shows the trend of the environmental risk in time, making it the best tool for interpretation of the monitoring results. A risk profile can be drawn from time-serial data. Quantitatively, the environmental risk is most simply described using the risk characterization ratio RCR (PEC/PNEC), which is the ratio of the predicted environmental concentration (PEC) to the predicted no-effect concentration (PNEC), i.e. the concentration in the environment that is not likely to impact adversely the whole of the ecosystem. The risk, expressed as RCR, changes in time due to changes of both the PEC and the PNEC. The PEC of the contaminant typically changes in the course of its transport by dilution and concentration (accumulation), mobilization or immobilization and other transport and fate processes. The PNEC changes along the same transport route, depending on the actual land uses and users. A runoff after reaching sensitive surface water (the habitat of a sensitive aquatic ecosystem) suddenly represents a much greater risk than along its transport pathway. The human risk of a subsurface plume far away from drinking water bases may be low, but it may represent a greater risk in the vicinity of a drinking water well.

Displaying a real picture of the risk in time and space, a risk map can be drawn and utilized for planning and managing environmental risk in an area, even subsequent to treatment. The risk can be estimated not only in space but also in time; the latter interpretation results in the *risk profile* used in this chapter to illustrate changes in time. Assuming a simple plume and linear transport, the profile is two dimensional (one spatial and the time dimension). To reflect the real transport process correctly, the transport should be mapped by a three-dimensional set of surfaces in time (both surface runoffs and subsurface plumes cover three spatial dimensions and the change in time).

After having detailed information about the area, the contaminant and its probable spatial and temporal distribution, an integrated risk model can be developed (see Figure 3.7). It is based on the *transport and fate model*, as well as land uses, land users and their *exposure model* from which the quantity of risk will be calculated. This kind of risk models are generally created for a certain time point and a certain locality, but with the help of the currently available information technology, it could even be a time map showing the quantity of risk and its changes spatially and temporally as a video. The risk manager should just read the location of the plume and the time point in question.

Such an integrated risk model should be the basis of all engineering activities in the environment – planning and implementing site assessment and monitoring, remediation, post-remediation activities, and future land uses. In the course of qualitative and quantitative risk assessment, increasingly more information can be utilized in a stepwise iterative methodology to refine the risk model and make the assessment procedure as efficient as possible.

The risk model should describe the site entirely: the contaminant, exposed receptor organisms, ecosystems and the interactions between all these participants. Exclusion of certain

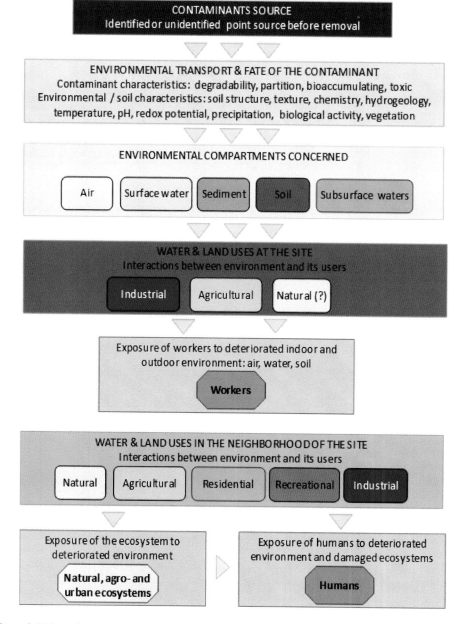

Figure 3.7 The risk model of an MNA site and the neighborhood of the site.

components of the model can be justified if a chemical substance, a pathway or a subsite does not contribute to the risk. All other details of the created risk model should be validated.

The integrated risk model is often referred to as the conceptual model because it should reflect the management concept. It is exceptional when natural attenuation is chosen as the primary remediation option. By this, a risk-based environmental management can be established and the autocratic managerial baseline information and judgment about the state of the art and priorities can be excluded. Priorities, e.g. the receptors that are taken into consideration, can be responsible for incorrect results. Risk models based only on human receptors may lead to ecosystem deterioration. Moreover, when considering only some protected species, most of the ecosystem may remain endangered, e.g. soil microbiota is often the last in line and not considered as a highly exposed receptor. The requirements of a good site assessment/environmental risk management are these:

1 To draw a risk model that is as objective as possible covering all possible environmental compartments and their users.
2 To create a management and decision-making scheme to set priorities from an environmental risk perspective.
3 Finally, to take socioeconomic, cultural, and political issues into account.

The site assessment and site and technology monitoring should not rely only on the results of a physicochemical analysis, having the main disadvantage of providing data only on a few, arbitrarily selected contaminants. They must also include the ecosystem, its observable deterioration, and the biological conditions of the soil. The living "cell factory" shall acquire a primary role, mainly in cases where natural attenuation is based on soil biology or where the soil ecosystem and function should be conserved. To measure the extent and type of environmental toxicity is an efficient means for environmental prevention, as information can be acquired on adverse effects posed by the contaminated environmental matrix (water or soil), even without (or before) identification of the contaminant(s). The actual adverse effects in soil or water caused by the pollutants may provide results that are contradictory to those of the physicochemical analytical methods, the reason for which may be one or more of the following:

– The analytical program does not include/identify all contaminants.
– Some contaminants at non-detectable low concentrations may have unacceptable high detrimental/adverse effects.
– The biological availability or non-availability of contaminants causes great differences compared to the toxicity of a pure substance or even to the predictable toxicity based on the chemically measured concentration (because the availability of the soil contaminants to the solvent used for chemical analysis differs from its biological availability).
– The picture gets more complicated due to the potential interaction of the contaminants with each other and with the soil matrix. These unpredictable complicated interactions may be strengthening, weakening, or even diminishing each other.

In summary, the quantitative environmental risk is

– Characterized by the RCR = PEC/PNEC ratio, which is determined from
– Modeled concentrations relying on calculated or measured data, and from
– Modeled effect results obtained from ecotoxicity tests and/or from default human toxicity data.

The risk profile is the variation of the quantitative risk in time substantiating the decisions made during the management of contaminated sites.

4.1 The risk model

The integrated risk model combines the transport and the exposure models (Figure 3.7). The transport model includes the pathway and the fate of the contaminant, starting from the source and leading to the environmental compartments likely to be reached by the contaminant. In the representation of a quantitative model the width of the arrows may be proportional to actual mass flows. In addition to the contaminant amount, the transport and fate models identify the physicochemical forms of the contaminant and its transformation/degradation. Transport and fate can be characterized by generic equations from available or newly measured data or can be based on measured data. The available software types are one-, two-, or multi-compartment models working generally predominantly with default data.

The exposure model (lower part of Figure 3.7) differentiates between exposures at the site and in its neighborhood. For instance, at an already identified contaminated site, it is mainly the workers that are exposed both to indoor and outdoor pollution. Exposures to and via the ecosystem are negligible, while in the neighborhood it is humans including children and pregnant woman and the members of the ecosystem that are targeted (potential receptors) by the air, water, and soil contaminants directly and *via* the food chain. The integrated risk model can map precisely the interactions between the environmental compartments (water-sediment, soil-air, etc.) and the receptors (e.g. food chains, human exposures *via* the ecosystem).

NA processes typically pose a risk to the following:

– Air – by the evaporation of volatile soil contaminants such as chlorinated solvents. Toxic soil vapor may intrude into indoor spaces and render toxic the air of residential buildings.
– Subsurface water – by soluble contaminants typically deriving from leaking containers or tubes disposed at the surface or the subsurface of the soil and transported by the groundwater in the form of plumes. A contaminant plume is the most common and best-known underground formation that may endanger water bases and residential areas.
– Surface waters – by runoffs contaminated by various dissolved soil contaminants and solid material. Acidic leachates containing toxic metals, industrial and mining wastes, pesticides, and nutrients from soil are the most common contaminants in runoffs.
– Soils – by all kinds of organic and inorganic contaminants from a primary or secondary source. Contaminants of the soil endanger/threaten an ecosystem and humans directly or indirectly *via* the food chain.

The site-specific risk model covers all the above-mentioned sources and transport routes and the local hydrogeological, meteorological, climate and contaminant data, receptors and interactions identified on the site. The water system (both surface and subsurface) and the characteristics of the soil are of major importance. The site-specific risk model refers to local industries, mining, and agriculture and the relevant occupational risks, as well as other, at highest risk the residential land uses, with focus on air quality, water uses, and the local population.

4.2   Integrated methodology for the assessment and monitoring of natural attenuation

An integrated methodology is the special combination of physicochemical, biological, and ecotoxicological methods (Gruiz, 2016a), also called the soil testing triad (STT, see more in Gruiz *et al.*, 2016a), providing complex information about the condition of the contaminated environment. The STT results cover the occurrence and the concentration of the contaminant, the characteristics of the environment, and the biota and its damage or potential damage, the latter demonstrated by the ecotoxicity of the environmental samples. Soil toxicity is in direct relation to environmental risk. One risk assessment approach that is worth applying to the monitoring of natural attenuation is the use of biological (e.g. biodegradation) and ecotoxicity results as priority information and all the other information on the contaminant, the environment and land uses can refine the results. Of course, the chemical data may have priority too, and in this case the biological and ecotoxicity data are supplemental and refine the chemical model. Applying either one or the other, the resulting integrated information is more comprehensive than merely the physicochemical analytical methods or the biological-ecotoxicological data. The STT methodology provides dynamic information by comparing the actual (contaminant) quantities and the produced effects. From time-serial data, the natural attenuating processes can be understood better, and a more correct forecast and decision can be made.

Sampling should fulfill the generally accepted sampling requirements (Gruiz *et al.*, 2016d) based on statistical patterns and site characteristics. One part of measuring and testing methods are the customized ones based on guidelines and protocols, while other parts are selected based on the first results on an iterative basis. Assessment and monitoring of natural attenuation may use *in situ*, real-time, on-site, or laboratory methods and modeling. For the evaluation, a wide range of statistical analyses is generally necessary, including an analysis of the trends.

The contaminants, hydrogeological environment, and microbiological activities are monitored in parallel by applying geological/geochemical survey methods, physicochemical analytical methods, biochemical, biological, as well as direct toxicological studies. The evaluation should be done in an integrated way. The interpretation of data differs in the different tiers of MNA management, but the conceptual risk model of the site always provides essential information.

Modeling is essential for mapping the contaminant and its spread at the site and the probabilities for reaching the receptors. Modeling is suitable to identify the dominating NA processes and the parameters that influence most their efficiency. The model can predict the most important parameter, the long-term transport and attenuation, as well as the time requirement of NA to reach the target (see also US EPA, 2011; US EPA Models, 2018; NAS, 2018).

4.2.1 *Physicochemical analytical methods*

The physicochemical analytical methods alone cannot satisfy the requirements of the management of natural attenuation because to correctly characterize beneficial biological and microbiological processes or the chemical time-bomb formation, a more complex consideration is necessary.

Physicochemical methods can characterize well the environment (geochemical characteristics, environmental parameters, i.e. pH, temperature, redox potential etc.) and the contaminant, if it has already been identified. In such cases, the physicochemical methods are the best for delineating the contamination and for monitoring the changes. If the contaminant is unknown, very complex (e.g. a mixture of different chemicals), or includes transformation products, the chemical methodology should be supplemented by tests for measuring the adverse effects.

What are the most frequently measured physicochemical parameters for the characterization of the contaminated environment and for monitoring natural processes? And what are the advantages and shortcomings? These questions will be answered in the following section.

4.2.1.1   Analysis of the contaminant

– Physicochemical analysis may determine the mass, flow rate, and concentration of the contaminant without (*in situ*) or with sample preparation (mainly in the lab).

  ○ Geological, geophysical, and hydrogeological measurements provide input parameters for modeling and predicting contaminant mass and flow rate.
  ○ Chemical analysis targeting the detection and the contaminant concentration can be carried out by direct analysis using selective sensors/detectors or after digestion/extraction of the sample and separation of the contaminant to analyze. In such a sample, the analyte is in a completely different state than it was in the environment.
  ○ Analytical methods work within a certain concentration range, which may require a largely different – more concentrated or diluted – analyte than in the natural sample. For example, in the case of an old mineral oil pollution instance, the already complex contaminant together with the transformation products may contain several hundred constituents. Whatever solvent is used to extract this complex contaminant, the extraction will be selective since none of the solvents can extract every contaminant to the same extent.
  ○ Most of the analytical methods are based on fractionation and/or separation (e.g. liquid- or gas-chromatography), which cause further selections in favor of the targeted analytes.
  ○ At the end of the chemical analytical procedure, a detector is used to determine the concentration by comparing the received signal to that of a reference material, which in environmental analytics is rarely identical to the environmental contaminant to be analyzed.

– Only those contaminants and components are analyzed that are foreseen and ordered from the lab. Typically, only the regulated contaminants are included in the analytical plan and many others are ignored. Neither are the degradation products always included; sometimes the degradation pathway is not completely known. This bad management practice endangers many soils all over the world.
– The chemical form (speciation) of a contaminant may be measured, but it is a labor-, time- and cost-intensive effort, and thus it is often not done. Even when it is done, the results cannot be always properly interpreted.
– Based on chemical analytical results only, the extrapolation from analytical data to the adverse effects on the ecosystem and humans involves high uncertainties because:

  ○ Not all chemical substances are taken into consideration; minor components and metabolites are often ignored;
  ○ The mobility and biological availability of the contaminants are not measured, in spite of their main influence on the realization of adverse effects (e.g. toxicity), so the actual toxicity cannot be judged correctly from the concentration-response function;
  ○ The interactions (antagonistic, additive, synergetic) between chemicals are not measured.

4.2.1.2   Characterization of the environment

Those environmental characteristics should be assessed that influence natural attenuation.

- For water/groundwater:
  - o   Depths, flow rates;
  - o   Temperature, pH;
  - o   Redox potential: dissolved oxygen, other oxygen supply, alternative electron acceptors such as nitrate, reduced Fe and Mn, sulfate, chloride, carbonate, or the presence of methane;
  - o   Ion content, inorganic and organic material contents, suspended solids, ash;
  - o   Material contents relevant to the element cycles and material balances: TOC, N forms, Ca, Mg, methane, etc.;
  - o   Salinity (can inhibit biological processes).

- For soil/sediment:
  - o   Soil type, soil texture;
  - o   Depths and layers;
  - o   Porosity, capillary head;
  - o   pH, redox potential (redox indicators, as for water);
  - o   Salt content, ion content and ion-binding capacity;
  - o   Humus content and composition;
  - o   Contents relevant to the element cycles and material balances.

These widespread measurement methods may characterize the entire environment at a specific point in time. However the dynamic nature of the complex, real, interaction-rich environment cannot be described properly by physical and chemical methods only, the biological/ecological and ecotoxicological responses are also needed, especially in the case of a contaminated environment.

4.2.2   Biological and ecological methods useful in natural attenuation monitoring

This group of assessment tools includes methods for the characterization of the living system of the environment (biota), for example the distribution of species in water and sediment ecosystems, the determination of the quantity and activity of soil-living organisms or the presence or lack of some typical/indicator species or genes. Molecular biology and DNA techniques gain increasing roles in characterizing the environment and its activities and health by indicator species or genes and the community with genomes and the metagenome (see more in Gruiz et al., 2016b). Most frequently measured biological characteristics:

- For water: microorganisms and algae counts, indicator species of aquatic animal and plant, species density, activities, diversity, specific communities such as micro- and mesoplankton, microorganisms associated with suspended solids, bioaccumulation in fish and other aquatic animals, food-chain effects, and biomagnifications. Enzyme activities, microbial degradation, and possible activation play a role in element cycles and contaminant transformations.

– For sediment: density and diversity of bottom sediment-dwelling organisms, density and diversity of micro-, meso-, and macrozoobenthos, indicator species (e.g. mussels, larvae), density and diversity of sediment-dwelling fish, bioaccumulation, density, diversity, and bioaccumulation of macroplants in bottom sediment, food-chain studies, biomagnification, microbial activities, degrading activity, and processes and activities playing roles in element cycling and contaminant transformations.

– For soil: density, diversity of soil microorganisms, indicators for microbial respiration, alternative types of respiration, degradation, activities in element cycles, and contaminant transformations, density, diversity and accumulation of plants, density and diversity of soil-living animals, accumulation, food chain, biomagnification, and indicator species.

Ecological monitoring follows the site-specific characteristics in time or, in the case of a missing time scale; the measured values are compared to a reference site. The assessment requires sophisticated statistical evaluation and separation of the effect of anthropogenic environmental damages from the natural trends and from adverse effects independent of the contamination, i.e. the damaging effect of pathogens independent of anthropogenic activity.

The available tool box is large, starting with the morphological identification of the entities and with determination of their percentage under a microscope and by biochemical methods to refined molecular biology methods. Molecular biology methods are fast and describe communities existing *in situ* in real time, compared to traditional assays based on incubation and propagation that never reflect the inherent situation. DNA and immune techniques provide information that is otherwise not available and that reflects most realistically the natural state of the ecosystem. Unfortunately, these methods are not yet standardized and are even as loaded by environmental heterogeneities and sampling bias as the conventional methods are.

DNA techniques for the monitoring of natural processes such as a biodegradation-based NA are possible and may give a highly refined picture of changes to the metagenome of the soil community responsible for the activities in the soil, but in practice the metagenome characteristics have not been translated properly yet into activities and soil health status. Nevertheless, these are promising techniques for following and understanding the response of nature to external conditions.

Both the biological and ecological characteristics should be evaluated in relation to the spatial direction and time scale of the presumed natural attenuation or for the validation of the modeled changes.

4.2.3 Measuring the effect – direct toxicity assessment (DTA)

An adequately designed test system can measure the present adverse effect (e.g. toxicity) of an environmental sample (i.e. contaminated water or soil) posed on certain, selected organisms. The extent of this measured effect is in direct relation to the magnitude of local environmental risk and makes predictions on expected damage or the necessary attenuation rate possible (see also Gruiz, 2016a). The test system may be applied/placed *in situ* into the environment to be studied, and alternatively it may be studied on site or in the lab after having taken the sample. Depending on the aim of the study, the test system can be a standard one or a site-specific one simulating actual conditions and processes using standard or site-specific test organisms. Based on the results, conclusions may be drawn on the effects and processes at the site where the sample derives from and by careful extrapolation to the whole of the relevant ecosystem.

Effects and end points in DTA

Direct toxicity testing of environmental water, sediment or soil samples means that the test system establishes/ensures contact between the sample and a controlled population of test organisms, generally one of the well-known, standardized species such as algae, daphnia, or fish for water and sediment-dwelling invertebrates, mussels, insect larvae, or bottom-fish, as well as plants rooted in sediment, and soil microorganisms, insects, and plants for the soil.

A more sophisticated concept is the use of a microcosm, i.e. a small-scale artificial eco-system, when a natural or similar artificial ecosystem is placed into the contaminated water, sediment, or soil under study and monitored. In such microcosms, complex natural processes can be simulated and detailed information can be obtained on natural attenuation in a con-taminated environmental matrix: processes, trends and rates can be measured. The impacts of the planned interventions can also be estimated based on the response of the small-scale artificial ecosystem (see more in Gruiz *et al.*, 2015c).

All of the usual ecotoxicity end points can be utilized for characterizing environmental toxicity and its possible attenuation such as a lethal effect on living organisms, a decrease in growth, a change in the number of individuals, reproduction inhibition, or an inhibitory effect on any life phenomenon, e.g. respiration, enzyme activities, movement, behavioral disturbances, mutagenicity, and teratogenicity.

From ecotoxicity results, direct conclusions can be drawn about the quality and quantity of the risk. The ecotoxicity results in comparison to the chemical analytical results provide refined conclusions on the dynamic behavior and fate of contaminants on the biological availability and interactions between contaminants and on any changes in time or on the presence of a chemical time bomb.

4.3 Evaluation and interpretation of natural attenuation monitoring data

Monitoring data on NA can be evaluated and interpreted for different management phases and for various purposes.

– To prepare decision making on whether NA, MNA, or ENA can be applied:

  o Exploration and identification of ongoing natural processes;
  o Identification of potential risk-reducing processes;
  o Estimation of the scale of NA and the mass balance;
  o Estimation of time requirement and residual risk;
  o Necessity of source removal or treatment;
  o Feasibility of NA for proper risk reduction at local and larger spatial scales and future land uses.

– Making the decision: selection of the best option such as these:

  o Use of MNA as the sole remediation;
  o Use of MNA as a component of a complex engineering solution involving the com-bination of different technologies;

- Rejection of MNA as it does not serve as a feasible solution and the selection of another, e.g. conventional technology.

– During implementation of remediation using MNA or ENA:

- Ongoing processes and velocities;
- Short- and long-term trends;
- Risk assessment;
- Validation of the prognosis.

– For aftercare and verification:

- Assessing long-term risks and the stability of environmental quality.

In addition to the environmental monitoring of energy consumption and cost assessment, it is also necessary to estimate and validate cost efficiency. Risk communication and educating stakeholders about natural attenuation are also important parts of the management.

The joint evaluation, the possible outcomes, and the interpretation of corresponding and contradicting physicochemical, biological, and ecotoxicity data will be discussed in the next two sections.

4.3.1   Corresponding monitoring results

Physicochemical analytical results (i.e. concentration) match the biological-ecotoxicological results.

1   Low concentration and low or no toxicity: Environmental risk is low.

For example, the soil did not show any toxicity according to the ecotoxicity tests performed with test organisms from three trophic levels. There is no visible sign of toxicity at the site and the performed chemical analytical tests are also negative. Therefore, it is very likely that the contaminant does not pose an unacceptable risk level to the site. However, this is not absolutely certain because undetected, unevenly distributed chemical substances, those missed by sampling and many others without acute effects missing from the analytical plan still can disprove the result of low or no risk. There might be adverse effects that can only be observed in the long term and by special tests, for example endocrine disruption. So, even if the situation looks clearly negative based on short-term monitoring, the negative results must be validated in long-term ecological monitoring and in biological or epidemiological studies on humans.

2   High concentration and significant toxicity: Environmental risk is high

The chemical analysis shows high lead concentration and the vegetation at the site is scarce and weak, while the toxicity of the soil samples is high. It is very likely that the adverse effect is already visible and the predicted damage is connected with the lead content. The certainty of this assumption may be checked with a multi-component analysis of metals to exclude the presence of other toxic metals. If e.g. cadmium or zinc cannot be excluded, the toxicity may be attributed to the joint effect of lead, cadmium, and zinc. In addition, undetected organic contaminants may also be in the soil.

4.3.2 Contradictory monitoring results

4.3.2.1 Analytically approved presence of a contaminant without adverse effects

1 The chemical substance has no adverse biological effect: Low-risk contaminated area

Some chemical substances are not referred to in the regulations and have no limiting values since they are not considered to be hazardous to the environment and humans *via* the environment. In such cases it is justified that the ecotoxicity study exhibits negative results. Some chemicals may pose adverse effects only in the long term or may have not been recognized or discovered yet to have hazardous effects such as the endocrine or immune disrupting effects. The presence of contaminants at chemically measurable quantities needs special attention and documentation because land users are not secured against risks emerging in the long term.

The lack of adverse effects due to low toxicity, low or missing bioavailability and the effects not reflected by the ecosystem should be differentiated. The explanation for the latter case can be the long-term adaptation of the biota of the contaminated site. Soil microorganisms, plants, and insects can adapt to certain contaminants. Ecotoxicological tests with standard species can circumvent this problem, but ecological assessment covering only a few species may lead to false negative results.

Some adverse effects may be of an aesthetical nature, showing disorder or mass, which should also be managed by different means using the tool box dedicated to reducing the ecological and human health risk of contaminants.

2 The contaminant is biologically non-accessible: Chemical time bomb! High risk!

Contaminants claimed to be non-hazardous in an unjustified way have to be fully separated from the intrinsically non-hazardous chemical substances. Seemingly non-hazardous contaminants may easily become hazardous under certain conditions. For example, those chemical substances that are hazardous via the atmosphere after evaporation are not risky at stable low temperatures, even if they are present at high concentrations or in bulk form. However, they are present and some risk of evaporation still remains (e.g. by lightning, fire, warming). Another example: undissolvable metal precipitates may be produced in the form of sulfides, e.g. from cadmium, zinc, or lead under anoxic conditions in a temporarily flooded three-phase soil or in bottom sediments. In the presence of pyrite (iron sulfide), arsenic, cadmium, copper, mercury, lead, and zinc co-precipitate with the pyrite. Metal sulfides are unavailable to biota, therefore the members of the ecosystem cannot perform uptake, and consequently the toxicity tests under anoxic condition fail to measure toxicity. But if such a soil/sediment gets dry and has the opportunity to come into contact with the atmosphere, the chemical form changes and the harmless substance becomes mobile, water-soluble, and biologically available. Stability of metal sulfides is strongly dependent on environmental conditions, as the "chemical time-bomb" easily explodes under oxygen-rich conditions. The mobilization of metals by the oxidation of sulfide minerals (or any other risk-increasing mobilizing affects) can be simulated or even provoked by "dynamized" simulation tests triggering the anticipated negative changes. Planning for these tests that show a realistic worst case for risk assessment purposes assumes good knowledge of the mechanisms influencing mobility/biological availability and the inclusion of appropriate negative and positive controls.

4.3.2.2   Adverse effects without analytically detected presence of the contaminant

Results of physicochemical analytical methods are negative, but visible or measurable adverse effects occur.

1   The contaminant has not been included into the analytical plan: Uncertainty! High or no risk

The first assessment plan of natural attenuation is mainly prepared based on historical information of former land uses, locally applied technologies, and on the results of previous assessments. It is typical for an abandoned industrial, mining, or disposal site that chemical substances with potential toxic effects may be missing from the analytical record, since some unknown/non-identified chemicals that no one is presently aware of could have been used on the site illegally. For this reason, stepwise assessment is important: repeated and refined integrated assessment is necessary until all questions are answered: in this case, toxic compounds or other agents with inhibitory effects should be identified.

2   Toxic transformation products are formed: Unknown scale of risk

During long-term interaction of the chemical substance with air, water, soil, and living organisms, toxic metabolites may be produced. In such cases, the results of ecotoxicity tests are the best decisive indicators. The measurement of toxicity by direct contact with environmental samples is a suitable risk assessment tool because the toxicity of the samples to living (test) organisms is associated with the risk, and a quantitative risk indicator, the risk characterization ratio (RCR = measured toxicity/no effect level), can be derived from the results. To identify the formation/presence of potential toxic metabolites for toxicity warning during the monitoring of NA processes, the mode of action, the biochemical mechanisms and the degradation pathway should be explored.

In summary, the integrated monitoring of natural attenuation and the comparative evaluation of the physicochemical, biological, and ecotoxicological results enable the identification of natural trends and the nature of attenuation, i.e. whether a real risk-reducing or just a concentration-decreasing (i.e. by dilution) process is in progress and what is the forecasted residual risk.

This integrated approach is extremely beneficial in the case of abandoned sites historically polluted by unknown, often complex contaminants. Using this approach, the shortcomings of the exclusive use of physicochemical models are avoided. On the other hand, the exclusive use of ecotoxicity tests, given the lack of chemical analytical results, enables quantification of the environmental risk (proportional to the scale of the adverse effect), setting of the target risk (no effect) and estimation for example of how many times the risk has exceeded the acceptable level or how many fold decrease is necessary to reach an acceptable no-effect level.

As MNA is always a long-term process, the model-based forecasts play a priority role. Using a transport model, e.g. to estimate the expansion or the final attenuation of a contaminant plume, the validation of the model can be done based on monitoring data. On an iterative basis, a realistic site-specific model can be developed from the original generic model. With the help of such transport models, planning and implementation provide information on:

- The scale of the necessary source reduction;
- The time until the plume stabilizes (does not continue to grow);
- The time of attenuation to the target concentration.

Modeling MNA in a contaminant plume can be based both on chemical concentrations and on the directly measured toxicity.

The chemical model evaluates the transport and fate of constituents to calculate an aggregated attenuation rate (see Figure 3.8), while the direct toxicity model is based on the measured (overall) toxicity attenuation (see Figure 3.9).

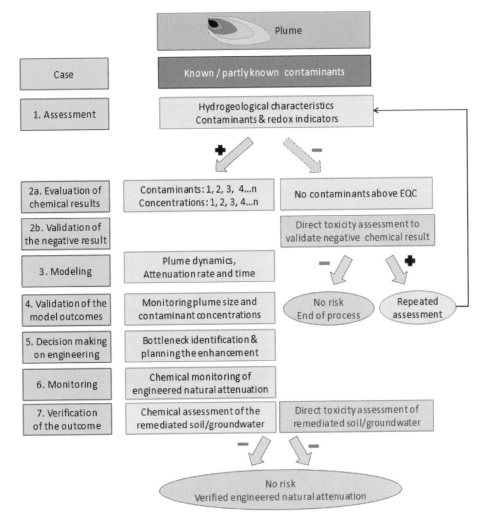

Figure 3.8 Chemical approach for the assessment, monitoring, and modeling of NA and decision making on its enhancement.

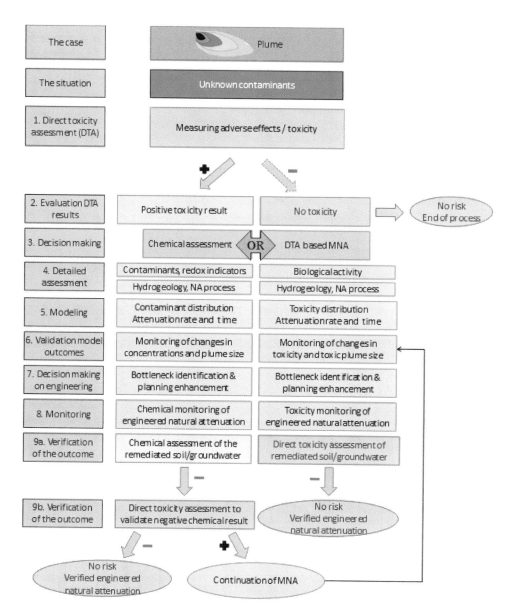

Figure 3.9 Direct toxicity assessment for the assessment, monitoring and modeling of NA, and decision making on its enhancement.

A well-established model such as the NAS (2018) requires the following data/information:

– Source data, plume length and width, thickness of the contaminated aquifer;
– Hydrogeologic and aquifer assessment data: hydraulic conductivity, hydraulic gradient, porosity, organic carbon content;

– Contaminant composition, concentration as a function of distance, redox indicators such as nitrate, sulfate, Fe(II), Mn(II), hydrogen, and their concentration as a function of distance.

From direct toxicity monitoring data as a function of the plume extension and time, similar modeling can be executed to determine the time requirement for natural attenuation to reach the no risk (no or acceptable adverse effects) situation. Ready-made software for directly measured toxicity-based modeling is not available. At unused sites and at those contaminated a long time ago, natural attenuation could already have eliminated the contaminants. For such cases, the DTA-based approach shown in Figure 3.9 offers an efficient solution involving no toxicity and no risk. This shortcut in the management process results in savings in costs and time. If the no risk state of the site cannot be verified, there are two possibilities:

1   Natural attenuation with DTA-based monitoring: this option is advantageous, when:

– Toxicity is not too high;
– Toxicity attenuation has been proven;
– The contaminant resupply is not significant;
– Enhancement is not necessary; or
– Biology-based enhancement is feasible.

2   Natural attenuation with conventional, concentration-based monitoring is advantageous if:

– The contaminants become known;
– All components can be monitored by chemical analysis;
– The degradation products are known and not toxic; or
– Chemical-related enhancement is planned.

4.3.3   Typical applications

The integrated physicochemical, biological and (eco)toxicological testing – i.e. the application of STT (Soil Testing Triad) – is best for those cases where the chemical model has limited use, i.e. where risk estimation is accompanied by high uncertainty from chemical analytical data such as in the case of the following:

– Abandoned land;
– Sites out of control;
– Sites with long-term contamination;
– The occurrence of more contaminants together;
– Unknown natural attenuation processes, unknown changes in the contaminants and their degradants;
– Complex environmental problems: several contaminants, several compartments, fast-changing situation.

The STT methodology is in particular useful for:

– Assessment and monitoring of abandoned contaminated land;
– Assessment and monitoring of natural attenuation;

- Monitoring of dynamic soil simulation tests to model natural attenuation;
- Monitoring of the enhancement of natural attenuation.

STT methodology supports the following:

- The quantitative, site-specific risk assessment;
- The quantitative, risk-based assessment of natural attenuation;
- Validation of the applied transport and fate models;
- Planning interventions into the natural attenuation;
- Controlling the technology to enhance natural attenuation;
- Controlling biological processes playing a role in natural attenuation;
- Controlling the residual risk of natural attenuation;
- Verifying the decision and the risk-reduction activity.

What cannot be controlled by the STT methodology is the uncertainty of sampling due to environmental heterogeneities. Spatial uncertainty can always jeopardize the environmental decisions.

4.4   Using the "risk profile" for planning, monitoring, and verification

The decision on the application of natural attenuation for reducing the risk at a contaminated site should be supported by technological, environmental, and economic efficiency assessment before (input for a feasibility study), during (input for MNA control), and after (verification of the remediation) NA implementation. A monitoring plan ensures the acquisition of those data that are necessary for the mass balance calculations, the estimation of time required and the enhancement of technological interventions, the trend and scale of risk reduction and the quantity of the residual risk. Monitoring provides data for cost estimation, which can then be compared to the costs of other remedial options or to the benefits from future land uses.

The risk profile, i.e. the change of the environmental risk in time, is the most important tool for engineering and managing natural attenuation. It is a forecast showing how the changes in the contaminants are likely to reduce or increase the extent of the predictable damage during a spontaneously occurring process in the environment.

The risk profile curve shows the quantitative variation of the risk in time. The risk is quantified in the simplest way by the ratio of the predicted environmental concentration and the predicted no-effect concentration (RCR = PEC/PNEC). PEC is a result of transport and fate modeling; PNEC is a result of extrapolation from a limited number of species to the whole ecosystem. In some cases, not the whole ecosystem but only some or one (largely protected, sensitive) species is targeted by NEC or NOEC values (since other ecosystem members are *ab ovo* protected when the target is calculated for the most sensitive one). Human risk *via* the environment is approached similarly. In this case, the predicted harmless (no-effect) concentration of an environmental contaminant will be calculated from available no-effect – likely harmless or acceptable – doses found in databases or in regulations. From the default limiting values (e.g. acceptable daily intake) and from the default intake rates (by inhalation, ingestion, or dermal exposure), the harmless environmental concentration can be calculated (see more in Gruiz *et al.*, 2015b).

In the planning and maintenance phases of natural attenuation, it is worthwhile to use default human and ecosystem limiting values, but for verification, the site-specific ecosystem and human population and land uses should also be considered (a safe threshold in a fishing lake is lower than in a large river, or the acceptable soil threshold should be much lower in hobby gardens, where vegetables are grown for long-term consumption by families/ children).

Evolution of the risk in time may be gradually or suddenly decreasing or increasing and changing as a function of time.

4.4.1   Risk profile of organic contaminants under natural impact

Natural biodegradation of organic contaminants and adequately engineered natural biodegradation usually lead to risk reduction. In the case of an ideally degrading microbiota, a continuously decreasing risk can be achieved during a relatively long term as shown in Figure 3.10. Biodegradation can be effectively enhanced by supplementation of the bottleneck substrate, which is most often oxygen in aerobic biodegradation and other electron acceptors in anoxic microbial processes. Due to toxic intermediaries or a toxic final product, a transitional risk increase or final residual risk may occur. In complex systems (several contaminants and several degradation products), the related risk increase is superimposed on the general decreasing trend.

Degradation, including biodegradation is not always well balanced. The lack of light/ sunshine can limit photodegradation, drought limits hydrolysis, and limited bioavailability and the lack of bottleneck substrates and co-substrates or the presence of toxic metabolites can complicate the simplified scenario of a seamlessly constant decrease.

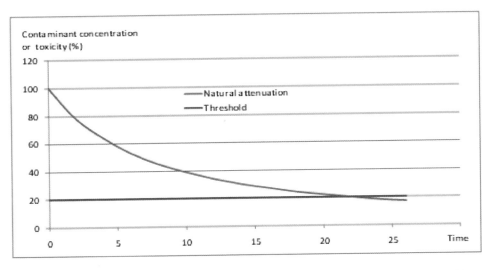

Figure 3.10 Risk profile of a well-balanced biodegradation-based natural attenuation process without the resupply of contaminants.

Table 3.1 Natural processes and their outcome for organic contaminants.

NA process for organic contaminants	Outcome of the process
Well-balanced natural degradation, including photodegradation, hydrolysis and biodegradation	Unequivocal risk reduction
Dilution by groundwater	Decrease in the source/secondary sources but increase in the contaminated volume or area. Local risk may be smaller, but dispersion may reach more sensitive receptors, e.g. water bases or conservation areas. Diffuse pollution and high background concentration are generated
Evaporation of volatile contaminants	Increased risk in atmosphere from an endless source
Partitioning from soil to groundwater or surface water	Increased risk in water, while negligible decrease in soil, so contaminated soil can be considered an endless source
Accumulation in high sorption capacity solid medium (e.g. in bed sediment) and in living organisms	Increased risk in the targeted medium and in accumulating organisms. Food chain/web effect: biomagnification
Partial or selective degradation	Persistent residue, residual risk
Limited biological availability	Delayed biodegradation and risk reduction
Toxic metabolites	Temporarily increased risk
Asynchronous mobilization and biodegradation	Temporarily increased risk
Enhancement of contaminants' availability	Increased mobility, faster leaching, faster biodegradation, temporarily increased but finally more rapid risk reduction
Stabilization by incorporation into soil's humus	Low mobility, low bioavailability, latent presence until humus degradation during soil deterioration
Enhancement of biological activity: ensuring optimal conditions for degrading/ transforming microbiota by optimal redox potential, suitable electron acceptor, pH, temperature, cometabolites.	More intensive biotransformation, biodegradation, bioleaching, biostabilization, biosurfactant production. Faster risk reduction

The natural attenuation processes involving organic contaminants and the result of these processes are summarized in Table 3.1.

4.4.2 Risk profile of toxic metals under natural impact

Some pollution sources containing toxic metals – such as abandoned mine wastes, hard rock and waste ore disposal sites, sediments disposed after dredging or by flood – give a typical increasing risk profile because weathering of these solid waste materials gradually increases the mobility of the contained toxic metals, which are then spread by leachates and runoffs. The decrease of the metal concentration in the source is accompanied by an

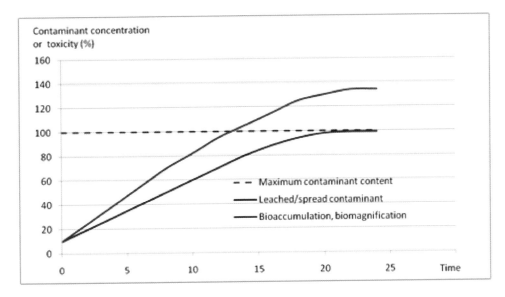

Figure 3.11 Risk profile downstream of a weathered metal-bearing mine waste.

increase in the vicinity, mainly downstream of the source. In such cases, the trend of the risk curve – characteristic for the downstream vicinity of the source – is increasing as shown in Figure 3.11. The risk increases until the full toxic metal content from the source becomes mobile and bioavailable (a chemical time bomb). Extrapolating the process in time, toxic metal content will spread from the contaminated area (functioning this time as a secondary source), continuously diluting and finally reaching a level that is considered not risky. But we cannot be happy because the background concentration has increased with the amount spread.

Figure 3.11 also shows the biomagnification risk profile indicating further risk increase in addition to the maximum risk in the soil. This surplus risk is posed to the ecosystem *via* bioaccumulation of interdependent members of the food chain.

The monitoring, the required technological intervention and the admissible land use may be planned according to the variation of the risk profile. The risk profile can also be used for the estimation of the benefits from any interventions such as monitoring, control, and enhancement.

Natural processes that reduce or increase risk with regard to metals and other inorganic contaminants are listed in Table 3.2.

4.4.3   Risk model of source management

Source management plays an essential role when using NA for remediation and also when enhancing it. Source removal is the only efficient solution to diminish the risk of non-degradable contaminants and significantly enhance risk reduction for the degradable ones. Large sources (deposits, leaking subsurface tanks and pipe systems, etc.) may release contaminants to

Table 3.2 Natural processes and their outcomes for metals.

NA process for contaminating toxic metals	Outcome of the process
Leaching of (not isolated) sulfide rock and ore or sulfur-containing material/waste	Increased risk in downstream water and soil, while negligible decrease in the source
Dilution of the contaminants by groundwater or erosion	Decrease in the source/secondary sources but increasing contaminated volume or area. The risk may be smaller but also greater if more sensitive receptors are reached. Diffuse pollution is produced with high background concentrations
Changes, e.g. increase in the redox potential	Increased metal mobility and increased risk in soil and groundwater
Changes, e.g. increase or decrease in pH	Increased metal mobility and increased risk
Changes in the chemical species of the metal in the presence of organic matter, iron hydroxides, carbonate minerals, H_2S or pyrite to form insoluble sulfides, silicates to incorporate metals	If mobility increases, the risk increases
	If mobility decreases, the risk decreases. Stability of the metal species is the key: irreversible stabilization results in long-term risk reduction
Biological methylation of some metals, e.g. Hg, As, Sb, Se, etc.	Highly increased risk, as the methylated forms are volatile and much more toxic than just the metal
Changed partitioning between physical soil phases	If mobility increases, the risk increases. If mobility decreases, the risk decreases
Weathering of metal-containing waste or rock	Increased mobility, increased risk
Immobilization/stabilization of metals in deeper soil and sediment layers	The risk of the water phase decreases, but the risk of the accumulating soil layer or sediment increases, creating a chemical time bomb
Adaptation of the ecosystem to contaminants	Smaller risk to adapted indigenous biota, but the same or increased risk, e.g. to humans, as the site looks healthy, or to the food chain in case of bioaccumulative contaminants
Revegetation of contaminated land	Risk due to lower metal mobility decreases, but a food-chain effect may significantly increase the risk
Flood	Deposition of contaminated sediment on soil: oxidation, mobilization and increased bioavailability increase the risk
Bioaccumulation	In plants and animals, including human nutrition
Biomagnification	Multiplied bioaccumulation along food chains and in food webs. Increased risk, mainly for predators and top predators

groundwater for decades. Small and declining contaminant sources may be liquidated by NA without removal, but it may increase the time requirement for NA-based remediation. For new sources, no matter whether it is degradable or not, whether small or large, the best management method is removal. The risk profiles of degradable and non-degradable contaminants are shown in Figure 3.12. In this figure, the risk is characterized by the risk characterization ratio RCR. When using the chemical model, RCR is interpreted as the ratio of the environmental concentration (a model prediction) to the no-effect concentration (default fixed by regulations or tested locally). If the toxicity approach is used, RCR is calculated as

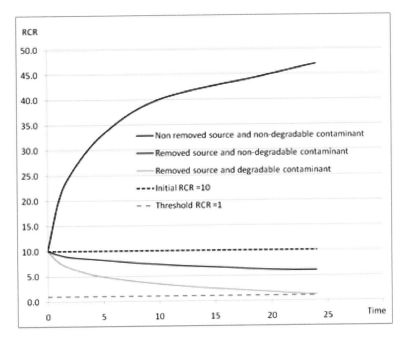

Figure 3.12 Impact of source removal on the risk profile of degradable and non-degradable contaminants.

sample toxicity/reference toxicity. Reference toxicity is often the non-toxic threshold (see also Gruiz *et al.*, 2015b; Gruiz, 2016a). According to the risk profiles:

1   Non-degradable contaminants in non-removed sources result in a continuously grow-ing amount of risky discharges, leachates and runoffs, growing subsurface contaminant plumes and a larger area exposed to adverse impacts.
2   Degradable contaminants of non-removed sources generate stable, shrinking or growing contaminant plume, depending on the flux ratio of degradation and addition from the source.
3   Removed sources of non-degradable contaminants – this case should be handled by sep-aration or immobilization to achieve smaller contaminant flux, shrinking underground plumes and decreasing risk.
4   Removal of the source of degradable contaminants results in the fastest contaminant flow reduction, shrinking plumes and the shortest time requirement.

MNA as the only remedial intervention is generally accepted for naturally decreasing contaminant flows, e.g. shrinking plumes. In the case of stable and growing fluxes, continu-ous release is very likely, so, source treatment or removal is necessary.

–   Case A: Removed source and degradable contaminant: the original 100% decreased to 18%, below the 20% threshold.
–   Case B: Non-removed source and non-degradable contaminant: constantly growing risk due to an increasing amount released.
–   Case C: Removed source and non-degradable contaminant: the original 100% decreased to 60% in the long term by dispersion, dilution, and other transport processes.

Degradable contaminants have a good chance to be eliminated from the soil and groundwater, but the final success depends on contaminant resupply from non-removed sources. The mass balance gives the answer to the question of whether or not degradation-based natural attenuation can overcome the continuous contaminant input (input into the soil but release from the source).

According to this relation, three cases may occur, as shown by Figure 3.13A.

NA can be enhanced in any of these three cases by source removal. Figure 3.13B shows the change of the risk curves after source removal at time point 8.

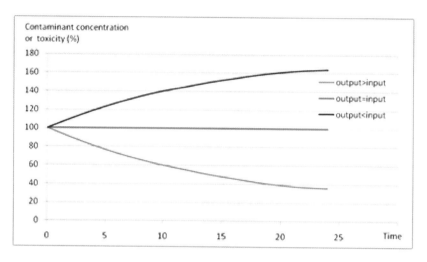

Figure 3.13A Non-removed source of degradable contaminant. Green: the original 100% decreased to 40%; reddish-orange: stable 100% – the input compensates the output; red: increase to 160%, in spite of degradation.

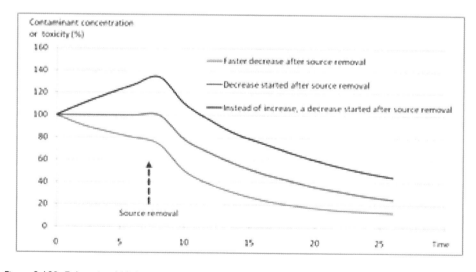

Figure 3.13B Enhancing NA by source removal of degradable contaminants at time point 7. Green: faster decrease below threshold; reddish-orange: stable 100% turns to decrease; brown: increasing risk (concentration or toxicity) turns to decrease.

5   NATURAL ATTENUATION IN ENVIRONMENTAL REMEDIATION

The assessment of natural attenuation, the monitoring data as well the prognosticated risk profile of a contaminated site, provide support for risk management, and for choosing between natural attenuation and engineered natural attenuation as the best possible and most efficient remediation option. Monitoring provides information for all management steps when using NA as a remedial option. The management steps are the following:

—   Studying the situation: stepwise site characterization and evaluation.
—   Conceptual risk model of the site.
—   Preliminary evaluation of evidence supporting NA:

    o   Do NA processes occur? Do they reduce risk? Partially or fully?
    o   Which kind of processes: sorption, dilution, degradation, partition, accumulation?

—   Treatability study: Only MNA? Source removal and MNA? Primary treatment and MNA? ENA and soft engineering? Other kind of engineering?
—   Modeling and refining the NA model with the stepwise expanding information and long-term prediction on time requirement and residual risk.
—   Feasibility study: Is NA efficient enough to serve as basis for remediation? Is it efficient enough compared to other options? Forecasting technological efficiency, environmental efficiency, costs and socioeconomic efficiency.
—   Planning the monitoring of NA to follow processes and validate the result of the model used.
—   Following the attenuation by problem-specific monitoring, characterizing environmental risk and checking the trends.
—   Validation and correction of the remediation plan and the conceptual model of NA.
—   Verification of the NA-based remediation: performance, efficiencies (technological, environmental, socioeconomic, cultural). Did it fulfill the planned objectives? Does it fulfill regulatory requirements in the long term?
—   Post-monitoring, long-term verification, sustainability: proving that the risk will further decrease or not increase again, if it has decreased once.

5.1   The natural attenuation management scheme

The preparatory and decision-making procedure for the application of natural attenuation to contaminated site remediation is the same as in any other remediation activity, except for an additional requirement to justify that NA fulfills remedial objectives in the long term (see Figure 3.14).

To control and verify natural attenuation at a contaminated site, thorough monitoring is necessary, including not only that of the contaminant in one soil phase but the soil conditions, the transformation (also the intermediate) products and the possible adverse effects. Especially important is to monitor the spread of the contaminant and the transformation products in such cases when the mobile contaminants can reach new areas and volumes with sensitive users. Monitoring, prevention, and control of such kinds of processes is of utmost importance. There are three typical cases for this type:

—   Volatile groundwater contaminants such as chlorinated solvents migrate from the surface of the groundwater upwards through the vadose zone and, depending on the depth

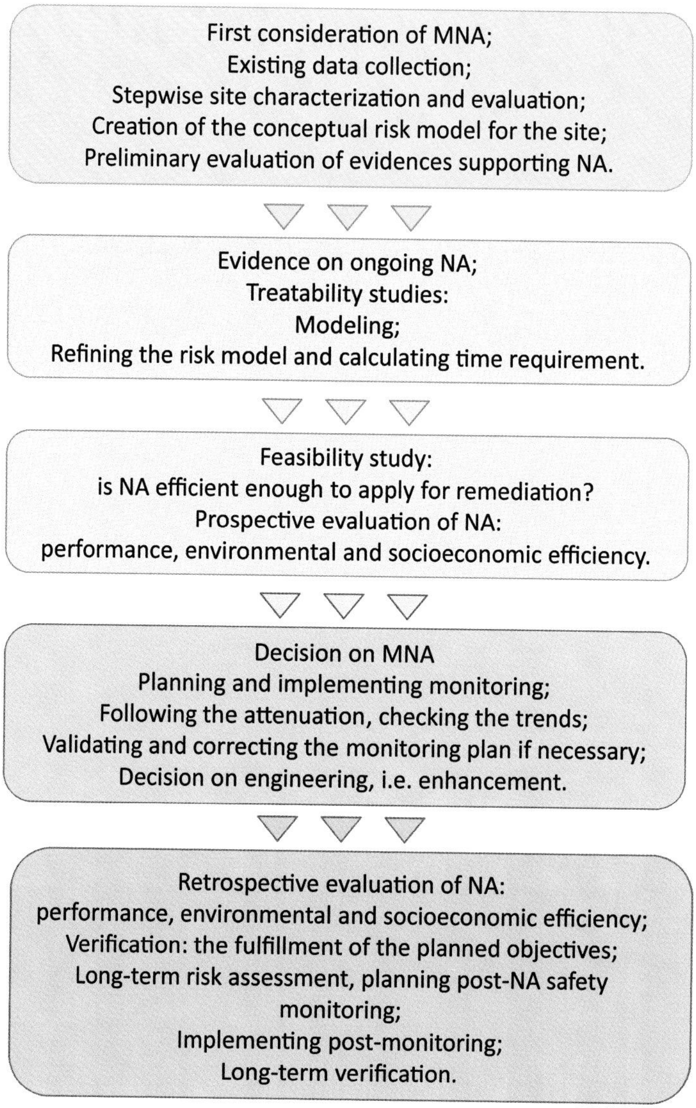

First consideration of MNA;
Existing data collection;
Stepwise site characterization and evaluation;
Creation of the conceptual risk model for the site;
Preliminary evaluation of evidences supporting NA.

▽ ▽ ▽

Evidence on ongoing NA;
Treatability studies:
Modeling;
Refining the risk model and calculating time requirement.

▽ ▽ ▽

Feasibility study:
is NA efficient enough to apply for remediation?
Prospective evaluation of NA:
performance, environmental and socioeconomic efficiency.

▽ ▽ ▽

Decision on MNA
Planning and implementing monitoring;
Following the attenuation, checking the trends;
Validating and correcting the monitoring plan if necessary;
Decision on engineering, i.e. enhancement.

▽ ▽ ▽

Retrospective evaluation of NA:
performance, environmental and socioeconomic efficiency;
Verification: the fulfillment of the planned objectives;
Long-term risk assessment, planning post-NA safety
monitoring;
Implementing post-monitoring;
Long-term verification.

Figure 3.14 Management of natural attenuation-based remediation: stepwise procedure.

of the plume and the permeability of the soil layer above it, can reach the atmosphere or enter buildings, if present, endangering people living, working, or spending time there.

– Water-soluble metal species from solid waste may contaminate huge volumes of water; subsurface water if buried and runoffs when located on the surface. In the latter case, additional solid transport worsens the situation.

– Solid-bound contaminants, which may form secondary pollution sources after random deposition and diffuse pollution in entire watersheds.

5.1.1   Pollution cases where monitored natural attenuation (MNA) is the best option

- In most of the abandoned and long-term contaminated sites, where nature found the way to heal itself without additional risks.
- Reduction of the risk of contaminated subsurface waters without any disturbance on the surface, e.g. at a residential area.
- Treatment of the groundwater and the soil without disturbing the surface in areas (e.g. at a natural conservation area) where disturbances are not desired.
- Large contaminated areas with contaminants at low concentration where the application of conventional remedial technologies such as excavation, pumping and treating extracted water, or placing and maintaining any machinery and operation is not costly.
- In combination with other remediation technologies, e.g. residual contamination after removal of the contaminant source or of the non-liquid phase from the surface of surface water or groundwater.

5.1.2   Pollution cases where engineered natural attenuation (ENA) is the best option

All cases where soft interventions can enhance the beneficial natural processes or reverse the seemingly disadvantageous natural processes are good candidates for engineered natural attenuation (ENA).

- Reversion of the natural process by physical technologies:
  - When a natural process appears to be disadvantageous because – as in the case of toxic metal mobilization – the result of leaching endangers the surrounding environment, isolation and leachate control can help;
  - Lowering or increasing water table;
  - Recycling soil air;
  - Recycling groundwater;
  - Controlling (directing, changing) flow direction of a plume by controlling hydraulic pressure and its spatial distribution;
  - Reversion of the natural process by physicochemical tools;
  - Collection and treatment of runoffs, leachates, eroded solid;
  - Supplementation of natural transport with subsurface filters, PRBs, wetlands, reactive soil zones, or other soft engineering tools to maintain its benefits but compensate for the disadvantageous natural processes.
- Supplementation and enhancement by biological tools:
  - Optimization of the natural conditions, i.e. pH, oxygen and nutrient supply, redox potential (by adding alternative electron acceptors), moisture content, additives to enhance or repel mobility, adjust redox potential and other "technological parameters";
  - Identification of the bottlenecks, i.e. lack of oxygen, nutrients, micronutrients, accessibility/availability of the contaminant, attenuation of cometabolic activity, unsuccessful adaptation of the microbiota;
  - Removal of the bottlenecks by adding nutrients, air or alternative electron acceptors, co-substrates (energy substrate) for cometabolic degradation/decomposition, increasing the redox potential with aeration/ventilation, decreasing the redox

potential (e.g. by temporary induction of aerobe processes for the consumption of atmospheric oxygen) or increasing the availability by the addition of mobility-enhancing agents or by the application of starter cultures (bioaugmentation).

Typical contaminants that natural attenuation has been applied to the following:

–   BTEX: main constituents of fuel hydrocarbons: benzene, toluene, ethylbenzene, and xylene;
–   CAH: chlorinated aliphatic hydrocarbons, with various degrees of volatility (VOC);
–   Chlorinated aromatic compounds;
–   Explosives such as trinitrotoluene (TNT), dinitrotoluene (DNT), or hexahydro-trinitro-triazine (RDX);
–   PAHs: polycyclic aromatic hydrocarbons;
–   Petroleum hydrocarbons, mineral oil derivatives;
–   Some pesticides;
–   Toxic metals and radionuclides;
–   Other inorganic contaminants such as nitrate, various salts, perchlorate.

5.2   Natural biodegradation and its enhancement

Remedial technologies relying on the activity of the natural microbiota may target the vadose zone, the saturated soil zone, or just the groundwater. Even if the groundwater is the targeted compartment, the solid phase will play a significant role in the implementation and control, while the efficiency of the natural attenuation-based technology will also be considered. Soil depth, pore conditions, pH and redox conditions, buffering capacities, habitat function, and the scale of heterogeneity will strongly influence the efficiency of the MNA or ENA and the uncertainties regarding the final risk.

5.2.1   The role of soil microorganisms in natural soil processes and in NA-based remediation

Once the soil gets contaminated, the soil microbiota reacts immediately to the new condition, i.e. the presence of the alien substance. In the case of a biodegradable contaminant, the species capable of degrading will proliferate better and those that can produce adaptive enzymes required for the degradation of the contaminant become activated, and the proportion of these species will increase within the microbial community. For example, diesel oil is an easily degradable contaminant. As a result of diesel oil contamination, the number of hydrocarbon-degrading cells in the soil increases in a few days from 10^2–10^3 cells/g of soil to 10^5–10^7 cells/g of soil. Behind the variations in species diversity in favor of the contaminant, changes of the metagenome (the totality of the genes in a community) can be identified. The flexibility of the metagenome makes these adaptive changes reversible, ensuring the health and the balanced functions and services of the soil – to a certain extent.

The huge genetic and biochemical potential of the soil and its continuous adaptation to contaminants is insured in the environment by various processes at population and gene level:

–   Change in the species distribution;
–   Switching to adaptive genes, and activating adaptive enzymes, enzyme systems;
–   Natural mutations and subsequent selections;
–   Increase of the mutation rate, i.e. due to the mutagenic effect of the contaminant;
–   Horizontal gene transfer, the direct transfer of useful genes from one individual to another, and their dispersion in the population and community. This process may occur

primarily in environmental compartments where the cells can meet and connect with each other, i.e. in biofilms.

The extent of natural biodegradation depends to a great extent on the soil type, soil air conditions, and nutrient supply. These characteristics may easily become limiting factors during spontaneous degradation, but in case they are available at the right time and in an adequate amount and the metagenome has optimally adapted to the contaminant, degrading microorganisms will completely degrade it to CO_2 and water under aerobic conditions. If the contaminant has a complex chemical structure and the degradation is only partial, degradation products can be formed. Adaptation of the microbiota to a mixture of contaminants cannot be perfect: in an optimal case, a multistep adaptation and degradation process can remediate the soil. In a worse case, significant residual contamination should be expected. Several degradation pathways require lower redox potential than aerobic degradation: environmental conditions/technological parameters should be adapted to the effective biochemical pathway.

Natural attenuation in saturated soils and groundwaters is a thoroughly studied process because it is one of the most frequent problems, as it poses a high risk to the environment for endangering water bases. Therefore, advanced technologies are available for monitoring and modeling it. Chemical analysis of the pumped groundwater provides data on contaminant concentration decrease and the redox potential or alternatively the quality and quantity of the electron acceptors (O_2, NO_3^-, Fe^{3+}, Mn^{2+}, SO_4^{2-}, etc.). The situation is certainly more complicated than that indicated by the analysis of the water samples, but the system may be modeled and handled as a quasi-packed flow through reactor. However, refined models can distinguish the processes occurring at the interfaces. Biodegradation in the saturated soil is relatively slow under anoxic or anaerobic conditions, so the extent of contaminant supply and the rate of biodegradation may be comparable in scale. When the scale of biodegradation compensates for or slightly over-compensates for the contaminant input, the polluted groundwater is under biological control, and the contaminant plume will not continue to grow/spread. But when the biodegradation is slower than the input supply, the plume will progress further. That is why the removal of the source is an essential task in contaminated land management in general and in the case of NA-based remediation, when the decision makers have to give NA a chance to heal our soils and groundwaters in the long term. A further limitation is that biodegradation occurs only on the surface of the plume, which is a relatively small volume compared to the total volume of the plume. The processes may be intensified, in addition to source removal, with the injection of additives (primarily electron acceptors, nutrients, and bioavailability-increasing agents) into the groundwater. The pH and redox potential can be monitored and controlled based on the information provided on the process status and on the biodegradation potential of the saturated soil.

The situation is different in unsaturated soil where the oxygen comes from the soil air and the speed of the microbiological degradation processes is an order of magnitude higher than that of the groundwater interface. The transport of the contaminant is also different. Instead of groundwater transport as a plume, diffusion and gravitational transport is typical from the surface or from a shallow location to a deep one by a separate phase or by water. The dissolved and mobile constituents of all three soil phases, including the mobile and less mobile fractions of the humus substances, play an important role in the C, N, and P metabolism of the unsaturated soil. The monitoring should cover not only the contaminants, serving as a carbon source for the microorganisms but the whole carbon cycle to get valuable information to evaluate and verify remediation by natural biodegradation. If the contaminant is a more complex molecule, e.g. with high chlorine (Cl) content, the related element cycle should also be monitored.

5.3 Engineered natural attenuation

5.3.1 A knowledge-based tool

Engineered natural attenuation (ENA) suggests a soft intervention to transform a natural process into a remedial technology with acceptable technological, economic, environmental, and social efficiencies. Soft interventions are compatible with the ecosystem, do not cause disturbances in ecosystem functioning, and represent a mixture of natural and engineering tools. The removal of a separable contaminant source is in most cases compatible with NA and a desired criterion. In general, reasonable isolation and separation of processes and mass flows may be able to change even risky processes to beneficial ones such as the supplementation of natural leaching with leachate collection and treatment. Groundwater extraction or recycling may restrain the contaminated groundwater plume from water bases by a properly established depression cone. Targeted depression cones can change the flow direction of the plume toward a natural sink or a filtration system. At the same time, depression lowers the water table and enhances aerobic biodegradation in the capillary fringe. A small or medium power blower or vacuum pump can exchange the soil air (soil venting) in a rather large soil volume and increase the redox potential for the benefit of aerobic biodegradation. Biodegradation of petroleum hydrocarbons may require a supplemental N source because the carbon surplus from the contaminant disturbs the proper C:N ratio, and the interventions aiming to increase bioavailability of hydrophobic contaminants result in natural biodegradation enhancement.

5.3.2 Enhancement of biodegradation-based natural attenuation

Enhancement of natural attenuation is the most efficient solution for living soils and biodegradable contaminants. It is reasonable to base the intervention on the reactor approach (see Chapter 2) by considering the impacted soil volume as a reactor and the biodegrading microbiota as the catalyst responsible for the rate of the transformation/degradation. Engineering ensures optimal conditions for the transforming/degrading microbiota (redox, pH, temperature, nutrient supply, etc.).

How can technologists optimize natural biodegradation? The technology should consider all three parts of the process: environmental conditions, contaminants, and the microbiota.

5.3.2.1 Setting optimal environmental conditions

The microbial community creates its food web, having the ability to utilize also the contaminants, but it has special needs determined by the aggregated individual nutrient requirements, pH, temperature, electron acceptors for energy production, etc. Those changes in environmental conditions that satisfy the needs of the original biota will trigger changes in the composition of the community that may or may not be ideal for the ultimate/final degradation of the contaminant. Engineered natural attenuation aims at strengthening that biota composition, which proves to be the best catalyst in the degradation of the contaminant that is present. If the contaminant is degradable by aerobic microorganisms, aeration, or an alternative oxygen source (e.g. an oxygen release compound) is necessary. If the best possible degradation takes place under a lower redox potential than the atmospheric value, then the conditions should be made more beneficial by lowering soil air oxygen (e.g. by isolating it from fresh air and letting it deplete oxygen levels by spontaneous aerobic respiration) and adding nitrate, Fe^{3+}, Mn^{4+} or other electron acceptors to ensure an optimal redox potential. If the degrading microbiota needs higher temperature, special nutrients, activators, or a cometabolite, the engineered NA enhancement should provide the optimal ranges or dosages. The most frequent interventions for NA enhancement are these:

- Aeration using bioventing;
- Nutrient and electron acceptor addition by injection;
- Flushing through existing extraction wells, push-pull tools or trenches, even by manual injection.

Groundwater flow and additive distribution can be adjusted in the soil at any rate and direction using air and water injection and extraction wells. The increasing scale of engineered intervention enables arbitrary 3D distributions to be designed and created. Thus multi-stage *in situ* processes or complex technologies can be implemented within the soil based on natural attenuation processes while maintaining microbiota viability and the habitat function of the soil (see more in Chapter 2).

5.3.2.2 Modifying contaminant availability

Environmental conditions not only strongly influence the microbiota but also determine the physicochemical state of the contaminant and the interaction between the microbiota and the contaminant. From the point of view of degradation, the mobility, and bioavailability of the contaminant is crucial, which can be influenced by physicochemical and biological tools.

To ensure the access of the catalyst and *the availability of the contaminant*, contaminant mobility as well as chemical and biological availability should be increased by desorption, volatilization, solubilization, shifting of partitioning toward mobile physical phases (soil air and water), plus chelate and complex formation. Thermal treatment, the application of cosolvents, surfactants, and chelating and complexing agents may lead to enhanced mobility and availability for physical, chemical, and biological transformation and degradation (see Chapter 2).

5.3.2.3 Bioaugmentation

If the microbiota is not capable of eliminating the contaminant(s) due to missing or inhibited degradation potential (missing genetics, low adaptive potential, low viability, etc.), but the contaminant is otherwise biodegradable, then bioaugmentation by living (micro)organisms may be necessary (see Chapter 2). Bioaugmentation works either by adding a great number of active microbes or ones with special degrading abilities. Products from living organisms specialized in direct degradation or increasing bioavailability of the contaminants can also be used as soil additives. These are typically degrading enzymes, which have *ab ovo* specific degrading abilities or biosurfactants, which are produced in great amount by some specialized cells.

The direct application of microbial preparations to soil raises a number of questions and issues. Our understanding of native, adaptive soil is limited, and often it is even less understood how the artificially produced supplemental microorganisms influence the soil community and its activities. Studying soil microbiota with conventional culturing methods, only a very small fraction of the microorganisms can be detected and isolated. Most of the cells, mainly the minor components of the community, remain hidden from the researchers.

Although today's DNA techniques allow the study of the complete genome – the metagenome – of the soil microbiota, the proper interpretation of the DNA results is often difficult.

The composition of a well-balanced and efficiently working soil community can hardly be imitated by artificially prepared bacterial mixtures, and this is justified from the many unsuccessful applications of inoculants made from some pure cultures of microorganisms. They are even more unsuitable for mixed pollutants, where a highly complex food web is behind the biodegradation. Their chance to remain active or even survive in the "wildness" of real soil is also questionable.

Laboratory study results on hydrocarbon mixtures and other biodegradable petroleum products prove that artificial preparations cannot surpass the adapted indigenous soil microbiota. However, the adapted microbiota can be promoted easily by satisfying their air and nutrient demand, so this way should be preferred if possible. On the other hand there are xenobiotics, which are completely non-degradable because the soil does not contain microbial species with the required genes of enzymes for degrading these alien molecules. Inoculation with a xenobiotic-degrading superbug in such cases may help. Even if the survival of the living cells is not ensured in the long term, interestingly their genes may "survive" and get incorporated into indigenous species and function until the xenobiotic is present and serving as energy substrate in the soil.

Ready-made inoculants, instead of accelerating the existing microbiota, may harm the well-balanced indigenous community, while aggressively consuming the easy-to-degrade components of a mixture, leaving behind a less degradable residual mixture, which will become degradable only after long-term adaptation. A laboratory microcosm study (unpublished results of the author) demonstrates the biodegradation of a high molecular weight, slowly degradable contaminant mixture of petroleum origin (given as extractable petroleum hydrocarbon = EPH) in soil without and with the addition of 1x and 5x inoculant concentrations. The inoculant was a self-developed mixture of isolated and in-vitro propagated microorganisms from the same soil that was inoculated. The 1x inoculant enhanced biodegradation evenly and reduced the final EPH content from 40% to 20% after 20 weeks, but the 5x amount caused a sudden decrease to 50% (consuming the easily degradable constituents). After this stage, the degradation stopped for a long while and the final degradation rate was smaller than without treatment (53% compared to 40% of the original 100%). The changes of the EPH content upon inoculation are shown in Figure 3.15. The increase in the initial phase is explained by a high

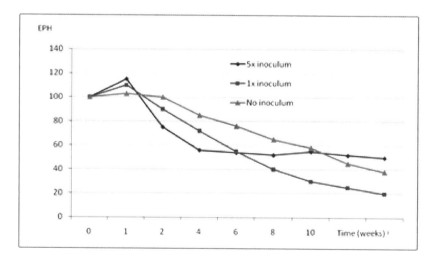

Figure 3.15 The effect of bioaugmentation by inoculant application in two concentrations.

mobilization rate (by biosurfactants), which overcomes the consumption rate. In the control soil, mobilization and consumption was equal during the first two weeks.

Bioaugmentation with artificial microbial inoculants –
applicability, advantages, risks

–   The soil inoculant should be specific both to the contaminant and the soil. Inoculants are not omnipotent.
–   It is recommended to differentiate nutrients, mobilizing additives (typically surfactants) and soil inoculants containing living microbes and the relevant effects. The commercially available products are generally a mixture of these.
–   The living cell count of the inoculant and the recommended dose give a picture of the likely effect when comparing with local or generic cell counts and microbial activities of soils.
–   The inoculant should be applied to contaminant biodegradation only in cases when:

   o   The indigenous microbiota is inactive and cannot be activated by providing convenient soil conditions for them;
   o   The fresh contamination is completely "unknown" to the local native microbiota;
   o   The contaminant is toxic or otherwise inhibits the local native biota.

–   The inoculant can be beneficial and its application is recommended in cases when:

   o   The contaminant is persistent, non-degradable under generic soil conditions, and the local microbiota have not adapted or could not adapt;
   o   Contaminant degradation requires special enzyme systems of specialized microorganisms;
   o   Cometabolic degradation should be applied, and the useful microbes are not present in the soil;
   o   Local adaptation of the microbiota would need too long of a time.

–   The application of alien microorganisms in the inoculum may have risks, given that viability of the inoculant microorganisms is questionable at the new place, and too many or too aggressive additional microorganisms can disturb the balance of the endogenous community by:

   o   Competition for available nutrients;
   o   Competition for the contaminant;
   o   Production of antibiotics or other substances with inhibitory effect;
   o   Consumption of easy-to-degrade nutrients and contaminants, which leaves a disadvantageous mixture behind;
   o   Partial degradation with toxic end products.

5.4   Summary of engineered natural attenuation

Risk reduction is typically the joint result of physicochemical and biological processes in soils contaminated with organic chemicals. Processes of fundamental importance are mobilization of contaminants with focus on partitioning between phases and biodegradation. The transport and dilution are always present as accompanying processes that have to be taken into account during modeling and calculations. The soil microbiota plays a primary role in the mobilization/immobilization and in the biotransformation/biodegradation processes.

Thorough assessment, modeling and planning as well as treatability studies are necessary before making the decision on the most suitable engineering tools to enhance natural attenuation. It is an increasingly popular solution that natural biological processes are combined with soft physicochemical treatment technologies such as the use of mild oxidizers or solubilizing agents together with biodegradation, *ex situ* water treatment with *in situ* biodegradation in the vadose zone, or combined chemical and phytostabilization.

The risks associated with *in situ* operations, the use of additives and the increased contaminant mobility should be continuously managed. Seemingly innocent processes, for example contaminant biodegradation, can lead to highly toxic intermediates of final products.

To manage engineered natural attenuation, the "reactor approach" (see Chapter 2) should be adopted by considering the soil volume affected by NA processes as a quasi-reactor and the operations necessary for monitoring and enhancement, as part of the remedial technology.

6 HISTORY AND STATE OF THE ART

Natural attenuation is an emerging concept all over the world. The huge number of contaminated and potentially contaminated sites and the finalized or likely numbers of yearly implementations of remediation show a deep gap. This justifies the use of NA for contaminated site remediation, even if it has a relatively long time requirement (between 3 and 15 years based on published results). In addition to limited human capacity, further benefits from NA are that costly technologies can be avoided and the site, its neighborhood, the ecosystem, and human land uses remain undisturbed or only very slightly disturbed. Of course, the long-term monitoring may generate expenditures (even as much as 30 years of monitoring may be required and this is equivalent to 100,000–300,000 USD). However, the costs of environmental monitoring as part of the aftercare of conventional remediation are at a similar range.

6.1 Growing utilization of natural attenuation for soil and groundwater remediation

Evidence has been presented on the existence and efficacy of natural attenuation already in the nineties in the US by the National Research Council (NRC, 1994) and in Canada by Lawrence Livermore National Laboratory (LLNL, 1995). These studies suggest that natural bioremediation is a general phenomenon at sites contaminated by petroleum hydrocarbons and related substances and is responsible for the attenuation of these pollutants in groundwater (Hinchee & Leeson, 1996; Bradley *et al.*, 1997; Landmeyer *et al.*, 1998, 2001; Wise, 2000; CIR, 2000a,b; Chapelle *et al.*, 2002a,b). The first technical protocol was prepared by the US Air Force Center for Environmental Excellence (US AFCEE) (Wiedemeier *et al.*, 1995d). Newell and his coworkers in 1996 worked out the BIOSCREEN (1997) decision support tool for natural attenuation and its feasibility as a risk reduction option (Newell *et al.*, 1996). Rifai *et al.* (1997) modeled contaminant transport by plume and developed software BIOPLUME (1997) that is a two-dimensional contaminant transport model for use under the influence of oxygen, nitrate, iron, sulfate, and methanogenic biodegradation.

A similar decision support system was created soon afterwards for dissolved chlorinated solvents (Aziz *et al.*, 1999), known as BIOCHLOR (2002). These software tools are still in use as screening models that simulate remediation through natural attenuation of petroleum and chlorinated hydrocarbons dissolved in groundwater under both aerobic and anaerobic conditions (see also US EPA Models, 2018).

In parallel to developing new monitoring methodologies and establishing the screening models to simulate remediation, the AFCEE Monitoring and Remediation Optimization System (MAROS) software has been developed for long-term monitoring optimization at contaminated groundwater sites in accordance with the Optimization Guide. This topic came into the fore because long-term monitoring (for process control, performance measurement, or compliance purposes) results in very high cumulative costs (several contaminants, attenuating activities, adverse effects, many monitoring points and long-term, i.e. years). The idea behind optimization is improving the efficiency of the monitoring system by optimally compiled, optimally timed *in situ* and real-time measuring techniques (Aziz *et al.*, 2000, 2003).

The US Environmental Protection Agency (US EPA, 1997, 1999a,b) incorporated MNA into the risk management of superfund and underground storage tank sites (OSWER Directive, 1999), dissolved fuel hydrocarbons (Wiedemeier *et al.*, 1998a; US EPA, 2011), and chlorinated solvents (Wiedemeier *et al.*, 1998b; Wiedemeier & Haas, 1999; Bradley *et al.*, 2000; US EPA, 1998a; Looney *et al.*, 2006; US EPA, 2007a). US EPA issued useful materials about natural biological processes and their assessment (US EPA, 1998b, 1999a,b) and the American Society for Testing Materials prepared a standard guide for groundwater remediation using NA (ASTM, 1998). The US Department of the Navy (USDN, 1998) and the Washington State Department of Ecology published technical guidelines on MNA for their own contaminated sites (WSDE, 2005). Later on, specific technical protocols were established for assessing biodegradation of organic contaminants (US EPA, 2008), especially for chlorinated organics (ITRC, 2008) and for chlorinated solvents (US EPA, 2007c, 2013). For volatile organics (US EPA, 2004, 2012), several field methods, protocols, guidelines, and strategies (e.g. US EPA, 2012) as well as projects were developed, such as the Wisconsin one for chlorinated solvents (WDNR, 2014). The United States Geological Survey (USGS, 2018a,b) – approaching practical and economic conditions – made calculations for the time requirement of remediation based on MNA (Chapelle *et al.*, 2003). The evaluation of MNA of metals and other inorganics (US EPA, 2007a,b; Ford *et al.*, 2007) and the decision framework for inorganic contaminants and radionuclides (ITRC, 2007, 2010; US EPA, 2015) all together established the basis for wider acceptance and application of MNA for contaminated land, mainly groundwater remediation (see also Clue-in, 2018b).

In spite of a significant amount of literature – books, studies, and protocols – and the growing number of applications, many questions still arise and doubts hinder wider acceptance, so it is still necessary to publish convincing overviews and reports on success cases and good practices.

Pachon (2011), in a presentation at the SNOWMAN Conference on MNA, summarized the US state of the art (Superfund, 2013):

– MNA is being used as part of long-term response actions for groundwater;
– MNA works as a finishing step for low levels of contamination, especially petroleum-related, chlorinated solvents and inorganics;
– Selecting MNA as a remedy shows growing tendency.

From the latest Superfund report, the application of MNA in percentages is shown here:

- 35% of Superfund sites with MNA combined it with some form of treatment;
- 21% of sites selected MNA alone with no additional form of source control or groundwater treatment.

A FAQ from the Environmental Security and Technology Certification Program (Adamson & Newell, 2014) provides a concise overview of current knowledge regarding the management of subsurface contaminant releases using monitored natural attenuation. Several research programs of the US Strategic Environmental Research and Development Program (SERDP, 2018) aim for better understanding, assessing, and monitoring as well as modeling of natural processes (Mass transfer, 2015; NA of TCE, 2015; CAH degradation, 2015; CAH diffusion, 2015; NMR for MNA, 2015; CSIA, 2015).

The early research by the USGS (Wiedemeier *et al.*, 1996b; Chapelle *et al.*, 2003) on the calculation of the necessary time for hydrocarbons NA, led to the development of the first version of the software (Widdowson *et al.*, 2005). Further research on biodegradation modeling of various organic compounds (Blum *et al.*, 2007) made the software more precise and generally useful. NAS is today a software package that provides a decision-making framework for determining the time needed to clean up groundwater contamination sites (NAS, 2018). The package has been upgraded several times and the newest version

- compares times of cleanup associated with monitored natural attenuation to pump and treat remediation;
- expands the kinds and numbers of contaminants considered and
- allows for concurrent consideration of solvents (chlorinated ethenes) and petroleum hydrocarbons (USGS, 2018c).

In Europe, Sinke and le Hencho (1999) prepared an overview on existing guidelines and protocols, and the following year, UK's Environment Agency published a relevant guidance document (Carey *et al.*, 2000). This guideline zooms the definition of natural attenuation for groundwater and risk-reducing natural processes. This narrowing may simplify the regulation and communication for practitioners working with small contaminated sites but does not support a more holistic approach, which requires the integrated evaluation of natural processes in soil, groundwater, air, surface water, and sediment and deals with larger spatial scales.

The Germany-based DECHEMA (2018), recognizing the importance of NA in environmental management, organized the first European Conference on Natural Attenuation in Heidelberg, which was followed by the second and third ones in 2005 and 2007 in Frankfurt am Main (NA Frankfurt, 2005, 2007).

One of the early projects including NA and MNA for groundwater remediation was the IN-CORE project (INCORE, 2003), which covered several aspects of NA, evaluated case studies on the possible use of NA as a remedial option, mainly for BTEX and PAH (Bockelman *et al.*, 2001a; INCORE Case Study 9, 2003). A methodology for the quantification of the mass flux and attenuation rate was first published by Bockelmann *et al.* (2001a,b,c; 2003). The development of monitoring and modeling methods did not stop with expanded knowledge but to the contrary: the new analytical, biotechnological, and molecular methods were immediately applied and transformed from laboratory-based techniques to *in situ* field methods (e.g. Berghoff *et al.*, 2014).

The European Network for Industrially Contaminated Land (NICOLE, 2018) prepared a cartoon booklet for better understanding (NICOLE, 2005) and put sustainability of NA in focus from 2007 (NICOLE, 2007).

The European SNOWMAN MNA (2018), being active from 2009, summarized the state-of-the-art information here based on a European survey in six countries (Declercq, 2011):

- MNA is a recent concept, a well-known remediation method in Europe;
- There is a certain level of heterogeneity across countries that will continue to exist, but nevertheless similarities between the different countries do exist;
- MNA is implemented in different ways such as stand-alone, parallel or follow-up action;
- Lack of "return on experience"!

To enhance dissemination and information transfer, the SNOWMAN Network organized a conference in 2011, the first demonstration of how the network wants to facilitate the exchange of knowledge on soil remediation between countries (SNOWMAN, 2018; MNA Conference, 2011). SNOWMAN stated that MNA presents a challenge in terms of technique and performance to harmonize with soil policy, the sociological impact, and financial and legal implications. They hope for a breakthrough via knowledge exchange and the inclusion of a broader audience. In this spirit, several general reviews and position papers were prepared for science and policy professionals (e.g. Rügner *et al.*, 2006; LABO, 2011).

An interesting view came from industry, showing openness toward MNA, saying that nature works for free and the existence of guidelines (in some countries) supports its application. At the same time, essential tasks have been marked (Jacquet, 2011) such as the following:

- Implementation on the market should be enhanced;
- Europe-wide inclusion into local and national legislations is necessary;
- Harmonization of the time frame for NA with administrative and management time frames;
- Relaxing the rigid implementation of protocols (from the side of authorities);
- The approach should fit into a risk-based management of contaminated sites;
- NA can be considered in any project;
- NA need not be considered as a stand-alone solution only;
- Long-term uncertainties must be reduced;
- Long-term sustainability is a prime target;
- Investment is needed in NA demonstrations;
- Changing ownership is a problem yet to be solved;
- Stakeholders must be practical when using NA as a remedial option.

Canada developed a study to assess evidence of NA at upstream oil and gas facilities, established a database and analyzed the site characterization and monitoring information for 124 sites (CAPP, 2002; Epp *et al.*, 2002).

Database review showed that the greatest amount of information was related to petroleum hydrocarbon contamination, particularly benzene, toluene, ethylbenzene, and total xylenes (BTEX). Measured data demonstrated that natural attenuation occurs for petroleum hydrocarbons: 47% of the plumes were interpreted to be stable, 26% as shrinking, and 6% as expanding.

No correlation was found between the hydrogeological factors such as geology, permeability, flow velocity, and plume classification. Plume lengths did not appear to correlate with groundwater velocity (data uncertainty was identified) but tended to be greater for inorganic and non-petroleum hydrocarbon plumes compared to petroleum hydrocarbons. Bacterial activity was identified in all monitoring wells except one with high salinity. Sulfate reduction was the typical biodegradation reaction. Estimated biodegradation rates tended to be lower than values reported in the literature, possible causes including cooler temperatures in Canada.

Published studies reflect the presence of biological and toxicity-based tools in the assessment and monitoring of NA in the early years. Biodegradation rates were measured to characterize NA by Suarez and Rifai (1999), ecotoxicity and genotoxicity in parallel with chemical analysis by Helma *et al*. (1998), a cell bioassay was performed by Schirmer *et al*. (2004a,b) for monitoring pollution and its attenuation.

The Australian Department of Environmental Protection (DEP, 2018) started in 2004 with MNA for groundwater remediation (DEP, 2004). After collecting data on background and realized cases (McLaughlan *et al*., 2006; Beck, 2010), CRC CARE published the national framework in 2013 on how MNA can be applied for groundwater contaminated with hydrocarbons and how to manage its risk to protect humans and the ecosystem (CRC CARE, 2013; CRC CARE, 2018). The guidance recommends a tiered methodology for MNA:

1   Preliminary assessment from technological feasibility and economic, social, and legal aspects (time frame, liability, sustainability, etc.).
2   Initial evaluation of NA and technological indicators.
3   Detailed characterization, evidence for an efficient NA.
4   Comprehensive monitoring of the performance and verification of the technology.
5   Documented closure and demonstration that the requirements set have been met.

Some MNA or ENA cases of the last 30 years from around the world will be discussed in the next section. In addition to the thoroughly studied and published cases, several sites have remediated/healed themselves spontaneously, without any human intervention. These sites were definitely polluted during former industrial, agricultural or mining activities, or inadequate land use and showed significant presence of the contaminant in the first assessment rounds, the reason why they were put on the contaminated site list. Nonetheless, if the time was long enough between the preliminary and detailed assessments, their detailed assessment showed that in time they became slightly contaminated or non-contaminated. Unfortunately, these cases are not always properly documented, so the uncertainties are significant. Assessment failures and overestimation can also play a role.

6.2   Some examples worldwide of MNA and ENA

In Table 3.3 some typical case studies are collected from the last 25 years to demonstrate the viability and feasibility of natural attenuation as a remedial option for soil and groundwater. The list includes the most frequently targeted contaminants worldwide. The development of the concept is also nicely reflected by the cases. The early applications were designed to prove that NA works at all. Then the benefits were compared to conventional engineered technologies, and, ultimately, they were acknowledged as a full-fledged sustainable technological solution. The acceptance has grown significantly and it resulted in a better understanding and cooperation between engineers and nature.

Table 3.3 Collection of published applications of MNA for environmental management – case studies.

Contaminants	Site	Problem	Way of application MNA only or together with others	Reference
TPH and other organic chemicals	Denver, Colorado, US	Contaminated groundwater	MNA + ENA based on intrinsic biodegradation	Wiedemeier et al., 1993
TPH: fuel hydrocarbons	Airport, US	Contaminated groundwater	MNA + ENA based on intrinsic biodegradation	Wiedemeier et al., 1994, 1995b, c
TPH: fuel hydrocarbons	Mining area, Tatabánya, Hungary, EU	Diesel-contaminated soil	ENA based on intrinsic biodegradation enhanced by soil venting in the vadose soil zone	Gruiz and Kriston, 1995
BTEX	Hill Air Force Base, Utah, US	BTEX plume and light NAPL (jet fuel)	NAPL recovery + MNA of dissolved BTEX	Wiedemeier et al., 1995d
BTEX	Patrick Air Force Base, Florida, US	BTEX plume	MNA: intrinsic bioremediation with long-term monitoring	Wiedemeier et al., 1995d
TPH	Jet-fuel tank farm, Hanahan, South Carolina, US	Contaminated groundwater	MNA and ENA: natural and engineered remediation	Vroblesky et al., 1996, 1997
TPH: fuel hydrocarbons BTEX	Four US air force bases: Hill AFB, Utah, Elmendorf, Alaska, Travis, California Langley, Virginia, US	BTEX plume	MNA: technology demonstration	Air Force Center, 1999
BTEX	Traverse City Coast Guard Base, Michigan, US	BTEX plume	Extensive NA of BTEX was measured Source removal by pump and treat + MNA of the plume	CIR, 2000b Wilson et al., 1990
BTEX	Bassendean Sands in the Perth basin, Western Australia	BTEX plume	MNA: natural degradation rates were obtained from a groundwater tracer test	Thierrin et al., 1993
BTEX and MTBE	Vandenberg Air Force Base, California, US	Fuel leak. Persistent MTBE in a Fuel Spill	MNA: While BTEX attenuated rapidly, MTBE did not	Durrant et al., 1999 CIR, 2000b
Benzene from gasoline	Swan Coastal Plain, Western Australia	BTEX plume	MNA: intrinsic benzene mineralization	Franzmann et al., 1999

(Continued)

Table 3.3 (Continued)

Contaminants	Site	Problem	Way of application MNA only or together with others	Reference
BTEX from gasoline	Swan Coastal Plain, Western Australia	BTEX plume	MNA: intrinsic sulfate reduction of toluene	Franzmann et al., 2002
PAH: fuel hydrocarbons and lubricant oil	Agricultural machine station, Hungary, EU	Contaminated soil and groundwater	Intrinsic biodegradation occurs, enhancement by soil venting and by mobilization/solubilization using cyclodextrins	Leitgib et al., 2008 Leitgib et al., 2005 Gruiz et al., 1996
BTEX	124 upstream oil and gas facilities in Alberta, Canada	BTEX plumes, other toxicants in some cases	MNA results: 47% stable plumes 26% shrinking plumes 25% growing plumes	CAPP, 2002 Cross, 2002 Corseuil et al., 2000
BTEX	NICOLE demonstration site, EU	BTEX plume	MNA feasibility study: NA is active, remediation can be based on it	NICOLE MNA, 2005
BTEX and mineral oil	NICOLE demonstration site, EU	Contaminant plume	NA occurs, and can be protective and efficient enough	NICOLE MNA, 2005
TPH, VOC, toxic metals	Trollberget, an abandoned dump site outside the city of Hanko, Southern Finland, EU	TPH plume and metals in the soil	MNA: biodegradation-based NA has been proven by an integrated assessment. Residual risk is acceptable for recreational land use	EU Life Project, 2006
TPH	Népliget, Budapest, Hungary, EU	Inhibited transformer oil both in soil and groundwater	ENA with soil venting and availability enhancement. Comparison with pump and treat	Molnár et al., 2009 Gruiz et al., 2008 Gruiz et al., 1996
TPH	Australia	TPH plume and capillary fringe smear	Source removal + MNA based on intrinsic biodegradation	Naidu et al., 2012
PAHs	South Glens Falls, New York, US	Coal tar remaining in groundwater from former coal tar production	Source removal + MNA based on microbiological transformation and chemical immobilization in humus	Taylor et al., 1996 Wilson and Madsen, 1996 Madsen et al., 1991

Contaminant	Site	Contamination	MNA/ENA description	Reference
BTEX, TPH, PAH	NICOLE demonstration site, EU	Inherited industrial site	MNA is highly efficient by itself	NICOLE MNA, 2005
PAH, mazout (heavy fraction of petroleum)	Mazout storage ponds, Budapest, Hungary, EU	Inherited industrial site with mazout-contaminated soil and diesel-contaminated groundwater	Source removal + MNA based on intrinsic biodegradation (bioavailable PAHs) and parallel immobilization of the large-molecule PAHs by humidification	Molnár et al., 2002, 2005; Molnár, 2006
PAH, creosote (coal tar)	Wood preservation plant, Hungary, EU	Creosote-contaminated soil	MNA: NA potential based on biodegradation under aerobic or anoxic conditions. 10–50% residual PAH do not pose significant toxicity due to immobilization	Molnár et al., 2007
BTEX	40-acre site in a coastal region of New Jersey, US	BTEX plume	ENA: surficial application of sulfate salts (gypsum, Epsom salts) over a groundwater plume. Infiltration process provides elevated sulfate concentrations and facilitates ENA: 4 years monitoring demonstrated the viability of surficial sulfate delivery	Kolhatkar and Schnobrich, 2017; Kolhatkar, 2015
Explosives: trinitrotoluene and other nitrogen-rich explosives	Near Hawthorne, Nevada, US	Shallow aquifer contaminated by nitrogen-rich explosives	MNA: natural microbiological and geochemical processes degraded dissolved explosives leaving nitrate behind	Van Denburgh et al., 1996
Explosives: dinitrotoluene (DNT)	Near Hawthorne, Nevada, US	Shallow aquifer contaminated with dissolved DNT	MNA: intrinsic mineralization of DNT in groundwater has been proven	Bradley et al., 1997
2,4,6-trinitrotoluene (TNT) 1,3,5-trinitro-1,3,5-hexahydrotriazine (RDX) degradation products	Louisiana Army Ammunition Plant, Minden, LA, US Joliet Army Ammunition Plant, Joliet, IL, US	Explosives in groundwater	Source removal + MNA: primary treatment: soil excavation and incineration MNA based on intrinsic biodegradation	Pennington et al., 1999a,b

(Continued)

Table 3.3 (Continued)

Contaminants	Site	Problem	Way of application MNA only or together with others	Reference
Munitions	Louisiana Army Ammunition Plant site, active 1942–94, near Doyline, Louisiana, US	Groundwater contaminated from production wastewater stored in a "pink lagoon"	Source removal + MNA-based bioremediation	Toze et al., 2003; Zakikhani et al., 2003; Pennington et al., 1999a,b
CAH: bromoform, carbon tetrachloride; PCE;1,2-dichlorobenzene; hexachloroethane	Borden Air Force Base, Canada	Mixed plume	MNA: partial degradation: mass of bromoform, dichlorobenzene and hexachloroethane decreased, but that of carbon tetrachloride and PCE did not	CIR, 2000b; Criddle et al., 1986
CAH: TCE	St. Joseph, Michigan, US	Contaminated groundwater	MNA: extensive anaerobic NA of CAH due to coincidental occurrence of organic pollution in the groundwater, which satisfied the high organic demand of the anaerobic biodegradants present	CIR, 2000b; Lendvay, 1998; Dolan and McCarty, 1995; Semprini et al., 1995; Haston et al., 1994; Wilson et al., 1994; McCarty and Wilson, 1992
AH	South Municipal Water Supply Well Superfund site in Peterborough, New Hampshire, US	2,200-foot long plume of dissolved contaminant	Source pump & treat + MNA of the dissolved CAH in the plume	Turley and Rawnsley, 1996
CAH: trichloroethene; cis-1,2-dichloroethene; vinyl chloride; and BTEX	Plattsburgh, Air Force Base, New York, US	Soil and groundwater of a fire training area	MNA: significant natural biodegradation by iron(III) reduction and methanogenesis. BTEX degradation resulted in oxygen, nitrate, and sulfate depletion	Wiedemeier et al., 1996a, c
CAH: solvents and degradation products	14 different US air force bases, US	DNAPL plume	MNA: anaerobic dechlorination at 13 out of 14 sites. Enhancement proved to be feasible	AFC, 1999

Contaminant	Site	Type	Description	Reference
CAH: trichloroethene	Picatinny Arsenal site in Morris County, New Jersey, US	TCE plume	MNA: based on intrinsic biodegradation and compared to pump and treat	Smith et al., 1999; Imbrigiotta and Ehlke, 1999
CAH: solvents	Twin Cities Army Ammunition Plant, Site A, MA, US	DNAPL plume	MNA: anaerobic dechlorination proved to be efficient after source removal	Ferrey et al., 2000
CAH: di- and trichloroethene, vinyl chloride, dichloropropane and degradation products	Brooklawn site Petro-Processors Inc. (PPI), Baton Rouge, LO, US	Solvent production	MNA: anaerobic biodegradation alone fulfills criteria	Clement et al., 2002
CAH chlorobenzene and BTEX	Contaminated megasite Bitterfeld/Wolfen, Eastern Germany, EU	Contaminated groundwater	MNA in a strictly anaerobic environment based on methanogenesis, sulfate and iron reduction, reductive dechlorination	Heidrich et al., 2004; Wycisk et al., 2003
Mixed: CAH: VOC, BTEX, mineral oil, styrene	NICOLE demonstration site, a harbor area, EU	Inherited industrial site	NA of VOC occurs, MNA is a feasible option	NICOLE MNA, 2005
Mixed: CAH: VOC, TPH, BTEX, styrene	NICOLE demonstration site in EU	Inherited industrial site	MNA occurs and proved to be protective for all contaminants. A buffer zone is recommended	NICOLE MNA, 2005
CAH: VOC	NICOLE demonstration site in EU	Inherited industrial site	NA is weak by itself, the redox potential is not supportive. ENA is recommended by redox modification	NICOLE MNA, 2005
CAH: chlorinated ethenes	Fairbanks, Alaska, US	Groundwater plume, cold environment	MNA + ENA with HRC additions for reductive dechlorination	USGS, 2016d; Bradley et al., 2005; Bradley and Chapelle, 2004

(Continued)

Table 3.3 (Continued)

Contaminants	Site	Problem	Way of application MNA only or together with others	Reference
CAH: trichloroethene (TCE)	Offutt AFB in eastern Nebraska close to the city of Omaha, US	TCE contaminated groundwater plume	ENA: biowall from mulch-sand mixture. Slow degradation of (organic) mulch provides substrates to generate hydrogen, the electron donor in reductive dechlorination	Early, 2007
CAH: dichloroethene (DCE) and vinyl chloride (VC)	Several TCE contaminated sites, US	No or less accumulation of DCE and VC under anoxic conditions, compared to the calculated value	MNA based on aerobic mechanisms contributes to DCE mineralization even as low as 0.01 mg/L dissolved oxygen!	Bradley, 2012
CAH: tetrachloroethene, trichloroethene, and their degradation products.	Medley Farm Superfund Site, South Carolina, US	A former waste solvent dump	ERD by lactate solution injection into GW	O'Steen and Howard, 2014
CAH: chlorinated ethenes; chlorinated benzenes	Naval Air Station Cecil Field, Jacksonville, Florida, US	Disposal pit for spent fuels, oils, solvents, paints	MNA + in situ air sparging	USGS, 2016b
Chlorinated benzene	Naval Air Station in Pensacola, Florida – demonstration site, US	Groundwater plume	In situ oxidation by passive curtain of ORC + MNA	USGS, 2016c
CAH: chlorinated ethenes	Textron Realty Operations Inc., Niagara Falls, New York, US	TCE plume from a neutralization pond	Source remediation + MNA in the plume	USGS, 2016e
Tetrachloroethene and vinyl chloride	U.S. Naval Submarine Base Kings Bay, Georgia, US	Leaking landfill	MNA + chemical oxidation	USGS, 2016a et al., 2007 Chapelle and Bradley, 1998, 1999

Chloroethenes and sulfuric acid together	Naval Air Station in Pensacola, Florida – demonstration site, US	MNA feasibility under acidic conditions: NA is not precluded, so it can be combined by source treatment causing acidification	Bradley et al., 2007
TCE and breakdown products	Boeing, Auburn, 20 locations Washington, US	263 active sampling points, stable and shrinking TCE plumes, MNA is a feasible solution for TCE and breakdown products	LAI, 2016
56 organic compounds, e.g. chloroform 1,2-dibromo-3-chloropropane 1,2-dichloropropane 1,3-dichloropropane 1,2,3-trichloropropane ethylene dibromide, dinoseb	Brown & Bryant Superfund Site Arvin, California, US	Pesticides, herbicides, fumigant formulation A Superfund site	Source treatment + MNA	USACE, 2012
Perchlorate	Naval Surface Warfare Center, Indian Head, MD, US	Perchlorate in groundwater	MNA based on natural reductive biodegradation; Performance and cost evaluation	ESTPC, 2010
Volatile organic contaminants (VOCs)	Aberdeen Proving Ground – West Bank Canal Creek, US	Volatile contaminants	ENA with a reactive mat that sorbs VOCs and enhances their biodegradation by allowing longer processing times for microbial breakdown and by VOCs phytovolatilization and phytodegradation	Majcher et al., 2009 Wein et al., 2007
PCBs	The Hudson River, Manhattan, US	PCBs in sediment due to discharges from a PCB manufacturing facility	Incomplete NA: combined aerobic-anaerobic biodegradation process, very slow and stops without enhancement, so ENA is recommended	Bedard and Quensen, 1995; McNulty, 1997 Quensen et al., 1988

(Continued)

Table 3.3 (Continued)

Contaminants	Site	Problem	Way of application MNA only or together with others	Reference
PCB and dioxins	Several places around the world	PCBs and dioxins in soil and groundwater	Slow and partial degradation. Enhancement is necessary by surfactants, nutrients, electron acceptors and/or donors (alternating aerobic and anaerobic conditions) and co-substrates. Evidence mainly from microcosms. Field degradation rate is generally not significant or very small, long-term MNA is needed	Billings, 2014 Nelson et al., 2014 Viisimaa et al., 2013 Wang and Oyaizu, 2011 Krumins et al., 2009 Ahn et al., 2008 Fava et al., 2002, 2003 Kao et al., 2001 CIR, 2000b Juhasz and Naidu, 2000
PCBs	Santa Susana Field Laboratory, California, US	PCB-contaminated aerobic soil	MNA based on biodegradation of PCBs unlikely, also because no anaerobic conditions exist at the site	Nelson et al., 2014
Dioxins	Santa Susana Field Laboratory, California, US	Dioxins contaminated aerobic soil	MNA based on biodegradation is slow, time requirement is proportional to the concentration: estimated as 1–50 years under aerobic conditions	Nelson et al., 2014
Red mud: Alkalinity and Na ion	Ajka, Hungary, EU, red mud spill	Na-ions and pH in flooded soil	NA of Na-ions were monitored for two years: fast attenuation was observed in the flooded soil	Gruiz et al., 2012 Rékási et al., 2013
Red mud Metals: Al, As, Co, Cr, Mo, V	Ajka, Hungary, EU, red mud spill	Metals, Na-ions and alkalinity in water and sediment	Metals accumulate after partition/ sorption in sediment. Gypsum addition lowered metal release	Mayes et al., 2011 Gruiz et al., 2012 Renforth et al., 2012 Klebercz et al., 2012 Lehoux et al., 2013 Anton et al., 2014
Metals: arsenic	Cape Canaveral Fire Training Area, US	Arsenic in groundwater	Arsenic MNA: adsorption to the sediment particles from the plume	Reisinger et al., 2005

Contaminant	Location	Contamination	Remediation approach	References
Metals: copper, cobalt, nickel, zinc	Pinal Creek Basin, US	Copper, cobalt, nickel, zinc in a groundwater plume	NA: multiple physical, chemical, microbiological processes ensure NA within 2 km of the plume. ENA: pH increase by carbonate enhances precipitation and sorption	Brown et al., 1997 Stollenwerk, 1994 Brown and Harvey, 1996 CIR, 2000b
Metals: Fe, Zn, Pb and sulfate	Mining area Murcia, Spain, EU	Metals in sediment and soil	Sulfate-reducing bacteria (SRB) produce sulfide, which precipitates ionic metals under anaerobic conditions. Metals bond to sulfide and carbonate	Jong and Parry, 2006 Van Roy et al., 2006 Bernardes et al., 2007
Metals: As, Cd, Zn and Pb	Abandoned mining area, Gyöngyösoroszi, Hungary, EU	Metals in sediment and soil	Source removal/isolation + enhanced immobilization of metals by adding steel shot, carbonate rock, and fly ash for sorption to iron hydroxides, carbonate minerals and formation of insoluble silicate minerals. Planting crops results in humus formation, and sorption on organic matter	Gruiz et al., 2006, 2009a,b Feigl et al., 2007, 2009, 2010
Chromium VI	100 Area, Hanford, US	Widespread Cr VI in groundwater endangers drinking water and Columbia river	Cr VI reduction occurs in two-phase soil and sediment. Sediment's reductive capacity is active in the presence of dissolved oxygen	Truex et al., 2015
Radionuclides: uranium	Hanford, US	Uranium contamination	Source removal + NA	Ford et al., 2009 Waichler and Yabusaki, 2005 Yabusaki et al., 2008 CIR, 2000b US EPA, MNA, 2018

(Continued)

Table 3.3 (Continued)

Contaminants	Site	Problem	Way of application MNA only or together with others	Reference
Radionuclides: [137]Cs	Chernobyl, Ukraine	[137]Cs contamination in water, soil, and ecosystem	NA processes: Radioactive decay Surface runoff Vertical migration in soil Immobilization in sediment and organic soil Biogeochemical migration	Chernobyl, Ukraine, 2018 Konoplev, 2017 Grygolinska and Dolin, 2003 Shestopalov et al., 2003 Sobotovich et al., 2003 Davydov et al., 2002 Ivanov and Dolin, 2001 Ivanov and Kashparov, 2003 Ivanov, 2006 Panin et al., 2001 Sadolko et al., 2000 Kudelsky et al., 1996
Radiocesium	Fukushima, Japan	[137]Cs contamination in water, soil, and ecosystem	NA processes: radioactive decay, surface runoff and flood, vertical migration in soil immobilization in sediment, organic soil and flooded soil, biogeochemical migration NA is faster compared to Chernobyl due to 3x times higher yearly precipitation and flood	Konoplev, 2017 Konoplev, 2016a Konoplev et al., 2016b Konoplev et al., 2016c Al-Masri et al., 2015
Multiple contaminants: tar oil, crude oil, gasoline, smoldered lignite-tar oil, and sulfuric acid, caustic soda, phenol, dimethylformamide	Tank farm of a former hydrogenation plant (lignite-based gasoline production) at megasite Zeitz, Germany, EU	Multi-contaminant plume	Verified intrinsic sulfate reduction and methanogenic process. NA with long-term monitoring is feasible for regional water protection	Schirmer et al., 2004a, 2006 Gödeke et al., 2004
CWA VX: O-ethyl S-(2-[diisopropylamino]ethyl) methylphosphonothioate	Indoor surfaces at any place	Attenuation time after application	99% NA within 10 days, on various materials under different environmental conditions	Oudejans et al., 2016 Columbus et al., 2012 US EPA, 2010 Munro et al., 1999

MNA: monitored natural attenuation; ENA: engineered/enhanced natural attenuation
TPH: total petroleum hydrocarbon; BTEX: benzene, toluene, ethylbenzene, xylenes; PAH: polycyclic aromatic hydrocarbons
CAH: chlorinated aliphatic hydrocarbons; TCE: trichloroethene; CWA: chemical warfare agent; VOC: volatile organics
ORC: oxygen release compound; HRC: hydrogen release compound; ERD: enhanced reductive dechlorination.

6.3   Examples from the authors' practice

In the early 1990s, the withdrawal of the Soviet army and the changes in basic ownership by privatization of state property in Central and Eastern Europe revealed an enormous number of contaminated sites. These sites could not be remediated because of the lack of financial resources, preparedness, and capacity, but several site assessments have been carried out mainly for valuation of the necessary remediation and compensation between owners. These assessments opened the door for research and development on the affected sites. Authors of this chapter also had the opportunity to assess and monitor former industrial and mining sites, and to test several new engineering tools and implement innovative technologies, including MNA and ENA, as well as the combination of MNA with conventional engineered technologies (Gruiz, 2002, 2009).

6.3.1   Engineered natural attenuation of soil and groundwater contaminated by organic substances

In this section, examples are introduced for monitored and engineered natural attenuation from the author's practice. The examples include sites polluted with organic and inorganic (metals) contaminants, environmental compartments such as soil groundwater and surface water, and examples of combined technologies. All the examples are related to actual contaminated sites and pilot or full-scale field applications.

6.3.1.1   Spontaneous biodegradation and its enhancement in the vadose zone

In relation to natural attenuation, generally groundwater plumes are meant although soil surfaces, runoffs and leachates, surface waters and the inside of the three-phase soil are also highly exposed to natural transport and fate processes, and spontaneous contaminant attenuation with various consequences. In this section, several cases are introduced, where the ongoing natural processes have been utilized for remediation of the three-phase soil, or the three-phase soil and the groundwater together.

Positive impact of soil venting and nutrient addition – a success story

Diesel oil, a biodegradable organic contaminant (i.e. hydrocarbon) mixture was identified at 0–3 m depths at a former coal mine area at Tatabánya, Hungary. The site is characterized by good quality soil and active intrinsic biodegradation with a soil microbiota well-adapted to the mixture of contaminants. The remediation of the abandoned site had started spontaneously prior to the first assessment. According to microcosm study results, the microbial degradation could be enhanced just by aeration, so a soil venting technology had been established in the field.

A rather low air flow rate – replacing pore air only two to five times daily – was enough to provide an adequate amount of atmospheric air for aerobic biodegradation and to remove the accumulated carbon dioxide. Figure 3.16 shows the soil venting scheme and the position of the soil gas extraction wells (middle row) connected to the blower and the atmospheric air "injection" wells (outer two rows), which consisted of perforated tubes (or just boreholes) for passive air input into the deeper soil layers. Alternatively, fresh air could be supplied

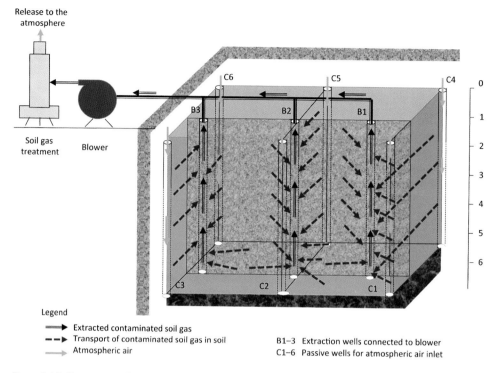

Release to the
atmosphere

Soil gas
treatment   Blower

Legend

⟹ Extracted contaminated soil gas
- - ▶ Transport of contaminated soil gas in soil
→ Atmospheric air

B1–3 Extraction wells connected to blower
C1–6 Passive wells for atmospheric air inlet

Figure 3.16 Bioventing scheme.

from the soil surface or through trenches instead of passive wells/boreholes. The air collected from the extraction wells by the connected blower may be subject to the gas treatment
depending on the contaminant. In this demonstration case, no volatile contaminant components were present. Intermittent nutrient supply (commercial nitrate-containing fertilizer) was
applied twice.

The gas chromatographic composition of the contaminant did not show significant
selection amongst the components of the extractable hydrocarbon fraction after biodegradation, compared to the initial. This indicates that the soil microbiota was well prepared to mineralize all (extractable) components of the contaminant mixture at a more or less similar rate.

Table 3.4 data – gas chromatographic (GC) contaminant concentration of the hexane-
acetone extract – reflects effective degradation of the 1000–20,000 mg/kg diesel oil within
2 and 3 months (Gruiz & Kriston, 1995). Soil reached the quality criteria (1000 mg/kg for
this industrial site) after 3 months, and the environmental risk of the site also decreased to
the acceptable level as shown by the risk profile in Figure 3.17.

Failed enhancement at a long-time contaminated site

Another area of the Tatabánya, Hungary, coal mine, filled up with gravel and cinders,
was contaminated with diesel oil and engine oil at 0–1.5 m depths forming a poor-
quality soil-like medium. The hydrocarbon content was high (30,000 mg/kg), but the

Table 3.4 Contaminant concentration (GC) in soil during the bioventing field experiment.

Depth (cm)	Sampling point 1 (mg/kg)			Sampling point 2 (mg/kg)		Sampling point 3 (mg/kg)	
	Start	8 weeks	12 weeks	Start	8 weeks	Start	8 weeks
0–50	3,600	420	320	6700	520	650	320
50–100	8600	1500	300	19,500	210	920	310
100–150	12,700	3,360	580	19,200	540	990	280
150–200	8800	1,790	400	18,700	840	960	370
200–250	6000	1,010	260	15,000	460	860	110
300–350	3300	990	200	400	ND	240	80
350–400	9600	800	200	100	ND	650	ND

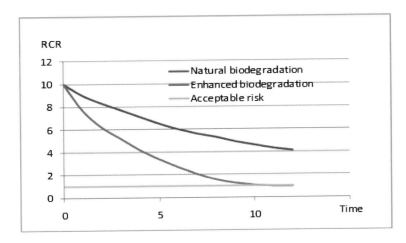

Figure 3.17 Risk profile of natural attenuation and of engineered natural attenuation at a diesel-contaminated site.

concentration of oil-degrading cells and the biodegradation rate was very low. Aeration and nutrient supply in treatability studies showed a low biodegradation rate. Gas chromatographic analysis of the contaminant explained the failure: the high share of non-paraffin isomers and large-molecule cyclic hydrocarbons. These are typical residues of an unbalanced biodegradation where the easily biodegradable components were consumed by the microbes and thus the amount of less degradable/available components increased. The main reason identified for poor degradation and unsuccessful activation by aeration and nutrients was the poor soil quality (bare gravel-like material), providing an improper habitat for soil microbes. Organic waste and compost as well as a self-developed inoculant prepared from a well-adapted soil were added to the soil, which accelerated biodegradation in microcosm studies (Figure 3.18). In spite of this possibility, the decision was taken for the soil not to be remediated *in situ*, but rather extracted and its place refilled with new soil.

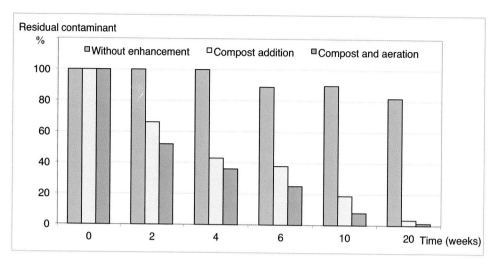

Figure 3.18 Biodegradation-based remediation after soil organic amendment and aeration (microcosm study for the Tatabánya, Hungary, coal mine contamination).

6.3.1.2 Enhancement of MNA with bioaugmentation

Natural bioremediation has limitations such as the bioavailability and degradability of the contaminants. From the catalyst perspective, it raises the question on whether the soil microbiota is able to produce biosurfactants and degrading enzymes in an adequate quantity and quality or to intensify the biodegradative capacity of the soil biota to detoxify and/or eliminate the contaminant.

The example of a transformer station at the Hejőcsaba, Hungary, cement works demonstrates that the degrading capacity of the soil biota may show high heterogeneities depending on soil conditions. The pollution at the transformer station originated from the product of TO-40, a PCB-free inhibited transformer oil, a mixture of C15–C50 hydrocarbons, stored and used at the site. The contaminant concentration in the soil showed a heterogeneous distribution with a maximum of 78,000 mg/kg (see Table 3.5) in the 1–1.5 m deep surface layer. The water table has not been affected.

Treatability studies proved that the biota could be activated significantly.

The most contaminated 20×20 m subarea was selected for demonstration purposes. The compacted soil was loosened, nutrients and water were injected by manual tools, as shown in Figure 3.19. Frequent soil sampling and analyses showed a sudden and significant decrease in contaminant concentration to 500–800 mg/kg in most parts of the site, but at some places the decrease failed to go below 13,000–20,000 mg/kg concentration. Looking for the reason, one of these troublesome spots was explored. Hard clayey aggregates completely impenetrable to air were found at a depth of 1.5 meters. Aerobic biodegradation was slow in these aggregates, but the anoxic degradation process was active. Therefore, two artificial inoculants were applied to enhance remediation: a self-developed mixture of bacteria isolated from the clayey aggregate and propagated in the lab and a commercially available "microaerophilic superbug". The two inoculants could reach similar results. The contaminant depletion in the soil is shown in Table 3.5: the clayey aggregates resisted enhancement with only aeration, but inoculation with adapted microorganisms brought success at the end (reddish brown highlights).

Table 3.5 Enhanced natural biodegradation-based soil remediation at Hejőcsaba, Hungary, cement works.

Soil sample	Contaminant content – initial (mg/kg)	Contaminant content – after 6 weeks (mg/kg)	Treated by inoculant – after 6 months
1	3200	500	80
2	6600	600	100
3	10,000	800	250
5	25 000	4000	90
6	25,000	7300	890
7	78,000	5300	600
8	28,000	19,000	90
9	n.d.	13,000	700
10	n.d.	20,000	880

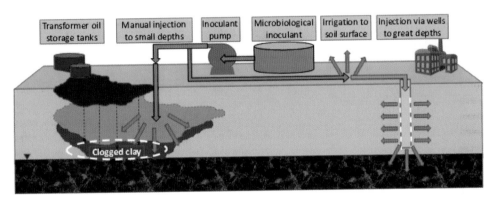

Figure 3.19 Inoculant delivery options: spraying onto the surface or into trenches for spontaneous infiltration, using injection wells or a handheld direct-push tool with an injector.

6.3.1.3   Natural attenuation enhanced by mobilization and bioventing

A mixture of diesel and engine oil at the Kutricamajor, Hungary, former agricultural machine yard with gas stations polluted both the soil and groundwater. The volume of the subsurface soil (a heterogeneous refill) contaminated with 1000 mg/kg average concentration of extractable hydrocarbons (EPH) was 280 m³. The highly contaminated central core with an average of 10,000 mg/kg contaminant was 20 m³ in volume. The highest contaminant concentration was 28,800 mg/kg at a 1.5 m depth.

The groundwater level was 3±1 m, and the EPH content 0.1–145 mg/L groundwater. The volume of the contaminated groundwater was estimated at about 600 m³.

The microbiota was active and could easily be further activated. The hydrocarbon-degrading cell concentration was 10–1000 CFU/g soil and 1–100 CFU/mL groundwater.

Biodegradation was enhanced by soil venting (one central active and five passive wells around) and by contaminant mobilization/solubilization using RAMEB = randomly methylated β-cyclodextrin (see also Fenyvesi *et al.*, 2016). The aim of the field study was to compare the efficiency of only biodegradation-based soil remediation with parallel application of pump and treat technology for the groundwater. The solubilizing agent was introduced into the three-phase soil intermittently by impulsive flushing and extracted together with the groundwater. For this purpose and to limit contaminant transport, a slight depression was established by water extraction (5 m³/day) parallel to soil aeration and contaminant mobilization.

Technology monitoring included continuous or daily analyses of CO_2 in extracted soil gas and the EPH content in the extracted groundwater. Soil sampling and analysis was done before and after remediation in order not to destroy the soil gas and groundwater flow pattern by additional drilling.

Figure 3.20 shows the increase of extractable contaminants in the water phase after RAMEB treatment: significant increase in CO_2 production (mainly in the vadose zone), in parallel to contaminant depletion in the water phase as a function of time and repeated treatments. Aeration in itself (for 18 weeks) did not increase the degradation rate, but aeration together with cyclodextrin application significantly enhanced biodegradation. It is typical for aged, strongly bound, large-molecule contaminants, similar to those at Kutricamajor that these strongly bound soil contaminants need to be mobilized to become available for biodegradation (Leitgib *et al.*, 2008; Gruiz *et al.*, 1996).

The repeated application of RAMEB confirmed the feasibility of the combined enhancement of a natural biological process, and the stepwise decreasing mobilizable amount of pollution of petroleum origin. The trends are clear in spite of an unexpectedly occurring alien contaminant (in groundwater flow), most likely stemming from uncontrolled fertilizer release independently of the study site.

The material balance is based on the mass of contaminants in soil and groundwater, its transfer from soil to groundwater by flushing and its elimination by biodegradation in the

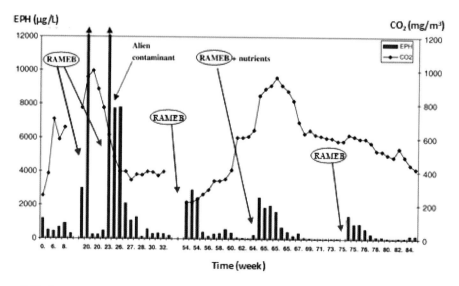

Figure 3.20 Contaminant mobilization and degradation enhanced by aeration and cyclodextrin addition to soil at the Kutricamajor, Hungary, former agricultural machine yard.

form of CO_2. Together with the contaminant, a certain part of soil organic matter was also degraded, so the amount calculated from the formed CO_2 was shared between the contaminant and the soil organic carbon, based on soil TOC and contaminant quantitative analyses: 408 kg of contaminant was eliminated from the 300 m³ soil and only 7 kg from the groundwater according to the calculated values; and 5 kg from the 7 kg contaminant derives from the three-phase soil (flushed into the groundwater after temporary flooding) and only 2 kg from the groundwater.

This type of mobilization aided by a combination of aeration and cyclodextrin addition with or without water treatment was applied for several demonstration cases in Hungary, as shown in Table 3.3. It was a general observation that the removed contaminant mass dissolved in groundwater was a minimum two orders of magnitude smaller than the contaminant mass from soil removed by biodegradation. This also proves that pump and treat alone cannot be a feasible solution if both soil and groundwater are contaminated.

The risk profiles in Figure 3.21 illustrate the biodegradation-based natural attenuation of bioavailable contaminants (green curve) and of the less available ones, which should be mobilized from time to time to artificially increase the size of the bioavailable fraction, or otherwise degradation would slow down. From the risk curves it is also clear that the mobilization of toxic contaminants results in a temporary risk increase, which should be managed.

6.3.1.4   Combination of soil ENA with physicochemical groundwater treatment

Soil and groundwater pollution was found at a former transformer station in Budapest, Hungary. The contaminant was an inhibited transformer oil, a PCB-free insulating oil, a mixture of C15–C50 hydrocarbons. The transformer oil pollution derived from a leaking spare transformer and reached the groundwater. The concentration of the contaminant in the soil was

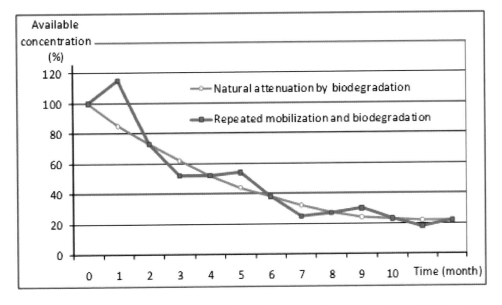

Figure 3.21 Engineered natural attenuation of bioavailable and artificially mobilized contaminants.

about 20,000 mg/kg (site-specific quality criteria: 500 mg/kg) and close to 1 mg/L in the groundwater (site-specific limit value: 0.5 mg/L). Intrinsic bioremediation has been proven, and enhancement was necessary because the biodegradation was slow. Soil biota could be activated by aeration and nutrient addition.

Based on treatability studies in microcosms, a complex technology was designed and applied for the remediation of a more-or less isolated subsite at the large-scale contaminated transformer station.

The 50 m³ soil and 1000 m³ groundwater at the subsite was treated for 47 weeks including a winter break of 20 weeks. The soil and the groundwater were treated under the heavy spare transformer without its removal as shown in Figure 3.22.

The *in situ* bioventing of the unsaturated soil was combined with groundwater recirculation: extraction and desiccation in gravel-filled trenches, ensuring the moisture supplement and the continuous flushing of the highly contaminated unsaturated soil. A combined well was used for pumping out groundwater and exhausting soil gas. Nutrients (N, P, K) were added three times at the 9th, 13th, and 21st weeks, and 2 × 10 kg RAMEB (to mobilize the transformer oil) was added to the soil on the 9th and 13th weeks.

Technology monitoring included continuous *in situ* soil gas analyses and frequent (minimum two daily) groundwater sampling and laboratory analyses. The solid soil was analyzed using core samples before and after remediation.

Biodegradation is demonstrated in real time by the increased CO_2 and decreased O_2 content of the constantly extracted soil gas after the first and second RAMEB and nutrient addition (Figure 3.23). The extracted water was partly recycled and partly treated by *ex situ* stripping.

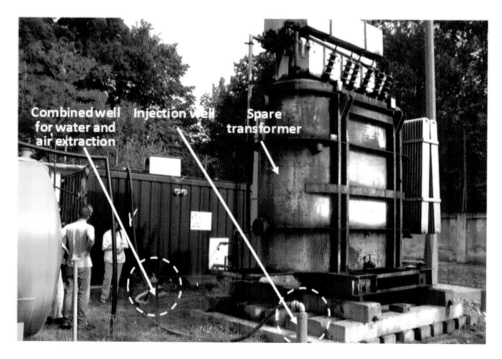

Figure 3.22 Bioventing of a soil volume underneath an unused transformer.

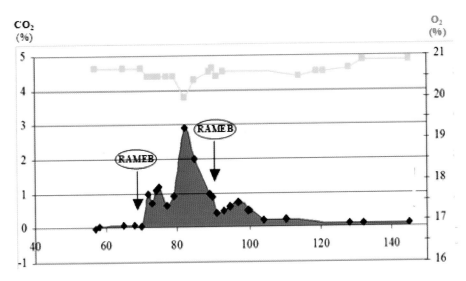

Figure 3.23 CO_2 (dark blue) and O_2 (green) content of the extracted soil gas measured *in situ* and in real time.

The soil and groundwater cleaning was completed in 47 weeks: 99% of the transformer oil was removed and the toxicity tests measured no toxicity both in the soil and in the groundwater.

This field application demonstrated that engineered natural attenuation can be applied in combination with groundwater extraction and *ex situ* water treatment (pump and treat) (Molnár *et al.*, 2009; Gruiz *et al.*, 1996, 2008). The verification and sustainability assessment of the complete case is introduced in Chapters 5 and 11.

6.3.1.5   Engineered natural attenuation of mazout-contaminated soil

Abandoned mazout storage ponds without isolation have been used for years to store heavy heating oil for industrial purposes. The mazout is rather dense oil; its transport in the soil's pore volume is strongly limited: it has a kind of self-isolating character. This was the argument for its storage in non-insulated ponds. Moreover, it would not have spread to the extent we experienced in the course of the assessment if it had not been regularly diluted with diesel oil in the winter cold to be able to be pumped out of the reservoirs. At this point, the transport of the diesel-dissolved mazout escalated and directly polluted a large soil volume and the groundwater. The groundwater has already been treated using pump and treat for many years without any success when our team investigated the site and recommended source removal and source treatment.

With regard to intrinsic biodegradation, mazout in the soil showed limited bioavailability, so its treatability was tested in laboratory microcosms by the addition of the mobilizing agent cyclodextrin (RAMEB), besides the implementation of aeration and nutrients supply. RAMEB increased the mobility and biological availability of the high-molecular weight and strongly bound contaminant. Aeration and nutrient supply resulted in 51% and 34% removal

Table 3.6 Effect of RAMEB on mazout removal from biovented soils.

Applied RAMEB (%)	Removed mazout (%)	
	Soil 1	Soil 2
0	51	34
0.3	72	20
0.5	73	54
0.7	53	40
1.0	69	45

from the soil contaminated with 13 and 21 g mazout/kg soil, and cyclodextrin could enhance this removal to 73% and 54%, respectively, in the best case as shown in Table 3.6. Gravimetrically measured, the initial mazout content in soil 1 was 12,780 mg/kg and in soil 2 it was 21,040 mg/kg.

The concentration of mazout-degrading bacterial cells in the soil increased 10–100 times upon addition of RAMEB. The relatively low concentration and the long biodegradation half-life of RAMEB (one year in soil) (Fenyvesi *et al.*, 2005) justified the conclusion that the increased bacterial cell number was the consequence of increased contaminant bioavailability. Interestingly, the toxicity of all soils was negative; it also demonstrated the limited bioavailability of mazout in soil – similar to the degrading microorganisms, the contaminants were not available for the test organisms either.

Environmental risk of the *in situ* application of solubilizing and bioavailability-increasing substances needs careful monitoring because higher mobility may endanger subsurface waters. An optimal concentration and dosage may prevent overdone mobilization but still allow for a proper increase in bioavailability (Molnár *et al.*, 2002; Molnár, 2006).

6.3.1.6 Coal tar creosote-contaminated soil: enhancement by bioaugmentation

The large molecular weight PAHs that make up coal tar creosote and coal tar oil were partly biodegraded and partly built into the matrix of the clayey humic soil of a former railway repair facility and waste disposal site in Hungary. Both the unsaturated and the saturated soil zones were highly polluted, but the groundwater only slightly, so the dominant part of the contaminant was sorbed on the solid phase of the clayey humic soil. The water table was at 1.5–2.5 meters. Seemingly, the risk decreased during decades of natural attenuation to an acceptable level; no signals of toxicity were observed. More detailed assessment of the site showed high creosote contents – 8,000 at subsite A and 133,000 mg/kg at subsite B.

Enhancement of aerobic biodegradation by aeration in treatability studies at first increased the extractable PAH content and the toxicity of the soil, but after this temporary increase (likely due to intrinsic biosurfactant production), the PAH content decreased to only 7000 mg/kg in the less contaminated site and to 65,000 mg/kg in the site with high creosote content within 4 weeks. Figure 3.24 shows the risk profile based on direct toxicity assessments. With a high level of pollution, the slow degradation could not be enhanced in soil A contrary to soil B. It is

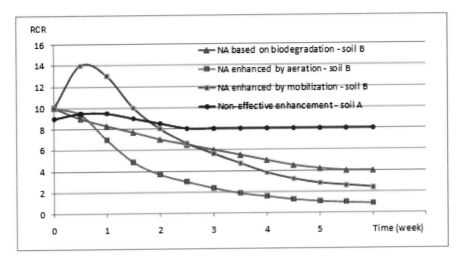

Figure 3.24 Risk profile of NA and ENA of coal tar creosote in the vadose zone: orange: slow NA; blue: NA enhanced by aeration; green: NA enhanced by solubilization and aeration; red: toxic residue inhibits NA.

Table 3.7 Summary of the most important measured data at the creosote-contaminated subsites at a former railway repair facility in Hungary.

Subsite Characteristics	A Before treatment	A After treatment	B Before treatment	B After treatment
Soil type, ingredients	Dark clayey soil; humus: 1.3% N: 0.1 mg/kg; P: 77 mg/kg		Black muddy filling; humus: 3.6% N: 0.8 mg/kg; P: 109 mg/kg	
TE (mg/kg)	20,000		165,000	
EPH (mg/kg)	8000	7000	134,000	65,000
Creosote-degrading cell (CFU per g soil)	4.6×10^5	4.7×10^5	4.6×10^5	4.7×10^7
Toxicity measured by four soil organisms	Toxic	Toxic	Very toxic	Slightly toxic

also interesting that the gravimetrically extracted content (total extract = TE) compared to the gas-chromatographically determined hydrocarbon content (EPH) shows large differences for the two subsites: the 8000 mg/kg EPH derives from 20,000 mg/kg TE (40%) at site A, while the 133,000 at the site B from 165,000 mg/kg (80%) (see Table 3.7).

The pollution at site A was likely the residue of long-term NA with selective biodegradation and stabilization, while the pollution at site B was of a later origin, where NA had just started. This theory was supported by the oil-degrading cell concentrations and the treatability study, applying aeration.

By aeration, the degradation in soil A did not show significant increase, e.g. the produced CO_2 or the creosote-degrading cell number had not increased significantly. While in soil B the CO_2 production showed a ten-fold increase, for the creosote-degrading cell concentration, it was 100-fold.

The average toxicity measured by two bacteria (*Azotobacter agile* and *Pseudomonas fluorescence*), plant (*Sinapis alba*), and animal (*Folsomia candida*) test organisms also indicated the very likely presence of a persistent and toxic residue in soil A, while the justified high toxicity of soil B decreased to a slightly toxic value after four weeks. Using an extremely sensitive toxicity test (*Aliivibrio fischeri* luminescence), both soils showed toxicity, and, due to bioavailability increase during aeration, the toxicity of soil A increased to 3x and that of soil B to 1.5x of the original.

The anoxic method of contaminant degradation was also tested because the sites were temporarily muddy due to excess water. Treatability was studied with and without bioaugmentation. Both the commercially available microaerophilic inoculant and the self-made one (separately propagated mixture of indigenous bacteria) significantly accelerated the initial degradation, but after 10–12 weeks, the difference was not so expressive between inoculated and non-inoculated parallel microcosms. The final degradation rates of soil A was 77% (without inoculant), 81% (commercial inoculant), and 87% (self-made inoculant), so the 8000 mg/kg creosote decreased in the best case to 1040 mg/kg and the degradation rate of soil B was 48%, 56%, and 63%, respectively, so 133,000 mg/kg decreased in the best case to 49,000 mg/kg within 10 weeks. The toxicity decreased significantly on both subsites, and both soils became classifiable as non-toxic. Even the most sensitive bioluminescence test was negative after 10 weeks of treatment.

The biodegradation of such high hydrocarbon concentrations in soil is often stated by professionals to be impossible, but the present example of creosote and some other examples from the author's experience refute such a statement (Molnár *et al.*, 2007).

The residual risk in this creosote case is attributed to the creosote fraction covalently bound to the stable part of the humus, where they are no longer bioavailable for the hydrocarbon-degrading biota. Under controlled circumstances, the natural immobilization of PAHs may reduce the risks posed to subsurface waters or to plant uptake, but other exposure routes such as dermal contact and direct soil ingestion still remain. A soil organic matter that binds strongly or has incorporated toxic PAHs in the humus should be considered a potential chemical time bomb because humus degradation may come to the point where PAHs became available again. Nevertheless, natural stabilization during humus formation can be a good solution for many low-toxicity PAHs and other large molecules or nanoparticles.

6.3.2   Natural attenuation of metals

Two examples are introduced where the authors actively assessed natural processes and studied the risk profile and the natural attenuation-based engineered solutions:

–   An abandoned zinc-lead mine in Gyöngyösoroszi with several non-isolated surface sources of sulfidic mine wastes with acid rock drainage (ARD) and acidic mine drainage (AMD).
–   The red mud flooded soil after the Ajka red mud disaster in 2010 resulting in high alkalinity and Na ion concentration in agricultural soil.

6.3.2.1   The former zinc-lead mine

Natural attenuation of sources containing toxic metals focuses on physical transport and dispersion, and chemical transformations, mainly dissolution and acidification. These transport and fate processes result in growing contaminant fluxes at watershed level and lead to unacceptably high contaminant concentrations in surface waters, in sediments and in soils in a diffuse form. The extensive diffuse contamination elevated the background concentrations in the long term.

The Toka Valley case is a typical example of toxic metals pollution of abandoned sulfide ore mine origin. In mines of this type, various contaminant sources can be found with more or less uniform characteristics all over the world (Younger *et al.*, 2002a, 2002b). These sources and the related transport and fate processes are summarized later. Most of these risk sources have been eliminated in the last 10 years, but prior to the complete rehabilitation of the site, the following natural processes and risk sources have been identified.

Acid mine drainage (AMD): the origin of AMD is the mine, where sulfides are exposed to air and water, thus the chemical and biological oxidation of metal sulfides generates sulfuric acid and dissolved toxic metals. Underground mining often takes place below the water table, so the acidic mine water must be constantly pumped out of the mine. When a mine is abandoned, the pumping ceases, and water floods the mine. Depending on the hydrogeology of the mine, acidity and metal content of the AMD will or will not be reduced by this. The fate of AMD without engineered water treatment is dilution and proportional pH increase, and metal precipitation and accumulation in the sediment. Intensive translocation of sediments to soil by floods is typical for mining in hilly areas with heavy runoffs. The displacement of the anoxic or anaerobic bottom sediment to the soil surface causes secondary acidification and metal mobilization.

Acid rock drainage (ARD): tailings piles and ponds, mine waste rock heaps or just the pyrite-containing rocks close to the surface are also a major source of acid drainage and dissolved metals. At the Toka Valley study site, hundreds of waste rock heaps were abandoned (some were more than 100 years old) without remediation, reclamation, or isolation. Natural bioleaching occurs on the heaps. The surface of the oldest heaps is completely leached, and in some cases spontaneous revegetation has started. Acidity and metal content of the leachate is attenuated by infiltration into the surrounding forest soil or by flowing into the surface water system where it is diluted. It results in increasingly extended diffuse pollution and creates chemical time bombs by insidious accumulations.

Contaminated sediment: Sediment may derive from acid rock and other mine wastes, having *ab ovo* high metal contents. Due to its fine texture, it has high metal sorption. Metal precipitation from these sediments is pH dependent. Liming of acid leachate could lower the risk posed to the water but may result in significant metal accumulation in sediments along the surface water system. Contaminated sediment may pose a high risk *via* site-specific land uses such as bathing in the creek and reservoirs, irrigation of agricultural land by sediment slurry, or sediment translocation by floods.

Soil contaminated by sediment: due to floods and irrigation, soil contamination was typical at the study site. Land users have not been aware of the hazards and risks of the sediment for decades. After floods, the sediment was transported by the creek to the soil surface, and the formerly immobile metals started to be mobilized naturally, becoming available for plant uptake in hobby gardens. Regular floods continuously increased the metal content of the soil. Vegetable and fruits grown by the owners to substitute market purchases for the family made the situation extremely risky.

Tailings pond: the residue of the flotation technology had extremely high toxic metal concentrations. The tailings pond was covered improperly by a thin soil layer (15–20 cm) without isolation. This erosion-prone soil cover was in direct contact with the tailings as shown in Figure 3.25.

Figure 3.25 Improper soil cover on top of tailings.

The tailings were less permeable for water than the cover soil. Infiltrated precipitation and aerobic conditions slowly mobilized the metals in the tailings and these metals were immediately transported by the capillaries and plant roots into the vegetation (Dobler *et al.*, 2001; Sipter *et al.*, 2002, 2005; Gruiz *et al.*, 2006, 2009b).

The first warning was provided by metal concentrations accumulating in plants, both in natural vegetation (Table 3.8) and in home-grown vegetables from the regularly flooded hobby gardens: in the Toka Valley it was usual to have metal concentrations that were 50- to 100-fold of the quality criteria. The toxicity assessment of the flotation tailings from the tailings pond and the cover soil explained plausibly the slow weathering even of the relatively stable mine waste. The tailings itself was not toxic (containing stable, not bioavailable toxic metals), but the cover layer and the mixture of the cover soil and the tailings were highly toxic. The chemically measured metal contents were contradictory to the toxicity: total metal content was extremely high in the tailings but relatively low in the cover soil (Table 3.8). The water-soluble metal concentrations correlate better with the toxicities.

The vegetation grown in the improper soil cover showed high metal tolerance and accumulation capacity: some plants had over 1100 mg/kg Zn and 20 mg/kg Cd concentration.

The natural mobilization of toxic metals at the interface of the tailings and the cover soil without isolation significantly increased the risk.

The soils of the same flooded gardens are highly polluted, and a decreasing gradient can be observed with increasing distance from the creek.

In summary, uncontrolled natural processes (i.e. leaching, acidification, runoff, and erosion) lead to enormous pollution of large areas, and in this way the whole catchment area may become endangered in a relatively short time. The source is practically infinite and the transport from primary sources results in an increasing number of diffusely distributed secondary sources. The early control of such sources may prevent expansion of the transport. The best approach was to remove the sources of manageable volume and to isolate the non-removable ones and to mitigate contaminant discharge and reduce the risk. Other solutions would be to control the risk somewhere along the transport pathway (runoff and leachate collection and treatment, erosion control) and the less-efficient solutions would be to remediate the already existing diffuse pollution, which

Table 3.8 Metal contents of natural vegetation on the improperly covered flotation tailings pond in 1998 (courtesy of Brandl, H. & Bachofen, W., University of Zurich, unpublished data).

Species/Sample	Location	Metal content (mg per kg dry weight)		
Uncontaminated site		Cd	Zn	Cu
Achillea millefolium	Dam, upper sediment reservoir	1.2	29	6
Robimia pseudoacacia	Károly tunnel	0.2	42	8
Silene vulgaris	Bagolyirtás	0.4	27	4
Contaminated sites				
Achillea millefolium	Tailings pond	2.4	255	17
Agrostis sp.	Tailings pond (Northwest)	6.3	410	32
Carex sp	Tailings pond (Northwest)	3.0	355	55
Echium vulgare	Tailings pond (Northwest)	5.0	608	45
Equisetum palustre	Tailings pond	0.7	755	8
Phalaris canadiensis	Tailings pond	0.5	145	4
Phragmites australis	Tailings pond (Northwest)	0.7	768	41
Populus sp.	Tailings pond (Northwest)	20	1160	13
Robinia pseudoacacia	Tailings pond	0.9	160	13
Robinia pseudoacaci	Tailings pond (top)	0.7	213	9
Silene alba	Tailings pond (Northwest)	2.6	694	50
Silene vulgaris (cucubulus)	Tailings pond (Northwest)	4.6	506	21
Solamum dulcamare	Tailings pond (Northwest)	10.7	32	52
Tussilago farifara	Tailings pond (Northwest)	8.8	569	39

Table 3.9 Toxicity and water-soluble (ws) metals content of the flotation tailings and the cover soil.

Test method Sample	Azotobacter agile dehydrogenase	Sinapis alba root and shoot	Vibrio fischeri luminescence	pH	Zn (mg/kg) total/ws	Pb (mg/kg) total/ws	Cu (mg/kg) total/ws
Black soil layer	Very toxic	Toxic	Very toxic	4.7	603/42	186/1.9	72/0.5
Gray flotation tailings	Non-toxic	Slightly toxic	Non-toxic	7.0	31,850/3.4	4970/1.2	2450/0.6

makes no sense in cases where primary and secondary sources still exist. Without reducing the exposure, the risk would constantly increase: the exposed area keeps growing and the pollution may reach highly sensitive localities and receptor organisms, amongst other children, pregnant women, or protected species as well as the food chains, posing a magnified risk relative to the risk at the original mining area. The risk profile of such cases includes a constantly growing curve both in time and space due to increasing concentrations and exposed areas (Figure 3.26). Another reason of the increase is the superposition of the risks due to the food-chain effect, or new land uses and receptors as well as other risky events and processes crossing the risk profile of metal pollution of mining origin. The risk in the source area is not decreasing either due to acidification and low-grade end products, but even if some decrease would be detectable in the source area, it would be absolutely negligible compared to that of the target area.

Table 3.10 Metal content of plants from the regularly flooded hobby gardens (selected data from an assessment campaign of the Eötvös Loránd University, Budapest, by Záray, Gy. – unpublished data, 1991).

Vegetable/Fruit	Cd (mg/kg dw)	Pb (mg/kg dw)	Zn (mg/kg dw)	Cu (mg/kg dw)
Beetroot	21	217	35,300	2800
Carrot	142	470	32,000	3280
Garlic	104	148	14,700	1530
Green beans	12	168	12,900	2790
Green salad	35–220	200–560	4500–94,000	600–3800
Horse-radish	80–50	170–280	9700–52,000	900–2100
Kohlrabi	433	325	67,000	4270
Onion	100–200	150–850	33,000–38,000	250–9000
Parsnips	130	427	16,600	2040
Peach	6	75	3590	550
Plum	7	100	600–1850	1100–2100
Pumpkin	5	79	1410	231
Rhubarb	101	199	6700	854
Strawberry	80	130	2280	830
Tomato	10	88	1850	790
30 flooded gardens maximum	1–6 25	200–600 1500	1000–2000 3600	50–200 400

dw = dry vegetable mass

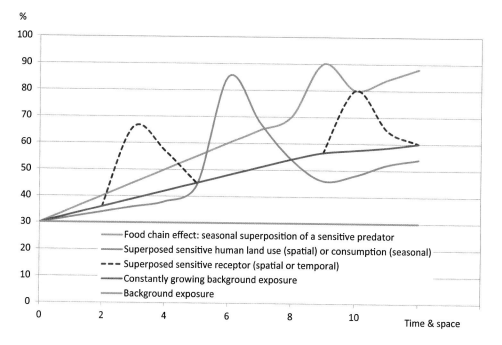

Food chain effect: seasonal superposition of a sensitive predator
Superposed sensitive human land use (spatial) or consumption (seasonal)
Superposed sensitive receptor (spatial or temporal)
Constantly growing background exposure
Background exposure

Figure 3.26 Risk profile of the target area.

Natural isolation could also be observed in this abandoned area, for example on the surface of the tailings pond where a new soil layer had formed from the dead organic matter of the surrounding forest. This was a humic carpet on the surface of the low-lying tailings after desiccation of the supernatant water on the top of the tailings pond. The strong organic "carpet" completely and continuously isolated and physically stabilized the risky tailings prone to erosion, separating it from the terrestrial ecosystem. Interestingly, the 15–20 cm thick layer could be removed as a real carpet because the roots avoided the tailing material, so that the "carpet" was not bound/crosslinked to the tailings material. Nevertheless, the risk *via* the food web (due to capillary transport and plant uptake) is still high (Gruiz, 2005; Gruiz *et al.*, 2009a).

6.3.2.2   Ajka, agricultural soils flooded by red mud

The dam of a red mud storage pond broke at the alumina production facility of MAL in Ajka, Hungary, in October 2010: 800,000 m³ of red mud of high alkalinity (pH = 13) rushed with high velocity, sweeping bridges and cars, and unfortunately led to human casualties – 10 people died and 60 were injured. A highly polluted residential area, seriously damaged surface waters (a 10-km long section of the Torna Valley) and more than 1000 hectares flooded agricultural land are the tragic consequences (Figure 3.27). Several studies were carried

Figure 3.27A The border of the red mud flood on agricultural land.

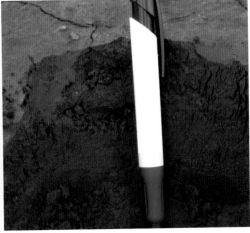

Figure 3.27B Flooded landscape from a higher point and a location covered by a 2–5 cm red mud layer.

out to estimate the short- and long-term damage and a detailed quantitative risk assessment was made by Gruiz *et al.* (2012) for all possible human and environmental risks related to soil uses.

The initially suspected hazards such as dusting and toxic metals concentrations did not reach an unacceptably risky level. Neither was the mobilization of the natural Se and As content of the soils (due to alkaline conditions) considered to be of a level requiring intervention or restrictions. In summary, a comprehensive risk assessment showed that the risks of the red mud to agricultural soil mainly derive from the presence of Na ions and alkalinity, as well as from fine particles that clog the pores (Gruiz *et al.*, 2012). The Na content of the soils – even in soils set free from red mud – was significantly higher than the reference. Removal of the solid phase red mud from the soil surface (after several weeks of the accident) could not invert adverse effects due to alkaline infiltrate into soil pores, which immediately impacted the soil after the red mud flood. Another problem was that the 20 or 30 cm red mud layers could be removed by earthwork machinery, but the removal of the 2–5 cm thin layers was hardly feasible. Heterogeneities related to an uneven surface and distance from the source and the flow path. The density of the red mud suspension further complicated the situation. Alkalinity has mostly leached from the thin red mud layer and infiltrated into the soil in a few days, resulting in a risk from high sodium ion content. The main risk of the desiccated red solid stems from its fine particulate consistency, which caused the clogging of soil pores, sealing the soil from atmospheric air and causing the soil biota to perish.

The yearly rate of attenuation of the Na concentration was calculated based on three months of monitoring and proved to be 3.5 for the upper and 2.0 for the deeper soil layer and only 0.8 for the red mud cover layer on the top of the soil. The half-life time estimate for Na ions in the soil is 3 months. From this value, the risk of four scenarios was calculated after 7 months:

1 Leaving red mud on the top of the soil.
2 Removing red mud completely.
3 Leaving a thin layer of residual red mud on the top and incorporating 5% of red mud into the soil.
4 Leaving a thin layer of residual red mud on the top and incorporating 10% of red mud into the soil.

The calculated risk characterization ratios (RCR) are shown in Table 3.11. To calculate the RCR values, a pH-based risk scale – created specifically for the impacted agricultural soil – and the risk of toxicity to soil-living organisms were considered, including pore occlusion and limited soil aeration in the top layer.

The recommended remediation based on the risk assessment is to remove red mud if its amount on the soil surface exceeds 10% and incorporate it into the upper 50 cm of soil if it is less than 10%. This view is supported by the RCR values: red mud incorporation under 10% is acceptable, but close to 10%, depending on soil type, risk may occur, so increased monitoring and specific measures may be necessary.

Environmental toxicity of field samples measured by soil-living organisms are shown in Table 3.12. Plants selected for cultivation (e.g. maize or energy woods) are much less sensitive for alkaline and high Na ion conditions, as found by the *Sinapis alba* (white mustard) test.

Table 3.11 Soil's Na content, its predicted attenuation and the risk of sodification (Gruiz et al., 2012).

Scenario	Na (mg/kg) in 7 months	RCR in 7 months	Verbal risk characterization	7 months after removal
Red mud on top of the soil	**3100**	**RCR$_{Na}$=3.4**	High	Not acceptable, remove
Complete removal of RM	200	RCR$_{Na}$=0.1	Negligible	Unlimited use but technical difficulties
Incorporation of 5% RM	420	RCR$_{Na}$=0.2	Negligible	Unlimited use, simple or deep plowing
Incorporation of 10% RM	800	RCR$_{Na}$=0.8	Moderate	Usable with conditional restrictions, plowing
Incorporation of 10% RM Assuming lowest attenuation	**1600**	**RCR$_{Na}$=1.6**	Significant	Usable with specific plants. Increased monitoring and control

Legend: RM = red mud; normal letter: measured value, **bold: estimated value**

Table 3.12 Inhibitory effect of red mud on soil ecosystem members (Gruiz et al., 2012).

Test	% red mud in soil causing 10% inhibition	% red mud in soil causing 20% inhibition	% red mud in soil causing 50% inhibition
Soil microorganisms	30	35	40
Seed germination	13	18	25
Test plant shoot growth	5	8	18
Test plant root growth	6	8	15
Collembolan lethality	15	20	25

Field assessments indicated that there is a good chance for long-term attenuation of Na ions due to their mobility and the high groundwater flow rate. As groundwater has not been utilized in the Torna Valley and the surface waters' flux is high enough to immediately dilute alkaline- and Na-containing inputs, the media representing long-term risk are not the groundwater, but soil and sediment.

6.3.2.3 Potential use of desiccated red mud for contaminated soil

The thorough study of the behavior of red mud in soil contributed to our previous research into risk reduction of zinc-, cadmium-, and lead-polluted soils in the Gyöngyösoroszi mining area using various stabilizing additives (see Section 6.3.2.1). In addition to metal-contaminated soils, other degraded soils such as sandy acid soils, unproductive waste soils, and barren materials were investigated. Figure 3.28 shows the results of a microcosm study on the soil of the mining site: a decrease in the water-soluble Cd and Zn concentrations and the enhanced plant growth upon red mud addition of 2% and 5%.

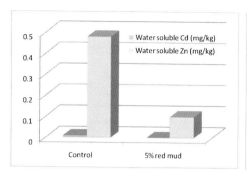 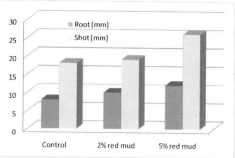

Figure 3.28 The effect of red mud addition to a cadmium- and zinc-contaminated soil: water-soluble metal contents and plant growth – microcosm experiment.

7   EVALUATION OF NA-BASED SOIL REMEDIATION – ADVANTAGES AND SUSTAINABILITY

Natural attenuation and other natural processes play an increasing role in environmental management. The following reasons should be considered in this respect:

1   The lack of unlimited financial resources and still available cheap building sites in many countries give greenfield investments priority, so brownfield sites that require costly remediation are avoided by developers and investors.
2   After the first large remediation wave, the most hazardous point sources have been remediated. A significant number of technologies have been tested and demonstrated. Hundreds of thousands of less hazardous, long-term abandoned contaminated sites remained untreated, where nature and natural processes dominate the situation, without or with monitoring or other engineering interventions. If such sites get into the horizon of spatial planning, it often turns out that the sites have spontaneously recovered or only a mild intervention is needed to reach an acceptable quality and utilizable state of the site.
3   Maintaining ecological value and ecosystem services of agricultural and natural land is especially important, thus in these cases only soft interventions are recommended that are in harmony but a minimum one, not in rough contrast to the natural environment and its quality. This agrees with the requirement of many natural conservation and ecological initiatives and legal requirements.

The requirements for eco-friendly and sustainable remediation have established an endless list of applicable technologies. However, owners, decision makers and managers do not trust these soft remedial methods because, based on their experience, they are only impressed by technologies that implement heavy machinery and spectacular gadgets and handle large quantities of materials. A longer period of time is needed to learn and appreciate "invisible" engineering that relies on knowledge-based soft interventions at the right time and the right place.

From the MNA of contaminated sites, there is an endless variability in potential engineering solutions through keeping agricultural soils in a healthy state until generic management of our entire environment, including ecosystem, is instated. This kind of fine-tuned environmental management requires continuous information about the environment, the monitoring of selected "health" indicators reflecting health or deterioration as well as technological tools suited to the level of the problem. Environmental management today concentrates largely on endangered environments and on those where the economic interest triggers it. Very rough visual or physicochemical indicators are used, such as pollutant concentrations in residential, industrial, and contaminated sites or the density or extinction of species at natural conservation sites. The early indication of unhealthy changes in the diversity has yet to be solved (except for some special cases), and even if the metagenome were monitored, the interpretation of the measured data would not be applicable to decision making and intervention.

7.1   Advantages and disadvantages of natural attenuation-based remediation

After decades of application, advantages and disadvantages of MNA and engineered NA, and other soft/ecoengineering technologies based on natural processes will be summarized here.

Advantages

Environmental and ecological advantages

- The full life cycle of the contaminant is isolated (enclosed inside the soil) and its components (apart from volatilization) return to the soil element cycle.
- The contaminant may be neutralized without harmful end products.
- The soil life is disturbed only to a minimum extent.
- The soil ecosystem and its function are preserved.
- It results in a relatively low environmental footprint.

Technological advantages

- Having information on the fate and behavior as well as the full life cycle of a contaminant, the suitable NA can then be designed and modeled.
- Observing/measuring only adverse effects, an effect-based management system should be applied.
- The technology can be implemented easily and inexpensively.
- The monitoring of the engineering interventions gives the possibility for control and corrective actions to operate at or close to optimum.
- The monitored process can be evaluated and verified continuously.

Economic advantages

- NA and MNA are low-cost technologies compared to conventional highly engineered solutions (excavation, soil washing, thermal methods) and the costs of MNA are generally much lower.
- Spontaneous physicochemical processes do not need engineering manipulation, or simple isolation may bring the result.
- Natural biochemical/biological processes use the contaminant as a substrate for transformation or degradation. Manipulation of the redox conditions may be necessary.

 – MNA is applicable both under aerobic and anoxic/anaerobic environments.
 – The operating environment is biological, so high temperature, pressure, and concentrations can be avoided.
 – Low environmental and economic risk.
 – NA can be applied in combination with other technologies such as source removal, treatment, and enhancement under biological conditions.

Disadvantages

 – Long-term process, and the site cannot always be utilized during the NA.
 – Long-term monitoring activity may be necessary: it needs generally a longer time compared to conventional highly engineered technologies.
 – High uncertainties in the predictions regarding:

 o Transport and fate of contaminants;
 o Changes in site conditions;
 o Interactions with the soil matrix;
 o Dispersion of the pollution;
 o Future land uses;
 o Attenuation rate and duration until achieving remedial goals.

However, these uncertainties can be lowered by continuous monitoring and validation of the prognosticated values.

 – Residual contamination and residual risk associated with MNA is likely (similar to other *in situ* remediations).
 – Migratory contaminants may reach larger areas, increasing background concentrations.
 – On their transport path, contaminants may reach more sensitive land and may interact with other contaminants, increasing the risk without a greater mass or concentration.
 – Our understanding on soil microbiota functioning and natural processes in the soil, in spite of significant development, is still poor, so the design and evaluation of NA processes (similar to all other processes in the environment) includes uncertainties.
 – Acceptance of NA, MNA, and other *in situ* technologies as well as the necessary communication needs to be increased.

7.2 NA and MNA sustainability

Measuring sustainability of soil remediation has developed a lot in the last 15 years, and the principles have been clarified to date, but the methodology is still under permanent change. There are good approaches, guidelines, and international standards (ASTM E2876–13, 2013; ISO, 18504, 2017) available, but a ready-made, uniform sustainability assessment method has not yet been developed. The actual sustainability assessment method should be compiled as part of the contaminated land management, considering the aim and scope and the site-specific conditions. Therefore, the published sustainability assessments show a large variety and several subjective or technology- and site-related features.

The essence of sustainability is that the risk-benefit balance should cover all possible ecological/environmental, human health, and social impacts in the widest sense, including all ecosystem services (provisioning, regulating habitat, and cultural services, see Gruiz,

2014). The actual quantitative parameter of the negative and positive impacts and the costs should be compared to each other as far as possible. The short-term balance can differ from the long-term one. Many of the components of sustainability include uncertainties and subjective judgments, so information should be gathered from the widest circle of stakeholders and the community.

Technological aspects of natural attenuation, "sustainability" of the natural process itself, meaning whether the natural remedial process indeed takes place and will be completed in the long run, and time requirement of the complete process play a major role. The sustainability of natural attenuation and enhanced NA are mainly driven by these technological (ongoing processes and time requirement) and economic (necessary costs and comparative economic benefit) aspects. The holistic sustainability approach has not been specialized for NA/MNA, but the same methodology has been used that was recommended for all other remediation technologies (see also Chapter 11).

More and more definitions and quantitative metrics are coming up within the comprehensive holistic-system approach that can characterize social, environmental, and economic aspects of sustainable soil remediation (see Chapter 11). An innovative solution has been developed by US EPA-ORD called "triple value" (3V) framework that helps capture the dynamic interactions among industrial, societal, and ecological systems (Fiksel *et al.*, 2012). There is an effective choice of indicators, which are classified into four major categories that are applicable as follows:

- Adverse Outcome – indicates destruction of value due to impacts upon individuals, communities, business enterprises or the natural environment.
- Resource Flow – indicates pressures associated with the rate of consumption of resources, including materials, energy, water, land, or biota.
- System Condition – indicates the state of the systems in question, i.e. individuals, communities, business enterprises, or the natural environment.
- Value Creation – indicates creation of value (both economic and well-being) through enrichment of individuals, communities, business enterprises, or the natural environment.

Natural environment and costs are in the focus when MNA sustainability is evaluated. The innovative Battelle EcoVal™ "Environmental Quality Evaluation System" tool may fit well to MNA sustainability assessments. EcoVal™ addresses the need for an integrative tool for the evaluation of human impacts on ecosystems and their services. The weighted environmental quality parameters are these:

- Ecosystem integrity;
- Intrinsic value;
- Ecosystem services value.

The intrinsic value is expressed as maintenance of rare and endangered species. Ecosystem service value is a monetized value. The aforementioned three ecosystem quality parameters are further divided into components (e.g. various ecosystem services such as groundwater uses and agricultural production) and the relevant measurable parameters (drinking water quality criteria, soil quality criteria, etc.) are evaluated.

Generic sustainability assessment justifies that the remediation protects human health and the environment while maximizing environmental, social, and economic benefits throughout the entire life cycle. In addition, there are some natural attenuation-specific aspects and indicators that can prove that the core natural processes facilitate reaching the remediation objectives of the site.

There are some generic key issues that could be responsible for NA and MNA sustainability. The following conditions should be thoroughly assessed and based on the acquired data model calculations carried out to decide whether or not the attenuating process will go on until complete decontamination. This can be considered as a basic criterion for the technological sustainability of MNA that determines the time requirement and the residual risk, two important components in MNA sustainability.

– The presence of the physicochemical or biological process that can be responsible for the attenuation and the presence of a satisfactory amount of driving force to maintain the urgent process for the required length of time.
– Physicochemical condition in the soils: water fluxes, recharges, contaminant fluxes, and transport pathways such as dissolution, partitioning, and abiotic degradation cycles and rates.
– Biological conditions, diversity of the biota, cycles, and biochemical pathways of transformation or degradation processes.
– Quantitative mass balance, energy balance, mass fluxes including degradations.
– Modeling mass balance and estimating the time required for contaminants to dissolve/disperse/degrade/biodegrade in the subsurface (duration of remediation).
– Short-term and long-term sustainability.

Biodegradation-based natural attenuation is sustainable from the technological point of view when the energy balance and the electron acceptor mass balance do not indicate a shortage in the energy source or in the available electron acceptors (Chapelle *et al.*, 2007). Specifically, the energy needs of the biochemical reaction for complete biodegradation (to harmless by-products) should be calculated and then compared to the available energy (used by microorganisms from organic substances). In addition to the energy balance, the density of the degrading microbiota (the catalyst) and the environmental conditions (primarily the redox potential and the responsible electron acceptors), are also basic requirements to be fulfilled. If energy and electron acceptors are available, and there is no other bottleneck, natural attenuation will go on until complete degradation and the process of biodegradation is sustainable. The progress and the completion of the natural process should be proven by monitoring. If monitoring data do not justify the predicted progress, enhancement or other interventions are needed based on thorough diagnosis. If the natural attenuation slows down, the cost of the monitoring will increase, and it may cause other remedial options to be more suitable. If the rate of the biodegradation-based NA decreases due to a shortage in the energy source, nutrients, or electron acceptors, then these should be supplied for the degrading microbiota and the expected duration recalculated.

The sustainability of the NA process (whether and when it is completed) and of the whole remediation based on NA (positive balance of the aggregated environmental, social, and economic risks and benefits) is often blurred. Although it could be differentiated properly if we consider NA-based remediation similar to any other remedial technology, considering NA as a controlled engineering activity and the affected soil/groundwater as a reactor that accommodates the technology. The engineer's task is to achieve optimal operation and process completion in time by a self-sustainable natural process such as biodegradation or bioleaching.

On the other hand, NA-based technologies pose various impacts on the natural and social environment that should be assessed similar to any other remediation (see also Chapter 11).

Maintaining a technology in a natural environment is beneficial because the processes are mainly performed by nature and the emission is partly excluded (mobilization, dissolution, and volatilization still exist). It is true however that the uncertainties are much higher compared to an isolated and completely controlled (engineering) technology. Implementing MNA in homogenous isotropic soils results in limited uncertainty with regard to performance. When dealing with heterogeneous anisotropic conditions, however, the uncertainty can be significant and may result in stakeholders being reluctant to accept MNA.

The risk from volatile emission and vapor migration of free or residual toxic substances is a critical issue, an example is chlorinated hydrocarbon plumes. The risk of chlorinated volatiles escaping from the soil may be extremely high, e.g. when residential homes, schools, or other public buildings may be affected. The chemical species, its mass loading and physical form (free, dissolved, or sorbed) in the soil as well as soil structure and type between the plume and the surface, are crucial in vapor intrusion.

Other risks and uncertainties are connected to water-soluble contaminants, high and low density non-aqueous phase liquids, the smeared zone (which is a potential long-term secondary source), and the degradation rate and its dependency on environmental conditions. Constant mass loading, long-term secondary sources and unstable biodegradation may necessitate monitoring over an extended period. Efficient MNA can be achieved only after the source mass loading has ceased.

In summary, even in cases when the occurrence of NA can be proven and the technical aspects clarified, extensive documentation on the evidence is required. Monitoring data should demonstrate the progress of the natural attenuation processes and validate the model setup and the time requirement estimated at the beginning (Lebrón *et al.*, 2013).

REFERENCES

Adamson, D.T. & Newell, C.J. (2014) *Frequently Asked Questions About Monitored Natural Attenuation in Groundwater*. ESTCP Project ER-201211. Environmental Security and Technology Certification Program, Arlington, VA, USA.

AFC (1999) *Natural Attenuation of Chlorinated Solvents: Performance and Cost Results From Multiple Air Force Demonstration Sites*. Technology Demonstration Technical Summary Report OMB No. 0704-0188. Parsons Engineering Science Inc. for Air Force Center for Environmental Excellence Technology Transfer Division. 101 pp. Available from: www.dtic.mil/dtic/tr/fulltext/u2/a425022.pdf. [Accessed 22nd May 2018].

Ahn, Y.-B., Liu, F., Fennell, D.E. & Häggblom, M.M. (2008) Biostimulation and bioaugmentation to enhance dechlorination of polychlorinated dibenzo-p-dioxins in contaminated sediments. *FEMS Microbiology Ecology*, 66(2), 271–281.

Al-Masri, M.S., Falck, E.W., Read, D. & Konoplev, A. (2015) *Applicability of Monitored Natural Attenuation at Radioactively Contaminated Sites*. IAEA, Technical Report Series, No. 445. Available from: www.researchgate.net/publication/284859160_Applicability_of_monitored_natural_attenuation_at_radioactively_contaminated_sites. [Accessed 22nd May 2018].

Alvarez, P.J.J. & Illman, W.A. (2005) *Bioremediation and Natural Attenuation: Process Fundamentals and Mathematical Models*. John Wiley & Sons, Inc., Published Online December 2005. doi:10.1002/047173862X.

Anton, A.D., Klebercz, O., Magyar, A., Burke, I.T., Jarvis, A.P., Gruiz, K. & Mayer, W.M. (2014) Geochemical recovery of the Torna-Marcal river system after the Ajka red mud spill, Hungary. *Environmental Science: Processes and Impacts,* 16(12), 2677–2685. doi:10.1039/c4em00452c.

ASTM (1998) *Standard Guide for Remediation of Ground Water by Natural Attenuation at Petroleum Release Sites*. E 1943–98, American Society for Testing Materials, ASTM International, West Conshohocken. Available from: www.astm.org/Standards/E1943.htm. [Accessed 7th July 2018].

ASTM E2876–13 (2013) *Standard Guide for Integrating Sustainable Objectives into Cleanup*. Available from: www.astm.org/Standards/E2876.htm. [Accessed 7th July 2018].

Aziz, C.E., Newell, C.J., Gonzales, J.R., Haas, P., Clement, T.P. & Sun, Y. (1999) *BIOCHLOR Natural Attenuation Decision Support System – User's Manual*. Air Force Center for Environmental Excellence, Brooks AFB, San Antonio, TX, USA.

Aziz, J.J., Ling, M., Rifai, H.S., Newell, C.J. & Gonzales, J.R. (2003) MAROS: A decision support system for optimizing monitoring plans. *Ground Water*, 41(3), 355–367. doi:10.1111/j.1745-6584.2003.tb02605.x.

Aziz, J.J., Newell, C.J., Rifai, H.S., Ling, M. & Gonzales, J.R. (2000) *Monitoring and Remediation Optimization System (MAROS): Software User's Guide*. Available from: www.environmentalrestoration.wiki/images/8/8d/Aziz-2000-Monitoring_and_Remed._Opt._Syst._Guide.pdf. [Accessed 7th July 2018].

Bayer-Raich, M., Jarsjö, J., Liedl, R., Ptak, T. & Teutsch, G. (2004) Average contaminant concentration and mass flow in aquifers from temporal pumping well data: Analytical framework. *Water Resources Research*, 40(8), W08303. doi:10.1029/2004WR003095.

Bayer-Raich, M., Jarsjö, J., Liedl, R., Ptak, T. & Teutsch, G. (2006) Integral pumping test analyses of linearly sorbed groundwater contaminants using multiple wells: Inferring mass flows and natural attenuation rates. *Water Resources Research*, 42, W08411. doi:10.1029/2005WR004244.

Beck, P. (2010) Use of monitored natural attenuation in management of risk form petroleum hydrocarbons to human and environmental receptors. *19th World Congress of Soil Science, Soil Solutions for a Changing World, 1–6 August 2010, Brisbane, Australia*. Published on DVD.

Bedard, D.L. & Quensen III, J.F. (1995) Microbial reductive dechlorination of polychlorinated biphenyls. In: Young, L.Y. & Cerniglia, C.E. (eds.) *Microbial Transformation and Degradation of Toxic Organic Chemicals*. Wiley-Liss Inc., New York, NY, USA. pp. 127–216.

Berghoff, A., Berning, A., Wortmann, C., Möller, A. & Mahro, B. (2014) Comparative assessment of laboratory and field based methods to monitor natural attenuation processes in the contaminated groundwater of a former coking plant site. *Environmental Engineering & Management Journal*, 13(3), 583–596.

Bernardes, F., Silva, D., Pérez, A.B. & Muñoz, J. (2007) *Natural Attenuation of Soils Contaminated With Heavy Metals*. Universidad Complutense de Madrid. Available from: pendientedemigracion.ucm.es/info/ . ./Informe_Felipe%20Duarte.pdf. [Accessed 7th July 2018].

Billings, M.L. (2014) *Microcosm Study of Natural Attenuation, Biostimulation, and Bioaugmentation of Soils Contaminated With PCBs, Dioxins, PAHs, and Petroleum Hydrocarbons*. Thesis, California Polytechnic State University, San Luis Obispo, CA, USA. doi:10.15368/theses.2014.171. Available from: http://digitalcommons.calpoly.edu/theses/1319. [Accessed 7th July 2018].

BIOCHLOR (2002) *Natural Attenuation Decision Support System for Chlorinated Solvent Release Sites, ver 2.2*. US EPA. Available from: www.epa.gov/water-research/biochlor-natural-attenuation-decision-support-system. [Accessed 7th July 2018].

BIOPLUME (1997) *BIOPLUME III: A Two-Dimensional Contaminant Transport Model for NA*. US EPA. Available from: www.epa.gov/water-research/bioplume-iii. [Accessed 7th July 2018].

BIOSCREEN (1997) *Natural Attenuation Decision Support System, ver 1.4*. US EPA. Available from: www.epa.gov/water-research/bioscreen-natural-attenuation-decision-support-system. [Accessed 7th July 2018].

Blum, P., Hunkeler, D., Weede, M., Beyer, C., Grathwohl, P. & Morasch, B. (2007) Quantification of biodegradation for various organic compounds using first-order, Michaelis-Menten kinetics and stable carbon isotopes. *Geophysical Research*, 9, 07285.

Bockelmann, A., Ptak, T., Liedl, R. & Teutsch, G. (2001b) Mass flux, transport and natural attenuation of organic contaminants at a former urban gasworks site. In: *Prospects and Limits of Natural*

Attenuation at Tar Oil Contaminated Sites. Dechema e.V. Texte, Frankfurt am Main, Germany. pp. 325–336. ISBN: 3-89746-029-7.

Bockelmann, A., Ptak, T., Liedl, R. & Teutsch, G. (2001c) Transport and natural attenuation of selected organic contaminants at a former urban gasworks site. *International Workshop 'Prospects and Limits of Natural Attenuation at Tar Oil Contaminated Sites', Dechema e.V., March 22–23, 2001, TU Dresden*. INCORE Publication P7. Available from: www.eugris.info/displayresource.aspx?r=66. [Accessed 7th July 2018].

Bockelmann, A., Ptak, T. & Teutsch, G. (2001a) An analytical quantification of mass fluxes and natural attenuation rate constants at a former gasworks site. *Journal Contaminant Hydrology*, 53(3–4), 429–453.

Bockelmann, A., Zamfirescu, D., Ptak, T., Grathwohl, P. & Teutsch, G. (2003) Quantification of mass fluxes and natural attenuation rates at an industrial site with a limited monitoring network: A case study. *Journal of Contaminant Hydrology*, 60(1–2), 97–121.

Bradley, P.M. (2012) *Microbial Mineralization of Cis-Dichloroethene and Vinyl Chloride as a Component of Natural Attenuation of Chloroethene Contaminants Under Conditions Identified in the Field as Anoxic*. U.S. Geological Survey Scientific Investigations Report 2012–5032. p. 30. Available from: https://pubs.usgs.gov/sir/2012/5032/ [Accessed 7th July 2018].

Bradley, P.M. & Chapelle, F.H. (2004) Chloroethene biodegradation potential. ADOT/PF Peger Road Maintenance Facility, Fairbanks, AK. U.S. Geological Survey Open-File Report 2004–1428.

Bradley, P.M., Chapelle, F.H., Landmeyer, J.E. & Schumacher, J.G. (1997) Potential for intrinsic bioremediation of a DNT-contaminated aquifer. *Ground Water*, 35(1), 12–17.

Bradley, P.M., Dinicola, R.S. & Landmeyer, J.E. (2000) Natural attenuation of 1,2-diochloroethane by aquifer microorganisms under Mn(IV)-reducing conditions. In: Wickramanayake, G.B., Gavaskar, A.R. & Kelley, M.E. (eds.) *Remediation of Chlorinated and Recalcitrant Compounds: Natural Attenuation Considerations and Case Studies*. Battelle Press, Columbus, OH, USA. pp. 169–174.

Bradley, P.M., Richmond, S. & Chapelle, F.H. (2005) Chloroethene Biodegradation in Sediments at 4°C. *Applied and Environmental Microbiology*, 71(10), 6414–6417. doi:10.1128/AEM.71.10.6414-6417.

Bradley, P.M., Singletary, M.A. & Chapelle, F.H. (2007) Chloroethene dechlorination in acidic groundwater – Implications for combining Fenton's treatment with natural attenuation. *Remediation Journal*, 18(1), 7–19. doi:10.1002/rem.20149.

Brown, J.G., Brew, R. & Harvey, J.W. (1997) Research on acidic metal contaminants in Pinal Creek Basin Near Globe, Arizona. FS-005-97. U.S. Geological Survey, Reston, VA, USA.

Brown, J.G. & Harvey, J.W. (1996) Hydrologic and geochemical factors affecting metal contaminant transport in Pinal Creek basin near Globe, Arizona. In: Morganwalp, D.W. & Aronson, D.A. (eds.) *U.S. Geological Survey Toxic Substances Hydrology Program: Proceedings of the Technical Meeting, 20–24 September 1993*, Colorado Springs, CO, USA. Geological Survey Water-Resources Investigations Report 94–4015. Vol. 2. USGS, Tallahassee, FL, USA. pp. 1035–1042. DOI: doi.org/10.3133/wri944015

CAH degradation (2015) *Biologically Mediated Abiotic Degradation of Chlorinated Ethenes: A New Conceptual Framework*. SERDP, ER-2532. Available from: www.serdp-estcp.org/Program-Areas/Environmental-Restoration/Contaminated-Groundwater/Persistent-Contamination/ER-2532.

CAH diffusion (2015) *A Field Method to Quantify Chlorinated Solvent Diffusion, Sorption, Abiotic and Biotic Degradation in Low Permeability Zones*. SERDP, ER-2533. Available from: www.serdp-estcp.org/Program-Areas/Environmental-Restoration/Contaminated-Groundwater/Persistent-Contamination/ER-2533. [Accessed 7th July 2018].

CAPP (2002) *Assessment of Monitored Natural Attenuation at Upstream Oil & Gas Facilities in Alberta: Final Report. Canadian Association of Petroleum Producers*. Available from: www.ptac.org/attachments/762/download. [Accessed 7th July 2018].

Carey, M.A., Finnamore, J.R., Morrey, M.J. & Marsland, P.A. (2000) Guidance on the assessment and monitoring of natural attenuation of contaminants in groundwater. Environment Agency, R&D Publication 95.

Chapelle, F.H. (2004) A mass balance approach to monitored natural attenuation. *Proceedings of the Fourth International Conference on Remediation of Chlorinated and Recalcitrant Compounds, 24–28 May 2004, Monterey, CA*. Battelle Press, Columbus, OH, USA. (Contact The Conference Group Inc. for information on the availability of the proceedings).

Chapelle, F.H. & Bradley, P.M. (1998) Selecting remediation goals by assessing the natural attenuation capacity of groundwater systems. *Bioremediation Journal*, 2(3–4), 227–238.

Chapelle, F.H. & Bradley, P.M. (1999) Selecting remediation goals by assessing the natural attenuation capacity of groundwater systems. In: Morganwalp, D.W. & Buxton, H.T. (eds.) *U.S. Geological Survey Toxic Substances Hydrology Program: Proceedings of the Technical Meeting, 8–12 March 1999, Charleston, SC*. Volume 3 of 3: Subsurface Contamination from Point Sources: U.S. Geological Survey Water-Resources Investigations Report 99–4018C. USGS, Tallahassee, FL,USA. pp. 7–19.

Chapelle, F.H., Bradley, P.M., Lovley, D.R., O'Neill, K. & Landmeyer, J.E. (2002a) Rapid evolution of redox processes in petroleum hydrocarbon-contaminated aquifer. *Ground Water*, 40, 353–360.

Chapelle, F.H., Landmeyer, J.E. & Bradley, P.M. (2002b) Identifying the distribution of terminal electron-accepting processes (TEAPS) in ground-water systems. *Workshop on Monitoring Oxidation-Reduction Processes for Groundwater Restoration. Workshop Summary, 25–27 April 2000*, Dallas, TX, USA. EPA/600/R-02/002.

Chapelle, F.H., Novak, J., Parker, J., Campbell, B.G. & Widdowson, M.A. (2007) *A Framework for Assessing the Sustainability of Monitored Natural Attenuation*. U.S. Department of the Interior and U.S. Geological Survey. Available from: https://pubs.usgs.gov/circ/circ1303/pdf/circ1303.pdf. [Accessed 7th July 2018].

Chapelle, F.H., Widdowson, M.A., Brauner, J.S., Mendez III, E. & Casey, C.C. (2003) *Methodology for Estimating Times of Remediation Associated With Monitored Natural Attenuation*. USGS Water-Resources Investigations, Report 03–4057, Columbia, SC, USA. Available from: https:// pubs.usgs.gov/wri/wri034057/pdf/wrir03-4057.pdf. [Accessed 7th July 2018].

Chernobyl, Ukraine (2018) *Chernobyl*. Available from: www.natural-analogues.com/nawg-library/na . . ./195-chernobyl/file. [Accessed 7th July 2018].

CIR (2000a) *Natural Attenuation for Groundwater Remediation*. Chapter 3: Scientific Basis for Natural Attenuation. Committee of Intrinsic remediation. National Academy Press, Washington, DC, USA. Available from: www.nap.edu/read/9792/chapter/5#66. [Accessed 7th July 2018].

CIR (2000b) *Natural Attenuation for Groundwater Remediation*. Committee on Intrinsic Remediation, Water Science and Technology Board, Board on Radioactive Waste Management, National Research Council. National Academy Press, Washington, DC, USA. Available from: www.nap.edu/ catalog/9792.html. [Accessed 7th July 2018].

Clement, T.P., Truex, M.J. & Lee, P. (2002) A case study for demonstrating the application of U.S. EPA's moniored natural attenuation screening protocol at a hazardous waste site. *Journal of Contaminant Hydrology*, 59, 133–162. Available from: www.eng.auburn.edu/~clemept/publsihed_pdf/ npcpaper.pdf. [Accessed 7th July 2018].

Clue-in (2018a) *Contaminated Site Clean-Up Information*. Available from: https://clu-in.org/tech focus/default.focus/sec/Natural%5FAttenuation/cat/Overview. [Accessed 7th July 2018].

Clue-in (2018b) *Natural Attenuation*. Available from: https://clu-in.org/techfocus/default.focus/sec/ Natural_Attenuation/cat/Overview. [Accessed 7th July 2018].

Columbus, I., Waysbort, D., Marcovitch, I., Yehezkel, L. & Mizrahi, D.M. (2012) VX Fate on Common Matrices: Evaporation versus Degradation. *Environmental Science & Technology*, 46, 3921–3927.

Corseuil, H.X., Fernandes, M. do Rosário, M. & Seabra, P.N. (2000) Results of a natural attenuation field experiment for an ethanol-blended gasoline spill. *Proceedings of the 2000 Petroleum Hydrocarbons and Organic Chemicals in Ground Water: Prevention, Detection, and Remediation, 14–17 November 2000*, Anaheim, CA, USA. pp. 24–31.

CRC CARE (2013) Defining the philosophy, context and principles of the national framework for remediation and management of contaminated sites in Australia. Cooperative Research Centre for Contamination Assessment and Remediation of the Environment, Technical Report Series, 27.

CRC CARE (2018) *Natural Attenuation: A Scoping Review*. CRC CARE Technical Report 03. Available from: www.crccare.com/4CABA9D0-F973-11E2-B4EA005056B60026   AUSTRALIA. [Accessed 7th July 2018].

Criddle, C., Elliott, C., McCarty, P.L. & Barker, J.F. (1986) Reduction of Hexachloroethane to Tetrachloroethylene in Groundwater. *Journal of Contaminant Hydrology*, 1(1–2), 133–142.

Cross, K.M. (2002) *Natural Attenuation at Upstream Oil and Gas Sites in Western Canada*. MSc Thesis, Department of Civil and Environmental Engineering, University of Alberta.

CSIA (2015) *Extending the Applicability of Compound-Specific Isotope Analysis to Low Concentrations of 1,4-Dioxane. ER-2535*. Available from: www.serdp-estcp.org/Program-Areas/ Environmental-Restoration/Contaminated-Groundwater/Emerging-Issues/ER-2535/(language)/ eng-US.

Davydov, Y., Voronik, N., Shatilo, N. & Davydov, D. (2002) Radionuclide speciation in soils contaminated by the Chernobyl accident. *Radiochemistry*, 44(3), 307–310.

DECHEMA (2018) *The Expert Network for Chemical Engineering and Biotechnology in Germany*. Available from: http://dechema.de/ [Accessed 7th July 2018].

Declercq, I. (2011) The use of monitored natural attenuation (MNA) in the SNOWMAN partner countries. *SNOWMAN MNA Conference, November 7th 2011*, Paris, France.

DEP (2004) *Use of Monitored Natural Attenuation for Groundwater Remediation, Land and Water Quality*. Branch, Environmental Regulation Division, Department of Environmental Protection, Perth, Australia. Available from: www.scribd.com/document/88228675/Use-of-Monitored-Natural-Attenuation-for-Groundwater-Remediation-WA. [Accessed 7th July 2018].

DEP (2018) *Australian Department of Environmental Protection*. Available from: www.environment. gov.au/ [Accessed 7th July 2018].

Dobler, R., Burri, P., Gruiz, K., Brandl, H. & Bachofen, R. (2001) Variability in microbial population in soil highly polluted with heavy metals on the basis of substrate utilization pattern analysis. *Journal of Soils and Sediments*, 1(3), 151–158.

Dolan, M.E. & McCarty, P.L. (1995) Small column microcosm for assessing methane-stimulated vinyl-chloride transformation in aquifer samples. *Environmental Science and Technology*, 29(8), 1892–1897.

Durrant, G.C., Schirmer, M., Einarson, M.D., Wilson, R.D. & Mackay, D.M. (1999) Assessment of the dissolution of gasoline containing MTBE at LUST Site 60, Vandenberg Air Force Base, California. *Proceedings of the 1999 Conference on Petroleum Hydrocarbons and Organic Chemicals in Ground Water: Prevention, Detection, and Remediation, 17–19 November, Houston, TX*. National Ground Water Association, Westerville, OH, USA.

Early, T.O. (ed) (2007) *Enhancements to Natural Attenuation: Selected Case Studies*. SRNL for the U.S. Department of Energy. Available from: https://energy.gov/lm/downloads/enhancements-natural-attenuation-selected-case-studies. [Accessed 7th July 2018].

Epp, T., Armstrong, J. & Biggar, K. (2002) *MNA Guideline Development in Alberta*. Available from: www.esaa.org/wp-content/uploads/2015/06/02-Epp.pdf. [Accessed 7th July 2018].

ESTPC (2010) *Monitored Natural Attenuation of Perchlorate in Groundwater: Cost and Performance Report*. Environmental Security Technology Certification Program (ESTPC), U.S. Department of Defense, Project ER-200428. Available from: https://clu-in.org/download/contaminantfocus/ perchlorate/Perchlorate-ER-200428-C&P.pdf. [Accessed 7th July 2018].

EU Life project (2006) *Demonstration of the Use of Monitored Natural Attenuation as a Remediation Technology*. DEMO MNA. Project Report of LIFE03 ENV/FIN/000250. Available from: www.environment.fi/syke/demo-mna and http://ec.europa.eu/environment/life/project/ Projects/index.cfm?fuseaction=search.dspPage&n_proj_id=2561#PD. [Accessed 7th July 2018].

Fava, F., Bertin, L., Fedi, S. & Zannoni, D. (2003) Methyl-β-cyclodextrin enhanced solubilization and aerobic biodegradation of polychlorinated biphenyls in two aged-contaminated soils. *Biotechnology and Bioengineeering*, 81, 384–390. doi:10.1002/bit.10579.

Fava, F., Di Gioia, D., Marchetti, L., Fenyvesi, E. & Szejtli, J. (2002) Randomly methylated β-cyclodextrins (RAMEB) enhance the aerobic biodegradation of polychlorinated biphenyls in aged-contaminated soils. *Journal of Inclusion Phenomenon and Molecular Recognition Chemistry*, 44, 417–421. doi:10.1023/A:1023019903194.

Feigl, V., Anton, A. & Gruiz, K. (2009) Combined chemical and phytostabilisation: Field application. *Land Contamination and Reclamation*, 17(3–4), 577–584.

Feigl, V., Atkári, Á., Anton, A. & Gruiz, K. (2007) Chemical stabilisation combined with phytostabilisation applied to mine waste contaminated soils in Hungary. *Advanced Materials Research*, 20–21, 315–318.

Feigl, V., Gruiz, K. & Anton, A. (2010) Remediation of metal ore mine waste using combined chemical- and phytostabilisation. *Periodica Polytechnica*, 54(2), 71–80. doi:10.3311/pp.ch.2010-2.03.

Fenyvesi, É., Gruiz, K., Verstichel, S., De Vilde, B., Leitgib, L., Csabai, K. & Szaniszló, N. (2005) Biodegradation of cyclodextrins in soil. *Chemosphere*, 60, 1001–1008.

Fenyvesi, É., Hajdu, C. & Gruiz, K. (2016) Potential of cyclodextrins in risk assessment and monitoring of organic contaminants. In: Gruiz, K., Meggyes, T. & Fenyvesi, É. (eds.) *Engineering Tools for Environmental Risk Management III*. Site Assessment and Monitoring Tools. CRC Press, Boca Raton, FL, USA, London, UK, New York, NY, USA & Leiden, The Netherlands. pp. 403–424.

Ferrey, M., Estuesta, P., Wilson, J. & Kampbell, D.H. (2000) *Evaluation of the Protocol for the Natural Attenuation of Chlorinated Solvents: Case Study at the Twin Cities Army Ammunition Plant, Site A*. Report No: EPA 600-R-01-025. 49 pp. Available from: www.pca.state.mn.us/sites/default/files/na-casestudy-tcaap.pdf. [Accessed 7th July 2018].

Fiksel, J., Eason, T. & Frederickson, H. (2012) *Framework for Sustainability Indicators at EPA*. US EPA, Office of Research and Development. Available from: https://cfpub.epa.gov/si/si_public_record_report.cfm?dirEntryId=254270. [Accessed 7th July 2018].

Ford, R.G., Miller, G., Fimmen, R. & Strauss, P. (2009) Monitored natural attenuation case study evaluations. Presented at *Interstate Technology and Regulatory Council Meeting*, Cincinnati, OH, USA. Available from: https://cfpub.epa.gov/si/si_public_record_report.cfm?dirEntryId=209509. [Accessed 7th July 2018].

Ford, R.G., Wilkin, R.T. & Puls, R.W. (2007) Monitored natural attenuation of inorganic contaminants in ground water. Volume 2: Assessment for non-radionuclides including Arsenic, Cadmium, Chromium, Copper, Lead, Nickel, Nitrate, Perchlorate, and Selenium. US EPA.

Franzmann, P.D., Robertson, W., Zappia, L. & Davis, G.B. (2002) The role of microbial populations in the containment of aromatic hydrocarbons in the subsurface. *Biodegradation*, 13, 65–78. doi:10.1023/A:1016318706753.

Franzmann, P.D., Zappia, L.R., Power, T.R., Davis, G.B. & Patterson, B.M.N. (1999) Microbial mineralisation of benzene and characterisation of microbial biomass in soil above hydrocarbon-contaminated groundwater. *FEMS Microbiology Ecology*, 30, 67–76.

Gödeke, S., Weiß, H., Geistlinger, H., Fischer, A., Richnow, H.H. & Schirmer, M. (2004) Strömungs- und Tracer-Transportmodellierung am Natural Attenuation Standort Zeitz. *Grundwasser*, 9(1), 3–11.

Griebler, C., Safinowski, M., Vieth, A., Richnow, H.H. & Meckenstock, R.U. (2004) Combined application of stable isotope analysis and specific metabolites determination for assessing *in situ* degradation of aromatic hydro-carbons in a tar oil contaminated aquifer. *Environmental Science and Technology*, 38(2), 617–631.

Gruiz, K. (2002) Relation of natural attenuation to environmental risk. *Book of Abstracts, European Conference on Natural Attenuation, October 2002*, Heidelberg, Germany. pp. 68–71.

Gruiz, K. (2005) Biological tools for soil ecotoxicity evaluation: Soil testing triad and the interactive ecotoxicity tests for contaminated soil. In: Fava, F. & Canepa, P. (eds.) *Soil Remediation Series No. 6*. INCA, Venice, Italy. pp. 45–70.

Gruiz, K. (2009) Soil bioremediation: A bioengineering tool. *Land Contamination and Reclamation*, 17(3–4), 543–551.

Gruiz, K. (2014) Ecosystem and man: Ecosystem services. In: Gruiz, K., Meggyes, T. & Fenyvesi, É. (eds.) *Engineering Tools for Environmental Risk Management, Volume 1. Environmental Contamination and Deterioration.* CRC Press, Boca Raton, FL, USA, London, UK, New York, NY, USA & Leiden, The Netherlands. pp. 16–22.

Gruiz, K. (2016a) Integrated and efficient characterization of contaminated sites. In: Gruiz, K., Meggyes, T. & Fenyvesi, É. (eds.) *Engineering Tools for Environmental Risk Management, Volume 3. Site Assessment and Monitoring Tools.* CRC Press, Boca Raton, FL, USA, London, UK, New York, NY, USA & Leiden, The Netherlands. pp. 1–98.

Gruiz, K. (2016b) Monitoring and early warning in environmental management – 5.2. Indicators in biomonitoring. In: Gruiz, K., Meggyes, T. & Fenyvesi, É. (eds.) *Engineering Tools for Environmental Risk Management, Volume 3. Site Assessment and Monitoring Tools.* CRC Press, Boca Raton, FL, USA, London, UK, New York, NY, USA & Leiden, The Netherlands. pp. 143–155.

Gruiz, K., Fekete-Kertész, I., Kunglné-Nagy, Z., Hajdu, C., Feigl, V., Vaszita, E. & Molnár, M. (2016b) Direct toxicity assessment – Methods, evaluation, interpretation. *Science of the Total Environment,* 563–564, 803–812. doi:10.1016/j.scitotenv.2016.01.007.

Gruiz, K., Fenyvesi, E., Kriston, E., Molnar, M. & Horvath, B. (1996) Potential use of cyclodextrins in soil bioremediation. *Journal of Inclusion Phenomena and Molecular Recognition in Chemistry,* 25(1–3), 233–236.

Gruiz, K., Fenyvesi, É., Molnár, M., Feigl, V., Vaszita, E. & Tolner, M. (2016c) *In situ* and real time measurements for effective soil and contaminated site management. In: Gruiz, K., Meggyes, T. & Fenyvesi, É. (eds.) *Engineering Tools for Environmental Risk Management, Volume 3. Site Assessment and Monitoring Tools.* CRC Press, Boca Raton, FL, USA, London, UK, New York, NY, USA & Leiden, The Netherlands. pp. 245–342.

Gruiz, K., Fenyvesi, É., Molnár, M., Feigl, V., Vaszita, E. & Tolner, M. (2016d) *In situ* and real time measurements for effective soil and contaminated site management – Sampling. In: Gruiz, K., Meggyes, T. & Fenyvesi, É. (eds.) *Engineering Tools for Environmental Risk Management, Volume 3. Site Assessment and Monitoring Tools.* CRC Press, Boca Raton, FL, USA, London, UK, New York, NY, USA & Leiden, The Netherlands. pp. 312–322.

Gruiz, K., Hajdu, C. & Meggyes, T. (2015b) Data evaluation and interpretation in environmental toxicology. In: Gruiz, K., Meggyes, T. & Fenyvesi, É. (eds.) *Engineering Tools for Environmental Risk Management. Volume 2. Environmental Toxicology.* CRC Press, Boca Raton, FL, USA. pp. 445–544.

Gruiz, K. & Kriston, E. (1995) *In situ* bioremediation of hydrocarbon in soil. *Journal of Soil Contamination,* 4(2), 163–173. doi:10.1080/15320389509383490.

Gruiz, K., Meggyes, T. & Fenyvesi, É. (eds.) (2015a) *Engineering Tools for Environmental Risk Management, Volume 2. Environmental Toxicology.* CRC Press, Boca Raton, FL, USA, London, UK, New York, NY, USA & Leiden, The Netherlands.

Gruiz, K., Meggyes, T. & Fenyvesi, É. (eds.) (2016a) *Engineering Tools for Environmental Risk Management, Volume 3. Site Assessment and Monitoring Tools.* CRC Press, Boca Raton, FL, USA, London, UK, New York, NY, USA & Leiden, The Netherlands.

Gruiz, K., Molnár, M. & Feigl, V. (2009a) Measuring adverse effects of contaminated soil using interactive and dynamic test methods. *Land Contamination & Reclamation,* 17(3–4), 443–462.

Gruiz, K., Molnár, M., Feigl, V., Vaszita, E. & Klebercz, O. (2015c) Microcosm models and technological experiments. In: Gruiz, K., Meggyes, T. & Fenyvesi, E. (eds.) *Engineering Tools for Environmental Risk Management, Volume 2. Environmental Toxicology.* CRC Press, Boca Raton, FL, USA. pp. 401–444.

Gruiz, K., Molnár, M. & Fenyvesi, É. (2008) Evaluation and verification of soil remediation. In: Kurladze, G.V. (ed) *Environmental Microbiology Research Trends.* Nova Science Publishers, Inc., New York, NY, USA. pp. 1–57. ISBN: 1-60021-939-X. Available from: www.novapublishers.com/catalog/product_info.php?products_id=9423. [Accessed 10th July 2018].

Gruiz, K., Vaszita, E., Feigl, V., Klebercz, O. & Anton, A. (2012) Environmental risk assessment of red-mud contaminated land in Hungary. In: Hryciw, R.D., Athanasopoulos-Zekkos, A. & Yesiller,

N. (eds.) *GeoCongress 2012 – State of the Art and Practice in Geotechnical Engineering.* Geotechnical Special Publication No. 225, *Proceedings of the Annual GeoCongress of the Geo-Institute of ASCE,* Oakland, CA, USA. pp. 4156–4165. ISBN: 978-0-7844-1212-1.

Gruiz, K., Vaszita, E. & Síki, Z. (2006) Regional scale environmental risk assessment of point and diffuse pollution of mining origin. *Methodological Baselines for Risk Based Inventories of Mining Sites, CD of the EC Workshop in Krakow, 24–25 November.* EUR 22515 EN.

Gruiz, K., Vaszita, E. & Siki, Z. (2009b) Environmental Risk Management of diffuse pollution of mining origin. In: Sarsby, R.W. & Meggyes, T. (eds.) *Construction for a Sustainable Environment.* CRC Press/Balkema, Leiden, The Netherlands. pp. 219–228.

Grygolinska, N.M. & Dolin, V.V. (2003) Self-clearing and natural attenuation rates in radioactive contaminated ecosystem. *Sixth International Symposium and Exhibition on Environmental Contamination in Central and Eastern Europe, at Prague.* Available from: www.researchgate.net/ publication/270620018_SELF-CLEAR-NG_AND_NATURAL_ATTENUATION_RATES_IN_ RADIOACTIVE_CONTAMINATED_ECOSYSTEM. [Accessed 7th July 2018].

Haack, S.K. & Bekins, B.A. (2000) Microbial populations in contaminant plumes. *Hydrogeology Journal,* 8(1), 63–76. doi:10.1007/s100400050008.

Haack, S.K., Fogarty, L.R., West, T.G., Alm, E.W., McGuire, J.T., Long, D.T., Hyndman, D.W. & Forney, L.J. (2004) Spatial and temporal changes in microbial community structure associated with recharge-influenced chemical gradients in a contaminated aquifer. *Environmental Microbiology,* 6(5), 438–448. doi:10.1111/j.1462-2920.2003.00563.x.

Haston, Z.C., Sharma, P.K., Black, J.N. & McCarty, P.L. (1994) Enhanced reductive dechlorination of chlorinated ethenes. *Symposium on Bioremediation of Hazardous Wastes: Research, Development, and Field Evaluation.* EPA/600/R-94/075. U.S. Environmental Protection Agency, Washington, DC, USA. pp. 11–14.

Heidrich, S., Weiss, H. & Kaschl, A. (2004) Attenuation reactions in a multiple contaminated aquifer in Bitter-feld (Germany). *Environmental Pollution,* 129, 277–288.

Helma, C., Eckl, P., Gottmann, E., Kassie, F., Rodinger, W., Steinkeller, H., Windpassinger, C. & Schulte-Hermann, R. (1998) Genotoxic and ecotoxic effects of groundwaters and their relation to routinely measured chemical parameters. *Environmental Science and Technology,* 32, 1799–1805.

Hinchee, R.E. & Leeson, A. (1996) *Soil Bioventing: Principles and Practice.* Lewis Publishers, Boca Raton, FL, USA

Hoffmann, R. & Dietrich, P. (2004) Geoelektrische Messungen zur Bestimmung von Grundwasser-fließrichtungen und -geschwindigkeiten. *Grundwasser,* 9, 194–203.

Hunkeler, D., Meckenstock, R.U., Sherwood Lollar, B., Schmidt, T.C. & Wilson, J.T. (2008) *A Guide for Assessing Biodegradation and Source Identification of Organic Groundwater Contaminants Using Compound Specific Isotope Analysis (CSIA).* EPA/600/R-08/148. U.S. Environmental Protection Agency, Washington, DC, USA. Available from: www.environmentalrestoration.wiki/ images/a/a9/Hunkeler-2008-A_Guide.pdf. [Accessed 7th July 2018].

Imbrigiotta, T.E. & Ehlke, T.A. (1999) Relative importance of natural attenuation processes in a tri-chloroe-thene plume and comparison to pump-and-treat remediation at Picatinny Arsenal, New Jersey. In: Morganwalp, D.W. & Buxton, H.T. (eds.) *Proceedings of the Technical Meeting of U.S. Geological Survey Toxic Substances Hydrology Program, 8–12 March 1999, Charleston, SC.* Volume 3 of 3: Subsurface Contamination from Point Sources. USGS Water-Resources Investigations Report 99–4018C,USGS, Tallahassee, FL, USA. pp. 615–624.

INCORE (2003) *Integrated Concept for Groundwater Remediation: INCORE, Final Report.* European Project. UW Umweltwirtschaft [CD-ROM], Stuttgart, Germany.

INCORE Case Study CS9 (2003) *Evaluation of the Field Study Kraftwerk Ost: Kohlebandbruecke with Respect to the Use of NA for BTEX and PAH.* Available from: www.eugris.info/displayresource. aspx?r=58. [Accessed 7th July 2018].

ISO 18504 (2017) *Soil Quality: Sustainable Remediation.* Available from: www.iso.org/standard/62688. html. [Accessed 7th July 2018].

ITRC (2007) *A Decision Flowchart for the Use of Monitored Natural Attenuation and Enhanced Attenuation at Sites with Chlorinated Organic Plumes.* Interstate Technology and Regulatory Council. Available from: www.itrcweb.org/Documents/EACODecisionFlowchart_v1.pdf. [Accessed 7th July 2018].

ITRC (2008) *Enhanced Attenuation: Chlorinated Organics. Technical and Regulatory Guidance.* The Interstate Technology & Regulatory Council, Enhanced Attenuation: Chlorinated Organics Team. Available from: www.itrcweb.org/Team/Public?teamID=31 and http://itrcweb.org/Guidance/GetDocument?documentID=28. [Accessed 7th July 2018].

ITRC (2010) *A Decision Framework for Applying Monitored Natural Attenuation Processes to Metals and Radionuclides in Groundwater.* APMR-1. Interstate Technology & Regulatory Council, Attenuation Processes for Metals and Radionuclides Team, Washington, DC, USA. Available from: www.itrcweb.org/GuidanceDocuments/APMR1.pdf. [Accessed 7th July 2018].

Ivanov, Y.U. & Dolin, V. (eds.) (2001) *Autorehabilitation Processes in Chernobyl Exclusion Zone Ecosystems.* Kiev, Ukraine.

Ivanov, Y.A. (2006) Autorehabilitation processes in ecosystems of abandoned areas and taking into account of these ones for planning of remediation measures. *Proceedings of the International Scientific Seminar "Radioecology of Chernobyl Zone", 27–29 September,* Slavutych, Ukraine.

Ivanov ,Y.A. & Kashparov, V.A. (2003) Long-term dynamics of radioecological situation in terrestrial ecosystems on the territory of exclusion zone. *Environmental Science and Pollution Research,* 1, 13–20.

Jacquet, R. (2011) *The MNA Concept, a View From Industry.* Solvay. Available from: http://snowman-network.com/wp-content/uploads/MNA-concept-a-view-from-industry-Roger-Jacquet-Solvay.pdf. [Accessed 7th July 2018].

Jong, T. & Parry, D.L. (2006) Microbial sulfate reduction under sequentially acidic conditions in an upflow an-aerobic packed bed bioreactor. *Water Research,* 40, 2561–2571.

Juhasz, A.L. & Naidu, R. (2000) Bioremediation of high molecular weight polycyclic aromatic hydrocarbons: A review of the microbial degradation of benzo[a]pyrene. *International Biodeterioration & Biodegradation,* 45(1), 57–88.

Kao, C.M., Chen, S.C., Liu, J.K. & Wu, M.J. (2001) Evaluation of TCDD biodegradability under different redox conditions. *Chemosphere,* 44(6), 1447–1454.

Klebercz, O., Mayes, W.M., Anton, Á.D., Feigl, V., Jarvis, A.P. & Gruiz, K. (2012) Ecotoxicity of fluvial sediments downstream of the Ajka red mud spill, Hungary. *Journal of Environmental Monitoring,* 14(8), 2063–2071. doi:10.1039/c2em30155e.

Kolhatkar, R.V. (2015) In situ bioremediation of contaminated groundwater using electron acceptor salts. US Patent 8986545.

Kolhatkar, R.V. & Schnobrich, M. (2017) Land application of sulfate salts for enhanced natural attenuation of benzene in groundwater: A case study. *Groundwater Monitoring and Remediation,* 37, 43–57. doi:10.1111/gwmr.12209.

Komex (1997) *Effectiveness of Intrinsic Bioremediation at Alberta Sour Gas Plants.* Final Report. Komex International Ltd. Public Works and Government Services Canada. Contract No. 23440–6–1011/002/SQ.

Konoplev, A. (2016a) *Fukushima and Chernobyl: Similarities and Differences in Radiocesium Behavior.* Radioactivity After Nuclear Explosions and Accidents: Consequences and Countermeasures. pp. 202–218. Available from: https://researchmap.jp/?action=cv_download_main&upload_id=132975. [Accessed 7th July 2018].

Konoplev, A. (2017) Fate and transport of radionuclides in soil-water environment. Review. EGU General Assembly. *Geophysical Research Abstracts,* 19, EGU 2017–3231.

Konoplev, A.,Golosov, A., Nanba, V., Omine, K., Onda, Y., Takase, T., Wada, T., Wakiyama, Y., Yoschenko, V., Zheleznyak, M. & Kivva, S. (2016c) Radiocesium solid-liquid distribution and migration in contaminated areas after the accident at Fukushima Dai-ichi Nuclear Power Plant. *Proceedings of International Conference on Environmental Radioactivity ENVIRA 2015, Thessaloniki, 21–29 September, New Challenges With New Analytical Technologies.* pp. 54–58.

Konoplev, A.V., Golosov, V.N., Yoschenko, V.I., Nanba, K., Onda, Y., Takase, T. & Wakiyama, Y. (2016b) Vertical distribution of radiocesium in soils of the area affected by the Fukushima Dai-ichi nuclear power plant accident. *Eurasian Soil Science*, 49(5), 570–580.

Kopinke, F.D., Voskamp, M., Georgi, A. & Richnow, H.H. (2005) Carbon isotope fractionation of organic contaminants due to retardation on humic substances – Implications for natural attenuation studies in aquifers. *Environmental Science & Technology*, 39, 6052–6062.

Krumins, V., Park, J.-W., Son, E.-K., Rodenburg, L.A., Kerkhof, L.J., Haeggblom, M.M. & Fennell, D.E. (2009) PCB dechlorination enhancement in Anacostia River sediment microcosms. *Water Research*, 43(18), 4549–4558.

Kudelsky, A., Smith, J., Ovsiannikova, S. & Hilton, J. (1996) Mobility of Chernobyl-derived 137Cs in a peatbot system within the catchment of the Pripyat River, Belarus. *Science of the Total Environment*, 188(2–3), 101–133.

Kuder, T., Philp, P., van Breukelen, B., Thouement, H., Vanderford, M. & Newell, C. (2014) *Integrated Stable Isotope-Reactive Transport Model Approach for Assessment of Chlorinated Solvent Degradation*. SERDP, ESTPC, ER-201029. Available from: www.serdp-estcp.org/Program-Areas/ Environmental-Restoration/Contaminated-Groundwater/ER-201029/ER-201029. [Accessed 7th July 2018].

Kueper, B.H., Stroo, H.F., Vogel, C.M. & Ward, C.H. (2014) *Chlorinated Solvent Source Zone Remediation*. Springer, New York, NY, USA.

LABO (2011) Consideration of Natural Attenuation in Remediating Contaminated Sites – Position Paper of 10/12/2009. Germany, German Environment Agency (Umweltbundesamt).

LAI (2016) *2016 Natural Attenuation Assessment: Work Plan, Boeing Auburn Facility, Auburn, Washington*. Landau Associates, Inc. Available from: https://fortress.wa.gov/ecy/gsp/DocViewer. ashx?did=56476. [Accessed 7th July 2018].

Landmeyer, J.E., Chapelle, F.H., Herlong, H.H. & Bradley, P.M. (2001) Methyl tert-butyl ether biodegradation by indigenous aquifer microorganisms under natural and artificial oxic conditions. *Environmental Science & Technology*, 35, 1118–1126.

Landmeyer, J.E., Chapelle, F.H., Petkewich, M.D. & Bradley, P.M. (1998) Assessment of natural attenuation of aromatic hydrocarbons in ground water near a former manufactured gas plant, South Carolina, USA. *Environmental Geology*, 34, 279–292.

Lebrón, C.A., Chapelle, F.H., Widdowson, M.A., Novak, J.T., Parker, J.C. & Singletary, M.A. (2013) *Verification of Methods for Assessing the Sustainability of Monitored Natural Attenuation*. US EPA. Available from: https://clu-in.org/download/techfocus/na/MNA-ER-200824.pdf. [Accessed 7th July 2018].

Lehoux, A.P., Lockwood, C.L., Mayes, W.M., Stewart, D.I., Mortimer, R.J.G., Gruiz, K. & Burke, I.T. (2013) Gypsum addition to soils contaminated by red mud: Implications for aluminium, arsenic, molybdenum and vanadium solubility. *Environmental Geochemistry and Health*, 35(5), 643–656. doi:10.1007/s10653-013-9547-6.

Leitgib, L., Gruiz, K., Fenyvesi, É., Balogh, G. & Murányi, A. (2008) Development of an innovative soil remediation: 'Cyclodextrin-enhanced combined technology'. *Science of the Total Environment*, 392, 12–21. doi:10.1016/j.scitotenv.2007.10.055.

Leitgib, L., Gruiz, K., Molnár, M. & Fenyvesi, É. (2005) Intensification of *in situ* bioremediation of hydrocar-bon-contaminated soils. *Proceedings of the 2nd European Conference on Natural Attenuation, Soil and Groundwater. Risk Management, 18–20 May*. Dechema, Frankfurt am Main, Germany.

Lendvay, J.M., Dean, S.M. & Adrianes, P. (1998) Temporal and spatial trends in biogeochemical conditions at a groundwater-surface water interface: Implications for natural attenuation. *Environmental Science and Technology*, 32(22), 3472–3478.

LLNL (1995) Rice, D.W., Dooher, B.P., Cullen, S.J., Everett, L.G., Kastenberg, W.E., Grose, R.D. & Marino, M.A.: *Recommendations to Improve the Cleanup Process for California's Leaking Underground Fuel Tanks (LUFTs)*. Lawrence Livermore National Laboratory, University of California,

Livermore, CA. Submitted to the California State Water Resources Control Board and the Senate Bill 1764 Leaking Underground Fuel Tank Advisory Committee, UCRL-AR-121762.

Looney, B.B., Early, T.O., Gilmore, T., Chapelle, F.H., Cutshall, N.H., Ross, J., Ankeny, M., Heitkamp, M., Major, D., Newell, C.J., Waugh, W.J., Wein, G., Vangelas, K.M., Adams, K.M. & Sink, C.H. (2006) *Advancing the Science of Natural and Enhanced Attenuation for Chlorinated Solvents.* Final Technical Document for the Monitored Natural Attenuation and Enhanced Attenuation for Chlorinated Solvents Project: Washington Savannah River Company WSRC-STI-2006-00377. Available from: www.researchgate.net/publication/236442501_Advancing_the_science_of_natural_and_enhanced_attenuation_for_chlorinated_solvents. [Accessed 7th July 2018].

Madsen, E.L., Sinclair, J.L. & Ghiorse, W.C. (1991) *In situ* biodegradation: Microbiological patterns in a contaminated aquifer. *Science*, 252, 830–833.

Majcher, E.H., Lorah, M.M., Phelan, D.J. & McGinty, A.L (2009) *Design and Performance of an Enhanced Bioremediation Pilot Test in a Tidal Wetland Seep, West Branch Canal Creek, Aberdeen Proving Ground, Maryland.* Scientific Investigations Report No 5112. U.S. Department of the Interior and U.S. Geological Survey. Available from: https://pubs.usgs.gov/sir/2009/5112/pdf/sir2009-5112.pdf. [Accessed 7th July 2018].

Mancini, S.A., Ulrich, A.C., Lacrampe-Couloume, G., Sleep, B., Edwards, E.A. & Lollar, S.B. (2003) Carbon and hydrogen isotopic fractionation during anaerobic biodegradation of benzene. *Applied Environmental Microbiology*, 69(1), 191–198.

Mass Transfer (2015) *Estimating Mobile-Immobile Mass Transfer Parameters Using Direct Push Tools.* SERDP, ER-2529. Available from: www.serdp-estcp.org/Program-Areas/Environmental-Restoration/Contaminated-Groundwater/Persistent-Contamination/ER-2529/(language)/eng-US. [Accessed 7th July 2018].

Mayes, W.M., Jarvis, A.P., Burke, I.T., Walton, M., Feigl, V., Klebercz, O. & Gruiz, K. (2011) Dispersal and attenuation of trace contaminants downstream of the Ajka bauxite residue (red mud) depository failure, Hungary. *Environmental Science & Technology*, 45(12), 5147–5155. doi:10.1021/es200850y.

McCarty, P.L. & Wilson, J.T. (1992) Natural anaerobic treatment of a TCE plume, St. Joseph, Michigan, NPL site. *Bioremediation of Hazardous Wastes.* pp. 47–50. EPA/600/R-92/126. U.S. Environmental Protection Agency Center for Environmental Research Information, Cincinnati, OH, USA.

McLaughlan, R.G., Merrick, N.P. & Davis, G.B. (2006) *Natural Attenuation: A Scoping Review.* CRC CARE, Adelaide, Australia.

McNulty, A.K. (1997) *In situ Anaerobic Dechlorination of PCBs in Hudson River Sediments.* Master's Thesis, Rensselaer Polytechnic Institute.

Meckenstock, R.U., Morasch, B., Griebler, C. & Richnow, H.H. (2004) Analysis of stable isotope fractionation as a tool to monitor biodegradation in contaminated aquifers. *Journal of Contaminant Hydrology*, 75, 215–255.

MNA Conference (2011) *Monitored Natural Attenuation Conference in Paris.* Available from: http://snowmannetwork.com/?page_id=266. [Accessed 7th July 2018].

MNA-DNA (2011) *Guidance Protocol: Application of Nucleic Acid-Based Tools for Monitoring Monitored Natural Attenuation (MNA), Biostimulation, and Bioaugmentation at Chlorinated Solvent Sites.* ESTCP Guidance Protocol. Environmental Restoration Project ER-0518. Available from: https://clu-in.org/download/ . . . /Bio-MNA-ER-200518-Guidance.pdf. [Accessed 7th July 2018].

Molnár, M. (2006) *Intesified Bioremediation of Contaminated Soil With Cyclodextrins: From the Lab to the Field.* PhD Thesis, Budapest University of Technology and Economics, Budapest, Hungary.

Molnár, M., Fenyvesi, É., Gruiz, K., Leitgib, L., Balogh, G., Murányi, A. & Szejtli, J. (2002) Effects of RAMEB on Bioremediation of Different Soils Contaminated with Hydrocarbons. *Journal of Inclusion Phenomena and Macrocyclic Chemistry*, 44, 447–452.

Molnár, M., Gruiz, K. & Halász, M. (2007) Integrated methodology to evaluate bioremediation potential of creosote-contaminated soils. *Periodica Polytechnica, Chemical Engineering*, 51(1), 23–32.

Molnár, M., Leitgib, L., Fenyvesi, É. & Gruiz, K. (2009) Development of cyclodextrin enhanced soil bioremediation: From laboratory to field. *Land Contamination & Reclamation*, 17(3–4), 601–612.

Molnár, M., Leitgib, L., Gruiz, K., Fenyvesi, É., Szaniszló, N., Szejtli, J. & Fava, F. (2005) Enhanced biodegradation of transformer oil in soils with cyclodextrin: From the laboratory to the field. *Biodegradation*, 16, 159–168. doi:10.1007/s10532-004-4873-0.

Morasch, B., Richnow, H.H., Schink, B. & Meckenstock, R.U. (2001) Stable hydrogen and carbon isotope fractionation during microbial toluene degradation: Mechanistic and environmental aspects. *Applied Environmental Microbiology*, 67, 4842–4849.

Morasch, B., Richnow, H.H., Schink, B., Vieth, A. & Meckenstock, R.U. (2002) Carbon and hydrogen stable isotope fractionation during aerobic bacterial degradation of aromatic hydrocarbons. *Applied Environmental Microbiology*, 68, 5191–5194.

Munro, N., Talmage, S., Griffin, G., Waters, L., Watson, A., King, J. & Hauschild, V. (1999) The sources, fate, and toxicity of chemical warfare agent degradation products. *Environmental Health Perspectives*, 107(12), 933–974.

NA Frankfurt (2005) *Second European Conference on Natural Attenuation, Soil and Groundwater Risk Management, Frankfurt am Main (Germany) 2005*. Available from: http://dechema.de/en/3_+ European+Conference+on+Natural+Attenuation+and+In_Situ+Remediation-p-20039960.html. [Accessed 7th July 2018].

NA Frankfurt (2007) *3. European Conference on Natural Attenuation and in situ Remediation, 19–21 November 2007, Frankfurt am Main*. Available from: http://dechema.de/en/Europ_+ Conf_+Natural+Attenuation-p-20044185.html. [Accessed 7th July 2018].

NA of TCE (2015) *Biogeochemical Processes that Control Natural Attenuation of Trichloroethylene in Low Permeability Zones*. SERDP, ER-2530. Available from: www.serdp-estcp.org/Program-Areas/Environmental-Restoration/Contaminated-Groundwater/Persistent-Contamination/ ER-2530.

Naidu, R., Nandy, S., Megharaj, M., Kumar, R.P., Chadalavada, S., Chen, Z. & Bowman, M. (2012) Monitored natural attenuation of a long-term petroleum hydrocarbon contaminated sites: A case study. *Biodegradation*, 23(6), 881–895. doi:10.1007/s10532-012-9580-7.

NAS (2018) *Natural Attenuation Software*. Available from: www.nas.cee.vt.edu/index.php. [Accessed 7th July 2018].

Nelson, Y., Croyle, K., Billings, M., Caughey, A., Poltorak, M., Donald, A. & Johnson, N. (2014) *Feasibility of Natural Attenuation for the Remediation of Soil Contaminants at the Santa Susana Field Laboratory*. Department of Civil and Environmental Engineering, California Polytechnic State University. Prepared for the US Dept of Energy. Available from: www.dtsc-ssfl.com/ files/lib_doe_area_iv/soiltreatstudies/evaluation_report/66902_SSFL_AreaIV_STS_Natural AttenuationPhase1_report.pdf. [Accessed 7th July 2018].

Nemeskéri, R.L., Neuhaus, M. & Pusztai, J. (2016) Dynamic site characterization for brownfield risk management. In: Gruiz, K., Meggyes, T. & Fenyvesi, É. (eds.) *Engineering Tools for Environmental Risk Management 2. Site Assessment and Monitoring Tools*. CRC Press, Boca Raton, FL, USA, London, UK, New York, NY, USA & Leiden, The Netherlands. pp. 343–360.

Newell, C.J., McLeod, R.K. & Gonzales, J.R. (1996) *BIOSCREEN Natural Attenuation Decision Support System*. User's Manual (Ver. 1.3), R.S. Kerr Environmental Research Center, Ada, OK: EPA/600/R-967-087.

NICOLE (2005) *Natural Attenuation, Cartoon Booklet, NICOLE Publication. SUSTAINABILITY*. Available from: www.nicole.org/uploadedfiles/2005-Natural-Attenuation-Cartoon-booklet.pdf. [Accessed 7th July 2018].

NICOLE (2007) Sustainability of natural attenuation of aromatics (BTEX). NICOLE Project Final Report.

NICOLE (2018) *Network for Industrially Contaminated Land in Europe*. Available from: www.nicole. org/ [Accessed 7th July 2018].

NICOLE MNA (2005) *Monitored Natural Attenuation/Demonstration & Review of the Applicability of MNA at 8 Field Sites*. NICOLE Project Final Report. Available from: www.nicole.org/ uploadedfiles/2005-Monitored-Natural-Attenuation-finalreport.pdf. [Accessed 7th July 2018].

NMR for MNA (2015) *NMR-based Sensors for in situ Monitoring of Changes in Groundwater Chemistry*. SERDP, ER-2534. Available from: www.serdp-estcp.org/Program-Areas/Environmental-Restoration/Contaminated-Groundwater/Monitoring/ER-2534. [Accessed 7th July 2018].

NRC (1994) Alternatives for groundwater clean-up. National Research Council (NRC), National Academy Press, Washington, DC, USA.

O'Steen, W.N. & Howard, R.O. Jr. (2014) *Challenges in Planning for Groundwater Remedy Transition at a Complex Site, U.S. EPA Region 4, Atlanta, Georgia.* Available from: https://clu-in.org/download/techfocus/bio/Challenges-in-Planning-for-Groundwater-Remedy-Transition-at-a-Complex-Site-Feb-2015.pdf. [Accessed 8th July 2018].

OSWER Directive (1999) *Directive 9200.4–17: Use of Monitored Natural Attenuation at Superfund, RCRA Corrective Action, and Underground Storage Tank Sites.* Available from: www.gpo.gov/fdsys/granule/FR-1999-05-10/99-11712 and www.epa.gov/sites/production/files/2014-02/ . ./d9200.4-17.pdf. [Accessed 7th July 2018].

Oudejans, L., Lemieux, P., Kaelin, L., Young, C., Fitzsimmons, Ch. & Englert, B. (2016) *Natural Attenuation of Persistent Chemical Warfare Agent VX on Selected Interior Building Surfaces.* EPA/600/R-16/110. U.S. Environmental Protection Agency, Office of Research and Development National Homeland Security Research Center.

Pachon, C. (2011) Groundwater monitored natural attenuation at cleanup sites in the United States. Office of Superfund Remediation & Technology Innovation. *SNOWMAN MNA Conference, 7 November 2011*, Paris.

Panin, A., Walling, D. & Golosov, V. (2001) The role of soil erosion and fluvial processes in the post-fallout redistribution of Chernobyl-derived 137Cs: A case study of the Lapki catchment, Central Russia. *Geomorphology*, 40(3–4), 185–204.

Pennington, J.C., Zakikhanf, M. & Harrelson, D.W. (1999a) Monitored natural attenuation of explosives in groundwater. Environmental Security Technology Certification Program Completion Report. Technical Report EL-99-7, ER-199518.

Pennington, J.C., Zakikhani, M., Harrelson, D.W. & Allen, D.S. (1999b) *Monitored Natural Attenuation of Explosives in Groundwater: Cost and Performance.* Technical Report EL-99–7, U.S. Army Engineer Waterways Experiment Station, Vicksburg, MS. Environmental Security Technology Certification Program, U.S. Department of Defense. Available from: www.dtic.mil/dtic/tr/fulltext/u2/a607329.pdf. [Accessed 7th July 2018].

Quensen III, J.F., Tiedje, J.M. & Boyd, S.A. (1988) Reductive dechlorination of polychlorinated biphenyls by anaerobic microorganisms from sediments. *Science*, 242, 752–754.

Reisinger, H.J., Burris, D.R. & Hering, J.G. (2005) Remediating subsurface arsenic contamination with monitored natural attenuation. *Environmental Science & Technology*, 39(22), 458A–464A.

Renforth, P., Mayes, W.M., Jarvis, A.P., Burke, I.T., Manning, D.A.C. & Gruiz, K. (2012) Contaminant mobility and carbon sequestration downstream of the Ajka (Hungary) red mud spill: The effects of gypsum dosing. *Science of the Total Environment*, 421–422, 253–259.

Rékási, M., Feigl, V., Uzinger, N., Gruiz, K., Makó, A. & Anton, A. (2013) The effects of leaching from alkaline red mud on soil biota: Modelling the conditions after the Hungarian red mud disaster. *Chemistry and Ecology*, 29(8), 709–723. doi:10.1080/02757540.2013.817568.

Richnow, H.H., Annweiler, E., Michaelis, W. & Meckenstock, R.U. (2003) Microbial *in situ* degradation of aromatic hydrocarbons in a contaminated aquifer monitored by carbon isotope fractionation. *Journal of Contaminant Hydrology*, 65, 101–120.

Rifai, H.S., Newell, C.J., Gonzales, J.R., Dendrou, S., Kennedy, L. & Wilson, J.T. (1997) *BIOPLUME III Natural Attenuation Decision Support System.* Version 1.0 User's Manual. Air Force Center for Environmental Excellence, Brooks AFB, San Antonio, TX, USA.

RT3D (2018) *Multi-Species Reactive Transport Simulation Software for Groundwater Systems.* ONNL. Available from: http://bioprocess.pnnl.gov. [Accessed 7th July 2018].

Rügner, H., Finkel, M., Kaschl, A. & Bittens, B. (2006) Application of monitored natural attenuation in contaminated land management – A review and recommended approach for Europe. *Environmental Science & Policy*, 9, 568–576.

Sadolko, I.V., Demchenko, L.V. & Lavrov, P.I. (2000) Effect of 137Cs vertical migration process in grounds of natural landscapes for reduction of external exposure doze. *Collection of Treatises of State. Scientific Center for Environmental Radiogeochemistry*, 1, 59–77.

Schirmer, K., Bopp, S., Russold, S. & Popp, P. (2004a) Dioxin-ähnliche Wirkungen durch Grundwasser am Indust-riestandort Zeitz: Erfassung und Ableitungen für Sanierungsstrategien. *Grundwasser*, 9(1), 33–42.

Schirmer, K., Dayeh, V.R., Bopp, S.K., Russold, S. & Bols, N.C. (2004b) Applying whole water samples to cell bioassays for detecting dioxin-like compounds at contaminated sites. *Toxicology*, 205, 211–221.

Schirmer, M., Dahmke, A., Dietrich, P., Dietze, M., Godeke, S., Richnow, H.H., Schirmer, K., Weiss, H. & Teutsch, G. (2006) Natural attenuation research at the contaminated megasite Zeitz. *Journal of Hydrology*, 328(3–4), 393–407.

Semprini, L., Kitanidis, P.K., Kampbell, D.H. & Wilson, J.T. (1995) Anaerobic transformation of chlorinated aliphatic hydrocarbons in a sand aquifer based on spatial chemical distributions. *Water Resources Research*, 31, 1051–1062.

SERDP (2018) *Monitoring of NA: Strategic Environmental Research and Development Program (SERDP)*. Available from: www.serdp-estcp.org/News-and-Events/Blog/Natural-Attenuation-Processes-A-New-Understanding. [Accessed 7th July 2018].

Shestopalov, V., Kashparov, V. & Ivanov, Y. (2003) Radionuclide migration into the geological environment and biota after the Chernobyl accident. *Environmental Science & Pollution Research, Special Issue*, 39–47.

Shukla, G. & Varma, A. (eds.) (2011) *Soil Enzymology, Soil Biology 22*. Springer-Verlag, Berlin, Germany & Heidelberg, Germany. doi: 10.1007/978-3-642-14225-3.

Sims, G.K. (2013) Current trends in bioremediation and biodegradation: Stable isotope probing. *Journal of Bioremediation & Biodegradation*, 4, e134. doi:10.4172/2155-6199.1000e134. Available from: www.omicsonline.org/current-trends-in-bioremediation-and-biodegradation-stable-isotope-probing-2155-6199.1000e134.php?aid=14580. [Accessed 7th July 2018].

Sinke, A. & Hencho, I. (1999) Monitored natural attenuation: Review of existing guidelines and protocols. 2864, TNO Institute of Environmental Sciences, Energy Research and Process Innovation, Apledoorn, the Netherlands.

Sipter, E., Auerbach, R. & Gruiz, K. (2005) Ecotoxicological testing and risk assessment of a heavy metal contaminated site. *Toxicology Letters*, 158(1), 253–254.

Sipter, E., Menczel, I., Ferwagner, A. & Gruiz, K. (2002) Natural processes in a toxic metal polluted site as potential risk source. *Book of Abstracts, European Conference on Natural Attenuation, October 2002*, Heidelberg, Germany. pp. 195–197.

Smith, J.A., Katchmark, W., Choi, J.-W. & Tillman, Jr, F.D. (1999) Unsaturated-zone air flow at Picatinny Arsenal, New Jersey: Implications for natural remediation of the trichloroethylene-contaminated aquifer. In: Morganwalp, D.W. & Buxton, H.T. (eds.) *U.S. Geological Survey Toxic Substances Hydrology Program: Proceedings of the Technical Meeting, 8–12 March 1999, Charleston, SC*. Volume 3: Subsurface Contamination from Point Sources. USGS Water-Resources Investigations Report 99–4018C, USGS, USA. pp. 625–633.

SNOWMAN (2018) *Knowledge for Sustainable Soils*. Available from: http://snowmannetwork.com. [Accessed 7th July 2018].

SNOWMAN MNA (2018) *Monitored Natural Attenuation*. Available from: http://snowmannetwork.com/?page_id=266. [Accessed 7th July 2018].

Sobotovich, E., Bondarenko, G. & Dolin, V. (2003) Biogenic and abiogenic migration of 90Sr and 137Cs of Chernobyl origin in terrestrial and aqueous ecosystems. *Environmental Science & Pollution Research, Special Issue*, 1, 31–38.

Stollenwerk, K.G. (1994) Geochemical interactions between constituents in acidic groundwater and alluvium in an aquifer near Globe, Arizona. *Applied Geochemistry*, 9(4), 353–369.

Suarez, M.P. & Rifai, H.S. (1999) Biodegradation rates for fuel hydrocarbons and chlorinated solvents in groundwater. *Bioremediation Journal*, 34(4), 337–362.

Superfund (2013) *Superfund Remedy Report*, Fourteenth Edition. EPA 542-R-13-016. Office of Solid Waste and Emergency Response, United States, Environmental Protection Agency (5203P), Washington. Available from: https://clu-in.org/asr/ and https://clu-in.org/download/remed/asr/14/SRR_14th_2013Nov.pdf. [Accessed 7th July 2018].

Taylor, B., Mauro, D., Foxwell, J., Ripp, J. & Taylor, T. (1996) Characterization and monitoring before and after source removal at a former Manufactured Gas Plant (MGP) disposal site. EPRI TR-105921. Electric Power Research Institute, Palo Alto, CA.

Thierrin, J., Davis, G.B., Barber, C., Patterson, B.M., Pribac, F., Power, T.R. & Lambert, M. (1993) Natural degradation rates of BTEX compounds and naphthalene in a sulphate reducing groundwater environment. *Hydrological Sciences Journal*, 38(4), 309–322. Published online in 2009. Available from: www.tandfonline.com/doi/abs/10.1080/02626669309492677. [Accessed 7th July 2018].

Toze, S., Zappia, L. & Davis, G.B. (2003) Determination of the potential for natural and enhanced biotransformation of munition compounds contaminating groundwater in a fractured basalt aquifer. *Land Contamination & Reclamation*, 8, 225–232.

Truex, M.J., Szecsody, J.E., Qafoku, N.P., Sahajpal, R., Zhong, L., Lawter, A.R. & Lee, B.D. (2015) *Assessment of Hexavalent Chromium Natural Attenuation for the Hanford Site 100 Area*. US Dept. of Energy. Available from: www.pnnl.gov/main/publications/external/ . . . /PNNL-24705.pdf. [Accessed 7th July 2018].

Turley, W.L. & Rawnsley, A. (1996) Applying natural attenuation of chlorinated organics in conjunction with ground-water extraction for aquifer restoration. *Proceedings of the Symposium on Natural Attenuation of Chlorinated Organics in Ground Water, September 11–13*, Dallas, TX, USA. 174 pp. Hull & Associates Engineering, Inc., Austin, TX, USA. Available from: https://cfpub.epa.gov/si/si_public_file_download.cfm?p_download_id=438427. [Accessed 7th July 2018].

USACE (2012) *Site-Specific Work Plan [Monitored Natural Attenuation]: Brown & Bryant Superfund Site, 600 South Derby Street, Arvin, California*. U.S. Army Corps of Engineers (USACE). Available from: https://clu-in.org/download/techfocus/na/BB-Site-MNA-Plan.pdf. [Accessed 7th July 2018].

US EPA (1997) Use of monitored natural attenuation at superfund, RCRA corrective action, and underground storage tank sites. Directive 9200.4–17.

US EPA (1998a) *Technical Protocol for Evaluating Natural Attenuation of Chlorinated Solvents in Ground Water*. EPA/600/R-98/128. U.S. Environmental Protection Agency, Office of Research and Development, Washington, DC, USA. Available from: www.epa.gov/superfund/health/conmedia/gwdocs/protocol.html. [Accessed 7th July 2018].

US EPA (1998b) *Monitoring and Assessment of in-situ Biocontainment of Petroleum Contaminated Ground-Water Plumes*. EPA/600/R-98/020. Available from: http://nepis.epa.gov/Simple.html. [Accessed 7th July 2018].

US EPA (1999a) *Ground Water Issue, Microbial Processes Affecting Monitored Natural Attenuation of Contaminants in the Subsurface*. Available from: http://nepis.epa.gov/Adobe/PDF/10002E30.pdf 12. [Accessed 7th July 2018].

US EPA (1999b) Use of monitored natural attenuation at superfund, RCRA corrective action, and underground storage tank sites. OSWER. April 21. OSWER Directive No. 9200.4–17P. U.S. Environmental Protection Agency, Office of Solid Waste and Emergency Response, Washington, DC, USA.

US EPA (2004) *Performance Monitoring of MNA Remedies for VOCs in Groundwater*. Available from: http://nepis.epa.gov/Adobe/PDF/10004FKY.pdf. [Accessed 7th July 2018].

US EPA (2007a) *Monitored Natural Attenuation of Inorganic Contaminants in Ground Water, Volume 1: Technical Basis for Assessment*. EPA/600/R-07/139. Available from: https://nepis.epa.gov/Exe/ZyPURL.cgi?Dockey=60000N4K.TXT. [Accessed 7th July 2018].

US EPA (2007b) *Monitored Natural Attenuation of Inorganic Contaminants in Ground Water, Volume 2: Assessment for Non-Radionuclides Including Arsenic, Cadmium, Chromium, Copper, Lead, Nickel, Nitrate, Perchlorate, and Selenium*. EPA/600/R-07/140. Available from: https://cfpub.epa.gov/si/si_public_record_report.cfm?dirEntryId=187248. [Accessed 7th July 2018].

US EPA (2007c) *Monitored Natural Attenuation of Inorganic Contaminants in Ground Water, Volume 3: Assessment for Radionuclides Including Tritium, Radon, Strontium, Technetium, Uranium, Iodine, Radium, Thorium, Cesium, and Plutonium-Americium.* Available from: http://nepis.epa.gov/Adobe/PDF/P100EBXW.pdf. [Accessed 7th July 2018].

US EPA (2008) *A Guide for Assessing Biodegradation and Source Identification of Organic Ground Water Contaminants using Compound Specific Isotope Analysis (CSIA).* Available from: http://nepis.epa.gov/Adobe/PDF/P1002VAI.pdf. [Accessed 7th July 2018].

US EPA (2010) *Decontamination of Residual VX on Indoor Surfaces Using Liquid Commercial Cleaners.* EPA/600/R-09/159. U.S. Environmental Protection Agency, Office of Research and Development, Washington, DC, USA. Available from: https://nepis.epa.gov/Exe/ZyPURL.cgi?Dockey=P100RBD6.txt. [Accessed 7th July 2018].

US EPA (2011) *An Approach for Evaluating the Progress of Natural Attenuation in Groundwater.* Available from: http://nepis.epa.gov/Exe/ZyPDF.cgi?Dockey=P100DPOE.pdf; www.epa.gov/ada/gw/mna.html and www.clu-in.org/techfocus/default.focus/sec/natural_attenuation/cat/guidance. [Accessed 7th July 2018].

US EPA (2012) *Framework for Site Characterization for Monitored Natural Attenuation of Volatile Organic Compounds in Ground Water.* EPA 600/R-12/712. Available from: http://nepis.epa.gov/Adobe/PDF/P100HYBY.pdf. [Accessed 7th July 2018].

US EPA (2013) *Ground Water Issue Paper: Synthesis Report on State of Understanding of Chlorinated Solvent Transformation.* Available from: https://clu-in.org/download/techfocus/na/NA-600-R-13-237.pdf. [Accessed 7th July 2018].

US EPA (2015) *Use of Monitored Natural Attenuation for Inorganic Contaminants in Groundwater at Superfund Sites.* Directive 9283.1–36, August 2015, United States Office of Solid Waste and Monitored Natural Attenuation of Inorganic Contaminants in Ground Water. Available from: ne-pis.epa.gov/Exe/ZyPURL.cgi?Dockey=P100ORN0. [Accessed 7th July 2018].

US EPA, MNA (2018) *Monitored Natural Attenuation for Radionuclides in Ground Water: Technical Issues.* EPA Science Inventory. Available from: https://cfpub.epa.gov/si/si_public_record_report.cfm?dirEntryId=205697. [Accessed 7th July 2018].

US EPA, Models (2018) *Models Website.* www2.epa.gov/water-research/methods-models-tools-and-databases-water-research. [Accessed 7th July 2018].

USDN (1998) Technical guidelines for evaluating monitored natural attenuation of petroleum hydrocarbons and chlorinated solvents in ground water at naval and marine corps facilities. US Department of the Navy.

USGS (2016a) *A Landfill Leaking Chlorinated Ethenes.* U.S. Naval Submarine Base Kings Bay, Georgia. Available from: https://toxics.usgs.gov/topics/rem_act/solvent_plume.html. [Accessed 7th July 2018].

USGS (2016b) *Biodegradation of Chlorinated Ethenes and Benzenes, Site 3, Naval Air Station (NAS) Cecil Field, Jacksonville, Florida.* Available from: https://toxics.usgs.gov/highlights/nas_2.2.0/case_study5.html. [Accessed 7th July 2018].

USGS (2016c) *Natural Attenuation of a Chlorinated Benzene Plume, U.S. Naval Air Station in Pensacola, Florida.* Available from: https://toxics.usgs.gov/topics/rem_act/benzene_plume.html. [Accessed 7th July 2018].

USGS (2016d) *Natural Attenuation of Chlorinated Ethenes in a Seasonally Cold Environment, Peger Road Operations and Maintenance Facility, Fairbanks, Alaska.* Available from: https://toxics.usgs.gov/topics/rem_act/natatten_ethenes.html. [Accessed 7th July 2018].

USGS (2016e) *Biodegradation of Chlorinated Ethenes in Fractured Dolomite, Textron Realty Operations Incorporated, Niagara Falls, New York.* Available from: https://toxics.usgs.gov/highlights/nas_2.2.0/case_study2.html. [Accessed 7th July 2018].

USGS (2018a) *U.S. Geological Survey Website.* www.usgs.gov/ [Accessed 7th July 2018].

USGS (2018b) *Environmental Health: Toxic Substances Hydrology Program.* Available from: https://toxics.usgs.gov/highlights/nas_2.2.0/ [Accessed 7th July 2018].

USGS (2018c) *Software Provides Estimates of How Long It Will Take for Remediation Efforts to Achieve their Goals.* Available from: https://toxics.usgs.gov/highlights/remediation_software.html. [Accessed 7th July 2018].

USGS DNA (2018) *Application of Molecular Methods in Microbial Ecology to Understand the Natural Attenuation of Chlorinated Solvents. USGS, Environmental Health: Toxic Substances.* Available from: https://toxics.usgs.gov/sites/molecular_methods.html. [Accessed 7th July 2018].

Van Denburgh, A.S., Goerlitz, D.F. & Godsy, E.M. (1996) Depletion of nitrogen-bearing explosives wastes in a shallow ground-water plume near Hawthorne, Nevada. In: Morganwalp, D.W. & Aronson, D.A. (eds.) *U.S. Geological Survey Toxic Substances Hydrology Program: Proceedings of the Technical Meeting, 20–24 September 1993, Colorado Springs, CO.* USGS Water-Resources Investigations Report, 94-4015, 2, USGS, USA. pp. 895–904.

Van Roy, S., Vanbroekhoven, K., Dejonghe, W. & Diels, L. (2006) Immobilization of heavy metals in the saturated zone by sorption and *in situ* processes. *Hydrometallurgy*, 83, 195–203.

Vieth, A., Fischer, A., Kästner, M., Gehre, M., Knoeller, K., Wachter, T., Dahmke, A. & Richnow, H.H. (2003) Isotope fractionation processes used to characterise natural attenuation of organic contaminants. In: Annokkee, G.J., Arendt, F. & Uhlmann, O. (eds.) *CONSOIL 2003, Proceedings of the 8th International FZK/TNO Conference on Contaminated Soil, 12–16 May 2003*, ICC Gent, Belgium. pp. 2384–2391.

Vieth, A., Voskamp, M., Kopinke, F.-D., Meckenstock, R.U. & Richnow, H.H. (2002) Isotope fractionation as a tool to document the in-situ biodegradation of organic pollutants in contaminated aquifers. In: *Third International Conference on Water Resources and Environment Research, Vol. II.* Eigenverlag des Forums für Abfallwirtschaft und Altlasten e.V., Dresden, Germany. pp. 198–202.

Viisimaa, M., Karpenko, O., Novikov, V., Trapido, M. & Goi, A. (2013) Influence of biosurfactant on combined chemical-biological treatment of PCB-contaminated soil. *Chemical Engineering Journal*, 220, 352–359.

Vroblesky, D.A., Bradley, P.M. & Chapelle, F.H. (1996) Influence of electron donor on the minimum sulfate concentration required for sulfate reduction in a petroleum hydrocarbon contaminated aquifer. *Environmental Science and Technology*, 30(4), 1377–1381.

Vroblesky, D.A., Robertson, J.F., Petkewich, M.D., Chapelle, F.H., Bradley, P.M. & Landmeyer, J.E. (1997) *Remediation of Petroleum Hydrocarbon-Contaminated Ground Water in the Vicinity of a Jet-Fuel Tank Farm, Hanahan, South Carolina.* U.S. Geological Survey, Branch of Information Services [distributor]. Water-Resources Investigations Report 96–4251, vii. 61 p.

Wachter, T., Dethlefsen, F., Gödeke, S. & Dahmke, A. (2000) Räumlich-statistische Charakterisierung der Hydrogeochemie einer BTEX-Grundwasserkontamination am Standort "RETZINA/Zeitz". *Grundwasser*, 9(1), S21–S32.

Waichler, S.R. & Yabusaki, S.B. (2005) *Flow and Transport in the Hanford 300 Area Vadose Zone-Aquifer-River System.* PNNL-15125, Pacific Northwest National Laboratory, Richland, WA, USA. Available from: www.hanford.gov/docs/gpp/library/programdocs-300/PNNL-15125.pdf. [Accessed 7th July 2018].

Wang, Y. & Oyaizu, H. (2011) Enhanced remediation of dioxins-spiked soil by a plant-microbe system using a dibenzofuran-degrading *Comamonas sp.* and *Trifolium repens L. Chemosphere*, 85(7), 1109–1114.

WDNR (2014) *Understanding Chlorinated Hydrocarbon Behavior in Groundwater: Guidance on the Investigation, Assessment and Limitations of Monitored Natural Attenuation.* Wisconsin Department of Natural Resources, RR-699. Available from: www.environmentalrestoration.wiki/images/f/f6/WIS-DNR-2014-Understanding_Chlorinated_Hydrocarbon_Behavior_In_GW.pdf. [Accessed 7th July 2018].

Wein, G., Michelle Lorah, M. & Majche, E. (2007) Wetland enhancement – Reactive mat at the Aberdeen proving ground – West Bank canal creek. In: Early, T.O. (ed) *Enhancements to Natural Attenuation: Selected Case Studies.* SRNL for the U.S. Department of Energy, Aiken, SC, USA.

Widdowson, M.A., Mendez III, E., Chapelle, F.H. & Casey, C.C. (2005) *Natural Attenuation Software (NAS) User's Manual Version 2*. Available from: www.environmentalrestoration.wiki/images/1/16/Widdowson2005-NAS_Users_Guide.pdf. [Accessed 7th July 2018].

Wiedemeier, T.H., Benson, L.A., Wilson, J.T., Kampbell, D.H., Hansen, J.E. & Miknis, R. (1996a) Patterns of natural attenuation of chlorinated aliphatic hydrocarbons at Plattsburgh, Air Force Base, New York. *Platform Abstract of the Conference on Intrinsic Remediation of Chlorinated Solvents, Salt Lake City, UT.*

Wiedemeier, T.H., Blicker, B. & Guest, P.R. (1994) Risk-based approach to bioremediation of fuel hydrocarbons at a major airport. *Federal Environmental Restoration III & Waste Minimization Conference & Exhibition, New Orleans, Louisiana.*

Wiedemeier, T.H., Guest, P.R., Henry, R.L. & Keith, C.B. (1993) The use of BIOPLUME to support regulatory negotiations at a fuel spill site near Denver, Colorado. *Proceedings of the Petroleum Hydrocarbons and Organic Chemicals in Ground Water: Prevention, Detection, and Restoration Conference, NWWA/API.* pp. 445–459.

Wiedemeier, T.H. & Haas, P. (1999) Designing monitoring programs to effectively evaluate the performance of natural attenuation. *Natural Attenuation of Chlorinated Solvents, Petroleum Hydrocarbons, and Other Organic Compounds*, 5(1), 313–323.

Wiedemeier, T.H., Miller, R.N., Wilson, J.T. & Kampbell, J.W. (1995a) Significance of anaerobic processes for the intrinsic bioremediation of fuel hydrocarbons. *Proceedings of the 1995 Petroleum Hydrocarbons Conference, November 29 to December 1*, Houston, TX, USA.

Wiedemeier, T.H., Swanson, M.A., Moutoux, D.E., Gordon, E.K., Wilson, J.T., Wilson, B.H., Kampbell, D.H., Haas, P.E., Miller, R.N., Hansen, J.E. & Chapelle, F.H. (1998b) *Technical Protocol for Evaluating Natural Attenuation of Chlorinated Solvents in Ground Water*. National Risk Management Research Laboratory Office of Research and Development, U. S. Environmental Protection Agency, Cincinnati, OH, USA, 45268.

Wiedemeier, T.H., Swanson, M.A., Wilson, J.T., Kampbell, D.H., Miller, R.N. & Hansen, J.E. (1995b) Patterns of intrinsic bioremediation at two United States Air Force Bases. In: Hinchee, R.E., Wilson, J.T. & Downey, D.C. (eds.) *Intrinsic Bioremediation*. Battelle Press, Columbus, OH, USA.

Wiedemeier, T.H., Swanson, M.A., Wilson, J.T., Kampbell, D.H., Miller, R.N. & Hansen, J.E. (1996b) Approximation of biodegradation rate constants for monoaromatic hydrocarbons (BTEX) in ground water. *Ground Water Monitoring & Remediation*, 16(3), 186–194.

Wiedemeier, T.H., Wilson, J.T. & Kampbell, D.H. (1996c) Natural attenuation of chlorinated aliphatic hydrocarbons at Plattsburgh Air Force Base, New York. *Proceedings of the Symposium on Natural Attenuation of Chlorinated Organics in Ground Water, September 11–13*, Dallas, TX, USA. pp. 76–84. Available from: https://cfpub.epa.gov/si/si_public_file_download.cfm?p_download_id=438427. [Accessed 7th July 2018].

Wiedemeier, T.H., Wilson, J.T., Kampbell, D.H., Miller, R.N. & Hansen, J.E. (1995d) *Technical Protocol for Implementing Intrinsic Remediation with Long-Term Monitoring for Natural Attenuation of Fuel Contamination Dissolved in Groundwater*. U.S. Air Force Center for Environmental Excellence, San Antonio, TX, USA. Vol. 2, ADA324247. Available from: www.dtic.mil/cgi-bin/GetTRDoc?Location=U2&doc=GetTRDoc.pdf&AD=ADA324247. [Accessed 7th July 2018].

Wiedemeier, T.H., Wilson, J.T., Kampbell, D.H., Miller, R.N. & Hansen, J.E. (1998a) *Technical Protocol for Implementing Intrinsic Remediation with Long Term Monitoring for Natural Attenuation of Fuel Contamination Dissolved in Groundwater*. EPA/600/R-98/128, US EPA. Available from: https://nepis.epa.gov/EPA/html/DLwait.htm?url=/Exe/ZyPDF.cgi/30003ONO.PDF?Dockey=30003ONO.PDF. [Accessed 7th July 2018].

Wiedemeier, T.H., Wilson, J.T. & Miller, R.N. (1995c) Significance of anaerobic processes for the intrinsic bio-remediation of fuel hydrocarbons. *Proceedings of the Petroleum Hydrocarbons and Organic Chemicals in Ground Water: Prevention, Detection, and Restoration Conference: NWWA/API.*

Wilson, B.H., Wilson, J.T., Kampbell, D.H., Bledsoe, B.E. & Armstrong, J.M. (1990) Biotransformation of monooaromatic and chlorinated hydrocarbons at an aviation gasoline spill site. *Geomicrobiology Journal*, 8, 225–240.

Wilson, J.T., Weaver, J.W. & Kampbell, D.H. (1994) Intrinsic bioremediation of TCE in ground water at an NPL Site in St. Joseph, Michigan. *Symposium on Intrinsic Bioremediation of Ground Water. EPA/540/R-94/515.* Environmental Protection Agency Office of Research and Development, Washington, DC, USA.

Wilson, M.S. & Madsen, E.L. (1996) Field extraction of a unique intermediary metabolite indicative of real time *in situ* pollutant biodegradation. *Environmental Science and Technology*, 30, 2099–2103.

Wise, D.L. (2000) *Bioremediation of Contaminated Soils.* Marcel Dekker Inc., New York, NY, USA.

WSDE (2005) Guidance on remediation of petroleum-contaminated ground water by natural attenuation. Publication No. 05-09-091 (Version 1.0), Washington State Department of Ecology, Olympia. Washington State Department of Ecology.

Wycisk, P., Weiss, H., Kaschl, A., Heidrich, S. & Sommerwerk, K. (2003) Groundwater pollution and remediation options for multisource contaminated aquifers (Bitterfeld/Wolfen, Germany). *Toxicology Letters*, 140–141, 343–351.

Yabusaki, S.B., Fang, Y. & Waichler, S.R. (2008) Building conceptual models of field-scale uranium reactive transport in a dynamic vadose zone-aquifer-river system. *Water Resources Research*, 44, W12403. doi:10.1029/2007WR006617.

Younger, P.L. & Banwart, S.A. (2002b) Time-scale issues in the remediation of pervasively contaminated groundwaters at abandoned mines sites. In: Oswald, S.E. & Thornton, S.F. (eds.) *Groundwater Quality: Natural and Enhanced Restoration of Groundwater Pollution.* IAHS Special Publ, Sheffield, UK. No 275. pp. 585–591.

Younger, P.L., Banwart, S.A. & Hedin, R.S. (2002a) *Mine Water: Hydrology, Pollution, Remediation.* Kluwer Academic Publishers, Dordrecht, The Netherlands.

Zakikhani, M., Pennington, J.C., Harrelson, D.W. & Gunnison, D. (2003) Monitored natural attenuation of explosives at the Louisiana Army Ammunition Plant. *Land Contamination and Reclamation*, 70, 233–239.

Chapter 4

Ecoengineering tools: Passive artificial ecosystems

R. Kovács[1], N. Szilágyi[1], I. Kenyeres[1*] & K. Gruiz[2]

[1]Organica, Budapest Hungary
[2]Budapest University of Technology and Economics
[*]Biopolus Institute, 1117 Budapest, Infopark, Hungary

ABSTRACT

Ecological engineering or ecotechnology, the exciting mixture of applied ecology, environmental engineering, biotechnology, systems control and complexity sciences has been matured recently into a well-established discipline with its own methodology. Its application fields range from natural habitat preservation and restoration, sustainable resource management to treatment of waste streams of high variety.

While the basis of ecoengineering is the ecological design – i.e. the principle which mimics efficient natural processes – the "reactor," the construction and the technology show a wide variety. The solutions vary from the simplest natural lakes with minimal engineering (e.g. controlled inflow, outflow, water level) to highly engineered artificial reactor systems placed on the surface or built under the surface combined with greenhouses and/or fisheries. The most common solutions use for surface and subsurface waste streams treatment (i) *in situ* placed operations such as aeration, or water recycling, (ii) artificially constructed wetlands with deep and shallow pools, marshes, various water flows and vegetation, or (iii) reactive subsurface constructions.

This chapter gives an overview of ecoengineering tools and technologies, covering passive systems, conventional reactor-based solutions and new developments in the field of advanced artificial ecosystems along with their main performance figures and characteristic application areas.

1 INTRODUCTION

Engineering science took a long time to reach the level of development where industrial and agricultural biotechnologies and their application in environmental technologies (such as biological wastewater treatment, composting, soil bioremediation, phytoremediation, and other phytotechnologies were transformed into proper engineering tools. After several decades of parallel development in the field of wastewater, solid waste, contaminated land, and natural habitat rehabilitation, the knowledge and experience became integrated into innovative applications such as integrated water resources management, closing water cycles, restoration ecology and habitat reconstruction, renewable energy production, waste utilization and solids recycling systems, green architecture and integrated building techniques, modern city planning, and in many other new disciplines.

Once the ecological processes and the ecosystem became the core compartment of engineering, the kinetics and the mass balance of basic processes were recognized and put into the focus of planning, implementation, monitoring, control and regulation of the ecotechnology, similar to other engineered processes (see more in Moo-Young *et al.*, 1996; Jördening & Winter, 2006; Ratledge & Kristiansen, 2006).

Environmental technologies based on living systems have the same problems as their advantages: a complex ecosystem plays the role of catalyst; thus engineering should ensure the optimum conditions in support of the best possible functioning/operation of that complex system. Moreover "the optimum" is a dynamic phenomenon, as a result of the combination of natural (spontaneous) and artificial (engineered) processes. The task of ecoengineering is to support the useful and to compensate the undesirable natural processes and supplement the ecological control system when needed. Thus engineering control should cover the monitoring of the natural processes by well selected indicators and comparing these to their required optimum or maximum. Technological operations may supplement the system with additives and/or with the appropriate environmental conditions such as flow rates, temperature, redox conditions, ion-concentrations, etc. To achieve the expectations, an ecotechnology should ensure that the continuously changing inflows are transformed with constant efficiency by the complex system of natural or artificial ecosystems.

When designing and planning a relatively simple biotechnological process, reactors and technological parameters, optimum operating conditions must be provided for the biocatalysis performed by a living and effectively working organism. Ecoengineering works with a much more complex *catalyst*, a living community and possibly its ideal *habitat*, which is a mixture of environmental and artificial establishments. Great uncertainty is involved in modeling the processes and implementing the operations/technologies as well as concerning the effect of the interventions. Ecoengineers should manage the uncertainties due to the complexity of both the catalyst ecosystem and the habitat matrix. Some tools are available to reduce these uncertainties such as laboratory studies, microcosms, and pilot mesocosms which provide information for planning and forecasting efficiency. Healthy and efficiently working, and in particular self-sustaining natural ecosystems such as lakes, wetlands, marshland, floodplains, mangrove swamps, and surface or subsurface soils can serve as good examples. Based on these complex systems the actively working community, the natural environmental gradients, the seasonal changes, and many other regularities of natural systems can be studied, understood, and copied.

Benyus (2002) defines this approach of "biomimicry" as a new discipline that studies nature's best ideas and then imitates these designs and processes to solve human problems by "innovation inspired by nature." Biomimicry relies on three key principles:

- Nature as model: study nature's models and emulate these forms, processes, systems, and strategies to solve human problems.
- Nature as measure: use an ecological standard to judge the sustainability of our innovations.
- Nature as mentor: view and value nature not based on what we can extract from the natural world, but what we can learn from it.

By understanding material flows, functions and control within healthy/self-sustaining natural systems and the way how they heal themselves and regain their balance, the ecoengineer can design a technology that mimics positive natural tendencies while utilizing an as complex as possible catalyst ecosystem of high diversity.

The engineering tools called "ecotechnologies" have mainly been applied to water treatment – including natural streams, lakes, runoff waters, effluents, wastewaters, and subsurface waters – but the field of ecoengineering expands in the direction of complex systems involving natural conservation, agriculture, forestry, and waste treatment technologies, e.g.

sustainable water systems (full recycling, ecological sanitation, blackwater management, composting toilet, etc.). Biological soil treatment should be considered as a special case of ecoengineering, where the soil ecosystem plays the role of catalyst. This ecosystem consists of the soil microbiota and macro-animals and the plants in the surface soil. The soil ecosystem is increasingly considered as the core factor, e.g. in environmental remediation, that should be supported by the technology. Unfortunately many soil technologies and management concepts still do not care about the health and functionality of the soil ecosystem, this primary environmental resource.

The tendency for expanding ecotechnologies to become part of everyday life is slow, but unstoppable. In the best case it can contribute to long-term sustainable use of our environment and to humans becoming an integral part of natural ecosystems.

Artificial ecosystems became increasingly complex, and today they can be characterized by high diversity and naturally forming hydrogeochemical cycles (with a shorter cycle time compared to natural systems). While a healthy natural ecosystem is self-sustaining, the artificial ecosystem should be controlled and supported by man to become sustainable over the long term. High species diversity can increase stability, and this has to be ensured by the man-made technology in complex, tiered artificial ecosystems.

Todd and Josephson (1996) and Todd *et al.* (2003) define the ecosystem-centered environmental technologies as "ecomachine technologies" and characterizes them as technologies which "reintegrate human and natural ecosystems to become mutually beneficial toward one another." The "ecomachines," the artificial living systems try to copy natural ecosystems by ensuring

- High species and genetic diversity;
- Long food chains and complex food webs;
- The use of sunlight as the primary energy source;
- Ecological succession;
- Stabilized state in hydrogeochemical cycling;
- High sustainability in technological terms (long-term stability and activity).

There are other characteristics in which ecotechnologies definitely differ from natural ones:

- Ecotechnologies are pragmatic with definite goals;
- They are more fragile compared to the robust natural systems;
- A higher value from ecosystem services is expected (productivity, purification, etc.);
- Ecotechnologies serve the exploitation of the ecosystem by man using refined tools to keep the "service provider" alive and in good health. It is important to find the ethically acceptable scale of intervention/interference.

In summary, the core catalyst of ecotechnologies is a partly artificial ecosystem, very similar to and not contradicting natural ecosystems and is controlled by humans. It mimics a natural ecosystem but generally is less complex and has a lower genetic diversity. Ecotechnologies are essential participants in a circular economy and regenerative design (see more in Braungart & McDonoug, 2002; Stahel, 2010).

The artificial ecosystem may have concrete technological goals (e.g. water purification, wastewater treatment, cultivation of plants or animals), but it can function as part of the residential environment in cities (parks, gardens, zoos, artificial lakes, aquaria, green belts, etc.), and even cities themselves are artificial ecosystems. In the absence of human interference, a man-made ecosystem fails to thrive. Ecoengineering should integrate environment and society based on equal opportunities.

1.1   Ecological engineering

The term "ecological engineering" used in its recent meaning can be found first in Odum's publications (Mitsch, 1997). The synonym "ecotechnology" is also widely used for denoting the scientific/technical field that can be defined *as the design of sustainable systems, consistent with ecological principles, which integrate human society with its natural environment for the benefit of both* (Bergen *et al.*, 2001).

The original definitions emphasized the self-designed ecosystems, the subsequent savings of non-renewable resources, the conservation ethic for ecosystems and an integrative rather than reductionist approach to ecology as the key characteristics of ecological engineering (Mitsch, 1997).

Odum (1962) defines ecological engineering as environmental manipulation by man using small amounts of supplementary energy to control systems in which the main energy drives are still coming from natural sources. The utilization of biological species, communities and ecosystems distinguishes ecotechnology from the traditional remedial technologies, which rely on devices and facilities to remove, transform or contain pollutants but which do not consider direct manipulation of ecosystems.

The goals of ecological engineering according to Bergen *et al.* (2001) are (i) to provide human welfare while simultaneously protecting the natural environment from which goods and services are drawn; (ii) to harness the self-design or self-organizational properties of natural systems; (iii) to establish systems where humans do not need to add matter or energy to maintain a particular ecosystem state.

To define ecological engineering, comparing with similar fields such as ecology, environmental engineering, biotechnology, and the traditional restoration-oriented sciences and the identification of the main differences, may also help. Both basic and applied ecology provide fundamentals to ecological engineering, but they do not define it completely. Environmental engineering applies scientific principles to solve pollution problems, but the concepts usually involve energy and resource intensive operations such as scrubbers, filters, chemical precipitators, etc., instead of natural solutions. Traditional biotechnology focuses usually on one or a few selected organisms for pollutants/waste treatment. Modern biotechnologies have opened up to the possibility of genetic manipulation to produce new strains and organisms to carry out specific functions. Conventional restoration-oriented applications usually lack the most important aspect of ecological engineering namely the self-sustainment of the designed ecosystem.

One questionable point with regard to traditional ecological engineering is its economic efficiency (Ierland & Man, 1996). The traditional economic standpoint cannot properly quantify the efficiency of ecotechnologies; socioeconomic cost and benefit assessment and life-cycle analysis are necessary in order to include all environmental, health, social and cultural aspects over the long term.

Some basic principles of ecoengineering are summarized here:

– Ecoengineering is ecologically based;
– Incorporates and adapts engineering and technology-based solutions;

- Is based on sustainable development;
- Is mutually beneficial to humans and the environment;
- Represents integrated systems;
- Focuses on reuse and recovery of waste;
- It is based on stakeholder participation during the design process.

1.2   Advantages and disadvantages

The most important among the ecological engineering advantages is that it strives to achieve sustainable solutions that it takes into account not only the interests of the actual project but the interests of the broader environment as well. Exploiting the potentials in complex ecosystems leads to better solutions in terms of (i) better removal performances and efficiencies in waste treatment processes; (ii) more robust solutions can be achieved owing to exploiting the diversity in the complex ecosystems.

Enhanced diversity in ecologically engineered systems means better removal performance and better system resilience. Holling (1996) makes a clear difference between engineering resilience and ecological resilience. Engineering resilience measures the degree to which a system resists moving away from its equilibrium point and how quickly it returns after a perturbation. In general, this is what is taken into account during usual design tasks. Ecological resilience reflects how large a disturbance an ecosystem can absorb before it changes its structure and function by changing the underlying variables and processes that control behavior. Diverse systems are more ecologically resilient and able to persist and evolve. Diversity can be measured as (i) the number of species present in the system; (ii) genetic variation within species; and (iii) functional diversity (redundancy), where a certain number of species or processes in the system can perform similar functions.

The biggest disadvantage of the approach is that it is a method without established routine processes for design. However, the reason for this is that the whole concept is quite new. Established processes are to be formed when more and more users become familiar with the discipline. It is true although that since ecological engineering deals with complex systems, properly working design methods will presumably require complex design tools.

1.3   Application fields

The literature uses the following categories for the identification of main application fields of ecological engineering:

- Ecological engineering is used to reduce or solve pollution problems that otherwise would be harmful to the ecosystems. This is probably the best known wide application area of the discipline: typical applications include wastewater or groundwater treatment using either constructed wetlands, or ecologically based "green machines" with indoor greenhouse-based applications (Guterstam & Todd, 1990). In these terms the design of ecological systems acts as an alternative to man-made/energy intensive systems to meet the respective human needs.
- Ecosystems are imitated, modified, or copied in an ecologically sound way to reduce or solve a resource problem.
- The tools of ecotechnology support the recovery of an ecosystem after significant disturbance (good examples are detailed e.g. in Luo et al., 2009 or in Jones & Hanna, 2004).

- The management, utilization and conservation of natural resources (Bergen *et al.*, 2001).
- Integration of society and ecosystems in built environments – for example in landscape architecture, urban planning and urban horticulture applications.

The components in artificial ecosystems cover almost the whole range of the natural system components. The exact composition of the respective artificial ecosystem is of course very much depending on the targeted application, however in general it is true that in these systems prokaryotic microorganisms, eukaryotic simple organisms, plants and higher order animals (arthropoda, worms, etc.) play always significant roles. The role of the plants is especially crucial since in addition to taking part in the metabolic processes occurring in the ecosystem, they serve as habitats for other organisms as well (biofilm formation in the root zones, humidity and shading control for other species, etc.) and, in many modern ecologically engineered systems, this habitat effect has utmost importance.

2   HISTORY AND DEVELOPMENT OF ECOENGINEERING-DRIVEN TECHNOLOGIES

Ecological engineering has its parallel roots in China and in the US emerging from the 1960s on. Initially this interdisciplinary field had quite a strong philosophical side as the definition in Section 2.1 also suggests; its scientific fundamentals were established only later, in the 1970–1980s, with the appropriate evolution of its most important boundary disciplines: applied ecology, biotechnology, and environmental engineering.

2.1   Traditional approaches

The roots of the discipline are more or less based on the activity of people coming from the ecological/social fields. According to this, initial publications are organized around the new approach of bringing together ecological principles and the results of science/engineering.

The first and to date the strongest school of ecological engineering deals with pollution removal processes, enhancing the traditional, biology-based engineered solutions using ecological principles. According to this kind of ecosystem-based design concept, the first experiments were conducted for wastewater treatment in small, environment-oriented communities. A few similar communities are prospering even now (e.g. the Findhorn ecovillage in Scotland), applying such traditional ecological engineering solutions.

From these experiments a logical way leads to the "living machines," that are actually small wastewater treatment plants installed under greenhouses, usually established in reactor cascades, with high numbers of higher organisms, most notably plants involved in the treatment process, installed on the top of traditional biological reactors.

In addition to the early communities the development of these technologies was performed at special research institutions. The most notable one was the New Alchemy Institute between 1971 and 1991. Recently it is called the Green Center and its main profile is education and supporting projects that demonstrate ecologically derived forms of energy, agriculture, aquaculture, housing, and landscapes, and living in harmony with the nature (The Green Center, 2010). John Todd, an emblematic figure of the early ecological engineering was among its founders. Later he founded Ocean Arks, a non-profit research and educational organization dedicated to the creation and the dissemination of the technologies for a sustainable future.

It is important to note that these early approaches had very strong ideological roots, and looking for philosophically sound solutions was equally important to create technically appropriate results.

2.2 Modern engineered artificial ecosystems

The most important difference between the traditional approaches and modern engineered artificial ecosystems is that modern methods mostly abandoned the philosophical background and take from the ecological side only those aspects that are usable and technically advantageous, after rigorous evaluation.

Another characteristic of these systems is that, originally, they are more or less engineered processes and have recently been developed into ecological systems.

The solutions falling into this group are the following: (i) constructed wetlands and other natural treatment processes in their modern form, with solid scientific background and (ii) various reactor-based solutions that aim at combining the advantages of traditional ecological systems with state-of-the-art biological technologies. It is important to note that, independently from these efforts, recently introduced biological processes for pollution prevention have very strong ecological bases. These are typically new multi-reactor designs that rely very much on interactions among more and more different species with various activities, as in the technologies of enhanced biological phosphorus removal (EBPR), simultaneous nitrification/denitrification (SND), Anammox process, DEMON process, and various biofilm-based systems.

3 TOOLS OF ECOLOGICAL ENGINEERING

3.1 Basic concepts

Special concepts of ecological engineering are outlined usually as follows (Mitsch, 1997): one of the central concepts of ecotechnology is "self-design." This means that the design and construction relies upon the adaptation capacity of ecosystems to the changing environment and this adaptation or self-organization is a central aspect in operation as well. The second essential characteristic of ecological engineering is the effort to construct systems that are as self-sustaining as possible. In practice this means reducing energy consumption and including solar power as much as possible.

3.2 Design principles

Ecological engineering has a few ecological design principles that are discipline specific and have an important effect on the designed systems (Bergen et al., 2001). Ecosystems have the capacity to self-organize, and this capacity of ecosystems that ecological engineering recognizes as a significant feature because it allows nature to do some of the engineering (Mitsch & Jorgensen, 1989).

The tools that ecological engineering uses are quite diverse and cover the following: knowledge and use of cultivated microorganisms, aquatic plant purification processes, constructed wetlands, stabilization lagoons, hydraulic engineering, biological membranes, architectural aspects (e.g. application of eco-concrete), multi-natural riverways, combined biotechnology, mathematical modeling/numerical methods, complex systems/system identification and control theories, architecture, and urban planning.

4   PASSIVE SYSTEMS, CONSTRUCTED WETLANDS, AQUATIC PLANT SYSTEMS

4.1   Constructed wetlands

Constructed wetlands are often recognized as natural treatment technologies. Natural treatment may be defined as treatment systems in which specific ecological environments found in nature are utilized for the treatment of wastewater (Tchobanoglous, 1997). The design of aquatic systems must be based on the contaminants of concern and knowledge of the processes that affect them.

The two principal categories of constructed wetlands are free water surface and subsurface flow wetlands. Free water surface wetlands consist of basins with channels containing a suitable medium to support the growth of emergent vegetation, open areas, and water flowing at relatively shallow depths (Figure 4.1). The open surface is exposed to atmospheric impacts and poses a risk to humans and ecosystem members through digestion and direct contact. Beside emergent plants, an increasing number of floating plants are applied in the free surface systems as detailed in Section 2.

A subsurface flow wetland (Figures 4.2 and 4.3) is constructed as a bed or channel containing coarse rock, gravel, sand, and soil to ensure a free flow of water to be treated and a growth medium for emergent plants. The bed medium is typically planted with marshland

Figure 4.1 Surface flow constructed wetland – horizontal flow, free water surface.

Figure 4.2 Subsurface flow constructed wetland – horizontal flow, solid medium on the surface.

Figure 4.3 Subsurface flow constructed wetland – vertical flow, solid medium on the surface.

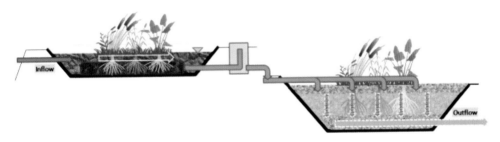

Figure 4.4 Hybrid of a horizontal surface flow and a vertical subsurface flow constructed wetland.

vegetation. In the subsurface flow wetlands water can flow horizontally or vertically (Figures 4.2 and 4.3). The water surface is below the surface of the bed medium so the appearance is just soil overgrown by plants. The advantages of a subsurface flow are prevention of mosquitoes and odors as well as stopping human contact with the contaminated waste stream. Horizontal and vertical water flows are the two options, and problem-specific combinations are also widespread.

Three major factors play a role in the design of wetlands:

– Selection and design of the most appropriate plant community: (i) emergent wetland; (ii) mixed wetland: emergent and forest; (iii) emergent and pond combination;
– Hydrological design: groundwater, surface runoff and solid flows and their control;
– Design of the landscape: linear design, basin, cascade of basins, and combination.

Surface and subsurface as well as horizontal and vertical constructed wetlands can be applied in combination generally known as a hybrid (see Figure 4.4).

The combination of wetlands with other types of wastewater treatment technologies are the "integrated systems." These may represent the combination of constructed wetland with compartments of traditional wastewater treatment (Figure 4.5) and/or with advanced ecomachine technologies or other reactor-based artificial ecosystems (Figure 4.6).

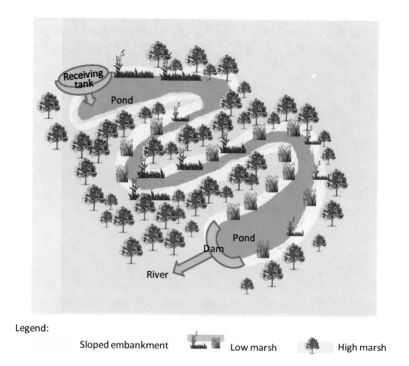

Legend:

        Sloped embankment   Low marsh   High marsh

Figure 4.5 Integrated constructed wetland with ponds and marshes.

Figure 4.6 Scheme of a constructed ecomachines containing wetland.

The integrated system in Figure 4.5 is a conventional open-air basin system placed between the natural or anthropogenic waste source and the receiving natural water, a river or lake, or possibly groundwater. Pumps are often used to feed the wetlands though gravity and automatic siphons are also used. The layout in Figure 4.5 represents a mixed design: ponds (primary/receiving tank for water level control, deep treatment ponds, and a detention pool at the end) and marshes (low marsh and high marsh) with emergent and forest plants.

Constructed wetlands can be installed indoors and partly underground as part of an advanced ecomachine such as the one in Figure 4.6. The scheme shows constructed ecomachines containing a wetland: (i) settlement tank for solid removal; (ii) buffer tank for quantity control; (iii) anoxic or anaerobic reactor for microbiological wastewater treatment; (iv) free surface constructed wetland with horizontal flow for water treatment; (v) aerated pond or lagoon for tertiary treatment; and (vi) sand filter before release to surface water or other recipient water body.

Table 4.1 Design guidelines for artificial ecosystems for water treatment with different setup.

Parameter	Constructed wetlands	Traditional living machines	Conventional biological WWT	Advanced artificial ecologies
Hydraulic detention time (h)	48–380	48–168	24–32	8–16
Water depth (m)	0.1–1	3–5	4–5	4–5
Specific footprint (m^2/m^3/day)	20–40	2.4–4.0	0.6–0.8	0.4–0.5

A comparison of typical design parameters for constructed wetlands and a few competitive solutions – such as traditional activated sludge-based wastewater treatment (WWT), living machines and advanced artificial ecologies – are included in Table 4.1.

Constructed wetlands are primarily for biological oxygen demand (BOD), suspended solids and nitrogen removal. Pathogenic microorganisms, metals and trace organics are also removed, but to a lesser extent. The basic removal processes are biological conversion (microbial mineralization and plant uptake), sedimentation, chemical precipitation, and sorption.

Constructed wetlands have extensive literary and regulatory backgrounds; the first worldwide initiative was the World Convention on wetlands (RAMSAR Convention, 1971). The US regulated the construction of wetlands for different purposes such as agriculture and municipal wastewater, animal waste, etc. from 1991 (USDA, 1991; US EPA, 1993–2000; 1993a,b, 1995, 1998, 1999a,b, 2000a,b, 2001a,b, 2004). In addition to lake eutrophication control, integrated constructed wetlands became widespread in many European countries for water treatment and pollution including diffuse pollution control: e.g. in the Netherlands (Harrington *et al.*, 2005; Scholz, 2006), in Denmark (Brix *et al.*, 2007), in Sweden (Hedmark & Jonsson, 2008), in Ireland (Babatunde *et al.*, 2008), and all over the world (e.g. Wu *et al.*, 2008). Design and performance evaluation guidelines were laid down (e.g. Vymazal, 2007; Carty *et al.*, 2008; Molle *et al.*, 2008; Kayranli *et al.*, 2010; Scholz, 2011), thus today the constructed wetland technology can efficiently combine the built and natural remedial potential and become an integral part of the innovative ecological technologies. There are several websites and books containing up-to-date information on new applications and results: Centre for Advancement of Water and Wastewater Technologies (2018); US EPA – Constructed Wetlands (2018); UNEP – Phytotechnology (2018); US EPA – Wetlands (2018); Wetlands for the Treatment of Mine Drainage (2018); Dotro *et al.* (2017); Higgins *et al.* (2017); Nagabhatla and Metcalfe (2017); Jørgensen *et al.* (2014); Green (2010); Kadlec and Knight (2004); Zedler (2000).

A topic often discussed is the exact function of macrophytes (larger aquatic plants). It is commonly agreed that most important functions are the physical effects: they stabilize the surface of the beds, provide good conditions for physical filtration, prevent vertical flow systems from clogging, insulate the surface against frost during winter and provide huge surface area for microbial growth (Brix, 1997; Brix *et al.*, 2007; Hoffmann *et al.*, 2011).

It is commonly assumed that the plant nutrient uptake is significant; however, this is only true in low loaded systems. The uptake capacity of emergent macrophytes and thus the amount that can be removed if the biomass is harvested is roughly in the range of 30–150 kg P/ha year and 200–1500 kg N/ha year (Gumbricht, 1993; Brix, 1994). If these values are compared to typical load values of low loaded constructed wetlands (approximately 300 kg P/ha year, 2700 kg N/ha year), it can be seen that the plants' contribution is significant in such cases, however it does not cover the entire degraded load.

An important contribution of plants is the macrophyte mediated transfer of oxygen to the rhizosphere by leakage from roots. This contributes to aerobic degradation of organic matter and nitrification in the rhizosphere. Although this transfer slows down during wintertime due to slower metabolic activities at lower temperatures, this comes together with oxygen solubility which has an overall result of equalized oxygen transfer conditions during the whole year. The plant mediated oxygen transfer rates are in the range of 10–160 ng O_2/cm^2 root surface/min (Brix, 1997), which – taken into account typical low-loaded systems – is a very significant amount even at the lower domains of this range.

An advantageous feature of the plants is that they are able to excrete a small amount of various organic compounds on the surface of the roots which in certain situations can act as food source for the biofilm during the low-load periods. This way the plants contribute to bacterial survival. Consequently, there will be a biomass in the system when the wastewater load is re-established. This is one source of the greater flexibility compared to conventional ones.

Another typical example is the excretion of antibiotics. As Seidel (1964) reported in case of bulrushes it is a usual phenomenon that has a very remarkable effect: a range of bacteria (Coliforms, Salmonella, Enterococci) disappear from polluted water by passing through a vegetation of bulrushes.

4.2   Aquatic plant systems

Aquatic plant systems are typically shallow ponds of channels with floating or submerged aquatic plants. Submersed and emerging plants are typical wetland and marsh vegetation such as edges, bulrushes and cattails, as well as shrubs and trees. Two major floating plant types – water hyacinths (*Eichhornia crassipes*) and duckweed (*Lemna spp.*) are used in general, both in aerated and non-aerated variations (Figure 4.7). A significant characteristic of these systems is that periodic harvesting is required for proper operation. Aquatic plant systems are effective in the removal of carbon and nitrogen both from dissolved and solid biodegradable matter.

Naturally floating plants have less developed root system, so the rhizosphere-based clean-up processes are less efficient compared to emergent plant species, on the other hand the plant material can easily be removed.

Tchobanoglous (1997) defines the following goals that current research with regard to constructed wetlands and aquatic systems has to address: (i) understanding how plants function, (ii) operate new configurations/improve operating techniques, and (iii) engineered application of plants in the system.

The classical example of the fields requiring more attention is the fact that a number of aquatic plants are able to transport oxygen from the atmosphere to the root zone – what is

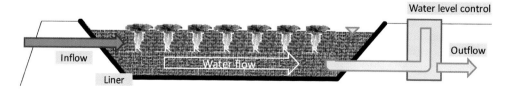

Figure 4.7 Horizontal surface flow constructed wetland with floating plants.

unknown is how to optimize the rate of oxygen transport to the root zone selectively. Currently, fully empirical approaches are used for the design so one of the most important goals here is to establish systematic design methods with solid scientific basis.

One way to innovate is to combine root-zone activity with aqueous phase processes and eliminate the construction and maintenance problems of a self-sustaining constructed bed. The result of these kinds of development are the floating constructed wetlands or floating islands which function similarly to traditionally constructed wetlands that foster vegetation to harvest nutrient or other pollutants from water. Vegetation is planted into soil or soil-like growth medium placed into the planting holes/wells formed/configured on the floating platform. The platform is constructed from a very light and porous fibrous material with a loose grid structure which supports but still lets the roots grow down to the water.

4.3   Constructed floating islands

Floating islands utilize terrestrial and wetland species grown in floating beds on the surface of stormwater sediment ponds and wetlands, eutrophic lakes and wastewater lagoons to treat the water. This kind of floating biofiltration restores ecological integrity to the aquatic ecosystem. Floating islands are embedded into the surrounding aquatic and terrestrial ecosystem mopping up nutrients and pollutants. They help tertiary treatment of wastewater through absorption and utilization of nitrogen, phosphorus, and suspended solids by plants and removal of colloidally suspended heavy metals out of the water body and into the rhizosphere (root zone). In addition, they allow denitrification and release of nitrogen gas to the atmosphere (Reinsel, 2018).

Wetland plants grown on top of the water can effectively purify water and can fulfill stringent quality standards (Duncan, 2012).

The world's largest aquaponics project in China's third largest aquaculture lake (Lake Taihu in the Yangtze Delta plain) experienced its worst ever algal bloom in 2004, pushing researchers to find innovative solutions. A large (4 acres) aquaponics system was designed to remove nutrients that fuel algal blooms. Experiments growing rice on fish ponds provided a foundation for scaling-up to lakes and larger water bodies for water repair (Duncan, 2014).

Sun et al. (2017) monitored the efficiency of water spinach and sticky rice in hydroponic floating bed systems for high-salinity tidal delta lakes. The total nitrogen removal rates were 89.7, 92.3, 85.1 (spinach) and 81.2 and 78.9% (rice), while total phosphorus removal rates were 94.4, 96.4, 93.5 (spinach) and 75.7 and 80.0% (rice) under different salinities. The results indicate that the two-plant species significantly contributed to the remediation of a polluted tidal river.

Floating islands of different size (from as small as 1.5 × 1.5 or 5 × 5 square feet to 20 × 20 square feet or even to 200 × 200 square feet) are increasingly popular around the world, from the Far East (where floating gardens represent a traditional practice such as the ones on the Inle lake in Myanmar) to the US, Australia, and Europe. The smaller ones are available as ready-made commercial products with complete planting pocket kits of growth medium and plant species. The platform itself is generally prepared from light natural plant material or from low-risk recycled foamed polymers. The starter plant compilation can be supplemented by local native species.

The floating roots have direct contact with the water to be treated and serve as a solid carrier which the degrading microorganisms bind to form a food net for pollutant utilization. Floating wetlands/islands can be used both for constructed wetlands, natural water bodies, agricultural and fishery ponds, wastewater lagoons, and stormwater and retention basins

Figure 4.8 Floating island developed on a solid foam-type foundation.

with shallow or deep-water levels. Floating islands treat polluted water bodies directly and may become part of the local ecosystem, performing widespread aquatic and terrestrial ecosystem functions such as fish refuge and providing a spawning habitat as well as a bird breeding ground (Figure 4.8).

Benefits of floating islands:

–   Good clean-up efficiency in BOD, N, P and suspended solid removal;
–   Improved water quality;
–   Adaptability: can fit into the native ecosystem;
–   Environmental sustainability: low energy requirement;
–   An inexpensive option;
–   Can be produced in any size and shape;
–   Applies to existing lagoon and pond systems;
–   No additional land is required;
–   Can be expanded, modified, or combined with other water quality tools (e.g. aeration);
–   Protecting aquatic ecosystems in streams, lakes, ponds, fish ponds, water retention, and storage ponds;
–   Efficient in erosion control and shoreline protection;
–   Can be used for cultivation of useful (e.g. edible) cultivars;
–   Contributes to natural beauty (e.g. formal gardens with flowers);
–   It can be moved to another place or removed if required;
–   It has a worldwide potential to improve surface water quality.

The reviews of Chen *et al.* (2016) and Pavlineri *et al.* (2017) give a good overview on the use of floating wetlands for wastewater treatment and on the design, operation and management aspects of floating wetlands. Several case studies and good examples are published, some of them are listed here for further study of the topic.

- *Bega, New South Wels (AU)* – Stormwater wetlands were installed in **2007** using Aqua Biofilter™ reed beds. The dense root mass of the 200 m² floating reedbed was located directly in the stormwater stream intercepting the polluted flows and mopped up the majority of the nutrients, suspended solids, metals, and toxicants. Carex was the most successful plant. After 14 months the small 30-cm plants had grown to 2.2 m in length, including roots. Root tissue lab analysis revealed a significant uptake of pollutants, nutrients, metals, and toxicants (Duncan, 2012).
- *Drinking water reservoir remediation in China in 2009* involved a 67-hectare reservoir with toxicants, health risks, toxic algal blooms, and wind-wave action resulting in sediment resuspension. Floating reed beds were placed near the shore to prevent wave action stirring up sediment and large-scale floating reed beds were deployed across the reservoir to treat the toxicants in the water and reduce nutrients. The project is being replicated across China (Duncan, 2009).
- *Coastal Restoration in Louisiana and Florida (US)* – Human-made floating islands have been established to support plant life in coastal areas since *2009* (BioHaven® Living Shorelines, 2013). The root zones of these plants extend below the solid growing medium of the island and take up nitrogen and phosphorous directly from the water (Coastal Restoration, 2018).
- *McLean's Pit Landfill, Greymouth (New Zealand) (2009)* – Landfill leachate containing BOD, nitrogen and suspended solid was treated by floating wetland technology in 2009 as a demonstration. The first stage of the development proved to be effective in suspended solid and color removal (about two-fold removal compared to preceding conventional treatment). BOD and total nitrogen removal increased by 1.5-fold (McLean's Pit Landfill, 2011). The performance of the Greymouth Landfill leachate treating floating wetland was compared in a study (Comparative Studies, 2009) to the performance of several hundred conventional wetlands used for the same purposes (Vymazal, 2009). The comparative results are shown in Table 4.2.
- *Rehberg Ranch Residential Subdivision, Billings, Montana (US) (2009)* is an early example of floating wetland application in an aerated wastewater lagoon. A 214-m² floating wetland was established covering 6.4% of the 3345 m² lagoon (Rehberg Ranch, 2010). The combination resulted in intensive nutrient and BOD removal. The increase in BOD removal was 38%, suspended solid removal: 27%, N and P removal: 9–9%, compared to the simply aerated control lake. The cost was reduced by 50%.

Table 4.2 Comparison between the performance of floating and conventional wetlands in landfill leachate treatment.

Parameter	Floating wetland	Conventional wetland
BOD	46%	33%
Total N	40%	33%
Suspended solid	89%	55%
Size	288 m²	872 m²

Table 4.3 Comparison between the performance of floating and conventional wetlands in municipal wastewater treatment.

Parameter	Floating wetland	Conventional wetland
BOD	89%	81%
Total N	69%	39%
Ammonium-N	83%	21%
Suspended solid	53%	68%
Size	214 m²	1197 m²

The performance of the Rehberg floating wetland system was compared to the average of several hundred conventional wetlands (Vymazal, 2009) for municipal/household wastewater. The main results of the comparative study are shown in Table 4.3 in percentage of removal (Comparative Studies, 2009).

– *Fish pond surplus nutrient (2009)* was reduced by planting aquatic vegetables on artificial floating beds in concrete fish ponds in China. After 120 days of treatment, 30.6% of total nitrogen and 18.2% of total phosphorus was removed by a 6-m² floating bed planted with *Ipomoea aquatica* (water spinach) (Li & Li, 2009).

– *Mermaid Pool, Somerset County, New Jersey (US) (2010)* – Phosphorus reduction was stable and effective with 83-m² passive floating wetlands treatment, 2% of the one-acre waterway. The platform was prepared from post-consumer polymer fibers. The 0.1–0.25 mg/L total phosphorous content of the inlet decreased to 0.05–0.065 mg/L in the outlet (Mermaid, 2011).

– *Lake Taihu, China(2010)* – An integrated ecological floating bed was constructed and used employing plant, freshwater clam and biofilm carrier. Mesocosm experiments were carried out to determine the optimal water exchange period (Li *et al.*, 2010).

– *Louisiana wastewater facility (US) (2011)* – Nutrient removal with passive floating islands. BioHaven floating treatment wetlands were installed (145 m² in total) in the wastewater treatment plant in 2011 and monitored during a 17-month period by monthly sampling. Chemical oxygen demand (COD), ammonia and phosphate were reduced to consistently manageable levels to keep the facility in compliance (Louisiana Wastewater Facility, 2011).

– *Sydney Olympic Park (2012)* – Authority owned floating islands habitat for migratory birds: the world's first marine species of saltmarsh floating islands was established in 2012 for a migratory water bird habitat. Two islands of 120 m² in total were installed. Black swiftlets, black swans, and migratory birds use the floating islands for a habitat (Ecoplan Australia, 2018).

– *Fish Fry Lake (2014)* – A dying lake in Montana (US) containing high levels of nitrogen and phosphorous and animal waste was remediated using rafts planted with marshland vegetation. The floating island helped the 6.5acre lake to recover, and regular fish harvesting contributed to the nutrient removal of the fish biomass. Fish growth increased to 120–135% compared to the control lake (Fish Fry Lake, 2014, 2018; Andresena, 2017).

– *Hayden Lake (2017)* in Idaho (US) is polluted, primarily with phosphorus, which has led to algae blooms and the proliferation of invasive aquatic plants. The combined pressures of increasing water temperatures and dropping dissolved oxygen levels are threatening the cold-water habitat. Experiments since 2011 confirmed the feasibility and efficiency of floating islands so seven installations were established in 2017. The floating fundament is made from recycled plastic bottles and natural wetland vegetation was planted on top. The floating wetlands solved the problem of nutrient pollution and the long-term stormwater runoff treatment. The effectiveness of 23 m² of floating wetland is equivalent to half a hectare of natural wetland surface area (Hayden Lake, 2017).

– *Marton (New Zealand) (2013)* – Treatment of an anaerobic lake, eliminating odors by using the floating wetland as a blanket to exclude odor transport and to form a low-rate anaerobic digestor for energy-efficient BOD removal from the wastewater. In September 2013 (after more than 3 years in operation), the system still performed optimally (Marton, 2013).

– *Urban river with a high nutrient and metal* content in China: the water in the floating wetland area became more transparent and showed efficient removal of nutrients and chemical and biological oxygen demand as well as metals content. However, the remediation area showed higher concentrations of organic carbon, total nitrogen, nitrate, sulfate, Fe, Cu, Pb, and Zn in the sediment compared to the control area. Similar to conventional wetland accumulation the same problem should be managed in relation to floating islands (Ning *et al.*, 2014).

– *Yangtze River Delta, China (2016)* – Three floating wetlands were constructed and planted to investigate the influence of seasonal change on contaminant removal and harvesting strategy. The best removal efficiency was achieved by *Thalia dealbata* (canna; a wetland plant): 71% for chemical oxygen demand (COD), 70% for total nitrogen, and 82% for total phosphorus. The COD and total N concentrations in effluent water were under 20 and 1.5 mg/L (Ge *et al.*, 2016).

– *High-salinity river in China* – Water spinach and sticky rice was used in hydroponic floating bed systems (Sun *et al.*, 2017).

– *Royal Botanic Garden (2018) in Melbourne (AU)* – By installing floating treatment wetlands water quality improved in the lakes, blue-green algal blooms were reduced in warmer months, as well as improvements were observed in wetland habitat for aquatic fauna and effective stormwater purification. Due to quality improvement, the water can be utilized for irrigating the botanic garden (Figure 4.9) (Royal Botanic Garden, 2018).

Figure 4.9 Guilfoyle's Volcano, the lake with floating wetlands in the Melbourne Royal Botanical Garden (Guilfoyle was an Australian landscape gardener and botanist).

- *Floating islands and floating riverbanks* are constructed by Biomatrix (2018a) using 2D, 3D, and 4D modular, thermo-fused, tough float systems. The modular float system has been deployed in the UK for food industries and also for residential developments in Scotland. The product is suitable both for industrial (including food and beverages industries) and municipal wastewaters. The floats are durable and UV resistant and the attached growth medium supports aquatic plants and the connected ecosystem. The construction is rapid (some weeks), the design is flexible, and the system can also be used for upgrading existing treatment plants (Biomatrix, 2018b).
- *Floating Island International (2018)* creates sustainable ecosystems for healthy natural microbial and plant processes resulting in better water quality and a diverse habitat for pollinators, wildlife, and fish.
- *Midwest Floating Islands* (2018) published many different applications on their website advertising several applications that closely mirror the fields of applications of floating islands in general:

  o   Stormwater treatment;
  o   New and retrofit for stormwater ponds;
  o   Wastewater and gray water lagoons and stabilization ponds;
  o   Aquaculture;
  o   Algae reduction;
  o   Improving water quality in lakes and rivers;
  o   Wetland restoration;
  o   Erosion control and wave dampening;
  o   Fish spawning platforms;
  o   Wildlife habitat restoration and promotion, e.g. establishing nesting bird habitat;
  o   Carbon sequestration;
  o   Aesthetic waterscape.

5   REACTOR BASED SOLUTIONS – LIVING SYSTEMS AND ECOMACHINES

5.1   Conventional living machines

The maximum degradation rate achievable with natural systems is determined by the environmental conditions (in most cases the temperature and/or dissolved oxygen concentration). In order to eliminate or diminish the limiting effects of these factors, the natural environment is imitated under controlled conditions within bioreactors. An important difference between traditional biotechnologies and the solutions of ecological engineering is that, in the latter, owing to the active presence of higher organisms (including plants), not only do the classical tanks act as reactors but their closed environment is involved in the treatment as well. For being able to control this environment, the reactors are covered by a greenhouse with temperate climate zones.

The application of reactors makes significantly higher process rates available. Table 4.1 gives an overview about comparable process rates achievable with natural treatment approaches and various reactor-based systems. As it turns out, the hydraulic detention time data show that it at least doubles the process rate that can be achieved with traditional living systems compared to constructed wetlands, and this still can be increased by a factor of ten using modern advanced artificial ecosystems.

Obviously, the process rate increase comes with an increase in capital and operational costs. A general comparison of capital expenditure (CAPEX) could be an important part of the overall technology comparison, however, such an analysis can be misleading as the different technologies can achieve different effluent limits with almost the same CAPEX. There is a very special difference however, in operational costs in the case of the ecologically enhanced systems. The cost of heating the greenhouse is a cost group that no competitive technologies have. Table 4.4 gives an estimation of these costs for different plant sizes. It is worth noting that this is around 10% of the total energy consumption of such treatment plants.

Ecomachines are always multi-reactor systems where each reactor is designed for hosting its own specialized ecosystem as comprehensively as possible. The reactors are halfway between natural habitats and conventional wastewater treatment tanks operated under anaerobic, anoxic or aerobic conditions. A typical reactor scheme of an ecomachine plant is illustrated in Figure 4.10 as published by Organica® (2018). The scheme visualized the different reactors with plants on their surface.

Table 4.5 gives an overview of the applied cell types in living systems. The combination and respective numbers and volumes of these cells depend on the respective influent characteristics and effluent limit requirements.

Table 4.4 Energy consumption for greenhouse heating of ecomachine installations at different sizes (0°C external average temperature during wintertime).

Plant capacity (m³/day)	Energy consumption (kWh/year)
500	35,770
2000	76,600
5000	140,500

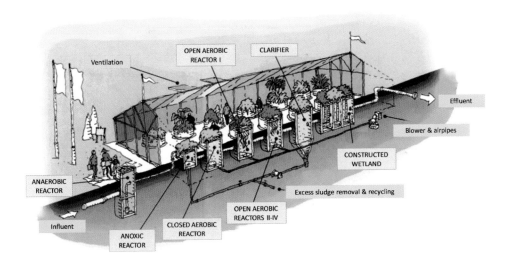

Figure 4.10 Typical reactor scheme of a multi-cellular living system plant.

Table 4.5 Reactor types in ecomachine technologies and in enhanced artificial ecosystems.

Type	Function	Conditions	Sensors installed*
Closed aerobic	First step of organic matter removal with odor control	DO $\geq$ 2 mg/L	DO, pH, COD/BOD
High-load aerobic (I)	Organic matter removal	DO < 1.0 mg/L usually covered by shallow biofilters	DO, pH, COD/BOD
Low-load aerobic (II–IV)	Simultaneous nitrification, denitrification; excess sludge reduction, biological phosphorus removal	DO $\approx$ 2.0 mg/L	DO, pH, NH_4^+-N, NO_3-N
High-load anoxic	Denitrification/organic matter removal	DO $\approx$ 0 mg/L	ORP, COD/BOD, NO_3-N
Low-load anoxic	Denitrification/organic matter removal	DO $\approx$ 0 mg/L	ORP, COD/BOD, NO_3-N
Anaerobic	Anaerobic fermentation, biological phosphorus removal	DO $\approx$ 0 mg/L & NO_3-N $\approx$ 0 mg/L usually covered by shallow biofilters	ORP, COD/BOD
Clarifier/phase separator	Suspended solids removal		
Ecological fluidized bed	Suspended solids/nutrient removal, effluent polishing	DO $\approx$ 4–5 mg/L	DO, pH

*Traditional living machines are usually operated with a minimum number of sensors. However, derivative technologies applying the same cell types rely very much on additional process control possibilities provided by the instrumentation, hence it is logical to include the sensors here.

The arrangement of a typical reactor cell can be seen in Figure 4.11. As the scheme shows, the plants are installed on top of the reactors with their roots dangling 1–1.5 m into the upper part of the basin. The root zone – just like in the natural systems – provides a healthy habitat for bacteria and a whole range of other organisms such as algae, protozoa, zooplankton, and worms. The clarification is followed by post-filtration treatment steps, usually in the form of ecological fluidized beds that act as living habitats for higher organisms, typically snails and fish (see also Laylin, 2010).

The functions of the plants are slightly different to that which was summarized in Section 4.2. In these reactor-based, high-load systems, the primary function of plant roots is being present as biofilm carriers. In addition to the carrier function, the oxygen transfer taking place across the root tissues is an important advantage of such systems. The oxygen transfer rates merely from plant roots can cover 1.2–10 g O_2/m^3 reactor volume/h respiration rates in planted reactors which is a significant contribution to the oxygen balance in the system.

As outlined in Section 4.2, the amount of nutrients absorbed by the plants (nitrogen and phosphorous) is very small compared to the amounts removed by the microorganisms, so it is usually neglected during the design process.

In recent decades the original living machine concept was further developed. The main development directions – as outlined in Section 5.2 – focus on combining the advantages of

Figure 4.11 A typical reactor in an ecomachine installation.

the ecological design with the beneficial properties of conventional biological wastewater treatment technologies. The most notable representative of this new generation is the Organica technology. In certain installations the technology combines the suspended biomass in the system (similar to conventional activated sludge) with fixed biomass present in two special forms in the reactors. First, the biomass lives on the surface of the plant root system. In order to optimize the aeration efficiency, an additional significant part of the active biomass resides on artificial fixed film carriers placed in the lower region of the reactors.

Todd and Josephson (1996) formulated a few design and operation criteria for living machine plants and introduced the term "ecomachine".

–   *Steep gradients.* A waste stream will benefit from passing through a series of stages that have different oxygen regimes, redox potentials, pH, temperature, etc. This principle is mostly about diversifying the habitats present in the treatment system so in this way fostering the development of highly diverse ecosystems.
–   *High exchange rates.* The goal is to maximize the surface area of living material to which the waste stream is exposed.
–   *Periodic and random pulsed exchange.* According to Todd and Josephson (1996), not only spatial variations but gradients in time can also contribute strongly to an increase in the treatment efficiency.
–   *Cellular design and structure of the system.* As discussed previously, compartmentalization contributes to achieving extended biodiversities, maximizing the

number of sub-ecosystems, so contributing to system stability and increasing treatment performance.

- *Solar based photosynthetic foundations.* Algae and higher plants are seen in conventional civil engineering as nuisance organisms to be eliminated physically and chemically from the treatment process. In the case of ecomachines these are integral parts of the treatment pathway.
- *Diversity of higher organisms.* Higher organisms in living systems consume the excess microbial biomass that develops on the degraded waste, so contributing to excess sludge reduction.
- *Biological exchanges beyond the system.* Operation of traditional living machines relies on periodic re-inoculation of the reactors so providing constant linkage with larger natural systems. The function of re-inoculation maintains or even enhances diversity in the systems.

5.2   Advanced artificial ecologies

This technology group emerged from the merger of conventional living machines and conventional activated sludge systems. These systems inherited the reactor layout of living machines and the process knowledge, equipment and instrumentation of activated sludge. After the initial combination of these two approaches, significant development led to the current state-of-the-art of these systems. One of the most notable solutions with the origins outlined previously is the Organica technology.

Artificial ecological solutions are positioned as a competitor of mainstream biological wastewater treatment technologies like activated sludge, membrane bioreactors (MBRs), moving bed biofilm reactors (MBBR) or sequencing batch reactors (SBRs). They are highly competitive both in capital and operational costs as well as footprint or manpower requirements. The beneficial features originate from the presence of artificial ecosystems in the reactors similar to living machines: there are plants in the system, and the plant roots act as biofilm carriers. The interior view of Organica technology-based waste treatment plants can be seen in Figure 4.12. The activity of biofilm organisms is added to that of the activated

Figure 4.12 Interior view of a wastewater treatment plant applying the Organica ecomachine technology.

sludge. This quantitative effect results in the increase of treatment efficiency. Higher animals with longer lifecycles (unicellular cilia, worms, arthropods) establish themselves on the biofilm. The formation of such a complex ecosystem enables a wider spectrum of nutritive material to be broken down. This qualitative effect leads to a lower organic matter content of the effluent and to higher efficiency of ammonia removal.

The specific surface area of plant roots is comparable to modern artificial carrier media, but plant roots are not susceptible to clogging. The oxygen transfer of plant roots contributes to lower aeration requirements.

In addition to the natural biofilm carriers, artificial "root systems" are installed in the reactors that contribute to the available biofilm carrier surface. The application of artificial biofilm carriers in combination with the plant roots has several advantages. The plant roots are protected from the shearing force of the air bubbles and a large surface area is provided for additional biofilm growth. Even when the plant roots have not grown (typical situation after system start-up), biomass can attach to these artificial carriers: the efficiency of the treatment is good, even just after the planting period.

Figure 4.13 shows the micro- and nanofiber structure of the artificial biofilm carrier which provides an active surface for microbial adhesion. The specialty fibrous product serves as a fix bed in flow-through bioreactors. The micro- and nanofibrous net becomes covered by a thick biofilm, as shown by Figure 4.14. It is also apparent in the pictures that the two biofilms from two different bioreactors significantly differ from each other.

The combination of natural and artificial root systems (see Figure 4.15) improves the wastewater treatment process by leveraging a fixed-bed biofilm (attached growth and not floating suspended in the water) that grows on both natural (plant) and engineered (biofiber media) root structures in the reactors allowing a much greater quantity and diversity of organisms to thrive in the same physical space compared to conventional activated sludge technologies. The larger and more diverse biomass per unit of reactor volume results in reduced construction costs as well as improved stability, increased nutrient removal performance, and reduced energy consumption. The reactors are placed into greenhouses, appearing as an odorless botanical garden, so it can be placed directly into the source of the wastewater.

Figure 4.13 Micro- and nanofibers function as an active surface for microbial adhesion. The specialty product serves as a fixed-bed in flow-through bioreactors.

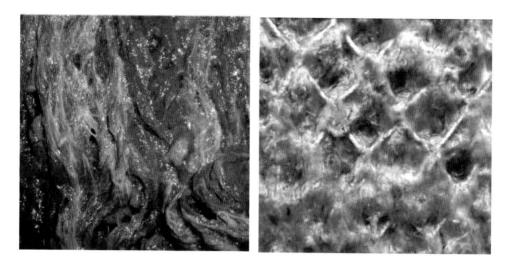

Figure 4.14 Micro and nanofiber net covered by a thick biofilm, derived from two different reactors.

Figure 4.15 Schematic representation of natural and artificial root system in the fixed-bed biofilm technology: artificial enlargement of plant roots increases microbial density in the reactor. Higher microbial density results in enhanced mineralization and an accelerated wastewater treatment process.

5.3   Living machines, ecomachines, ecomachine systems – case studies

The first living machine projects are dated from the 1990s, 30 years after the birth of the term "ecoengineering," when the design of sustainable ecosystems was called Odum in 1962 (Odum, 1962; Odum et al. 1963). Living machine- or ecomachine-type technologies use integrated wetland-based technologies, the main difference being that while wetlands are open air technologies, ecomachines use reactors placed inside. Ecomachines can integrate any type of wetlands, resulting in horizontal flow ecomachines, vertical flow ecomachines, tidal flow or hydroponic ecomachine systems, wastewater gardens, or even reactor-based floating ecomachines built on a boat or raft (the reactor-version of floating islands). Living machine systems may vary largely in size, design and capacity and the time needed for the development of a self-sustaining community (1–3 years). Ecomachine systems are typically combined with botanical gardens and teaching facilities in higher education. Ecomachines and ecomachine systems can solve local household wastewater/sewage treatment and water reuse and can serve as the basis for sustainable cities and sustainable agriculture and husbandry. Ecomachine systems utilize all nutrients and energy from wastes converting it to biomass and products that can join the normal biogeochemical cycles.

First developers of living- or ecomachines and systems:

– US-based John Todd Ecological Design (Todd, 2018);
– Ocean Arks International (Ocean Arks, 2018);
– Worrell Water Technologies (Worrell, 2018);
– The Scottish Biomatrix Water (Biomatrix, 2018a);
– Organica in Hungary (Organica, 2018).

The most popular fields of application are these:

– Replacement of small size wastewater treatment plants with ecoengineered living systems, named as living machines, ecomachines or food-chain reactors (a series of various rectors) and their systems, which are placed into botanical garden looking greenhouses.
– Supplemental application of the ecotechnology to large capacity traditional wastewater treatment plants: these kinds of facilities adequately insert the ecotechnologies and the greenhouses into existing conventional technological steps.
– Surface water restoration by using known restorer systems, which can be placed in-stream, directly into the water, as a floating reactor system (anchored or built on a boat) or off-stream; in this case a side-stream will flow through the treatment unit. Re-inoculation plays an important role in both cases.
– Recycling and reusing water to reduce primary water consumption in arid areas or in any case of water shortage (cf. global warming).
– Recycling and reuse of water in cases of activities with high water requirement, e.g. sports fields, zoo parks and botanical gardens, and industrial technologies where the water is used for washing/cleaning.
– Recycling and reuse of water in green buildings: water treatment locally by using the aesthetic greenhouse-based technology and reuse the treated water for non-potable purposes.
– Living factories: a kind of aggregate of various bio and ecotechnologies in multifunctional buildings in the urban environment. The aim is to close the metabolic loop by

integrating biological production and waste and wastewater recycling in adequate architectural forms fusing public spaces with bio-manufacturing and water reuse (e.g. recreation, urban farming, education and research) (Kenyeres, 2015).

– The previous approach is supported by those technologies which provide simultaneous wastewater treatment and electricity generation by the combination of an upflow constructed wetland and a microbial fuel cell (Oon et al., 2015; Xu et al., 2015; Song et al., 2017). The oxidation reduction potential (ORP) and dissolved oxygen (DO) profile in the study of Oon et al. (2015) showed that the anaerobic and aerobic regions were well developed in the lower and upper beds and the removal efficiencies of COD, NO_3^-, and NH_4^+ were 100%, 40%, and 91%, respectively. A maximum power density of 6.12 mW/m^2 was achieved

Some living machine and ecomachine-type ecotechnologies and systems are introduced as examples in this section, starting with the early installations and showing some of the recent ones. Several variations have been developed and combined with other biological and ecological technologies for today after the initial trials brought success, and the US EPA published an evaluation of performance and system costs in 2001 clearly demonstrating the advantages (US EPA, 2001b).

Findhorn Ecovillage (1995) in Scotland was the very first settlement in Europe that established an ecologically engineered sewage treatment plant for wastewater from about 500 residents. This cost-effective and aesthetic ecological water treatment system utilizes a set of sequenced ecologies built up in tanks in a visually appealing greenhouse. The tanks contain biofilters with complete ecological systems made up of diverse communities of bacteria, algae, members of the micro-fauna, and numerous plant (aquatic and mash vegetation, including trees) and animal species (snails and fish). The quality of the outflow from the series of tanks meets environmental standards without chemical additives or disinfectants. Biomatrix (2018a) company operates the Findhorn ecovillage today simultaneously providing a research and educational facility.

South Burlington (1995) municipal ecomachine, in Vermont, US, demonstrated high performance in the treatment of 80,000 gallon/day wastewater from 1200 residents in cold temperatures. Sewage flowed to a greenhouse with two treatment trains, each with five aerobic reactors, a clarifier and three ecological fluidized beds. The aesthetic beauty of the greenhouse and the surroundings as well as the lack of offensive odor (compared to previous wastewater treatment plant) makes the system compatible with the residential environment and can be used as an educational center for local schools.

Darrow School ecomachine (1998) in New Lebanon, New York, was an early example of a cost-effective, environmentally friendly solution to repair an inadequately functioning wastewater treatment plant. The Darrow school ecomachine treats 8500 gallon/day wastewater from campus buildings before returning the water to the Hudson River watershed. The diversity of microorganisms, snails, fish, and higher plants break down and mineralize organic pollutants. The ecomachine is placed into a botanical garden like greenhouse which provides a setting for educational activities: monitoring water quality and ecosystem activities and receiving more than 500 guests per year.

Oberlin College (2000) in Ohio, US, between Toledo and Cleveland, represents a complex eco-friendly solution which includes an ecomachine setting among other things. All the materials within the Adam Joseph Lewis Center for Environmental Studies are recycled, reused, or sustainably grown and harvested. Lighting, plumbing, and air-conditioning all represent energy-efficient and passive solutions. The variety of native ecosystems constructed not only provide food, but also solve stormwater management and storage, and wastewater treatment and reuse by living machine technology. The living machine contains two underground anaerobic tanks, two covered aerobic reactors, a series of hydroponic indoor reactors hosting aquatic plants, a clarifier, a vertical-flow constructed wetland and a reuse tank with disinfected water. The reactors and the reactor-based wetlands are all placed inside a greenhouse which also provides educational purposes for the college.

Emmen Zoo living machine (2002) in the Netherlands. These works have been operating since 2002 and treat 830,000 m³/day wastewater in the Noorder Zoo. A closed water cycle has been established and the effluent of the living machine is treated by ultrafiltration before it goes back in the reuse cycle. The living machine is placed in a heated building to create a constant climate and to optimize the biological processes. The system consists of three small and one large wastewater cycle. The large cycle combines the residues of the three small cycles from zoo animals which is mixed with the restaurant wastewater, the toilets and with the dirty pond water. Rainwater from the pavements is also collected and reused. Tap water use has been reduced from 180,000 m³ to 30,000 m³ per year and the system produces less waste. Total investment of the water treatment system was 10 million Euro. The total system can be visited by visitors to the zoo to learn more about water and aquatic ecosystems.

Telki (2004) is a rapidly growing village near Budapest, Hungary. Due to the rate of development the wastewater treatment plant reached its maximum capacity in 2002, so it required a complete upgrade. The municipal government selected the Organica system (a living machine-type ecotechnology) to solve the problem. The facility was constructed during 2004–2005 for the treatment of 800 m³/day municipal wastewater. The technological system consists of several closed and open aerobic and anoxic reactors to efficiently degrade organic components and wastewater nutrients. The microbiologically prepared water flows through the enhanced artificial ecosystems contain constructed wetland type reactors (called FCR = food chain reactors) to remove residual organics and nutrients as shown in Figure 4.10. The highly versatile Organica system and the Telki greenhouse garden (Figure 4.16) have been providing a home for several research projects over the past years so their operation was followed by intensive monitoring; the main water quality parameters are shown in Table 4.6.

Furman University (2006) Liberal Arts college in Greenville, South Carolina, US, selected the living machine technology as the most energy-efficient option with the lowest

Figure 4.16 The Organica ecomachine system in Telki, Hungary.

Table 4.6 Average input and output water characteristics of the Telki Organica system.

Measured parameter	Average influent (mg/L)	Average effluent (mg/L)	Standard quality criteria (mg/L)
Chemical oxygen demand (COD)	890	52	100
Biological oxygen demand (BOD)	430	9	30
Ammonium-nitrogen (NH₄-N)	64	3	10
Total nitrogen (TN)	87	10	35
Total phosphorous (TP)	21	1	5
Total suspended solid (TSS)	426	9	50

Table 4.7 Average input and output water characteristics of the Etyek (Hungary) Organica system.

Measured parameter	Average influent (mg/L)	Average effluent (mg/L)	Standard quality criteria (mg/L)
Chemical oxygen demand (COD)	630	43	100
Biological oxygen demand (BOD)	325	7	30
Ammonium-nitrogen (NH₄-N)	79	4	10
Total nitrogen (TN)	99	17	35
Total phosphorous (TP)	10	3	5
Total suspended solid (TSS)	283	11	50

life cycle costs: 5000 gallons of wastewater from the campus sewer is collected in an underground primary settling tank and pumped into the living machine®. A tidal flow wetland is integrated into the living machine system mimicking coastal wetlands with fill and drain cycles and intensive oxygenation. After treatment and disinfection by UV radiation, the water is reused for toilet flushing.

Etyek (2007), a small village just outside Budapest, was chosen by Hollywood producers to build the world-class Korda film studio. The increased wastewater production could not be handled by the existing wastewater treatment plant, thus its capacity had to be increased to the predicted demand of the studio while considering the seasonally fluctuating water demand of the wine country. The Organica system was built in a greenhouse with the capacity of 1000 m³/day (Table 4.7).

Guilford County Schools (2007) Greensboro, North Carolina, US, invested in the living machine® system after they had been told that the cost just to connect to the nearest municipal wastewater treatment plant would be US$ 4 million. Guilford County decided to invest in a "green" onsite wastewater treatment system, so they chose to feature a living machine system. The hybrid wetland type living machine designed for Guilford is comprised of horizontal flow and tidal flow wetland cells and disinfection and ultrafiltration units for the outflow of the wetlands. The main benefit of the system is a significant water saving (2 million gallon/year) and savings in the construction of sewer lines to reach the conventional wastewater treatment plant.

The **Omega Center (2009)** in Rhinebeck, New York, started to establish an ecomachine system in 2003 as an alternative for the failing field system of wastewater treatment. The ecomachine treats up to 52,000 gallon/day of wastewater for irrigation to a leach field. The technology contains (i) septic and equalization tanks, (ii) anoxic tanks, (iii) aerated aquatic cells, (iv) an outdoor wetland, and (v) a recirculating sand filter. The reactors are placed in a greenhouse that is also used as a classroom and event space.

The stepwise structure of the ecomachine: (i) sewage is collected and pumped to underground septic tanks where the anaerobic microbiological degrading process starts spontaneously and where the solid phase is separated from water. (ii) Equalization tanks regulate the uneven inflow by distributing it into two anoxic reactors where nitrate and phosphate is partly removed. (iii) Four constructed wetlands receive the effluent of the anoxic reactors. Here the root zone microorganisms mineralize organic matter providing nutrients for plants and convert nitrate surplus into nitrogen gas. (iv) The ecomachine lagoon within the greenhouse contains a complex ecosystem of high species diversity for polishing the outflow of the wetlands. The artificial ecosystem includes microscopic algae, fungi, bacteria, protozoa, snails, fish, and zooplankton. Plants, shrubs and trees are grown on floating racks within the system. Air compressors are installed to maintain oxygen levels. (v) The final cleansing process is filtering by a recirculating sand filter. The water leaving the filter is usable for any non-potable purposes. The water is utilized for irrigation, for flushing toilets and for water gardens or fish ponds.

San Francisco (2010) Public Utilities Commission, California, US, decided to use a sustainable design when planning a huge administration building in downtown San Francisco (525 Golden Gate Avenue). A living machine® system was established to treat and reuse the wastewater from the building. The living machine integrated the lobby, the front walk, and the sidewalk of the building. The technology consists of four stages:

–   Primary treatment after sewage collection in a subsurface storage tank;
–   Another subsurface tank-system for equalization and recirculation of the water;
–   Tidal flow wetland for mimicking the filling and draining cycle 12 times a day;
–   The vertical flow wetland cell for polishing the treated water. Both wetlands are planted and function as indoor and outdoor gardens.
–   The water treated in the wetlands is filtered and disinfected in underground reactors and reused for flushing toilets.

The benefits of the San Francisco living machine system are savings of 750,000 gallons water per year in the building itself and providing 900,000 gallons per year for off-site uses. The attractive foliage also contributes to a pleasant public space in the town.

Port of Portland (2010) Headquarters, in Portland, OR, US, uses a tidal flow wetland-type living machine®. The wetlands are situated in the lobby and along the exterior front walkway of Port of Portland administrative office building. The wastewater goes through:

–   Primary treatment after collection;
–   Equalization;
–   Six tidal flow living machines for secondary and tertiary wastewater treatment;
–   A vertical flow cell for polishing;
–   The treated effluent is filtered and disinfected with ultraviolet radiation and chlorine.

Benefits from this living machine technology are: savings in water use, reuse of treated wastewater for the building's cooling system, and the planted wetlands function as gardens for recreation. The project attained a LEED Platinum certification by the US Green Building Council.

The *Fisherville Mill Canal (2012)* restorer is a hybrid of in-stream and side-stream technologies, combined into a circulating loop. The ecomachine provides a healthy ecology for re-seeding the contaminated canal. The complex system consists of interlinking technologies: a greenhouse with solar aquatic tanks, a mycelial loop, and a floating plant raft anchored in the canal. Water passes from the canal through the greenhouse and back into the canal in a continuous loop. The system provides a large number of beneficial organisms to the canal on a year-round basis. Since the installation of the system, the canal water quality has improved significantly, and amphibians and turtles have returned.

Moskito Island (2015) ecomachine on the British Virgin Islands where water is a limited resource was designed and constructed for 29,000 gallon/day domestic sewage with the aim of reusing the treated water for non-potable irrigation purposes. The terraced configuration allows the ecomachine system to be sited on a steep hillside as a hillside garden and utilizes the power of gravity.

- Wastewater is received by underground septic storage tanks where primary treatment takes place and the settled solid phase is removed.
- The wetland setting consists of a deep aerated wetland chamber and a vertical- and a horizontal-flow wetland. The wetland zones of aeration and mixing are followed by anoxic areas to efficiently remove nitrogen.
- Gravity continues to move water from the lowest wetland into a series of six deep aquatic cells working as small aerobic ponds with a high species diversity to consume any remaining nutrients.
- A series of cartridge filters and ultraviolet disinfection provide final polishing of the treated effluent.

South Pest (2012) wastewater treatment plant was built in 1966, but by the 2000s it could no longer comply with the increased needs of Budapest, Hungary. Serious capacity, operational and odor problems as well as intensive foaming had arisen, mainly due to the increased contaminant concentrations. The facility of 80,000 m^3/day capacity (wastewater from 300,000 residents and numerous local industries), was completely redesigned and the following technological steps developed:

1 Screen to remove grit and a primary clarifier;
2 Distribution chamber;
3 Two anoxic reactors;
4 Five aerobic reactors with fine bubble aeration;
5 Secondary clarification;
6 BIOFOR system for nitrogen removal.

The BIOFOR technology needs methanol to efficiently remove nitrogen and its capacity is limited by the costs. The methanol consumption can be reduced by the inserted Organica ecomachine-type reactors (iii) (iv) (v), which are housed in three huge greenhouse systems at the South Pest facility (see Google map in Figure 4.17). The Organica reactor system is vegetated and additionally supplied by the artificial root system (see Figure 4.15) which, together with the natural root system and the adherent microorganisms, ensures the removal of all residual organics and nutrients resulting in a good quality effluent water with low chemical and biological oxygen demand, low nutrient and metal content, reduced odors and no foaming. The effluent fulfills even the most stringent standards as it is shown in Table 4.8.

In summary, the result of the upgrade in South Pest was more than a 50% reduction in air requirements compared to conventional solutions and more than 40% increase in loading capacity without increased footprint and operational savings mainly due to savings in nitrification and denitrification.

Table 4.8 Water quality parameters of the effluent water of the South Pest water treatment facility.

Measured parameter	Effluent water (mg/L)	Quality standard (mg/L)
COD	<10	80
BOD	<10	25
pH	7.4	6.5–9.0
Extractable by organic solvent	< 2	5
TSS	<10	35
TN average cold season	3.5	20
NH_4-N average cold season	2.54	4
TP	0.04	1.8
Phenol-index	<0.05	0.1
Total Cd	<0.001	0.005
Total Cr	<0.002	0.2
Total Pb	<0.002	0.05
Total Cu	<0.01	0.5
Total Ni	0.003	0.5
Total Hg	<0.0001	0.001

(Source: FCSM, 2018)

Figure 4.17 Bird's eye view of the South Pest WWTP: the combination of the primary and secondary clarifier and the greenhouses with the complex living system.

In *Tianshan, China (2018),* similar upgrading to South Pest is introduced by the Organica technology where a poorly working, 70,000 m³/day capacity wastewater treatment plant was "repaired." The Chinese WWTP will produce an effluent complying with the standard while maintaining its capacity without additional reactors. The upgraded plant is odorless and has the visual appearance of a series of botanical gardens. Upgrading the wastewater treatment makes real-estate development in the area possible.

6   PROCESS MONITORING AND CONTROL POSSIBILITIES

6.1   Process monitoring and instrumentation

Process monitoring and instrumentation in natural systems are usually kept at the lowest possible level, mostly containing periodical manual inspection only. Practically the same approach is applied in conventional living machines. Extensive instrumentation is only used in modern reactor-based artificial ecologies because they originate from conventional wastewater treatment technologies like activated sludge.

The instruments and sensors are similar to those applied in traditional wastewater treatment plants. The most frequently monitored parameters are dissolved oxygen, temperature, pH, and, in some situations, turbidity (these are the low-price probes that are easily available are robust and work reliably and, in addition, provide useful information about the ongoing biological processes). However, with the decreasing price of various in-line probes the constant monitoring of ammonia, nitrate, suspended solids, and organic matter (the latter two in the forms of chemical oxygen demand or biochemical oxygen demand) is often used.

In addition to monitoring and control of biological processes, the conditions within the greenhouse (temperature, humidity) and those environmental parameters that determine the greenhouse operation (like wind speed and rainfall) are constantly monitored and used in greenhouse operation control (operation of internal humidification, temperature control with shading, heating or opening the windows, etc.)

Irrespective of how sophisticated instrumentation is applied, periodical manual inspection is unavoidable. The most important area of manual intervention is plant nursing (removal of dry floral parts, excessive growth, etc.).

6.2   Process control considerations

Just like sophisticated process monitoring, automated process control emerged with the advanced reactor-based artificial ecologies. The signals coming from the sensors are mostly used in traditional wastewater treatment plant algorithms (e.g. dissolved oxygen control by modifying the blower frequency using a Proportional-Integral-Derivative = PID algorithm). In addition to the traditional process control methodologies, the complexity of the ecosystems applied and the elaborate instrumentation open up possibilities for different multi-input/multi-output (MIMO) control algorithms. In these control situations artificial neural networks or genetic algorithms combined with detailed dynamic system models are used.

7   SUMMARY

Ecological engineering started with very strong philosophical motivations in the 1960s. Since then, with the progress in disciplines like applied ecology, biotechnology, and environmental engineering, it gradually moved toward a scientifically solid alternative to conventional methods in the field of waste treatment, habitat reconstruction, urban planning, or resource control. Artificial ecosystems brought new advantages when they were combined with conventional technologies that helped in establishing more sustainable solutions.

Taking a look at the current trends, the development direction of both ecological and conventional engineering shows a path where the difference between the two fields gradually fades away.

Attracting and maintaining wider biological diversity in the waste stream treatment processes gives wider biodegradation ranges and better removal performances along with lower excess biomass production thus contributing to a more economical operation.

REFERENCES

Andresena, A. (2017) *Nature's Water Purifiers Help Clean Up Lakes*. Available from: www.linkedin.com/pulse/natures-water-purifiers-help-clean-up-lakes-aan-andresena. [Accessed 16th March 2018].

Babatunde, A.O., Zhao, Y.Q., O'Neill, M. & O'Sullivan, B. (2008) Constructed wetlands for environmental pollution control: A review of developments, research and practice in Ireland. *Environment International*, 34, 116–126.

Benyus, J.M. (2002) *Biomimicry. Innovation Inspired by Nature*. Perennial, Harper Collins Publisher, New York, NY, USA.

Bergen, S.D., Bolton, S.M. & Frindley, J.L. (2001) Design principles for ecological engineering. *Ecological Engineering*, 18, 201–210.

BioHaven® Living Shorelines (2013) *BioHaven® Living Shorelines, BioHaven® Floating Breakwaters*. Available from: www.floatingislandinternational.com/wp-content/plugins/fii/casestudies/35.pdf. [Accessed 16th March 2018].

Biomatrix (2018a) *Biomatrix Water Company*. Available from: www.biomatrixwater.com. [Accessed 16th March 2018].

Biomatrix (2018b) *Helix Flow Reactor*. Available from: www.biomatrixwater.com/helix-flow-reactor. [Accessed 16th March 2018].

Braungart, M. & McDonoug, W. (2002) *Cradle to Cradle: Remaking the Way We Make Things*. North Point Press, New York, NY, USA.

Brix, H. (1994) Functions of macrophytes in constructed wetlands. *Water Science and Technology*, 29(4), 71–78.

Brix, H. (1997) Do macrophytes play a role in constructed treatment wetlands? *Water Science and Technology*, 35(5), 11–17.

Brix, H., Schierup, H.-H. & Arias, C.A. (2007) Twenty years experience with constructed wetland systems in Denmark – What did we learn? *Water Science and Technology*, 56, 63–68.

Carty, A., Scholz, M., Heal, K., Gouriveau, F. & Mustafa, A. (2008) The universal design, operation and maintenance guidelines for farm constructed wetlands (FCW) in temperate climates. *Bioresource Technology*, 99, 6780–6792.

Centre for Advancement of Water and Wastewater Technologies (2018) Available from: https://cawt.ca and https://cawt.ca/centre/publications-and-presentations/ [Accessed 28th May 2018].

Chen, Z., Cuervo, D.P., Müller, J.A., Wiessner, A., Köser, H., Vymazal, J., Kästner, M. & Kuschk, P. (2016) Hydroponic root mats for wastewater treatment – A review. *Environmental Science and Pollution Research*, 23(16), 15911–15928.

Coastal Restoration (2018) Available from: https://capecodgreenguide.wordpress.com/floating-wetlands and www.floatingislandinternational.com/research/case-studies/#cs_4. [Accessed 16th March 2018].

Comparative Studies (2009) *Floating Islands Outperform Constructed Wetlands*. Available from: www.floatingislandinternational.com/wp-content/plugins/fii/casestudies/3.pdf. [Accessed 16th March 2018].

Darrow School (1998) *Darrow School Eco-Machine*. Available from: www.toddecological.com/index. php?id=projects and www.toddecological.com/data/uploads/casestudies/jtedcasestudy_darrow. pdf. [Accessed 16th March 2018].

Dotro, G., Langergraber, G., Molle, P., Nivala, J., Puigagut Juárez, J., Stein, O.R. & von Sperling, M. (2017) *Treatment Wetlands*. Volume 7. Biological Wastewater Treatment Series. IWA Publishing, London, UK. ISBN: 9781780408767. OCLC 984563578.

Duncan, T. (2009) *Floating New Ideas*. Available from: www.aquabiofilter.com/Tom%20Duncan%20 Aqua%20BiofilterTM%20October%202009%20WME%20Magazine.pdf. [Accessed 16th March 2018].

Duncan, T. (2012) Floating reedbeds biofilter performance in urban stormwater treatment wetlands (Ecoplan Consulting Pty Ltd and Aqua Biofilter™). *National Conference: Stormwater Australia, 2012*. Available from: www.aquabiofilter.com/Tom%20Duncan%20Stormwater%2009%20Floating% 20Biofilter%20Conference%20Paper%20V1.pdf. [Accessed 16th March 2018].

Duncan, T. (2014) *Lake Taihu: World's Largest Aquaponics Project, in China's Third Largest Aquaculture Lake*. Available from: https://permaculturenews.org/2014/10/14/worlds-largest-aquaponics-project-chinas-third-largest-aquaculture-lake. [Accessed 16th March 2018].

Ecoplan Australia (2018) Available from: www.ecoplan.net.au. [Accessed 16th March 2018].

Emmen Zoo living machine (2002) *Northern Zoo, Emmen, the Netherlands*. Available from: www. livingmachines.com/Portfolio/Zoos-Animal-Shelters/Norther-Zoo,-Emmen,-Netherlands.aspx and www.urbangreenbluegrids.com/projects/water-factory-in-emmen-zoo-the-netherlands. [Accessed 16th March 2018].

Etyek (2007) *The Organica® System in Etyek, Hungary: A Case Study*. Available from: www. organicawater.com/case-study/ and www.organicawater.com/case-study/organica-detailed-case-study-etyek-cs/ [Accessed 16th March 2018].

FCSM (2018) *Water Quality Parameters of 28. 02. 2018*. Available from: www.fcsm.hu/szolgaltatasok/ szennyviztisztitas/delpesti_szennyviztisztito_telep. [Accessed 16th March 2018].

Findhorn Ecovillage (1995) *Biomatrix Operates Living Machine in Findhorn*. Available from: www. biomatrixwater.com/portfolio-posts/biomatrix-operates-living-machine-in-findhorn. [Accessed 16th March 2018].

Fisherville Mill Canal (2012) *The Fisherville Mill Canal Restoration Pilot*. Todd Ecological. Available from: www.toddecological.com/index.php?id=projects and www.toddecological.com/data/ uploads/casestudies/jtedcasestudy_grafton.pdf. [Accessed 16th March 2018].

Fish Fry Lake (2014) *Transforming Fish Fry Lake From a Eutrophic Pond to a Productive Fishery*. Available from: www.floatingislandinternational.com/research/case-studies/ and www.floatingisland international.com/wp-content/plugins/fii/casestudies/40.pdf. [Accessed 16th March 2018].

Fish Fry Lake (2018) Available from: www.bbc.com/future/story/20120925-natures-water-purifiers? goback=.gde_94811_member_169923730. [Accessed 16th March 2018].

Floating Island International (2018) Available from: www.floatingislandinternational.com. Accessed 16th March 2018].

Furman University (2006) *Living Machine® System, Worrell Water Technologies*. Available from: www. livingmachines.com/Portfolio/Schools-Universities/Furman-University.aspx. [Accessed 16th March 2018].

Ge, Z., Feng, C., Wang, X. & Zhang, J. (2016) Seasonal applicability of three vegetation constructed floating treatment wetlands for nutrient removal and harvesting strategy in urban stormwater retention ponds. *International Biodeterioration & Biodegradation*, 112, 80–87.

The Green Center (2010) Available from: http://thegreencenter.net. [Accessed 16th March 2018].

Green, J. (2010) Integrated constructed wetlands for treating domestic wastewater. Uniting the built and natural environment. *The Dirt*. American Society of Landscape Architects. Available from: https://dirt.asla.org/2010/09/10/using-constructed-wetlands-for-wastewater-treatment. [Accessed 16th March 2018].

Guilford County Schools (2007) *Living Machine® System, Worrell Water Technologies.* Available from: www.livingmachines.com/Portfolio/Schools-Universities/Guilford-County-Schools,-Greensboro,-NC.aspx. [Accessed 16th March 2018].

Gumbricht, T. (1993) Nutrient removal processes in freshwater submersed macrophyte systems. *Ecological Engineering,* 2, 1–30.

Guterstam, B. & Todd, J. (1990) Ecological engineering for wastewater treatment and its application in New England and Sweden. *Ambio,* 19, 173–175.

Harrington, R., Dunne, E.J., Carroll, P., Keohane, J. & Ryder, C. (2005) The concept, design and performance of integrated constructed wetlands for the treatment of farmyard dirty water. In: Dunne, E.L., Reddy, K.R. & Carton, O.T. (eds.) *Nutrient Management in Agricultural Watersheds: A Wetlands Solution.* Wageningen Academic Publishers, Wageningen, The Netherlands, 179–188.

Hayden Lake (2017) Available from: http://stormwater.wef.org/2015/07/idaho-groups-utilize-monitor-floating-wetlands. [Accessed 16th March 2018].

Hedmark, Å. & Jonsson, M. (2008) Treatment of log yard runoff in a couch grass infiltration wetland in Sweden. *International Journal of Environmental Studies,* 65, 273–278.

Higgins, J., Mattes, A., Stiebel, W. & Wootton, B. (2017) *Eco-Engineered Bioreactors: Advanced Natural Wastewater Treatment.* CRC Press. Taylor & Francis Group, Boca Raton, FL, USA.

Hoffmann, H., Platzer, Ch., Winker, M. & von Muench, E. (2011) *Technology Review of Constructed Wetlands. Subsurface Flow Constructed Wetlands for Greywater and Domestic Wastewater Treatment. Sustainable Sanitation – Ecosan Program.* Deutsche Gesellschaft für Internationale Zusammenarbeit (GIZ) GmbH, Eschborn, Germany. Available From: https://www.susana.org/en/knowledge-hub/resources-and-publications/library/details/930. [Accessed 8th October 2018].

Holling, C.S. (1996) Engineering resilience versus ecological resilience. In: Schulze, P.C. (ed) *Engineering Within Ecological Constraints.* National Academy Press, Washington, DC, USA. pp. 66–80.

Ierland, E.C. & Man, N.Y.H. (1996) Ecological engineering: First step towards economic analysis. *Ecological Engineering,* 7, 351–371.

Jones, K. & Hanna, E. (2004) Design and implementation of an ecological engineering approach to coastal restoration at Loyola Beach, Kleberg County, Texas. *Ecological Engineering,* 22, 249–261.

Jördening, H.-J. & Winter, J. (2006) *Environmental Biotechnology: Concepts and Applications.* Wiley-VCH Verlag GmbH, Weinheim, Germany.

Jørgensen, S.E., Ni-Bin, C. & Xu, F. (eds.) (2014) *Ecological Modelling and Engineering of Lakes and Wetlands.* Elsevier, New York, NY, USA.

Kadlec, R.H. & Knight, R.L. (2004) *Treatment Wetland.* Lewis Publishers, Boca Raton, FL, USA.

Kayranli, B., Scholz, M., Mustafa, A., Hofmann, O. & Harrington, R. (2010) Performance evaluation of integrated constructed wetlands treating domestic wastewater. *Water, Air & Soil Pollution,* 210, 435–451.

Kenyeres, I. (2015) *BIOPOLUS Introduction: From Living Machines to Living Factories.* Available from: www.kazincbarcika.hu/_mellekletek/s_ptalalkozo20150817_Biopolus.pdf and www.biopolus.org/technology. [Accessed 16th March 2018].

Lake Taihu, China (2010) see Li *et al.* (2010).

Laylin, T. (2010) The living machine: An ecological approach to poo. *Ecologist,* 8th June 2010. Available from: https://theecologist.org/2010/jun/08/living-machine-ecological-approach-poo. [Accessed 16th March 2018].

Li, W. & Li, Z. (2009) In situ nutrient removal from aquaculture wastewater by aquatic vegetable *Ipomoea aquatica* on floating beds. *Water Science and Technology,* 59, 1937–1943.

Li, X.N., Song, H.-L., Li, W., Lu, X.-W. & Nishimura, O. (2010) An integrated ecological floating-bed employing plant, freshwater clam and biofilm carrier for purification of eutrophic water. *Ecological Engineering,* 36, 382–390.

Louisiana Wastewater Facility (2011) *Nutrient Removal With Passive Floating Treatment Wetlands.* Available from: www.floatingislandinternational.com/research/case-studies/#cs_4 and www.floatingislandinternational.com/wp-content/plugins/fii/casestudies/32.pdf. [Accessed 16th March 2018].

Luo, H., Huang, G., Wu, X., Peng, J., Fu, X. & Luo, L. (2009) Ecological engineering analysis and eco-hydrodynamic simulation of tidal rivers in Shenzhen City of China. *Ecological Engineering*, 35, 1129–1137.

Marton (2013) *Using the Floating Wetland as a Blanket to Exclude Odor Transport and to Form a Low-Rate Anaerobic Digestor. Eliminating Odors Using BioHaven Technology*. Available from: www.floatingislandinternational.com/research/case-studies/ and www.floatingislandinternational.com/wp-content/plugins/fii/casestudies/24.pdf. [Accessed 16th March 2018].

McLean's Pit Landfill (2011) *Demonstrating Treatment of Landfill Leachate Using Floating Treatment Wetland Technology*. Available from: http://www.floatingislandinternational.com/research/#case_studies and www.floatingwetlandsolutions.com/pdfs/McLeans-Pit-CS-032513.pdf. [Accessed 6th October 2018].

Mermaid (2011) *Phosphorus Reduction With Passive Floating Treatment Wetlands*. Available from: www.floatingislandinternational.com/research/case-studies/#cs_4 and www.floatingisland international.com/wp-content/plugins/fii/casestudies/27.pdf. [Accessed 16th March 2018].

Midwest Floating Island (2018) Available from: www.floatingislandinternational.com/wp-content/plugins/fii/news/54.pdf. [Accessed 16th March 2018].

Mitsch, W.J. (1997) Ecological Engineering: The roots and rationale of a new ecological paradigm. In: Etnier, C. & Guterstam, B. (eds.) *Ecological Engineering for Wastewater Treatment*. 2nd ed. CRC Press, Boca Raton, FL, USA.

Mitsch, W.J. & Jorgensen, S.E. (1989) Introduction to ecological engineering. In: Mitsch, W.J. & Jorgensen, S.E. (eds.) *Ecological Engineering: An Introduction to Ecotechnology*. Wiley, New York, NY, USA. pp. 79–101.

Molle, P., Prost-Boucle, S. & Lienard, A. (2008) Potential for total nitrogen removal by combining vertical flow and horizontal flow constructed wetlands: A full-scale experiment study. *Ecological Engineering*, 34, 23–29.

Moo-Young, M., Anderson, W.A. & Chakrabarty, A.M. (1996) *Environmental Biotechnology: Principles and Applications*. Springer, Dordrecht, The Netherlands.

Moskito Island (2015) *Moskito Island Eco-Machine*. Todd Ecological. Available from: www.todd ecological.com/index.php?id=projects and www.toddecological.com/data/uploads/casestudies/jtedcasestudy_moskito.pdf. [Accessed 16th March 2018].

Nagabhatla, N. & Metcalfe, C. (eds.) (2017) *Multifunctional Wetlands*. Springer, New York, NY, USA.

Ning, D., Huang, Y., Pan, R., Wang, F. & Wang, H. (2014) Effect of eco-remediation using planted floating bed system on nutrients and heavy metals in urban river water and sediment: A field study in China. *Science of the Total Environment*, 485–486, 596–603.

Oberlin College (2000) *Oberlin College: Setting a Sustainable Example in Ohio*. Available from: https://inhabitat.com/oberlin-college-setting-a-sustainable-example-in-ohio/ and www.youtube.com/watch?v=fmdsBlWwSTU. [Accessed 16th March 2018].

Ocean Arks (2018) *Ocean Arks International*. Available from: www.oceanarksint.org.

Odum, H.T. (1962) Man in the ecosystem. In: Proc. Lockwood Conference on the Suburban Forest and Ecology. *Connecticut Agricultural Experiment Station Bulletin*, 652, 57–75.

Odum, H.T., Siler, W.L., Beyers, R.J. & Armstrong, N. (1963) Experiments with engineering of marine ecosystems. *Publications of the Institute of Marine Science (University of Texas)*, 9, 374–403.

Omega Center (2009) *Omega Center for Sustainable Living Eco-Machine®*. Ocean Ark. Available from: www.toddecological.com/index.php?id=projects and www.toddecological.com/data/uploads/casestudies/jtedcasestudy_omega.pdf. [Accessed 16th March 2018].

Oon, Y.-L., Ong, S.-A., Ho, L.-N., Wong, Y.-S., Oon, Y.-S., Lehl, H.K. & Thung, W.-L. (2015) Hybrid system up-flow constructed wetland integrated with microbial fuel cell for simultaneous wastewater treatment and electricity generation. *Bioresource Technology*, 186, 270–275. doi:10.1016/j.biortech.2015.03.014.

Organica (2018) Available from: www.organicawater.com. [Accessed 16th March 2018].

Pavlineri, N., Skoulikidis, N.T. & Tsihrintzis, V.A. (2017) Constructed floating wetlands: A review of research, design, operation and management aspects, and data meta-analysis. *Chemical Engineering Journal*, 308, 1120–1132.

Port of Portland (2010) *Port of Portland Headquarters Living Machine® System*. Worrell Water Technologies. Available from: www.livingmachines.com/Portfolio/Municipal-Government/Port-of-Portland-Headquarters,-Portland,-OR.aspx and http://sustainablewater.com/wp-content/uploads/2013/07/POP-Case-Study-070213.pdf and https://urbanecologycmu.wordpress.com/2016/10/14/port-of-portland-headquarters. [Accessed 16th March 2018].

RAMSAR Convention (1971) *World Convention on Wetlands*. Available from: www.ramsar.org/sites/default/files/documents/library/original_1971_convention_e.pdf. [Accessed 16th March 2018].

Ratledge, C. & Kristiansen, B. (2006) *Basic Biotechnology*. Cambridge University Press, Cambridge, UK.

Rehberg Ranch (2010) *Achieving Significant Nutrient Removal in Aerated Wastewater Lagoons Using floating Island Technology*. Available from: www.floatingislandinternational.com/research/case-studies/#cs_4 and www.floatingislandinternational.com/wp-content/plugins/fii/casestudies/5.pdf. [Accessed 16th March 2018].

Reinsel, M.A. (2018) Floating wetlands help boost nitrogen removal in lagoons. *WaterWorld*. Available from: www.waterworld.com/articles/print/volume-28/issue-6/editorial-features/floating-wetlands-help-boost-nitrogen-removal-in-lagoons.html. [Accessed 16th March 2018].

Royal Botanic Garden (2018) Available from: http://spel.com.au/news/greening-royal-botanic-gardens-melbourne. [Accessed 16th March 2018].

San Francisco (2010) *Public Utilities Commission Living Machine® System*. Worrell Water Technologies. Available from: www.livingmachines.com/Portfolio/Municipal-Government/San-Francisco-Public-Utilities-Commission,-San-Fra.aspx and https://sfwater.org/index.aspx?page=1156. [Accessed 16th March 2018].

Scholz, M. (2006) *Wetland Systems to Control Urban Runoff*. Elsevier, Amsterdam, The Netherlands.

Scholz, M. (2011) *Wetland Systems*. Springer, Dordrecht, The Netherlands.

Seidel, K. (1964) Abbau von Bacterium coli durch höhere Wasserpflänzen. *Naturwissenschaft*, 51, 395.

Song, H., Zhang, S., Long, X., Yang, X., Li, H. & Xiang, W. (2017) Optimization of bioelectricity generation in Constructed wetland-coupled microbial fuel cell systems. *Water*, 9(3), 185. doi:10.3390/w9030185.

South Burlington (1995) *South Burlington Municipal Eco-Machine*. Designed by Living Technologies, Inc. and John Todd with Ocean Arks. Available from: www.toddecological.com/data/uploads/casestudies/jtedcasestudy_southburlington.pdf. [Accessed 16th March 2018].

South Pest (2012) *Upgrade South Pest, Hungary, Organica Case Study*. Available from: www.organicawater.com/case-study/ and www.organicawater.com/case-study/south-pest-upgrade-cs/and www.budapestvideo.hu/attachments/917_viz-vilagnapja-nyilt-nap.mp4. [Accessed 16th March 2018].

Stahel, W.R. (2010) *The Performance Economy*. 2nd ed. Palgrave MacMillan, London, UK.

Sun, S., Sheng, Y., Zhao, G., Li, Z. & Yang, J. (2017) Feasibility assessment: Application of ecological floating beds for polluted tidal river remediation. *Environmental Monitoring and Assessment*, 189, 609. doi:10.1007/s10661-017-6339-y.

Sydney Olympic Park (2012) see Ecoplan Australia (2018).

Tchobanoglous, G. (1997) Land-based systems, constructed wetlands, and aquatic plant systems in the United States: an overview. In: Eitner, C. & Guterstam, B. (eds.) *Ecological Engineering for Wastewater Treatment*. 2nd ed. CRC Press, Boca Raton, FL, USA.

Telki (2004) *The Organica® System in Telki, Hungary: A Case Study*. Available from: www.organicawater.com/case-study/ and www.organicawater.com/case-study/organica-detailed-case-study-telki-cs. [Accessed 16th March 2018].

Tianshan (2018) *Real Estate Development in Tianshan, China*. Organica Case Study. Available from: https://www.organicawater.com/case-study/real-estate-development-potential-cs/ and https://d3o1jlvchszf6h.cloudfront.net/wp-content/uploads/RealEstateDevelopment_Tianshan.pdf. [Accessed 16th March 2018].

Todd, J. (2018) *John Todd Ecological Design*. Available from: www.toddecological.com/ [Accessed 16th March 2018].

Todd, J., Brown, E.J.G. & Wells, E. (2003) Ecological design applied. *Ecological Engineering*, 2003, 421–440.

Todd, J. & Josephson, B. (1996) The design of living technologies for waste treatment. *Ecological Engineering*, 6, 109–136.

UNEP – Phytotechnology (2018) *Examples of Environmental Applications of Phytotechnology. C. Constructed Wetlands*. Freshwater Management Series No. 7. Available from: www.unep.or.jp/ietc/ Publications/Freshwater/FMS7/14.asp. [Accessed 16th March 2018].

USDA (1991) *A Guide to Creating Wetlands for Agricultural Wastewater, Domestic Wastewater, Coal Mine Drainage Stormwater. Volume 1, General Considerations*. United States Department of Agriculture.

US EPA (1993–2000) *Constructed Wetlands Handbooks (Volumes 1–5): A Guide to Creating Wetlands for Agricultural Wastewater, Domestic Wastewater, Coal Mine Drainage and Stormwater in the Mid-Atlantic Region)*. United States Environmental Protection Agency. Available from: www.epa. gov/owow/wetlands/pdf/hand.pdf. [Accessed 16th March 2018].

US EPA (1993a) *A Handbook of Constructed Wetlands: A Guide to Creating Wetlands For: Agricultural Wastewater Domestic Wastewater Coal Mine Drainage Stormwater*. Available from: www. epa.gov/wetlands/handbook-constructed-wetlands. [Accessed 16th March 2018].

US EPA (1993b) *Constructed Wetlands for Wastewater Treatment and Wildlife Habitat: 17 Case Studies* by Spenser, E., Port Blakely Mill Company, EPA 832/R-93/005. Available from: www.epa. gov/wetlands/constructed-wetlands-wastewater-treatment-and-wildlife-habitat-17-case-studies. [Accessed 16th March 2018].

US EPA (1995) *Constructed Wetlands for Animal Waste Treatment: A Manual on Performance, Design, and Operation with Case Histories*. United States of America Environmental Protection Agency (USEPA) Number: 855B97001. Available from: https://nepis.epa.gov/Exe/ZyPURL. cgi?Dockey=200054US.TXT. [Accessed 16th March 2018].

US EPA (1998) *Constructed Wetlands and Aquatic Plant Systems for Municipal Wastewater Treatment*. EPA/625/1–88/022. EPA, Office of Research and Development, Cincinnati, OH, USA.

US EPA (1999a) *Free Water Surface Wetlands for Wastewater Treatment: A Technology Assessment*. https://nepis.epa.gov/Exe/ZyPDF.cgi/9100JFW4.PDF?Dockey=9100JFW4.PDF.

US EPA (1999b) *Subsurface Flow Constructed Wetlands for Wastewater Treatment: A Technology Assessment*. Available from: http://nepis.epa.gov/Exe/ZyPDF.cgi/2000475V.PDF?Dockey=2000475V. PDF. [Accessed 16th March 2018].

US EPA (2000a) *Design Manual: Constructed Wetlands Treatment of Municipal Wastewater*. EPA/625/R-99/010. Available from: https://nepis.epa.gov/Exe/ZyPDF.cgi/30004TBD.PDF?Dockey= 30004TBD.PDF. [Accessed 16th March 2018].

US EPA (2000b) *Guiding Principles for Constructed Treatment Wetlands: Providing for Water Quality and Wildlife Habitat*. EPA 843-B-00-003. Available from: www.epa.gov/owow/wetlands/constructed/ guide.html and http://nepis.epa.gov/Exe/ZyPDF.cgi/2000536S.PDF?Dockey=2000536S.PDF. [Accessed 16th March 2018].

US EPA (2001a) *Constructed Wetlands: Passive Systems for Wastewater Treatment by Lorion, R.* Available from: www.epa.gov/wetlands/constructed-wetlands. [Accessed 16th March 2018].

US EPA (2001b) *The Living Machine (R) Wastewater Treatment Technology: An Evaluation of Performance and System Cost*. National Service Center for Environmental Publications (NSCEP). Available from: https://nepis.epa.gov/EPA/html/DLwait.htm?url=/Exe/ZyPDF.cgi/91022RS0. PDF?Dockey=91022RS0.PDF. [Accessed 16th March 2018].

US EPA (2004) *Constructed Treatment Wetlands*. EPA 843-F-03-013. Available from: http://nepis. epa.gov/Exe/ZyPDF.cgi/30005UPS.PDF?Dockey=30005UPS.PDF. [Accessed 16th March 2018].

US EPA – Constructed Wetlands (2018) *The Science of Wetlands: Constructed Wetlands.* www.epa. gov/wetlands/constructed-wetlands and http://nepis.epa.gov/Exe/ZyPDF.cgi/30005UPS.PDF? Dockey=30005UPS.PDF.

US EPA – Wetlands (2018) *Constructed Wetlands.* Available from: www.epa.gov/wetlands and www. epa.gov/wetlands/constructed-wetlands and www.epa.gov/wetlands/wetlands-restoration-definitions-and-distinctions. [Accessed 16th March 2018].

Vymazal, J. (2007) Removal of nutrients in various types of constructed wetlands. *Science of the Total Environment*, 380, 48–65.

Vymazal, J. (2009) The use of constructed wetlands with horizontal sub-surface flow for various types of wastewater. *Ecological Engineering*, 35, 117.

Wetlands for the Treatment of Mine Drainage (2018) Available from: http://technology.infomine.com/ enviromine/wetlands/welcome.htm#examples_constr. [Accessed 16th March 2018].

Worrell (2018) *Worrell Water Technologies Living Machine Systems.* Available from: www.living machines.com/Company/Worrell-Water-Technologies.aspx. [Accessed 16th March 2018].

Wu, Y., Chung, A., Tama, N.F.Y., Pia, N. & Wong, M.H. (2008) Constructed mangrove wetland as secondary treatment system for municipal wastewater. *Ecological Engineering*, 34, 137–146.

Xu, B., Ge, Z. & He, Z. (2015) Sediment microbial fuel cells for wastewater treatment: Challenges and opportunities. *Environmental Science: Water Research & Technology*, 1, 279. doi:10.1039/ c5ew00020c.

Zedler, J.B. (2000) *Handbook for Restoring Tidal Wetlands.* CRC Press, Boca Raton, FL, USA.

Chapter 5

Biodegradation-based remediation – overview and case studies

M. Molnár[1], K. Gruiz[1] & É. Fenyvesi[2]

[1]Department of Applied Biotechnology and Food Science, Budapest University of Technology and Economics, Budapest, Hungary
[2]CycloLab Cyclodextrin Research & Development Laboratory Ltd, Budapest, Hungary

ABSTRACT

Biodegradation of hazardous contaminants in soil may lead to complete mineralization and elimination from soil under biological (compatible with life) conditions. Biodegradation-based *in situ* remediation belongs to the group of green and eco-friendly soft technologies characterized by high efficiency in clean-up, low secondary impacts on the surrounding environment, and low ecological footprint. In spite of biodegradation's excellent performance, more aggressive technologies using heavy machinery and conditions far from those of nature's still often get priority. Energy-, material-, water-, and labor-intensive "pump and treat," "dig and dump" or "dig and treat" technologies are still practiced, even in cases when high risk or economic interest does not justify urgency.

Bioremediation, natural attenuation, and remediation based on biodegradation has long been involved in research and development and thousands of papers have been published in this field and the number of technology demonstrations and patents are also significant. In the practice biodegradation-based *in situ* soil remediation is less popular due to the lack of knowledge from the engineering side and the lack of trust from the managers and owners. Biodegradation-based natural attenuation has a pioneering role in this field – mainly through the approved spontaneous biodegradation on long-term abandoned sites – proven that the soil biota is capable of cleaning up the groundwater and the soil in many cases and they deserve the support by engineering tools, which is needed to work at maximum efficiency.

Environmental technology verification and sustainability assessment also became a requirement for remediation. Sustainability assessment is applied to remediation both as a prospective evaluation tool in support of decision making and as a retrospective one for the validation of the predictions in technology efficiencies and approval site (soil) quality (see more in Chapter 11). Sustainability assessment clearly presents the advantages and benefits of biodegradation-based *in situ* technologies in an absolute sense and in comparison to conventional geotechnical and physicochemical solutions. Biodegradation-based soil remediation can be considered as an ecotechnology, because the soil biota, this special community work according to ecological rules and their remedial function is integral part of the global biogeochemical cycles.

This chapter focuses on the following:

- Main principles of microbial degradation (of natural petroleum hydrocarbons and xenobiotics);
- Some key reactions of common aerobic and anaerobic biotransformation mechanisms;
- Major factors influencing the rate of biodegradation, among others bioavailability the main barrier of biodegradation; and, finally,

– CDT technology, the innovation of the authors, is introduced from the laboratory experiments to the field demonstration. CDT is the cyclodextrin-enhanced *in situ* bioremediation, a combination of bioventing and periodic flushing using cyclodextrin for bioavailability enhancement. The CDT case used CDT in combination with groundwater extraction to keep water table low (increasing the aerobic zone depths) and preventing contaminated groundwater from escaping from the site.

I INTRODUCTION

Bioremediation is a biotechnology based on the biological transformation, primarily the breakdown or biodegradation of contaminating compounds in environmental matrices. During biodegradation, microbes utilize chemical contaminants in the soil as an energy source – or cometabolize them with a food source – and, through oxidation-reduction reactions, metabolize the target contaminant into useable energy for microbes.

Biodegradation can be defined as the biologically catalyzed reduction in complexity of chemicals, usually by microorganisms. Complete degradation is often termed "mineralization." *Mineralization* is a form of biodegradation that results in conversion of an organic compound into its inorganic constituents (e.g. CO_2, CH_4, H_2O, SO_4^{2-}, and PO_4^{3-}) or mineral salts and new microbial cellular constituents (biomass). *Ultimate biodegradation* is a term sometimes used as a synonym for mineralization. *Biotransformation*, often used synonymously with *bioconversion*, usually refers to a single step in a biochemical pathway, in which a molecule is catalytically converted into a different molecule. The *cometabolic degradation* of organic pollutants has also the potential to achieve required cleanup goals. Cometabolism, which is a non-growth-linked biological process catalyzed by microorganisms, is the metabolic transformation of a substance which is not utilized for growth or energy while a second substance serves as primary energy or carbon source.

2 BIODEGRADATION-BASED REMEDIATION: MAIN PRINCIPLES AND KEY REACTIONS

Microbial processes have long been known to be important in the environment for the degradation of a large number of organic compounds. In spite of extensive research carried out to study microbial activities and microorganism-mediated processes, the details of many metabolic pathways are still not understood.

However, some basic factors and major determinants of biodegradation processes are identified (Alvarez & Illman, 2006; Swartjes, 2011). The following factors must be taken into account for successful bioremediation:

– The existence of microorganisms with the potential to degrade the target contaminant(s);
– The presence of a substrate that can be utilized as an energy and carbon source;
– The presence of optimal environmental conditions (e.g. water content, pH, temperature);
– The redox potential, the availability of an appropriate electron acceptor;
– The availability of target pollutants to the microorganisms;
– The presence of nutrients necessary to support the microbial cell growth and enzyme production.

2.1   Microorganisms capable of degrading organic pollutants

A great number of bacterial and fungal genera possess the capability of degrading organic pollutants. Microorganisms that can degrade organic compounds in the environment include species of Gram-negative and Gram-positive bacteria, as well as fungi. Biodegradation is most commonly associated with bacteria. However, fungi are capable of degrading a wide range of environmental pollutants, including recalcitrant compounds such as polycyclic aromatic hydrocarbons, dichlorodiphenyltrichloroethane, trinitrotoluene, and pentachlorophenol.

The predominant aerobic bacteria in polluted soils belong to a spectrum of genera and species such as *Pseudomonas, Acinetobacter, Alcaligenes, Flavobacterium, Bacillus, Arthrobacter, Nocardia, Corynebacterium, Mycobacterium*. Further important degraders of organic pollutants can be found within the genera *Comamonas, Burkholderia, and Xanthomonas*. Anaerobic biodegradation of organic compounds by *Acetobacterium, Rhodobacter, Desulfobacterium, Clostridium, Desulfococcus, Desulfovibrio, Syntrophus* species was found in the environment (Crawford, 2002; Schink, 2005). There are many species of fungi – primarily white rot and mycorrhizal fungi – which are particularly efficient at degrading xenobiotics: *Phanerochaete chrysosporium, Agrocybe aegarita, Coriolopsis gallica, Lenzites betulina, Nematoloma frowardii, Morchella conica, Agarius bisporus, Aspergillus sp., Penicillium sp.* etc. (Crawford, 2002).

After a microbial consortium has been exposed to a new chemical stress, there is a characteristic lag time before it adapts and begins to degrade the compound and grow and reproduce actively. This phenomenon is known as *adaptation*, and it refers to adaptive changes that occur at the individual, bacterial population and microbial community levels that increase the rate of biodegradation. The adaptation may involve inter- and intraspecific genetic exchange. Catabolic genes can be exchanged among cells of the same genus or different genera. The likelihood of genetic exchange and the compatibility of the transferred genes are often genetically predetermined. Many plasmid-containing genes coding for catabolic enzymes are transmissible. Genetic rearrangements and mutations can also be important adaption mechanisms that contribute to the evolution of novel catabolic pathways (Alvarez & Illman, 2006).

Recombinant DNA techniques have been studied intensively to improve the degradation of hazardous wastes. Recombinant bacteria can be obtained by genetic engineering techniques or by natural genetic exchange between bacteria. Applications for genetically engineered microorganisms (GEM) in bioremediation have received great attention, but have largely been confined to the laboratory.

The use of genetically engineered microorganisms has been applied to bioremediation process monitoring, strain monitoring, stress response, end point analysis, and toxicity assessment. The range of tested contaminants includes chlorinated compounds, aromatic hydrocarbons, heavy metals, and non-polar toxicants, etc. (Menn *et al.*, 2008).

Biodegradation of organic contaminants in the environment is always the result of the complex activity of a microbial consortium. There are very few examples for contaminants whose degradation depends on one single bacterial strain. Cooperation between different microbial species is necessary for efficient biodegradation of organic compounds. Major examples of such beneficial microbial interaction are commensalism, syntrophism, and interspecies hydrogen transfer (Alvarez & Illman, 2006).

Microorganisms have limits of tolerance for particular environmental conditions, and the success rate of microbial biodegradation depends on nutrient availability, moisture content, pH, redox potential, and temperature of the habitat where the microorganisms live (soil, sediment, water). Inorganic nutrients involved, but not limited to, are nitrogen, phosphorus, and meso- and micronutrients. Well-balanced nutrient supply is necessary for microbial activity and cell growth, even in cases when the contaminant satisfies energy demand. All soil microorganisms require moisture for cell growth and function. Availability of water affects the transport of water and soluble nutrients into and out of microorganism cells. However, excess moisture, such as in saturated soil, may be undesirable for aerobic microorganisms limiting molecular oxygen for aerobic respiration, but may be beneficial for anoxic or a must for anaerobic microorganisms. Soil pH is also important because most microbial species can survive only within a certain pH range. Furthermore, soil pH can affect availability of nutrients. Temperature influences the rate of biodegradation by controlling the rate of enzymatic reactions within microorganisms.

Commercially available microbiological cultures may be even more sensitive for environmental conditions compared to native soil microbiota and have limited chance to optimally cooperate with the native consortium. Thus for soils, where the native microbiota cannot fulfill its role in the biogeochemical cycling, thoroughly selected and artificially grown microorganisms may help with their density, constituents of special biodegradation potency, or resistance. When using microbiological inoculants, their impact on the still existing beneficial microbiota and the locally existing healthy diversity is necessary to learn and to be taken into consideration during the planning of the proportion and composition of the inoculant, otherwise their application may increase the imbalance in degraded, deteriorated soils.

Enzyme technologies use enzymes, the products of micro- or larger organisms can also be applied to soil remediation. Enzymes, used as additives for the transformation of contaminants, similar to naturally occurring soil enzymes (inside and outside of microorganisms, closely bonded to soil matrices) can act as specific degrading agents. Several enzyme products have been developed and are marketed as commercial bioremediation agents. Among biological agents, enzymes have a great potential to effectively transform and detoxify soil pollutants. The use of enzymes may represent a technological option/a soft technological tool for overcoming most of the disadvantages caused by living microorganisms as additives. Enzymes can be applied under extreme conditions that limit microbial activity and are also effective at low pollutant concentrations. Enzymatic methods generally have low energy requirements, are easy to control, and have minimal environmental impact (Piotrowska-Długosz, 2017).

2.2   Electron acceptors and redox transformations

Contaminant degradation by oxidation may result in complete mineralization, partial degradation, and several types of chemical transformations. Mineralization and degradation by oxidation assumes aerobic or anoxic respiration and the presence of sufficient quantity and the appropriate quality of electron acceptor molecules.

Chemically, the degradation of organic contaminants mostly involves *redox transformations*. In soil, the oxidation-reduction (redox) potential (*Eh*) – measurable in Volt [V] – represents the sum of numerous oxidation-reduction process pairs. The *Eh* values of these pairs differ in magnitude: from the strongest reductant CO_2/CH_2O (–0.43 V) to the strongest oxidant O_2/H_2O (0.82 V) (Paul & Clark, 1996). The redox potential is of great significance in determining the soil chemical and biological phenomenon. In the microbial oxidation of a substrate (e.g. a contaminant molecule) in an aerobic/oxic soil, O_2 is the electron acceptor. In anoxic soils, NO_3^-, Fe^{3+}, Mn^{4+}, SO_4^{2-}, CO_2, and organic intermediators can act as electron acceptors (Paul & Clark, 1996). Degradation of organic matter in the absence of oxygen can be coupled to the reduction of these alternative electron acceptors following a certain sequence that appears to be determined by the respective redox potentials. The availability of either high- or low-potential electron acceptors may also influence the biochemistry of degradation processes.

In groundwater systems the biodegradation potential not only depends on the relative energy yields of the different reactions, but also on the availability of organic substrates, electron acceptors, and active microbial populations. The identification of different redox zones is of fundamental importance to assess the overall potential for degradation of most organic contaminants in groundwater (Swartjes, 2011). In fact, attenuation of specific organic contaminants is strongly influenced by redox zonation. This concept is valid not only for contaminants degrading preferentially (and/or most rapidly) in an oxidizing environment (e.g. petroleum hydrocarbons (PH, BTEX) or PAHs), but also for contaminants (e.g. chlorinated solvents) that are typically reduced under anaerobic conditions (Wiedemeier et al., 1999). Consequently the presence and role of electron acceptors and electron donors is a key point in biodegradation-based remediation (Middeldorp et al., 2002; Swartjes, 2011; Gruiz, Chapter 3). As oxidation and reduction should go hand in hand, (i) oxidation and mineralization of organic substances needs the parallel reduction of an electron acceptor (typically O_2, NO_3, SO_4, Fe(III), Mn(IV) or other organic substances when they are reduced to CH_4), or (ii) if the substance is degraded by reduction, the counterpart of the reaction should be the oxidation of sulfide to sulfate, Mn(II) to Mn(IV), Fe(II) to Fe(III), N_2 to NO_3 or an organic substance to CO_2 and water (see Figure 5.1). Biodegradation rates and dominant degradation

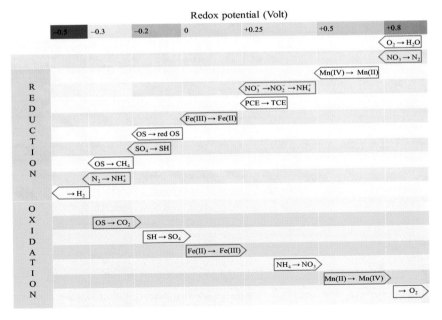

Figure 5.1 Reduction and oxidation processes in soil as functions of the redox potential.

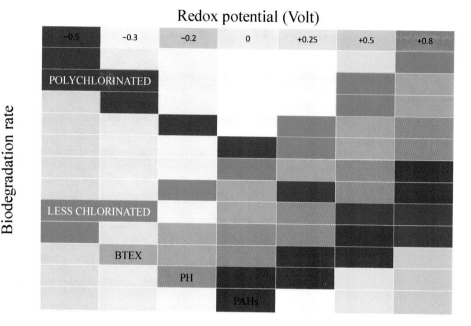

Figure 5.2 Biodegradation rates of typical organic contaminants as function of redox potential.

mechanisms for different classes of contaminants depend on redox conditions as is demonstrated in Figure 5.2.

2.3   Redox manipulations in soil remediation

The environmental risk of contaminants in subsurface soils and sediments can be mitigated by *in situ* manipulation of natural processes to reduce the mobility or the form of contaminants by an *In situ* Redox Manipulation (ISRM) technology. ISRM is targeted to remediate groundwater containing chemically reducible metallic and organic contaminants (DOE, 2000). The ISRM approach involves creating a permeable treatment zone downstream of a contaminant plume or contaminant source by injecting a chemical-reducing agent and/ or microbial nutrients to alter the oxidation/reduction state of the groundwater and sediments (DOE, 2000; Fruchter *et al.*, 1994, 2000; Vermeul *et al.*, 2003, 2004; Faybishenko *et al.*, 2009). Appropriate manipulation of the redox potential can result in the destruction of organic contaminants (King *et al.*, 2012).

3   BIODEGRADABILITY, BIOAVAILABILITY

Biodegradation of a contaminant in soil depends on the contaminants characteristics, on the density and composition of the microbiota and on environmental conditions.

Biodegradability and bioavailability concern the contaminating chemical substance. Having access, having the capability to take up and degrade – these are meant for the microbiota. The two is connected by the contact and the interaction between the two: the contaminant and the microorganism, see Figure 5.3.

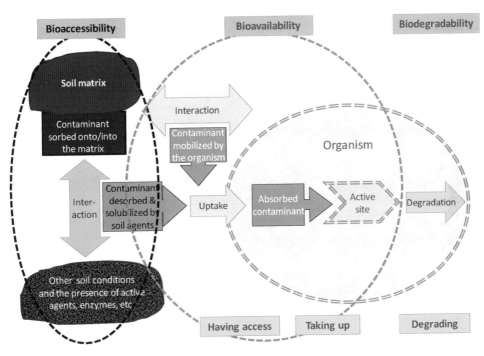

Figure 5.3 The scheme of contaminant availability and degradability related to microbial access and degradation (modified from Hajdu and Gruiz, 2015).

3.1   Biodegradability

The chemical structure determines biodegradability; it has been widely studied and information has been gathered in databases. Based on these data and the relationship between structure and biodegradability, mathematical models have been created for several chemical substance groups to predict biodegradability. The physical and chemical characteristics of a contaminant suggested to predict biodegradability include molecular mass, shape, and size of the molecule, water solubility, octanol–water partition coefficient (K_{ow}), the occurrence of branching, the presence of ester bonds, aromatic and heteroaromatic rings, substitution by halogens etc. (Alexander, 1994; Calow, 1993; Crawford, 2002). The Quantitative Structure-Activity Relationship (QSAR) models are increasingly used to estimate biodegradability of chemicals (Pavan & Worth, 2006; Gruiz *et al.*, 2015).

Biodegradability of the substance is an essential, but not the only parameter, that determines biodegradation in real soil. Biodegradability of the substance is the potency to become degraded, but if the soil conditions and the microorganisms present can actually trigger biodegradation, may be questionable.

Actual biodegradation in real soil equally depends on the presence of a capable microbial consortium and environmental conditions supportive to the degrading microorganisms. If the community of the soil is not ready to degrade the freshly occurred contaminant (the necessary enzymes are absent), the adaptive capacity (activation of the suitable genes present) of the microbiota is responsible for biodegradation. The adaptive capacity positively correlates with species diversity, the metagenome size, and diversity.

Beneficial environmental conditions involve the optimal conditions for and maximal supply of the microbial community with nutrients and bioactive microelements.

3.2   Bioavailability and its enhancement

Actual biodegradation of a biodegradable substance can be restricted or facilitated by its accessibility and availability to degrading organisms. The organic contaminants, which are in prolonged contact of the soil, are bound to the soil particles and display reduced accessibility and availability towards biodegradation (Haritash & Kaushik, 2009). A clear definition is found in the publication of Hajdu and Gruiz (2015), differentiating the following:

- Biological accessibility of the contaminant: being at a place, in a form that allows access for microorganisms (e.g. not strongly bound or built into the solid matrix).
- Biological availability of the contaminant: being in a form, which can interact with the surface of the microorganism making uptake possible. Bioavailability determines the potential of a chemical substance to be absorbed by an organism.
- Biodegradability of the contaminant: having a physicochemical structure that is compatible with the enzyme system of the microorganism.
- Access of the microorganism to the contaminants assume the presence of the microorganisms in the right place at the right time and the interaction between the contaminant and the microorganism.
- Uptake and transformation of the contaminant by the microorganism means that the enzymes of the microorganism catalyze the actual reaction after having access to the contaminant.

Interaction between a soil pollutant and the soil biota depends on the actual bioavailability of the substance, which is associated with the Kp value (soil-solid–soil-water partition coefficient = concentration in solid/concentration in water) in the soil, and with the interaction between the soil and the test organism (Hajdu *et al.*, 2009). The prerequisite of any interaction is that the living organisms should have access to the contaminant.

If a toxic organic pollutant in soil is sufficiently bioavailable for biodegradation, it can also exert toxic effects to organisms. However, degrading bacteria may have protective biochemical tools against toxicity, but the real solution is that they have specific mechanisms to mobilize substances before taking them up (Valentín *et al.*, 2013). But the biological mobilization applies only to a small proportion of the contaminant, which is also taken up and degraded after mobilization. This way the bioavailable proportion of the pollutant results in relatively low effective concentrations and do not harm the microbiota as much as the completely dissolved pollutants or drastically solubilized ones by additives. This is why technologists are advised to apply a stepwise addition to mobilize contaminants. It is the requirement of safety too: contaminants mobilized this way cannot escape from treatment volumes during *in situ* remediation.

Recognizing the biodegradation-limiting effect of substance bioavailability, several methods have been developed to overcome this disadvantage. Bioavailability of organic pollutants is controlled by parameters such as the physical state of the pollutant, its solubility in water, and its tendency to adsorb or bind to soil or sediment particles (Fetzner, 2009; Gruiz et al., 2015).

Some typical additive groups are listed here which suitable to increase the bioavailability of organic contaminants (Burmeier, 1995):

– Surfactants, preferentially anionic and/or non-ionic;
– Biosurfactants, optionally produced by the indigenous microflora;
– Long-chain alcohols;
– Substances capable of forming hydrophilic complexes with the non-polar contaminant;
– Inorganic compounds such as pyrophosphate.

Each of these options can have its own problems that need to be addressed during technology application. For example, most of these agents can act as a carbon source, thereby the contaminant utilization rate may decrease while the added compound is being oxidized. On the other hand, the new additive could also act as a co-substrate in cases where the microbiota cannot yield enough energy from the utilization of the contaminant. Another problem is that the *bioavailability-enhancing* substance must be biodegradable itself but not too fast compared to the "host-molecule" (Burmeier, 1995).

4 MOBILIZING AND SOLUBILIZING AGENTS TO INCREASE CONTAMINANT BIOAVAILABILITY

The main bottleneck in soil remediation is typically the availability of pollutants for the transformation process whether it is based on a physical, chemical, or biological process, each of them reaches proper conversion rate when the contaminant is in solubilized form. Solubilizing agents – solvents, surfactants, emulsifiers, complex-forming and micelle-forming agents, cyclodextrins, dendrimers, and several other nanomaterials – being man-made or biological and acting in various ways, thus all yield the same: an increase of the contaminants' proportion in water compared to solid.

4.1 The use of surfactants in soil remediation

Surfactant molecules have a hydrophobic (lipophilic) and a hydrophilic part (amphiphilic), so they have affinity both to water and oil like non-polar molecules. Surfactants or surface-active agents can modify the water:solid partition by shifting the distribution rate of the contaminant toward the mobile physical phase, to the soil water in our case. Hydrophobic contaminants can be transferred into the aqueous phase by different mechanisms and in various forms.

– Wetting effect of the surfactants leads to the disposal of sorbed hydrocarbons by water and this way the partition of the contaminant is shifted toward the aqueous phase, resulting contaminant mobilization. The contaminant is in dissolved form in the aqueous phase.
– Micelle formation occurs when using higher surfactant concentration (above the critical micelle concentration of the surfactant). The product of this type of solubilization is a micelle: the hydrophobic contaminant in the middle, completely surrounded by the

surfactant molecules. The process is also called micellar solubilization and the contaminants are this way "solubilized" (not dissolved). The term tries to differentiate this state of the molecules from real solutions and also refers to the uncertainty in considering the solubilized hydrophobic molecules as dissolved, as a colloidal solution/dispersion or as an emulsified separate phase or all these together.

– An emulsion is produced by large-molecule surfactants from two immiscible liquids: many of the petroleum hydrocarbon type contaminants form emulsion in water on the effect of surfactants. In this case, small droplets of the oil are surrounded by relatively large surfactant molecules. The large size micelles in water show opalescence, so emulsions cannot be considered as real solutions.

"Solubilized" in the context of soil contaminants is used in the widest sense: for every contaminant which becomes more mobile in water and more available for water-related processes during soil remediation.

Efficient removal or *in situ* transformation of a soil contaminant relates to the aqueous phase as follows: (i) only solubilized contaminants can be removed by water extraction, (ii) chemical reactions are much faster in water, compared to solid, and (iii) solubilized substances are much more bioavailable than the solid/sorbed or gaseous forms. Water can transfer contaminants to the location of treatment, e.g. to a subsurface reactor, to a collector system or just into an extraction well for surface treatment.

4.2   Synthetic surfactants

The most widely used types of synthetic surfactants are anionic surfactants, i.e. sulfate, sulfonate, and phosphate esters (e.g. ammonium and sodium lauryl sulfate, alcohol ethoxysulfates, linear alkylbenzene sulfonates) and carboxylates (including the toxic perfluorononanoate and perfluorooctanoate). Another high-volume surfactant type is the non-ionic surfactant such as long-chain alcohols, alcohol ethoxylates, and polyethylene glycol alkyl ethers. Cationic and zwitterionic surfactants (having both cationic and anionic centers in the same molecule) are produced and used in lesser amount.

Synthetic surfactants are applied as detergents, wetting agents, emulsifiers, and dispersants. All these activities may play a role in soil remediation enhanced by the mobilization mainly of organic, but also inorganic, contaminants.

The most popular applications of surfactants in soil remediation are the surfactant-enhanced removal of contaminants, oil spill dispersion, surfactant-enhanced bioremediation, and surfactant-enhanced oxidation or reduction. The presence of surfactants in soil results in weakening sorption of contaminants and improving water's wetting and permeability properties.

Some developments aim at contaminant-specific surfactant formulations for soil remediation. Patented non-ionic surfactants are provided for example by Ivey International (Ivey, 2017) to selectively desorb and mobilize sorbed petroleum hydrocarbons, chlorinated solvents, and certain toxic metals from soil. The introduced applications show improved mass recovery.

Many of the synthetic surfactants, mainly the cationic ones, are toxic to the ecosystem, so their environmental risk should be managed. Anionic and non-ionic surfactants are mildly toxic or non-toxic, so their risk may be acceptable in those cases when highly toxic contaminants cannot be removed from soil otherwise and when the contaminant in the soil in direct contact with the ecosystem and humans poses higher risk than the surfactant application. The use of surfactants can also be confirmed by their ability to accelerate remediation and shortened treatment duration and as a result reduce the risk of emissions from the technology. Of course, surfactants with low or no hazard can be applied more widely without strict restrictions.

4.3 Biosurfactants

Biosurfactants, the amphiphilic molecules of microbial origin pose a much smaller risk. Biosurfactants are produced extracellularly or as part of the cell membrane by a combination of yeast, bacteria, and filamentous fungi. Several soil microorganisms produce biologically harmless surfactants, it is their main tool in supplying themselves with nutrients and energy sources sorbed to the solid phase in soil. There are some demonstrative experimental results proving that bioavailability (mobility) enhancement precedes biodegradation (Hajdu and Gruiz, 2015). The concentration of an old, strongly bound contaminant significantly increased in the beginning of the remediation parallel to the increase of cell concentration and microbial activity, but it started to decrease only in a second step (Molnár *et al.*, 2002). Monitoring bioremediation at frequent time intervals indicates that this biologically controlled initial increase and accumulation of a more mobile and available fraction of the pollutant can always be observed when the pollutant is aged, partly degraded, consists of large molecules, and is not readily biodegradable. The role of the naturally occurring biosurfactants is extremely important in nature. Good surfactant-producing microorganisms are more competitive as they can utilize molecules not available to others. Biosurfactants are widely used in cosmetics and the food industry, and nowadays they are increasingly researched and recommended as solubilizing, bioavailability-enhancing agents in soil remediation (Molnár *et al.*, 1998; Santanu, 2008; Pacwa-Płociniczak *et al.*, 2011; Soberón-Chávez & Maier, 2011; Mulligan *et al.*, 2014; Eruke and Udoh, 2015; Karlapudi *et al.*, 2018).

Biosurfactant molecules have affinity both to water and oil like non-polar molecules as other surfactants. This character plays a role in microbial uptake of the lipophilic organic substances (Fritsche & Hofrichter, 2008). Biosurfactants are characterized by proper biodegradability, low toxicity, high effectiveness in enhancing solubilization, and biodegradation of low solubility compounds (Mulligan, 2005). Most of them are specific or narrow-spectrum biosurfactants, more efficient in solubilization, so the necessary amount is smaller, compared to the synthetic ones. Biosurfactants tolerate a wide range of pH and temperature. The most active biosurfactants can lower the surface tension of water from 72 to 30 mN/m and interfacial tension between water and *n*-hexadecane from 40 to 1 mN/m.

Bioavailability- and biodegradation-enhancing biosurfactants were studied by Cameotra and Singh (2009), and their electron microscopic study brought the evidence on the direct uptake of biosurfactant-coated hydrocarbon droplets by the degrading microorganisms: living cells "internalized" the surfactants layered oil droplets using the uptake mechanism of pinocytosis.

The number of surfactant-producing microorganisms are extremely large, the types and amount of synthesized surfactants also show a large variety. Microbial surfactants involve various glycolipids (mainly rhamno-, trehalo-, and sophorolipids) or lipoproteins and are produced by the microorganism extracellularly. The best know lipoprotein biosurfactants are surfactin (a cyclic lipopeptide) from *B. subtilis* and lichenysin from *Bacillus licheniformis*. Many long-chain fatty acids, phospholipids, and neutral lipids of microorganisms also have surface-active characteristics. Polymeric biosurfactants include emulsan, liposan, alasan, lipomanan, and other polysaccharide-protein complexes consisting of three or four repeating sugars with fatty acids attached to the sugars. Emulsan is a complex extracellular acylated polysaccharide synthesized by *Acinetobacter calcoaceticus* with an average molecular weight of about 1000 KD. Its industrial applications as emulsifier are widespread (Gorkovenko *et al.*, 1999; Mulligan, 2005; Rahman & Gakpe, 2008). Some microorganisms produce extracellular vesicles to help hydrocarbon uptake by forming microemulsion. Sometimes the whole bacterial cell itself can function as surfactant, having strong affinity for hydrocarbon–water and air–water interfaces (Karanth *et al.*, 1999).

Many of the biosurfactant-producing microorganisms are hydrocarbon degraders (Franzetti *et al.*, 2010; Vijayakumar & Saravanan, 2015; Cai *et al.*, 2015). Yeasts and other fungi are also active surfactants producers. Shete *et al.* screened the patents in 2006 and found 35 surfactant production patents with *Acinetobacter*, 30 with *Bacillus*, 29 with *Pseudomonas*, 80 with other bacteria, and 59 patents with fungal species.

Pacwa-Płociniczak *et al.* (2011) gave a good summary on biosurfactants and their application for environmental remediation, and Mulligan *et al.* (2014) on biosurfactant research trends and applications in general. Main applications of biosurfactants involve bioremediation of toxic and persistent contaminants, dispersion of **oil spills**, enhanced contaminant removal by physicochemical and biological methods (Zhang & Miller, 1992; Nayak *et al.*, 2009; Yin *et al.*, 2009; Pacwa-Płociniczak *et al.*, 2011).

Some environmental ***applications of biosurfactants*** are shown here as examples:

Enhancement of physicochemical removal/washing of organic contaminants:

 – The studies presented by Franzetti *et al.* (2008, 2009) showed that the BS29 bioemulsans from *Gordonia* species are promising washing agents for remediation of hydrocarbon-contaminated soils.
 – Kildisas *et al.* (2003) and Baskys *et al.* (2004) applied surfactant-enhanced washing in a combined technology for extremely high concentration organic pollutants: 1000 m³ of contaminated soil was treated, with an initial oil concentration of 180–270 g/kg. The washed soil with reduced contaminant concentration (34–59 g/kg) was treated using common bioremediation and phytoremediation. The pollutant concentrations dropped to 3.2–7.3 g oil per kg of soil after biodegradation.
 – Marcos *et al.* (2015) published 70–90% removal of motor oil from contaminated sand using biosurfactants, while only 55–80% by synthetic surfactants, both within 5–10 minutes.

Enhancement of physicochemical removal of inorganic contaminants:

 – Herman *et al.* (1995) used a rhamnolipid biosurfactant (produced by Pseudomonas species) for cadmium, lead and zinc removal from soil.
 – Hong *et al.* (2002) studied saponin, a plant-derived biosurfactant for recovery of heavy metals from contaminated soils using washing. Saponin was effective in removing the exchangeable and carbonated fractions of heavy metals from soil. The limits of Japanese leaching test were met for all of the soil residues after saponin treatment.
 – Biologically produced surfactants like surfactin, rhamnolipids, and sophorolipids were tested for removal of heavy metals such as Cu, Zn, Cd, and Ni from

contaminated sites (Wang & Mulligan, 2004a). Rhamnolipid was studied for its metal removal capacity both in liquid and foam forms (Wang & Mulligan, 2004b; Mulligan & Wang, 2006). Rhamnolipid type I and type II was found to be suitable for heavy metal removal.

- Arsenic mobilization from mine tailings (Wang & Mulligan, 2009) increased greatly (20x compared to water) after biosurfactant application. Mass ratio was 10 mg surfactant per g mine tailings at pH = 11.

Enhancement of biodegradation-based soil remediation both by bioavailability enhancement and biodegradation intensification:

- Wang and Mulligan (2004b) evaluated the surfactant foam technology in remediation of contaminated soil and concluded that it can be used to remove contaminants and act as augmentation for other technologies such as washing, leaching or bioremediation to improve contaminant removal efficiency and cost effectiveness.
- Rhamnolipid biosurfactants produced by *Pseudomonas aeruginosa* DS10–129 showed significant applications in the bioremediation of hydrocarbons in gasoline contaminated soil and petroleum oily sludge. Rhamnolipid biosurfactant enhanced the bioremediation process by releasing the weathered oil from the soil matrices and enhanced the bioavailability of hydrocarbons for microbial degradation. Biosurfactants are also used in the bioremediation of sites contaminated with toxic heavy metals like uranium, cadmium, and lead (Mulligan & Wang, 2006).
- Whang *et al.* (2008) applied rhamnolipid and surfactin for the enhanced biodegradation of diesel-contaminated water and soil.
- For bioavailability enhancement trehalolipids were used by Franzetti (2010).
- Marcos *et al.* (2015) conducted degradation experiments in motor oil-contaminated sand. The presence of the biosurfactants increased the degradation rate by 10–20%, especially during the first 45 days, indicating that biosurfactants acted as efficient enhancers for hydrocarbon biodegradation.

Enhancement of bioleaching and bioleaching-based soil remediation:

Bioleaching of metals from a low-grade mining ore and bioleaching of contaminating metals from soil or rock may use the same technology, i.e. a microbiologically catalyzed leaching process. (Bioleaching typically applies to sulfide ores: it is a two in one process in which microorganism produce acids for metal dissolution, in parallel to metal oxidation for getting energy.)

- Singh and Cameotra (2004) applied biosurfactants to the enhancement of metal-contaminated soil bioremediation.
- Wang and Mulligan (2009) studied arsenic mobilization from mine tailings in the presence of a biosurfactant.

Oil recovery from dredged material:

- Pesce (2002) patented a biotechnological method for the regeneration of hydrocarbons from dregs and oily muds, using a sophorolipid type biosurfactant.

– Baviere *et al.* (1994) patented a technology using sophoroside solution for washing oil containing drill material. Sophorolipids are produced by Torulopsis yeast species with high productivity.

In addition to the application of separately produced biosurfactants (from the market), the inoculation by microorganisms producing surfactants is also a useful concept to overcome availability-limited biodegradation in soil and groundwater remediation.

4.4   Nanoparticles in environmental remediation

The high surface area of nanoparticles makes them excellent sorbents for both organic and inorganic contaminants. Once sorbed, a pollutant may be mobilized, immobilized, removed, or destroyed depending on the pollutant, the nanoparticle, and the environmental matrices. Nanoparticles may increase the bioavailability of pollutants (Lewinski, 2008). Solutions with solid phase particles (nano- and microparticles) are extracting agents for the removal of PAHs from soil (Lau *et al.*, 2014). Nanoparticles as sorbents for scavenging contaminants are SAMMS™ (nanoporous ceramics coated with a monolayer of functional groups), nanotubes, ferritin, dendrimers, metalloporphyrinogens, organically modified silica (SOMS), etc. SAMMS are useful for the sorption of radionuclides, mercury, chromate, arsenate, and selenite (Fryxell *et al.*, 2007). Carbon nanotubes (CNTs) can be used to immobilize toxic metals (Pb^{2+}, Cu^{2+}, Ni^{2+}, and Zn^{2+}) and thus increasing the soil adsorption capacity (Matos *et al.*, 2017). Ferritin, an iron storage protein is photocatalytic in reduction of water-soluble hexavalent chromium to the less toxic water-insoluble trivalent chromium (US EPA, 2008). Dendrimers are hyper-branched polymers with functional groups. Polyamidoamine (PAMAM) dendrimers are effective chelators of copper (Diallo & Savage, 2005; Xu *et al.*, 2005; Castillo *et al.*, 2014). Crosslinked organosilica particles (trade name: Ecotreat) swell and absorb small organic compounds such as PCE, TCE, and PCBs (ABS Materials, 2017). The dechlorination reaction takes place within these particles with the embedded zero-valent iron (ZVI) catalyst.

Nanobioremediation is the combination of an abiotic degradation using nanoparticles such as nZVI for dechlorination of PCBs or Pd/Fe nanoparticles for dechlorination of dioxins and lindane, and subsequently applied bioremediation for further bacterial dechlorination (Le *et al.*, 2015; Bokare *et al.*, 2012; Singh *et al.*, 2013). The abiotic step decreases the concentration of the contaminant to a level non-toxic to the microbiota. Nanomaterials are injected into the treatment zone through injection wells, piezometers or monitoring wells using direct push, pressure pulse technology, liquid atomization injection, pneumatic fracturing, and hydraulic fracturing (US EPA, 2008).

5   CYCLODEXTRINS IN BIODEGRADATION-BASED REMEDIATION

The non-toxic, biodegradable cyclodextrins (CDs) can solubilize poorly soluble organic contaminants such as PAHs, PCBs, chlorinated solvents, etc. *via* molecular encapsulation, as shown in Figure 5.4 (Fenyvesi *et al.*, 2009). CDs can be used for bioavailability enhancement either for aerobic, anaerobic or cometabolic biodegradation in soil or in other solid matrices.

The biodegradation of PCBs, especially those with lower degree of chlorination can be enhanced by improving the bioavailability using randomly methylated beta-cyclodextrin

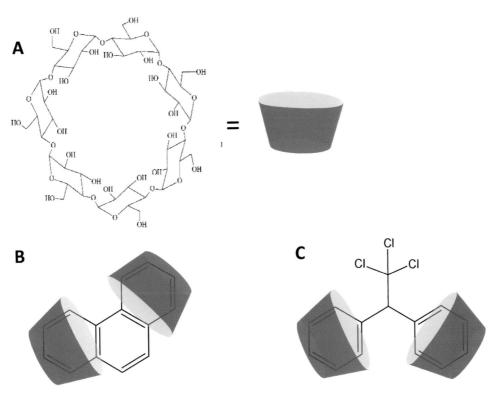

Figure 5.4 Structure and symbol of beta-cyclodextrin (A), and its complexes with phenanthrene (B) and DDT (C).

(RAMEB) (Fava *et al.*, 2002; Hu *et al.*, 2016). Similarly, the biodegradation of diesel oil, transformer oil, and mazout was enhanced by RAMEB due to the enhanced bioavailability (Gruiz *et al.*, 2011; Molnár *et al.*, 2009). In the case of contaminants with extremely high affinity toward CD complexation, biodegradation can be hindered by the strong binding of the contaminant to the CD as it was observed for dichlorodiphenyltrichloroethane (DDT) and 2-hydroxypropyl-beta-cyclodextrin (HPBCD) (Gao *et al.*, 2015). Figure 5.4C shows the CD-DDT complex. Most of the originally hydrophobic and biologically non-available organic compounds become more bioavailable by getting a hydrophobic outer surface after cyclodextrin complexation (Figure 5.4B), but there are exemptions, such as DDT, which bind so strongly to HPBCD that they together form a much larger and less bioavailable complex molecule at high HPBCD concentration. In such cases it is important to find the optimal concentration of the complexing agent which improves bioavailability.

5.1 Microbial degradation under aerobic conditions

Application of cyclodextrins in soil remediation technologies was first reviewed in the end of the 1990s (Wang *et al.*, 1998). Later, several applications were published for water, soil, and air treatment and for environmental analyses, mainly in Europe and in Japan (Fenyvesi *et al.*, 2009; Gruiz *et al.*, 2011; Fenyvesi *et al.*, 2011; Nagy *et al.*, 2013). CDs improve biodegradation by enhancing the bioavailability of organic pollutants. Some successful applications are shortly introduced in the following:

- Phenanthrene aerobic biodegradation was accelerated by HPBCD (Wang *et al.*, 1998).
- Fluorene biodegradation in a slurry soil that was augmented by fungi (such as *Absidia cylindrospora*) was enhanced by maltosyl beta-cyclodextrin (Garon *et al.*, 2004).
- Microbial decomposition of PAHs, phenols and biphenyl was intensified by the addition of HPBCD (and nutrients) (Allan *et al.*, 2007; Yoshii *et al.*, 2001).
- Amendment with HPBCD at as low concentration as 14 mg/kg soil enhanced the biodegradation of hexadecane and phenanthrene (Stroud *et al.*, 2009).
- The unsubstituted BCD, which usually forms complexes of low solubility, have beneficial effect on biodegradation, as the availability of contaminants that are molecularly encapsulated is higher compared to those adsorbed to soil. The rate of degradation of aliphatic and aromatic compounds was significantly enhanced in the presence of BCD (Bardi *et al.*, 2000) and was increased with increasing BCD concentration (Steffan *et al.*, 2003). BCD was beneficial for biomass production, ensuring an additional carbon source for the microbes. In addition to the type and concentration of the CD, the soil properties were also found to have an influence on biodegradation: the rate followed the order: clay < loamy soil < sand (Steffan *et al.*, 2001).
- Alpha-CD was inefficient but gamma-CD improved slightly the biodegradation of hexadecane (Sivaraman *et al.*, 2009) as well as of polychlorinated biphenyls (Fava *et al.*, 1998).
- In an *in situ* field experiment, a hydrocarbon-polluted soil (total petroleum hydrocarbon (TPH) content 310–660 mg/kg) was treated with 1 g BCD/m^2, N and P nutrients, and augmented with an adapted culture of indigenous microbiota (Bardi *et al.*, 2003). Three months later, the hydrocarbons were fully degraded (TPH reduced to 5–23 mg/kg) and practically disappeared.
- Randomly methylated beta-cyclodextrin (RAMEB) was used to enhance the biodegradation of diesel and transformer oil as well as of mazout (black oil, the residue of petroleum distillation) in spiked soils and also in real-site soils with aged pollution (Gruiz *et al.*, 1996; Molnár *et al.*, 2002). RAMEB was selected because it has a higher solubilizing potential than HPBCD (Fenyvesi *et al.*, 1996; Balogh *et al.*, 2007) and can form inclusion complexes of higher stability with the typical hydrocarbons (Szaniszló *et al.*, 2005) compared to other cyclodextrins. RAMEB has a beneficial effect on the soil pore structure, thus improving the habitat of the microbes (Jozefaciuk *et al.*, 2003). RAMEB is less biodegradable than HPBCD, and its 1.0–1.5 year half-life time makes it a suitable additive to enhance bioremediation, without being degraded in the course of technology application (Fenyvesi *et al.*, 2005, Yuan & Jin, 2007).
- Adding RAMEB to the soil led to a clear increase in the bioavailability of the contaminants at the beginning of the treatment. In the presence of RAMEB, biodegradation started earlier in soils contaminated with diesel or transformer oil, resulting in more efficient removal of contaminants depending on the RAMEB concentration (Molnár *et al.*, 2002). In laboratory experiments, 0.7%, 0.5%, and 0.2% RAMEB solutions were found to give optimum results in clay, sandy, and loamy soils contaminated with transformer oil, respectively. Even mazout, which is difficult to biodegrade, became partly bioavailable in the presence of RAMEB: the number of specific degrading bacteria was

increased by one to two orders of magnitude, and 0.7% RAMEB yielded a 40% decrease in mazout concentration compared with <10% achieved without using RAMEB.

– The optimum RAMEB concentration was 1–3% for PCB-contaminated soil, resulting in an enhanced specific PCB-degrading biomass and improved biodegradation, first of all in a slurry-phase treatment (Fava *et al.*, 2003).

– A combined cyclodextrin technology (CDT, ventilation, CD flushing, nitrogen, and phosphorus) was applied at a transformer station (see later subsection for more detail) in Hungary. The soil was flushed with a RAMEB solution from time to time through the injection well, and, after a few days' pause, the groundwater was continuously pumped out from the extractor wells on the other side of the transformer (Molnár *et al.*, 2005). The mass balance proved that most of the contaminants (>98%) were removed as a consequence of the enhanced microbial activity (Gruiz *et al.*, 2008).

– A similar combined technology was used at a former fuel tank station on an agricultural site in Hungary contaminated with aged diesel and engine oil from leaking underground tanks (Leitgib *et al.*, 2008). In this case, however, the injection and extraction were performed alternately in the same well, i.e. a combination of push-pull and drive-through techniques was applied: half of the additives (RAMEB and nutrients) were applied through the combined injection – extraction well, and the other half through the five additional injection wells arranged in a circle around the combined well. The combined technology included the following: (i) intensification of spontaneous biodegradation by *in situ* bioventilation in the unsaturated soil zone by the addition of nutrients and RAMEB as additives to enhance bioavailability; (ii) *ex situ* physicochemical treatment of the pumped groundwater; and (iii) impulsive flushing of the unsaturated zone by a RAMEB solution. After the addition of RAMEB, the hydrocarbon concentration in the groundwater increased 10–40-fold and the specific oil-degrading bacteria 2–10-fold. The contaminant concentrations decreased significantly from 30,000 mg/kg to 3500 mg/kg and from >1000 mg/L to <200 mg/L in soil and water, respectively, at the end of the treatment. The mass balance based on soil gas (volatiles, CO_2, and O_2), groundwater (contaminant and cyclodextrin content) and soil (contaminant, microbiology, and toxicity) monitoring data showed that over 98% of the contaminant was removed by biodegradation, and only less than 2% by groundwater treatment. The specific feature of this combined technology is that it applies the same additive (RAMEB) for the enhancement of both soil flushing and bioremediation. The depressed level of groundwater and its *ex situ* treatment ensure that the contaminants with enhanced solubility cannot spread into the surrounding environment.

5.2 Cyclodextrin-enhanced biodegradation under Fe(III)-reducing conditions

The previous examples show the efficacy of cyclodextrins in enhancing aerobic biodegradation. The anaerobic processes also require enhancement of contaminant availability.

The Fe(III) ion is the most abundant electron acceptor in soils; it plays an important role in natural attenuation of PAH contamination. HPBCD application under Fe(III)-reducing conditions to improve the bioavailability of PAHs increased the initial phenanthrene mineralization rate with increasing HPBCD concentration (Ramsay *et al.*, 2005). At a high (5 g/L) HPBCD concentration, however, the biodegradation ceased after 25 days. A possible explanation is that too much CD protects the contaminants included in their cavity from the microbes, in a similar way to the surfactants at concentrations above the critical micelle concentration (CMC).

5.3   Cometabolites and cyclodextrins

Amending soil with biphenyl can enhance PCB degradation. Further minor enhancement can be achieved by jointly applying biphenyl and cyclodextrins, e.g. HPBCD and GCD (Fava *et al.*, 1998), and RAMEB (Fava *et al.*, 2003). Other studies showed that there is competition between biphenyl and the PCBs for the cyclodextrin cavity resulting in decreased depletion of PCBs compared with HPBCD alone, especially in the low-chlorinated region (Luo *et al.*, 2007).

Considerable synergism was observed when cometabolites of plant origin (naringin, cumarin, limonene, isoprene, and carvone) were applied together with HPBCD: higher PCB depletion was found in the presence of carvone and HPBCD in soil with a low organic carbon content (55–85%) compared to carvone alone (Luo *et al.*, 2007). HPBCD improved the availability (solubility) of these secondary plant metabolites by the formation of complexes. It was also observed that HPBCD treatment changed the phospholipid fatty acid (PLFA) composition: several lipids indicative of Gram-negative bacteria were found to be enriched, and lipids suggesting the presence of Gram-positive bacteria were found to be depleted.

6   ASPECTS AND KEY REACTIONS OF AEROBIC AND ANAEROBIC BIOTRANSFORMATION MECHANISMS

In this section the most common biodegradation pathways will be introduced, which play a role in soil and groundwater remediation: aerobic, anaerobic, and cometabolic biotransformation/degradation of organic contaminants.

6.1   Aerobic biotransformation pathways

The most important classes of organic pollutants in the environment are constituents of mineral oils and petroleum products, industrial, and agricultural chemicals such as solvents and pesticides or various halogenated petrochemicals. Deposition from the atmosphere as well as solid waste and sewage sludge deposition also significantly contribute to soil contamination and deterioration. Although a wide array of organic contaminants can be metabolized by microorganisms, some are more readily degraded than others. Evolutionary processes have resulted in the development of enzymes and pathways for the metabolism of naturally occurring compounds (petroleum hydrocarbons, for example). These pathways are found in diverse species of microorganisms that are widespread and indigenous to most environments. The biodegradation of the majority of hydrocarbons proceeds by aerobic pathway mechanisms. The main principle of aerobic degradation of hydrocarbons is shown in Figure 5.5.

Compounds such as polycyclic aromatic hydrocarbons, halogenated aliphatic and aromatic hydrocarbons, nitroaromatic and azo compounds, s-triazines, organic sulfonic acids, dioxins and some synthetic polymers – to which microorganisms have not been exposed during evolution over geologic time scale (*xenobiotics*), often are supposed to be resistant to bioremediation. But many examples have been reported, where xenobiotics such as haloaromatic compounds undergo biotransformations (Alexander, 1994; Alvarez & Illman, 2006; Jördening & Winter, 2005; Schink, 2005; Das & Chandran, 2010).

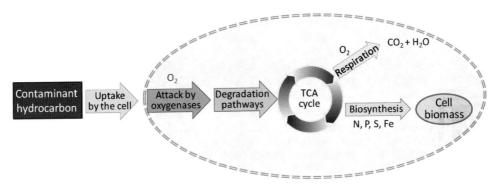

Figure 5.5 Main principle of aerobic degradation of hydrocarbons by microorganisms.

Any of the aerobic oxidation, anaerobic oxidation, anaerobic reduction, fermentation, and cometabolism can lead to the degradation of organic pollutants. Numerous mechanisms and pathways have been described for the biodegradation of organic compounds. The main reaction mechanisms are dehalogenation, deamination, desulfuration, metoxilation, hydroxilation, oxidation by oxidases, hydrolysis, acetogenesis, and methanogenesis. The EWAG Biocatalysis/Biodegradation Database (EAWAG, 2018) contains information on microbial biocatalytic reactions and biodegradation pathways for chemical compounds, primarily for xenobiotics with 219 pathways, 1503 reactions, 1396 compounds, and 993 enzymes. The goal of the EWAG is to provide information on microbial enzyme-catalyzed reactions that are important for biotechnology. All metabolic reactions are mediated by enzymes such as oxidoreductases, hydrolases, transferases, isomerases, ligases etc.

The enzymes playing a role in biodegradation have also been isolated and studied and extensive variability has been found in terms of their effect mechanism, the catalyzed reaction type: oxidative, reductive, hydrolytic, and synthetic processes as well as the substrates of the enzymes. Enzymes may accept not only natural molecules, but also compatible contaminants (with similar structure to natural substrates), and that makes soil decontamination possible.

Petroleum hydrocarbons comprise diverse groups of compounds including alkanes, alkenes, aromatic constituents, and heterocyclic compounds. Among petroleum hydrocarbons n-alkanes are degraded faster than branched alkanes and aromatics compounds. Long-chain n-alkanes (C_{10}–C_{24}) are degraded most rapidly. Short-chain alkanes (less than C_9) are toxic to many microorganisms, but they evaporate rapidly from petroleum-contaminated sites. Oxidation of alkanes is classified as being terminal or diterminal, the monoterminal oxidation being the main pathway. It proceeds via the formation of the corresponding alcohol, aldehyde, and fatty acid (see Figure 5.6). Beta-oxidation of the fatty acids results in the formation of acetyl-CoA (Fritsche & Hofrichter, 2008). Cyclic alkanes representing minor components of mineral oil are relatively resistant to microbial attack. The absence of an exposed terminal methyl group complicates the primary attack. A few species are able to use cyclohexane as sole carbon source; more common is its cometabolism (see later) with mixed cultures.

Aromatic hydrocarbons, e.g. benzene, toluene, ethylbenzene, and xylenes (BTEX compounds), and naphthalene belong to the large volume of petrochemicals widely used as fuels

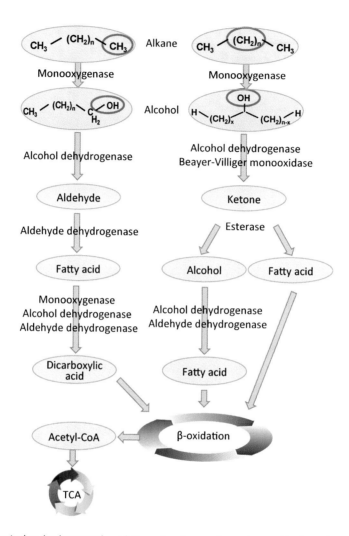

Figure 5.6 Terminal and subterminal oxidation of n-alkanes: the main aerobic degradation pathways.

and industrial solvents. These aromatic organic compounds are classified by the presence of two to six aromatic rings (e.g. fused benzene ring) with or without alkyl substituents arranged in linear, cluster, or angular configurations. Polycyclic aromatic hydrocarbons (PAH) are of major concern to both public and environmental health due to their acute and chronic toxicity, as well as their mutagenic and carcinogenic properties.

The aerobic degradation of aromatic hydrocarbons by bacteria and fungi is mainly accomplished via three different pathways (see Figure 5.7).

The reaction is catalyzed by aromatic hydrocarbon ring hydroxylating enzymes (ARHDs) and form cis-dihydrodiols by bacteria. The product, a diol, is then converted to catechol by a dehydrogenase. These initial reactions, i.e. hydroxylation and dehydrogenation, are common to pathways of bacterial degradation of aromatic hydrocarbons. Figure 5.7

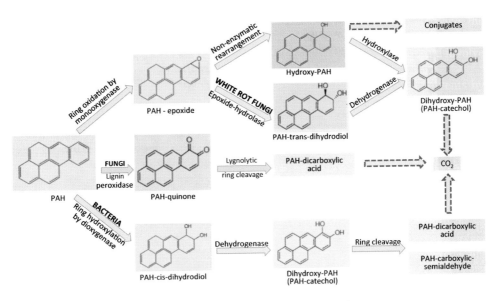

Figure 5.7 Aromatic hydrocarbon breakdown pathways in bacteria and fungi.

shows the bacterial degradation pathways of the oxygenolytic ring cleavage to intermediates of the central metabolism. At the branch point catechol either is oxidized by the intradiol *o*-cleavage or the extradiol *m*-cleavage. Both ring cleavage reactions are catalyzed by specific dioxygenases (see Figure 5.8) (Fritsche & Hofrichter, 2008).

Two different metabolic pathways are involved during the fungal biodegradation of PAH. Non-ligninolytic fungi deal with PAHs using the cytochrome P450 monooxygenase pathway where PAHs are oxidized to arene oxides via the incorporation of a single oxygen atom into the ring of the substrate. In contrast, white rot fungi (a ligninolytic fungus) mineralize PAHs using a soluble extracellular ligninolytic enzyme such as manganese peroxidase, laccases, and lignin peroxidase (Koshlaf & Ball, 2017).

Polycyclic aromatic hydrocarbons (PAHs) are also widely distributed environmental contaminants that have harmful biological effects including acute and chronic toxicity, mutagenicity, and carcinogenicity. In the case of these contaminants the effect of chemical structure on biodegradability in soil is also obvious. Anthracene, phenanthrene, and acenaphthylene contain three rings and are biodegraded at reasonable rates when oxygen is present. Other compounds with four rings in contrast are highly persistent (Alexander, 1994). For example, anthracene biodegradation by Gram-negative, Gram-positive bacteria, and numerous fungi has been reported. Bacteria initiate anthracene degradation by hydroxylation of the aromatic ring to yield *cis*-1,2-dihydroanthracene-1,2-diol. This intermediate is converted to anthracene-1,2-diol, which is cleaved at the meta position to yield 4-(2-hydroxynaph-3-yl)-2-oxobut-3-enoate (see Figure 5.9). This compound may spontaneously rearrange to form 6,7-benzocoumarin or be converted to 3-hydroxy-2-naphthoate from which degradation proceeds through 2,3-dihydroxynaphthalene to phthalate. This pathway has been identified in different bacteria such as *Pseudomonas, Sphingomonas, Rhodococcus, Mycobacterium*, and *Bacillus* sp. (EAWAG, 2018).

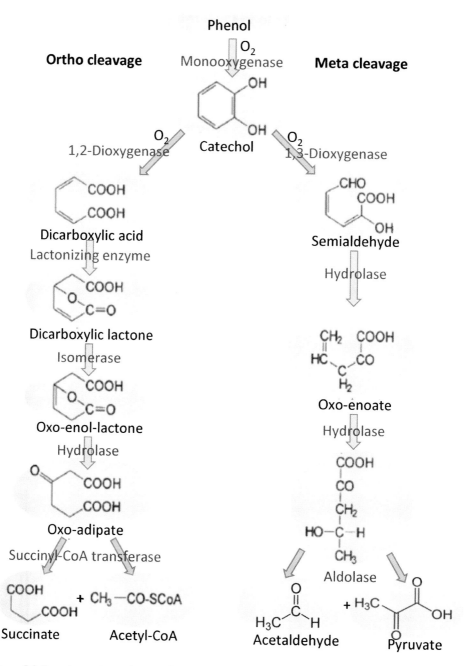

Figure 5.8 Two alternative pathways of aerobic degradation of aromatic compounds (after Fritsche & Hofrichter, 2008).

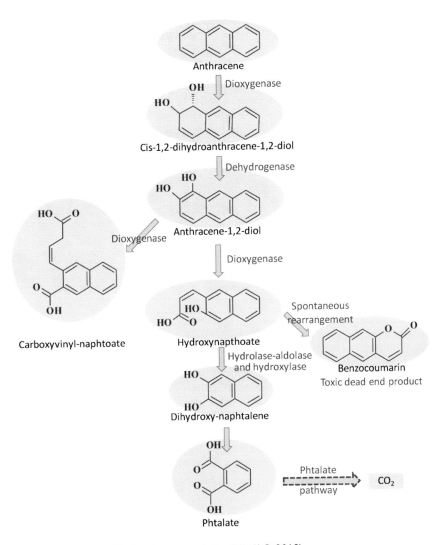

Figure 5.9 Anthracene graphical pathway map (after EAWAG, 2018).

6.2 Anaerobic biotransformation pathways

Anaerobic degradation pathways are utilized by microorganisms that can grow in the absence of oxygen and employ some other electron acceptor in their respiration processes. Anaerobic systems proved to be successful in bioremediation for such recalcitrant compounds as chlorinated solvents, pesticides, polychlorinated biphenyls (Alexander, 1994; Alvarez & Illman, 2006; Jördening & Winter, 2005; King *et al.*, 1992; Schink, 2005).

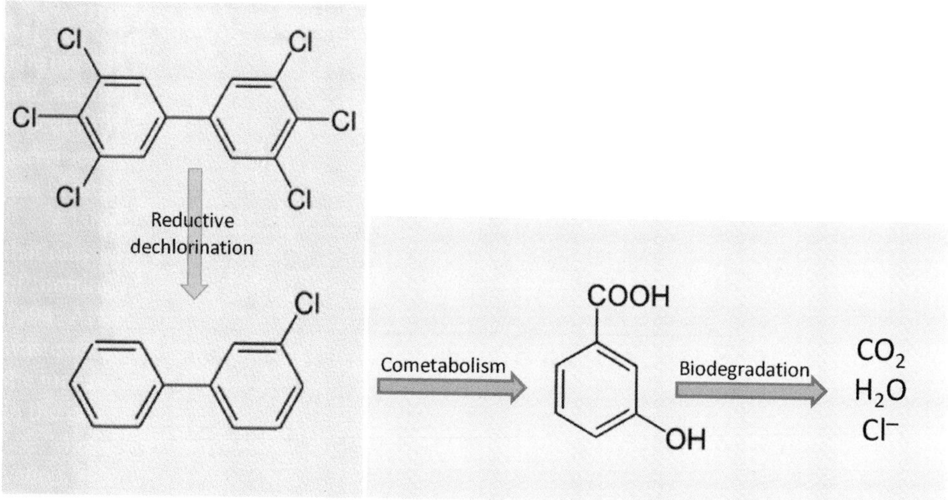

Figure 5.10 Sequence of anaerobic/aerobic degradation of highly chlorinated PCB.

Mononuclear aromatic compounds can be degraded anaerobically rather efficiently if they carry at least one carboxy, hydroxy, methoxy, amino, or methyl substituent, and four major degradation pathways have been identified in the recent past which differ basically from the well-known aerobic oxygenase-dependent pathways. The degradation kinetics differs considerably, depending on the sites of substitution. Halogenated aliphatics and aromatics are reductively dehalogenated more efficiently and better than by aerobes, the higher the degree of halogenation (Schink, 2005).

The sequence of anaerobic and aerobic bacterial activities for the mineralization of chlorinated xenobiotics has also been observed. An example for sequential anaerobic/aerobic degradation of a compound is the mineralization of highly chlorinated PCBs (see Figure 5.10) or dioxins (Alvarez & Illman, 2006). Reductive dehalogenation, the first step in PCBs degradation, requires anaerobic conditions and organic substrates acting as electron donors. Anaerobic dechlorination is always incomplete; the products are di- and monochlorinated biphenyls. These products can be further metabolized by aerobic microbes (Fritsche & Hofrichter, 2008).

6.3 Cometabolic degradation of organic contaminants

Cometabolism of a chemical substance without nutritional benefit in the presence of a growth substrate is a common phenomenon in microorganisms. *Cometabolism* is the metabolic transformation of a substance which is not utilized for growth or energy while a second substance serves as primary energy or carbon source.

Chlorinated solvents (e.g. TCE), chlorinated pesticides, polychlorinated biphenyls (PCBs), and alkylbenzene sulfonates (ABS) are chemicals that are often transformed through cometabolism (Alexander, 1994; Wackett, 1995; Schink, 2005; Alvarez & Illman, 2006; Arp *et al.*, 2001). Wide range of conversions, reaction types and products are associated with cometabolism. Among cometabolic conversions that appear to involve single enzyme, the reaction may be hydroxilations, oxidations, denitrations, deaminations, hydrolyses, acylations, or cleavages of ether linkages, but many of the conversions are complex and involve several enzymes (Alexander, 1994; Fritsche & Hofrichter, 2008).

A special type of cometabolism involves the oxidation of chlorinated solvents by oxygenases which is known as cooxidation (Alvarez-Cohen & Speitel, 2001). The phenomenon of cometabolism is often called *gratuitous* or *fortuitous metabolism* (Crawford, 2002).

The prerequisites for cometabolic transformations are the enzymes of the growing cells and the synthesis of cofactors necessary for enzymatic reactions, e.g. of hydrogen donors [reducing equivalents, NAD(P)H] for oxygenases. For example, methanotrophic bacteria can use methane or other chlorine compounds as a sole source of carbon and energy. They oxidize methane to carbon dioxide via methanol, formaldehyde, and formate. The assimilation requires special pathways, and formaldehyde is the intermediate that is assimilated. The first step in methane oxidation is catalyzed by methane monooxygenase, which attacks the inert methane. Methane monooxygenase is unspecific and also oxidizes various other compounds, e.g. alkanes, aromatic compounds, and trichloroethylene (TCE). Toluene as growth substrate for cometabolic transformation of TCE can also be used (Suttinun *et al.*, 2013), the first step in this case being toluene oxidation catalyzed by toluene monooxygenase (see Figure 5.11).

Consequently, methane or toluene might be injected in an effort to stimulate the cometabolism of trichloroethylene.

Cometabolic degradation can be limited by a low substrate utilization rate, depletion of the growth substrate, enzyme inhibition or inactivation, and cytotoxicity of the degradation products. Monitoring should cover these conditions to run the process at optimum rate.

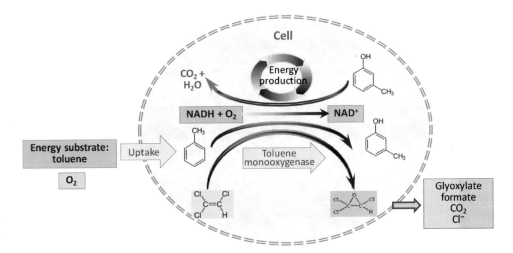

Figure 5.11 Aerobic cometabolic degradation of trichloroethylene (TCE) by the toluene monooxygenase system.

6.4   Harmful degradation products/hazardous metabolites

The most important role of microorganisms in biodegradation-based bioremediation, i.e. the transformation of organic contaminants to less harmful product. But the most undesirable aspect of microbial transformations in the environment is the formation of products with increased toxicity, enhanced mobility, enhanced lipophilicity leading to bioconcentration, and bioaccumulation.

A large number of chemicals that are themselves harmless often are converted to products that may be toxic and harmful to humans, animals, plants, or microorganisms. This process of forming harmful products from innocuous precursors is known as activation (Alexander, 1994; Alvarez & Illman, 2006). Activation can occur in soil, groundwater, surface waters, wastewater, and other environments where microorganisms are active. This process is one of the main reasons to study the pathways and products of breakdown of organic contaminants in the environment by integrated chemical-biological-ecotoxicological methods.

Many different pathways, mechanisms and enzymes are associated with environmental activation. Vinyl chloride, which is a potent carcinogen, is formed during the microbial metabolism of perchloroethene (PCE) and trichloroethylene (TCE) (see Figure 5.12) under anaerobic conditions. TCE has been widely used and now represents a major contaminant of many aquifers.

In the case of pesticides numerous cases have been documented of metabolites with greater bioactivity (Krieger, 2001). The epoxidation of the pesticide aldrin to dieldrin under aerobic conditions is also a transformation that increases the toxic nature of the compound. Other examples of toxic metabolites include the formation of 1,1-dichloro-2,2-bis (p-chlorophenyl)ethylene (DDE), which is apparently the form most responsible causing thin egg shells in birds that have accumulated [1,1,1-trichloro-2,2-bis(p-chlorophenyl) ethane] DDT or DDE from their prey; and 1,1-dichloro-2,2-bis(p-chlorophenyl)ethane (DDD), which can persist for years in some soil and water systems: S-oxidation of aldicarb (and some other N-methylcarbamates) to the more water-soluble compound and in some cases more persistent sulfoxide and sulfone forms.

Microorganisms can convert a number of secondary amines found in common household products and pesticides to the N-nitroso compounds. The activation is the N-nitrosation of the secondary amine to give the highly toxic N-nitroso compound.

Some bacteria can O-methylate chloroguaiacols; the products of such a methylation are toxic to fish (Alexander, 1994). A number of other types of hazardous metabolites can be formed during bioremediation in environment theoretically and in model experiments, but only a few were detected during remediation. Nevertheless, these known harmful products

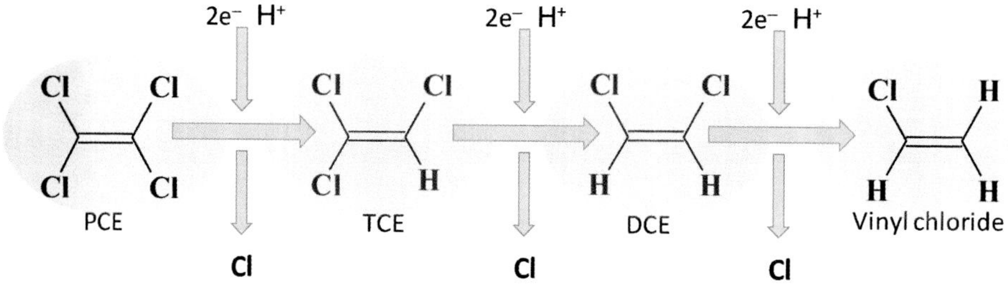

Figure 5.12 Reductive dechlorination pathway of PCE degradation.

should be considered when assessing risks associated with natural attenuation and enhanced biodegradation and they must be purposefully monitored. Direct toxicity testing of the remediated soil can draw attention to residual soil toxicity, also in a case when chemical analysis proves no presence of the original contaminant.

7 BIOREMEDIATION – A CASE STUDY

Bioremediation is a biotechnology using and supporting microorganisms to eliminate the contaminant and thus clean the soil. Soil microbes utilize chemical contaminants as an energy source, or cometabolize them together with a food source. For hydrocarbons, for example, the most common technology applied currently are natural attenuation, bioaugmentation, and biostimulation (Koshlaf & Ball, 2017), for chlorinated hydrocarbons anoxic or alternating aerobic-anoxic degradation is used (see more in Chapter 3).

Soil bioremediation based on natural biological processes is often viewed with mistrust and a lack of understanding. *In situ* technologies are still something mysterious and suspicious since the equipment and the machinery is not demonstratively large and noisy to make them convincing.

Biodegradation-based *natural attenuation* is a process by which the indigenous microorganisms can eliminate or detoxify, e.g. hydrocarbon pollutants hazardous to human health and/or the environment into less toxic forms. It might be the simplest bioremediation in terms of technology, but the knowledge for selecting, planning, monitoring, and verifying natural attenuation is actually more demanding than just an engineering method.

The bioremediation technology is based on the stimulation of microbial degradation of organic pollutants in contaminated sites by optimizing various factors such as nutrient concentrations, water content, temperature, redox potential, pH, oxygen supply, and availability of contaminants to microorganisms and temperature (*biostimulation*). Bioremediation can involve not only the indigenous microbial population but exogenous organisms, using them as an inoculant for soil, which is called *bioaugmentation*. Microbiological inoculants, enzyme additives, and nutrient additives that significantly increase the rate of biodegradation are defined as bioremediation agents by the US EPA (Nichols, 2001).

7.1 Cyclodextrin-enhanced bioremediation technology (CDT)

In situ cyclodextrin-enhanced bioremediation technology (CDT) has been developed to increase the efficiency of a natural biodegradation-based soil remediation technology that is environmentally friendly and ensures sustainable land and soil risk management.

Randomly methylated beta-cyclodextrin (RAMEB) was found to significantly enhance the bioremediation and detoxification of transformer oil-contaminated soil by increasing the bioavailability of the pollutants and the activity of indigenous microorganisms. The feasibility of the innovative cyclodextrin-enhanced bioremediation (CDT) was demonstrated in a field application.

7.1.1 Site description

Transformer oil contamination is a frequent and serious problem all over the world, also in Hungary, mainly at transformer stations. At present times most of the pollution at transformer stations derives from PCB-free transformer oils. Successful bioremediation of transformer oil-contaminated soils has been demonstrated by Molnár *et al.* (2005). It was found that the process is adversely affected by the poor bioavailability of the transformer oil. The use of CD may increase the bioavailability of oily pollutants and in turn accelerate soil microbiota and intensify soil bioremediation.

The site selected for CDT implementation was a legacy contaminated site, experiencing long-term pollution derived from leaking subsurface oil distribution pipelines and a spare transformer at a transformer station in Budapest, Hungary. The site where the technology was demonstrated was a relatively small separate part of the transformer station with 30 m³ (50 t) contaminated soil for the demonstration of the innovative CDT technology. The transformer oil (TO 40A) had already reached the groundwater with an average depth of 2.6 m. The chemical analysis of the environmental samples showed a high level of contamination in the soil: i.e. 20,000–30,000 mg/kg of GC-measured extractable petroleum hydrocarbon (GC EPH) content and 20,000–44,000 mg/kg solvent-extractable organic material (SEM) content in the upper 1–2 m, just under the transformer. The oil content in the groundwater was 0.99 mg/L. The concentration of the volatile fraction in the soil gas was under the detectable limit. Both saturated and unsaturated zones had to be decontaminated. The biological characterization of the contaminated soil indicated that the microflora has adapted and is capable of biodegrading the contaminants (see more information in Chapter 11).

7.1.2   Technology development – from the laboratory to the field

The primary aim of this technology development was to prove that CDT is a potential option for reducing the hazards posed to the environment by hydrocarbon-contaminated sites compared to other less efficient biotechnologies.

The second objective was to compile and test a high-quality and generally applicable, integrated methodology for site assessment, and for planning, development, and monitoring of remediation.

A scale-up concept was applied for the technology development, which included four steps from the laboratory to the field. The scientific background of CDT, using randomly methylated ß-cyclodextrin (RAMEB), was established in small- and large-scale laboratory experiments. Various organic contaminants of different biodegradability (such as diesel oil, transformer oil, polycyclic aromatic hydrocarbons, mazout) and cyclodextrins with different modes of actions were tested in various combinations in small-scale laboratory biodegradation experiments. The influence of the soil characteristics, type and concentration of cyclodextrins, addition of nutrients, aeration and soil moisture content was tested in technological experiments. The effect of cyclodextrins at a concentration range of 0–1% by weight was tested in soil microcosms. Open static soil microcosms and flow-through soil microcosms were used to study the progress of biodegradation and the effects of technological parameters. Some parameters such as cyclodextrin type and concentration, necessary level of aeration and nutrients (NPK supply) have been optimized for larger scale application at this phase of scale-up. The influence of soil properties was also studied by using three different texture types of pristine soils (sandy, clayey, humic-loamy) in laboratory technological experiments. To assess the efficiency of cyclodextrin-enhanced remediation on long-term pilot-scale laboratory experiments were carried out prior to field demonstration: 60-litre solid-phase soil reactors were set up for a 10-month pilot-scale laboratory experiment. Brown forest soil was spiked with 30,000 mg/kg of transformer oil. RAMEB was applied at concentrations of 0, 0.5, and 1.0% by weight.

Based on the results of the laboratory experiments, field demonstration was designed and performed for the remediation of a site contaminated with transformer oil (see Figure 5.13).

A site-specific complex technology was implemented for the remediation of the contaminated soil based on the results of the site assessment and the previous laboratory-scale technological experiments:

– *In situ* bioventing by air extraction and air inlet through passive wells.
– Biodegradation enhancement by using cyclodextrin (RAMEB) to mobilize/solubilize the solid-bound contaminants and nutrients (N, P, and K) for better microbial growth and activity.
– Temporary flushing the unsaturated and saturated zones by cyclodextrin solution and recycled water.
– *Ex situ* treatment of the extracted groundwater by phase separation, filtration through a sand filter and adsorption on activated carbon.

Figure 5.14 shows the scheme of the complex technology: the combined well for air and water extraction, the infiltration trench for water and additives delivery. The passive wells used for atmospheric air introduction and the groundwater treatment unit are not shown.

The photos in Figure 5.15 show the combined well and the trench, demonstrating that the *in situ* cyclodextrin-enhanced bioventing does not cause significant disturbances on the surface, and that inaccessible soil volumes such as the volume under the transformer can also be reached by this treatment. The site where the technology was demonstrated – was a relatively small separate part of the transformer station with 30 m³ (50 t) contaminated soil to demonstrate the innovative CDT technology. After the successful demonstration the whole site has been remediated.

In order to characterize the biodegradation processes in the contaminated soil and to evaluate the effect of the treatment, an integrated monitoring was used in both the laboratory and field experiments. This technique combined physical and chemical analyses with biological monitoring and direct toxicity testing. After the technology was demonstrated on the contaminated site, an evaluation and verification of the CDT was performed (see in Chapter 11).

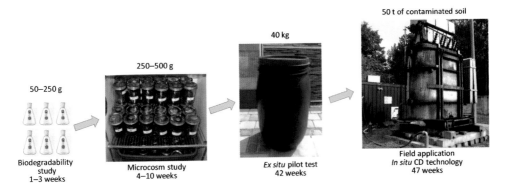

Figure 5.13 Scaling-up of CDT.

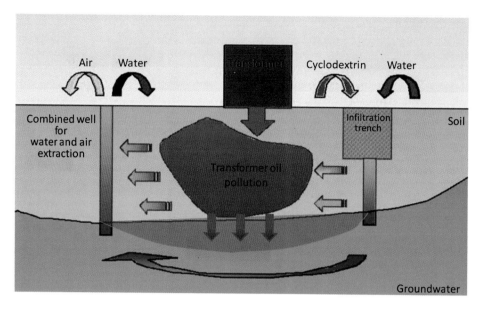

Figure 5.14 The scheme of the *in situ* CDT technology at the transformer station.

Figure 5.15 The combined well for air and water extraction and the infiltration trench.

7.1.3   Monitoring tool box for CDT and the surrounding environment

The technology was monitored by the Soil Testing Triad, including physico-chemical, biological, and ecotoxicological methods (Gruiz, 2005) (see Figure 5.16). The parallel chemical analyses and direct toxicity assessment made possible the differentiation between mobilized and still fixed forms of the contaminants. The results of the monitoring were evaluated in an integrated way (Chapter 3).

Soil samples were analyzed and characterized for the quality and quantity of the contaminant and the microbiota as well as for the toxicity of the soil using the integrated methodology. Physicochemical characterization of the soil and the groundwater included the

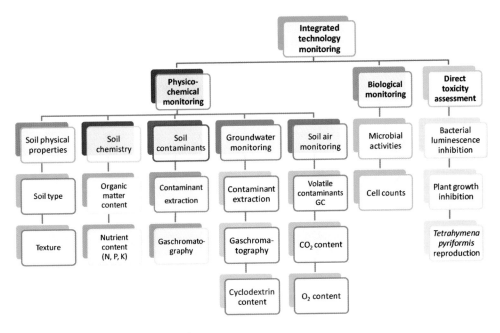

Figure 5.16 Monitoring plan for CDT.

following measurements: pH, redox conditions, grain-size distribution, humus content, and nutrient supply.

Chemical analytical methods were used for characterization of the contaminants in the soil and groundwater. The extractable organic material (EOM) content of the soil was determined after hexane-acetone (2:1) extraction using gravimetry. The gas-chromatographed extractable petroleum hydrocarbon (GC-EPH) content of the soil was analyzed from the same extract by employing gas chromatography (GC) using a flame-ionization detector. The total petroleum hydrocarbon (TPH) content of the groundwater was determined by gas chromatography after extraction by n-pentane. Volatile hydrocarbons in soil gas were also measured by gas chromatography. As the hydrocarbon content was negligible in the samples, this analysis was performed only at the start of the technology.

Soil biological parameters such as the aerobic heterotrophic bacterial cell concentration, the number of pollutant-degrading cells, and soil respiration were used to monitor biodegradation. The aerobic heterotrophic bacterial cell concentration was determined by colony counting after cultivation of microorganisms occurring in soil suspensions in water on peptone-glucose-meat-extract (PGM) agar plates in petri dishes (CFU/g soil). The population density of the hydrocarbon-degrading cells was measured after cultivation in tubes of liquid nutrient medium. For growing the oil-degrading cells a dilution series of contaminated soils were used containing transformer oil as the only carbon source.

The applied liquid medium was supplemented with an inorganic salt solution, trace elements and an artificial electron acceptor 2-(p-iodophenyl)-3-(p-nitrophenyl)-5-phenyl tetrazolium chloride (INT). The most probable number (MPN) was calculated by using probability tables.

The production of carbon dioxide by soil microbes was measured to characterize soil respiration. The CO_2 production of the soil microorganisms at laboratory scale was measured by a self-designed respirometer. CO_2 was absorbed in NaOH and determined by HCl titration. In the field experiments CO_2 and O_2 content of the soil gas was measured by an Oldham Mx21 Gas Analysator (Oldham Gas Detection Ltd. UK).

Direct contact soil ecotoxicity tests were applied using test organisms of three trophic levels. Ecotoxicity testing serves the safety and gives information on bioavailability of the contaminants. In the case of *in situ* remediation, ecotoxicity testing may play a role in the monitoring of emissions from the technology. The applied ecotoxicity methods were self-developed tests based on similar Hungarian, German, and European standard methods used for wastewaters or hazardous waste materials. *Aliivibrio fischeri* bioluminescence test, *Sinapis alba* root and shoot elongation test together with *Folsomia candida* mortality test were used to follow up the adverse effect of the soil during the laboratory and field experiments. The end points used for the plant and animal tests were ED_{20} (LD_{20}) or ED_{50} (LD_{50}) values where ED_{20} (LD_{20}) and ED_{50} (LD_{50}) mean soil doses that caused 20% and 50% inhibition (lethality) in *Aliivibrio fischeri* bioluminescence test.

The results of the monitoring were evaluated in an integrated way as described in Chapter 3, Section 4.3).

7.1.4   Bench-scale and pilot tests

A large number of treatability studies were performed in microcosms in order to find the differences, trends, and optimum parameters with regard to the use of RAMEB for the intensification of biodegradation-based remediation. The results of the laboratory experiments proved the bioavailability-enhancing effect of RAMEB: increased biodegradation rate, higher microbial activity, and decreased toxicity were observed as a consequence of an enhanced bioavailability (Molnár *et al.*, 2003, 2005; Molnár, 2006). The main results and conclusions of the bench-scale and pilot-scale technological experiments are summarized in Figures 5.17 and 5.18.

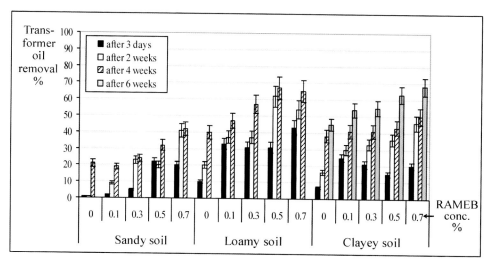

Figure 5.17 Effect of RAMEB on transformer oil degradation in different soils in microcosm experiments (the soils were initially spiked with 30,000 mg/kg transformer oil) (Gruiz *et al.*, 2008) Error bars represent standard deviation (SD ± 10% or less).

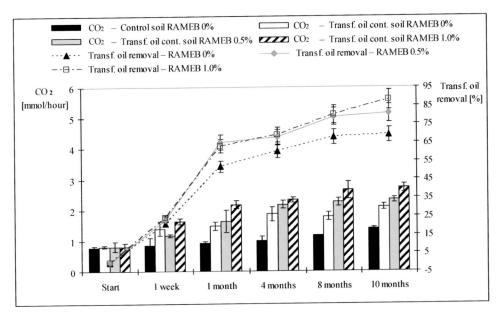

Figure 5.18 Transformer oil removal and CO_2 production rates throughout the CDT pilot experiment (Gruiz *et al.*, 2008) (values are the means of three replicates. Error bars represent standard deviations).

The study carried out on three-phase sandy, clayey and humic-loamy soils proved enhanced transformer oil degradation in all soils in the presence of RAMEB. The various soil characteristics resulted in various extents of transformer oil biodegradation in the sandy, humic-loamy and clayey soil (see Figure 5.17). The main outcome of these experiment series is that the application of 0.1–0.3% RAMEB was sufficient for a significant intensification of the biodegradation of transformer oil.

Addition of RAMEB was also effective both on microbial activity and contaminant removal in long-term pilot-scale experiments carried out on humic-loamy soil. Because of the enhanced microbial activity, a higher biodegradation rate (contaminant removal) and higher CO_2 production rate were observed using respirometry in CD-treated soils (see Figure 5.18). These results have obviously demonstrated how transformer oil biodegradation is enhanced by the addition of RAMEB. Higher RAMEB concentration did not increase transformer oil removal: there was no significant difference between soils treated with 0.5% and 1% RAMEB. This is an important result from the point of view of cost efficiency.

7.2   Field demonstration – full-scale remediation

The final phase of technology development was the field demonstration of CDT. On the basis of the bench-scale study results a field demonstration on the transformer oil-contaminated site was designed and performed.

The 6-month *ex situ* pilot study was successful; 0.3%–0.8% RAMEB resulted in a shortened adaptation phase for microbial degradation and remarkable differences in the decrease

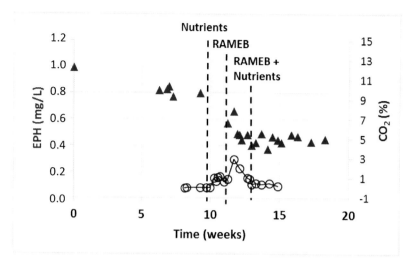

Figure 5.19 Changes in the EPH content in the groundwater (▲) and in the CO_2 content in soil gas (○) after treatment with CDT (RAMEB and nutrient additions are indicated with dashed lines) (Gruiz *et al.*, 2008).

of hydrocarbons compared to untreated. After the successful pilot test, the efficiency of RAMEB was demonstrated by *in situ* field application to the two- and three-phase soil contaminated with transformer oil (Molnár *et al.*, 2005).

Based on the detailed site assessment and preliminary studies, a site-specific combination of three technologies was planned and implemented for 47 weeks:

– Treatment of the unsaturated zone by bioventing and intensification of the biodegradation with nutrients (chemical fertilizers containing N, P and K) and RAMEB. Moisture is supplied continuously by slow infiltration of the treated water;
– Temporary flushing of the unsaturated and the saturated zones with cyclodextrin solution and the extracted water;
– Continuous extraction of the groundwater (ensuring depression to prevent contaminant transport from the site) and *ex situ* treatment of the extracted groundwater.

The complex technology applied at the field resulted in a decrease in groundwater contaminant concentration from 1 mg/L to 0.3 mg/L. Figure 5.19 shows the results for the groundwater in the first 20 weeks: the contaminant content decreased from 1 mg/L to 0.4 mg/L in this period. The contaminant concentration in the soil decreased from the initial 25,000 mg/kg to 240 mg/kg by the 47th week: this is approximately a 99% removal. Both water and soil meet quality criteria after the treatment.

Direct toxicity assessment confirmed the lack of adverse effects of the treated soil (see Table 11.6 in Chapter 11).

Water contamination showed a constant low value after the 20th week, but a minimum rate aeration and water extraction was continued for the winter period, and monitoring continued during spring time for greater certainty. The demonstration was terminated after 47 weeks.

Overall, the innovative cyclodextrin technology proved to be efficient for the transformer oil-contaminated soil, and the combination of bioventing and bioflushing with the solubilizing agent cyclodextrin could fulfill quality criteria within 47 weeks. The cyclodextrin-enhanced biological technology was combined this time with the *ex situ* treatment of the extracted groundwater, what was necessary in order to establish depression at the site, to prevent the surrounding area from contaminant dispersion.

The detailed verification of the cyclodextrin technology is described in Chapter 11.

REFERENCES

ABS Materials (2017) *EcoTreatTM in situ Injection Media*. Available from: http://abswastewater.com/wp-content/uploads/2016/08/Remediation-Solutions.pdf. [Accessed 26th July 2018].

Alexander, M. (1994) *Biodegradation and Bioremediation*. Academic Press, San Diego, CA, USA.

Allan, I.J., Semple, K.T., Hare, R. & Reid, B.J. (2007) Cyclodextrin enhanced biodegradation of polycyclic aromatic hydrocarbons and phenols in contaminated soil slurries. *Environmental Science & Technology*, 4, 5498–5504.

Alvarez, P.J.J. & Illman, W.A. (2006) *Bioremediation and Natural Attenuation*. Wiley-Interscience, Hoboken, NJ, USA.

Alvarez-Cohen, L. & Speitel, Jr. G.E. (2001) Kinetics of aerobic cometabolism of chlorinated solvents. *Biodegradation*, 12, 105–126.

Arp, D.J., Yeager, C.M. & Hyman, M.R. (2001) Molecular and cellular fundamentals of aerobic co-metabolism of trichloroethylene. *Biodegradation*, 12, 81–103.

Balogh, K., Szaniszló, N., Otta, K. & Fenyvesi, E. (2007) Can CDs really improve the selectivity of extraction of BTEX compounds? *Journal of Inclusion Phenomena and Macrocyclic Chemistry*, 57, 457–462.

Bardi, L., Mattei, A., Steffan, S. & Marzona, M. (2000) Hydrocarbon degradation by a soil microbial population with beta-cyclodextrin as surfactant to enhance bioavailability. *Enzyme and Microbial Technology*, 27, 709–713.

Bardi, L., Ricci, R. & Marzona, M. (2003) *In situ* bioremediation of a hydrocarbon polluted site with cyclodextrin as a coadjuvant to increase bioavailability. *Water, Air, & Soil Pollution: Focus*, 3, 15–23.

Baskys, E., Grigiskis, S., Levisauskas, D. & Kildisas, V. (2004) A new complex technology of cleanup of soil contaminated by oil pollutants. *Environmental Research, Engineering and Management*, 30, 78–81.

Baviere, M., Degouy, D. & Lecourtier, J. (1994) *Process for Washing Solid Particles Comprising a Sophoroside Solution*. U.S. Patent, 5,326,407.

Bokare, V., Murugesan, K., Kim, J.-H., Kim, E.-J. & Chang, Y.-S. (2012) Chang Integrated hybrid treatment for the remediation of 2,3,7,8-tetrachlorodibenzo-p-dioxin. *Science of the Total Environment*, 435–436, 563–566.

Burmeier, H. (1995) Bioremediation of soil. In: Alef, K. & Nannipieri, P. (eds.) *Methods in Applied Soil Microbiology and Biochemistry*. Academic Press, Cambridge, UK. doi:10.1016/B978-0-12-513840-6.X5014-9.

Cai, Q., Zhang, B., Chen, B., Song, X., Zhu, Z. & Cao, T. (2015) Environ Screening of biosurfactant-producing bacteria from offshore oil and gas platforms in North Atlantic Canada. *Environmental Monitoring and Assessment*, 187, 284. doi:10.1007/s10661-015-4490-x.

Calow, P. (1993) *Handbook of Ecotoxicology*. Blackwell Science Ltd, Hoboken, NJ, USA.

Cameotra, S.S. & Singh, P. (2009) Synthesis of rhamnolipid biosurfactant and mode of hexadecane uptake by *Pseudomonas* species. *Microbial Cell Factories*, 8, 1–7. doi:10.1186/1475-2859-8-16.

Castillo, V.A., Barakat, M.A., Ramadan, M.H., Woodcock, H.L. & Kuhn, J.N. (2014) Metal ion remediation by polyamidoamine dendrimers: A comparison of metal ion, oxidation state, and titania immobilization. *Environmental Science and Technology*, 11, 1497–1502. doi:10.1007/s13762-013-0346-5.

Crawford, R.L. (2002) Biotransformation and biodegradation. In: Hurst, C.J., Crawford, R.L., Knudsen, G.R., McInerney, M.J. & Stetzenbach, L.D. (eds.) *Manual of Environmental Microbiology*. 2nd ed. ASM Press, Washington, DC, USA.

Das, N. & Chandran, P. (2010) Microbial degradation of petroleum hydrocarbon contaminants: An overview. *Biotechnology Research International*, 2011, 941810. doi:10.4061/2011/941810. Epub 2010 September 13.

Diallo, M.S. & Savage, N. (2005) Nanoparticles and water quality. *Journal of Nanoparticle Research*, 7(4–5), 325–330.

DOE (2000) In situ redox manipulation subsurface contaminants focus area. U.S. Department of Energy, Office of Environmental Management, Office of Science and Technology. Innovative Technology Summary Reports.

EAWAG (2018) *Microbial Biocatalytic Reactions and Biodegradation Pathways*. Biocatalysis/Biodegradation Database. Available from: http://eawag-bbd.ethz.ch/ [Accessed 26th July 2018].

Eruke, O.S. & Udoh, A.J. (2015) Potentials for Biosurfactant enhanced bioremediation of hydrocarbon contaminated soil and water – A review. *Advances in Research*, 4(1), 1–14. Available from: www.journalrepository.org/media/journals/AIR_31/2014/Dec/Udoh412014AIR11933_1.pdf. [Accessed 26th July 2018].

Fava, F., Di Gioia, D. & Marchetti, L. (1998) Cyclodextrin effects on the ex-situ bioremediation of a chronically polychlorobiphenyl-contaminated soil. *Biotechnology and Bioengineering*, 58, 345–355.

Fava, F., Di Gioia, D., Marchetti, L., Fenyvesi, E. & Szejtli, J. (2002) Randomly methylated β-cyclodextrins (RAMEB) enhance the aerobic biodegradation of polychlorinated biphenyl in aged-contaminated soils. *Journal of Inclusion Phenomena and Macrocyclic Chemistry*, 44, 417–421.

Fava, F., Di Gioia, D., Marchetti, L., Fenyvesi, E. & Szejtli, J. (2003) Randomly methylated beta-cyclodextrins (RAMEB) enhance the aerobic biodegradation of polychlorinated biphenyl in soils. *Journal of Inclusion Phenomena and Macrocyclic Chemistry*, 44(1–4), 417–421.

Faybishenko, B., Hazen, T.C., Long, P.E., Brodie, E.L., Conra, M.E., Hubbard, S.S., Christensen, J.N., Joyner, D., Borglin, S.E., Chakraborty, R.W., Kenneth, H., Peterson, J.E., Chen, J., Brown, S.T., Tokunaga, T.K., Wan, J., Firestone, M., Newcomer, D.R., Resch, C.T., Cantrell, K.J., Willett, A. & Koenigsberg, S. (2009) *In situ Long-Term Reductive Bioimmobilization of Cr(VI) in Groundwater Using Hydrogen Release*. California Digital Library. Available from: https://escholarship.org/uc/item/1222c6sk. [Accessed 26th July 2018].

Fenyvesi, É., Balogh, K., Oláh, E., Bátai, B., Varga, E., Molnár, M. & Gruiz, K. (2011) Cyclodextrins for remediation of soils contaminated with chlorinated organics. *Journal of Inclusion Phenomena and Macrocyclic Chemistry*, 70(3–4), 291–297. doi:10.1007/s10847-010-9839-8.

Fenyvesi, E., Gruiz, K., Verstichel, S., De Wilde, B., Leitgib, L., Csabai, K. & Szaniszló, N. (2005) Biodegradation of cyclodextrins in soil. *Chemosphere*, 60, 1001–1008.

Fenyvesi, E., Molnar, M., Leitgib, L. & Gruiz, K. (2009) Cyclodextrin-enhanced soil-remediation technologies. *Land Contamination & Reclamation*, 17(3–4), 585–597.

Fenyvesi, E., Szeman, J. & Szejtli, J. (1996) Extraction of PAHs and pesticides from contaminated soils with aqueous CD solutions. *Journal of Inclusion Phenomena and Macrocyclic Chemistry*, 25, 229–232.

Fetzner, S. (2009) Biodegradation of xenobiotics. In: Doelle, H.W., Rokem, J.S. & Marin Berovic, M. (eds.) *Biotechnology – Vol X – Biodegradation of Xenobiotics*. EOLSS Publications Co. Ltd, Oxford, UK.

Franzetti, A., Bestetti, G., Caredda, P., Colla, La P. & Tamburini, E. (2008) Surface-active compounds and their role in the access to hydrocarbons in Gordonia strains. *FEMS Microbiology Ecology*, 63, 238–248.

Franzetti, A., Caredda, P., Ruggeri, C., Colla, La P., Tamburini, E., Papacchini, M. & Bestetti, G. (2009) Potential applications of surface active compounds by Gordonia sp. strain BS29 in soil remediation technologies. *Chemosphere*, 75, 801–807.

Franzetti, A., Gandolfi, I., Bestetti, G., Smyth, T.J. & Banat, I.M. (2010) Production and applications of trehalose lipid biosurfactants. *European Journal of Lipid Science and Technology*, 112, 617–627.

Fritsche, W. & Hofrichter, M. (2008) Aerobic Degradation by Microorganisms. In: Rehm, H.J. & Reed, G. (eds.) *Biotechnology: Environmental Processes II*. Volume 11b. 2nd ed. Wiley-VCH Verlag GmbH, Weinheim, Germany.

Fruchter, J.S., Cole, C.R., Williams, M.D., Vermeul, V.R., Amonette, J.E., Szecsody, J.E., Istok, J.D. & Humphrey, M.D. (2000) Creation of a subsurface permeable treatment barrier using *in situ* redox manipulation. *Groundwater Monitoring and Remediation*, 20(2), 66–77.

Fruchter, J.S., Spane, F.A., Fredrickson, J.K., Cole, C.R., Amonette, J.E., Templeton, J.C., Stevens, T.O., Holford, D.J., Eary, L.E., Bjornstad, B.N., Black, G.D., Zachara, J.M. & Vermeul, V.R. (1994) *Manipulation of Natural Subsurface Processes: Field Research and Validation*. PNL-10123. Pacific Northwest National Laboratory, Richland, WA, USA.

Fryxell, G., Mattigod, S., Yuehe, L., Hong, W., Fiskum, S., Parker, K., Feng, Z., Yantasee, W., Zemanian, T., Addleman, R., Jun, L., Kemner, K., Kelly, S. & Xiangdong, F. (2007) Design and synthesis of self-assembled monolayers on mesoporous supports (SAMMS™): The importance of ligand posture in functional nanomaterials. *Journal of Materials Chemistry*, 17, 2863–2874.

Gao, H., Gao, X., Cao, Y., Xu, L. & Jia, L. (2015) Influence of Hydroxypropyl-β-cyclodextrin on the extraction and biodegradation of p,p′-DDT, o,p′-DDT, p,p′-DDD, and p,p′-DDE in soils. *Water, Air, & Soil Pollution*, 226(7), 208–220.

Garon, D., Sage, L., Wouessidjewe, D. & Seigle-Murandi, F. (2004) Enhanced degradation of fluorene in soil slurry by *Absidia cylindrospora* and maltosyl-cyclodextrin. *Chemosphere*, 56, 159–166.

Gorkovenko, A., Zhang, J., Gross, R.A., Kaplan, D.L. & Allen, A.L. (1999) Control of unsaturated fatty acid substitutes in emulsans. *Carbohydrate Polymers*, 39, 79–84.

Gruiz, K. (2005) Soil testing triad and interactive ecotoxicological tests for contaminated soil. In: Fava, F. & Canepa, P. (eds.) *Soil Remediation Series No. 6*. INCA, Venice, Italy.

Gruiz, K., Fenyvesi, E., Kriston, E., Molnár, M. & Horváth, B. (1996) Potential use of cyclodextrins in soil bioremediation. *Journal of Inclusion Phenomena and Macrocyclic Chemistry*, 25, 233–236.

Gruiz, K., Molnár, M. & Fenyvesi, E. (2008) Evaluation and verification of soil remediation. In: Kurladze, V.G. (ed) *Environmental Microbiology Research Trends*. Nova Sci Publ, Inc, New York, NY, USA.

Gruiz, K., Molnár, M., Fenyvesi, É., Hajdu, C., Atkári, Á. & Barkács, K. (2011) Cyclodextrins in innovative engineering tools for risk-based environmental management. *Journal of Inclusion Phenomena and Macrocyclic Chemistry*, 70(3–4), 299–306. doi:10.1007/s10847-010-9909-y.

Gruiz, K., Molnár, M., Nagy, Z.M. & Hajdu, C. (2015) Fate and behavior of chemical substances in the environment. In: Guiz, K., Meggyes, T. & Fenyvesi, E. (eds.) *Engineering Tools for Environmental Risk Management: 2. Environmental Toxicology*. CRC Press, Boca Raton, FL, USA. pp. 72–124.

Hajdu, C. & Gruiz, K. (2015) Bioaccessibility and bioavailability in risk assessment. In: Gruiz, K., Meggyes, T. & Fenyvesi, É. (eds.) *Engineering Tools for Environmental Risk Management: 2. Environmental Toxicology*. CRC Press, Boca Raton, FL, USA. pp. 337–400.

Hajdu, C., Gruiz, K. & Fenyvesi, É. (2009) Bioavailability- and bioaccessibility-dependent mutagenicity of pentachlorophenol (PCP). *Land Contamination & Reclamation*, 17(3–4), 473–481.

Haritash, A.K. & Kaushik, C.P. (2009) Biodegradation aspects of polycyclic aromatic hydrocarbons (PAHs): A review. *Journal of Hazardous Materials*, 169, 1–15. doi:10.1016/j.jhazmat.2009.03.137.

Herman, D.C., Artiola, J.F. & Miller, R.M. (1995) Removal of cadmium, lead, and zinc from soil by a rhamnolipid biosurfactant. *Environmental Science & Technology*, 29, 2280–2285.

Hong, K.J., Tokunaga, S. & Kajiuchi, T. (2002) Evaluation of remediation process with plant-derived biosurfactant for recovery of heavy metals from contaminated soils. *Chemosphere*, 49(4), 379–387.

Hu, J., Wang, Y., Su, X., Yu, C., Qin, Z., Wang, H., Hashmic, M.Z., Shi, J. & Shen, C. (2016) Effects of RAMEB and/or mechanical mixing on the bioavailability and biodegradation of PCBs in soil/slurry. *Chemosphere*, 155, 479–487.

Ivey (2017) *Ivey International*. Available from: www.iveyinternational.com/products.php. [Accessed 26th July 2018].

Jördening, H.J. & Winter, J. (2005) *Environmental Biotechnology: Concepts and Application*. Wiley-VCH Verlag GmbH, Weinheim, Germany.

Jozefaciuk, G., Murányi, A. & Fenyvesi, E. (2003) Effect of randomly methylated beta-cyclodextrin on physical properties of soils. *Environmental Science & Technology*, 37, 3012–3017.

Karanth, N.G.K., Deo, P.G. & Veenanadig, N.K. (1999) Microbial production of biosurfactants and their importance. *Current Science*, 77, 116–123.

Karlapudi, A.P., Venkateswarulu, T.C., Tammineedi, J., Kanumuri, L., Ravuru, B.L., Dirisala, V. & Kodali, V.P. (2018) Role of biosurfactants in bioremediation of oil pollution: A review. *Petroleum*, in press, available online: 10 March 2018. doi:10.1016/j.petlm.2018.03.007.

Kildisas, V., Levisauskas, D. & Grigiskis, S. (2003) Development of clean-up complex technology of soil contaminated by oil pollutants based on cleaner production concepts. *Environmental Research, Engineering and Management*, 25, 87–93.

King, A., Jensen, V., Fogg, G.E. & Harter, T. (2012) Groundwater Remediation and Management for Nitrate. *Technical Report 5 in: Addressing Nitrate in California's Drinking Water with a Focus on Tulare Lake Basin and Salinas Valley Groundwater. Report for the State Water Resources Control Board Report to the Legislature.* Center for Watershed Sciences, University of California, Davis, CA, USA.

King, R.B., Long, G.M. & Sheldon, J.K. (1992) *Practical Environmental Bioremediation*. Lewis Publishers, Boca Raton, FL, USA.

King, A., Jensen, V., Fogg, G.E. & Harter, T. (2012) Groundwater Remediation and Management for Nitrate. *Technical Report 5 in: Addressing Nitrate in California's Drinking Water with a Focus on Tulare Lake Basin and Salinas Valley Groundwater. Report for the State Water Resources Control Board Report to the Legislature.* Center for Watershed Sciences, University of California, Davis, CA, USA.

Koshlaf, E. & Ball, A.S. (2017) Soil bioremediation approaches for petroleum hydrocarbon polluted environments. *AIMS Microbiology*, 3(1), 25–49. doi:10.3934/microbiol.2017.1.25.

Krieger, R. (2001) *Handbook of Pesticide Toxicology. Principles*. 2nd ed. Academic Press, San Diego, CA, USA.

Lau, E.V., Gan, S., Ng, H.K. & Poh, P.E. (2014) Extraction agents for the removal of polycyclic aromatic hydrocarbons (PAHs) from soil in soil washing technologies. *Environmental Pollution*, 184, 640–649.

Le, T.T., Nguyen, K.-H., Jeon, J.-R., Francis, A.J. & Chang, Y.-S. (2015) Nano/bio treatment of polychlorinated biphenyls with evaluation of comparative toxicity. *Journal of Hazardous Materials*, 287, 335–341.

Leitgib, L., Gruiz, K., Fenyvesi, E., Balogh, G. & Murányi, A. (2008) Development of an innovative soil remediation: 'Cyclodextrin-enhanced combined technology'. *Science of the Total Environment*, 392, 12–21.

Lewinski, N. (2008) *Nanotechnology for Waste Minimization and Pollution Prevention*. Available from: http://wp.vcu.edu/nalewinski/wp-content/uploads/sites/4986/2014/08/Lewinski_NNEMS_2008.pdf. [Accessed 26th July 2018].

Luo, W., D'Angelo, E.M. & Coyne, M.S. (2007) Plant secondary metabolites, biphenyl, and hydroxypropyl-beta-cyclodextrin effects on aerobic polychlorinated biphenyl removal and microbial community structure in soils. *Soil Biology & Biochemistry*, 39, 735–743.

Marcos, J.C., Ferreira, I.N.S., Correa, P.F., Rufino, R.D., Luna, J.M., Silva, E.J. & Sarubbo, L.A. (2015) Application of bacterial and yeast biosurfactants for enhanced removal and biodegradation of motor oil from contaminated sand. *Electronic Journal of Biotechnology*, 18(6), 471–479.

Matos, M.P.S.R., Correia, A.A.S. & Rasteiro, M.G. (2017) Application of carbon nanotubes to immobilize heavy metals in contaminated soils. *Journal of Nanoparticle Research*, 19, 126–137. doi:10.1007/s11051-017-3830-x R.

Menn, F.-M., Easter, J.P. & Sayler, G.S. (2008) Genetically engineered microorganisms and bioremediation. In: Rehm, H.J. & Reed, G. (eds.) *Biotechnology: Environmental Processes II*. Volume 11b. 2nd ed. Wiley-VCH Verlag GmbH, Weinheim, Germany.

Middeldorp, P., Langenhoff, A. *et al.* (2002) *In situ* clean-up technologies. In: Agathos, S. & Reineke, W. (eds.) *Focus on Biotechnology. Biotechnology for the Environment: Soil Remediation*. Series, Volume 3a. Kluwer Academic Publishers, Dordrecht, The Netherlands.

Molnár, M. (2006) *Intensification of Soil Bioremediation by Cyclodextrin: From the Laboratory to the Field*. PhD Thesis, Budapest University of Technology and Economics, Budapest, Hungary.

Molnár, M., Fenyvesi, E., Gruiz, K., Leitgib, L., Balogh, G., Murányi, A. & Szejtli, J. (2002) Effects of RAMEB on bioremediation of different soils contaminated with hydrocarbons. *Journal of Inclusion Phenomena and Macrocyclic Chemistry*, 44, 447–452.

Molnár, M., Fenyvesi, É., Gruiz, K., Leitgib, L., Balogh, G., Murányi, A. & Szejtli, J. (2003) Effects of RAMEB on bioremediation of different soils contaminated with hydrocarbons. *Journal of Inclusion Phenomena and Macrocyclic Chemistry*, 44, 447–452.

Molnár, M., Gruiz, K., Fenyvesi, É. & Szőnyi, M. (1998) Potential use of ß-cyclodextrin for bioremediation of contaminated soil. *Contaminated Soil'98* (Preprints of the International Conference on Contaminated Soil, Edinburgh, May 17–21, 1998). Thomas Telford, London, UK. pp. 1207–1208.

Molnár, M., Leitgib, L., Fenyvesi, E. & Gruiz, K. (2009) Development of cyclodextrin-enhanced soil remediation: From the laboratory to the field. *Land Contamination & Reclamation*, 17(3–4), 599–610.

Molnár, M., Leitgib, L., Gruiz, K., Fenyvesi, E., Szaniszló, N., Szejtli, J. & Fava, F. (2005) Enhanced biodegradation of transformer oil in soils with cyclodextrin: From the laboratory to the field. *Biodegradation*, 16, 159–168.

Mulligan, C.N. (2005) Environmental applications for biosurfactants. *Environmental Pollution*, 133(2), 183–198. doi:10.1016/j.envpol.2004.06.009.

Mulligan, C.N., Sharma, S.K. & Mudhoo, A. (eds.) (2014) *Biosurfactants: Research Trends and Applications*. CRC Press Taylor & Francis Group, Boca Raton, FL, USA, London, UK & New York, NY, USA.

Mulligan, C.N. & Wang, S. (2006) Remediation of a heavy metal contaminated soil by a rhamnolipid foam. *Engineering Geology*, 85(1–2), 75–81. doi:10.1016/j.enggeo.2005.09.029.

Nagy, Zs., Gruiz, K., Molnár, M., Fekete-Kertész, I., Fenyvesi, É. & Perlné Molnár, I. (2013) Removal of emerging micropollutants from water using Cyclodextrin. *Proceedings Aquaconsoil 2013, Theme C: Assessment and Monitoring*, Paper 2323.

Nayak, A.S., Vijaykumar, M.H. & Karegoudar, T.B. (2009) Characterization of biosurfactant produced by *Pseudoxanthomonas* sp. PNK-04 and its application in bioremediation. *International Biodeterioration and Biodegradation*, 63, 73–79.

Nichols, W.J. (2001) The U.S. Environmental Protect Agency: National oil and hazardous substances pollution contingency plan, Subpart J Product Schedule (40 CFR 300.900). *Proceedings of the International Oil Spill Conference, March 2001*. American Petroleum Institute, Washington, DC, USA. pp. 1479–1483.

Pacwa-Płociniczak, M., Płaza, G.A., Piotrowska-Seget, Z. & Cameotra, S.S. (2011) Environmental applications of biosurfactants: Recent advances. *International Journal of Molecular Sciences*, 12(1), 633–654. doi:10.3390/ijms12010633. Available from: www.ncbi.nlm.nih.gov/pmc/articles/PMC3039971/ [Accessed 26th July 2018].

Paul, E.A. & Clark, F.E. (1996) *Soil Microbiology and Biochemistry*. 2nd ed. Academic Press, San Diego, CA, USA.

Pavan, M. & Worth, A.P. (2006) Review of QSAR models for ready biodegradation. EUR 22355 EN Report. European Commission Directorate, General Joint Research Centre, Institute for Health and Consumer Protection, Ispra, Italy.

Pesce, L. (2002) A biotechnological method for the regeneration of hydrocarbons from dregs and muds, on the base of biosurfactants. 02/062,495. World Patent.

Piotrowska-Długosz, A. (2017) The use of enzymes in bioremediation of soil xenobiotics. In: Hashmi, M., Kumar, V. & Varma, A. (eds.) *Xenobiotics in the Soil Environment. Soil Biology*. Volume 49. Springer, Cham, Switzerland.

Rahman, K.S.M. & Gakpe, E. (2008) Production, characterization and applications of biosurfactants-Review. *Biotechnology*, 7(2), 360–370.

Ramsay, J.A., Robertson, K., van Loon, G., Acay, N. & Ramsay, B.A. (2005) Enhancement of PAH biomineralization rates by cyclodextrins under Fe(iii)-reducing conditions. *Chemosphere*, 61, 733–740.

Santanu, P. (2008) Surfactant-enhanced remediation of organic contaminated soil and water. *Advances in Colloid and Interface Science*, 138(1), 24–58. doi:10.1016/j.cis.2007.11.001.

Schink, B. (2005) Principles of anaerobic degradation of organic compounds. In: Jördening, H.J. & Winter, J. (eds.) *Environmental Biotechnology. Concepts and Application*. Wiley-VCH Verlag GmbH, Weinheim, Germany.

Shete, A.M., Wadhawa, G., Banat, I.M. & Chopade, B.A. (2006) Mapping of patents on bioemulsifier and biosurfactant: A review – NOPR. *Journal of Scientific & Industrial Research*, 65, 91–115.

Singh, P. & Cameotra, S.S. (2004) Enhancement of metal bioremediation by use of microbial surfactants. *Biochemical and Biophysical Research Communications*, 319(2), 291–297.

Singh, R., Manickam, N., Mudiam, M.K.R., Murthy, R.C. & Misrab, V. (2013) An integrated (nano-bio) technique for degradation of HCH contaminated soil. *Journal of Hazardous Materials*, 35, 258–259.

Sivaraman, C., Ganguly, A. & Mutnuri, S. (2009) Biodegradation of hydrocarbons in the presence of cyclodextrins. *World Journal of Microbiology & Biotechnology*, 26, 227–232.

Soberón-Chávez, G. & Maier, R.M. (2011) Biosurfactants: A general overview. In: Soberón-Chávez, G. (ed) *Biosurfactants*. Springer-Verlag, Berlin, Germany. pp. 1–11.

Steffan, S., Bardi, L. & Marzona, M. (2001) Biodegradation of hydrocarbon in polluted soils using beta-cyclodextrin as a coadjuvant. *Biological Journal of Armenia, Special Issue: Cyclodextrins*, 53, 218–225.

Steffan, S., Tantucci, P., Bardi, L. & Marzona, M. (2003) Effects of cyclodextrins on dodecane biodegradation. *Journal of Inclusion Phenomena and Macrocyclic Chemistry*, 44, 407–411.

Stroud, J.L., Tzima, M., Paton, G.I. & Semple, K.T. (2009) Influence of hydroxypropyl-beta-cyclodextrin on the biodegradation of C-14-phenanthrene and C-14-hexadecane in soil. *Environmental Pollution*, 157, 2678–2683.

Suttinun, O., Luepromchai, E. & Müller, R. (2013) Cometabolism of trichloroethylene: Concepts, limitations and available strategies for sustained biodegradation. *Reviews in Environmental Science and Bio/Technology*, 12, 99–114. doi:10.1007/s11157-012-9291-x.

Swartjes, F.A. (2011) *Dealing With Contaminated Sites: From Theory Towards Practical Application*. Springer, Dordrecht, The Netherlands.

Szaniszló, N., Fenyvesi, E. & Balla, J. (2005) Structure – Stability study of cyclodextrin complexes with selected volatile hydrocarbon contaminants of soils. *Journal of Inclusion Phenomena and Macrocyclic Chemistry*, 53, 241–248.

US EPA (2008) *Nanotechnology for Site Remediation: Fact Sheet*. Available from: https://nepis.epa.gov/Exe/ZyPDF.cgi/P1001JIB.PDF?Dockey=P1001JIB.PDF. [Accessed 26th July 2018].

Valentín, L., Nousiainen, A. & Mikkonen, A. (2013) Introduction to organic contaminants in soil: Concepts and risks. In: Vicent, T. *et al.* (eds.) *Emerging Organic Contaminants in Sludges: Analysis, Fate and Biological Treatment. The Handbook of Environmental Chemistry*. Volume 24. Springer, Berlin, Germany & Heidelberg, Germany. doi:10.1007/698_2012_208.

Vermeul, V.R., Rockhold, M.L., Bjornstad, B.N., Szecsody, J.E., Murray, C.J., Williams, M.D., Newcomer, D.R. & Xie, Y. (2004) *In situ Redox Manipulation Permeable Reactive Barrier Emplacement: Final Report, Frontier Hard Chrome Superfund Site, Vancouver, WA, Battelle*. Pacific Northwest Division, 99352, Report PNWD-3361, Richland, WA, USA. 93 p.

Vermeul, V.R., Williams, M.D., Szecsody, J.S., Fruchter, J.S. Cole, C.R. & Amonett, J.E. (2003) Creation of a subsurface permeable reactive barrier using in situ redox manipulation. Applications to

Radionuclides, Trace Metals, and Nutrients. In: Naftz, D.L. et al. (eds.) *Handbook of Groundwater Remediation using Permeable Reactive Barriers*. Academic Press, San Diego, CA, USA. doi: 10.1016/B978-012513563-4/50010-4

Vijayakumar, S. & Saravanan, V. (2015) Biosurfactants-Types, sources and applications. *Research Journal of Microbiology*, 10, 181–192. doi:10.3923/jm.2015.181.192.

Wackett, L.P. (1995) Bacterial co-metabolism of halogenated organic compounds. In: Young, L.Y. & Cerniglia, C. (eds.) *Microbial Transformation and Degradation of Toxic Organic Chemicals*. John Wiley & Sons, New York, NY, USA.

Wang, J.M., Marlowe, E.M., Miller-Maier, R.M. & Brusseau, M.L. (1998) Cyclodextrin-enhanced biodegradation of phenanthrene. *Environmental Science & Technology*, 32, 1907–1912.

Wang, S. & Mulligan, C. (2004a) Rhamnolipid foam enhanced remediation of cadmium and nickel contaminated soil. *Water Air and Soil Pollution*, 157(1), 315–330.

Wang, S. & Mulligan, C. (2004b) An Evaluation of surfactant foam technology in remediation of contaminated soil. *Chemosphere*, 57(9), 1079–1089. doi:10.1016/j.chemosphere.2004.08.019.

Wang, S. & Mulligan, C. (2009) Arsenic mobilization from mine tailings in the presence of a biosurfactant. *Applied Geochemistry*, 24(5), 928–935.

Whang, L.M., Liu, P.W.G., Ma, C.C. & Cheng, S.S. (2008) Application of biosurfactant, rhamnolipid, and surfactin, for enhanced biodegradation of diesel-contaminated water and soil. *Journal of Hazardous Materials*, 151, 155–163.

Wiedemeier, T.H., Rifai, H.S., Newell, C. & Wilson, J.T. (1999) *Natural Attenuation of Fuels and Chlorinated Solvents in the Subsurface*. Wiley, New York, NY, USA.

Xu, Y. & Zhao, D. (2005) Removal of copper from contaminated soil by use of poly(amidoamine) dendrimers. *Environmental Science and Technology*, 39, 2369–2375.

Yin, H., Qiang, J., Jia, Y., Ye, J., Peng, H. *et al.* (2009) Characteristics of biosurfactant produced by *Pseudomonas aeruginosa* S6 isolated from oil-containing wastewater. *Process Biochemistry*, 44, 302–308.

Yoshii, H., Furuta, T., Shimizu, J., Kugimoto, Y., Nakayasu, S., Arai, T. & Linko, P. (2001) Innovative approach for removal and biodegradation of contaminated compounds in soil by cyclodextrins. *Biological Journal of Armenia, Special Issue: Cyclodextrins*, 53, 226–236.

Yuan, C. & Jin, Z. (2007) Aerobic biodegradability of hydroxypropyl-beta-cyclodextrins in soil. *Journal of Inclusion Phenomena and Macrocyclic Chemistry*, 58, 345–351.

Zhang, Y. & Miller, R.M. (1992) Enhanced octadecane dispersion and biodegradation by a *Pseudomonas rhamnolipid* surfactant (biosurfactant). *Applied and Environmental Microbiology*, 58, 3276–3282.

Chapter 6

Traditional and innovative methods for physical and chemical remediation of soil contaminated with organic contaminants

É. Fenyvesi[1], K. Gruiz[2], E. Morillo[3] & J. Villaverde[3]

[1]*CycloLab Cyclodextrin R&D Lab. Ltd., Budapest, Hungary*
[2]*Budapest University of Technology and Economics, Budapest, Hungary*
[3]*Institute of Natural Resources and Agrobiology, Spanish National Research Council (CSIC), Seville, Spain*

ABSTRACT

This chapter provides an overview of the remediation technologies based on physicochemical processes. In addition to emerging technologies, innovative developments of traditional technologies are presented. There is a demonstration of the physicochemical methods useful for the remediation of soils contaminated with pesticides. A case study on full-scale application of *in situ* chemical oxidation on a site historically contaminated with chlorinated hydrocarbons is reported. Hydrogen peroxide combined with alternating well pumping ensured intensive mass transfer between the soil liquid phases (DNAPL lens and groundwater) and achieved significant contaminant removal.

1 INTRODUCTION

É. Fenyvesi and K. Gruiz

The classification of the technologies used in this book is primarily based on the characteristics of the contaminants and the contaminated phases of the soil to be treated (see Chapter 1 in this volume). A further consideration in sorting the large number of technologies is the dominant physical, chemical, biological, and ecological processes which play a role in decontamination and remediation of the environment.

The most frequent organic contaminants in the soil include fuel hydrocarbons, polynuclear aromatic hydrocarbons (PAHs), chlorinated aromatic compounds, polychlorinated biphenyls (PCBs), pesticides, insecticides, germicides, detergents, etc. Some of these are volatile and can be in the soil gas or vapor, whereas others are dissolved or emulsified in the aqueous phase or sorbed to the solid phase of the soil. The water-immiscible liquids partition into the gaseous, liquid, and solid phases of the soil and form separate phases above the groundwater (LNAPL) or below the aquifers (DNAPL). The solid contaminants can be mixed with or sorbed to the soil particles or bound to the matrix by physical or chemical bonds, e.g. built in the humus by covalent bonds (Gruiz, 2009). Partition of contaminants in soil involves all three physical phases and also the living part, the biota. One or two phases may dominate in contaminants-binding and others may play a negligibly small role.

This chapter introduces and evaluates the soil remediation technologies based on physicochemical processes even if they are used typically in combination with biological or ecological methods.

The ENFO database (2018), Remediation Technologies Screening Matrix and Reference database (FRTR, 2018a), Center for Public Environmental Oversight (CPEO, 2018), and EPA CLU-IN database (2018) were used to review the technologies, in addition to the generic literature search.

The physical and chemical remediation technologies for reducing the risk of organic contaminants are based on either mobilization or immobilization of the pollutants (Gruiz, 2009) as well as mobilization or immobilization of the contaminated medium. Mobilization means increasing the mobile (dissolved, gaseous, bioavailable) fraction of the contaminant and removing it together with the mobile physical phases. Immobilization, on the contrary, decreases this contaminant fraction and prevents the transport of the contaminant or the contaminated medium from its – possibly separated – location. Technologies based on mobilization remove the contaminants from the contaminated soil phases either physically (volatilization, extraction), chemically (destruction, transformation into less hazardous compounds), or biologically (extraction, stabilization, transformation by biological contribution, as well as partial or ultimate degradation). Biological technologies are summarized in Chapters 3–5 of this volume. In the case of technologies based on immobilization, the environmental risk of the contaminant is reduced by decreasing – as far as possible, irreversibly – the mobility, transport, and bioavailability of the contaminants without removing them from the contaminated medium. The classification of these technologies is covered in Chapter 1 of this volume.

Remedial technologies can usually be applied both *in situ* (without excavation of the soil) or *ex situ* by removing the soil and treating it on the surface on site, off site, at a dump site, in heaps, or in reactors. Landfilling the excavated soil or sludge is generally permitted only if a limited amount of free liquid is present and no or very little leaching of hazardous contaminants is expected. Landfill management should ensure the collection and treatment of the generated gas (landfill gas) and/or liquid (leachate).

The common physical and chemical methods can be used separately or combined with each other and/or with thermal, electrochemical, and biological methods. The combined technologies are usually unique, problem-specific solutions, and are thus often considered innovative. The traditional technologies enhanced by new additives/reagents are also classified as innovative. The cutting-edge technologies combine various methods, e.g. physicochemical remediation with bio- and ecotechnologies.

The *in situ* remediation technologies are usually more efficient in homogeneous and permeable soils compared to *ex situ* technologies. *Fracturing*, e.g. pneumatic or hydraulic fracturing, is an option for enhancing the transport processes in low-permeability soils by forming cracks and decreasing soil particle size. The transport of both the contaminants and the chemical reactants is improved, which is why fracturing can be useful to increase the efficiency of some kind of *in situ* treatments, e.g. extractions, augmentation, or application of additives. In addition to low permeability, high heterogeneity also hampers *in situ* physicochemical processes.

2 TECHNOLOGIES BASED ON MOBILIZATION

Physicochemical mobilization covers many technologies which are based on shifting the partition of the contaminants toward the more mobile soil phases, i.e. from liquid to gas (evaporation), from solid to liquid or gas (desorption), from any of the physical phases into the biological "phase" (biological uptake and/or accumulation). These treatments can be intensified by the use of mobilizing additives (surfactants, micelle-forming agents, chemical reagents, or enzymes resulting in smaller and more mobile degradants) or heat and electricity. Another group of technologies aims to "mobilize" the mobilizable physical phases, e.g. by groundwater flow modification, extraction of soil air, groundwater and leachates, as well as soil moisture collection.

The most frequently applied physical contaminant removal technologies are based on the extraction of contaminated soil gas, soil moisture, and soil water from the unsaturated (three-phase soil) or the groundwater from the two-phase soil, and their surface treatment.

2.1   Separation technologies

This section discusses the separation of the mobile soil phases or the contaminant (insofar as it represents a separate mobile physical phase), starting with the gases and vapors, followed by the liquid phases such as the contaminated soil waters and the liquid phase contaminants.

2.1.1   Soil venting

Soil venting, soil vapor extraction, and soil vacuum extraction (called also as soil vapor extraction or soil volatiles extraction, SVE) are used for the removal of contaminated soil gas and vapor from the soil's vadose zone. In addition to the removal of vapors, it promotes further desorption and evaporation in the three-phase soil as well as chemical oxidation and aerobic biodegradation. Soil venting involves the introduction of fresh air with high oxygen content (passively or actively by injection) into the soil while contaminated air with increased CO_2 and volatile contaminant contents is extracted, i.e. an enhanced air exchange is established in the soil, similar to any ventilation. The difference to normal ventilation is that the air space is limited to the pore volume of the vadose zone (Figure 6.1).

Passive wells are simple, open-top wells, screened at the depth of contaminant occurrence. They allow in fresh atmospheric air as close as possible to the contaminant.

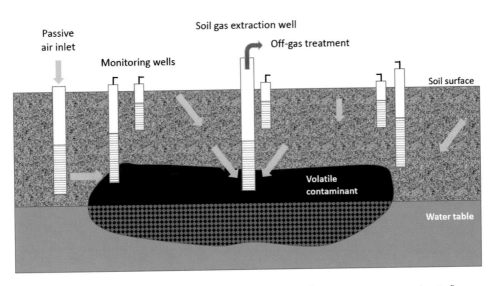

Figure 6.1 Soil venting with passive atmospheric air inlet. The yellow arrows represent the air flow.

Active wells on the other hand inject air into the soil with pressure. Air extraction with blowers can be connected to both. A more sophisticated system is the directed soil air circulation system (SAC). Two screens – an upper and a lower one – are built into the borehole and separated from each other. Both are connected to a surface blower. This allows for soil gas extraction from either the upper or the lower segment of the well or from both at the same time. By doing so, horizontal and vertical flow circulation cells are generated in the soil surrounding the extraction well. The circulation direction is reversible and can be adjusted according to the contaminant distribution in the soil (IEG Technologie, 2018).

Optimal air flow depends on the pore volume and the volatility of the contaminant. Soil venting by suction is generally more appropriate, as the contaminated air can be collected and better controlled than in pressurized systems where the overpressurized soil gas easily escapes. In contrast, the depression in the soil gas pressure moves the gas flux in the desired direction. The collected contaminated soil gas is treated by the most appropriate technology – using separators, sorbents, washers, catalytic oxidizers, combustors, or incinerators, etc. – depending on the quantity and quality of the contaminants in the exhausted gas. Thermal destruction (e.g. direct flame thermal oxidation, catalytic oxidizers), adsorption (e.g. granular activated carbon, zeolites, polymers), biofiltration, photolytic/photocatalytic destruction, non-thermal plasma destruction, membrane separation, gas absorption, and vapor condensation are used. The most common methods are thermal oxidation and granular activated carbon adsorption.

Some relevant concerns are these:

– Soil vapor extraction can be used both *in situ* and *ex situ* – the latter for soil heap treatment.
– Increased aerobic biological activities are expected due to SVE, even if the technology is planned and used as a single physical treatment. For biodegradable soil contaminants soil venting is used especially for the activation of the biodegrading soil biota, which is prepared to degrade the organic contaminants. This application is referred to as bioventing (see Chapters 1 and 5 in this volume).
– Air permeability of the soil has an inverse relationship with soil moisture content.
– Increased soil gas/soil air flux promotes water evaporation and causes soil desiccation.

2.1.2   Air stripping/sparging

Air stripping, also called air sparging or volatilization, forces the contaminants to transfer from waters, typically from groundwater into the sparged air bubbles (Loden, 1992). To remove volatile and semivolatile organic contaminants (VOCs and SVOCs) characterized by high Henry's law constant (high volatility), contaminant-free air is injected into the water. The extracted vapor can be treated by any known gas cleanup technology, e.g. gas separation, sorption, catalytic oxidation, incineration, or biofiltration.

The process can be implemented *in situ* in the groundwater or in groundwater extraction wells, and in reactors placed on the soil surface. Such a reactor is shown in Figure 6.2. (The packed towers or tray towers are operated with countercurrent flow of the extracted groundwater and air to enhance the contact time between these two phases.)

For *in situ* air stripping, contaminant-free air is injected into the saturated soil zone (two-phase soil) and contaminant-containing soil air/soil vapor is extracted from the unsaturated

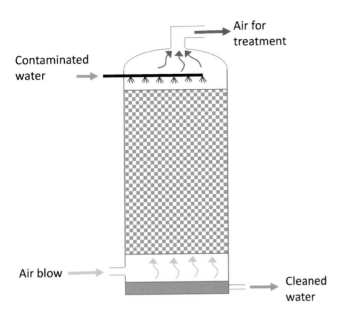

Figure 6.2 Packed-column air stripper for *ex situ* treatment of volatiles-contaminated groundwater.

(three-phase) soil. The pressure gradient resulting from creating a vacuum in the extraction wells pulls the vapors, that is, the contaminated soil air into the extraction wells.

The highly engineered version of air stripping is carried out in the well itself. A well constructed for such purpose can be considered as a subsurface-placed, *in situ* operating reactor or reactive well (Figure 6.3). The well is under depression (vacuum blower) and is screened in two depths, one under the water table at the level of contaminant occurrence, and the other one directly above the water table. The negative pressure in the well is adjusted to lift the water from the deeper to the upper screen (without extraction to the surface), from where the water is forced to leave, causing a circulation in the surrounding of the well between the screens. The groundwater flow direction through the well can be upward or downward depending on site-specific conditions. Pressurized ambient air is injected to the bottom of the active well, where air bubbles meet the contaminated water and strip the volatile contaminants. Contaminant-containing bubbles leave the water at the depths of the upper screen, from where the vapors are extracted and transported to the vapor treatment facility by the vacuum blower (SITE, 1993; FRTR, 2018a,b).

This technology can enhance the contact between the phases of the soil. A circulation flow cell develops in the saturated soil surrounding the wells, with water entering at the base of the well and leaving through the upper screened segment or vice versa depending on the desired flow direction (Hirschberger, 1998).

Another implementation of the *in situ* air stripping is performed by horizontal wells (Figure 6.4). The full-scale demonstration of horizontal wells at a military site contaminated with trichloroethylene (TCE) and PCE reduced the initial 10–1031 µg/L and 3–124 µg/L TCE and PCE concentration to 0.67–6.29 µg/L and 0.44–1.05 µg/L, respectively, in 139 days (US Department of Energy, 1995).

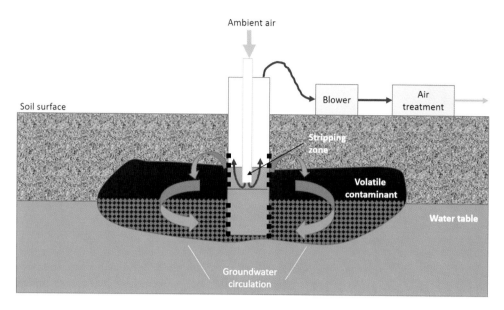

Figure 6.3 *In situ* in-well air stripping.

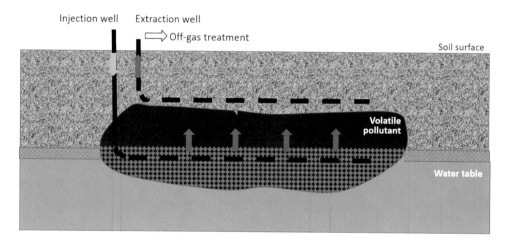

Figure 6.4 Scheme of *in situ* air stripping technology applying horizontal wells.

Pulsed mode soil gas extraction can improve contaminant removal. Pulsing air injection often induces the collapse of the air channels, and the new channels in each cycle ensure more efficient removal of the contaminants (Abdel-Moghny *et al.*, 2012).

Another method of intensification is periodical injection of an ozone/air mixture in conjunction with a pulsing pump (Kerfoottech, 2018). In addition to air stripping for extracting dissolved volatile organic compounds (VOCs) out of contaminated groundwater, the ozone reacts rapidly with the volatile compounds to decompose them into end products such as carbon dioxide, dilute hydrochloric acid, and water (see also in Section 2.2.1 in this chapter).

By removing soil air, not only contaminant vapor but also gaseous metabolites such as CO_2 are replaced by the fresh air stimulating the microbiota. In the special combination of *in situ* air stripping and *in situ* biodegradation the vapor from the groundwater is transported into the soil pore spaces of the unsaturated soil where they can be biodegraded by the activated microbiota. When applied in combination with bioremediation, additives (nutrients, etc.) to enhance biodegradation can be injected into the stripping well (see Chapter 5 in this volume).

2.1.2.1 Heat-enhanced soil gas extraction

Soil volatile extraction (SVE) can easily be enhanced by hot air injection: 10–20°C increase may duplicate the efficacy of desorption and volatiles removal, especially that of the heavier (less volatile) fractions. Heat intensifies the processes of both desorption and volatilization (Henry's law constants increase significantly with increasing temperature). Microbiological degradation can also be accelerated by increasing temperature to 30°C. Thermally enhanced vapor extraction system (TEVES) uses preheated air or steam to be injected into the soil. The three-phase soil can be heated up by other means suitable for energy transfer such as hot air/steam injection, electrical resistance heating, electromagnetic heating, fiber optic/radio frequency heating, and their combinations (TEVES, 2018).

Hot air or steam is generally injected below the contaminated soil zone to heat up the contaminated volume, deliberate the volatiles and semivolatiles from the soil matrix and transfer the vapors by the upward movement of the heated soil gas to the surface. The contaminant-containing soil gas should be collected and treated. The collector system can be placed into the soil (perforated tube or well system under negative pressure) or on the surface (collector system on the surface with local exhaust ventilation (LEV) hood and gas treatment).

The mobilization, evaporation, and removal of volatile and semivolatile contaminants (VOC and SVOC) by using heat can be applied both *in situ* and *ex situ*.

Temperature increase in soil has a complex impact on soil physics and chemistry as well as on microbial life and activity. In the vadose soil zone water also evaporates in addition to contaminant mobilization. It may promote further evaporation of the volatile contaminants. The vapor pressure and mobility (higher diffusivity, lower viscosity) of the volatile or semivolatile contaminants are increased. Soil moisture is heated up locally, which functions as a steam source mobilizing and stripping the steam-volatile contaminants. Permeability of the dry soil increases after evaporation of the soil moisture, and thus the vapor extraction can be more efficient.

The activity of the microbiota increases with increasing temperature until a certain limit. Higher temperature than the usual 8–15°C in the subsurface causes changes in the diversity of the microbiota: heat-tolerant species gain advantage and the heat-sensitive will be pushed back. When planning soil heating, it should be decided if the technology will be built upon microbial degradation or not. If biological activity is to be kept, soil moisture supply, aeration, and supplemental nutrients may be necessary and the temperature should not exceed a certain level, e.g. 30°C (such a critical level may vary depending on climate, season, and the spatial and temporal characteristics of the environment).

If soil vapor extraction is the basic process for contaminant removal and therefore not a requirement to sustain biological activity, the temperature can go up to 105°C (in the case of high moisture content) or even to 300°C and higher. These high-temperature soil treatment technologies are characteristic of thermal desorption, as any increase in temperature enhances desorption. Even higher temperatures can be used for *in situ* or reactor-based thermal destruction and/or stabilization of soil/sediment contaminants. In such technologies, even if the main goal is destruction or stabilization, several mobilizable soil components and contaminants will be volatilized and products of thermal destruction will also increase the "must be treated" volatile fraction.

The efficiency of air sparging and the removal of the heavier fractions in particular can be enhanced by increasing soil temperature: heat intensifies the processes of both desorption and volatilization. Higher temperature may be beneficial for biodegradation as well. A thermally enhanced vapor extraction system (TEVES) uses preheated air or steam to be injected into the soil. The temperature should be optimized for maximum efficiency of contaminant removal, i.e. in reducing the risk of soil contamination and the minimum deterioration of soil biota and soil structure. In the case of high-risk contaminants the survival of the soil biota will be downgraded, but soil structure can still be rescued. For instance, preheating the air to 45°C at the inlet resulted in an increase in recovery efficiency of between 70% and 90% after 5 hours of *in situ* air stripping of benzene-contaminated soil and groundwater (Shah *et al.*, 1995).

Steam-enhanced remediation (SER) in combination with dual phase vacuum extraction (DPVE) and air stripping was used to remove predominantly creosote contamination from the soil of a former timber treatment facility in Burton on Trent (UK). The temperature was increased to 60–90°C and the volatile components of the DNAPL were removed by DPVE from contaminated soil and groundwater; 4500 kg of contaminants was removed from an area of 1100 m² over a period of 7 months (Churngold, 2009). In a recent application soil and groundwater contaminated with diesel oil in Rotterdam Port were heated up to 105°C using plant steam and vacuum to remove the free oil phase, vapors, and dissolved phase in groundwater. As a final polishing step the temperature in the vadose soil zone was increased to 120°C. The remediation was successfully completed in 6 months (Churngold, 2018).

In addition to hot air/steam injection, other technologies such as electrical resistance heating, electromagnetic heating, fiber optic/radio frequency heating, and their combinations can be used for increasing the temperature of the soil (TEVES, 2018). ***Electrical resistance heating*** uses an electrical current generated by low-frequency electricity to heat up the soil. It applies to low permeability soils, such as clays or clayey sediment and it works in high moisture content soils, too, and is excellent for steam-volatile contaminants. The vapors are extracted and treated. Electrodes are placed directly into the soil in such a way that the contaminated volume is located between the electrodes so the current passes through the contaminated soil and heats it up. The evaporation of the soil water is a kind of *in situ* steam source inducing the evaporation of the contaminant together with the water vapor. Water content of the soil continuously decreases, the heat causes desiccation and disintegration of the soil structure, making it less cohesive and more permeable.

A special *in situ* solution called ***six-phase soil heating*** (SPSH) has been developed and was demonstrated first in 1994 by Gauglitz *et al.* Low-frequency electricity was used to heat the soil by six electrodes in a hexagonal array for better heat distribution. The vapor extraction well, which removes the contaminants, air, and steam from the subsurface, is located in the center of the hexagon.

Radio frequency heating (RFH) uses electromagnetic energy to heat soil and enhance evaporation of volatile and semivolatile soil contaminants to be extracted. Three rows of vertical electrodes are placed into the soil: the soil volume to be heated is between the two outer rows of ground electrodes. The third row of electrodes (between the two) receives electricity. The three rows act as a buried triplate capacitor and can heat up soil to over 300°C. The soil continuously heats and dries until the current stops flowing.

2.1.2.2   Effect of surfactants and complexing agents

Surfactant-enhanced air stripping (SEAS) coupled with soil vapor extraction (SVE) is useful in the second phase of remediation when the removal rate with conventional air stripping using SVE is slowed down. By reducing the surface tension surfactants help the migration of NAPLs such as chlorobenzene from the small pores of the soil (Qin *et al.*, 2013) and promote air flow (Kim *et al.*, 2004). For instance, the presence of trace quantities of the surfactant Triton X100 leads to uniform small diameter bubbles ensuring higher contact surfaces between the phases and enhanced performance of the technology (Burns & Zhang, 2001).

Both surfactants and complexing agents decrease the Henry's law constant and reduce the volatility of VOCs, thus resulting in the decreased efficiency of air stripping (Anderson, 1992; Kashiyama & Boving, 2004).

2.1.3   Soil washing/chemical extraction for the removal of the soluble contaminants

Ex situ treatment of soil washing aims to remove the contaminants from the solid phase of the soil/sediment, which use either water (soil washing) or organic solvent (chemical extraction) (EPA CLU-IN, 2018; CL:AIRE, 2007) (see also Chapters 1 and 2 in this volume). The process separates the fine soil particles containing most of the contaminants from bulk soil (Figure 6.5).

The wash-water may contain a basic leaching agent, surfactant, or chelating/complexing agent or additives to adjust pH to help remove organics and heavy metals. Soils and

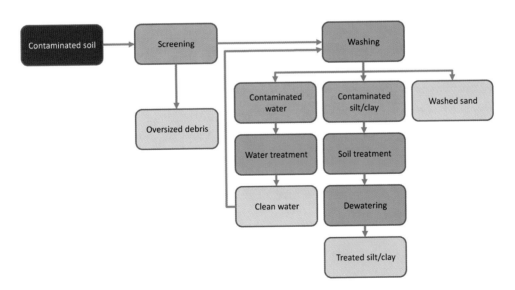

Figure 6.5 Flowchart of soil washing.

wash-water are mixed in a tank or other treatment unit and water is passed through sieves, mixing blades, or high-pressure water sprays on site using mobile, transportable units, or at soil treatment centers. With sizing and classification separated fractions are obtained: coarse gravel and debris, a coarse silt fraction, clay, and fine silt fraction. In many cases, the coarse gravel fraction is relatively a contaminant-free and considered non-hazardous material which can be recycled on the site or used on another site as fill or as valuable construction material (sand and gravel). If the contaminant content in this coarse fraction is above the target requirement, it is crushed and reprocessed. The silt and clay containing the most of the contaminants are treated separately, either by repeated soil washing or by bioremediation, catalytic oxidation, etc. Depending on the K_{ow} (octanol–water partition coefficient) of the contaminant, a large volume of polluted wash-water is formed. This water can be separated from the soil fractions using gravity settling, magnetic separators, or flotation and then cleaned up by air stripping, adsorption, chemical oxidation, or biodegradation technologies usually used for high concentration wastewater. The clean water then can be reused or discharged.

The technology may be inefficient because hydrocarbon retention after soil washing may be influenced by a number of factors unrelated to particle size, such as soil humic acids, metal oxide coatings, shape, and porosity of the soil particles, and clay type (Dove *et al.*, 1992).

When the soil washing is performed using an organic solvent instead of water or aqueous solution of solubilizing additives, the technology is called ***chemical extraction***. It can be implemented as a separate technology or as a component of the complex soil washing plant. It is used *ex situ*, if possible, *on site* to avoid transporting the soil to a different place of treatment. The sediments, sludge, and soil contaminated with organic pollutants are mixed with the solvent in an extractor to desorb and dissolve the contaminants. The solvent is then recovered from the extracted solution and recycled, while the contaminants drawn off are reused or disposed of. Soils of high moisture content need to be dried before treatment to achieve high removal rates. Multi-stage continuous extraction and countercurrent extraction improve the efficiency of contaminant removal (Li *et al.*, 2012).

Individual solvent or a solvent mixture (ethanol, propanol, acetone, ethyl acetate, liquefied gases such as propane, dimethyl ether or carbon dioxide) is used as extracting chemical. For instance, an ethyl acetate-acetone-water mixture was successfully applied for the removal of petroleum hydrocarbons (Silva *et al.*, 2005). Traces of solvent may remain within the treated soil matrix, so the toxicity of the solvent is an important consideration. Fire and explosion hazard should also be taken into account and preventive measures implemented.

The potential of *supercritical carbon dioxide* as solvent for soil extraction has been reviewed by Anitescu and Tavlarides (2006). This innovative green technology combined with catalytic oxidation is useful for the quick remediation of heavily polluted hotspots, for the removal of the less strongly bound, biodegradable fraction of PAHs, petroleum hydrocarbons, creosote, etc. and to reduce the most severe risk. Solvent extraction is inefficient in the removal of both high molecular weight organic and very hydrophilic substances.

Soil washing can be made more efficient by using solubilizing agents such as surfactants and complex-forming additives, similarly to *in situ* soil flushing (see Section 2.1.4). For heavy oil recovery, for instance, water with an added surfactant is the most common processing solvent.

2.1.4   In situ *soil flushing*

Soil flushing means *in situ* washing of the contaminated soil. Most often the soil between two wells or through ditches is washed by injecting water/aqueous solution/solvent mixtures into the contaminated zone; this may be within the vadose zone, the saturated zone, or both. The water table is raised in the vicinity of the injection well so that the solution flows through the contaminated zone, and the hydraulic gradient between the enhanced water level in the injection well and the depressed water level of the extracting well helps the contaminants to be desorbed from the soil and leached into the groundwater, which is extracted and treated on the surface. The dissolved organics are removed *ex situ* either partitioning into vapor phase (stripping) and/or into solid phase (adsorption) or by destroying them (oxidation, UV treatment) or by biodegradation-based decontamination. Process sludge and residual solids such as spent carbon and spent ion exchange resin deriving from the treatment of the recovered washing fluids must be appropriately treated before disposal. Air emissions of volatile contaminants from recovered flushing fluids should be collected and treated to meet applicable regulatory standards. The recovered fluids are reused in the flushing process: reinjected back into the contaminated soil, circulated, re-extracted, and re-purified. Low-permeability or heterogeneous soils are difficult to treat (fracturing can help). Pollutants which sorb strongly to the soil surface require many flushing cycles.

Usually vertical wells are used, but angled wells can also be applied below buildings or other structures that do not allow access for vertical wells to the area. Usually the flushing solution is injected in the source zone and extracted down-gradient using one or more wells (*line drive treatment*). In another implementation the same well is used both for injection and after a reaction phase, when the contaminants are solubilized by the flushing solution for extraction (*push-pull technique*) (Boving et al., 2008). An effective collection system is required to prevent migration of contaminants and potentially toxic extraction fluids to uncontaminated areas. Recharge is typically controlled by maintaining the water level in injection wells or drains or by pumping at specified rates.

A decrease in contaminant concentrations at wells may indicate success of the remediation, but it may also indicate dispersion or contaminant transport to down-gradient areas. Increased or constant, but elevated, concentrations in a well may indicate the presence of a continuing source of contamination from the unsaturated zone or from NAPL. If such sources are not removed or treated, the technology will likely operate endlessly.

Soil flushing can be combined with bioremediation. In this case the applied water is generally amended by nutrients to stimulate *microbial degradation*. The biological conversion can be

- Aerobic: aeration is combined with periodic flooding in the three-phase soil; or
- Anoxic: groundwater aeration, ORC or alternative electron acceptors addition is used to establish optimal redox potential for the active microbiota in the saturated soil/groundwater.

Elevated temperature is favorable for desorption and dissolution of the contaminants due to their enhanced solubility and for volatilization of liquid or sorbed contaminants. Steam injection is used to combine soil flushing with steam distillation and stripping of volatile and semivolatile contaminants (Davis, 1998). Hot water can also be utilized for a similar effect.

For instance, TCE removal was increased by a factor of two when hot water flushing was used to increase the temperature from 5°C to 40°C prior to air sparging (Imhoff *et al.*, 1997).

Pulsed hydraulic pressure can be used to enhance the recovery rate of LNAPL generating intense mixing so that liquids flow into and out of pore throats (Davidson, 2018). Pressure pulse technology (PPT) is applied so that the liquids repeatedly flow into and out of the wellbore and pore throats, ensuring accelerated fluid flow and pore-scale mixing, and enhancing the contact area between the remedial fluid and porous media.

Solubilizing agents such as cosolvents, surfactants, biosurfactants, and complexing agents can enhance efficacy. Recently, nano-sulfonated graphene was used as a washing agent for the removal of PAHs from a coking plant soil: <80% of PAHs was removed in four washing cycles, thus outperforming Tween 80 and RAMEB (Gan *et al.*, 2017).

The first two groups – cosolvents and most of the industrial surfactants – may represent an environmental hazard. Residual flushing additives in the soil may be a concern and should be evaluated on a site-specific basis.

Cosolvent flushing involves injecting a solvent mixture (e.g., water plus a miscible organic solvent such as alcohol) into either unsaturated or saturated soil zones, or both, to extract organic contaminants. Cosolvent flushing can be applied to soils to dissolve both the source and the contaminant plume originating from it. The cosolvent mixture is injected into the injection wells, and the solvent with dissolved contaminants is extracted from the extraction wells and then treated *ex situ*. For example, ethanol (95%) in water was applied for the removal of PCE at a former dry cleaner site in Florida (EPA CLU-IN, 1998). When selecting the cosolvent its effect on microbiota should be considered.

2.1.4.1   Surfactant-enhanced soil flushing

Surfactants can mobilize the organic contaminants: their amphiphilic molecules form micelles and the contaminants with large K_{ow} are solubilized in the core of these micelles. Numerous studies indicated that surfactants enhance the removal rate of NAPLs (Mulligan *et al.*, 2001). The surfactants applied should be biodegradable, non-toxic, and soluble in water (the solubility can be enhanced by a cosolvent such as isopropanol). The anionic and non-ionic surfactants are less likely adsorbed by the soil than the cationic surfactants, which can adhere to soil and reduce effective soil porosity. They are also generally less toxic than cationic surfactants.

The non-toxic, biodegradable biosurfactants such as rhamnolipids, trehalose lipids, and sophorose lipids produced by bacteria and yeast were found to effectively enhance both solubility and biodegradation of contaminants such as PAHs (Noordman *et al.*, 1998; Lau *et al.*, 2014).

A synergy of surfactants and cosolvents (Tween 80 and ethanol) was observed in TCE extraction in lab-scale experiments (Table 6.1). It should be noted that the surfactant solutions showed foaming, thus causing technological problems in field applications. Foaming was particularly strong when in-well pressure pulse technology was modeled by periodically mixing the vessels to intensify the contact of the phases.

Although emulsification can enhance the mobility of the contaminants to a high extent, such mobilization holds a threat of downward or horizontal movement of contaminants and subsequent contamination of groundwater. This can be prevented by applying a hydraulic barrier or permeable reactive barrier (PRB). The other difficulty of surfactant-enhanced soil flushing lies in the high cost of the separation of surfactants from recovered flushing fluid.

Table 6.1 Solubilized TCE in groundwater amended by additives.

Solubilized TCE	Without additives	4% Tween 80	20% ethanol	4% Tween 80 + 20% ethanol
Intensive mixing (mg/L)	1000	55,000	1500	105,000
Without mixing (mg/L)	500	11,000	1000	20,000
Enhancement ratio (mixed)	1	55	1.5	105
Enhancement ratio (non-mixed)	1	22	2	40

2.1.4.2 Complex-forming agents for enhancement of soil flushing

Cyclodextrins (CDs) are typical complex-forming agents. These cyclic oligosaccharides consisting of 6, 7, or 8 glucose units form complexes with most of the organic contaminants by including them into their hydrophobic cavity. The complexed contaminants show enhanced solubility especially if CD derivatives with high solubilizing effect are used. Hydroxypropyl and randomly methylated beta-cyclodextrins (HPBCD and RAMEB) are such solubilizers produced industrially. They reduce the K_{ow} and the soil–water partition coefficient (K_D) of the contaminants (Fenyvesi *et al.*, 2009; Ko & Yoo, 2003).

Compared to the often toxic surfactants, the cyclodextrins, e.g. HPBCD are non-toxic. They can efficiently remove various organic contaminants (Berselli *et al.*, 2004; Zeng *et al.*, 2006; Guo *et al.*, 2010; Sánchez-Trujillo *et al.*, 2013; Wan *et al.*, 2010). In a contaminant mixture, the components compete for access to the CD cavities, and therefore the extraction efficiency is lower than that for the single components (Badr *et al.*, 2004). The efficiency also depends on the organic-matter content of the soil: the higher is the content, the slower the desorption rate. Humic acids can be complexed by CDs, and they also compete for the cavity (Ishiwata & Kamiya, 1999). While surfactants are also competitors (it makes no sense to combine surfactants with CDs), synergic effect of methyl BCD, and 30% ethanol as cosolvent was observed in extracting hexachlorobenzene from soil (Wan *et al.*, 2009).

Soil flushing with HPBCD solution ("sugar flushing") (Boving & Brusseau, 2000) is a potential technology for the removal of crude oil containing normal alkanes (nC_{15}–nC_{35}) and polyaromatic hydrocarbons (PAHs) from porous media (Gao *et al.*, 2012). The beneficial effect of CD extraction decreases with increasing *n*-alkane chain length. CD significantly enhanced PAH extraction from sand, and the enhancement effect increased in the order of parent compounds < C-1 substituted < C-2 substituted < C-3 substituted for most PAHs tested (Gao *et al.*, 2012).

Carboxymethyl BCD (CMBCD) was found to be effective for washing soils containing mixed (organic and inorganic) contaminants (Brusseau *et al.*, 1997; Chatain *et al.*, 2004; Skold *et al.*, 2009). The mechanism includes salt formation with metals and inclusion complex formation with organic pollutants. Especially good results were obtained when RAMEB and EDTA were applied together to a soil contaminated with metals and PCBs. By repeating the extraction three times, as much as 76% and almost 100% of the PCB and mobile metal content (Cd, Cu, Mn, and Pb), respectively, could be removed (Ehsan *et al.*, 2007). Good removal efficiencies (>80%) for PAHs, toxic metals, and fluorine were obtained by applying CMBCD and carboxymethyl chitosan (CMC) solution for washing a soil from an abandoned metallurgic plant (Ye *et al.*, 2014).

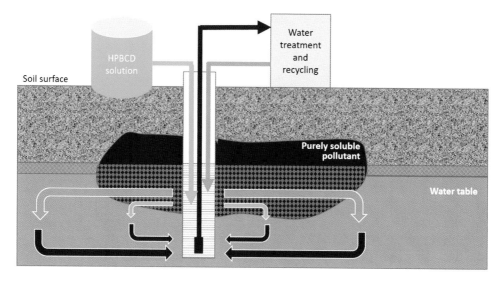

Figure 6.6 Scheme of the push-pull soil flushing technology using complex-forming cyclodextrin (HPBCD) (modified after Boving *et al.*, 2008).

The efficiency of the "sugar flushing" technology using aqueous HPBCD solution has been demonstrated in various field experiments on a waste dump site (McCray & Brusseau, 1998, 1999) and on military sites (Tick *et al.*, 2003; Blanford *et al.*, 2000) contaminated with PCE or TCE as DNAPLs. The push-pull flushing system (applying the same well periodically for injection and extraction) proved to be more efficient than the line-drive system (separate injection and extraction wells) (Figure 6.6) (Boving *et al.*, 2008). The HPBCD solution injected to the well was extracted after a reaction time and the contaminant was separated by stripping, distillation, or sorption on activated carbon. The regenerated HPBCD solution was pumped back for the next cycle of soil flushing.

Cost efficiency analysis showed that the total implementation costs for the CD-enhanced and surfactant-enhanced soil flushing technologies (CDEF and SEF) are comparable. The time needed to complete remediation can be reduced significantly by both CDEF and SEF technologies (Blanford *et al.*, 2006). CDEF also offers the ecological benefit of only introducing a non-toxic and degradable material into the subsurface. Regeneration of the CD solution makes the technology more economical (Boving *et al.*, 2007). Recent studies have revealed that the HPBCD remained in the subsurface after cessation of the active remediation was utilized as co-substrate in anaerobic degradation of the chlorinated hydrocarbons (Hinrichs, 2004; Blanford *et al.*, 2018). In this spontaneous process also nitrate and sulfate concentrations were reduced due to their role as terminal electron acceptors in the anaerobic biodegradation, while no significant change was observed in the concentration of dissolved oxygen and total iron at the site.

Soil flushing with CD solution can be combined with *in situ* bioremediation utilizing the benefits of the enhanced bioavailability of the CD-solubilized contaminants (Gruiz *et al.*, 2011; Leitgib *et al.*, 2007; Molnár *et al.*, 2005) (for more details, see Chapter 5 in this volume).

The soil effluents deriving either from soil washing or soil flushing can be decontaminated in various ways, but the effects of the complex-forming agents or surfactants on the selected technology should be considered:

- Stripping (the residence time in the stripper should be increased as the volatility of the contaminants is usually decreased in the presence of both CDs and surfactants) (Kashiyama & Boving, 2004);
- Sorption on activated carbon (the sorbent should be carefully selected to avoid the sorption of the additives without reducing the sorption of the organic pollutants) (Sniegowski *et al.*, 2014);
- Advanced oxidation processes (AOPs) (see Section 2.2.1 in this chapter);
- Liquid/liquid extraction by using vegetable oil (colza oil) (Petitgirard *et al.*, 2009);
- Electrochemical treatment (Gomez *et al.*, 2010);
- Biological treatments (the most economical technologies for regeneration of the CD solutions containing the soil contaminants using either the indigenous microflora or activated sludge) (Berselli *et al.*, 2004; Yoshii *et al.*, 2001; Gruiz *et al.*, 2008). CDs are non-toxic to the microbes, they catalyze the decomposition processes and are eventually biodegraded themselves (Fenyvesi *et al.*, 2005; Molnár *et al.*, 2005; Verstichel *et al.*, 2004).

In summary, the non-toxic, biodegradable cyclodextrins (CDs) can solubilize poorly soluble organic contaminants such as PAHs, PCBs, chlorinated solvents, etc. (Fenyvesi *et al.*, 2009). The biodegradation of PCBs, especially those with a lower degree of chlorination, can be enhanced by improving bioavailability using randomly methylated beta-cyclodextrin (RAMEB) (Fava *et al.*, 2002; Hu *et al.*, 2016). Similarly, the biodegradation of diesel oil, transformer oil, and mazout was enhanced by RAMEB due to enhanced bioavailability (Gruiz *et al.*, 2011; Molnár *et al.*, 2009). In the case of contaminants with extremely high affinity toward CD complexation, biodegradation can be hindered by the strong binding of the contaminant to the CD as it was observed for dichlorodiphenyltrichloroethane (DDT) and HPBCD (Gao *et al.*, 2015).

2.1.4.3 *Electrokinetic extraction*

The electrokinetic extraction process removes metals and polar organic contaminants from low-permeability soil, mud, sludge, and marine dredging by applying electric current. This *in situ* technology based on electrochemical and electrokinetic processes is primarily a separation and removal technique for extracting contaminants from soils with an electric charge.

Applying direct current to the soil increases its temperature, which helps to desorb the organic contaminants and volatize the VOCs. Solubilizing agents (surfactants, CDs, cosolvents) can enhance the electromigration of the contaminants and enhance the removal efficiency (Huang *et al.*, 2012; Ko *et al.*, 2000; Yuan *et al.*, 2006). Charged cyclodextrins such as CMBCD help the removal of both inorganic and organic pollutants (Jiradecha *et al.*, 2006).

A more detailed evaluation of this technology is in Chapter 9 of this volume.

2.2   Destructive physicochemical technologies – chemical degradation

During the degradation of the contaminants dissolved in water or sorbed to the solid phase of the soil, hazardous compounds are converted to non-hazardous or less toxic ones using reduction/oxidation (redox), acid-base reactions, hydrolysis, or other dissociation-based (e.g. dehalogenation) reactions. The chemical reactions are carried out during the *ex situ* treatment of the extracted air and groundwater or by contacting the solid phase (usually suspended in slurry) with the reagents. The *ex situ* treatment can be carried out *on site* using mobile, transportable units or at soil treatment centers by mixing the reagents into excavated soil and pumped groundwater, then separated and recycled.

In situ chemical treatment of soil and groundwater has become widespread through the development of solutions for reagent delivery and PRBs. For groundwater treatment the reagents can be injected by direct push injectors or via wells as well as filled into permeable reactive barrier (PRB) placed in such a way that the contaminated groundwater flows through. The solid phase can be treated either with gas-phase reagents such as ozone or reagents dissolved in water which permanently or periodically floods and flushes the contaminated soil volume. Only the two-phase soil can be treated efficiently with *in situ* oxidation and reduction technologies since water as reaction medium is required. Chemical reactions-based destruction in the three-phase soil has the drawback of a relatively small area of influence. The solutions of reagents are delivered to the contaminated zone through wells and trenches or by injection either by gravity or by using vacuum wells. Pressurized injection of liquids or gases, either through the screen of a well or the probe of a direct push rig, force the reagent into the soil or aquifer. The direct push rig offers a cost-effective way of delivering the reagent.

These technologies can be applied alone or in combination with other technologies, e.g. in the technology chain of soil extraction and air stripping of the extracted groundwater plus catalytic oxidation of VOCs in the effluent gases. Another example combines *in situ* chemical oxidation (ISCO) or reduction (ISCR) with subsequent aerobic/anaerobic biodegradation of the remaining contaminants. In the latter cases ISCO and ISCR aim to degrade the contaminants of low biodegradability to compounds more easily degraded by the soil microflora.

2.2.1   Technologies based on redox reactions

The most often used chemical reactions are redox reactions. The oxidizing agents most commonly applied are ozone, hydrogen peroxide, sodium persulfate, potassium permanganate, hypochlorites, chlorine, chlorine dioxide, ferrate [Fe(VI)], while the reducing agents are zero-valent metals, preferably iron (Fe^0 or Fe^{2+}), sodium bisulfite, sodium hydrosulfite, dithionite, hydrogen, etc. The technologies can be used for an extensive variety of organic pollutants, such as VOCs and SVOCs, fuel hydrocarbons, PAHs, micropollutants, and pesticides both *ex situ* and *in situ*. Reaction rates can be accelerated by use of a catalyst and by adding some form of energy (radiation, solar, heat, electrical, kinetic). The ***advanced oxidation processes*** (AOPs) use the oxidizing agents in the presence of the catalyst and/or UV light.

When planning the technology, the soil oxidant demand (SOD) arising from the consumption of oxidant due to organic carbon contained in soils should be considered (Brown, 2003). When chemical oxidation is designed the oxidant necessary for the remediation should be calculated based on the experimentally determined concentrations (Haselow *et al.*, 2003):

- Dissolved phase contaminant;
- Sorbed phase contaminant;
- Free phase contaminant;
- Dissolved phase reduced minerals;
- Solid phase (or sorbed phase) reduced minerals;
- Dissolved and sorbed phase natural organic matter (NOM).

Similar to SOD, the soil reductant demand (SRD) can also be determined but it is rarely used for soils contaminated with organics (McGrath *et al.*, 2007).

2.2.1.1 Technologies based on oxidation

Table 6.2 is a comparison of the most frequently used oxidants. These oxidants produce radicals (hydroxyl OH·, superoxide O_2·, perhydroxyl HO_2·) which are responsible for oxidation (Cheng *et al.*, 2016). Highly reactive hydroxyl radicals can be generated *in situ* also by sonication or hydraulic cavitation of water (Lehr, 2004). When a catalyst is used the technology is considered an advanced oxidation process (AOP).

Sodium/potassium permanganate ($NaMnO_4$ and $KMnO_4$) is an easy-to-handle and stable oxidant both in solid form and in solution. Figure 6.7 shows the scheme of *in situ* technology implementation using permanganate aqueous solution.

In a field demonstration, PCE and TCE were degraded *in situ* using potassium permanganate solution (Schnarr *et al.*, 1998). In another successful case, horizontal wells were used for injection of sodium permanganate solution (Siegrist *et al.*, 2001). Hrapovic *et al.* (2005) observed that permanganates may be inhibitory to *Dehalococcoides ethenogenes*,

Table 6.2 Comparison of the most frequently used oxidants.

	Sodium or potassium permanganate	Hydrogen peroxide	Sodium persulfate	Sodium percarbonate ($Na_2CO_3+H_2O_2$)	Ozone
Standard oxidation potential	low	medium	high	high	highest
Soil oxidant demand (SOD)	high	not high	low	not high	low
pH	wide range	3–5	wide range	wide range	wide range
Catalyst	no	ferrous iron	ferrous iron	ferrous iron	no
Precipitate	manganese oxide	ferric iron	ferric iron	ferric iron	no
Effect on soil biota	provisional decrease in activity			enhanced activity	enhanced activity (oxygen-rich environment)

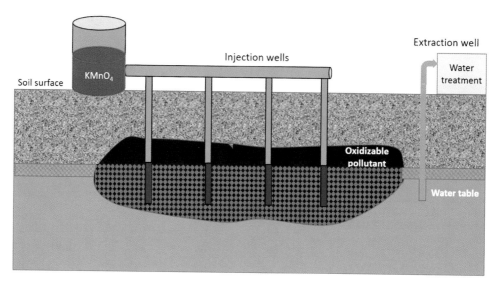

Figure 6.7 Scheme of ISCO with permanganate.

the microbial species that completely dechlorinates PCE and TCE. Permanganate oxidation induces a shift from a microbial consortium composed predominantly of aerobic and anaerobic heterotrophs, nitrate- and sulfate-reducing bacteria and methanogens, to primarily aerobic heterotrophs (Klens *et al.*, 2001).

Hydrogen peroxide is typically applied combined with an iron catalyst and is one of the most efficient oxidants (EPA, 1998). In many cases, however, there is sufficient iron or other transition metals in the subsurface with no additional ferrous sulfate. The hydroxyl radicals (OH·) formed can oxidize complex organic compounds. Residual hydrogen peroxide decomposes to water and oxygen in the subsurface. The process requires acidic pH and can be applied both *in situ* (Fekete-Kertész *et al.*, 2013) and *ex situ*. A case study on ISCO with hydrogen peroxide is described in Sections 5 and 6 of this chapter.

Light irradiation during oxidation by hydrogen peroxide can significantly increase oxidation efficiency (Ruppert *et al.*, 1993). Light is typically applied *ex situ* for extracted groundwater.

Hydrogen peroxide can be continuously generated in the soil on a suitable cathode with the addition of iron catalyst (Brillas *et al.*, 2009; Nidheesh & Gandhimathi, 2012). The enhanced rate of degradation is due to the continuous regeneration of Fe^{2+} on the cathode.

Hydrogen peroxide is toxic to microorganisms (it is a commonly used antiseptic agent in medical practice). When applied to soil a concentration-dependent reduction of both microbial numbers and diversity is observed (Waddell & Mayer, 2003).

Persulfate has to be activated (turned into sulfate radicals) by heat, ultraviolet light, high pH, hydrogen peroxide, or transition metals. For instance, enhanced temperature (50–70°C) and simultaneous application of nano zero-valent iron (nZVI) resulted in fast removal of PAHs from sediment in *ex situ* treatment (Chen *et al.*, 2014). At temperatures above 40°C, persulfate becomes especially reactive and can degrade most organics (Block *et al.*, 2004). Explosives and PCBs were successfully removed from the soil of a burning site by persulfate oxidation (Waisner *et al.*, 2008). Alkaline (NaOH)-activated persulfate oxidation effectively reduced the phenanthrene and TPH concentrations in a few days' time in an aged, contaminated soil originated from a railway maintenance facility (Lominchar *et al.*, 2018). Persulfate oxidation eliminated a part of PAHs and transferred another part of them into more bioavailable forms to be treated by bioremediation (Medina *et al.*, 2018). As with other oxidants, persulfate also has a negative effect on the soil microbiota, which can be restored by amendments (Satapanajaru *et al.*, 2017).

Sodium percarbonate is claimed to have a long-term effect (up to 30 days) and to oxidize even chlorinated ethanes and PCBs which are degraded by other oxidants only in a low efficiency (Regenesis, 2018).

Ozone can oxidize contaminants directly or through forming hydroxyl radicals. When applied together with hydrogen peroxide, enhanced transformation to hydroxyl radicals of high reactivity improves the efficiency of pollutant degradation compared to the efficiency of ozone alone (Peroxone, 2018). The o*zone sparging process* employs nano- to micro-sized bubbles of air and ozone mixtures pressured through soil and groundwater (Kerfoottech, 2018). Pulsing of air-ozone mixtures enhances the process scrubbing capacity and pushes the bubbles through the soil capillary pores (Figure 6.8). The technology can be applied to a wide variety of contaminants, such as petroleum hydrocarbons, chlorinated solvents, and other chlorinated compounds, BTEX, PAHs, PCBs, and pesticides.

The application of any of these strong oxidants destroys the contaminants in a relatively short period of time. However, rebound of dissolved contaminant concentrations can occur if the source is not treated. Real remediation can usually only be achieved by multiple chemical oxidant addition followed by enhanced or accelerated bioremediation. Chemical oxidation can be combined with bioremediation only under specific circumstances, e.g. at low contaminant concentration and if the microbiota is resistant to the oxidizing reagent.

To avoid the adverse effects of oxidants on intrinsic microbes, carefully designed technology should be applied, e.g. low oxidant concentrations are recommended by Sutton *et al.*

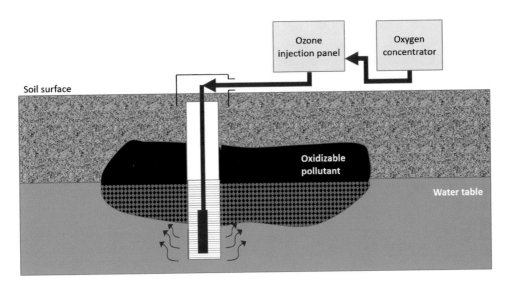

Figure 6.8 Scheme of ISCO with ozone.

(2011). Chen *et al.* (2015) found that hydrogen peroxide and permanganate were less harmful to the biota than persulfate.

The combined technology is less costly than the chemical technology alone and takes less time than bioremediation alone (Huang *et al.*, 2017). For instance, Hu *et al.* (2018) combined chemical oxidation using H_2O_2 + Fe^{2+} with fungal treatment (*Phanerochaete chrysosporium*) for the removal of bisphenol A from river sediment. Higher degradation efficiency was achieved compared to either of these technologies applied separately. The organic acids produced by the fungi were beneficial for the hydrogen peroxide oxidation ensuring acidic pH. On the other hand, the ferrous iron stimulated the growth of fungi.

Chemical oxidation can be coupled with the simultaneous or subsequent stimulation of bioremediation, e.g. by the addition of oxygen release compounds (ORC) to treat the low-concentration or remaining contaminants in the saturated soil. ORC is an oxygen-providing agent that enhances aerobic biodegradation in groundwater and in saturated soil. Usually it is phosphate-intercalated magnesium peroxide or calcium oxy-hydroxide that, when hydrated, provides a controlled release of molecular oxygen for periods of up to 12 months on a single application (Regenesis, 2018).

The contaminants (VOCs) in air streams resulting from remedial operations, such as SVE and air stripping, are degraded by catalytic oxidation. In the presence of a catalyst, typically metal oxide, oxidation is accelerated by adsorbing both the oxygen and the contaminant on the catalyst surface where they react and form CO_2, H_2O, and HCl at much lower temperatures than required by conventional thermal oxidation (see Chapter 1 in this volume).

UV light can oxidize organic contaminants in air and water, which can be enhanced by a catalyst such as TiO_2 or ozone oxidation. This combination is more effective than UV or ozone applied alone. Similarly, the efficiency of an H_2O_2/Fe^{2+} system can be enhanced by UV irradiation. Both ozone and H_2O_2/Fe^{2+} systems produce highly reactive hydroxyl radicals (OH·). The main advantage of this system is that sunlight is appropriate (eliminates need for expensive UV lamps). The disadvantage is that low pH is required to avoid the precipitation of iron. In some experiments the use of only endogenous iron also resulted in effective treatment: de S. e Silva *et al.* (2008) reported about approximately 76% removal of PAHs in sandy soil.

The UV technology is not widely used because of the following:

– UV decomposition does not usually reach complete mineralization of the organic contaminants.
– The technology is applicable for *ex situ* treatment of soil vapor and contaminated groundwater extracted by either soil washing or flushing (Lee *et al.*, 2016; Mishra and Chun, 2015; Schneider *et al.*, 2014), but it is not feasible for the contaminated soil, as the UV light cannot penetrate deep into the soil (thin layers should be treated). The technology is also not energy-efficient.

Figure 6.9 shows the scheme of a UV treatment of soil vapor containing VOCs (Hishimoto *et al.*, 2005). The volatile organic compounds such as TCE can be removed from the soil by heating (e.g. by mixing with calcium oxide) and simultaneous photocatalytic decomposition using a sheet made of a specially impregnated paper containing both activated carbon and TiO_2.

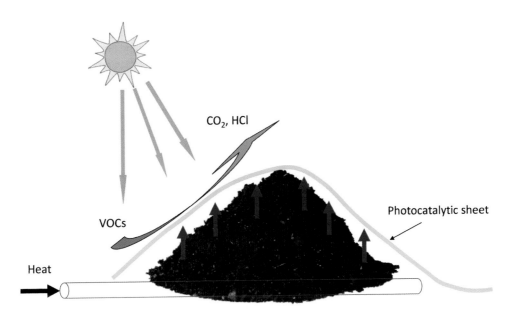

Figure 6.9 TCE is volatilized and removed from the soil by heating and subsequently undergoes UV decomposition using a photocatalytic sheet that contains TiO_2 catalyst (redrawn from Hishimoto *et al.*, 2005).

2.2.1.2 Technologies based on chemical reduction

The aim of chemical reduction is to degrade toxic organic compounds in the soil and groundwater to less toxic compounds. The most-often-used reducing agent is zero-valent iron (ZVI). It is usually applied as filler in the *permeable reactive barriers* (PRBs) for *in situ* reduction of halogenated organic compounds dissolved in groundwater (see Chapter 10 in this volume for a detailed description of the technology) (Figure 6.10).

PRB affects only the water proportion flowing through, and thus source treatment may be necessary in addition. ZVI is delivered into contaminant zones using pneumatic or hydraulic injection, e.g. direct push rigs. The emulsified and micronized forms are injectable. *In situ* reduction enables not only the soluble, but also the sorbed-phase and free-phase halogenated hydrocarbons to be reduced to target levels (US EPA, 2002).

Emulsified zero-valent iron (EZVI) is composed of surfactant, biodegradable vegetable oil, water, and ZVI particles (either nano- or micro-scale iron). EZVI forms emulsion particles that contain the ZVI in water surrounded by an oil/liquid membrane. EZVI combines directly with the target contaminants until the oil membrane is consumed by biological activity. In addition to the abiotic degradation associated with the ZVI, EZVI injection contributes to enhanced biodegradation of dissolved chlorinated hydrocarbons because the vegetable oil and surfactant act as electron donors to promote anaerobic biodegradation processes (ESTCP Project CU-0431, 2006).

Bimetallic nanoscale particle (BNP) technology consists of submicron particles of ZVI with a trace coating of palladium that acts as a catalyst. Due to the extremely small size of

Soil surface

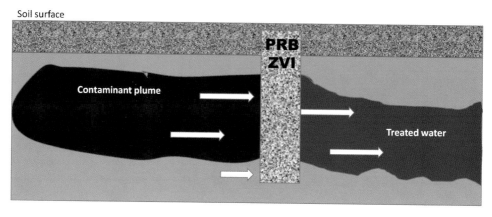

Figure 6.10 Treatment of contaminated groundwater with vertical PRB made of ZVI.

the particles (in the order of 10–100 nm) they can be transported by groundwater to establish *in situ* treatment zones, thus addressing not only dissolved contaminant plumes but also highly concentrated, dissolved contaminants within source areas. BNP can be used to treat contaminant areas that generally are inaccessible to conventional technologies, e.g., beneath buildings and in deep aquifers (US Navy, 2003). Both EZVI and BNP technologies can be further improved by applying organically modified silica that swells and captures small organic compounds in close proximity to the embedded ZVI (Clue in, 2018). Further details are discussed in Chapter 10 in this volume.

Sodium dithionite ($Na_2S_2O_4$) is used for chemical reduction of Fe(III) present in the soil or sediment to Fe(II), providing reducing conditions. The lower redox potential favors both abiotic and biotic reduction of contaminants such as explosives and chlorinated solvents, e.g. TCE (Szecsody *et al.*, 2004). For the treatment of soils not rich in iron, the combined injection of ferrous iron with sodium dithionate is beneficial.

Solvated electron treatment (SET) process is a special reduction technology which applies alkaline earth metals (e.g. sodium, calcium, lithium, and potassium) dissolved in ammonia to form metal ions and free electrons (SET, 2018). These free electrons produce a strong reducing agent that removes halogens (primarily chlorine) from organic molecules and reduces other contaminants. The process is used *ex situ*: the soil is excavated, dried, and treated in anhydrous ammonia slurry, then separated. The technology is not widely applied because of the high risk associated with the reactants. The main application is the remediation of PCB-contaminated soils.

Catalytic reductive dehalogenation (CRD) uses dissolved hydrogen as a reducing agent, in the presence of a palladium-on-alumina catalyst to chemically transform compounds such as TCE and PCE into environmentally benign ethene without the accumulation of intermediate transformation products such as vinyl chloride. Because of its rapid reaction rates (removal efficiencies for most of the chlorinated hydrocarbons are greater than 99% within minutes), a treatment unit system can be placed in a dual-screened well, allowing contaminated groundwater to flow through (Berg, 2000). The extra cost of the palladium catalyst is justified when groundwater has cocontaminants that are safer to leave in the ground, such as radionuclides. The system was developed for *in situ* application, in which case TCE was destroyed without pumping groundwater to the surface.

Hydrogen release compounds (HRC) can be used for the anaerobic degradation of residual low-concentration chlorinated compounds in groundwater. The injectable form can be delivered directly to the saturated soil. It can also be used as filler in PRBs (US EPA, 2009). It is a polylactate ester (typically with glycerol) which is decomposed by soil microbes resulting in controlled release of hydrogen for periods of up to 6–24 months on a single application. For more details, see Chapter 5 in this volume.

2.2.1.3   Enhancing redox technologies by surfactants and complex-forming agents

Surfactants and complex-forming agents can enhance the efficiency of the technologies based on redox reactions by ensuring increased concentrations (availability) of the poorly soluble contaminants. In addition to enhancing the solubility of the contaminants, CDs can solubilize also the catalyst [Fe(II)] and are non-toxic Fe-chelating agents compared to the toxic EDTA. CDs can accelerate the oxidation reactions due to the simultaneous complexation of both the contaminants (including them into the cavity) and the metal or metal oxide catalyst (forming outer sphere complexes). In this way the reactants get into close proximity to each other due to their interaction with CDs, resulting in two- to five-fold enhancement of oxidation compared to the CD-free treatments of soil slurry or groundwater.

It should be taken into account, however, that not only the contaminants but also both CDs and surfactants can be oxidized to a certain degree, and it is therefore necessary to find an optimum concentration which suffices for solubilization but does not consume too much of the oxidant (Chen et al., 2017b; Fenyvesi et al., 2011; Hanna et al., 2005, ABS Materials, 2017).

Some examples for the beneficial applications of CDs or surfactants are listed in Table 6.3.

In the literature both positive, negative, and even neutral impact of non-ionic surfactants such as Brij 35 and Tween 80 on TCE photodegradation was reported by Chu and Choy (2010, 2001a,b). The technology should therefore be carefully planned, because the catalytic effect is usually realized only at lower concentrations of both CDs and surfactants, but at high concentrations these additives can inhibit degradation, as was shown for the photodegradation of various contaminants such as phenanthrene, bisphenol Z, pentachlorophenol, pyrene, or TCE (Zhang et al., 2011; Wang et al., 2007; Hanna et al., 2004; Petitgirard et al., 2009; Fenyvesi et al., 2011). For instance, UV protection of TCE by surfactants results in increased degradation half-life (Table 6.4). Only the cosolvent of ethanol had a catalytic effect on TCE photodegradation.

The data in Table 6.4 shows that the solubilizing agents enhance the solubility of contaminants (washing solution containing both surfactant and cosolvent exhibited the highest initial TCE concentration), and this determines the efficiency of the whole soil washing and photodegradation technology: the higher the contaminant concentration in the washing solution/soil effluent, the higher amount of contaminant is removed.

Concerning the technologies based on reduction, Ayoub et al. (2008) claimed that no positive effect was achieved in the treatment of TCE by ZVI or by mobilizing agents such as surfactants, starch, and BCD. Two factors can hinder the process:

1   CD binding to the iron can decrease the effective surface.
2   TCE complexed by CD can react with iron at a lower rate than the uncomplexed one (protecting effect of complexation).

Table 6.3 Examples of the beneficiary application of cyclodextrins or surfactants on soil remediation technologies based on redox reactions.

Technology	Contaminant	CD/surfactant	Reference
Fe-catalyzed hydrogen peroxide	pentachlorophenol	HPBCD[a]	Hanna et al., 2005
	PAHs and PCBs	CMBCD[b]	Lindsey et al., 2003; Zheng and Tarr, 2004
	hexachlorobenzene	BCD[c]	Oonnittan et al., 2009
	trinitrotoluene (TNT)	CMBCD	Wei and Tarr, 2003; Matta et al., 2008
	TCE	RAMEB[d]	Fenyvesi et al., 2011
	bisphenol-A	CMBCD and BCD	Chen et al., 2017b
photo-, electrochemical generation of H_2O_2+ Fe	TNT	RAMEB	Yardin and Chiron, 2006; Murati et al., 2009
FE-activated persulfate	TCE and PCE	HPBCD	Liang et al., 2007; Liang and Lee, 2008
	PAHs, PCBs and polybrominated diphenyl ethers	HPBCD	Chen et al., 2017a
Ozone and persulfate and H_2O_2	PAHs	HPBCD	Enchem, 2010; Dettmer et al., 2017
TiO₂-catalyzed photodegradation	Phenanthrene	Triton X100[e]	Zhang et al., 2011
photodegradation	TNT	RAMEB	Yardin and Chiron, 2006
	Norflurazol	RAMEB	Villaverde et al., 2007
	Chlorobenzene	CD+fullerene	Ali and Sandya, 2014

[a]hydroxypropyl-beta-CD, [b]carboxymethyl beta-CD, [c]beta-CD, [d]random methylated beta-CD, [e]non-ionic surfactant, polyethylene glycol p-(1,1,3,3-tetramethylbutyl)-phenyl ether

Table 6.4 Results of UV irradiation of dissolved/emulsified TCE in groundwater amended by additives (Fenyvesi et al., 2011).

	Without additive	4% Tween 80	5% RAMEB	20% ethanol	4% Tween 80 + 20% ethanol
Initial TCE (mg/L)	1300	50,000	3100	1000	100,000
$t_{1/2}$ (min)	60	300	102	17	250
Removed TCE compared to the initial in 15 min (%)	28	17	21	55	9
Removed TCE in 15 min (mg/L)	400	8200	700	450	9000

Enhanced contaminant solubility can counterbalance these negative effects by using a properly selected CD derivative such as carboxymethyl BCD. CMBCD has a high degree of substitution with low iron binding affinity and a medium protecting effect, but still being capable of solubilizing TCE (Shirin *et al.*, 2004).

2.2.2 Chemical dehalogenation

Dehalogenation is the process of removing covalently bound halogen from the contaminants. The main targets of the technologies are PCBs, polychlorinated-dibenzo-p-dioxins (PCDDs), polychlorinated dibenzofurans (PCFs), polychlorinated terphenyls (PCTs), and some halogenated pesticides. High organic content and water content in the soil decreases the efficiency. Two *ex situ* technologies are used; both apply base catalysis and enhanced temperature.

Base-catalyzed decomposition is a simple technology using sodium hydroxide or sodium bicarbonate mixed with contaminated soil (FRTR, 2018a). The mixture is heated to above 330°C in a reactor to partially decompose and volatilize the contaminants.

The *base-catalyzed glycolite dehalogenation* technology is used also only *ex situ* and applies alkaline polyethylene glycol or tetraethylene glycol reagents to convert aromatic halogenated contaminants to non-toxic glycol ethers by replacing chlorine atoms in a nucleophilic substitution reaction at 150°C (FRTR, 2018a).

Both treatments leave the soil alkaline. It should be neutralized and excess glycol reagents washed out. Both the high temperature and the alkaline pH are detrimental for the soil microbiota.

2.2.3 Mechanochemical remediation (ball milling)

Mechanochemistry is the chemical and physicochemical transformation of substances during aggregation caused by mechanical energy. Mechanochemical remediation is a reactor-based technology used for soils contaminated with both organic and inorganic contaminants. Inorganic contaminants are simply immobilized (for details, see Chapter 8 in this volume), while in the case of organic contaminants the kinetic energy induced by balls breaks the bonds of contaminants through exothermic reactions (mechanochemical destruction technology) (Figure 6.11). The *dehalogenation by mechanochemical reaction* (Birke *et al.*, 2004) has several advantages over chemical dehalogenation technologies:

– Works at ambient conditions;
– Short reaction times;
– Both soil and slurry can be treated;
– No pretreatment is needed;
– Concentration of the halogenated contaminant can range from ppb to pure contaminant;
– Low energy, equipment, personal, and reagent costs.

The general scheme of reductive dehalogenation is shown by the following equation:

$$R\text{-}Cl + Me + H \rightarrow R\text{-}H + MeCl_2$$

Alcohols, polyethers, amines, etc. can be used as hydrogen (H) donor. For PCBs magnesium is required as metal (Me) (Birke, 2018).

Small mobile plants have been developed (Tribochem, 2018). Ball milling has been successfully used for the degradation of hexachlorobenzene (Mulas *et al.*, 1997),

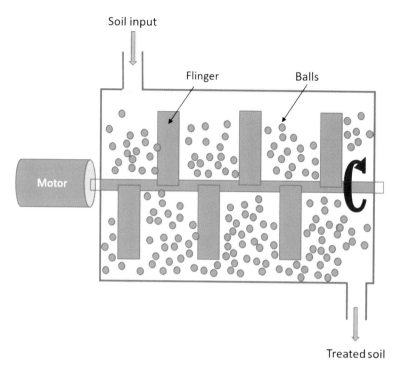

Figure 6.11 Scheme of ball mill (Anwar, 2011).

hexabromobenzene (Zhang *et al.*, 2002), PCBs (Birke *et al.*, 2004), other organohalogenated compounds (Monagheddu *et al.*, 1999), and sulfonated compounds (Caschili *et al.*, 2006). The mechanochemical treatment of dry soil and metal oxide catalysts such as ferrihydrite achieved dechlorination of both 4-chloroaniline and PCP to a greater extent than in the batch experiment with soil slurry (Pizzigallo *et al.*, 2004). Mechanochemical treatment (ball milling) of DDT using calcium oxide results in a graphitic product through a series of reactions and various intermediates (Hall *et al.*, 1996).

High water content decreases the efficiency; water content below 20% yields acceptable destruction rates for organic contaminants such as naphthalene and diesel oil (Anwar, 2011). Soils rich in organic material need extended milling time. The treatment is detrimental for the soil microbiota because of the heat but it can be revitalized by inoculation. Also the soil structure is destroyed by first decreasing and later increasing the particle size (aggregation), thus changing also porosity and water retention. Mixing the treated soil with unmilled soil makes it suitable for growing plants.

3   TECHNOLOGIES BASED ON IMMOBILIZATION

Irreversible immobilization aims to prevent or reduce the migration of contaminants in soil, water, and the atmosphere. The risk from the contaminants can be diminished by reducing or completely preventing their transport on their own or together with air, water and solid phases. Immobilization of the contaminant is generally associated with the immobile, non-volatile, and non-soluble chemical form, typically a stable precipitate. Another concept to reduce the transport of contaminants in soil is by strong bonding to a highly stable solid phase. If the soil solid itself is dispersive or tends to erode, it is worth stabilizing the soil and the contaminants together.

Various strategies can be applied:

− Disperse stabilization by physicochemical processes;
− Mass stabilization (solidification by block formation);
− Thermal stabilization.

The main physicochemical processes useful for immobilization of organic contaminants in soil, as given in Chapter 1, are sorption, chemical condensation, polymerization, oxidation, and thermal stabilization. Strong sorption may operate in the soil in a dispersed form keeping the contaminant in place, or in a separate volume, when contaminated water is treated, e.g. by a permeable barrier or a filter, in a filled column, or in a batch reactor. Compared to sorption the organic contaminants can be transformed to more stable, non-soluble, immobile chemical forms by condensation or polymerization reactions. The products are chemically immobilized, stable molecules dispersed in the contaminated solid matrix or precipitated from liquid media. Stabilization of contaminants may also be combined by partial or full chemical destruction. The stabilization can be performed both *in situ* and *ex situ* and the stabilized matrix can be compact (block of concrete or ceramic) or of granular nature dispersed in the soil. The *disperse stabilization* may not be harmful for the microbiota of the soil, while the *solidification and block formation* are usually performed under drastic conditions, e.g. on high temperature and physicochemical conditions detrimental for the microorganisms.

The risk of the inappropriate texture of soils, e.g. too loose mechanical composition, consequently too low resistance against erosion, landslide, soil desiccation, and nutrient leaching, can also be improved by stabilization thus extending the technological palette from deep soil mixing with heavy machinery to agricultural and ecological soil stabilizing technologies. *Soil stabilization* technology is most widely used in geotechnical applications, i.e. rendering of backfill, highway capping, slope stabilization and foundation improvement, and waste and contaminant containment. These applications are typically connected to the remediation of polluted sites and the treatment of contaminated soil and sludge when geotechnical and environmental goals should be combined, e.g. brownfields rehabilitation and utilization of abandoned sites, when large amounts of polluted material will remain in place in parallel to long-term risk reduction.

Mass stabilization (*stabilization/solidification*, S/S) is a method to improve soft or contaminated soils by adding binders. Mass stabilization includes the stabilization of the contaminants and solidification of the soft soil (clay, sludge, peat, silt, or sediments) allowing it to be used on the area for construction works in wet areas and areas with soft soils at reasonable costs without expensive piling or huge mass replacements. These processes involve encapsulation of the contaminant by forming a solid phase material from the contaminated medium. Many of the solidifying additives such as cement, gypsum, bitumen, or polymers simultaneously immobilize the contaminant. Contaminant migration from the solid phase blocks is restricted both by the immobile form and the largely reduced surface area potentially exposed to leaching.

S/S technologies neither remove nor destroy the contaminants present, but hinder their migration and transport, thus prevent their contact with the biota and humans, so they cannot exert their adverse effect.

Coating/encapsulation of the contaminated solid medium, e.g. soil, solid waste or sludge with low-permeability materials (a clay layer or polymer-based liners) results in complete isolation of the hazardous contaminant from the environment (see *containment* later). The technology can be accomplished by mechanical processes (physical stabilization by block formation) or by a chemical reaction between a contaminant and binding reagents such as cement kiln dust (by-product of cement production), lime, or fly ash as well as polymers in order to reduce the solubility and leachability of contaminants (chemical stabilization). Environmental conditions such as pH and redox potential may have less influence on organic contaminants compared to metals.

Table 6.5 gives an overview of the technological measures to achieve immobilization, stabilization and solidification of contaminants and contaminated soil. The technological intervention should be selected based on the nature and hazards of the contaminant, the characteristics (the geochemical composition) of the soil, as well as the sensitivity of the surrounding environment.

The remobilization risk of the immobilized contaminant upon changed conditions and their transport to surface- and groundwater should always be considered, given that the hazardous contaminant is still present in the environment, but showing no effects, and posing low risk. However this situation represents a chemical time bomb (Stigliani, 1991; Gruiz, 2000), that needs continuous monitoring and safety management.

Here only "cold" stabilization is discussed, thermostabilization (e.g. vitrification) is included in thermal remediation technologies in Chapter 1 of this volume.

Biological stabilization technologies are summarized in Chapter 5 of this volume.

3.1   Immobilization of the contaminants by sorption

Disperse stabilization of organic contaminants in soil can be achieved by strong sorbents which cause a shift in the contaminants' partition toward the solid phase resulting in low concentrations in the aqueous and gaseous phases. Therefore sorption of the contaminants can be used for their removal from soil air and soil water, and for decreasing biological availability. The risk in the mobile soil phases can be reduced by treatment in separate reactors after soil gas or water extraction using subsurface reactive barriers (with passive or active groundwater flow), or the addition of the sorbent to the soil where the sorption capacity of the solid phase will be increased and the partition of contaminants shifted from the water to the solid phase.

Activated carbon (AC) is usually used to adsorb VOCs from contaminated air (Russel *et al.*, 1992). The effluent gases in air stripping can be cleaned in this way. AC and synthetic resins are also utilized for the adsorption of dissolved organic contaminants from the extracted groundwater, soil flushing solutions and waste water. The two most common reactor configurations for carbon adsorption systems are the fixed bed and the pulsed or moving bed reactors (Cheremisinoff & Davletshin, 2015). The fixed-bed configuration is the most widely used type for adsorption from liquids. Depending on the type, concentration, and volume, AC should be regenerated or properly treated and/or disposed of in the case of explosives or metal contaminants. For remediating sediments, AC is applied either by direct mixing, as a component of a geotextile mat, or by being incorporated into a cover that forms a physical and adsorptive barrier between the sediment and water (Koehlert, 2018). Nowadays *biochar* (produced by pyrolysis of biomass) has been increasingly used as a sorbent for soil remediation (Beesley *et al.*, 2011; Zhang *et al.*, 2013). Compared to activated

Table 6.5 Summary table of technological measures to achieve immobilization, stabilization, and solidification of contaminants and contaminated soils.

	Contaminant immobilization	Soil stabilization	Immobilization & stabilization of contaminant & soil together	Mass stabilization, solidification/block formation
Soft physicochemical treatment, even in combination with biological processes	Addition of contaminant-specific sorbent or sorption enhancing agent	Erosion prevention by mechanical staples, stakes, nets, etc.	Sorbents + erosion prevention	Not relevant
	Creation of more inclusive contaminant forms by agents modifying pH and redox potential, oxidizing, reducing, forming large molecules etc.	Enhancement of humus and clay mineral formation by additives to achieve more stable soil texture	Contaminant incorporation into the dispersedly formed humus and clay minerals	Not relevant
	Chemical reagent addition	Erosion prevention by stabilizing additives (clay, humus, cement, fly ash, etc.)	Coincidental contaminant immobilization and soil microstructure stabilization	Large amount of solidifying additives, e.g. cement in slurry form to produce solid blocks from soil
	Biofiltration and/or accumulation	Soil erosion prevention by plant roots, and by special plants	Biologically facilitated immobilization combined with soil stabilization	Not relevant
Treatment with chemical reagents	Oxidizing or reducing agents for precipitation, condensation, polymerization	pH and redox potential setting agents, which let the contaminant immobilize in addition to soil texture modification	Contaminant immobilizing and soil texture stabilizing agents together	Stabilization of dispersive or contaminated soils by cement-like and polymer-based additives

(Continued)

Table 6.5 (Continued)

	Contaminant immobilization	Soil stabilization	Immobilization & stabilization of contaminant & soil together	Mass stabilization, solidification/block formation
Thermal treatment	Enhanced condensation, polymerization of organic contaminants due to mild heating	Not relevant	Condensation- and polymerization-based incorporation of organic contaminants into humus	
	High temperature enhanced condensation and polymerization	High temperature vitrification of the soil to stabilize soil texture	High temperature vitrification of the contaminated soil to stabilize soil texture	High temperature vitrification of the mineral soil to form glass like blocks
Application of heat & reagents together	Enhanced immobilization of the contaminant of larger size or less soluble chemical form	Not relevant?	Heat enhanced incorporation of the modified contaminant into minerals or humus or the organomineral matrix of the soil	Heat and reagent enhanced block formation
Isolation and containment	Isolation of contaminant lens or buried, solid pollution within the soil	Not relevant	In situ isolation of the contaminated soil by geotextile, impermeable walls or injected stabilizing additives (lime or cement)	In situ isolation by block formation with solidifying agents (injection or mixing in slurry form). The created block remains in the original place of the contaminated soil volume

carbon, biochar has the advantage of high cation exchange capacity (CEC) due to residual carboxylic acid functionalities ensuring improved sorption of toxic metals. The sorption capacity, mechanical hardness, and bulk density of biochar are lower than those of AC, but the improved water retention of the biochar-amended soils is accompanied with reduced leaching of the nutrients which is an advantage in the technologies combined with phyto- or bioremediation (García-Delgado *et al.*, 2015; Koltowski *et al.*, 2016). See some examples in Section 4.3.2 of this chapter.

Various *clay minerals* can be used for the immobilization by sorption of both organic and inorganic contaminants from contaminated groundwater. For example, modified (organophilic, hydrophobic) clays show enhanced adsorption of the organic pollutants such as 2,4-dichlorophenol (Pernyeszi *et al.*, 2006). Bandosz *et al.* (1992) described the "molecular engineering" of clay for specific organic substances. Modification of clay minerals by exchange, intercalation, calcination, and imbibition of organic groups, followed by their polymerization and carbonization, changes the surface properties of clays significantly. Carbonization of organic material in the interlayer space of the clay minerals leads to a microporous activated carbon that exhibits unique properties as a sorbent. Bowman studied the use of modified zeolites as a filling in permeable reactive barriers (PRBs) to sorb tetrachloroethylene (PCE) and also as sorbent for the removal of petroleum hydrocarbons such as BTEX from oilfield wastewater (Bowman, 2002).

Nanomaterials are being increasingly developed and applied for environmental remediation as sorbents or catalysts. Carbon nanotubes, dendritic polymers, functionalized ceramics, tuneable biopolymers are all promising sorbents for water, mainly for surface water purification and disinfection. Nanoparticles may be used as sorbents or as reactive agents (photocatalysts or redox agents) for organic contaminants as well as for nanofiltration in membranes. In spite of many open questions – such as the fate of nanoparticles in the environment and the long-term hazard, e.g. human health effects and environmental toxicity – the use of nanomaterials have started and more than 70 full-scale applications have been described up to 2017 all over the world (Araújo *et al.*, 2015; Clue In, 2018, Diallo, 2005, Lofrano *et al.*, 2017; Theron *et al.*, 2008; Watlington, 2005).

Carbon nanotubes have been recognized for their ability to adsorb highly toxic organic contaminants, such as dioxins, much greater than traditional activated carbon (Long & Yang, 2001). SAMMS™ materials consist of a nanoporous ceramic substrate coated with a monolayer of functional groups tailored to preferentially bind to the target contaminant. The functional molecules bind covalently to the silica surface, and other functional groups serve as adsorbents to scavenge specific contaminants from waste streams. It has good chemical and thermal stability and can be readily reused or restored (Fryxell *et al.*, 2007).

3.2 Chemical stabilization of contaminants

Chemical stabilization aims at transforming the contaminants into a less mobile form by using chemical reactions such as changing pH (lime, $CaCO_3$), oxidation (ozone, hydrogen peroxide), hydrolysis, condensation, and polymerization. The irreversibly immobilized chemical forms ensure low risk even if they are dispersed.

DCR (dispersion by chemical reaction) is a group of patented waste treatment technologies using CaO (quicklime/burnt lime) for the immobilization of heavily oiled sludge, oil-contaminated soils, acid tars, and toxic metals in $Ca(OH)_2$ and $CaCO_3$ matrices (Marion *et al.*, 1997).

Lime treatment (alkaline hydrolysis) has achieved contaminant removal as high as 80–98% in soils contaminated with petroleum hydrocarbons and on a burning site polluted with explosives, PCBs, TCE, and heavy metals (Ko *et al.*, 2010; Schifano & Thurston, 2007; Waisner *et al.*, 2008). Liming reduced the concentration of the explosive RDX by alkaline hydrolysis at a military site (Miller & Foran, 2012). Liming can also increase soil pH, cation exchange capacity (CEC) and be beneficial for the ecosystem. As an example, the diversity and abundance of forest tree species was increased after 25 years of dolomite lime application in a mining region in Canada (Nkongolo *et al.*, 2016).

3.2.1   Simultaneous contaminant immobilization and soil microstructure stabilization by using binders

Contaminant immobilization and the stabilization of the contaminated soil should go hand in hand to efficiently hinder further contaminant spread. The contaminant should be kept in the solid phase and pollution of atmosphere and surface waters should be prevented. The prevention is more efficient when the contaminant itself is in an immobile form, i.e. not being able to evaporate, dissolve, or otherwise become solubilized in water. Strong bonding of the contaminant to the soil matrix or any of its components requires the contaminants to become integral, and the soil texture and structure stable at the same time.

Before field application, bench-scale testing is necessary to evaluate how effectively a potential binder system is at reducing the leachability of the contaminants. The nature of contaminants may vary across a contaminated site, which means that more than one binder formulation may be required during remediation of an area (Bone *et al.*, 2005).

The binders can be thermoplastic polymers (asphalt bitumen, paraffin, polyethylene), thermosetting polymers (polyvinyl esters, urea formaldehyde, epoxy polymers), and other proprietary additives. The thermoplastic polymers operate at 120–180°C therefore they are used for soils of low moisture content, while the thermosetting polymers work at ambient temperature but are not as efficient in dry soils. The latter require the combination of several liquid ingredients such as monomers, catalysts, initiators, etc.

Geopolymers based on aluminosilicate have a similar stabilizing effect as cement. They can be prepared from waste products such as fly ash, using an alkaline activator. The aluminosilicate swells in the soil moisture and forms a gel which coats soil particles (Du *et al.*, 2016; Alhomair *et al.*, 2017).

Biopolymers, most often polysaccharides such as cellulose, starch, chitosan, etc. are also used. They form hydrogel when wetted and have a similar strengthening effect on soil to that of cement (Chang *et al.*, 2016).

3.3   Soil stabilization

Stability of the soil structure may show a wide range from non-stable through stable enough to serve the (agro)ecosystem and the healthy microbial life ensuring element cycling to an extremely stable, solidified state, which is not suitable for life but fulfills a zero or close to zero emission. Soft solutions can stabilize degraded soils which are exposed to water and wind erosion or have a texture unable to keep water and nutrients for plant life. Soil texture of degraded soils, such as sandy acid soil with low humus content, can be improved by the addition of synthetic or biopolymers, thus most of the agricultural by-products and wastes (shells, stalks, stems, husks, cobs, waste wood, etc.). These are eco-friendly additives having the ability to amend soil texture and structure and beneficially impact the biogeochemical carbon cycle.

Soft solutions preventing soil loss and soil movement keep the habitat functions of the soil and can provide ecosystem services. In *disperse stabilization* the textural and structural changes take place in the microstructure of the soil and are homogeneously distributed. The forming aggregates ensure a loose but still stable structure. Soil stabilization uses the whole range from humus and clay mineral formation and loose aggregation to heat-enhanced melting and block formation from the clayey aggregates.

In situ stabilization of the soil in itself or together with the contaminant involves both the disperse stabilization of the three-phase soil including natural surfaces and agricultural soils (see also Chapter 8) and treatment of the upper layers to immobilize the contaminants by chemicals such as sorbents, oxidizing agents, or block formation by the addition of binding agents (see mass stabilization later).

Fly ash and lime are used for disperse stabilization of soil to enhance strength properties (compressive and shearing strength), to reduce soil moisture content, improve compaction, and control shrink swell properties. They can be used separately or combined. Fly ash is also a sorbent for toxic metals and organic contaminants.

Soil loss due to wind and water erosion is one of the most serious soil degradation problems, and many traditional and creative new ideas have been established to prevent soil migration. Most of the *erosion prevention* methods can be applied for stabilization of contaminated soil, too.

There is a plethora of *mechanical soil immobilization* methods: retaining walls and terraces, geotextile application, other cover by net from plant material (pl. coco or palm), spikes, vetiver plant (plant with extremely long roots to keep firm the upper layer of the soil), etc. Some products marketed for temporary and permanent soil erosion protection are listed here:

– Aspen Excelsior Logs, biodegradable (Western Excelsior, 2018);
– Erosion control blankets (ECB) made of straw and/or coconut, biodegradable (ECB Verdyol, 2018);
– Turf reinforcement mats (TRMs), synthetic, non-biodegradable, stabilized against UV degradation, provide long-term erosion resistance (Contech, 2018);
– High performance turf reinforcement mats (HPTRMs) for slope control (Propex, 2018a);
– ArmorMax: HPTRMs and engineered earth anchors (Propex, 2018b);
– Terra-Tubes fiber filtration tubes (FFTs) engineered composites of wood fibers, man-made fibers and performance-enhancing polymers (Profiles, 2018a);
– Geocell cellular confinement system, an engineered, high-strength network of interconnected cells for stabilizing soil against erosion (Presto, 2018);
– Gabion baskets, Reno mattresses, and Terramesh, hexagonal woven galvanized steel wire mesh compacted baskets filled with stone or soil (Geo-Solutions, 2018a);
– Geotube® Dewatering Technology (Geo-Solutions, 2018b);
– HydroTurf™ (AOF, 2018a);
– Native Grass Sod (GeoSolutions, 2018c);
– Flexterra Flexible Growth Medium (FGM) (AOF, 2018b);
– ProMatrix™ Engineered Fiber Matrix (EFM) (Profiles, 2018b);
– Sod Staples for proper anchoring the erosion products (blankets, etc.).

3.3.1 Combination of contaminant sorption and stabilization/ solidification of the soil

The efficiency of stabilization treatment of organic contaminants can be improved by using sorbents either incorporated as additives in the solidifying mix or used as a pretreatment prior to conventional mass stabilization. Many of these additives are waste products from industrial processes. Additives such as activated carbon, shredded tire particles, and organoclays can increase the chemical stabilization of the contaminants. Silica fume and fly ash can improve the physical stabilization of organic compounds by reducing the porosity and permeability of the soil.

Cement formulations containing activated carbon were used for stabilization of organic contaminants – such as creosote, dioxins, and pentachlorophenol (PCP) – at the American Creosote site in Jackson, TN (Bates *et al.*, 2000). The results show reduced leaching of organics. Organophilic clays can act as successful sorbents for organic contaminants and enable them to be treated by cement-based stabilization (Sora *et al.*, 2002). Organoclay is usually a smectite clay (e.g. bentonite, hectorite) that has been modified to be hydrophobic by exchanging the naturally occurring cations (e.g. Na^+, K^+) with organic cations (Reible, 2005). In a clean-up conducted at West Drayton, UK, soil contaminated with high levels of hydrocarbons was treated successfully on a commercial scale using cement containing organophilic clay additives (Al-Tabbaa & Perera, 2003).

Immobilization of organic compounds in a cement matrix, with or without sorbent, is mainly a result of physical entrapment. Degradative stabilization combines the immobilization and degradation of contaminants transforming them to less hazardous compounds and decreasing their mobility simultaneously. Cement slurries containing Fe(II) have been tested on PCE, effectively reducing the chlorinated compound to non-chlorinated by-products (Hwang & Batchelor, 2000).

Clay or nanoclay with synthetic or natural polymers represents special combinations of sorbents and stabilizers. The slowly biodegradable stabilizing nanocomposite has good sorption, water and nutrients retention capacity from the biopolymer and improved (compared to the biopolymer in itself) mechanical properties originating from the clay component. These nanocomposites fulfill carbon sequestration and other eco-requirements such as reducing greenhouse gas emission and sustaining healthy geobiochemical cycles. The combined sorption and stabilization is active during the length of (bio)polymers life span.

Nanoclay can be combined with several organic materials such as rubber, natural fibers, resins, wood-polymers, synthetic polymers, and nanocarbon products. Many of these composites can be applied for environmental technologies including stabilization of degraded and contaminated soil (Syakir *et al.*, 2016).

3.3.2 Solidification/stabilization by producing blocks

For the purpose of *block formation*, the mixture of soil, water, and stabilizing agent is repeatedly mixed in place. *In situ* stabilization is used for wet sites or contaminated and/or degraded sediment disposal as well as for foundations. Specialized cementing agents (Portland cement, lime, or lime-fly ash mixture), and clays (attapulgite, bentonite, modified organophyllic clays), as well as contaminant-specific chemical reagents are used for soil remediation. Mixing itself may have the goal of physicochemical stabilization but also the enhancement of chemical reactions (typically for decontamination, e.g. by *in situ* chemical oxidation) in slurries and soils.

Four types of geotechnical applications are used for *in situ* soil stabilization:

- Injection through push-pull tools: fluid materials (suspension or solution) are injected into the subsurface soil or rock (grouting), to (i) decrease permeability, (ii) increase shear strength, and (iii) decrease compressibility.
- Shallow soil mixing works with a crane-mounted mixing system, using a mixing head, e.g. a bucket mixer or a rotary blender. The mixers are generally 1–4 m in diameter and driven by a high torque turntable. The up and down moving mixing head can be enclosed in an open-bottomed cylinder or used without a cylinder. After completion of one mixing unit, the cylinder is removed and placed adjacent overlapping with the previous one. Shallow mixing is utilized to lose soils and sludges to depths up to 9–10 m. It mechanically mixes soil with solidifying agents (lime, iron, clay minerals, compost, etc.) in powder form or injected as a slurry (Raito, 2018; Jasperse, 2018). Also, large diameter shallow soil mixing can be applied, e.g. by Geo-Solutions (2018a), for the purpose of hazardous waste remediation.
- Deep soil mixing equipment is a crane-supported special auger, a set of leads which guides a series of one to four hydraulically driven, overlapping mixing paddles and auger flights (see Figure 8.3 in Chapter 8). The auger flights are 0.6–0.9 m in diameter. After penetration, the slurry is injected into the soil through the tip of the hollow-stemmed augers. The auger flights loosen the soil and lift it to the mixing paddles which blend the slurry and soil. As much as 60–80% of the slurry is injected during downward penetration and the remainder during the withdrawal so that the mixing process is repeated on the way out (Jasperse, 2018). Nowadays large flighted rotary augers capable of injecting slurry chemicals and water through the auger flights are used for *in situ* stabilization. The auger bores and mixes a large-diameter "plug" of the contaminated soil (USACE, 2003; Geo-Solutions, 2018a).

  Application of deep mixing in addition to environmental remediation is typically used for building a hydraulic barrier or retaining wall systems, support and strengthen foundations of constructions, ensuring a support system for excavation as well as damage mitigation for landslide, erosion, and seismic deterioration. These applications are good examples of geotechnical and environmental technologies being inseparable and mutually beneficial (Porbaha *et al.*, 2005).
- Backhoe stabilization can be applied for surface soil treatment. In this case a backhoe-operated rotary drum soil mixing tool is used for geotechnical soil improvement (Geo-Solution, 2018b).

3.4   Mass stabilization

The mass stabilization method can be applied for contaminated dredged sediments, contaminated soils and waste sludges at places where river, channel, or port dredging is going on and large amounts of contaminated sediments need to be treated to reduce toxicity and mobility. Dredged sediment is normally stored in contained disposal facilities where the mass stabilization technology can be applied and the treated solid material reused. In many cases the sediment, sludge, or contaminated soil is part of a large area to restore. There are three options depending on spatial planning and the risk posed by the contaminants (Ringeling, 1998):

- The contaminated medium remains in place after stabilization/solidification and becomes part of the subsurface environment/construction isolated from surface activities. This

solution is acceptable when the risk according to the future land use can be reduced to an acceptable level.
- Stabilization off site after removal, transportation, and treatment in a soil treatment plant (if available) and reuse it as construction material. Demand for reuse is a key factor to consider when planning for *ex situ* stabilization (Makusa, 2012; PIANC, 2009). For instance, the dredged material is fractionated and the fine clay fraction is utilized for ceramics production. The method is useful for incorporating toxic metals in a solid form which are resistant to chemical degradation and leaching (Harrington & Smith, 2013).
- Stabilization off site and disposing at a disposal site.

The selection of the best possible solution for handling dredged material is primarily not a technological, but rather a managerial issue, as the decision is based on the locally available environmental, economic and engineering options, their costs and benefits. It is worth studying the experience of the large dredging projects such as the POSW II (1992–1996) or the PIANC (2009) projects, which evaluate dredged sediment treatment and reuse options and constraints:

- Geotechnical uses: construction materials, isolation, flood and coastal protection, land improvement, placement on waterways banks (Liu *et al.*, 2011).
- Environmental utilizations: habitat creation and enhancement, aquaculture, agriculture, recreation, water quality improvement, sustainable relocation, filling of deep borrow pits.
- Constraints: often higher costs of treatment and use compared with disposal; difficulty of finding suitable locations to use dredged materials; lack of markets for products as secondary raw materials; lacking standards for the products; dredged material is often classified as hazardous waste; negative public perception, low acceptance.

Many opportunities exist for beneficial reuse of dredged material such as expanded clay grains, isolation layers, and protection layers on landfills, brownfields and remediation sites, road construction and sintering (Foged *et al.*, 2018), bricks, and ceramics (Alvarez-Guerra *et al.*, 2008; Hamer & Karius, 2002; Hill & Haber, 2012; Rodriguez & Brebbia, 2015; Sly & Hart, 1987; Tangprasert *et al.*, 2015).

Although a variety of technologies have been developed for *ex situ* stabilization in addition to ceramic processing, cement-based processes are still frequently used both *ex situ* and *in situ*, and many different additives (lime, fly ash, slag, clay, and gypsum) and proprietary blends are applied to improve the performance and reduce cost (Paria & Yuet, 2006).

3.5   Containment

The aim of containment is to prevent or reduce the migration of contaminants in soil and water (surface and subsurface waters) and the atmosphere. Passive containment does not decrease toxicity or mobility of contaminants, it only mitigates their migration and transport. This risk reduction option is selected for irremovable or very toxic contaminants when no other remediation technologies can be applied at a reasonable cost. *In situ* containment is of moderate cost because the soil is not removed, groundwater is not extracted, and no soil phase treatment is performed. The cost of periodical monitoring for potential leaching and for other emissions and discharges should be taken into account. However, engineered containment may have significant costs covering transport, isolation by linings and barriers, and monitoring the performance. Containment is often applied in combination with physical and/or chemical stabilization.

The containment aims at the following:

–   Physical separation (isolation) of the risky material possibly in all directions (covering, isolating by vertical barriers and bottom linings);
–   Exclusion of water and control of leaching;
–   Gas extraction if necessary.

Landfill capping (isolation from the surface) is a common form of implementation (FRTR, 2018a). Landfill caps can be a vegetated soil layer, asphalt, concrete, or multiple layers consisting of geotextile and soils. Geosynthetics/geomembranes are synthetic polymers of low permeability, such as polyethylenes of various densities, polypropylenes, polyvinyl chloride (PVC). High-permeability soil/rock/gravel or synthetic materials are used for a constructed drainage layer to conduct water away from the watertight surface and low permeability soil (clay) layers or synthetic liners are used to hinder water infiltration.

The organic contaminants of low volatility and solubility such as PCBs, DDT (dichlorodiphenyltrichloroethane) and large PAH molecules are immobile enough to be isolated by landfill capping at low risk. Vertical walls are required to prevent the horizontal flow of groundwater and also a bottom barrier layer is necessary if the groundwater level can reach the contaminated solids. Base layer drainage needs thorough monitoring to control the integrity of the liners. Gas migration is controlled by providing migration pathways and off-gas treatment if necessary.

4   PHYSICOCHEMICAL TECHNOLOGIES FOR REMEDIATION OF PESTICIDES IN SOIL

E. Morillo & J. Villaverde

Remediation of pesticide-contaminated soils can be a complicated process, as most of such soils contain mixtures of different compounds rather than a single contaminant. The majority of pesticides applied in agriculture are organic compounds, and their physicochemical remediation will be introduced in this section. As for the rest of organic contaminants, technologies to treat pesticide-contaminated soils fall into two categories: containment immobilization or treatment based on mobilization, and the latter technologies fall into two more categories: separation and destruction. Some technologies accomplish two or more of these functions depending on the type of both contamination (point source or diffuse) and contaminant. Remediation of organic pesticides can be done using any of the techniques developed for other organic pollutants with similar characteristics; however, some specific techniques are really used for pesticides remediation, and only these will be discussed in this chapter.

This subchapter, containing the remedial technologies of pesticide-contaminated soil remediation repeats the preceding, more general methods for organically contaminated soil; main differences being that pesticides are generally large, non-volatile, and not even semi-volatile molecules, and, depending on the chemical family to which they belong, they have limited tendency to be degraded. It is also specific that when the origin of contamination comes from the application, pesticides are mostly sorbed onto solid soil particles in the surface layer of the soils. Thus remediation suitable for pesticide-contaminated soil and groundwater is a subset of the physicochemical methods and is typically applied *in situ*. In these cases, the applicable remediation is dominated by biological and ecological or other soft/green methods compatible with the soil ecosystem and agricultural production. In the case of a point-source contamination of the soil by pesticides arising from their manufacture, handling, and disposal, a more drastic remediation is needed. Depending on the method selected, it will be carried out *in situ* or *ex situ*.

4.1   Separation technologies

The remediation of many pesticide-contaminated soils is often limited by their persistence and resistance, decreasing their availability to be treated using different destruction methods. In those cases a process that removes the contaminant from the host medium is needed.

4.1.1   Soil washing

Soil washing using different kinds of extractants to separate the pesticides from the soil has been extensively used in Europe but only has limited use in the US. In general, soil washing is most appropriate for soils that contain at least 50% sand and gravel. Pesticides are extracted from soil material by application of different extractants such as organic compounds, acids, surfactants, etc. Percolating liquid may be collected and treated for more degradation or landfilled.

Ye *et al.* (2013) assess the potential applicability of organic solvent washing as a remediation technique for organochlorine (OCP)-contaminated soils examining five organic solvents (ethanol, 1-propanol, and three different petroleum ether fractions) and studied the effect of certain factors such as solvent concentration, washing time and temperature, mixing speed or solution to soil ratio at a laboratory scale. Among these factors the quality of solvent (petroleum ether of 60–90°C boiling point gave the best results) and the solution to soil ratio (at least 10:1) were the most significant parameters. The thermal desorption techniques and solvent washing approaches using n-alcohols and surfactants were compared by Gao *et al.* (2013) to remediate OCPs-contaminated soils. About 87% of OCPs was removed from the soils using ethanol, and more than 90% of ethanol used could be recovered. The ethanol extraction was more efficient than aqueous surfactant solutions but less efficient than thermal desorption resulting in >99.9% removal of OCPs by treatment at 500°C for 30 min. However, there are several factors that may limit the applicability of the extraction process such as the extraction of organically bound metals together with pesticides, the low efficacy of solvent extraction on very high molecular weight pesticides, or the toxicity of the solvent to soil microbial population (Pavel & Gavrilescu, 2008).

4.1.1.1   Soil washing/chemical extraction enhanced by surfactants and complex-forming agents

Advances in the development of surfactants capable of extracting a greater variety of organic pesticides have made washing with water a feasible alternative for their remediation in soils. Industrially synthesized surfactants include, among others, chemically ethoxylated alcohols, sulfonates, Triton, Brij 35, and sodium dodecyl sulfate (SDS). Although the washing process using surfactants can be efficient, the contaminants are not destroyed as in solvent extraction, so that further treatment is necessary to treat target compounds in soil and wastewater.

Villa *et al.* (2010) chose the non-ionic surfactant TX-100 to remediate a soil contaminated by long-term exposure to DDT and DDE, and the wastewaters obtained from washing experiments were submitted to photodegradation combined with hydrogen peroxide oxidation. Removal efficiencies of 66% (DDT) and 80% (DDE) were achieved for three sequential washings and 99% and 95% of wastewater degradation efficiencies. Bandala *et al.* (2013) also carried out the restoration of a soil contaminated with 2,4-D using surfactant-enhanced soil washing followed by the application of advanced oxidation processes (Fe-catalyzed hydrogen peroxide and UV treatment). Dos Santos *et al.* (2015) also used the surfactant SDS for the removal of atrazine from soil, but the resulting washing waste was treated by electrolysis with a boron-doped diamond electrode. Surfactants are also used to increase the bioavailability of pesticides to be degraded by soil microorganisms. Torres *et al.* (2012) have studied the removal of methyl parathion using anionic, non-ionic, cationic, and natural surfactants, and wastewater was biologically treated. However, surfactants may have disadvantages as they could be used as a substrate for microorganisms or have toxic effects when present at high concentrations (Laha *et al.*, 2009). Betancurt-Corredor *et al.* (2015) observed a reduced DDT biodegradation when using Tween 80 plus biostimulation compared to the experiment when only biostimulation was used.

Due to the environmental risk posed by the chemosynthetic surfactants, an alternative way is the use of biogenic compounds such as biosurfactants, more ecologically acceptable, in bioremediation of pesticide-contaminated soils (Mulligan, 2009). A diverse range of bacteria capable of producing biosurfactants has been proposed. In particular, the feasibility of rhamnolipid biosurfactants, mainly produced by *Pseudomonas aeruginosa*, to remove pesticides from soil has been studied. Wan *et al.* (2015) observed the simultaneous removal of lindane, lead, and cadmium from soils by rhamnolipids combined with citric acid. The effect of three different biosurfactants, rhamnolipid, sophorolipid, and trehalose-containing lipid, on solubilization and biodegradation of four HCH isomers in a soil bound matrix was studied by Manickam *et al.* (2012), indicating that sophorolipid offered the highest solubilization and enhanced degradation. Odukkathil and Vasudevan (2015) evaluated the capability of biosurfactant-producing bacterial strains in enhancing the bioavailability of endosulfan and its metabolite.

In recent years CDs have gained much attention as pesticide complexation agents. Many of these studies only determined the increasing solubility and complexation constants between specific pesticides with a variety of CDs (Orgoványi *et al.*, 2009; Yáñez *et al.*, 2012), but increasingly more studies can be found that indicate the capacity of CDs for soil remediation based on their extractant ability and subsequent soil washing of a wide variety of pesticides such as the herbicide norflurazon (Villaverde *et al.*, 2005, 2006), hexachlorobenzenes (Wan *et al.*, 2009), organochlorine pesticides (Wong & Bidleman, 2010; Mao *et al.*, 2013; Ye *et al.*, 2014a), 2,4-D, acetochlor, alachlor, dicamba, dimethenamid, metolachlor, and propanil (Flaherty *et al.*, 2013). The enhanced extraction from soils facilitates the subsequent remediation by phytoremediation (Ye *et al.*, 2014b), oxidation (Villaverde *et al.*, 2007), or biodegradation (Villaverde *et al.*, 2012, 2013, 2017, 2018; Gao *et al.*, 2015; Zhao *et al.*, 2015; Rubio-Bellido *et al.*, 2015, 2016).

4.1.2 Soil flushing

Soil flushing has seldom been used for pesticide remediation in the traditional mode. The remediation of phosalone and atrazine was studied on an artificially contaminated synthetic soil by flushing using an aqueous solution of ethanol (Di Palma, 2003; Di Palma et al., 2003). The extraction solutions were submitted to hydrolysis treatment in the case of phosalone and to oxidation by hydrogen peroxide in the case of atrazine. However, in recent years, electrokinetic soil flushing (EKSF) is becoming one of the most promising technologies for pesticide-contaminated soil remediation, especially when the pesticide consists of ionic species. For detailed description of this technology, see Chapter 9 in this volume.

In relation to pesticides, the removal of the anionic herbicide 2,4-D has been widely studied as a model ionic pollutant. Souza et al. (2016) observed 90% removal of the pesticide 2,4-D using electrokinetic soil flushing, and Risco et al. (2016) studied the influence of the arrangement of electrodes in the soil to obtain increasing amounts of 2,4-D removed. Ribeiro et al. (2011) observed that electrokinetic process was able to efficiently remove molinate and bentazone from soils and it was related to the differential pH between both electrodes. In a study with kaolinite spiked with diuron, results showed that in the presence of humic acids, the removal efficiency of diuron decreased from 90% to 35% (Polcaro et al., 2007). These results have important implications for application in real soils, since most pesticides tend to strongly adsorb on the soil organic matter. In such cases, their extraction has to be carried out using enhanced electrokinetic techniques (Karagunduz, 2009; Gomes et al., 2012). Bocos et al. (2015) used electrokinetic oxidation with hydrogen peroxide to remediate a soil polluted with the pesticide pyrimethanil and PAHs, also evaluating the effect of several complexing agents. Surfactants have been used to bring DDT into soil solution (Karagunduz et al., 2007). Vieira dos Santos et al. (2016) studied the application of electrokinetic soil flushing to a soil spiked with four herbicides (2,4-D, oxyfluorfen, chlorsulfuron and atrazine) demonstrating that efficiency depended on the chemical characteristics of the pesticide used. 2,4-D was the pesticide most efficiently removed (95% of removal), while chlorsulfuron was more resilient to the treatment. Volatilization was found to be a process of major significance in the application of electrokinetic techniques to soil polluted with herbicides.

4.2 Destruction technologies

4.2.1 Application of chemical oxidation for pesticide-contaminated soil

The Fe-catalyzed oxidation with hydrogen peroxide is the most common process in the remediation of pesticides in soil. One of the main advantages of the process is the possibility of *in situ* remediation, eliminating the need for excavating contaminated soils and thus allowing the rapid treatment of refractory contaminants. DDT, diuron, 2,4-dichlorophenol (2,4-DCP), pentachlorophenol (PCP), and numbers of other pesticides in the soils can be effectively degraded by applying hydrogen peroxide (Usman et al., 2014; Rosas et al., 2014). However, there are collateral effects with the process. Villa et al. (2008) observed that, despite the high percentages of DDT degradation in soil, there was also an increase of dissolved organic carbon (from 80 to 880 mg/L) and 80% of the natural organic matter was degraded; a significant increase in metal concentrations was also observed in the slurry filtrate due to the acidic pH required by the technology. Direct application of Fe-catalyzed oxidation by hydrogen peroxide is very aggressive to the soil and can also be a disaster to the microbes (Brillas *et al.*, 2009).

Different pesticides have been successfully decomposed in soils using photocatalytic degradation with TiO_2 catalyst under UV irradiation or solar light. Xu *et al.* (2011) observed that only a small amount of catalyst (0.5%) was needed when glyphosate content in soil was low. Similar results were observed for diuron (Higarashi & Jardim, 2002). Xu *et al.* (2011) studied glyphosate and Sharma *et al.* (2015) found that with imidacloprid the photocatalytic reaction mainly occurs in the surface soil layer and the degradation efficiency decreases as the soil layer becomes thicker.

Ozonation is one of the most promising processes which can be applied *in situ* or *on site*. The pesticide can be directly decomposed by O_3 or can react indirectly with the OH· radical which has been generated by the O_3 decomposition. O_3 could decompose on soil active surfaces (metal oxides, soil organic matter) to generate OH·. As ozonation is a relatively new technique for pesticide remediation, the majority of the publications deal with aqueous media. Ozonation of organic contaminants in soil is difficult due to strong limitations of the diffusion process in this heterogenic matrix. Ozone penetration within the soil layer has been determined to be only about 15–45 mm. Balawejder *et al.* (2014, 2016) developed an effective remediation method of DDT-contaminated soil. The preliminary results of investigations at a laboratory scale were exploited to scale up the technology to pilot scale. It is based on the utilization of hydroxyl radicals that were generated from water aerosol and ozone in a fluidized bed reactor. During the remediation procedure the level of soil contamination was reduced by 80%. No degradation products were detected and the total organic carbon was preserved.

In addition to the most frequently applied oxidizing agents, hydrogen peroxide and persulfate, a new and least studied oxidant was also used for *in situ* chemical oxidation (ISCO) of propachlor accelerated by metal ion activation (Cu^{2+} and Fe^{2+}) (Liu *et al.*, 2012). Another two persulfate-based oxidants, peroxydisulfate ($S_2O_8^{2-}$, PDS) and peroxymonosulfate (HSO_5^-, PMS) were used by Waclawek *et al.* (2016) for remediation of HCHs activated by electrochemical processes.

Surfactant application and the combined utilization with bioremediation or soil washing are also documented. 95% of DDT was removed by Fe-catalized oxidation with hydrogen peroxide after soil washing using a Triton X-100 solution (Villa *et al.*, 2010). These researchers have also investigated the combined application of photodegradation and oxidation by hydrogen peroxide to degrade DDT and DDE in soil, which were decomposed after a 6-hour treatment. Laboratory data showed that 99% degradation for chlorimuron-ethyl was achieved in 10 minutes by using a combined technology, however, only 68% degradation was observed after 30-min Fe-catalyzed oxidation by hydrogen peroxide without using light (Gozzi *et al.*, 2012). Souza *et al.* (2013) studied the degradation of glyphosate by the same combined technology.

Research in advanced oxidation processes is still at an early stage. Most studies to date have been conducted at the lab scale and many improvements are required before the technology can be scaled up to bench and pilot plant levels in terms of assessing the cost and operational conditions. According to Rodrigo *et al.* (2014), most of the studies published in the literature have focused on pesticide removal at concentrations that are significantly higher than typical pollutant concentrations in a soil and which are only found in industrial wastes. However, these studies are the only way of determining the efficiency of the technology.

4.2.1.1 Application of chemical reduction

Some pesticides that are persistent in aerobic environments are more readily degraded under reducing conditions. Thus generating a reducing environment in soils, sediments, and aquifers has become a popular remediation strategy. One application of this technique uses zero-valent iron (ZVI) as a chemical reductant (see more details in Chapter 10 in this volume). ZVI can be used to remediate soils contaminated with pesticides (Shea *et al.*, 2004). In these studies, degradation was enhanced by adding $Al_2(SO_4)_3$ to ZVI. $Al_2(SO_4)_3$ increased acidity and promoted the formation of green rust [mixed Fe(II)/Fe(III) double hydroxides], which facilitated metolachlor dechlorination (Satapanajaru *et al.*, 2003).

Degradation can be maximized by combining chemical treatment (e.g. ZVI) with biological degradation. Boparai *et al.* (2008) reduced the concentration of metolachlor in a contaminated soil after chemical treatment with ZVI and aluminum sulfate but atrazine and nitrate remained. Decreases in atrazine and nitrate concentrations were obtained after adding sucrose, resulting in an enhanced microbial activity in the presence of readily available carbon.

Nanoscale zero-valent iron (nZVI) has been conceived for cost-efficient degradation of chlorinated pollutants in soil as an alternative to permeable reactive barriers or excavation. Treatment with ZVI promotes rapid abiotic degradation of chlorinated contaminants *via* reductive dechlorination (Cong *et al.*, 2010; Han *et al.*, 2016) (see more in Chapter 10).

4.3 Containment and immobilization technologies

4.3.1 Containment technologies

Pesticide contamination of soils may result not only from agricultural processes but also from manufacturing, and improper storage, handling, and disposal of pesticides. With the signing of the Stockholm Convention and the development of global monitoring programs, many OCP-contaminated sites are left by hundreds of abandoned OCP production factories close to cities. Currently, most of these contaminated sites face urgent land use conversion and redevelopment for commerce (Li *et al.*, 2008; Ye *et al.*, 2013) and likely pose a threat to residents and the environment. As POPs such as OCPs are only poorly soluble in water, they tend to remain as a highly persistent contamination in the soil (Caliman *et al.*, 2011). Owing to specific properties of such pesticides, macroscale remediation efforts using containment immobilization technologies both for removed contaminated soils and for *in situ* immobilization. Containment immobilization technologies include capping to limit infiltration and leaching and subgrade barriers to limit lateral plume migration. These processes are considered as a type of waste disposal because no real remediation of the polluted site is reached comprising only a provisional solution. In spite of this, they have great application potential.

The pesticides lindane and DDT and their metabolites and isomers are some of the compounds included as POPs in the Stockholm Convention. Technical hexachlorocyclohexane (HCH) and the separated γ isomer (lindane) were among the most extensively used organochlorine pesticides.

At most factories the production of such chlorinated organic wastes was generally dumped, usually in an uncontrolled manner, in landfills near the production facilities (Gałuszka *et al.*, 2011). This practice has resulted in numerous contaminated sites around the world. A multilayered landfill cover system with solid household waste was deposited on top of a HCH/lindane landfill in Hamburg in order to "immobilize" or rather cover the hazardous wastes (Götz et al., 2013), but leaching from the landfill remains a challenge for groundwater management. Sometimes a containment cement or concrete wall is constructed around the contaminated area to prevent the dispersal of the pollutants (Weber & Varbelow, 2013). In many cases, despite the encapsulation of the sites, traces of the contaminants were detected in the surrounding soil and in groundwater (Usman et al., 2014). These past experiences with such disposal practices highlights their insustainability due to the risks of contamination of ecosystems, the food chain, together with ground and drinking water supplies.

4.3.2 Sorption technologies

Adsorption is the first process that takes place when pesticides are in contact with soil and it affects other processes such as chemical transport, leaching, bioavailability, or ecotoxicological impacts on non-target organisms. Organic amendments can be used as an immobilization technology increasing adsorption of pesticides to soil and, consequently, reducing their availability.

According to Khorram et al. (2016), the use of organic amendments to reduce bioavailability has less disruptive effects and can be used in pesticide-contaminated soil by:

1   Binding pesticides to reduce their potential mobility into water resources and living organisms.
2   Providing nutrients to promote plant growth and stimulate ecological restoration.

As organic materials originate from biological matter, they require minimal pretreatment before application to soil.

Carbonaceous materials (CM), often referred to as "black carbon," represent a continuum of heterogeneous materials ranging from partly charred plant material to char and charcoal to soot and graphite. It has been shown that biochar and other CM addition to soils and sediments enhances sorption of hydrophobic organic compounds (HOC), thereby potentially playing a crucial role in controlling transport, bioavailability, fate, and health risk of organic contaminants (Kupryianchyk et al., 2016). The sorption strength and mechanism controlling pollutant sequestration by CM is dependent upon molecular (carbonized and non-carbonized fractions) and structural (surface, pore, and bulk properties) characteristics. This in turn depends on the CM starting material (feedstock) and pyrolytic conditions.

Biochar has attracted attention for its powerful ability to reduce the bioavailability of hydrophobic contaminants. The effect of biochar on the retention of pesticides applied to agricultural soils has received little study, although research dealing with this issue has increased in recent years. There are published studies on the impact of biochar on sorption and dissipation of pesticides such as atrazine, terbuthylazine, pyrimethanil, or bentazone (Cao et al., 2009; Wang et al., 2010; Yu et al., 2010; Cabrera et al., 2014). The adsorption capacity of biochar for pesticides depends on its physicochemical properties such as organic carbon content, specific surface area, and porous structure (Khorram et al., 2016). Several studies have demonstrated that biochar amendment can lead to irreversible adsorption of pesticides (Yu et al., 2010). It includes surface-specific adsorption, entrapment into micropores and partitioning into condensed structures. Sopeña et al. (2012) reported that the adsorption capacity for isoproturon on a soil amended with 2% (W/W) biochar from *Eucalyptus dunni* was nearly five times higher for amended soil than for the unamended one.

Cabrera *et al.* (2011) observed a drastic increase in the adsorption of fluometuron and MCPA (4-chloro-2-methylphenoxyacetic acid) in soil amended with 2% (w/w) of biochars made from different feedstocks. Adsorption was specially increased using wood pellets (about 30 times higher for fluometuron and up to 50 times higher for MCPA). However, they observed an enhanced leaching of both herbicides after amendment with biochars containing soluble organic compounds, concluding that the amount and composition of the organic carbon content, especially the soluble part, can play an important role in the immobilization of these herbicides. Martin *et al.* (2012) also observed that soils freshly amended with biochars showed a two- to five-fold increase in sorption of herbicides as compared with that in the unamended soil, and desorption hysteresis was prominent in the soil amended with fresh biochars. In contrast, the soil containing aged biochars over a period of 32 months exhibited sorption – desorption properties similar to that of the control soil, indicating that aging of biochars in the soil reduced their sorption capacity.

Organic green wastes and compost from various origins as soil amendment has been increasingly used globally. This land management is accepted as an ecological method for the disposal of such wastes while it simultaneously increases soil fertility and organic matter, preventing losses of pesticides from runoff or leaching due to an increase in pesticide immobilization by sorption. Additionally, beneficial effects on soil biochemical properties and microbiological parameters are reached, accelerating the dissipation of pesticides in soil (García-Jaramillo *et al.*, 2016). López-Piñeiro *et al.* (2013) studied the sorption and degradation of MCPA in soils amended with freshly composted and aged olive mill waste, observing that the higher the amount and maturity of wastes applied, the higher the retention of MCPA. However, they also observed that MCPA could be easily desorbed if the amendment was not aged or composted. Also, different organic residues from olive oil production were used by García-Jaramillo *et al.* (2014) to study their effect on the immobilization and leaching of bentazone and tricyclazole. Tricyclazole was not detected in any of the leachates in the amended soil. The authors concluded that the sorption of dissolved organic matter from the amendments changed the physicochemical properties of the soil surface. Centofanti *et al.* (2016) used aged dairy manure and biosolids compost amendment for *in situ* risk mitigation of aged DDT, DDE, and dieldrin residues in an old orchard soil. The addition to soil of the waste material, spent mushroom substrate (SMS), have also been demonstrated to be a promising strategy to optimize pesticide immobilization and control water pollution (Marín-Benito *et al.*, 2009, 2013; Álvarez-Martín *et al.*, 2016).

5   ISCO COMBINED WITH INTER-WELL AGITATION FOR CHLORINATED HYDROCARBONS – A CASE STUDY

É. Fenyvesi and K. Gruiz

In situ chemical oxidation was applied to chlorinated hydrocarbons at a historically contaminated industrial site in Mezőlak, Western Hungary. The risk management involved site investigation and site-specific risk assessment as well as risk reduction by *in situ* chemical oxidation (ISCO) in combination with intensive inter-well agitation. ISCO was selected based on previous experience and on small-scale treatability studies. As a large number of extraction wells had already been established on the site, and the location of the contaminant sources could not be precisely identified, alternating injection into and extraction from the wells was applied to enhance interaction between the injected reagent and hidden lenses of high-density chlorinated contaminants. The effectiveness of agitation was proved by laboratory treatability studies and by a field pilot study. The successful field trial between two wells was followed by field application of the ISCO combined with inter-well agitation technology specifically developed for the site. The innovative element of this *in situ* remediation was a directed water transport between the extraction and injection wells by alternatingly operating the injection and extraction wells. By so doing the water was actively moved between the wells and the interaction with the contaminant provided optimum performance of the oxidizing agent. The technology utilized extensive pump and treat (by stripping) on the site, which had been going on for several years but was completely ineffective due to constant contaminant recharge from unidentified subsurface sources.

As a first step additional site assessment and pumping tests were performed, and an attempt was made to localize the subsurface contaminants sources. Some of the sources were identified in the neighboring lot but were inaccessible during the project. After selecting and planning the operation, the monitoring plan was prepared in agreement with the verification plan (see Chapter 11). The ISCO technology using hydrogen peroxide in combination with intensive groundwater agitation was introduced and the original pump and treat technology was modified. Alternating pumping of the wells ensured intensive mass transfer between soil solid and groundwater as well as between groundwater and DNAPL lenses. The technology was developed within the Hungarian MOKKA project (2004–2008) with participation from the Weprot company (2018).

5.1 Site description, history

The site selected for pilot experiments and technology demonstration is located near the village of Mezőlak. For a long time it has been an industrial site where trichloroethylene (TCE) was applied for metal degreasing. The soil and groundwater had been contaminated for decades.

TCE was identified as the main contaminant by the first site assessment. Some other chlorinated ethylenes such as tetrachloroethylene (PCE) and dichloroethylene isomers (DCE) and chlorinated ethanes were detected at lower concentrations. In some wells, vinyl chloride (VC) was also found, suggesting microbiological decomposition.

The owner was obliged to take remedial action. The site-specific regulatory quality criteria in the groundwater for each chlorinated hydrocarbon (CHC) have been set at 150 μg/L (multifunctional quality criterion: 10 μg/L) and for total volatile hydrocarbons at 300 μg/L. The site-specific quality criteria for TCE, the main contaminant found in the soil at that time was established as 0.57 mg/kg (multifunctional quality criterion: 0.1 mg/kg). The most contaminated soil was landfilled and a groundwater decontamination system (extraction and monitoring wells, stripping tower) was established and started to operate in 2004, the same year.

The technology was based on groundwater extraction by pumping, collecting in a container, and treating it on site (pump and treat technology, see Section 2.1.2) and using air stripping. The volatile chlorinated compounds were removed from the extracted groundwater before it was discharged.

This technology was run for 3 years, but it failed to permanently reduce the TCE content below the limit in the wells. The reason was that the subsurface sources (two main and several smaller sources were suspected) could not be identified and eliminated and they contaminated the groundwater constantly. It is very probable that in addition to the two identified larger lenses several smaller ones remained unidentified close to the footwall in the liquid sand. In addition, there might be TCE lenses under the buildings and the neighboring ground providing a steady TCE supply in spite of the successful local clean-up.

5.2   Step-by-step risk management

The management concept is illustrated in Figure 6.12.

The first management step was the creation of the site conceptual risk model (CRM). CRM specifies the pollution sources, transport routes, land uses, and receptors (see Section 5.1.2), as well as the size of risk represented by the width of the arrows (see later in Figure 6.15).

Historical data and current groundwater sampling and analysis results were used to delineate the contaminated area and divide it into two subsites: a highly contaminated sub-site near to the supposed sources and a less contaminated subsite somewhat farther from the source (see map in Figure 6.13). The flow direction of the groundwater is from southwest to northeast, which means that contaminated groundwater flows continuously from the more contaminated area to the less contaminated one.

The next step was the development of substitute technology to replace the ongoing inefficient pump and treat. The remediation running on the site was evaluated, the reasons of failure were identified, and conclusions drawn (see Section 5.2.1).

The *technology options* were analyzed and the most fitting ones studied in bench-scale experiments. The conclusions of the treatability studies enabled the design of the field pilot study and selection of the full-scale technology in the field. An integrated methodology (including physicochemical measurement and analyses, biological tests, and direct soil and water toxicity assessment) was designed for monitoring the implementations during the stepwise scaling-up to satisfy the planned technology evaluation.

The selected technology was demonstrated in *full scale* at two subsites. Monitoring had two main goals: to control technological parameters and observe environmental quality; changes in the soil and groundwater, emission from the *in situ* reagents applied and the mobilized contaminants as well as the quality of the treated and released water.

The *verification* of the technology was carried out by using a four-staged verification system, consisting of the assessment of (i) technological efficiency through the mass balance, (ii) environmental efficiency through assessment of risk posed on the ecosystem and humans, (iii) cost-benefit assessment, and (iv) a SWOT to investigate all the other impacts and non-quantitative characteristics (Gruiz *et al.*, 2014). Only mass balance is discussed here.

5.2.1   *Site assessment and mapping*

Sampling wells (26) were drilled for detailed site investigation and remediation in 2004. The contaminated plume was delineated and its size was found to be 2000 m^2 within the 12,000 m^2 industrial area shown in the map in Figure 6.13. The soil profile based on geotechnical assessment comprises 1–1.5 m mixed soil backfill followed by very diverse natural layers (mainly sand and gravel and sandy loam and clayey layers) at 5.3–5.5 m. Beneath this layer silt, sandy and clayey silt extend to 10.3 m depth and sand with silt to 10.3–11.5 m. This layer is directly above the footwall and consists of quicksand lenses storing TCE in diffuse distribution under the whole of the identified main plume. The groundwater level varies between 2 and 2.5 m.

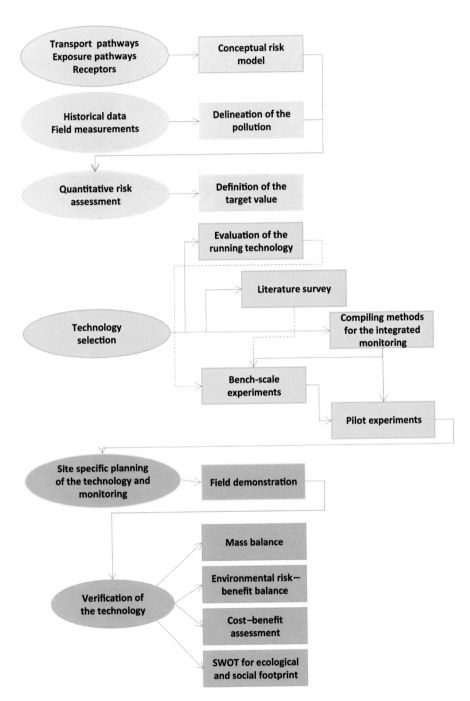

Figure 6.12 Environmental risk management methodology used on the site.

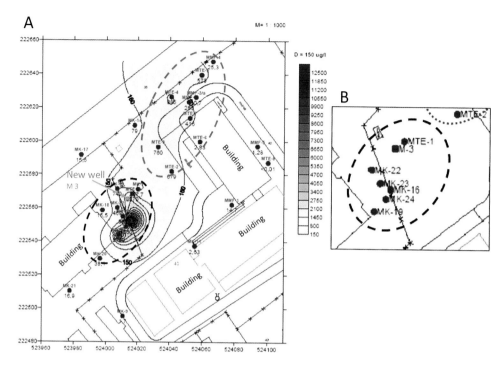

Figure 6.13 A) Trichloroethylene distribution in groundwater on the site. MK, MMF: groundwater monitoring wells, MTE: extraction wells. The new well (M 3) is marked by an arrow. The areas of the three experiments are marked (heavily contaminated area surrounded by black dashed line and a less contaminated area surrounded by brown dashed line).
B) Location of the wells in the heavily contaminated area.

The sampling was repeated every three months for three years from September 2007 when the ISCO demonstration project started. The soil and water samples were analyzed and tested using an integrated methodology, involving chemical, biological, and ecotoxicological methods. The results showed that the contaminant concentration periodically increased in certain wells near to the assumed subsurface sources (e.g. in the vicinity of the wells MTE-1 and MK-16) where particularly high concentrations of TCE were measured in the range of 1400 to 17,700 µg/L (the site-specific quality criteria is 150 µg/L). The concentration of perchloroethylene (PCE) (88–623 µg/L) was also above the regulatory limit (150 µg/L) while low concentrations of *cis*- and *trans*-dichloroethylene (DCEc and DCEt, respectively) were measured (1–30.3 µg/L and 1–13.9 µg/L). Vinyl chloride was detected only twice in this period suggesting limited biodegradation.

In addition to CHCs the concentrations of characteristic ions (chloride, sulfate, nitrate, ammonium, etc.) were also measured. Chloride ion content near to a suspected source (45 g/L in well MTE-1, August 2007) suggested biological dechlorination. Increased ammonium ion content (0.7 g/L in well MTE-1) compared to other wells (< 0.1 g/L) indicated that anaerobic processes were also going on.

In the following March a new well (M-3) was bored for the pilot tests between the new one and the most contaminated one. The TCE contents of the soil from the new drilling was

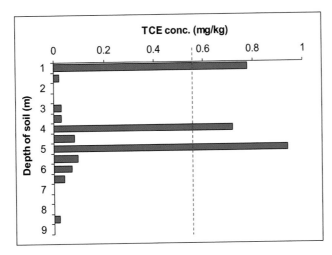

Figure 6.14 TCE concentration of soil in the drilling M-3 as a function of depth (measured by head space GC). The red line shows the regulatory limit 0.57 mg/kg.

analyzed by gas chromatography from the soil extracts. Figure 6.14 shows the TCE profile as a function of depth.

The upper (three-phase soil) layer at 0.8–1.2 m was also contaminated suggesting that the pollutant has been transported through the capillary fringe and the vadose zone to the upper soil section, and from the pores into the atmosphere. The other two contaminated layers are located in the saturated soil at 4 and 5 m depths.

5.2.2   Site-specific risk characterization

Contaminants: CHCs with TCE as the main component;
Henry's law constant of 985 Pa/m³/mol at 20°C;
Vapor pressure: 7.8 kPa at 20°C;
Water solubility of 1.1 g/L at 20°C;
$\log k_{ow}$ (octanol–water partition coefficient) 2.29;
k_{aw} (= Henry/RT) (air–water partition coefficient) 0.4.

The partitioning tendency of trichloroethylene in the environment generally develops as follows: air: 97.7%; water: 0.3%; soil: 0.004%; sediment: 0.004%. Based on this TCE tends to fully evaporate from soil to air and if a building is on the top of the contaminated soil and the vapor finds its way to the inside space where TCE vapor can mainly accumulate in indoor air. Human toxicity features of CHCs are: neurotoxic, immunotoxic, hepatic, developmental, and carcinogenic effects (TOXNET, 2018).

Figure 6.15 shows a conceptual risk model (CRM). This is the risk map of the site, showing sources, transport routes, the contaminated environmental compartments, and the uses of these compartments (atmosphere, water, soil) by the various receptors, mainly workers in this case. The first version of the CRM is quantified by archive and site investigation data. In this way CRM highlights the unacceptable risks.

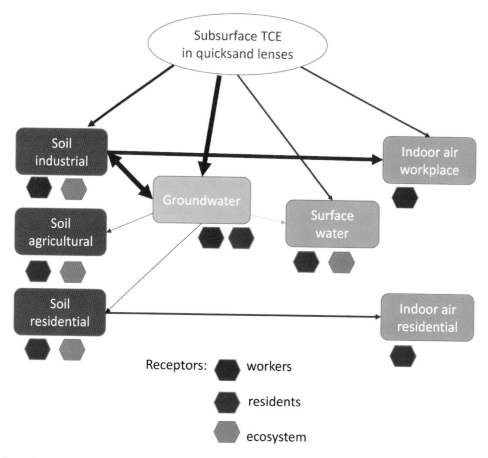

Figure 6.15 Conceptual risk model (CRM) of the site.

The CRM shown in Figure 6.15 was used for developing site risk management and monitoring.

As the CRM suggests, the land use within the fence is industrial and outside of the fence residential and agricultural. A small creek flows within 2 km. There is no residential area downstream of the plume. The industrial site was used by workers (an assembly plant was in operation at that time) and staff of the remediation activities. Ecological receptors have not been considered within the fence.

The TCE content of the unidentified subsurface sources polluting the groundwater represented the dominant risk. TCE can typically be detected in indoor air of buildings on a contaminated area. Industrial buildings are the target of vapor intrusion and workers working inside the buildings on the Mezőlak site. Thus, the main exposure route in this case is inhalation of indoor air. TCE together with the other halogenated compounds is especially harmful: it is carcinogenic, genotoxic and has several other organ-specific toxic effects.

Risk evaluation in 2007

Sources

Main sources: unidentified underground TCE (contained in quicksand lenses), contaminated groundwater, soil and soil air. Primary sources could not been identified in spite of attempting to determine the exact location of the DNAPL lenses. The suggestion of the professionals is that DNAPL is diffusely dispersed in quicksand above the footwall at about 10–11 m.

Transport routes

The main transport route leads *from the TCE-containing lenses to the groundwater*. The natural transport with groundwater would follow the groundwater flow, but the groundwater has been continuously extracted since 2004 and so established a hydraulic barrier. This is why the contaminated water does not leave the industrial site.

From groundwater to soil air: TCE is volatile and most of TCE is in vapor form in the environment. This tendency results in high TCE concentrations in soil air and transport by diffusion into atmospheric air. The process leads to vapor intrusion into buildings on the top of contaminated soil or above contaminated groundwater.

From contaminated subsoil to water and soil air: in addition to the deep lying lenses, the soil sorbs significant amounts of TCE as detected by the depth-dependent analyses (Figure 6.14). The contaminated subsoil may also function as a secondary source and supplement the water and soil air with contaminants.

Contaminated and potentially contaminated environmental compartments

Soil of the industrial site is contaminated; the level of contamination is above the site-specific quality criterion.

Groundwater of the industrial site: groundwater flow and contaminant transport with groundwater is restricted by the hydraulic barrier. If the continuous groundwater extraction is terminated, the contaminated plume can leave the site, reach the vicinity downstream or even more remote areas and the creek.

The surface water (the creek at a distance of approximately 2 km from the site) and its aquatic ecosystem were not endangered at that time because the calculated groundwater flux from the site toward the creek was negligible and the measured contaminant concentration away from the treated zone showed steep decrease. The small flux and TCE concentration are due to the hydraulic barrier, i.e. the depression caused by continuous extraction of the contaminated water to the air stripper. The Henry's law constant value of 985 Pa/m^3/mol at 20°C (ASTDR, 2018) suggests that TCE partitions rapidly to the soil gas from the groundwater or to soil gas and the atmosphere from the unsaturated topsoil.

Indoor air: transport limitation of groundwater also leads to limitations for TCE vapors, meaning that vapor intrusion may occur into industrial buildings on the site but not into residential homes.

Ambient air: the transport of TCE into ambient air and its further access to ambient air outside of the site cannot be limited just by the low contaminant concentrations in groundwater and soil.

Land and water uses

An industrial facility is still active at the site. The facility is a closed one. Neighboring land use is agricultural and residential and only agricultural land use occurs downstream of the plume.

The water extracted from the site has not been used for drinking or irrigation purposes. Outside of the industrial site the water wells do not serve as drinking water supply but water from local wells is used in some cases for irrigation.

Users of the land

The site itself is used by the employees of the plant, working in the buildings and by the staff or the company carrying out remedial work.

The residential area is used by average population including children, adults, and pregnant women. Some of the workers may live in the neighboring village (double exposure).

The workers in the buildings on the site may be chronically exposed to certain levels of TCE. The risk characterization ratio for respiration (RCR_{resp}) can be calculated by the following equation:

$$RCR_{resp} = IC/RfC$$

where IC = inhaled concentration (calculated from water concentrations = 400–5000 µg/m³) and RfC = reference concentration (40 µg/m³) (RfC, 2018), yielding RCR_{resp} = 10–125.

The CHCs have not reached the aquatic ecosystem (as the groundwater does not escape from the site) or terrestrial ecosystem, and the agroecosystem in the neighboring area was also not impacted at the time of the assessment due to preventive measures (hydraulic barrier). This is the reason why neither contaminated groundwater nor air pollutants were detected in the surrounding residential and agricultural area.

It should be noted that the risks at that time were low due to the hydraulic barrier but would significantly increase if the groundwater extraction ceased. If water extraction terminated without proper groundwater remediation, the pollutant could leave the industrial site and endanger residents, agroecosystem, and the nearby creek.

The site-specific quantitative risk is characterized by the risk characterization ratio RCR, which is the ratio of the actual concentration in air, water, or soil to the concentration which does not create adverse effects to the land users (Gruiz, 2016).

Various strategies were applied to determine the RCR by the following:

1   Comparing the measured concentration to site-specific regulatory quality criteria (e.g. given by the local authority).
2   Comparing the measured or calculated figures to quality criteria reflecting the multifunctional use of water and soil.

The acceptable RCR is RCR = 1 in case (i), in case (ii) the acceptable RCR will be determined with respect to actual land uses, e.g. RCR < 2 for residential uses, and RCR < 10 for industrial purposes. A different concept is used for managing air concentrations: the RCR for indoor air is calculated from the international recommendation of 40 µg/m³ acceptable concentration (ATSDR, 2018) and the actual air concentration either measured or calculated from the concentration of the water. When RCR >1–10 the risk is too high, remedial actions are necessary. Air concentration in this case was calculated from the concentration in the contaminated groundwater (partitioning) and from three-phase soil, both providing high TCE concentration in the soil gas and indoor atmosphere threatening workers.

Risk calculation – before remediation

RCR of the groundwater for industrial and residential (multipurpose) land uses:

Industrial use

Low-level contamination: 1440 µg/L
Strongly contaminated plume: 17,700 µg/L
Authority requirement for industrial land use: 150 µg/L

$RCR_{ind\ min}$: 9.6 $RCR_{ind\ max}$: 118

The yellow colored area on the map (Figure 6.14) with the borderline of 150 µg/L represents the area where $RCR > 1$.

Residential use

Regulatory quality criteria for CHC for multiuse purposes: 40 µg/L

$RCR_{res\ min}$: 36 $RCR_{res\ max}$: 442

RCR of the topsoil for residential and industrial land uses:

Industrial use

$RCR_{ind\ topsoil} = 0.8\ mg/kg\ /\ 0.57\ mg/kg = 1.4$

It is slightly risky for industrial land uses as the measured concentration is somewhat larger than the target values required by the authority.

Residential use

$RCR_{res\ topsoil} = 0.8\ mg/kg\ /\ 0.1\ mg/kg = 8$

Topsoil proved to be risky for human receptors (including children, pregnant women), based on the multipurpose quality criteria of 0.1 mg/kg for soil.

The map shows that two buildings are located above the TCE plume (PEC > 0.15 mg/L). Working in these buildings, especially in the one near to the source poses a high risk to humans.
Volatile TCE from the topsoil can enter buildings and contaminate indoor air. A large proportion of 0.8 mg/L TCE in the three-phase soil is in the pores, meaning that the diffusion of soil air (depending on geometry and diffusion rate) is the main transport route within the soil.
TCE vapor intrusion was calculated from the TCE concentration in the three-phase soil assuming geometrically proportional direct transport by diffusion from soil to indoor air.
The estimated indoor air contamination can be as high as 3000 µg/m³ (estimated average from topsoil under the building) in indoor air within a large building directly above the contaminated plume, representing an $RCR = 3000\ µg/m^3\ /\ 40\ µg/m^3 = 75$.
This figure may be lower assuming poor isolation of the windows. The industrial building is equipped with ventilation, so an intensified ventilation and frequent sampling and measuring volatile CHCs was recommended.

Table 6.6 Toxicity of soil samples from different depth measured by *Folsomia candida* acute mortality test (Fekete-Kertész *et al.*, 2013).

Depth	ED_{20} (g soil)	ED_{50} (g soil)	Toxicity	RCR
1.5–2.5 m	10.9	18	slightly toxic	RCR=2.5
2.0–3.5 m	>20.0	>20.0	non-toxic	RCR <1
3.5–5.0 m	12.9	>20.0	non-toxic	RCR <1
5.0–6.0 m	5.5	11.3	toxic	RCR=5

Ecosystem risk

The adverse effect of the soil itself was evaluated by direct toxicity assessment using soil-living test organisms in the study. The *Vibrio fischeri* (bacterial) bioluminescence-inhibition test, the plant root and shoot growth inhibition test (Gruiz & Molnár, 2015) and *Folsomia candida* (collembola, an insect) acute mortality test (Gruiz *et al.*, 2015) were used to study soil toxicity.

The soil samples were found to be non-toxic to *Vibrio fischeri* and *Sinapis alba* (white mustard). *Folsomia candida* (insect) was more sensitive to TCE exposure through inhalation. Direct toxicity results are shown by Table 6.6.

Special, TCE-degrading microbes were found in high concentrations in samples from the deeper layers suggesting anaerobic degradation.

5.2.3   *Evaluation of the former pump and treat technology*

Prior to the application of ISCO, the technology at the Mezőlak site was a pump and treat with *ex situ* air stripping and was without any results after 3 years of application.

In 40 months of operation, 25 kg TCE was removed by extracting and stripping 23,173 m³ groundwater. In spite of this huge effort high concentrations of TCE (> 10 mg/L) were measured in the monitoring wells every now and then.

It was concluded that the existing pump and treat technology was inefficient, time consuming and expensive. A water-insoluble contaminant residing under the aquifers in separate phases cannot be effectively removed by extracting the groundwater.

Mixing the strongly contaminated water with the water from less contaminated wells was also not economic because an increased volume of water had to be treated and discharged. Termination of this practice was advised.

A periodic increase of the CHCs content in the extracted water suggested that there were unidentified plumes in the area.

The following management measures were decided:

– *Approximation of the location of the source plumes.*
– *Assessing the microbial activity at the site to see whether microbial degradation of the CHC compounds was going on* and performing laboratory experiments to see if these microbiological processes could be enhanced using additives (nutrients, electron acceptors or donors, cometabolites, etc.).
– *A step-by-step selection of the technology* was based on (i) literature survey, (ii) bench-scale treatability studies, (iii) field pilot studies, and (iv) a full-scale demonstration.

5.2.4   Technology selection, comparative evaluation of the options

Four technologies were selected and compared based on published information and labora-tory treatability studies (Fekete-Kertész *et al.*, 2013): i) Pump and treat; (ii) intensified pump and treat; (iii) *in situ* anaerobic biodegradation; and (iv) *in situ* chemical oxidation.

Pump and treat together with *ex situ* air stripping (the technology used on the site for a long period of time without success) can be intensified by solubilizers (SEP&T). This may accelerate the clean-up, however, decomposition/degradation of TCE would be more advan-tageous from a waste production perspective given that SEP&T only transfers the contami-nant from one phase to another resulting in contaminated carbon at the end which should be incinerated. *Ex situ* UV degradation was not favored due to the high energy demand (low energy efficiency). The two *in situ* technologies: anaerobic biodegradation and ISCO are more beneficial concerning cost efficiency. ISCO using hydrogen peroxide was more effec-tive than potassium permanganate and persulfate (Fekete-Kertész *et al.*, 2013). Moreover, low concentration of hydrogen peroxide is not harmful to the soil biota, even increased activ-ity was observed, which can enhance the natural attenuation.

The comparative evaluation of the technologies based on the criteria set by Kukacka *et al.* (2010) can be seen in Table 6.7.

There is no need for further investment at the site to implement any of the technologies, the existing establishment can be used or modified as required. The lowest maintenance cost is for biodegradation but it is of low efficiency and is limited to deeper soil layers. The cost of the chemicals in SEP&T and ISCO is compensated by the reduced energy costs due to a shorter duration. Taking into account that the owner of the site was forced to achieve the remedial goals in a short time period, ISCO seemed to be the best available technology. The bench-scale ISCO experiments with the site soil revealed that the addition of iron did not improve the efficiency of the technology, showing that the intrinsic iron content in the soil is sufficient for the catalysis of oxidation by hydrogen peroxide (Fenyvesi *et al.*, 2011).

5.2.5   In-well treatability test

A new well (M-3) was drilled for the pilot study in the direction of the hydraulic gradi-ent between the assumed source and the extraction well MTE-1, in a short distance from well MTE-1 (see map in Figure 6.14). These two wells served for the *in situ* pilot study of the combination of ISCO with inter-well agitation by alternating injection and extraction. The agitation between the two wells due to pressure difference and intensive water flow facilitated the contact between hydrogen peroxide and groundwater, and groundwater and

Table 6.7 Comparative evaluation of the technologies: summary table.

	P&T	SEP&T	Biodegradation	ISCO
Effectiveness	low	**high**	low	**high**
Duration of remediation	long	medium	long	**short**
Further investment costs at the site	**no**	**no**	**no**	**no**
Maintenance costs (including chemicals, labor and energy)	medium	**low**	**low**	medium
Regulatory compliance	low	medium	**high**	**high**
Need of follow-up technologies	high	high	**low**	**low**

DNAPL-containing lenses and enhanced the *in situ* oxidation. In the pilot test, hydrogen peroxide and sulfuric acid were added to the extracted groundwater and injected back alternately into the two wells. The acidic pH was necessary for the iron-catalyzed reaction of hydrogen peroxide (see Section 2.2.3). Extra iron was not added, the technology was based on the intrinsic iron content of the soil (Fekete-Kertész *et al.*, 2013).

The groundwater level measured from the surface was 2 m in both wells. The volume of the wells MTE-1 and M-3 was 100 L and 60 L, respectively; 180 L groundwater was extracted from MTE-1, 30% hydrogen peroxide solution added (1 L) and the pH adjusted to 3 (in a container). This was the lowest effective peroxide concentration in laboratory studies. The water containing peroxide was reinjected. Then 120 L water was extracted from the M-3 well, 0.7 L hydrogen peroxide solution added and reinjected. This cycle of groundwater extraction, reinjection was repeated once more alternately in both wells, then in a third cycle four times more peroxide was added to each well. Sampling was done one day before the experiment (sample 0), directly before peroxide injection (sample 1) and between each treatment steps. (For routine application extraction and injection wells can be equipped with alternating pumps and peroxide dosing.)

The pH of the extracted groundwater remained unchanged (6.6–6.7) due to the buffering capacity of the soil. The electrical conductivity changed parallel with the sulfate concentration increasing gradually due to the sulfuric acid additions.

TCE concentration started to decrease due to pumping in well MTE-1 (from sample 0 to sample 1) and it was practically non-existent after addition of the oxidant (Figure 6.16). The concentration of PCE decreased stepwise after consecutive treatment steps. In the samples MTE-1/4–7 and M-3/6–7 the TCE concentration was below the remediation goal (0.15 mg/L).

The *in situ* chemical oxidation with alternating inter-well treatment proved to be efficient: the pollutants practically disappeared after one day. No TCE was detected in the water extracted at the last sampling of well MTE-1, while before the experiment 4.8 mg/L TCE was measured. 0.04 mg/L TCE was measured in the water extracted at the last sampling of M-3 well, while 3.5 mg/L TCE was determined at the start of the experiment.

The TCE mineralized to inorganic chloride was calculated from the increased chloride concentration: 24.1 and 24.8 mg/L from 15.1 and 19.5 mg/L at the start in wells MTE-1 and M-3, respectively.

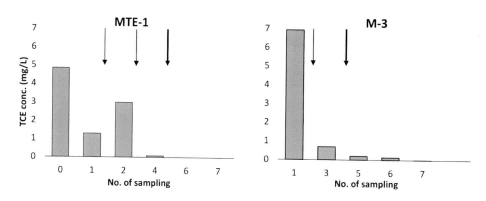

Figure 6.16 TCE concentration in MTE-1 and M-3 wells at sampling 0 to 7. The arrows show the addition of hydrogen peroxide (the thicker line shows the addition of four times higher amount of oxidant) (Fekete-Kertész *et al.*, 2013).

Since the pilot experiment was proved successful, the technology was designed for the entire site. The sources could not be identified, the most probable suggestion is that TCE occurs diffusely in the quicksand lenses. The distribution of these lenses depends on the unevenness of the footwall and the depths of the quicksand layer.

6 FULL-SCALE DEMONSTRATION OF THE REMEDIATION TECHNOLOGY

Two subsites were distinguished throughout the site (see Figure 6.13): the source zone contaminated with TCE lenses and the less-contaminated area, where the soil is contaminated by the groundwater arriving from the source zone. Three experiments (one in the source zone and two in the less contaminated area) were performed simultaneously in 2008.

In the most contaminated area, six wells were treated once a day with hydrogen peroxide mixed into the extracted groundwater, the pH was adjusted with the more environmentally friendly phosphoric acid instead of sulfuric acid and then injected back. One well was used for the extraction of the groundwater which was then treated on the surface by air stripping. The extraction was started in this well only after approximately 1 hour contact time after the groundwater with mixed reagent was reinjected into the six treated wells.

The wells MK-16, MK-19, MK-22, MK-23, MK-24, and M-3 were treated for 4 days (5 L 30% hydrogen peroxide and 0.4 L 85% phosphoric acid were added to the extracted 180 L water and injected back into each well). Sampling (extracting 180 L water from each treatment well) was continued for another 3 days. The well MTE-1 was used for extracting groundwater at a rate of 1.5 L/min for 7 days.

The treatments reduced the TCE concentration in each well (Figure 6.17). The most remarkable effect was observed in well MK-19 where the starting concentration of 13 mg/L TCE decreased to less than 1 mg/L after the first treatment. In well MK-22, the starting TCE level dropped to one tenth after the first addition of hydrogen peroxide. A remarkable decrease was achieved in the well (MK-16) with the highest TCE concentration (18 mg/L): the TCE concentration was first enhanced (from 18 to 38 mg/L) due to pumping then gradually decreased to the half of this outstanding concentration after one week.

All wells were sampled and the contaminant concentration measured, the results are shown in Figure 6.17. Some wells showed a proportional decrease with the number of treatments, but others showed a spontaneous increase in days 5–7, when only extraction and no treatment was applied. It shows the resupply from less available sources and indicates the need of longer treatment.

In extraction well MTE-1, an extremely high TCE concentration (36 mg/L) was measured after starting the pumping which decreased to approximately one tenth (<3 mg/L) due to the *in situ* oxidation in the other wells (Figure 6.18). Altogether 8.1 m³ water was extracted from this well during the 1-week experiment. The amount of TCE removed by water extraction was ~*70 g*. The TCE decomposed by chemical oxidation (calculated from the chloride content of the extracted water) was ~*84 g*. Altogether *154 g* TCE was degraded and removed during 1 week.

Comparing this result with the conventional pump and treat, which removed 25 kg (25,000 g) TCE from 23,000 m³ water by stripping (within 40 months), i.e. approximately 1 g/m³, the ISCO in combination with the intra-well agitation could remove 154 g/8 m³ = 19 times more than pump and treat. More than half of that amount was completely eliminated, i.e. degraded, thus there was no need for further treatment. Considering the degraded amount

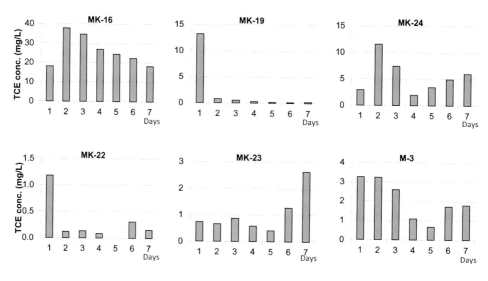

Figure 6.17 TCE concentration in the more contaminated area in the wells MK-16, MK-19, MK-22, MK-23, MK-24, and M-3 treated for the first 4 days with hydrogen peroxide.

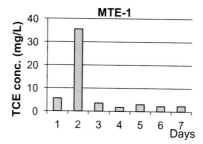

Figure 6.18 TCE concentration in the strongly contaminated area in the extraction well MTE-1 after the treatment of wells MK-16, MK-19, MK-22, MK-23, MK-24, and M-3 for the first 4 days with hydrogen peroxide.

only, the method was 10 times more efficient than pump and treat in removal and 100% more efficient from the point of view of residual wastes treatment. Conventional pump and treat achieved 25,000 g/1200 days = 20 g/day daily removal compared to 154 g/day, of the innovative solution which is a 7.7 times higher efficiency. On day 8, the samples were analyzed for other chlorinated aliphatics using the GCMS method. TCE mobilization by inter-well agitation caused a consistently high dissolved concentration in the groundwater in the source zone, meaning that the TCE could be mobilized from the lenses and become available for the dosed oxidant. In the samples from the most contaminated wells 18.1 and 2.1 mg/L (MK-16), 2.6 and 0.2 mg/L (MK-23), 6.0 and 0.3 mg/L (MK-24), and 1.8 and 0.2 mg/L (M-3) TCE were measured. Dichloroethylene, trichloroethane concentrations were all below the limit. Vinyl chloride and tetrachloroethane were not detected in any of the samples.

In the less-contaminated area two experiments for *in situ* treating and agitation were carried out. The locations of the wells are shown in Figure 6.13. For the wells MTE-2 and MTE-3 (two wells near each other), the alternating method was applied for the first 4 days then both wells were treated every day. The treatment included mixing 5 L hydrogen peroxide and phosphoric acid to the extracted 180 L groundwater and injected back. The extraction rate in the actual extraction well was 8 L/min. In the well MTE-2, the TCE concentration first increased then gradually decreased over the course of the alternating injection and extraction. In the well MTE-3 the decrease of TCE concentration started immediately after treatment began (Figure 6.19). The extracted groundwater (14.4 m³ in either well) removed 23.3 g TCE (13.4 g and 9.9 g TCE from wells MTE-2 and MTE-3).

In the less-contaminated area, wells MTE-4, MMF-3 and MMF-4 were treated and MTE-5 and MTE-7 were used as extraction wells.

At the end of the experiment the TCE concentration in wells MTE-2, MTE-5, MMF-3, and MMF-4 was below the official limit for the site (TCE: 0.15 mg/L, total chlorinated

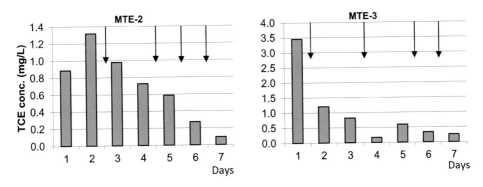

Figure 6.19 TCE concentration in the wells MTE-2 and MTE-3 alternately treated (the arrows indicate the hydrogen peroxide treatments).

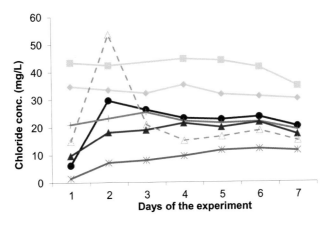

Figure 6.20 Typical chloride profiles in various wells from the source zone – MTE-1 (D), MK-23 (✳), MK-24 (●), M-3 (▲) – and from the less contaminated area – MMF-3 (■), MTE-5 (◆) and MTE-22 (+).

hydrocarbon: 0.3 mg/L). On the day eight the samples were analyzed for the other chlorinated aliphatics using the GCMS method. PCE, dichloroethylene, trichloroethane concentrations were all below the limit. Vinyl chloride and tetrachloroethane were not detected in any of the samples. The less contaminated part of the site downstream the source zone was cleaned up after 1 week of treatment, as the contaminant was resupplied in dissolved form.

The chloride concentration increased during the treatment in the strongly polluted area (source zone) (wells MK16, MK-23, and M-3). In well MTE-1 the chloride content showed an outstandingly high value on the second day similarly to the TCE concentration. Some typical chloride profiles are shown in Figure 6.20.

In the less-contaminated area the TCE decomposed by chemical oxidation (calculated from the enhancement of the chloride content in the extracted 86.4 m^3 water) was ~*131 g*. TCE removed by extraction and stripping was ~*50 g*. The highest and lowest chloride concentrations measured in the extraction wells were 25.4–18.9 mg/L, 34.5–20.9 mg/L, 35.3–30.0 mg/L, and 29.4–24.9 mg/L for MTE-2, MTE-3, MTE-5, and MTE-7, respectively. Based on this 1-week experiment, the technology proved the efficiency of the *in situ* chemical oxidation using hydrogen peroxide and phosphoric acid combined with intensive transfer of groundwater between the wells. After this demonstration experiment the site remediation was continued using this technology. As a result the contamination diminished in most of the wells in the less contaminated area. After post-monitoring in every 6 months for 2 years, this area was *declared to be non-risky* (the contamination decreased permanently below the regulatory limit). The extraction and injection wells were filled with soil. The monitoring wells have been continued to be sampled annually.

In summary, the ISCO combined with alternating pumping efficiently reduced the contaminant concentration:

- Intensified mixing of the soil phases is ensured and probably even the contaminant plumes are removed;
- The reagents in the applied concentration are not harmful for the soil life;
- The end products of the reaction (CO_2, chloride, and water) do not pollute the environment;
- In comparison to a conventional pump and treat technology the contaminant removal is 19 times better, complete degradation is 10 times more efficient than TCE removal by stripping;
- Time requirement is 7.7 times less than that of the pump and treat

REFERENCES

Abdel-Moghny, T., Mohamed, R.S.A., El-Sayed, E., Aly, S.M. & Snousy, M.G. (2012) Removing of hydrocarbon contaminated soil via air flushing enhanced by surfactant. *Applied Petrochemical Research*, 2(1), 51–59.

ABS Materials (2017) *Remediation Solutions*. Available from: http://abswastewater.com/wp-content/uploads/2016/08/Remediation-Solutions.pdf. [Accessed 1st May 2017].

Alhomair, S.A., Gorakhki, M.H. & Bareither, C.A. (2017) Hydraulic conductivity of fly ash-amended mine tailings. *Geotechnical and Geological Engineering*, 35(1), 243–261. doi:10.1007/s10706-016-0101-z.

Ali, M.M. & Sandya, K.Y. (2014) Visible light responsive titanium dioxide-cyclodextrin-fullerene composite with reduced charge recombination and enhanced photocatalytic activity. *Carbon*, 70, 249–257.

Al-Tabbaa, A. & Perera, A.S.R. (2003) Stabilisation/solidification binders and technologies: UK current practice and research needs. *Land Contamination & Reclamation*, 11, 71–79.

Alvarez-Guerra, M., Alvarez-Guerra, E., Alonso-Santurde, R., Andrés, A., Coz, A., Soto, J., Gómez-Arozamena, J. & Viguri, J.R. (2008) Sustainable management options and beneficial uses for contaminated sediments and dredged material. *Fresenius Environmental Bulletin*, 17, 1539–1553.

Álvarez-Martín, A., Rodríguez-Cruz, M.S., Andrades, M.S. & Sánchez-Martín, M.J. (2016) Application of a biosorbent to soil: A potential method for controlling water pollution by pesticides. *Environmental Science and Pollution Research*, 23, 9192–9203.

Anderson, M.A. (1992) Influence of surfactants on vapor-liquid partitioning. *Environmental Science & Technology*, 26, 2186–2191.

Anitescu, G. & Tavlarides, L.L. (2006) Supercritical extraction of contaminants from soils and sediments. *Journal of Supercritical Fluids*, 38, 167–180.

Anwar, A. (2011) *The Effect of Soil Type, Water and Organic Materials on the Mechanochemical Destruction of Organic Compounds*. Master thesis, Auckland University of Technology, Auckland, New Zealand.

AOF (2018a) *HydroTurf*. Available from: www.acfenvironmental.com/products/erosion-control/hard-armor/hydroturf/ [Accessed 16th February 2018].

AOF (2018b) *Introduction to Flexterra Flexible Growth Medium*. Available from: www.acfenviron mental.com/wp-content/uploads/2015/09/Hydraulic-Mulch-TMM.pdf. [Accessed 16th February 2018].

Araújo, R., Castro, A.C.M. & Fiúza, A. (2015) The use of nanoparticles in soil and water remediation processes. *Materials Today: Proceedings*, 2(1), 315–320.

ATSDR (2018) *Toxicological Profile for Trichloroethylene (TCE) Agency of Toxic Substances & Desaese Registry*. Toxic Substances Portal. Available from: www.atsdr.cdc.gov/ToxProfiles/tp.asp?id=173&tid=30. [Accessed 8th July 2018].

Ayoub, S.R.A., Uchiyama, H., Iwasaki, K., Doi, T. & Inaba, K. (2008) Effects of several surfactants and high-molecular-weight organic compounds on decomposition of trichloroethylene with zero valent iron powder. *Environmental Technology*, 29, 363–373.

Badr, T., Hanna, K. & de Brauer, C. (2004) Enhanced solubilization and removal of naphthalene and phenanthrene by cyclodextrins from two contaminated soils. *Journal of Hazardous Materials*, 112, 215–223.

Balawejder, M., Antos, P., Czyjt-Kuryło, S., Józefczyk, R. & Pieniazek, M. (2014) A novel method for degradation of DDT in contaminated soil. *Ozone: Science and Engineering*, 36, 166–173.

Balawejder, M., Józefczyk, R., Antos, P. & Pieniążek, M. (2016) Pilot-scale installation for remediation of DDT-contaminated soil. *Ozone: Science and Engineering*, 38, in press.

Bandala, E.R., Cossio, H., Sánchez-Lopez, A.D., Córdova, F., Peralta-Heránández, J.M. & Torres, L.G. (2013) Scaling-up parameters for site restoration process using surfactant-enhanced soil washing coupled with wastewater treatment by Fenton and Fenton-like processes. *Environmental Technology*, 34, 363–371.

Bandosz, T.J. Jagiello, J., Amankwah, K.A.G. & Schwarz, J.A. (1992) Chemical and structural properties of clay minerals modified by inorganic and organic material. *Clay Minerals*, 27, 435–444.

Bates, E.R., Sahle-Demessie, E. & Grosse, D.W. (2000) Solidification/stabilization for remediation of wood preserving sites: Treatment for dioxins, PCP, creosote, and metals. *Remediation Journal*, 10, 51–65.

Beesley, L., Moreno-Jiménez, E., Gomez-Eyles, J.L., Harris, E., Robinson, B. & Sizmur, T. (2011) A review of biochars' potential role in the remediation, revegetation and restoration of contaminated soils. *Environmental Pollution*, 159(12), 3269–3282.

Berg, L.L. (2000) *Explanation of Significant Differences for the Trailer 5475 Groundwater Remediation*. Lawrence Livermore National Laboratory, Livermore Site. Available from: www.erd.llnl.gov/library/AR-136189.pdf. [Accessed 14th June 2016].

Berselli, S., Milone, G., Canepa, P., di Gioia, D. & Fava, F. (2004) Effects of cyclodextrins, humic substances, and rhamnolipids on the washing of a historically contaminated soil and on the aerobic bioremediation of the resulting effluents. *Biotechnology and Bioengineering*, 88, 111–120.

Betancurt-Corredor, B., Pino, N.J., Cardona, S. & Peñuela, G.A. (2015) Evaluation of biostimulation and Tween 80 addition for the bioremediation of long-term DDT-contaminated soil. *Journal of Environmental Sciences*, 28, 101–109.

Birke, V. (2018) *Economic and Ecologically Favorable Detoxification of Polyhalogenated Pollutants Applying the DMCR Technology*. Available from: http://tribochem.com/downloads/download/feahi_eng.pdf. [Accessed 16th February 2018].

Birke, V., Mattik, J. & Runne, D. (2004) Mechanochemical reductive dehalogenation of hazardous polyhalogenated contaminants. *Journal of Materials Science*, 39(16), 5111–5116.

Blanford, W.J., Barackman, M., Boving, T.B., Klingel, E. & Brusseau, M. (2000) Cyclodextrin-enhanced vertical flushing of a trichloroethene contaminated aquifer. *Ground Water Monitoring and Remediation*, 21, 58–66.

Blanford, W.J., Boving, T. & Wade, R. (2006) Aquifer monitoring shows complex-sugar flushing increases potential for enhanced biodegradation. *Technology News and Trends*, EPA, 25, 3–4.

Blanford, W.J., Pecoraro, M.P., Heinrich, R. & Boving, T.B. (2018) Enhanced reductive de-chlorination of a solvent contaminated aquifer through addition and apparent fermentation of cyclodextrin. *Journal of Contaminant Hydrology*, 208, 68–78.

Block, P., Brown, R. & Robinson, D. (2004) Novel activation technologies for sodium persulfate *in situ* chemical oxidation. *4th International Conference of Chlorinated and Recalcitrant Compounds, Monterey, CA, 24–27 May 2004*, Monterey, CA, USA. Available from: www.peroxychem.com/media/22892/FMC_Peroxygen_Talk_2010-10_Persulfate_Oxidation_and_Reduction_Reactions.pdf. [Accessed 14th May 2017].

Bocos, E., Fernandez-Costas, C., Pazos, M. & Sanroman, M.T. (2015) Removal of PAHs and pesticides from polluted soils by enhanced electrokinetic-Fenton treatment. *Chemosphere*, 125, 168–174.

Bone, B.D., Barnard, L.H., Boardman, D.I., Carey, P.J., Hills, C.D., Jones, H.M., MacLeod, C.L. & Tyrer, M. (2005) *Review of Scientific Literature on the Use of Stabilisation/Solidification for the Treatment of Contaminated Soil, Solid Waste, and Sludges*. CL:AIRE. Guidance Bulletin, GB01, 2005, 8. Available from: www.claire.co.uk/information-centre/cl-aire-publications. [Accessed 14th May 2017].

Boparai, H.K., Comfort, S.D., Shea, P.J. & Szecsody, J.E. (2008) Remediating explosive-contaminated groundwater by *in situ* redox manipulation (ISRM) of aquifer sediments. *Chemosphere*, 71, 933–994.

Boving, T.B., Barnett, S.M., Perez, G., Blanford, W.J. & McCray, J.E. (2007) Remediation with cyclodextrin: Recovery of the remedial agent by membrane filtration. *Remediation Journal*, 17, 21–36. doi:10.1002/rem.20131.

Boving, T.B., Blanford, W.J., McCray, J.E., Divine, C.E. & Brusseau, M.L. (2008) Comparison of line-drive and push-pull flushing schemes. *Groundwater Monitoring and Remediation*, 28(1), 75–86.

Boving, T.B. & Brusseau, M.L. (2000) Solubilization and removal of residual trichloroethene from porous media: Comparison of several solubilization agents. *Journal of Contaminant Hydrology*, 42, 51–67.

Bowman, R.S. (2002) Applications of surfactant-modified zeolites to environmental remediation. *Microporous and Mesoporous Materials*, 61, 43–56.

Brillas, E., Sirés, I. & Oturan, M.A. (2009) Electro-Fenton process and related electrochemical technologies based on Fenton's reaction chemistry. *Chemical Reviews*, 109(12), 6570–6631.

Brown, R. (2003) *In situ* chemical oxidation: Performance, practice, and pitfalls. *2003 AFCEE Technology Transfer Workshop, 25 February 2003, San Antonio, TX*. Available from: https://clu-in.org/download/techfocus/chemox/4_brown.pdf. [Accessed 16th May 2018].

Brusseau, M.L., Wang, X. & Wang, W.-Z. (1997) Simultaneous elution of heavy metals and organic compounds from soil by cyclodextrin. *Environmental Science & Technology*, 31, 1087–1092.

Burns, S.E. & Zhang, M. (2001) Effects of system parameters on the physical characteristics of bubbles produce through air sparging. *Environmental Science & Technology*, 35, 204–208.

Cabrera, A., Cox, L., Spokas, K.A., Celis, R., Hermosín, M.C., Cornejo, J. & Koskinen, W.C. (2011) Comparative sorption and leaching study of the herbicides fluometuron and 4-chloro-2-methylphenoxyacetic acid (MCPA) in a soil amended with biochars and other sorbents. *Journal of Agricultural and Food Chemistry*, 59, 12550–12560.

Cabrera, A., Cox, L., Spokas, K.A., Hermosín, M.C., Cornejo, J. & Koskinen, W.C. (2014) Influence of biochar amendments on the sorption – Desorption of aminocyclopyrachlor, bentazone and pyraclostrobin pesticides to an agricultural soil. *Science of the Total Environment*, 470–471, 438–443.

Caliman, F.A., Robu, B.M., Smaranda, C., Pavel, C.L. & Gavrilescu, M. (2011) Soil and ground water clean-up: Benefits and limits of emerging technologies. *Clean Technologies and Environmental Policy*, 13, 241–268.

Cao, X., Ma, L., Gao, B. & Harris, W. (2009) Dairy-manure derived biochar effectively sorbs lead and atrazine. *Environmental Science and Technology*, 43, 3285–3291.

Caschili, S., Delogu, F., Concas, A., Pisu, M. & Cao, G. (2006) Mechanically induced self propagating reactions: Analysis of reactive substrates and degradation of aromatic sulfonic pollutants. *Chemosphere*, 63, 987–995.

Centofanti, T., McConnell, L.L., Chaney, R.L., Beyer, N.W., Andrade, N.A., Hapeman, C.J., Torrents, A., Nguyen, A., Anderson, M.O., Novak, J.M. & Jackson, D. (2016) Organic amendments for risk mitigation of organochlorine pesticide residues in old orchard soils. *Environmental Pollution*, 210, 182–191.

Chang, I., Im, J. & Cho, G.C. (2016) Introduction of microbial biopolymers in soil treatment for future environmentally-friendly and sustainable geotechnical engineering. *Sustainability*, 8, 251–274. doi:10.3390/su8030251.

Chatain, V., Hanna, K., de Brauer, C., Bayard, R. & Germain, P. (2004) Enhanced solubilization of arsenic and 2,3,4,6-tetrachlorophenol from soils by a cyclodextrin derivative. *Chemosphere*, 57, 197–206.

Chen, C.F., Binh, N.T., Chen, C.W. & Dong, C.D. (2014) Removal of polycyclic aromatic hydrocarbons from sediments using sodium persulfate activated by temperature and nanoscale zero-valent iron. *Journal of the Air & Waste Management Association*, 65(4), 375–383. doi:10.1080/1096224 7.2014.996266.

Chen, F., Luo, Z., Liu, G., Yang, Y., Zhang, S. & Ma, J. (2017a) Remediation of electronic waste polluted soil using a combination of persulfate oxidation and chemical washing. *Journal of Environmental Management*, 204, 170–178.

Chen, K.F., Chang, Y.C. & Chiou, W.T. (2015) Remediation of diesel-contaminated soil using In Situ Chemical Oxidation (ISCO) and the effects of common oxidants on the indigenous microbial community: A comparison study. *Chemical technology and Biotechnology*, 91, 1877–1888.

Chen, W., Zou, C., Liu, Y. & Li, X. (2017b) The experimental investigation of bisphenol a degradation by Fenton process with different types of cyclodextrins. *Journal of Industrial and Engineering Chemistry*, 56, 428–434.

Cheng, M., Zeng, G., Huang, D., Lai, C., Xu, P., Zhang, C. & Liu, Y. (2016) Hydroxyl radicals based Advanced Oxidation Processes (AOPs) for remediation of soils contaminated with organic compounds: A review. *Chemical Engineering Journal*, 284, 582–598.

Cheremisinoff, N.P. & Davletshin, A. (2015) *Hydraulic Fracturing Operations: Handbook of Environmental Management Practices*. John Wiley & Sons, New York, NY, USA.

Choy, W.K. & Chu, W. (2001a) The modelling of trichloroethene photodegradation in Brij 35 surfactant by two-stage reaction. *Chemosphere*, 44(2), 211–215.

Choy, W.K. & Chu, W. (2001b) The rate improvement and modeling of trichloroethene photodegradation by acetone sensitizer in surfactant solution. *Chemosphere*, 44(5), 943–947.

Chu, W. & Choy, W.K. (2000) The study of lag phase and rate improvement of TCE decay in UV/ surfactant systems. *Chemosphere*, 41(8), 1199–1204.

Churngold (2009) *Steam Enhanced Remediation of Creosote at Former Timber Treatment Facility*. Available from: www.churngold.com/case-studies/remediation/steam-enhanced-remediation-of-creosote-at-former-timber-treatment-facility.html. [Accessed 23th January 2018].

Churngold (2018) *Steam Enhanced Remediation*. Available from: www.churngold.com/remediation/ thermal-technologies/steam-enhanced-remediation.html. [Accessed 23th January 2018].

CL:AIRE (2007) *Understanding Soil Washing*. Technical Bulletin TB13, 2007. Available from: www.claire.co.uk/information-centre/cl-aire-publications. [Accessed 6th May 2018].

Clue in (2018) *Nanotechnology: Application for Environmental Remediation*. Available from: https://clu-in.org/techfocus/default.focus/sec/Nanotechnology:_Applications_for_Environmental_Remediation/cat/Application/#10. [Accessed 16th February 2018].

Cong, X., Xue, N., Wang, S., Li, K. & Li, F. (2010) Reductive dechlorination of organochlorine pesticides in soils from an abandoned manufacturing facility by zero-valent iron. *Science of the Total Environment*, 408, 3418–3423.

Contech (2018) Available from: www.conteches.com/products/erosion-control/temporary-and-permanent/turf-reinforcement-mats. [Accessed 6th March 2018].

CPEO (2018) *Center for Public Environmental Oversight*. Available from: www. cpeo.org. [Accessed 25th February 2018].

Davidson, B., Spanos, T. & Zschuppe, R. (2018) *Pressure Pulse Technology: An Enhanced Fluid Flow and Delivery Mechanism*. Available from: www.onthewavefront.com/inc/pdfs/tech/primawave/articles/davidson-chlorcon-paper.pdf. [Accessed 16th February 2018].

Davis, E.L. (1998) *Steam Injection for Soil and Aqufer Remediation*. EPA Ground Water Issue. Available from: https://nepis.epa.gov/Exe/ZyPDF.cgi/10002E1U.PDF?Dockey=10002E1U.PDF. [Accessed 16th July 2017].

de S. e Silva, P.T., Locatelli, M.A.F., Jardim, W.F., Neto, B.B., da Motta, M., de Castro, G.R. & da Silva, V.L. (2008) Endogenous iron as a photo-Fenton reaction catalyst for the degradation of Pah's in soils. *Journal of the Brazilian Chemical Society*, 19(2). Available from: http://dx.doi.org/10.1590/S0103-50532008000200020. [Accessed 16th May 2018].

Dettmer, A., Carroll, K.C., Schaub, T., Khan, N., Appuhamilage, N.S., Cruz, S., Ball, R., Boving, T.B. & Fernandez, C.A. (2017) Stabilization and prolonged reactivity of aqueous-phase ozone with cyclodextrin. *Journal of Contaminant Hydrology*, 196, 1–9.

Diallo, M.S., Christie, S., Swaminathan, P., Johnson, J. Jr. & Goddard III, W. (2005) Dendrimer enhance ultrafiltration. 1. Recovery Cu(II) from aqueous solutions using PAMAM dendrimers with ethylene diamine core and terminal NH_2 groups. *Environmental Science and Technology*, 39(5), 1366–1377.

Di Palma, L. (2003) Experimental assessment of a process for the remediation of organophosphorous pesticides contaminated soils through *in situ* soil flushing and hydrolysis. *Water, Air, & Soil Pollution*, 143, 301–314.

Di Palma, L., Ferrantelli, P. & Petrucci, E. (2003) Experimental study of the remediation of atrazine contaminated soils through soil extraction and subsequent peroxidation. *Journal of Hazardous Materials*, 99, 265–276.

Dos Santos, E.V., Sáez, C., Martínez-Huitle, C.A., Cañizares, P. & Rodrigo, M.A. (2015) Combined soil washing and CDEO for the removal of atrazine from soils. *Journal of Hazardous Materials*, 300, 129–134.

Dove, D., Bhandari, A. & Novak, J. (1992) Soil washing: Practical considerations and pitfalls. *Remediation Journal*, 3, 55–67.

Du, Y.J., Yu, B.W., Liu, K., Jiang, N.J. & Liu, M.D. (2016) Physical, hydraulic, and mechanical properties of clayey soil stabilized by lightweight alkali-activated slag geopolymer. *Journal of Materials in Civil Engineering*, 29(2). doi:10.1061/(asce)mt.1943-5533.0001743.

ECBVerdyol (2018) *Erosion Control Blanket*. Available from: www.erosioncontrolblanket.com/ [Accessed 6th March 2018].

Ehsan, S., Prasher, S.O. & Marshall, W.D. (2007) Simultaneous mobilization of heavy metals and Polychlorinated Biphenyl (PCB) compounds from soil with cyclodextrin and EDTA in admixture. *Chemosphere*, 68, 150–158.

Enchem (2010) *Advanced Mixed Oxidation and Inclusion Technology*. EPA Research Grant EPD10024. Enchem Engineering Inc. Available from: https://cfpub.epa.gov/ncer_abstracts/index.cfm/fuseaction/display.highlight/abstract/9085/report/F. [Accessed 14th May 2018].

ENFO Database (2018) Available from: www.körinfo.hu/drupal/en. [Accessed 14th March 2018].

EPA (1998) *Field Applications of in situ Remediation Technologies: Chemical Oxidation.* EPA 542-R-98-008. Available from: http://nepis.epa.gov/Exe/ZyPDF.cgi/1000305N.PDF?Dockey=1000305N. PDF. [Accessed 16th May 2017].

EPA CLU-IN (1998) *Cosolvent Flushing Pilot Test Report Former Sages Dry Cleaner.* Florida Department of Environmental Protection, Tallahassee, FL, USA. Available from: https://clu-in.org/download/remed/sages.pdf. [Accessed 16th May 2018].

EPA CLU-IN (2018) *A Citizen's Guide to in situ Soil Flushing.* Available from: https://clu-in.org/download/remed/soilflsh.pdf. [Accessed 14th July 2018].

EPA CLU-IN Database (2018) Available from: www.clu-in.org/ [Accessed 14th March 2018].

ESTCP CU-0431 (2006) *Final Laboratory Treatability Report For: Emulsified Zero Valent Iron Treatment of Chlorinated Solvent DNAPL Source Areas.* Available from: www.dtic.mil/cgi-bin/GetTRD oc?Location=U2&doc=GetTRDoc.pdf&AD=ADA451083. [Accessed 14th May 2018].

Fava, F., Di Gioia, D., Marchetti, L., Fenyvesi, E. & Szejtli, J. (2002) Randomly Methylated β-Cyclodextrins (RAMEB) enhance the aerobic biodegradation of polychlorinated biphenyl in aged-contaminated soils. *Journal of Inclusion Phenomena and Macrocyclic Chemistry*, 44, 417–421.

Fekete-Kertész, I., Molnár, M., Atkári, Á., Gruiz, K. & Fenyvesi, É. (2013) Hydrogen peroxide oxidation for *in situ* remediation of trichloroethylene: From the laboratory to the field. *Periodica Polytechnica: Chemical Engineering*, 57, 41–51.

Fenyvesi, E., Balogh, K., Olah, E., Batai, B., Varga, E., Molnár, M. & Gruiz, K. (2011) Cyclodextrins for remediation of soils contaminated with chlorinated organics. *Journal of Inclusion Phenomena and Macrocyclic Chemistry*, 70, 297–297.

Fenyvesi, E., Gruiz, K., Verstichel, S., De Wilde, B., Leitgib, L., Csabai, K. & Szaniszló, N. (2005) Biodegradation of cyclodextrins in soil. *Chemosphere*, 60, 1001–1008.

Fenyvesi, E., Molnár, M., Leitgib, L. & Gruiz, K. (2009) Cyclodextrin-enhanced soil-remediation technologies. *Land Contamination & Reclamation*, 17, 585–598.

Flaherty, R.J., Nshime, B., DeLaMarre, M., DeJong, S., Scott, P. & Lantz, A.W. (2013) Cyclodextrins as complexation and extraction agents for pesticides from contaminated soil. *Chemosphere*, 91, 912–920.

Foged, S., Duerinckx, L. & Vandekeybus, J. (2018) *An Innovative and Sustainable Solution for Sediment Disposal Problems.* Western Dredging Association. Available from: https://western dredging.org/index.php/woda-conference-presentations/category/73-session-9a-sediment-dewatering-treatment-and-disposal. [Accessed 15th March 2018].

FRTR (2018a) *Remediation Technologies Screening Matrix and Reference Guide.* Federal Remediation Technology Roundtable. Available from: nepis.epa.gov/Exe/ZyPURL.cgi?Dockey=2000KG7K. TXT. [Accessed 14th May 2018].

FRTR (2018b) *In-Well Air Stripping.* Federal Remediation Technologies Roundtable. Available from: https://frtr.gov/matrix2/section4/4-40.html. [Accessed 4th February 2018].

Fryxell, G.E., Mattigold, S.V., Lin, Y., Wu, H., Fiskum, S., Parker, K., Zheng, F., Yantasee, W., Zemanian, T.S., Addleman, R.S., Liu, J., Kemner, K., Kelly, S. & Feng, X. (2007) Design and synthesis of self-assembled monolayers on mesoporous supports (SAMMS): The importance of ligand posture in functional nanomaterials. *Journal of Materials Chemistry*, 17, 2863–2874. doi:10.1039/B702422C.

Gałuszka, A., Migaszewski, Z.M. & Manecki, P. (2011) Pesticide burial grounds in Poland: A review. *Environment International*, 37, 1265–1272.

Gan, X., Teng, Y., Ren, W., Ma, J., Christie, P. & Luo, Y. (2017) Optimization of ex-situ washing removal of polycyclic aromatic hydrocarbons from a contaminated soil using nano-sulfonated graphene. *Pedosphere*, 27(3), 527–536.

Gao, H., Gao, X., Cao, Y., Xu, L. & Jia, L. (2015) Influence of hydroxypropyl-β-cyclodextrin on the extraction and biodegradation of p,p'-DDT, o,p'-DDT, p,p'-DDD, and p,p'-DDE in soils. *Water, Air, & Soil Pollution*, 226, 208–213.

Gao, H., Miles, M.S., Meyer, B.M., Wong, R.L. & Overton, E.B. (2012) Assessment of cyclodextrin-enhanced extraction of crude oil from contaminated porous media. *Journal of Environmental Monitoring*, 14, 2164–2169.

Gao, Y.F., Yang, H., Zhan, X.H. & Zhou, L.X. (2013) Scavenging of BHCs and DDTs from soil by thermal desorption and solvent washing. *Environmental Science and Pollution Research*, 20, 1482–1492.

García-Delgado, C., Alfaro-Barta, I. & Eymar, E. (2015): Combination of biochar amendment and mycoremediation for polycyclic aromatic hydrocarbons immobilization and biodegradation in creosote – Contaminated soil. *Journal of Hazardous Materials*, 285, 259–266.

García-Jaramillo, M., Cox, L., Cornejo, J. & Hermosín, M.C. (2014) Effect of soil organic amendments on the behavior of bentazone and tricyclazole. *Science of the Total Environment*, 466–467, 906–913.

García-Jaramillo, M., Cox, L., Hermosín, M.C., Cerli, C. & Kalbitz, K. (2016) Influence of green waste compost on azimsulfuron dissipation and soil functions under oxic and anoxic conditions. *Science of the Total Environment*, 550, 760–767.

Gauglitz, P.A., Roberts, J.S., Bergsman, T.M., Caley, S.M., Heath, W.O., Miller, M.C., Moss, R.W., Schalla, R., Jarosch, T.R., Dilek, E.C.A & Looney, B.B. (1994) Six-Phase Soil Heating accelerates VOC extraction from clay soil. *Proceedings of Spectrum '94, August 1994*, Atlanta, GA, Georgia.

Geo-Solutions (2018a) *In situ Stabilization: Soil Mixing for Environmental Remediation and Geotechnical Application*. Available from: www.geo-solutions.com/soil-insitu-soil-stabilization-solidification. [Accessed 20th January 2018].

Geo-Solutions (2018b) *Backhoe Operated Soil Stabilization*. Available from: www.youtube.com/watch?v=UOgau92qfS4. [Accessed 20th January 2018].

Geo-Solutions (2018c) *Native Grass Sod*. Available from: www.geosolutionsinc.com/products/erosion-control-native-grass-sod.html. [Accessed 20th January 2018].

Gomes, H.I., Dias-Ferreira, C. & Ribeiro, A.B. (2012) Electrokinetic remediation of organochlorines in soil: Enhancement techniques and integration with other remediation technologies. *Chemosphere*, 87, 1077–1090.

Gomez, J., Alcantara, M.T., Pazos, M. & Sanroman, M.A. (2010) Soil washing using cyclodextrins and their recovery by application of electrochemical technology. *Chemical Engineering Journal (Amsterdam, Netherlands)*, 159, 53–57.

Götz, R., Sokollek, V. & Weber, R. (2013) The dioxin/POPs legacy of pesticide production in Hamburg: Part 2, Waste deposits and remediation of Georgswerder landfill. *Environmental Science and Pollution Research*, 20, 1925–1936.

Gozzi, F., Machulek Jr., A., Ferreira, V.S., Osugi, M.E., Santos, A.P.F., Nogueira, J.A., Dantas, R.F., Esplugas, S. & de Oliveira, S.C. (2012) Investigation of chlorimuron-ethyl degradation by Fenton, photo-Fenton and ozonation processes. *Chemical Engineering Journal*, 210, 444–450.

Gruiz, K. (2000) When the chemical time bomb explodes? – Chronic risk of toxic metals at a former mining site. *Proceedings of ConSoil 2000*, Leipzig, Germany. pp. 662–670.

Gruiz, K. (2009) Contaminated site remediation: Role and classification. *Land Contamination & Reclamation*, 17, 533–542.

Gruiz, K. (2016) Quantifying the risk of contaminated sites. Section 7 (Integrated and efficient characterization of contaminated sites. Chapter 1). In: Gruiz, K., Meggyes, T. & Fenyvesi, E. (eds.) *Engineering Tools for Environmental Risk Management. Volume 3. Site Assessment and Monitoring Tools*. CRC Press, Boca Raton, FL, USA. pp. 74–90.

Gruiz, K. & Molnár, M. (2015) Aquatic toxicology. Chapter 4. In: Gruiz, K., Meggyes, T. & Fenyvesi, E. (eds.) *Engineering Tools for Environmental Risk Management. Volume 2. Environmental Toxicology*. CRC Press, Boca Raton, FL, USA. pp. 171–228.

Gruiz, K., Molnár, M., Feigl, V., Hajdu, C., Nagy, Z.M., Klebercz, O., Fekete-Kertész, I., Ujaczki, É. & Tolner, M. (2015) Terrestrial toxicology. Chapter 5. In: Gruiz, K., Meggyes, T. & Fenyvesi, E. (eds.) *Engineering Tools for Environmental Risk Management. Volume 2. Environmental Toxicology*. CRC Press, Boca Raton, FL, USA. pp. 229–310.

Gruiz, K., Molnár, M. & Fenyvesi, E. (2008) Evaluation and verification of soil remediation. Kurladze, V.G. (ed) *Environmental Microbiology Research Trends*. Nova Science Publishers, Inc., New York, NY, USA. pp. 1–57.

Gruiz, K., Molnár, M., Fenyvesi, É., Hajdu, C., Atkári, Á. & Barkács, K. (2011) Cyclodextrins in innovative engineering tools for risk-based environmental management. *Journal of Inclusion Phenomena and Macrocyclic Chemistry*, 70, 299–306.

Gruiz, K., Vaszita, E. & Clement, O. (2014) Site-specific risk management of point and diffuse sources. In: Gruiz, K., Meggyes, T. & Fenyvesi, E. (eds.) *Engineering Tools for Environmental Risk Management. Volume 1. Environmental Deterioration and Contamination – Problems and Their Management*. CRC Press, Boca Raton, FL, USA.

Guo, H., Zhang, J., Liu, Z., Yang, S. & Sun, C. (2010) Effect of Tween80 and β-cyclodextrin on the distribution of herbicide mefenacet in soil – Water system. *Journal of Hazardous Materials*, 177, 1039–1045.

Hall, A.K., Harrowfield, J.M., Hart, R.J. & McCormick, P.G. (1996) Mechanochemical reaction of DDT with calcium oxide. *Environmental Science & Technology*, 30(12), 3401–3407.

Hamer, K. & Karius, V. (2002) Brick production with dredged harbour sediments. An industrial-scale experiment. *Waste Management*, 22, 521–530.

Han, Y., Shi, N., Wang, H., Pan, X., Fang, H. & Yu, Y. (2016) Nanoscale zerovalent iron-mediated degradation of DDT in soil. *Environmental Science and Pollution Research*, 23, 6253–6263.

Hanna, K., Chiron, S. & Oturan, M.A. (2005) Coupling enhanced water solubilization with cyclodextrin to indirect electrochemical treatment for pentachlorophenol contaminated soil remediation. *Water Research*, 39, 2763–2773.

Hanna, K., de Brauer, Ch., Germain, P., Chovelon, J.M. & Ferronato, C. (2004) Degradation of pentachlorophenol in cyclodextrin extraction effluent using a photocatalytic process. *Science of the Total Environment*, 332, 51–60.

Harrington, G. & Smith, J. (2013) *Guidance on the Beneficial Use of Dredge Material in Ireland*. Available from: www.epa.ie/pubs/reports/research/sss/Beneficial%20Use%20of%20Dredging%20Material.pdf. [Accessed 16th July 2018].

Haselow, J.S., Siegrist, R.L., Crimi, M. & Jarosch, T. (2003) Estimating the total oxidant demand for *In situ* chemical oxidation design. *Remediation Journal*, 13, 5–16.

Higarashi, M.M. & Jardim, W.F. (2002) Remediation of pesticide contaminated soil using TiO2 mediated by solar light. *Catalysis Today*, 76, 201–207.

Hill, K. & Haber, R.A. (2012) Use of Mid-Delaware river dredged sediment as a raw material in ceramic processing. In: Sundaram, S.K., Spearing, D.R. &'Vienna, J.D. (eds.) *Environmental Issues and Waste Management Technologies in the Ceramic and Nuclear Industries VIII*. John Wiley and Sons, Hoboken, NJ, USA.

Hinrichs, R.M. (2004) *Post Monitoring of a Cyclodextrin Remediated Chlorinated Solvent Contaminated Aquifer*. Master's Thesis, Luisana State University. Available from: http://digital commons.lsu.edu/cgi/viewcontent.cgi?article=2452&context=gradschool_theses. [Accessed 16th July 2018].

Hirschberger, F. (1998) *Remediation of a CHC Contamination in Viernheim, Germany*. Available from: http://aguas.igme.es/igme/publica/pdflib15/027.pdf. [Accessed 13th May 2018].

Hishimoto, K., Irie, H. & Fujishima, A. (2005) TiO$_2$ photocatalysis: A historical overview and future prospects. *Japanese Journal of Applied Physics*, 44, 8269–8285.

Hrapovic, L., Sleep, B.E., Major, D.J. & Hood, E.D. (2005) Laboratory study of treatment of trichloroethene by chemical oxidation followed by bioremediation. *Environmental Science & Technology*, 39, 2888–2897.

Hu, C.J., Huang, D.L., Zeng, G.M., Cheng, M., Gong, X., Wang, R.Z., Xue, W.J., Hu, Z. & Liu, Y. (2018) The combination of Fenton process and Phanerochaete chrysosporium for the removal of bisphenol A in river sediments: Mechanism related to extracellular enzyme, organic acid and iron. *Chemical Engineering Journal*, 338, 432–439.

Hu, J., Wang, Y., Su, X., Yu, C., Qin, Z., Wang, H., Hashmic, M.Z., Shi, J. & Shen, C. (2016) Effects of RAMEB and/or mechanical mixing on the bioavailability and biodegradation of PCBs in soil/slurry. *Chemosphere*, 155, 479–487.

Huang, D., Hu, C., Zeng, G., Cheng, M., Xu, P., Gong, X., Wang, R. & Xue, W. (2017) Combination of Fenton processes and biotreatment for wastewater treatment and soil remediation. *Science of the Total Environment*, 574, 1599–1610.

Huang, D., Xu, Q., Cheng, J., Lu, X. & Zhang, H. (2012) Electrokinetic remediation and its combined technologies for removal of organic pollutants from contaminated soils. *International Journal of Electrochemical Science*, 7, 4528–4544.

Hwang, I. & Batchelor, B. (2000) Reductive dechlorination of tetrachloroethylene by Fe(II) in cement slurries. *Environmental Science & Technology*, 34, 5017–5022.

IEG Technologie (2018) *Soil Air Circulation Systems.* Available from: www.ieg-technology.com/en/Soil-and-Groundwater-Remediation-Technologies/Vacuum-Vapour-Extraction/Soil-Air-Circulation-Vacuum-Vapour-Extraction.html. [Accessed 16th January 2018].

Imhoff, P.T., Frizzell, A. & Miller, C.T. (1997) Evaluation of thermal effects on the dissolution of a nonaqueous phase liquid in porous media. *Environmental Science & Technology*, 31(6), 1615–1622. doi:10.1021/es960292x.

Ishiwata, S. & Kamiya, M. (1999) Effects of humic acids on the inclusion complexation of cyclodextrins with organophosphorus pesticides. *Chemosphere*, 38, 2219–2226.

Jasperse, B.H. (2018) *In-situ Stabilization Using Shallow Soil Mixing and Deep Soil Mixing, Geo-con.* Available from: www.geocon.net/pdf/paper03.pdf. [Accessed 16th February 2018].

Jiradecha, C., Urgun-Demirtas, M. & Pagilla, K. (2006) Enhanced electrokinetic dissolution of naphthalene and 2,4-DNT from contaminated soils. *Journal of Hazardous Materials*, 136(1), 61–67.

Karagunduz, A. (2009) Electrokinetic transport of chlorinated organic pesticides. In: Reddy, K.R. & Cameselle, C. (eds.) *Electrochemical Remediation Technologies for Polluted Soils, Sediments and Groundwater.* John Wiley & Sons, Inc., Hoboken, NJ, USA. pp. 235–248.

Karagunduz, A., Gezer, A. & Karasuloglu, G. (2007) Surfactant enhanced electrokinetic remediation of DDT from soils. *Science of the Total Environment*, 385, 1–11.

Kashiyama, N. & Boving, T.B. (2004) Hindered gas-phase partitioning of trichloroethylene from aqueous cyclodextrin systems: Implications for treatment and analysis. *Environmental Science & Technology*, 38, 4439–4444.

Kerfoottech (2018) *C-SPARGER® "Genesis of Ozone Sparging".* Available from: www.kerfoottech.com/environmental-technology-products-c-sparger.asp. [Accessed 16th February 2018].

Khorram, M.S., Zhang, Q., Lin, D., Zheng, Y., Fang, H. & Yu, Y. (2016) Biochar: A review of its impact on pesticide behavior in soil environments and its potential applications. *Journal of Environmental Sciences*, 44, 269–279.

Kim, H., Soh, H.E., Annable, M.D. & Kim, D.J. (2004) Surfactant-enhanced air sparging in saturated sand. *Environmental Science & Technolology*, 38(4), 1170–1175.

Klens, J., Pohlmann, D., Scarborough, S. & Graves, D. (2001) The effects of permanganate oxidation on subsurface microbial populations. In: Leeson, A., Kelley, M.E., Rifai, H.S. & Magar, V.S. (eds.) *Natural Attenuation of Environmental Contaminants* 6(2). Andrea Battelle Press, Columbus, OH, USA. pp. 253–259.

Ko, J.H., Musson, S. & Townsend, T. (2010) Destruction of trichloroethylene during hydration of calcium oxide. *Journal of Hazardous Materials*, 174, 876–879.

Ko, S.-O., Schlautman, M.A. & Carraway, E. (2000) Cyclodextrin-enhanced electrokinetic removal of phenanthrene from a model clay soil. *Environmental Science & Technolology*, 34(8), 1535–1541.

Ko, S.O. & Yoo, H.C. (2003) Enhanced desorption of phenanthrene from soils using hydroxypropyl-beta-cyclodextrin: Experimental results and model predictions. *Journal of Environmental Science and Health Part B Pesticides, Food Contaminants, and Agricultural Wastes*, B38, 829–841.

Koehlert, K. (2018) Activated carbon: Fundamentals and new applications. *Chemical Engineering*, (7), 32–40. Available from: www.chemengonline.com/activated-carbon-fundamentals-new-applications. [Accessed 16th March 2018].

Koltowski, M., Hilber, I., Bucheli, T.D. & Oleszczuk, P. (2016): Effect of steam activated biochar application to industrially contaminated soils on bioavailability of polycyclic aromatic hydrocarbons and ecotoxicity of soils. *Science of the Total Environment*, 556–567, 1023–1031.

Kukacka, J., Vána, J. & Urban, O. (2010) Investigation and remediation of oil lagoons – Selected technological approaches. In: Sarby, R. & Meggyes, T. (eds.) *Construction for a Sustainable Environment*. CRC Press Francis & Taylor Group, London, UK. pp. 251–259.

Kupryianchyk, D., Hale, S., Zimmerman, A.R., Harvey, O., Rutherford, D., Abiven, S., Knicker, H., Schmidt, H.P., Rumpel, C. & Cornelissen, G. (2016) Sorption of hydrophobic organic compounds to a diverse suite of carbonaceous materials with emphasis on biochar. *Chemosphere*, 144, 879–887.

Laha, S., Tansel, B. & Ussawarujikulchai, A. (2009) Surfactant-soil interactions during surfactant-amended remediation of contaminated soils by hydrophobic organic compounds: A review. *Journal of Environmental Management*, 90, 95–100.

Lau, E.V., Gan, S., Ng, H.K. & Poh, P.E. (2014) Extraction agents for the removal of Polycyclic Aromatic Hydrocarbons (PAHs) from soil in soil washing technologies. *Environmental Pollution*, 184, 640–649.

Lee, K.M., Lai, C.W., Ngai, K.S. & Juan, J.C. (2016) Recent developments of zinc oxide based photocatalyst in water treatment technology: A review. *Water Research*, 88, 428–448.

Lehr, J.H. (2004) *Wiley's Remediation Technologies Handbook: Major Contaminant Chemicals and Chemical Groups*. Wiley Interscience, Hoboken, NJ, USA.

Leitgib, L., Gruiz, K., Fenyvesi, E., Balogh, G. & Murányi, A. (2007) Development of an innovative soil remediation: 'Cyclodextrin-enhanced combined technology'. *Science of Total Environment*, 392, 12–21.

Li, X., Du, Y., Wu, G., Li, Z., Li, H. & Sui, H. (2012) Solvent extraction for heavy crude oil removal from contaminated soils. *Chemosphere*, 88(2), 245–249. doi: 10.1016/j.chemosphere.2012.03.021.

Li, X.H., Wang, W., Wang, J., Cao, X.L., Wang, X.F., Liu, J.C., Liu, X.F., Xu, X.B. & Jiang, X.N. (2008) Contamination of soils with organochlorine pesticides in urban parks in Beijing, China. *Chemosphere*, 70, 1660–1668.

Liang, C.J., Huang, C.F., Mohanty, N., Lu, C.J. & Kurakalva, R.M. (2007) Hydroxypropyl-beta-cyclodextrin-mediated iron-activated persulfate oxidation of trichloroethylene and tetrachloro ethylene. *Industrial & Engineering Chemistry Research*, 46, 6466–6479.

Liang, C.J. & Lee, I.L. (2008) *In situ* iron activated persulfate oxidative fluid sparging treatment of TCE contamination – A proof of concept study. *Journal of Contaminant Hydrology*, 100, 91–100.

Lindsey, M.E., Xu, G., Lu, J. & Tarr, M.A. (2003) Enhanced Fenton degradation of hydrophobic organics by simultaneous iron and pollutant complexation with cyclodextrins. *Science of the Total Environment*, 307, 215–229.

Liu, C.S., Shih, K., Sun, C.X. & Wang, F. (2012) Oxidative degradation of propachlor by ferrous and copper ion activated persulfate. *Science of the Total Environment*, 416, 507–512.

Liu, Z., Jiao, S. & Guo, Y. (2011) The research of marine dredged sediment ceramic's application in pavement – The performance experiment of a low-carbon road construction material. *International Conference on Electrical and Control Engineering (ICECE)*. doi:10.1109/ICECENG.2011.6058155. Available from: http://ieeexplore.ieee.org/document/6058155/?reload=true. [Accessed 15th March 2018].

Loden, M.E. (1992) *A Technology Assessment of Soil Vapor Extraction and Air Sparging*. US EPA/600/R-92/173, US EPA Office of Research and Development. Available from: https://nepis.epa.gov/Exe/ZyPDF.cgi/30003W66.PDF?Dockey=30003W66.PDF. [Accessed 16th May 2017].

Lofrano, G., Libralato, G. & Brown, J. (2017) *Nanotechnologies for Environmental Remediation: Applications and Implications*. Springer, New York, NY, USA.

Lominchar, M.A., Santos, A., de Miguel, E. & Romero, A. (2018) Remediation of aged diesel contaminated soil by alkaline activated persulfate. *Science of the Total Environment*, 622–623, 41–48. doi:10.1016/j.scitotenv.2017.11.263.

Long, R.Q. & Yang, R.T. (2001) Carbon nanotubes as superior sorbent for dioxin removal. *Journal of the American Chemical Society*, 123(9), 2058–2059.

López-Piñeiro, A., Peña, D., Albarrán, A., Sánchez-Llerena, J. & Becerra, D. (2013) Behavior of MCPA in four intensive cropping soils amended with fresh, composted, and aged olive mill waste. *Journal of Contaminant Hydrology*, 152, 137–146.

Makusa, G.P. (2012) *Soil Stabilization Methods and Materials. State of the Art Review*. Luleå University of Technology, Luleå, Sweden.

Manickam, N., Bajaj, A., Saini, H.S. & Shanker, R. (2012) Surfactant mediated enhanced biodegradation of hexachlorocyclohexane (HCH) isomers by Sphingomonas sp. NM05. *Biodegradation*, 23, 673–682.

Mao, Y., Sun, M., Yang, X., Wei, H., Song, Y. & Xin, J. (2013) Remediation of Organochlorine Pesticides (OCPs) contaminated soil by successive hydroxypropyl-β-cyclodextrin and peanut oil enhanced soil washing-nutrient addition: A laboratory evaluation. *Journal of Soils and Sediments*, 13, 403–412.

Marin-Benito, J.M., Brown, C.D., Herrero-Hernández, E., Arienzo, M., Sánchez-Martín, M.J. & Rodríguez-Cruz, M.S. (2013) Use of raw or incubated organic wastes as amendments in reducing pesticide leaching through soil columns. *Science of the Total Environment*, 463–464, 589–599.

Marín-Benito, J.M., Sanchez-Martinez, M.J., Andrade, M.S., Perez-Clavijo, M. & Rodriguez-Cruz, M.S. (2009) Effect of spent mushroom substrate amendment of vineyard soils on the behavior of fungicides. 1. Adsorption – Desorption of penconazole and metalaxyl by soils and subsoils. *Journal of Agriculture and Food Chemistry*, 57, 9634–9642.

Marion, G.M., Payne, J.R. & Brar, G.S. (1997) *Site Remediation via Dispersion by Chemical Reaction (DCR)*. CRRL Special Report 97–18. Available from: http://permanent.access.gpo.gov/websites/armymil/www.crrel.usace.army.mil/techpub/CRREL_Reports/reports/sr97_18.pdf. [Accessed 13th June 2018].

Martin, S.M., Kookana, R.S., Van Zwieten, L. & Krull, E. (2012) Marked changes in herbicide sorption – Desorption upon ageing of biochars in soil. *Journal of Hazardous Materials*, 231–232, 70–78.

Matta, R., Hanna, K., Kone, T. & Chiron, S. (2008) Oxidation of 2,4,6-trinitrotoluene in the presence of different iron-bearing minerals at neutral pH. *Chemical Engineering Journal (Amsterdam, Netherlands)*, 144, 453–458.

McCray, J.E. & Brusseau, M.L. (1998) Enhanced *in situ* flushing of multiple-component immiscible organic liquid contamination at the field scale using cyclodextrin: Mass removal effectiveness. *Environmental Science & Technology*, 32, 1285–1293.

McCray, J.E. & Brusseau, M.L. (1999) Cyclodextrin-enhanced *in situ* flushing of multiple-component immiscible organic liquid contamination at the field scale: Analysis of dissolution behavior. *Environmental Science & Technology*, 33, 89–95.

McGrath, A., Oberle, D., Schroder, D., McInne, J. & Maxwell, C. (2007) Bench scale evaluation of ex-situ and in-situ Cr(VI) remedial methods. In: Guertin, J., Jacobs, J.A. & Avakian, C.P. (eds.) *Chromium(VI). Handbook Written by Independent Environmental Technical Evaluation Group (IETEG)*. CRC PRESS, Boca Raton, FL, USA, London, UK, New York, NY, USA & Washington, DC, USA. pp. 333–346. Available from: www.engr.uconn.edu/~baholmen/docs/ENVE290W/National%20Chromium%20Files%20From%20Luke/Cr(VI)%20Handbook/L1608_C09.pdf. [Accessed 9th June 2018].

Medina, R., Gara, P.M.D., Fernandez-Gonzalez, A.J., Rosso, J.A. & Del Panno, M.T. (2018) Remediation of soil chronically contaminated with hydrocarbon through persulfate oxidation and bioremediation. *Science of the Total Environment*, 618, 518–530. doi:10.1016/j.scitotenv.2017.10.326.

Miller, J. & Foran, C. (2012) Development of cleanup technologies for the management of US military installations. In: Rose, E.P.F. & Mather, J.D. (eds.) *Military Aspects of Hydrogeology*. Geolological Society of London, London, UK.

Mishra, M. & Chun, D.M. (2015) α-Fe_2O_3 as a photocatalytic material: A review. *Applied Catalysis A: General*, 498, 126–141.

MOKKA Project (2004–2008) *Innovative Decision Support Tools for Risk-based Environmental Management*. Available from: http://enfo.hu/mokka/index.php?lang=eng&body=mokka. [Accessed 24th July 2018].

Molnár, M., Leitgib, L., Fenyvesi, E. & Gruiz, K. (2009) Development of cyclodextrin-enhanced soil remediation: From the laboratory to the field. *Land Contamination & Reclamation*, 17(3–4), 599–610.

Molnár, M., Leitgib, L., Gruiz, K., Fenyvesi, E., Szaniszló, N., Szejtli, J. & Fava, F. (2005) Enhanced biodegradation of transformer oil in soils with cyclodextrin: From the laboratory to the field. *Biodegradation*, 16, 159–168.

Monagheddu, M., Mulas, G., Doppiu, S., Cocco, G. & Raccanelli, S. (1999) Reduction of polychlorinated dibenzodioxins and dibenzofurans in contaminated muds by mechanically induced combustion reactions. *Environmental Science & Technology*, 33, 2485–2488.

Mulas, G., Loiselle, S., Schiffini, L. & Cocco, G. (1997) The mechanochemical self-propagating reactions between hexachlorobenzene and calcium hydride. *Journal of Solid State Chemistry*, 129, 263–270.

Mulligan, C.N. (2009) Recent advances in the environmental applications of biosurfactants. *Journal of Colloid and Interface Science*, 14, 372–378.

Mulligan, C.N., Yong, R.N. & Gibbs, B.F. (2001) Surfactant enhanced remediation of contaminated soil – A review. *Engineering Geology*, 60, 371–380.

Murati, M., Oturan, N., van Hullebusch, E.D. & Oturan, M.A. (2009) Electro-Fenton treatment of TNT in aqueous media in presence of cyclodextrin. Application to ex-situ treatment of contaminated soil. *Journal of Advanced Oxidation Technologies*, 12(1), 29–36.

Nidheesh, P.V. & Gandhimathi, R. (2012) Trends in electro-Fenton process for water and wastewater treatment: An overview. *Desalination*, 299, 1–15.

Nkongolo, K.K., Michael, P., Theriault, G., Narendrula, R., Castilloux, P., Kalubi, K.N., Beckett, P. & Spiers, G. (2016) Assessing biological impacts of land reclamation in a mining region in Canada: Effects of dolomitic lime applications on forest ecosystems and microbial phospholipid fatty acid signatures. *Water, Air, & Soil Pollution*, 227, 104. doi:10.1007/s11270-016-2803-5.

Noordman, W.H., Ji, W., Brusseau, M.L. & Janssen, D.B. (1998) Effects of rhamnolipid biosurfactants on removal of phenanthrene from soil. *Environmental Science & Technolology*, 32(12), 1806–1812.

Odukkathil, G. & Vasudevan, N. (2015) Biodegradation of endosulfan isomers and its metabolite endosulfate by two biosurfactant producing bacterial strains of Bordetella petrii. *Journal of Environmental Science and Health, Part B. Pesticides, Food Contaminants, and Agricultural. Wastes*, 50, 81–89.

Oonnittan, A., Shrestha, R.A. & Sillanpaa, M. (2009) Removal of hexachlorobenzene from soil by electrokinetically enhanced chemical oxidation. *Journal of Hazardous Materials*, 162, 989–993.

Orgoványi, J., Oláh, E., H-Otta, K. & Fenyvesi, É. (2009) Dissolution properties of cypermethrin/cyclodextrin complexes. *Journal of Inclusion Phenomena and Macrocyclic Chemistry*, 63, 53–59.

Paria, S. & Yuet, P.K. (2006) Solidification/stabilization of organic and inorganic contaminants using portland cement: A literature review. *Environmental Reviews*, 14, 217–255.

Pavel, L.V. & Gavrilescu, M. (2008) Overview of *ex situ* decontamination techniques for soil cleanup. *Journal of Environmental Engineering*, 7, 815–834.

Pernyeszi, T., Kasteel, R., Witthuhn, B., Klahre, P., Vereecken, H. & Klumpp, E. (2006) Organoclays for soil remediation: Adsorption of 2,4-dichlorophenol on organoclay/aquifer material mixtures studied under static and flow conditions. *Applied Clay Science*, 32, 179–189.

Peroxone (2018) Available from: www.cpeo.org/techtree/ttdescript/peroxz.htm and www.frtr.gov/matrix2/section4/4_4.html. [Accessed 14th February 2018].

Petitgirard, A., Djehiche, M., Persello, J., Fievet, P. & Fatin-Rouge, N. (2009) PAH contaminated soil remediation by reusing an aqueous solution of cyclodextrins. *Chemosphere*, 75, 714–718.

PIANC (2009) *Dredged Material as a Resource: Option and Constraints*. Report No. 104–2009 Available from: www.cedaconferences.org/documents/dredgingconference/html_page/9/4-1-murray.pdf. [Accessed 29th January 20188].

Pizzigallo, M.D., Napola, A., Spagnuolo, M. & Ruggiero, P. (2004) Mechanochemical removal of organo-chlorinated compounds by inorganic components of soil. *Chemosphere*, 55(11), 1485–1492.

Polcaro, A.M., Vacca, A., Mascia, M. & Palmas, S. (2007) Electrokinetic removal of 2,6-dichlorophenol and diuron from kaolinite and humic acid – Clay system. *Journal of Hazardous Materials*, 148, 505–512.

Porbaha, A., Weatherby, D., Macnab, A., Lambrechts, J., Burke, G., Yang, D. & Puppala, A.J. (2005) Report: American practice of deep mixing technology. *Proceeding of the International Conference on Deep Mixing-Best Practice and Recent Advances, Stockholm*.

POSW II (1997) *Final Report Development Programme for Treatment Processes for Contaminated Sediments (1992–1996)*. RIZA Report nr 97.051, ISBN 90.369 5097 X, PO Box 17, 8200 AA Lelystad, The Netherlands Rogaar H. Available from: www.helpdeskwater.nl/secundaire-navigatie/english/sediment/@176390/treatment-for/ [Accessed 9th March 2018].

Presto (2018) *Geocells Cellular Confinement System*. Available from: www.prs-med.com/category/geocells/ [Accessed 6th March 2018].

Profiles (2018a) *Terra-Tubes® Fiber Filtration Tubes*. Available from: http://profilelibrary.info/Files/A013-025334_Terra-Tubes%20Broch.pdf. [Accessed 6th March 2018].

Profiles (2018b) *ProMatrix™ Engineered Fiber Matrix*. (EFM). Available from: www.profileevs.com/products/hydraulic-erosion-control/advanced-fiber-matrices/promatrix-efm. [Accessed 6th March 2018].

Propex (2018a) Available from: http://propexglobal.com/GeoSolutions/Product-Tour/PYRAMAT. [Accessed 6th March 2018].

Propex (2018b) *ARMORMAX® for Erosion Control & Slope Stabilization*. Available from: http://propexglobal.com/Geo-Solutions/Product-Tour/ArmorMax. [Accessed 6th March 2018].

Qin, C.Y., Zhao, Y.S., Su, Y. & Zheng, W. (2013) Remediation of nonaqueous phase liquid polluted sites using surfactant-enhanced air sparging and soil vapor extraction. *Water Environment Research*, 85(2), 133–140.

Raito (2018) *Shallow Soil Mixing. SCM Method. Rotary Blender & Bucket Mixing*. Available from: www.raito.co.jp/english/construction/pdf/scm.pdf. [Accessed 24th February 2018].

Regenesis (2018) *RegenOx® in-situ Chemical Oxidation (ISCO) and Oxygen Release Compound (ORC®)*. Available from: www.regenesis.com. [Accessed 14th February 2018].

Reible, D.D. (2005) *Organoclay Laboratory Study*. Oregon, McCormick and Baxter Creosoting Company Portland. Available from: www.clu-in.org/download/contaminantfocus/sediments/Organoclay LabStudy.pdf. [Accessed 14th May 2017].

RfC (2018) *Indoor Air Reference Levels*. Available from:www.canada.ca/en/health-canada/services/publications/healthy-living/indoor-air-reference-levels.html. Measured workplace air was in the lower range. [Accessed 8th May 2018].

Ribeiro, A.B., Mateus, E.P. & Rodríguez-Maroto, J.M. (2011) Removal of organic contaminants from soils by an electrokinetic process: The case of molinate and bentazone. Experimental and modeling. *Separation and Purification Technology*, 79, 193–203.

Ringeling, R.H.P. (1998) *Handling of Contaminated Dredged Material*. Delft University Press, Delft, The Netherlands.

Risco, C., Rodrigo, S., López-Vizcaíno, R., Sáez, C., Cañizares, P., Navarro, V. & Rodrigo, M.A. (2016) Electrokinetic flushing with surrounding electrode arrangements for the remediation of soils that are polluted with 2,4-D: A case study in a pilot plant. *Science of the Total Environment*, 545, 256–265.

Rodrigo, M.A., Oturan, N. & Oturan, M.A. (2014) Electrochemically assisted remediation of pesticides in soils and water: A review. *Chemical Reviews*, 114, 8720–8745.

Rodriguez, G.R. & Brebbia, C.A. (2015) *Coastal Cities and Their Sustainable Future*. WitPress, New Forest National Park, UK.

Rosas, J.M., Vicente, F., Saguillo, E.G., Santos, A. & Romero, A. (2014) Remediation of soil polluted with herbicides by Fenton-like reaction: Kinetic model of diuron degradation. *Applied Catalysis B: Environmental*, 144, 252–260.

Rubio-Bellido, M., Madrid, F., Morillo, E. & Villaverde, J. (2015) Assisted attenuation of a soil contaminated by diuron using hydroxypropyl-β-cyclodextrin and organic amendments. *Science of the Total Environment*, 502, 699–705.

Rubio-Bellido, M., Morillo, E. & Villaverde, J. (2016) Effect of addition of HPBCD on diuron adsorption – Desorption, transport and mineralization in soils with different properties. *Geoderma*, 265, 196–203. doi:10.1016/j.geoderma.2015.11.022.

Ruppert, G., Bauer, R. & Heisler, G. (1993) The photo-Fenton reaction – An effective photochemical wastewater treatment process. *Journal of Photochemistry and Photobiology A: Chemistry*, 73, 75–78.

Russel, H.H., Matthews, J.E. & Sewell, G.W. (1992) *TCE Removal From Contaminated Soil and Ground Water*. EPA Ground Water Issue EPA/540/S-92/002. Available from: www.epa.gov/sites/production/files/2015-06/documents/tce.pdf. [Accessed 16th May 2018].

Sánchez-Trujillo, M.A., Morillo, E., Villaverde, J. & Lacorte, S. (2013) Comparative effects of several cyclodextrins on the extraction of PAHs from an aged contaminated soil. *Environmental Pollution*, 178, 52–58.

Satapanajaru, T., Chokejaroenrat, C., Sakulthaew, C. & Yoo-iam, M. (2017) Remediation and restoration of petroleum hydrocarbon containing alcohol-contaminated soil by persulfate oxidation activated with soil minerals. *Water Air Soil Pollution*, 228, 345. doi:10.1007/s11270-017-3527-x.

Satapanajaru, T., Shea, P.J., Comfort, S.D. & Roh, Y. (2003) Green rust and iron oxide formation influences metolachlor dechlorination during zerovalent iron treatment. *Environmental Science & Technology*, 37, 5219–5227.

Schifano, V. & Thurston, N. (2007) Remediation of a clay contaminated with petroleum hydrocarbons using soil reagent mixing. *22nd Annual International Conference on Contaminated Soils, Sediments and Water, Association for Environmental Health and Sciences*. pp. 264–277. Available from: http://scholarworks.umass.edu/soilsproceedings/vol12/iss1/27/ [Accessed 16th May 2017].

Schnarr, M., Truax, C., Farquhar, G., Hood, E., Gonullu, T. & Stickney, B. (1998) Laboratory and controlled field experiments using potassium permanganate to remediate trichloroethylene and perchloroethylene DNAPLs in porous media. *Journal of Contaminant Hydrology*, 29, 205–224.

Schneider, J., Matsuoka, M., Takeuchi, M., Zhang, J., Horiuchi, Y., Anpo, M. & Bahnemann, D.W. (2014) Understanding TiO2 photocatalysis: Mechanisms and materials. *Chemical Reviews*, 114(19), 9919–9986.

SET (2018) *Solvated Electron Treatment*. Available from: www.cpeo.org/techtree/ttdescript/solvelectr.htm. [Accessed 16th February 2018].

Shah, F.H., Hadim, H.A. & Korfiatis, G.P. (1995) Laboratory studies of air stripping of VOC-contaminated soils. *Journal of Soil Contamination*, 4, 1–17.

Sharma, T., Toor, A.P. & Rajor, A. (2015) Photocatalytic degradation of imidacloprid in soil: Application of response surface methodology for the optimization of parameters. *RSC Advances*, 5, 25059–25065.

Shea, P.J., Machacek, T.A. & Comfort, S.D. (2004) Accelerated remediation of pesticide-contaminated soil with zerovalent iron. *Environmental Pollution*, 132, 183–188.

Shirin, S., Buncel, E. & vanLoon, G.W. (2004) Effect of cyclodextrins on iron-mediated dechlorination of trichloroethylene – A proposed new mechanism. *Canadian Journal of Chemistry*, 82, 1674–1685.

Siegrist, R.L., Urynowitz, M.A., West, O.R., Crimi, M.L. & Lowe, K.S. (2001) *Principles and Practices of in Situ Chemical Oxidation Using Permanganate*. Battelle Press, Columbus, OH, USA.

Silva, A., Delerue-Matos, C. & Fiuza, A. (2005) Use of solvent extraction to remediate soils contaminated with hydrocarbons. *Journal of Hazardous Materials*, B124, 224–229.

SITE (1993) *Superfund Innovative Technology Evaluation Program (SITE)*. Technology Profiles. US EPA. pp. 123–124.

Skold, M.E., Thyne, G.D., Drexler, J.W. & McCray, J.E. (2009) Solubility enhancement of seven metal contaminants using carboxymethyl-beta-cyclodextrin (CMCD). *Journal of Contaminant Hydrology*, 107, 108–113.

Sly, P.G. & Hart, B. (eds.) (1987) *Sediment/Water Interactions: Proceedings of the Fourth International Symposium*. Melbourne, Australia, February 16–20th, 1987. Springer, New York, NY, USA. Available from: www.springer.com/la/book/9780792302599. [Accessed 15th March 2018].

Sniegowski, K., Vanhecke, M., D'Huys, P.-J. & Braeken, L. (2014) Potential of activated carbon to recover randomly-methylated-β-cyclodextrin solution from washing water originating from *in situ* soil flushing. *Science of the Total Environment*, 485–486, 764–768.

Sopeña, F., Semple, K., Sohi, S. & Bending, G. (2012) Assessing the chemical and biological accessibility of the herbicide isoproturon in soil amended with biochar. *Chemosphere*, 88, 77–83.

Sora, I.N., Pelosato, R., Botta, D. & Dotelli, G. (2002) Chemistry and microstructure of cement pastes admixed with organic liquids. *Journal of the European Ceramic Society*, 22(9–10), 1463–1473.

Souza, D.R.D., Trovó, A.G., Filho, N.R.A., Silva, M.A.A. & Machado, A.E.H. (2013) Degradation of the commercial herbicide glyphosate by photo-fenton process: Evaluation of kinetic parameters and toxicity. *Journal of the Brazilian Chemical Society*, 24, 1451–1460.

Souza, F.L., Saéz, C., Llanos, J., Lanza, M.R.V., Cañizares, P. & Rodrigo, M.A. (2016) Solar-powered electrokinetic remediation for the treatment of soil polluted with the herbicide 2,4-D. *Electrochimica Acta*, 190, 371–377.

Stigliani, W.M. (ed) (1991) *Chemical Time Bombs: Definition, Concepts, and Examples.* Laxenburg, Austria, International Institute for Applied Systems Analysis. Available from: http://pure.iiasa.ac.at/3510/1/ER-91-016.pdf. [Accessed 7th May 2017].

Sutton, N.B., Grotenhuis, T.C., Langenhoff, A.A.M. & Rijnaarts, H.H.M. (2011) Efforts to improve coupled in situ chemical oxidation with bioremediation: A review of optimization strategies. *Journal of Soils and Sediments*, 11, 129–140.

Syakir, M.I., Nurin, N.A., Zafirah, N., Kassim, M.A. & Khalil, H.P.S.A. (2016) Nanoclay reinforced on biodegradable polymer composites: Potential as a soil stabilizer. In: Jawaid, M., Qaiss, A.K. & Bouhfid, R. (eds.) *Nanoclay Reinforced Polymer Composites: Nanocomposites and Bionanocomposite, Engineering Materials.* Springer, Berlin, Germany. pp. 329–356. doi:101007/978-981-10-1953-1.

Szecsody, J.E., Fruchter, J.S., Williams, M.D., Vermeul, V.R. & Sklarew, D. (2004) *In situ* chemical reduction of aquifer sediment: Enhancement of reactive iron phases and TCE dechlorination. *Environmental Science & Technology*, 38, 4656–4663.

Tangprasert, W., Jaikaew, S. & Supakata, N. (2015) Utilization of dredged sediments from Lumsai Canal with rice husks to produce bricks. *International Journal of Environmental Science and Development*, 6(3), 217–220. Available from: www.ijesd.org/vol6/593-M0028.pdf. [Accessed 15th March 2018].

TEVES (2018) *Thermal Enhanced Vapor Extraction System.* Available from: www.cpeo.org/techtree/ttdescript/thevapor.htm. [Accessed 25th February 2018].

Theron, J., Walker, J.A. & Cloete, T.E. (2008) Nanotechnology and water treatment: Applications and emerging opportunities. *Critical Reviews in Microbiology*, 34(1), 43–69. doi:10.1080/10408410701710442.

Tick, G.R., Lourenso, F., Wood, A.L. & Brusseau, M.L. (2003) Pilot-scale demonstration of cyclodextrin as a solubility-enhancement agent for remediation of a tetrachloroethene-contaminated aquifer. *Environmental Science & Technology*, 37, 5829–5834.

Torres, L.G., Ramos, F., Avila, M.A. & Ortiz, I. (2012) Removal of methyl parathion by surfactant-assisted soil washing and subsequent wastewater biological treatment. *Journal of Pest Science*, 37, 240–246.

TOXNET (2018) *Toxiclology Data on Trichloroethylene.* Available from: https://toxnet.nlm.nih.gov/cgi-bin/sis/search/a?dbs+hsdb:@term+@DOCNO+133. [Accessed 8th July 2018].

Tribochem (2018) *Mechanochemistry Comprising Mechanochemical Reductive Dehalogenation/Destruction of PCBs, DDT, HCH, TCE, Dioxins Using Ball Mills (Vibratory Mills, Vibrating Mills).* Available from: www.tribochem.com/projects/index.html. [Accessed 16th February 2018].

USACE (U.S. Army Corps of Engineers) (2003) *Safety and Health Aspects of HTRW Remediation Technologies.* EM 1110-1-4007, 4-1-4-12. Available from: www.publications.usace.army.mil/Portals/76/Publications/EngineerManuals/EM_1110-1-4007.pdf. [Accessed 16th May 2018].

US Department of Energy (1995) *In situ Air Stripping Using Horizontal Wells.* Innovative Technology Summary Report. Available from: www.dndkm.org/DOEKMDocuments/ITSR/SoilGroundWater/In_Situ_Air_Stripping_Using_Horizontal_Wells.pdf. [Accessed 19th May 2018].

US EPA (2002) *Field Sampling and Treatability Study for in-situ Remediation of PCB's and Leachable Lead With Iron Powder.* EIMS Metadata Report R825511C019. Available from: https://cfpub.epa.gov/si/si_public_record_Report.cfm?dirEntryID=57704. [Accessed 16th May 2018].

US EPA (2009) *Hydrogen Release Compound (HRC®) Barrier Application at the North of Basin F Site, Rocky Mountain Arsenal Innovative: Technology Evaluation Report.* EPA 540/R-09–004.

Usman, M., Tascone, O., Faure, P. & Hanna, K. (2014) Chemical oxidation of hexachlorocyclohexanes (HCHs) in contaminated soils. *Science of the Total Environment,* 476, 434–439.

US Navy (2003) *Pilot Test Final Report: Bimetallic Nanoscale Particle Treatment of Groundwater at Area I.* Available from: http://costperformance.org/pdf/20040618_346.pdf. [Accessed 14th May 2018].

Verstichel, S., De Wilde, B., Fenyvesi, E. & Szejtli, J. (2004) Investigation of the aerobic biodegradability of several types of cyclodextrins in a laboratory-controlled composting test. *Journal of Polymers and the Environment,* 12, 47–55.

Vieira dos Santos, E., Souza, F., Saez, C., Cañizares, P., Lanza, M.R.V., Martinez-Huitle, C.A. & Rodrigo, M.A. (2016) Application of electrokinetic soil flushing to four herbicides: A comparison. *Chemosphere,* 153, 205–211.

Villa, R.D., Trovó, A.G. & Nogueira, R.F.P. (2008) Environmental implications of soil remediation using the Fenton process. *Chemosphere,* 71, 43–50.

Villa, R.D., Trovó, A.G. & Pupo Nogueira, R.F. (2010) Soil remediation using a coupled process: Soil washing with surfactant followed by photo-Fenton oxidation. *Journal of Hazardous Materials,* 174, 770–775.

Villaverde, J., Maqueda, C. & Morillo, E. (2005) Improvement of the desorption of the herbicide norflurazon from soils via complexation with β-cyclodextrin. *Journal of Agricultural and Food Chemistry,* 53, 5366–5372.

Villaverde, J., Maqueda, C. & Morillo, E. (2006) Effect of the simultaneous addition of β-cyclodextrin and the herbicide norflurazon on its adsorption and movement in soils. *Journal of Agricultural and Food Chemistry,* 54, 4766–4772.

Villaverde, J., Maqueda, C., Undabeytia, T. & Morillo, E. (2007) Effect of various cyclodextrins on photodegradation of a hydrophobic herbicide in aqueous suspensions of different soil colloidal components. *Chemosphere,* 69, 575–584.

Villaverde, J., Posada-Baquero, R., Rubio-Bellido, M., Laiz, L., Saiz-Jimenez, C., Sanchez-Trujillo, M.A. & Morillo, E. (2012) Enhanced mineralization of diuron using a cyclodextrin-based bioremediation technology. *Journal of Agricultural and Food Chemistry,* 60, 9941–9947.

Villaverde, J., Posada-Baquero, R., Rubio-Bellido, M. & Morillo, E. (2013) Effect of hydroxypropyl-β-cyclodextrin on diuron desorption and mineralisation in soils. *Journal of Soils and Sediments,* 13, 1075–1083.

Villaverde, J., Rubio-Bellido, M., Lara-Moreno, A., Merchán, F. & Morillo, E. (2018) Combined use of microbial consortia isolated from different agricultural soils and cyclodextrin as a bioremediation technique for herbicide contaminated soils. *Chemosphere,* 193, 118–125. doi:10.1016/j.chemosphere.2017.10.172.

Villaverde, J., Rubio-Bellido, M., Merchán, F. & Morillo, E. (2017) Bioremediation of diuron contaminated soils by a novel degrading microbial consortium. *Journal of Environmental Management,* 188, 379–386. doi:10.1016/j.jenvman.2016.12.020.

Waclawek, S., Antos, V., Hrabak, P., Cernik, M. & Elliot, D. (2016) Remediation of hexachlorocyclohexanes by electrochemically activated persulfates. *Environmental Science and Pollution Research,* 23, 765–773.

Waddell, J.P. & Mayer, G.C. (2003) Effects of Fenton reagent and potassium permanganate applications on indigenous subsurface microbiota: A literature review. In: Hatcher, K.J. (ed) *Proceedings of the 2003 Georgia Water Resources Conference, 23–24 April 2003.* University of Georgia, Athens, Georgia. Available from: www2.usgs.gov/water/southatlantic/ga/projects/airforce/Waddell-GWRC2003.pdf. [Accessed 31st May 2018].

Waisner, S., Medina, V.F., Morrow, A.B. & Nestler, C.C. (2008) Evaluation of chemical treatments for a mixed contaminant soil. *Journal of Environmental Engineering*, 134, 743–749.

Wan, C.L., Yang, X., Du, M.A., Xing, D.F., Yu, C.G. & Yang, Q.L. (2010) Desorption of oil in naturally polluted soil promoted by beta-cyclodextrin. *Fresenius Environmental Bulletin*, 19, 1231–1237.

Wan, J., Meng, D., Long, T., Ying, R., Ye, M., Zhang, S., Li, Q., Zhou, Y. & Lin, Y. (2015) Simultaneous removal of Lindane, lead and cadmium from soils by rhamnolipids combined with citric acid. *Plos One*, 10(6), e0129978. doi:10.1371/journal.pone.0129978.

Wan, J., Yuan, S., Mak, K., Chen, J., Li, T., Lin, L. & Lu, X. (2009) Enhanced washing of HCB contaminated soils by methyl-beta- cyclodextrin combined with ethanol. *Chemosphere*, 75, 759–764.

Wang, G., Li, H., Yu, R. & Deng, N. (2007) Beta-cyclodextrin enhanced photodegradation of bisphenol C under UV light. *Fresenius Environmental Bulletin*, 16, 690–696.

Wang, H., Lin, K., Hou, Z., Richardson, B. & Gan, J. (2010) Sorption of the herbicide terbuthylazine in two New Zealand forest soils amended with biosolids and biochars. *Journal of Soils and Sediments*, 10, 283–289.

Watlington, K. (2005) *Emerging Nanotechnologies for Site Remediation and Wastewater Treatment.* Available from: www.clu-in.org/download/studentpapers/K_Watlington_Nanotech.pdf. [Accessed 16th January 2018].

Weber, R. & Varbelow, G. (2013) Dioxin/POPs legacy of pesticide production in Hamburg: Part 1 – Securing of the production area. *Environmental Science and Pollution Research*, 20, 1918–1924.

Wei, B. & Tarr, M.A. (2003) Role of cyclodextrins in Fenton remediation of 2,4,6- trinitrotoluene. *Abstracts of ACS National Meeting American Chemical Society, Division of Environmental Chemistry*, 43, 127–129.

Weprot Company (2018) Available from: www.weprot.hu. [Accessed 16th July 2018].

Western Excelsior (2018) Available from: www.westernexcelsior.com/products/erosion.html. [Accessed 6th March 2018].

Wong, F. & Bidleman, T.F. (2010) Hydroxypropyl-β-cyclodextrin as non-exhaustive extractant for organochlorine pesticides and polychlorinated biphenyls in muck soil. *Environmental Pollution*, 158, 1303–1310.

Xu, X., Ji, F., Fan, Z. & He, L. (2011) Degradation of glyphosate in soil photocatalyzed by Fe_3O_4/ SiO_2/TiO_2 under solar light. *International Journal of Environmental Research and Public Health*, 8, 1258–1270.

Yáñez, C., Cañete-Rosales, P., Castillo, J.P., Catalán, N., Undabeytia, T. & Morillo, E. (2012) Cyclodextrin inclusion complex to improve physicochemical properties of herbicide bentazon: Exploring better formulations. *PLoS ONE*, 7(8), e0041072. doi:10.1371/journal.pone.0041072.

Yardin, G. & Chiron, S. (2006) Photo-Fenton treatment of TNT contaminated soil extract solutions obtained by soil flushing with cyclodextrin. *Chemosphere*, 62, 1395–1402.

Ye, M., Sun, M., Hu, F., Kengara, F.O., Jiang, X., Luo, Y. & Yang, X. (2014a) Remediation of organochlorine pesticides (OCPs) contaminated site by successive methyl-β-cyclodextrin (MCD) and sunflower oil enhanced soil washing – Portulaca oleracea L. cultivation. *Chemosphere*, 105, 119–125.

Ye, M., Sun, M., Kengara, F.O., Wang, J., Ni, N., Wang, L., Song, Y., Yang, X., Li, H., Hu, F. & Jiang, X. (2014) Evaluation of soil washing process with carboxymethyl-beta-cyclodextrin and carboxymethyl chitosan for recovery of PAHs/heavy metals/fluorine from metallurgic plant site. *Journal of Environmental Science*, 26, 1661–1672.

Ye, M., Sun, M., Liu, Z., Ni, N., Chen, Y., Gu, C., Kengara, F.O., Li, H. & Jiang, X. (2014b) Evaluation of enhanced soil washing process and phytoremediation with maize oil, carboxymethyl-β-cyclodextrin, and vetiver grass for the recovery of organochlorine pesticides and heavy metals from a pesticide factory site. *Journal of Environmental Management*, 141, 161–168.

Ye, M., Yang, X.L., Sun, M.M., Bian, Y.R., Wang, F., Gu, C.G., Wei, H.J., Song, Y., Wang, L., Jin, X. & Jiang, X. (2013) Use of organic solvents to extract Organochlorine Pesticides (OCPs) from aged contaminated soils. *Pedosphere*, 23, 10–19.

Yoshii, H., Furuta, T., Shimizu, J., Kugimoto, Y., Nakayasu, S., Arai, T. & Linko, P. (2001) Innovative approach for removal and biodegradation of contaminated compounds in soil by cyclodextrins. *Biological Journal of Armenia, Special Issue: Cyclodextrins*, 53, 226–236.

Yu, X.Y., Pan, L.G., Ying, G.G. & Kookana, R.S. (2010) Enhanced and irreversible sorption of pesticide pyrimethanil by soil amended with biochars. *Journal of Environmental Science*, 22, 615–620.

Yuan, S., Tian, M. & Lu, X. (2006) Electrokinetic movement of hexachlorobenzene in clayed soils enhanced by Tween 80 and beta-cyclodextrin. *Journal of Hazardeous Materials*, 137(2), 1218–1225.

Zeng, Q.R., Tang, H.X., Liao, B.H., Zhong, T.F. & Tang, C. (2006) Solubilization and desorption of methyl-parathion from porous media: A comparison of hydroxypropyl-beta-cyclodextrin and two nonionic surfactants. *Water Research*, 40, 1351–1358.

Zhang, Q., Matsumoto, H., Saito, F. & Baron, M. (2002) Debromination of hexabromobenzene by its co-grinding with CaO. *Chemosphere*, 48, 787–793.

Zhang, X., Wang, H., He, L., Lu, K., Sarmah, A., Li, J., Bolan, N.S., Pei, J. & Huang, H. (2013) Using biochar for remediation of soils contaminated with heavy metals and organic pollutants. *Environmental Science and Pollution Research*, 20(12), 8472–8483.

Zhang, Y., Wong, J.W.C., Liu, P. & Yuan, M. (2011) Heterogeneous photocatalytic degradation of phenanthrene in surfactant solution containing TiO_2 particles. *Journal of Hazardous Materials*, 191, 136–143.

Zhao, J., Chi, Y., Liu, F., Jia, D. & Yao, K. (2015) Effects of two surfactants and beta-cyclodextrin on beta-cypermethrin degradation by Bacillus licheniformis B-1. *Journal of Agricultural and Food Chemistry*, 63, 10729–10735.

Zheng, W. & Tarr, M.A. (2004) Evidence for the existence of ternary complexes of iron, cyclodextrin, and hydrophobic guests in aqueous solution. *Journal of Physical Chemistry B*, 108, 10§172–10§176.

Chapter 7

Leaching, bioleaching, and acid mine drainage case study

H.M. Siebert[1], G. Florian[1], W. Sand[1], E. Vaszita[2], K. Gruiz[2], M. Csővári[3], G. Földing[3], Zs. Berta[3] & J.T. Árgyelán[4]

[1]Biofilm Centre, Aquatic Biotechnology, University of Duisburg-Essen, Essen, Germany
[2]Department of Applied Biotechnology and Food Science, Budapest University of Technology and Economics, Budapest, Hungary
[3]MECSEK-ÖKO Co., Pécs, Hungary
[4]Nitrokémia Zrt., Hungary

ABSTRACT

This chapter defines and discusses the leaching and bioleaching process in the context of mining and soil remediation. It focuses on the prevention, mitigation, and technological utilization of the leaching process and on the treatment of the acidic leachate.

The topic is discussed in three sections:

(i) The first section gives a general overview of the leaching process, on the history and principles of acid mine drainage treatment, on the technical solution and applicability of active and passive treatment systems such as oxic and anoxic ponds, channels, wetlands, and several other engineered reactor-based technologies and permeable reactive barriers. (ii) The second subchapter focuses on sulfidic ore bioleaching. The biochemistry of the process is presented and the activities of iron- and sulfur-oxidizing microorganisms are clarified. The biotechnological solutions in the monitoring, prevention, and treatment of the acidic leachate is in focus.

(iii) The third subchapter is also closely related to the topic of natural leaching and its consequences. A case study is introduced where sulfide-ore mining produced AMD and ARD. Ongoing mine closure, and complex site rehabilitation is aimed to overcome environmental damage and prevent any future damage. The study presents the engineering solution applied in the closure of the Gyöngyösoroszi metal mine in Hungary. It is currently under restoration and provides an opportunity to study and manage all the impacts of leaching on water, sediment, and soil. Leaching in the underground mine, leaching from various mine wastes and waste rocks resulted in hundreds of relatively well-defined contaminated sites in addition to the 15 km long watershed-scale diffuse pollution.

I LEACHING AND LEACHATE TREATMENT

E. Vaszita and K. Gruiz

1.1 Introduction – leaching in general and bioleaching in particular

Leaching in the widest context is a natural process by which soluble substances are washed out of soil, rock, solid waste, or any other solid material. Leaching in the environment is the effect of the interaction between precipitation, irrigation, runoff, or flood and a solid surface. Water, rainwater, and mainly acid rain may leach away nutrients (e.g. calcium), fertilizers, pesticides, or other toxic substances from the soil, rock, or hazardous wastes stockpiled/ deposited on the soil surface without proper insulation/lining. The leachate containing the released chemicals may cause typical pollution of surface and subsurface waters.

Leaching is a physicochemical process which starts with the infiltration of the water or of the diluted acidic leachant (rainwater), and involves dissolution and/or mobilization of the originally bound chemicals (typically minerals) and the transport of the leached substances by soil water in the form of solute (leachate). It can be modeled as a special kind of extraction process in a flow through packed column (the impacted solid volume) with a water-based extractant (leachant).

Leaching in soil science is a natural rock weathering process resulting in the release of minerals or organic matter into the surrounding environment. Although leaching occurs naturally, it may also be applied as a preparatory step for metal recovery (Bayat & Sari, 2010). Chemical leaching technologies apply the principle of natural leaching when using water or weak acids such as acetic acid to mobilize and leach minerals from the soil or other solid matrices. The metals recovered by the process can possibly be recycled.

Leaching is widely used in extractive metallurgy to convert metals into soluble salts in aqueous medium. Chemical leaching-based industrial processes may apply dump leaching (from ore taken directly from the mine), heap leaching (from fine-grained crushed ore or tailings), tank leaching (ore is filled into closed reactors and the leachant flows through) or *in situ* leaching (*in situ* recovery or solution mining through boreholes drilled into metal ore deposits).

The leaching process may also be used in the remediation of metals-contaminated soil to remove the metals from the soil by dissolution and washing with acids, alkali, or surfactants.

Natural leaching of sulfide minerals is a special case of leaching which may be enhanced by microbes by up to a hundred thousand times (Singer & Stumm, 1970). Without the microbes, natural leaching of sulfide minerals is a slow process. Bioleaching is a process whereby the leachant (the acidic and/or surfactant-containing extractant) is generated by microorganisms as a result of their catabolic (energy-producing oxidative procedure) and/ or biosurfactant-producing activities. Chemical and biological leaching generally occurs together in nature. Complex leaching is widely utilized as the technological basis for metal recovery and soil and sediment remediation.

The two most widespread technologies in the mining industry are bioleaching in heaps and in stirred tanks. The piles for bioleaching are constructed from crushed rock on impervious sloped pads which allow the percolated solute to flow by gravity into the collection system. Oxygen can be added by using low-pressure blowers at the base of the heap. Bioleaching in aerated and stirred tanks uses a mineral slurry. The large size of the reactors, stirring, and aeration generate high costs compared to heap leaching, but higher oxidation efficiency can be reached.

In addition to the traditional biomining of copper, gold, and uranium, the recovery of critical metals such as rare earth elements (REE) has become increasingly prominent in recent years. In addition to low-grade ore processing, mine wastes and tailings, phosphogypsum deposits, and REE-containing electronic wastes are also in the focus of interest. (For more on this topic read Chen *et al.*, 2018; Barmetler *et al.*, 2016; Zhuang *et al.*, 2015; Hennebel, 2015; Glombitza & Reichel, 2014; Brandl, 2008; Krebs, 1997.)

Besides the controlled utilization of the leaching process for metal recovery, the unintentional occurrence of the process in the environment is very widespread and diverse, ranging from acid rain leaching of the earth surface to the leaching of oxidizable wastes of mining or industrial origin. The presence of atmospheric oxygen and moisture is enough to oxidize what is oxidizable (typically reduced minerals such as pyrite), and the produced energy is immediately utilized by microorganisms, which grow and metabolize the energy until source depletion.

1.2 History and principles of AMD treatment – a general overview

Acid mine drainage (AMD) or acid rock drainage (ARD) is a strongly acidic wastewater containing high concentrations of dissolved ferrous and non-ferrous metal sulfates, and salts (Johnson & Hallberg, 2005; Macingova & Luptakova, 2012).

The process of AMD generation is extremely complex. It involves chemical, biological, and electrochemical reactions which vary with environmental conditions (Johnson & Hallberg, 2005; Simate & Ndlovu, 2014).

AMD is generated by the oxidation of sulfide minerals such as pyrite (FeS_2) as a result of exposure of these minerals to both oxygen and water (Skousen, 1995; Johnson, 2003; Johnson & Hallberg, 2005) and microorganisms (Schippers et al., 1996; Younger et al., 2002; Johnson & Hallberg, 2003; Dold, 2010).

The oxidation of pyrite can follow several pathways (Ali, 2011; Buzzi et al., 2013) involving surface interactions with dissolved O_2, Fe^{3+}, and other mineral catalysts. The primary oxidant involved in pyrite oxidation in most situations is ferric iron rather than molecular oxygen (Evangelou, 1995). The formation of AMD can be considered to occur in three major steps, as detailed by several studies (Kalin et al., 2006; Simate & Ndlovu, 2014) and in Section 2:

1   Oxidation of iron sulfide and enhanced oxidation of sulfide minerals by ferric iron.
2   Oxidation of ferrous iron.
3   Hydrolysis and precipitation of ferric iron and other minerals.

The activities of moderately and extremely acidophilic iron-oxidizing bacteria have a pivotal role in the genesis of acid mine drainage (Johnson & Hallberg, 2003; Sand et al., 2007).

Thiobacillus ferroxidans (Waksman, 1922; Leathen et al., 1953; Singer & Stumm, 1970) and several other microbial species widespread in the environment have been considered to be involved in pyrite weathering. *Thiobacillus ferroxidans* has been shown to increase iron conversion rate by a factor of hundreds to as much as a million (Singer & Stumm, 1970; Blowes et al., 2003, 2014).

AMD is considered one of the main pollutants in both operating and abandoned polymetallic sulfide mining sites (Johnson & Hallberg, 2005; Sheoran & Sheoran, 2006) – in tunnels, mine workings, open pits, waste rock piles, and tailings facilities (Blowes et al., 2014; Johnson & Hallberg, 2005; Sheoran & Sheoran, 2006), but the causes of AMD are not limited to only the mining industry. Any activity that disturbs mineralized materials can lead to AMD (highway and tunnel construction and other deep excavations where sulfide materials are exposed) (Skousen, 1995; Skousen, 2000; Skousen et al., 1998, 2000, 2002).

Untreated AMD can contaminate the soil, groundwater, and surface watercourses, damaging the health of plants (Jiwan & Kalamdhad, 2011; Yadav, 2010), humans (Garland, 2011), wildlife, and aquatic species (Gerhardt et al., 2004, Hansen et al., 2002; Schmidt et al., 2002; Soucek et al., 2000). The development of sustainable and cost-efficient remediation options for the AMD problem has been the subject of extensive research (Coulton et al., 2003; Gaikwad & Gupta, 2008; Johnson, 2000; Johnson & Hallberg, 2003; Kleinmann et al., 1998; Luptakova et al., 2010; Nordstrom, 2011; Younger, 2000; Younger et al., 2002, 2003). According to the US EPA's policy (US EPA, 2014) to reduce the environmental footprint of activities associated with the clean-up of contaminated sites, the focus is on reducing energy usage, air pollution, and impacts on water resources, improving waste management, and protecting ecosystem services. Developments cover both source control and migration control (Johnson & Hallberg, 2005). Source control techniques concentrate on the control of AMD formation at the source (Egiebor & Oni, 2007; Luptakova *et al.*, 2010) based on the removal of oxygen and/or water from the system (Johnson & Hallberg, 2005; Kuyucak, 2002; Skousen *et al.*, 2000) since oxygen and water are two of the three principal reactants (Skousen *et al.*, 2000). Migration control techniques aim to treat the resulting drainage (Egiebor & Oni, 2007; Luptakova *et al.*, 2010). These techniques can be classified in two broad categories – active and passive treatments (Johnson & Hallberg, 2005; Skousen *et al.*, 2000; Akcil & Koldas, 2006).

Both active and passive treatment methods combine and adopt physical, biological, and chemical approaches to treat AMD by raising pH, lowering dissolved metal concentrations, and sulfate content (EPA, 2014; RoyChowdhury *et al.*, 2015).

Active treatment generally involves the continuous application of alkaline materials to neutralize acidic mine waters and precipitate metals (Johnson & Hallberg, 2005). Younger *et al.* (2002) define "active" treatment as the improvement of water quality by methods which require ongoing inputs of artificial energy and/or (bio)chemical reagents. According to Younger *et al.* (2002), the most predominant "active" treatment method is "ODAS" – oxidation, dosing with alkali, and sedimentation, which is similar to the process applied in traditional wastewater treatment plants. Others traditional or "active" treatments common to wastewater treatment plants include: sulfidation, biosedimentation, sorption and ion exchange, and membrane processes like filtration and reverse osmosis (Younger *et al.*, 2002).

Passive treatment involves processes that do not require frequent human intervention, operation or maintenance, and that typically employ natural construction materials (e.g. soils, clays, broken rock), natural treatment media (e.g. plant residues such as straw, wood chips, manure, compost), and promote growth of natural vegetation (US EPA, 2014). Studies of these techniques more than 30 years ago (Wieder & Lang, 1982) and have continued to be researched ever since (Hedin *et al.*, 1994a; Hyman & Watzlaf, 1995; Skousen, 1996; Gusek, 1995, 1998; Younger *et al.*, 2002, 2003). Further to the review of various scientific papers, technical reports, reference guides authored by reputed US, Australian, and British scientists and engineers cited herein, a short description is provided in the next sections of the technical solution and applicability of some active and passive AMD treatment systems is provided hereinafter.

1.3 Technical solution and applicability of active and passive treatment systems for AMD

Active and passive treatment systems have been extensively used for AMD remediation (Coulton *et al.*, 2003; Gaikwad & Gupta, 2008; Johnson, 2000; Johnson & Hallberg, 2003; Kleinmann *et al.*, 1998; Nordstrom, 2011; Younger, 2000; Younger *et al.*, 2002, 2003).

As detailed by Taylor *et al.*, 2005, the key factors in the selection and design of active and passive AMD treatment systems are: water chemistry (including pH, metals, sulfate levels, and redox state) and influent AMD flow rate, and the objectives of AMD treatment (e.g. protection of site infrastructure, downstream aquatic ecosystems, or water resources). Other important factors include capital and operating costs, availability of suitable treatment reagents/materials and sludge management.

According to a review by Johnson and Hallberg (2005), both the active and the passive AMD treatment systems may be grouped in two categories:

1   Remediation technologies that rely on biological activities (biological remediation strategies).
2   Those which rely on abiotic processes (abiotic remediation strategies).

1.3.1   Active AMD treatment systems

In general, the active remediation systems utilize the following key processes for treatment of AMD (Taylor *et al.*, 2005):

–   pH control or precipitation;
–   Electrochemical concentration;
–   Biological mediation/redox control (sulfate reduction);
–   Ion exchange/absorption or adsorption/flocculation and filtration;
–   Crystallization.

Johnson and Hallberg (2005) divided the active AMD treatment systems into two groups: active abiotic remediation systems (based on lime addition and aeration) and active biological remediation systems (microbial bioreactors [offline sulfogenic bioreactors]).

1.3.1.1   Active abiotic remediation systems

The active treatment process requiring the addition of a chemical-neutralizing agent (Coulton *et al.*, 2003) is the most widespread method used to mitigate acidic effluents. According to Taylor *et al.* (2005), active treatment systems should accommodate any acidity, flow rate, and acidity load. Addition of an alkaline material to AMD will raise its pH, accelerate the rate of chemical oxidation of ferrous iron (for which active aeration, or addition of a chemical-oxidizing agent such as hydrogen peroxide is also necessary), and cause many of the metals present in solution to precipitate as hydroxides and carbonates, resulting an iron-rich sludge containing various other metals depending on the chemistry of the mine water treated.

A wide range of chemical agents such as limestone ($CaCO_3$), hydrated lime ($Ca(OH)_2$), caustic soda ($NaOH$), soda ash (Na_2CO_3), calcium oxide (CaO), anhydrous ammonia (NH_3), magnesium oxide (MgO) and magnesium hydroxide [$Mg(OH)_2$] are used during the active treatment of AMD water worldwide (Johnson & Hallberg, 2005; RoyChowdhury *et al.*, 2015; Skousen *et al.*, 2000).

Less common alkaline reagents include Lime Kiln Dust (LKD) and Cement Kiln Dust (CKD), fly ash, fluidized bed combustion ash, calcium peroxide, potassium hydroxide, and seawater-neutralized red mud (from bauxite processing) (Taylor *et al.*, 2005). The efficiency of each of the chemicals depends on the seasonal variation, daily AMD load, and metal concentration.

The addition of various acid-neutralizing and metal-precipitating chemical agents to AMD water is a common practice to meet the effluent discharge limits within a short time span (Taylor *et al.*, 2005).

The disadvantages of this active system are high operating costs and problems with disposal of the bulky sludge that is produced (Johnson & Hallberg, 2005). To improve the efficiency of the process and eliminate the problems associated with the large sludge amount produced, several refinements have been made (Aube & Payant, 1997; Coulton *et al.*, 2003; Taylor *et al.*, 2005).

As an example, the technical operation and applicability of some of the most common active treatment plants for AMD neutralization and/or metal recovery (low density sludge plant and high density sludge plant) is provided below based on a summary by Taylor *et al.*, 2005).

Low Density Sludge (LDS) plants are the most common fixed plants for AMD neutralization. Their operation includes pH control and precipitation in three main treatment stages as summarized by Taylor *et al.* (2005).

1   Reagent mixing and dosing stage: a solid neutralization reagent is mixed with water in a tank to produce a slurry. This slurry is then dosed into a reactor containing AMD.
2   Reaction stage: the solution is mixed with mechanical stirrers and aerated if necessary to oxidize any reduced metals (e.g. convert Fe^{2+} to Fe^{3+}). Mixing and/or aeration is continued as the solution flows through one or more reactors. The volume of reactor(s) needs to provide sufficient water retention capacity to allow complete oxidation and neutralization.
3   Flocculation and clarification stage: neutralized water from the reaction stage is transferred to a clarifier/thickener tank. The flow velocity is significantly reduced in the clarifier, and a flocculant may be added during this stage to facilitate sludge settling. Sludge from the clarifier base is removed and generally disposed of on site, while supernatant water is discharged from the plant.

High Density Sludge (HDS) plants also involve the three main treatment stages (i) reagent mixing and dosing stage, (ii) reaction stage, and (iii) flocculation and clarification stage. According to Taylor *et al.* (2005), the key difference associated with HDS plants is that a proportion of alkaline treatment sludge from the thickener underflow is recycled back through the plant to complete the first phase of neutralization. As a result, sludge density increases progressively – sometimes up to 40wt% solids – and the efficiency of reagent use improves. Sludge handling and disposal costs can be significantly reduced by the HDS process, and reagent costs can also be reduced as a result of the improved efficiency of reagent use.

The ***Alkaline barium calcium (ABC) desalination process*** is one of the best recent active treatment options of AMD (Mulopo, 2015). This technology is able to reduce both metals and sulfate concentrations below the toxic level and results in low levels of sludge to be disposed of after useful chemicals or metals are recovered. In general, the ABC desalination process has three major steps: neutralization for metal removal, sulfate removal, and sludge processing. The ABC technology needs further research to provide drinking water quality.

1.3.1.2   Active biological remediation – microbial reactor systems

Microbial reactor systems implement active biological remediation by utilizing the biogenic production of hydrogen sulfide to generate alkalinity and to remove metals as insoluble sulfides (De Vegt *et al.*, 1998). These systems consist of a sulfate-reducing bioreactor and metal sulfide precipitators. In the sulfate-reducing bioreactor, bacterial activity (sulfate-reducing bacteria [SRB]) reduces sulfate (SO_4^{2-}) to soluble H_2S (and HS^-) and produces HCO_3^-. In metal sulfide precipitators, HCO_3^- partially neutralizes the incoming water while H_2S and HS^- react with dissolved metals in the AMD, resulting in precipitation of the metals as sulfide minerals. The precipitation of sulfide minerals is an acid-consuming reaction. Partially treated AMD from the metal sulfide precipitators is then fed into a limestone reactor for further neutralization. Water flows from the limestone reactor at nearly neutral pH into the sulfate-reducing bioreactor where it provides a source of sulfate for the bacterial activity. Treated AMD is then discharged from the bioreactor while the H_2S, HS^-, and HCO_3^- produced in the bioreactor are circulated into the precipitator. These plants are best suited to treating AMD with a pH of 3.0–5.5 and ambient oxygen conditions (De Vegt et al., 1998). Commercial sulfidogenic plants for AMD treatment are rare since they rely on continued growth of sulfate-reducing bacteria (SRB), which require temperatures between 5°C and 40°C, pre-treated pH levels above 5.5 and redox potential (ORP) levels below +150 mV.

I.3.2 Passive AMD treatment systems

Passive treatment systems for AMD use the chemical, biological, and physical removal processes that occur naturally in the environment to modify the influent characteristics and ameliorate any associated environmental impacts. The major processes include the following:

– Chemical processes: oxidation, reduction, coagulation, adsorption, absorption, hydrolysis, precipitation.
– Physical processes: gravity, aeration, dilution.
– Biological processes: biosorption, biomineralization, bioreduction, alkalinity generation.

The design of passive systems must accommodate slow reaction rates; thus, passive treatment systems are best suited to AMD with low acidity (<800 mg $CaCO_3$/L), low flow rates (<50 L/sec), and therefore low acidity loads (<100–150 kg $CaCO_3$/day) (Taylor et al., 2005). The life expectancy of a passive treatment system depends on the mass of limestone and/or organic matter in the system. The available porosity within the limestone and organic matter can also affect life expectancy, as porosity determines the capacity to store treatment precipitates. Passive treatment system may become ineffective if the system gets blocked with treatment precipitates due to insufficient porosity within the limestone/organic matter layers (Ziemkiewicz et al., 1994; Hedin et al., 1994a,b).

Johnson and Hallberg (2005) divided the passive AMD treatment systems into two groups: passive abiotic remediation systems (anoxic limestone drains) and passive biological remediation systems (constructed wetland [aerobic and anaerobic], permeable reactive barriers).

I.3.2.1 Passive abiotic remediation systems

Anoxic limestone drains

Anoxic limestone drains (ALDs) are passive, abiotic remediation systems. ALDs are generally constructed of a plastic bottom liner and a clay cover containing limestone into which anoxic water is introduced (Figure 7.1). The dimensions of the drain vary from narrow (0.6–1.0 m) to wide (10–20 m) in diameter, typically ca. 1.5 m deep and ca. 30 m in length. The objective of these systems is to add alkali to AMD while maintaining the iron in its reduced form to avoid the oxidation of ferrous iron and precipitation of ferric hydroxide on the limestone. The limestone dissolves in AMD, and since CO_2 cannot escape, a buildup of

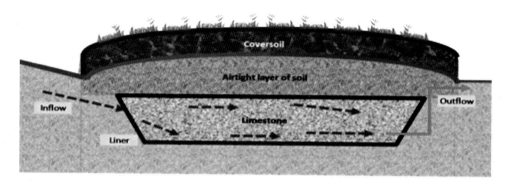

Figure 7.1 Anoxic limestone drain.

bicarbonate occurs, thus adding alkalinity (Watzlaf & Hedin, 1993). The pH of the ALD effluent is around 6.3, at which ferrous hydroxide will not precipitate, but ferric hydroxide and aluminum hydroxide do. For this reason it is important that the ALDs treat AMD that contains no O_2, Fe^{3+}, or Al^{3+} and the design should prevent oxygen from entering into, and carbon dioxide escaping from, the drain. Metal hydroxide precipitation within an ALD will retard water flow, leading to premature failure (US EPA, 2014). ALDs were first described by Turner and McCoy (1990) and Brodie *et al.* (1990), who found them useful to pre-treat acid water for wetland.

1.3.2.2 Passive biological remediation systems

Constructed wetlands

Constructed wetlands are built on the land surface using soil or crushed rock/media and wetland plants. Constructed wetlands can be designed as aerobic wetlands, anaerobic horizontal-flow wetlands, and vertical-flow ponds (vertical flow wetlands) (see also Chapter 4, Section 5). Constructed wetlands are designed to treat contaminants over a long period and can be used as the sole technology, where appropriate, or as part of a larger treatment approach. Contaminants can be removed by microbial transformation and/or plant uptake. Organic and inorganic contaminants are mobilized for plant uptake. The soil- and water-based microorganisms typically remove dissolved and suspended metals from AMD by (bio) sorption and/or uptake. The indirect impact of the biota is achieved by changing environmental conditions, thus the physicochemical characteristics of the contaminants, causing metal immobilization/precipitation and sorption of metals. Both aerobic and anaerobic types often use limestone as a base or as an additive. Key factors that need to be considered when determining the type, size, and cost of an appropriate wetland system include influent acidity loads, pH and redox state, water flow rates and retention times, and the area available for a wetland (Taylor *et al.*, 2005).

Aerobic or oxidizing wetlands are principally suited to near-neutral waters contaminated with Fe. The main remediative reaction that occurs within them is the oxidation of ferrous iron and subsequent hydrolysis of the ferric iron produced. In order to maintain oxidizing conditions, aerobic wetlands are relatively shallow systems (typically <30 cm) that operate by surface flow. Macrophytes (*Typha* (cattail), *Juncus* (rush), and *Scirpus* sp. (e.g. sedges)) are planted for aesthetic reasons to regulate water flow (e.g. to prevent channeling) and to filter and stabilize the accumulating ferric precipitates.

Oxygen infiltration is encouraged and metals precipitate as oxyhydroxides, hydroxides, and carbonates. Aerobic wetlands are designed to provide sufficient residence time to allow metal oxidation and hydrolysis, thereby causing precipitation and physical retention of Fe, Al, and Mn hydroxides. Brodie (1993) reported that wetlands receiving net alkaline AMD (pH range of 4.5–6.3, Fe <70 mg/L, Mn <17 mg/L, Al <30 mg/L) were capable of removing the metals effectively to discharge standards.

Anaerobic or reducing wetlands and compost bioreactors (Johnson & Hallberg, 2005) are built with organic-rich substrates containing oil, peat moss, spent mushroom compost, sawdust, straw/manure, hay bales, or other organic mixtures which provide reducing conditions and also contain limestone for acid neutralization. Often anaerobic wetlands are constructed underground and are devoid of vegetation since penetrating plant roots may cause the ingress of oxygen into the anaerobic zones, which is detrimental to reductive processes.

In this kind of a system, net acidity of AMD water is removed by the dissolution of limestone (Brodie *et al.*, 1990) and the metabolism of iron- and sulfate-reducing bacteria (Tuttle *et al.*, 1969; Gusek, 1998; Hedin & Nairn, 1990). In addition, metals are precipitated as sulfides, hydroxides, and/or carbonates (Henrot & Wieder, 1990). Like aerobic systems, anaerobic wetlands must have substantial residence time for the water; therefore, they require large areas to treat large volumes of strongly acidic AMD. In mountainous areas, wetlands have been most successful when applied to small AMD flows of moderate water quality (Wieder, 1993). Hedin *et al.* (1994b) suggest that anaerobic wetlands can be.

Reducing and alkalinity producing systems (RAPS) are important engineering versions of the basic compost bioreactors (Johnson & Hallberg, 2005; Younger *et al.*, 2003).

In the RAPS system, AMD first flows downwards through a layer of compost (to remove dissolved oxygen and facilitate the reduction of iron and sulfate). Subsequently, the AMD passes through a limestone and gravel bed (to add additional alkalinity, as in ALDs) (Figure 7.2). Usually, water draining from a RAPS flows into a sedimentation pond, and/or an aerobic wetland, to precipitate and retain iron hydroxides.

The successive alkalinity producing systems (SAPS) may include RAPS and additional settling ponds (Figure 7.3), oxidation ponds, wetlands, or multiple RAPS (Kepler & McCleary, 1994; Demchak *et al.*, 2001; US EPA, 2014).

Permeable reactive barriers (PRBs) are passive systems to treat a wide range of polluted groundwaters (Simon & Meggyes, 2000; Meggyes *et al.*, 2001; Simon *et al.*, 2001; Simon *et al.*, 2002; Smyth *et al.*, 2003; Roehl *et al.*, 2005; Meggyes *et al.*, 2009 and Chapter 4,

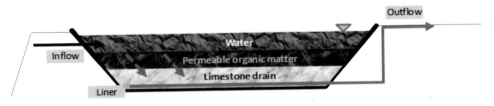

Figure 7.2 Reducing and alkalinity producing system (RAPS).

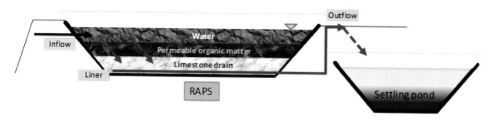

Figure 7.3 Successive alkalinity producing systems (SAPS).

Section 2 of this volume). Some of them are biology-based and designed to remediate AMD and operate on the same basic principles as anaerobic wetlands (Benner *et al.*, 1997; Gilbert *et al.*, 2002; Younger *et al.*, 2003; Johnson & Hallberg, 2005). Biology-based PRBs include buried layers of reactive material (a mixture of organic matter and possibly limestone, zero valent iron) to intercept groundwater plumes of AMD (Johnson & Hallberg, 2005; Taylor *et al.*, 2005). The organic material can encourage reductive microbiological processes, generating alkalinity which is further enhanced by dissolution of limestone and/or other basic minerals resulting in precipitation of metals (As, Cd, Cu, Fe, Ni, Pb, and Zn) as well as sulfides, hydroxides, and carbonates (Taylor *et al.*, 2005). A series of drainage pipes placed below the limestone layer carry the water to aerobic ponds where ferrous ions oxidize and precipitate (Johnson & Hallberg, 2005; Younger *et al.*, 2003).

According to Taylor *et al.* (2005), the key factors which may limit the lifetime of PRBs are the mass of available reactive material and the available volume of pore spaces (and permeability) of the barrier. Metal precipitation and substrate compaction can result in a decrease in porosity and permeability of the barrier. PRBs are ideally suited to cold climates as low soil temperatures (<5°C) inhibit bacterial activity.

1.4 Sustainability of the leachate treatment technologies

The sustainability of any remediation system is a factor that is becoming increasingly critical in decision making.

AMD and ARD treatment from the environmental point of view is highly beneficial for water, soil, and ecosystem at watershed level. Untreated sources and continuously formed AMD and ARD can diffusely contaminate

- the surface water systems with acidic leachates containing toxic metals,
- sediments with eroded acid rock, and
- soils with sediments disposed by floods.

In addition to diffuse pollution, the background values increase, initiating undesirable stress and genetic changes in the ecosystem. The impact of AMD and ARD can be reduced to a minimum by the controlled collection and treatment of the acidic effluents containing metals. The final solution is to remove, encapsulate or otherwise eliminate or inactivate the source of leachate production.

No generally accepted good practice has been developed yet for the utilization of leachate treatment residues. Similar to other water treatment technologies, the residue, i.e. the metal precipitate, must further be treated and/or contained. An example is an iron oxide sludge recovered from a drainage channel at an abandoned coal mine in Pennsylvania and used to manufacture burnt sienna pigment in a commercially successful venture. Base metals recovered by active biological treatment of AMD from metal mines provide some financial return on the investment and running costs of sulfidogenic bioreactors (Hedin, 2003).

The choice of which option to use to remediate AMD is dictated by a number of economic and environmental factors. Traditionally, large discharge volume mine waters have been treated by active chemical processing, particularly when the waters are acidic.

The necessary land surface area and topographic problems may rule out passive biological systems in some situations. However, the mining industry is becoming increasingly attracted to the latter as it seeks to avoid the high recurrent costs of lime addition and sludge disposal. In theory the land areas required for passive systems can be dramatically reduced by focusing on optimizing biological processes, for example, packed bed bioreactors for the removal of iron from acidic mine waters, which are far more effective than aerobic wetlands. It does need to be recognized that, in reality, none of the remediation systems described in this review are maintenance free; passive systems also require a certain amount of management and will eventually fill with accumulated ochre (aerobic wetlands) and sulfides (compost bioreactors).

The ecological and cost efficiency of AMD treatment can be upgraded by utilizing waste materials to fill the reactors. Jones and Cetin (2017) studied two different types of waste materials to remediate acid mine drainage (AMD): recycled concrete aggregates (RCAs) and fly ashes. Results of the column leach tests suggest that RCAs and the highly alkaline fly ash can effectively raise pH of the AMD and reduce Cr, Cu, Fe, Mn, and Zn concentrations in AMD. In addition, sulfate concentrations of AMD decreased significantly after being treated by RCAs while sulfate concentrations of the AMD increased when remediated by fly ashes.

Implementation of AMD/ARD treatment can be expected to result in significant improvement in human health and safety: pollution will be controlled, agricultural land, and surface water system protected for human uses, aesthetic and use value of the landscape increases, and the economic value of land increases in areas where natural acid rock or disposed waste is under control.

Similar to other natural attenuation-based remediation, source removal is always the preferred approach when the sources can be identified and delineated. Non-removed acid rock and waste producing leachate may "provide" unlimited sources, and the leachate treatment in such cases cannot be sustainable, as both the ecological and socioeconomic efficiency of the treatment will be poor due to the extremely long duration.

The extremely long-time requirement is demonstrated well by the leaching study performed by the authors. Acid rock containing Cd, Zn, and Pb was leached over the long term to provide site-specific quantitative characteristics of pH changes, leached metal amounts, and leaching efficiency under average annual rain and dry conditions (Gruiz *et al.*, 2009a,b; Vaszita *et al.*, 2009).

The time requirement of leachate collection and treatment was estimated based on the measured parameters. The pH was constant at pH $= 1.5$ for average rain conditions which decreased even to below pH $= 1$ during the dry period. The pH started to rise after 1400 days (almost 4 years) during the final depletion phase.

Under average rain conditions Cd supply for leaching lasts 8.5 years, 12 years for Zn, and 3300 years the for Pb; under dry conditions, the same values are 24, 26, and 25,000 years. These estimates and the large-scale destruction, ecosystem, and human health damage observable at abandoned mining sites clearly prohibit the disposal of any acid rock waste without controlling oxidation (acidification), leaching, and transport with water. Operators are well advised not to abandon AMD producing mines, rather undertake immediate and full rehabilitation of mines and mining sites.

2 BIOLEACHING AND LEACHATE REMEDIATION

H.M. Siebert, B. Florian, and W. Sand

2.1 Bioleaching

Bioleaching is used in industrial processes such as heap or tank leaching to extract valuable metals from low-grade ores. Bioleaching also occurs as an unwanted natural process called acid rock drainage/acid mine drainage (ARD/AMD): metal sulfides pyrite (FeS_2), chalcopyrite ($CuFeS_2$), or arsenopyrite (AsFeS) are dissolved by oxidation processes. In nature this may be accompanied by acidification of water bodies concomitantly with the release of (large amounts) of heavy metals (Barnes & Romberger, 1968; Gray, 1996). In recent decades a lot of work has been done to prevent ARD/AMD, reduce negative aspects of industrial leaching processes and remediate contaminated areas (Evangelou & Zhang, 1995; Johnson & Hallberg, 2005; Santomartino & Webb, 2007; Gibert et al., 2011).

2.1.1 Mechanism of bioleaching in sulfidic ores and wastes

Bioleaching is the conversion of insoluble metal sulfides into soluble ionic form using chemical and mainly bacterial oxidation processes. Metal sulfides are oxidized to metal ions and sulfate under aerobic conditions by acidophilic iron- and/or sulfur-oxidizing bacteria or archaea (Temple & Colmer, 1951; Kelly & Wood, 2000; Rohwerder et al., 2003; Vera et al., 2013). Molecular oxygen acts as a final electron acceptor for the overall biochemical pathway of the metal sulfide bioleaching process. Ferric ions are the relevant oxidizing agents in the chemical process where Fe(III) ions are reduced to Fe(II) ions. The sulfur moiety of a metal sulfide is oxidized to sulfate and various intermediate sulfur compounds such as elemental sulfur, polysulfide, thiosulfate, and polythionates (Schippers et al., 1996; Rohwerder et al., 2003). Metal sulfides can be classified as acid-soluble (e.g. chalcopyrite) and acid-insoluble (e.g. pyrite). Solubility is determined by electron configuration and is relevant for the determination of the pathways of dissolution. Acid-soluble metal sulfides have valance bands from the metal and sulfur atom orbitals and can be attacked by protons, whereas acid-insoluble metal sulfides have valance bands that are derived from orbitals of the metal atoms which make them non-reactive for protons (Tributsch & Bennett, 1981; Crundwell, 1988; Schippers & Sand, 1999; Edelbro et al., 2003).

Two dissolution pathways – named after the first sulfur intermediate generated in the process – are proposed (Figure 7.4). The polysulfide mechanism is concerned with the dissolution of acid-soluble metal sulfides and the thiosulfate mechanism with the dissolution of acid-insoluble metal sulfides (Schippers & Sand, 1999; Sand et al., 2001; Rohwerder et al., 2003).

Metal sulfides are generally dissolved into the oxidized ionic form of the metal and different ionic sulfur compounds. Fe(III) ions serve as main oxidants for metal sulfides. Electrons are extracted by the oxidation process, whereby Fe(III) ions are reduced to Fe(II) ions.

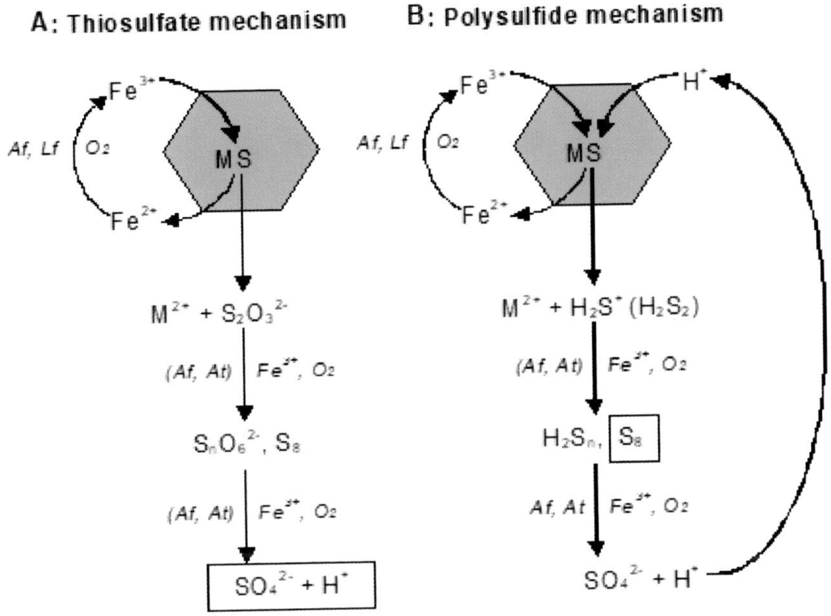

MS = metal sulfide, M = metal cation, A.f. = *Acidithiobacillus ferrooxidans*, L.f.= *Leptospirillum ferrooxidans*, A.t. = *Acidithiobacillus thiooxidans*

Figure 7.4 Schematic illustration of thiosulfate (A) and polysulfide (B) leaching pathways for metal sulfides with intermediates and final products.

The oxidation of a metal sulfide to elemental sulfur is described by the following equation (Schippers & Sand, 1999; Sand *et al.*, 2001):

Chemical oxidation: $MS + 2\ Fe^{3+} \leftrightarrow M^{2+} + 0.125\ S_8 + 2\ Fe^{2+}$

Iron-oxidizing bacteria and archaea oxidize Fe(II) to Fe(III) ions to maintain their metabolism and regenerate the oxidizing agent (Temple & Colmer, 1951; Kelly & Wood, 2000):

Microbial iron oxidation: $4\ Fe^{2+} + O_2 + 4\ H^+ \leftrightarrow 4\ Fe^{3+} + 2\ H_2O$

Sulfur compound-oxidizing bacteria or archaea generate energy by oxidizing sulfur/ sulfur compounds to sulfate and protons (sulfuric acid):

Microbial sulfur oxidation $2\ S^0 + 2\ H_2O + 3\ O_2 \leftrightarrow 4\ H^+ + 2\ SO_4^{2-}$

However, Fe(III) ions are the attacking agents for dissolving acid-insoluble ores such as pyrite. Therefore, Fe(II) ion-oxidizing microorganisms are required. Acid-soluble ores can be dissolved by protons released by sulfur-compound oxidizing organisms (Schippers *et al.*, 1996; Sand *et al.*, 2001).

2.1.2   Contact and non-contact bioleaching

Two mechanisms for microbial metal sulfide dissolution are proposed: the "contact" and the "non-contact" mechanism (Figure 7.5). It has previously been confirmed that bacterial leaching processes occur mainly in the region of contact between bacteria and the sulfide surface. This requires that microorganisms attach to the mineral surface. The non-contact mechanism assumes that microorganisms oxidize dissolved Fe(II) ions in the liquid phase (Sand et al., 2001; Rawlings, 2002).

Attachment of leaching bacteria to a substrate such as pyrite requires the secretion of specific extracellular polymeric substances (EPS) at the bacterial surface (Gehrke et al., 1998; Harneit et al., 2006). The organisms are positively charged by complexing Fe(III) ions in the EPS (Gehrke et al., 1998; Rohwerder et al., 2003), while, according to the ionic strength and pH, metal sulfide has a net negative charge in an aqueous solution (Bebié & Schoonen, 2000). Cell attachment is based on electrostatic interactions between a positively charged cell and a negatively charged surface. The EPS complexed Fe(III) ions are involved in the leaching process. Thus, the function of the complexed ions is similar to the role of free Fe(III) ions in the non-contact mechanism. Sand and Gehrke (2006) suggest that the amount of EPS-bound Fe(III) ions varies with the metabolic activity of the bacteria. Additionally, quorum sensing molecules, like N-acylhomoserine lactone with acyl chains of 12–14 carbon atoms, were described to enhance the biofilm formation of A. ferrooxidans on elementary sulfur and metal sulfides (González et al., 2012).

2.1.3   Microbial bioleaching community

The composition of leaching communities depends on environmental conditions such as temperature, humidity, heavy metal concentration and substrate conditions. Leaching communities consist of diverse acidophilic bacterial and archaeal species. Chemolithotrophic bacteria like Leptospirillum spp., Acidithiobacillus spp., and archaea such as Sulfolobus spp. or Metallosphaera spp. have been isolated from sites of natural mineral oxidation (Temple &

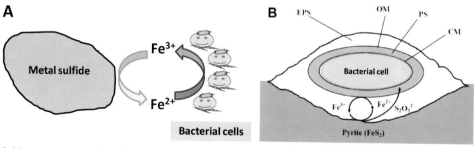

A: Non-contact mechanism: planktonic cells oxidize Fe(II) ions to Fe(III) ions. These Fe(III) ions attack a metal sulfide; the attack is limited by the diffusion rate.

B: Contact mechanism: cells are embedded in EPS and attached to the metal sulfide surface (e.g. pyrite). The Fe(III) ions complexed in the EPS facilitate the dissolution of the metal sulfide and can be directly reoxidized by the bacteria. OM = outer membrane, PS = periplasmic space, CM = cytoplasmic membrane (simplified from Rohwerder et al., 2003).

Figure 7.5 Schematic overview of the non-contact (A) and the contact (B) mechanisms for bioleaching.

Colmer, 1951; Sand *et al.*, 1992; Kelly & Wood, 2000; Schippers, 2007). Acid-tolerant chemoorganotrophic bacteria such as *Acidiphilum spp.* are also present in bioleaching communities (Harrison, 1984; Nancucheo & Johnson, 2009). Microorganisms have been subdivided into three groups according to preferred growth temperature ranges. Mesophilic organisms are those with optimum temperatures between 25°C and 40°C, e.g. *Acidithiobacillus ferrooxidans, Acidithiobacillus thiooxidans*, or *Leptospirillum ferrooxidans*. They cannot grow above 45°C (Rawlings, 1997). Moderate thermophilic microorganisms such as *Leptospirillum ferriphilum, Sulfobacillus thermosulfidooxidans*, or *Ferroplasma acidarmanus* can grow in mixed cultures with mesophiles or with thermophiles. The temperatures for moderate thermophiles range between 40°C and 55°C (Norris, 1997). Thermophilic archaea such as *Acidianus brierleyi* or *Sulfolobus metallicus* can grow at temperatures between 55°C and 80°C (Clark & Norris, 1996; Zhu *et al.*, 2011).

2.1.4 Natural bioleaching process and its industrial application

Acid rock drainage/acid mine drainage (ARD/AMD) is the result of the oxidation of metal sulfides (especially pyrite FeS_2) containing rock, ore, tailings or any solid material in contact with oxygen, water, and microorganisms further to their disturbance/displacement (mining, excavation, construction, waste disposal).

Recovery of valuable metals such as copper, zinc, nickel, or uranium via industrial bioleaching from sulfide ores employing microorganisms is an established biotechnology (Bosecker, 1997; Rawlings, 1997; Rohwerder *et al.*, 2002; Olsen *et al.*, 2003; Rawlings *et al.*, 2003; Brierley & Brierley, 2013). It is an environmentally friendly process with minimal process emissions or tailings. Process solutions are usually recycled. Associated possible problems are long-term release of contaminated acid mine drainage (AMD) effluents and accidental water pollution by dam failures, such as break of the technically undersized tailings ponds and dams. A new approach for recovery of valuable metals is the coupling of bioleaching and electrokinetics (Huang *et al.*, 2015).

However, unwanted leaching processes such as acid rock/acid mine drainage (ARD/AMD) are a great problem. They increasingly attract public attention because environmental awareness and legislation are forcing mining companies and state agencies to fight these problems effectively.

2.2 Acidic leachates and their treatment

The sulfides (typically pyrite) in rocks and in mine waste may be chemically or biologically oxidized when exposed to air and water. The mining activity itself evokes sudden access of air and water to the formerly isolated, underground rock formations. Subsurface mining under the water table needs continuous pumping of the water from the mine, giving atmospheric air exposure and time to oxidize subsurface sulfide rock. The combined *in situ* chemical and biological oxidation generates AMD in the mine water fed by seeps, sinkholes, or springs inside the mine.

Piled rock, coal, waste rock, and tailings without sealing are *ab ovo* exposed to air and water so ARD generation is very likely in such cases. The released acid gets into contact with sulfidic rock formations and mobilizes toxic elements such as arsenic, zinc, nickel, boron, copper, and others. Without control and treatment it ends up in groundwater and surface waters, thus endangering the aquatic ecosystems and the food chains which are based on them.

Surface and subsurface installations can be used for the treatment of the acidic leachate. The passive and active ARD/AMD treatment technologies were introduced in Section 7.1.3 in general; the following applications show water treatment examples for acid coal pile drainage. The origin of the acid in these leachates – similar to sulfide ores – is the elemental sulfur in the coal, which oxidizes after contacting air and water. Several overviews were prepared between 2000–2010 on this topic after the clarification of the possible chemical and biological processes (Kirby *et al.*, 1999; Dempsey *et al.*, 2001) and the results of the first full-scale applications were published (Watzlaf *et al.*, 2002; Younger *et al.*, 2002; Simmons *et al.*, 2002; Cravotta & Watzlaf, 2002; Garrett *et al.*, 2002; Ziemkiewicz, 2003; and the European project of PIRAMID, 2003).

2.2.1 Active and passive treatment of acidic coal pile drainage

Coal mine drainage is characterized by high amounts of acids, plus sulfate and metal ions, typically iron or aluminum. This polluted water must undergo a remediation process for further utilization or for discharge. The aim of acid mine drainage (AMD) treatment is to increase the pH to near-neutral values and remove metals and sulfates. Passive and active methods ranging from chemical to biological methods are useful for treating coal mine and coal pile drainage, as reviewed by Johnson and Hallberg in 2005. Limestone and aerobic conditions, for example, in open ponds, encourage the oxidation of ferrous ions and the precipitation of the metals as oxides or hydroxides. However, a further step is necessary to remove the iron or other metal precipitates. An innovative procedure for the removal of the suspended lime precipitate from the treated water is the use of surfactants-enhanced dissolved-air flotation technology. The adherence of air bubbles to suspended lime precipitate makes it float to the water surface from where it can be skimmed. Both synthetic and microbial surfactants can be used (Menezes *et al.*, 2011).

Another option is constructed wetlands, which generally apply close to neutral (in a preceding technology step already neutralized) leachates (see Chapter 4, Section 1.3). Design and dimensioning can control certain processes such as oxidation and metal precipitation, and complexation of metals with organic matter or anaerobic sulfide-generation from metal ions.

Anoxic limestone drains (ALD) for coal mine drainage and their long-term performance was studied by Watzlaf *et al.* (2000, 2004) and found that proper isolation from air and additional alkalinity can exclude hydroxy precipitation of metals on the long term. The efficiency of such a system is affected by the carbon dioxide partial pressure and calcite saturation index. The compounds formed *in situ* may coat the limestone surface, which results in a decreased limestone dissolution rate. In the case of low or medium Fe(III) and aluminum concentration this problem does not occur even over the long term, otherwise this problem can be eliminated by thorough design. Santomartino and Webb (2007) estimated the sufficient time of safe limestone bed application as a function of the surface area of the limestone (SLS), the proportion of Fe retained in the bed (f × Fe^{2+}) and the flow rate of the leachate (Q) as follows:

$$t = SLS \times 4.4 / f \times Fe^{2+} \times Q,$$

where

t   maximum time of safe use (year)
SLS   the surface area of the limestone (m^2)

f   fraction retained in the bed system (−)

$f \times Fe^{2+}$ concentration in limestone bed (mg/L)

Q   flow rate (L/h).

Authors recommend the use of this equation both to ALD, OLD, and also to open lime-stone channels, which generally retain higher fractions of iron in the system. If the necessary geometrics cannot be ensured, the limestone must be regenerated, i.e. the iron hydroxide coating removed by agitation from time to time.

Relative diluted coal mine AMDs can be treated by oxic limestone drains (OLD) as reported by Cravotta and Trahan (1999). The pH, alkalinity, and calcium concentration increased in their study, while acidity and iron and aluminum contents decreased by around 95% due to limestone treatment. Toxic metal content also decreased due to sorption and coprecipitation on metal oxides at values above pH = 5.

Pulsed limestone bed (PLB) reactor was tested both at laboratory and field scales to overcome the problems of limestone treatment, e.g. slow dissolution rates and precipitation of metals (Ziemkiewicz et al., 1994; Watten et al., 2005, 2007). This reactor used a carbon dioxide pretreatment step for increased limestone dissolution and a pulsed liquid addition. The method provided high alkalinity levels in the laboratory. In field trials on three AMD sources the pH was raised from pH = 3.2–4.8 to values between pH = 6.2 and 6.9, allowing ferric- and aluminum, but not ferrous or manganese, ions to precipitate.

Another possibility is the use of dispersed alkaline substrates (DAS). This was tested in a laboratory-scale experiment in columns fed with natural AMD at different flow rates with a pH of 2.3–3.5 (Rötting et al., 2008). In this study calcite sand was mixed with wooden chips. They showed 1200 mg/L net acidity removal rates within 70 weeks, which is 3–4.5 times greater compared to conventional passive treatment systems. Aluminum, iron, lead, and copper ions were almost completely removed in this system. An alternative to lime or limestone treatment was described by Petrik et al. (2005) and Vadapalli et al. (2008): they tested the application of fly ash for AMD neutralization. The pH increased from 2.7 to 11.5 and aluminum and iron ions were eliminated depending on the AMD-to-fly ash ratio. This was attributed to a precipitation and coprecipitation of metal hydroxides as well as an adsorption to the fly ash surface. The authors' opinion is that the material loop of coal mining can be closed by utilizing the fly ash of coal-fired power plants: in addition to neutralizing AMD, the solid residue of AMD treatment can be used for mine backfill (see Gyöngyösoroszi mine closure case study in Section 3 of this chapter). Macrocapsules with included phosphate buffer used to improve the pH in AMD waters at a laboratory scale. The pH increased from pH 2.5 to pH 6 over 8 days before a decrease occurred. In field trials an initial rise in pH to 6 was observed, followed by a decrease to pH 2.5 within 10 days (Aelion et al., 2009). In another laboratory study in sand columns, microbial denitrification increased the pH by two units to pH = 8 and potassium dihydrogen phosphate encapsulated in a pH-sensitive coating maintained it for four weeks. This approach can be used inside permeable reactive barriers where the phosphate ions move through the aquifer with the groundwater flow. The advantage is that the phosphate can act as a buffer when it is precipitated in strongly acidic soils (Rust et al., 2002).

RAPS and SAPS became widespread for coal mine AMD from the 2000s and several applications have been published. A full-scale passive treatment is presented by Matthies et al. (2010), composed of two parallel RAPS and one aerobic reed wetland. The system reduced iron and aluminum ion content and acidity by 83–87% compared to a system applied previously. The pH and alkalinity were increased constantly, and the removal rates were constant over five years of operation regardless of seasonal variations. Other authors, e.g. Bhattacharya *et al.* (2008) pointed out the disadvantages of such systems: regular monitoring and maintenance due to toxic metal precipitation that changes the substrate producing alkalinity. Koschorreck (2008) studied how sulfate reduction at pH values below 5 is influenced by electron donor addition and by the supply with carbon and energy sources as well as inhibitory factors, e.g. the metabolites H_2S and organic acids.

2.2.2 Mitigation of AMD production

An investigation of different waste piles in Germany showed an 85.0–99.98% reduction in sulfate and toxic metals release (Willscher *et al.*, 2010) due to revegetation. The revegetated soil cover above the heap reduced pyrite oxidation rates due to the presence of oxygen-consuming organic substances infiltrated from the top or transported by groundwater. The cell counts of sulfate-reducing bacteria did significantly increase in this environment in comparison to an uncovered heap.

Inhibition of AMD production by treatment of the parent mine waste using surfactants has been known for more than 20 years. The influence of detergents on planktonic leaching microorganisms, e.g. *Acidithiobacillus ferrooxidans* was investigated at laboratory scale. The application inhibited or enhanced microbial pyrite oxidation depending on detergent type and concentration used (Sand, 1985; Onysoko *et al.*, 1984; Dugan, 1986). In other laboratory experiments quaternary ammonium compounds or hydrocarbons for planktonic microorganisms were used (Sikkema *et al.*, 1995; Kourai *et al.*, 2006; Sumitomo *et al.*, 2006). Sand *et al.*, 2007 described the field-scale application of sodium dodecyl sulfate (SDS) and the biocide isothiazolinone in large percolators with a volume of 65 m³ containing acidic mining waste. The addition of 0.2 g SDS per kg waste material decreased the metabolic activity and the number of leaching microorganisms, but the bacteria were not killed. The SDS concentration ranged from 0 to 100 mg/L SDS in the effluent, which was explained by the adsorption or consumption of the surfactant. Application of the biocide reduced the release of toxic metals and sulfides to some extent, but a complete AMD inhibition was not achieved. This may be explained by an inefficient biocide concentration or the heterogeneous access of the additives. Another study where 0.3 or 3 g/L SDS was used on river sediment showed similar low AMD inhibition (Seidel *et al.*, 2000), and only 3.6% of the added SDS was detected in the first 2 hours after application. Even higher surfactant concentration showed an inhibition of bioleaching for a short time only. The authors' explanation was the strong sorption on solid particles immediately after application, and microbial degradation.

The lesson learned from the above studies is that it is important to adopt and combine various AMD mitigation and treatment methods depending on site conditions, acidity of the drainage as well as the amount of pollutant metal ions. The other generic rule for such complex chemically and biologically determined cases is the necessity to find optimal technological parameters in the preceding laboratory and small-scale field experiments. The results can serve as input for the thorough design of the type and amount of reagents and additives, the residence times and the consequent size of the *in situ* or *ex situ*, active or passive reactor.

2.3 Monitoring bioleaching, microbiological status, and leaching activity

The microorganism monitoring methods can be divided into cultivation- and molecular-biology-based techniques.

The classical approach is the cultivation-based technique to detect living microorganisms, which can be achieved by the most-probable-number technique or by cultivation on agar plates. It enables the cell counts of living microorganisms and their direct analysis in liquid or solid samples. The disadvantage of this method may be an underestimation of the microorganisms present due to nonculturable microorganisms or cells embedded in biofilms. However, this method can distinguish microorganisms with the help of metabolism, e.g. iron-oxidizing from sulfur-oxidizing, or aerobic from anaerobic microorganisms. Investigations on artificial coal spoils showed first heterotrophic and then autotrophic microorganisms that were detected by cultivation-based techniques within 76 weeks of cultivation (Harrison, 1978). The disadvantage of this method is the preference of particular microorganisms due to different nutrition requirements. However, monitoring bacterial biomass can reduce this disadvantage by measuring e.g. the lipid phosphate content (Porsch et al., 2009). Microcalorimetry is an independent cultivation approach which can determine metabolic activity in solid or liquid samples to calculate the concentration of microorganisms (Schippers et al., 1995; Schroeter & Sand, 1993). Total cell number can be determined from liquid or solid samples using light and phase contrast and fluorescence microscopy. In this case direct cell counting with the nucleic acid stains DAPI (4',6-diamidino-2-phenylindole) or SYBR Green for the detection of the microorganisms are performed. This method was used to count microorganisms in German mine tailings and in an active lignite mining area up to 10^9 cells per gram sand, and 10^8 cells per gram sand, respectively (Kock et al., 2007; Kock & Schippers, 2006; Siebert et al., 2009). The disadvantage of this method is the problem of differentiation between living and dead cells. To overcome this problem the Fluorescence In situ Hybridization technique (FISH) with specific fluorescence probes is used to detect, for example, the 16S ribosomal RNA genes (e.g. Urbieta et al., 2012). This technique also allows the differentiation between bacteria and archaea and between species. An improvement of this technique is the Catalyzed Reporter Deposition Fluorescence In situ Hybridization Technique (CARD-FISH) described by Schippers et al. (2007).

Quantitative Polymerase Chain Reaction (qPCR) offers another option for monitoring and quantifying microbial communities. The technique is based on the amplification of bacterial or archaeal 16S ribosomal RNA genes. This method is used after the extraction of nucleic acids from solid or liquid samples as described earlier (e.g. Kock & Schippers, 2008; Liu et al., 2006). A community study of different coal mine drainage treatment systems using cultivation-based and molecular-based qPCR techniques has indicated the predominance of heavy metal resistant fungal and bacterial strains. The fungi were identified using morphological characterization and the amplification of the 18S rRNA gene, 28S rRNA gene, and ITS1–5.8S rRNA-ITS2 region (Santelli et al., 2010). Another study that used Denaturing Gradient Gel Electrophoresis (DGGE) analysis of amplified 16S rRNA genes and sequencing showed that the microbial diversity in the oxic sediment zone of a constructed wetland is very low (Nicomrat et al., 2006). Iron- and sulfur-oxidizing acidophiles A. ferrooxidans and A. thiooxidans as well as the bacteria Alcaligenes sp. and Bordetella sp. were identified. Another study on samples from three layers of surface sediments from a mining lake showed high microbial diversity. Depending on the physicochemical properties of the samples, cultivation- and molecule-based techniques identified acidophilic iron-reducing or -oxidizing or neutrophilic iron-reducing bacteria (Porsch *et al.*, 2009).

Factors beyond the microbial community, e.g. physicochemical properties, must be measured to be able to properly monitor the remediation process. Not only the pH is a significant marker, but oxygen, carbon dioxide, sulfate, ferrous, and ferric ion contents or other metals are also important. Temperature, conductivity, or redox potential measurements can also be used for control. Various methods are available for mineralogical and chemical analysis of samples affected by AMD. The morphology and the arsenic specimen from a mining waste were analyzed by Environmental Scanning Electron Microscope (ESEM) (Recio-Vazquez *et al.*, 2010). This study qualitatively and quantitatively determined heavy metals by X-ray diffraction (XRD) and X-ray fluorescence (XRF) spectrometry, and Inductively Coupled Plasma Atomic Emission Spectrometry (ICP-AES) with a detection limit of 10 g/L arsenic in the samples. The measurement of pH, alkalinity, and other factors allows the development of a model for the prediction of the efficiency of different remediation treatments (Nixdorf *et al.*, 2010).

2.4   Impacts of inhibiting agents on bioleaching microorganisms

As described previously, the treatment of leaching microorganisms by surfactants can be effective at different concentrations. This can be explained by the growth behavior of these microorganisms. The increased effect of ethanol, H_2O_2 or SDS on planktonic cells of *Candida albicans* was shown in comparison to biofilm-grown cells (Nett *et al.*, 2008). A high cell density system with 10^{10} planktonic *A. thiooxidans* cells was treated with 0.5–10 g/L SDS, which resulted in the release of nucleic acids and a 25–75% decrease in cell numbers. The incomplete disappearance of cells even when treated with high surfactant concentrations suggests the existence of a lysis mechanism (Siebert *et al.*, 2011). The cell lysis in bacteria and the different ways of its induction have been known for more than 100 years (Rice & Bayles, 2008). This may explain the differences between laboratory and field studies using biocides for the inhibition of microorganisms. Since the acidophilic bacteria, e.g. *A. ferrooxidans* or *Leptospirillum ferrooxidans*, can grow planktonically and in a biofilm on a mineral surface, the inhibition concentration might differ because an efficient detergent concentration depends not only on surface adhesion and biodegradation but also on the type of microorganism grown in a biofilm and perhaps on the substrate. While relatively low concentrations inhibited planktonic microorganisms because of different accessibility, microorganisms grown in a biofilm were not vulnerable even at higher concentrations.

A novel approach to reduce pyrite dissolution was shown in 2008. In this study freshly milled mine waste was pre-cultivated with the heterotrophic acidophiles *Acidiphilium*, *Acidocella*, & *Acidobacterium*. Subsequently, cultures were incubated with the lithoautotrophic *A. ferrooxidans* for four weeks. This reduced the pyrite dissolution rate by 57–75% (Johnson *et al.*, 2008). However, co-incubation of the heterotrophs and autotrophs failed to reduce pyrite dissolution in comparison to the controls.

2.5   Conclusions

Demand for valuable metals such as copper, gold, or nickel is increasing constantly all over the world. This situation is aggravated by fast economic growth, e.g. in China, India, and Brazil, resulting in shortages of such metals. Bioleaching is an efficient, competitive, and environmentally friendly technology to extract metals also from low-grade ores. Research is still needed to increase its efficiency and to reduce environmental impacts of industrial leaching technology. In addition, technologies to minimize the impact of unwanted bioleaching such as acid rock drainage and methods for the remediation of mining-impacted areas, including forecast and monitoring techniques, also need to be considerably improved.

3   GENERAL OVERVIEW OF THE GYÖNGYÖSOROSZI MINING SITE, CONCEPT FOR THE REMEDIATION OF THE FORMER ZINC-LEAD MINES

Zs. Berta, G. Földing, J.T. Árgyelán & M. Csővári

3.1   Introduction

This case study presents an engineering solution applied to a mine closure in Hungary. The former Gyöngyösoroszi metal mine currently is under restoration. AMD originates from the abandoned and flooded mine, and ARD is generated from hundreds of solid mine waste dumps and waste rock heaps of various sizes and from flotation tailings ponds. Leaching in the underground mine, mine wastes and waste rocks caused watershed-scale diffuse pollution. The impacts of leaching on water, sediment, and soil can be observed at watershed level. Additionally, a series of industrial accidents took place at the site during ore mining and processing which resulted in serious contamination of the nearby *Toka Creek* and valley. This section gives a general overview of the environmental conditions and the remediation concept of the site and the first results of the remediation. The main steps of the remediation are remediation of the waste rock piles (WRP) and the flotation tailings dam, clean-up of the *Toka Creek* and the water reservoirs, and backfilling the highly pyritic *Mátraszentimre* mine.

A short overview is given of the geology of the site, the history of mining at this site, the acid mine drainage (AMD) and its treatment, the waste rock piles, the ore processing plant, and the tailings pond.

3.1.1   Mine and mining history

The first written information on metal mining at Gyöngyösoroszi, Eastern Hungary, dates back to the second half of the 18th century. Mining continued in the19th century, especially for lead and silver. Industrialized mining started in 1925, and the mining area became state owned in 1948. At the time the official strategy in Hungary was the development of heavy industry. In line with this, the Hungarian Government decided in 1951 to open a new zinc-lead ore mine at the Gyöngyösoroszi site with a mining capacity of 500 t/day and to build an ore processing plant of 200 t/day capacity. Some shafts built earlier (1926–1930) were dewatered and newly exploited. New shafts were built (Károly shaft in 1955 and Bányabérc shaft in 1964), one last one (*Mátraszentimre* shaft) being added in 1967. The total production of the mine was 3.6 million metric tons of ore from which 260,000 t of concentrates (sphalerite, galena and pyrite) were obtained by 1986 when ore mining activity was abandoned and the mine was placed on long-term stand-by (Soós, 1966; Vidacs, 1966a; Kun, 1966, 1980, 1998). Mine water treatment is being conducted by precipitating iron and toxic metals (zinc, cadmium, cobalt, nickel, lead) from the water using a lime milk process.

Initially the ore extracted in Gyöngyösoroszi was processed in Bulgaria and Germany. A processing plant was built on the site to reduce transportation costs. Technological water demand (2000 m^3/day) determined the location: the *Toka Creek* was capable of supplying the processing plant with water, and thus it was decided to build the plant and the water reservoir in the Toka Valley. The flotation process started in the plant in 1955.

A flotation tailings pond was constructed in the nearby valley (Száraz Valley). During the operation a few dam failures occurred when a large amount of process water from the pond and solid tailings entered the *Toka Creek*. The estimated total volume of tailings reaching the *Toka Creek* was a few thousand tons. One of the most serious remediation tasks is the clean-up of the *Toka Creek* valley of tailings.

The site and the related environmental risks have been studied by Fügedi *et al.* (2001), Fügedi (2004, 2006) and Gruiz *et al.* (2007). Based on measured data between 1991–2005 (Gruiz *et al.*, 2007) the metal concentration of the *Toka Creek* was, on average, As: 10 µg/L, Cd: 2 µg/L, Pb: 30 µg/L, and Zn: 800 µg/L. Gruiz and Vodicska (1993) found that total As and Pb concentrations in the soil decreased from 110 mg/kg to non-detectable, and from 462 mg/kg to 63 mg/kg, respectively, at a distance of 5–50 m from the *Toka Creek*. Similarly, Cd and Zn concentrations decreased from 7.5 mg/kg to 0.6 mg/kg, and from 1685 mg/kg to 208 mg/kg, respectively. The size of the Toka watershed affected by mining activity was 10 km^2, according to the catchment scale GIS assessment performed by Gruiz *et al.* (2007). Broad information on the Gyöngyösoroszi mining site was provided by MAFI (Hungarian Central Geological Survey) (Kun, 1998).

3.1.2 Geology of the site and development of the ore mineralization

The geology of the polymetallic ore deposit at Gyöngyösoroszi is presented in the publications of Vidacs (1966b); Kun *et al.* (1988); Fügedi (2006).

The area affected by ore mining at Gyöngyösoroszi is located in the western part of the Mátra Mountains, consisting of mainly Tertiary eruptive rocks. There is no direct information regarding the pre-Tertiary basement as the prospecting boreholes did not even reach the Oligocene underlying beds. The deepest borehole (−700 m) ended in the lower andesite formation (Hasznos andesite). The thickness of the *Mátra Andesite Formation* is 500–600 m, but it exceeds even 1000 m near *Mátraszentimre*.

The mineralized territory around Gyöngyösoroszi in the Western Mátra Mts. covers an area of about 30 km^2 between Gyöngyösoroszi, Mátrakeresztes and *Mátraszentimre*. Its structure is of the caldera type, i.e. late collapse structure, with connected *hydrothermal veins*. Hydrothermal alteration, silicification, and argillization of volcanic rocks can be observed regionally in the territory.

The two main components of the ore targeted by mining are Zn and Pb. Considering that it is a sulfidic ore, the main minerals are sphalerite and smaller amounts of wurtzite and galena. Fe is also present primarily in pyrite and in smaller amounts in marcasite, chalcopyrite and arsenopyrite. The length of the veins in the Gyöngyösoroszi ore deposit area is generally 500–1000 m; their thickness is 0.3–2.0 m. A detailed description of the mineralization can be found in publications by Vidacs (1966b); Kun *et al.* (1988). The characteristic metal composition is summed up in Table 7.1.

Thirty-one veins were found in the mineralized territory in the course of ore mining, and they were mined from 1955 to 1986. The lowest level of mining activity is the +100 m.a.s.l.

Table 7.1 Characteristic metal compositions of the Gyöngyösoroszi ore.

Characteristic parameters of Gyöngyösoroszi veins							
Level m.a.s.l.	Number of samples	Pb %	Zn %	Cu %	Fe %	Au g/t	Ag g/t
Veins of the central deposit							
+ 510	85	0.52	1.66	0.06	5.06	0.71	9.73
+ 460	590	1.32	3.42	0.20	4.31	1.21	19.39
+ 400	1657	1.39	3.70	0.08	4.27	1.22	14.41
+ 350	1297	1.55	4.31	0.16	4.77	0.59	15.32
+ 300	1138	1.65	4.93	0.21	4.96	0.55	17.44
+ 250	1188	1.62	4.34	0.22	5.46	0.18	18.73
+ 200	1052	1.59	4.31	0.25	7.52	0.22	21.85
+ 150	634	1.92	4.68	0.36	9.73	0.21	23.89
+ 100	396	1.85	3.55	0.38	9.57	0.04	26.30
Mátraszentimre vein		0.96	2.77	0.09	6.28	0.40	11.80

(meters above sea level) drift within the Károly vein. The highest point was the *Mátraszentimre* shaft at +656 m.a.s.l. height, which means that the Gyöngyösoroszi operations affected the 556 m thick interval of the ore deposit.

The deposit comprises a polymetallic ore with medium to high zinc and lead content. The ore also contains a large amount of pyrite and some traces of gold and silver. Most of the ore was mined from the central mining areas.

3.1.3 Summary of the mining activity

The last mining period lasted from 1949 to 1986, and mining and ore processing were terminated due to financial difficulties.

The mining area can be divided into three parts according to the groups of productive veins that were found on the site:

- *Central deep zone* (+100 to +400 m, situated below the main transporting adit. (Károly vein). The volume of the excavated rock is estimated to be 748,000 m³.
- *Central zone* and other veins above the main transporting adit (~+400 to +560 m, Bányabérc vein); the volume of mined-out rock is estimated to be 42,000 m³.
- *Northern mining area* (*Mátraszentimre* vein) was only included in the operation in 1967. Approximately 193,000 m³ rock was mined out.

Metal content of the ROM (run of mine ore: the unprocessed form) is presented in Table 7.2. Total ROM mass amounted to 3.634 million tons, which produced 2.944 million tons of tailings (Ötvös *et al.*, 2004).

The connection and location of the three mining areas are presented in the simplified cross-section in Figure 7.6. The main transporting adit (Adit) is also shown together with the two bulkheads (plugs) constructed in the Adit for enhancing the flooding of the *Mátraszen-timre mine* (V-2 bulkhead) and regulation of the mine water volume for the water treatment station (V-3 bulkhead). The details of the bulkheads are described by Bagdy and Sulyok (2000). However, after several months of operation, the V-2 plug *failed in compression of water* collected in the *Mátraszentimre mining* openings. The damage of the plug and its surroundings led to the outburst of large amounts of highly contaminated mine water from the *Mátraszentimre* openings. This water (100,000–300,000 m³) ran through the *Toka Creek* and extensively contaminated the creek bed.

Plug V-3 was built in 1989 to control the flow rate of the mine water to the water treatment plant. Both plugs have been removed recently as part of remediation.

The field still has ore reserves after the termination of mining. These reserves are presented in Table 7.3. It can be seen that the pyrite content in the *Mátraszentimre* vein is twice as high as in the whole area. The quantity of the available ore is estimated to be almost twice that of the already exploited amount, but the grade is lower.

Table 7.2 ROM quality.

Metals	Zn	Pb	Cu	Fe	Cd	Au	Ag	As
Content	2.85%	0.98%	0.147%	9.8%	150 g/t	0.74 g/t	47 g/t	350 g/t

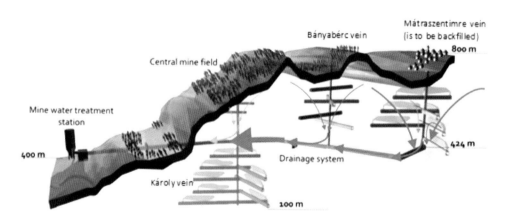

Figure 7.6 Simplified cross-section of the mining area.

Table 7.3 Ore reserves left in the *Mátraszentimre* site.

Item	Zn	Pb	FeS₂	Au	Ag	Quantity of the ore
	%	%	%	g/t	g/t	million t
Total geological reserves	3.71	1.39	13.78	0.15	0.39	5.761
Reserves in the *Mátraszentimre* site	2.75	0.96	11.72	0.09	0.24	1.027

3.2   AMD in the Gyöngyösoroszi mine

The volume of the created underground cavities exceeded 1.7 million m³ (Table 7.4) in the Gyöngyösoroszi mine during the industrial-scale mining activities.

Mine waters are part of the groundwater connected with the mine openings: 25–30% of all mine waters in the former Gyöngyösoroszi ore mine originated from the *Mátraszentimre* mining territory. The main vein mined contains a large amount of pyrite, the alteration of which in oxygen-rich conditions causes intense acidification and dissolves toxic elements into the mine water. This component of contaminated mine water has a significant effect on the overall mine water quality.

3.2.1   Quality of the mine water

Only limited reliable data is available about the quality of the discharged mine water from the 1970–1990 period. The most frequently measured parameters were electrical conductivity, pH and Zn concentration because of the nature of the mine water treatment technology. Flooding of the deep mining fields in 1988 had a significant effect: the pH increased and the Zn concentration decreased and stabilized at a high level (10–20 mg/L) (Figure 7.7). However, it was

Table 7.4 Main geometric parameters of the mine cavity.

Mine field	Total mine openings (m³)	Free face (m²)
Mátraszentimre	118,086	220,143
Bányabérc	30,060	
Central	1,585,756	3,521,798
Total	**1,733,902**	**3,741,941**

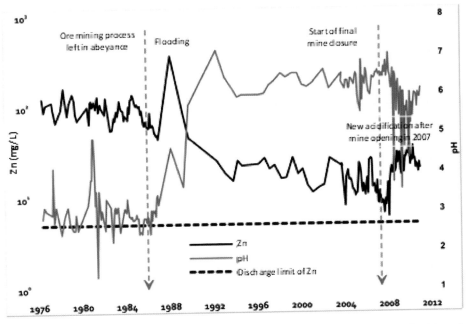

Figure 7.7 pH and Zn concentration time series of the mine water sampled at the mouth of the adit.

Table 7.5 Results of chemical analyses of the mine water sampled at the mouth of the adit.

Component	Unit of measurement	Water sample	
		Filtered	Non-filtered
Al	mg/L	2.48	22.18
Fe		49.38	68.83
Mn		4.67	4.84
As	µg/L	120.6	1111.2
Cd		62.8	122
Co		30.2	35.7
Cr		17.9	15.4
Cu		207.9	158.1
Ni		19.2	17.2
Pb		60.9	131.9
Zn		17,892	22,738
Na^+	mg/L	6.7	75
K^+		1.5	40
Ca^{2+}		18.4	405
Mg^{2+}		12.5	238
Cl^-		0.7	31
SO_4^{2-}		63	1,540
CO_3^{2-}		<10	<10
HCO_3^-		12	207
TDS		635	2,265

obvious at that time that the decreasing trend of Zn concentration would not result in water quality changes, albeit in the distant future and mine water treatment had to be continued.

There were 125 sampling events at the mouth of the adit during 2004–2011. Table 7.5 shows the statistical summary of the representative parameters including main ions and toxic elements. The analyses of toxic elements were carried out using ICP-OES and ICP-MS, and classic laboratory methods were implemented to analyze the main anions both from filtered and unfiltered water samples.

Fe, Zn, and the Cd concentrations have exhibited an increasing trend since the mine was reopened in 2007 (accompanied by the complex remediation and rehabilitation of the site) because of the aeration of the workings and the resulting AMD (Figure 7.7).

After reopening the *Mátraszentimre* mine field (2008), in parallel to an increasing sulfate trend, bicarbonate and the pH have shown a clear decrease as a result of acidification.

3.2.2 Quantity of AMD

The water was pumped from the deeper mining fields to the surface via the Károly shaft during mining operations because of dewatering the mining level below the elevation of 400 m.a.s.l (Figure 7.6). Dam V-3 controlled the drained mine water in this period.

After 25 years of intensive mining, from 1986 (the date of abeyance) to late 2004, the flow rate of the outflowing mine water ranged between 1800 and 3000 m³/day (Figure 7.8). The flow rate of the resulting AMD depended on the precipitation/rainfall at the mine field: the total average daily mine water rate was 3000 m³/day in 2007 and 2008, and more than 4000 m³/day in 2006, 2009, and 2010 (rather wet years), see Figure 7.9.

Dam V-3 controlled the mine water drainage flow before its failure at the end of 2007. At present, such a control is not possible. Thus, it was necessary to reconstruct and modernize the mine water treatment facilities. In addition to the underground buffer space, two surface buffering pools were built to control the hazardous situation. The mine water was pumped via the Károly shaft to the underground buffer space.

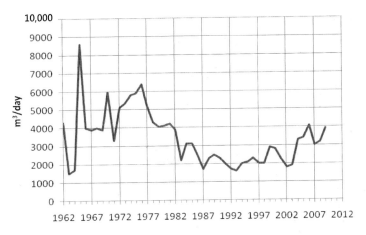

Figure 7.8 The average quantity of mine water measured at the mouth of the adit between 1962 and 2009.

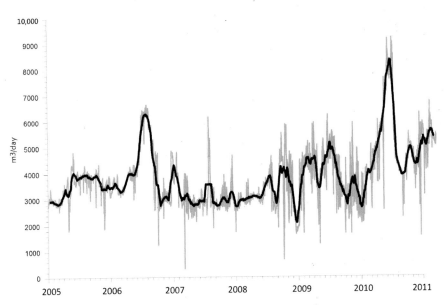

Figure 7.9 Daily quantity of total mine water measured at the mouth of the adit between 2005 and 2011. The black line shows the 30-day moving average.

3.2.3 AMD flow rate

The AMD flow rate was measured starting in 2009: 25–30% of the total mine water derived from the *Mátraszentimre* mine field, 5% from the northern part of the adit, and 20–25% from the Bányabérc and Béla cross headings, and the remaining 40–50% came from the central mine field (Figure 7.10). The most problematic mine field is the *Mátraszentimre* field, which supplies 25–30% of the total mine water but 50–85% of the toxic metals; 75% of the total contamination typically originates from the *Mátraszentimre* field.

3.2.4 Mine water quality at different mine fields

Three main mine water groups can be distinguished by the chemistry of the major elements.

Group 1: These are descending water inflows of meteoric origin, filtering into the mine workings via fissures and faults. However, the sulfate content indicates the influence of the host rocks, pervaded through by ore-bearing veins. The variability in their quality and

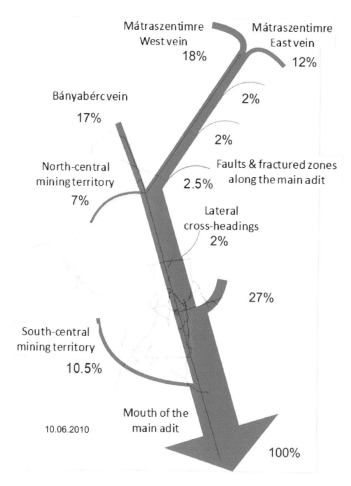

Figure 7.10 Distribution of the mine water quantity between the different mining territories according to the measurements on 10 June 2010.

quantity depends on the precipitation amount and detention time. The following additional subgroups can be distinguished:

1A: low or low-to-moderate TDS,
1B: slightly oxidative character,
1C: rainwater like, and
1D: neutral pH.

Group 2 does not differ significantly from group 1. It has higher TDS concentration, slightly lower pH values (4.6–7) compared to group 1, and an increasing sulfate anion concentration indicating a longer retention time in the fissures of the host rock traversed by ore-bearing hydrothermal veins. These waters are typical of the central mine field.

Group 3 differs significantly from the others. Waters in this group have moderate to high TDS content, they are Ca-(Mg)-SO$_4$ water type, often exhibiting strongly acidic chemistry. The prevailing anion is the sulfate which is present at high concentrations. These waters occur in the flooded deep mine openings ("old man"), where the water reaches the ore-bearing host rock, and in the airy mine openings where the oxidation is intensified (AMD). These waters come from non-flooded, oxygen-saturated mine openings, where an intensive alteration process takes place (see AMD), and are typical of the *Mátraszentimre* mine field. Table 7.6 summarizes the characteristics of typical mine waters.

The highest Al, Fe, Mn, Zn, and Cd concentrations can be found in the mine waters originated from the mine reopenings (Groups 2 and 3). Significantly high concentrations (especially Fe and As) are typical of the mine waters from the non-flooded *Mátraszentimre* mine field where the pyrite-rich *Mátraszentimre* vein is located (subgroup 3B).

In general, the toxic elements (As, B, Ba, Cd, Co, Cu, Cr, Ni, Pb) are present at low concentration in the water feeders (<0.1 mg/L), typical of Group 1. These elements occur at extremely high concentrations (0.1–10 mg/L) in the mine waters of Group 2, and Group 3 in particular. The most common toxic contaminant in these waters is the As. Further

Table 7.6 Statistical summary of the waters belonging to different mine water groups.

Parameter	Water types within the mine						
	1A	1B	1C	1D	2	3A	3B
TDS (mg/L)	377	454	612	763	994	2140	2696
pH	7.2	6.9	7.0	7.0	6.4	5.2	3.1
EH (mV)	143	138	100	87	157	176	447
SO$_4^{2-}$ (mg/L)	148	262	275	344	550	1266	1768
Al (mg/L)	1.0	0.1	0.06	0.06	0.7	7	37
As (µg/L)	33	76	13	75	40	647	3142
B (µg/L)	21	17	86	49	49	71	29
Cd (µg/L)	14.4	5.6	3.6	2.6	14	76	481
Fe (mg/L)	7.7	4.1	1.2	1.8	4.3	67	271
Mn (mg/L)	0.4	0.7	0.7	0.8	2.1	5.1	7.3
Zn (mg/L)	2.3	1.4	0.9	0.5	4.3	25	70

characteristic toxic elements are Cd, Cu, and Co, and Pb in the mine waters originating from *Mátraszentimre*. The distribution of these elements depends on the composition of the host rock: As and Co with Fe indicate pyrite-containing rock, Cd and Pb in addition to Zn reflect host rock containing sphalerite and galenite.

3.2.5 Summary of the AMD at the Gyöngyösoroszi mine

More than 1.7 million m^3 underground cavities were excavated during the time of non-ferrous metal mining in 1950–1986. In the period of 1960–2010 more than 64.5 million m^3 water was discharged from the system through the adit.

At present, 25–30% of the total mine water derives from the *Mátraszentimre* mine field, with approximately 5% from the northern part of the adit, 20–25% from the Bányabérc and Béla cross headings, and 40–50% from the central mine field.

The most problematic part of the mine is the *Mátraszentimre* field, which provides 25–30% of the total mine water flow and 50–85% of the total toxic metals discharged (75% in average).

The main vein located in the *Mátraszentimre* mine field contains predominantly pyrite, whose weathering under oxygen-rich conditions results in intensive acidification and leaching of toxic elements into the mine water, thus producing AMD. Mine operation has been suspended since 1986. One of the main remediation actions was acid mine water treatment. The highly pyritic part of the mine field had to be backfilled to avoid further damage and higher risk due to AMD.

The aims of backfilling were as follows: (i) avoid the accumulation of mine water and sludge, which may cause unpredictable pressure underground; (ii) backfill the entire mining space to principally eliminate pyrite oxidization; (iii) keep out water inflows from the cavity system and thus reduce mine water quantity; and (iv) prevent contamination, acidification and solubilization of toxic elements.

The aims of the drainage system are these: (i) ensure long-term free leakage of the water feeders which still exist after backfilling; (ii) ensure the proper treatment of mine water as long as necessary; and (iii) Rule out any unsafe or uncontrollable situation.

3.3 AMD mitigation by backfilling

The environmental impact of the previous mining activity has been evaluated by many experts and institutions. MECSEK-ÖKO Co. summarized all the relevant information during preparation of the remediation concept of the site, taking into account two main factors:

–   *Mátraszentimre* mine is located upstream of Gyöngyösoroszi village (see Figure 7.10). AMD flows through the 5-km long adit. If the adit is blocked, the strongly contaminated mine water accumulates in the mine openings, posing high risk to the residents of the village and the environment along the *Toka Creek*. Such an event already occurred in 1988 (when the V-2 plug failed), resulting in widespread contamination of the *Toka Creek* and its flood plain.

–   It is advisable to backfill the *Mátraszentimre* mine to avoid such risks. A proper backfill could prevent both mine water accumulation and air inflow into the reactive pyritic zone and suppress AMD formation.

Figure 7.11 In situ column experiment in Matraszentimre mine.

The risk assessment concluded that the *Mátraszentimre* mine should be backfilled using suitable fly ash with significant pozzolanic activity available at a nearby power station.

Several laboratory and *in situ* mine studies were conducted to assess the leaching behavior of the mine water when it seeps through the ash bed. Column experiments performed inside the mine simulated the exclusion of air. The photo in Figure 7.11 shows one of the column experiments in *Mátraszentimre*. Mine water is *seeping directly from the wall of the drift* and fly ash is used as filling in the column.

The more than 3-month long study concluded that three phases can be distinguished:

1   The ash behaves as a neutralizing agent (due to its CaO content) and some metals precipitate.
2   Transitional leaching occurs.
3   Ash functions as neutral backfill material with minimal chemical interactions with mine water but shows surface sorption effects.

Figure 7.12 shows the average heavy metal content of the mine water seeping from the wall of the tunnel (influent) and passing through the column (effluent) filled with thick fly ash slurry.

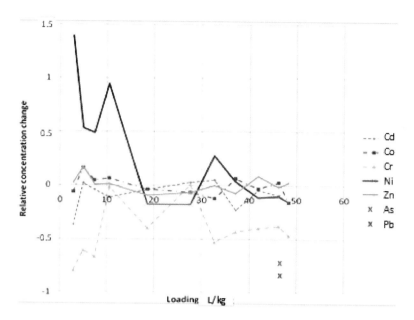

Figure 7.12 Relative concentration change of heavy metals during the mine water – fly ash interaction.

The duration of the experiment was 110 days, pH of the influent AMD was 3.4–3.7 (acidity: ~22,000 mg $CaCO_3$/L), pH of the effluent: 3.7–4.2.

The relative change of metal concentrations in the effluent:

$$[c_{effluent} - c_{influent}]/c_{influent}$$

versus the loading rate was plotted in Figure 7.12. Chromium decreased to 50%, arsenic to 30%, lead to 20% at the highest loading rate of 48L per kg of fly ash. The other metals do not show significant changes at this load when seeping through the column filled with fly ash.

Following the impact of loading, Cd, Co, Ni, and Zn did not show significant changes at loads higher than 18–20 L mine water per kg of ash. Ni was the only metal showing a relative increase at low loadings.

After repeated passes of the mine water through the fly ash backfill, the reactivity of the fly ash towards the mine water ceased and behaved almost like an inert material.

The neutralizing capacity of the backfill material (ash + CaO) increased when the admixture of ash and CaO (2%) was used. The interaction of the mine water and fly ash has been studied by MECSEK-ÖKO Co. (Csővári, 2007). The study led to the conclusion that the governing parameter was the pH of the AMD when considering the possible dissolution of heavy metals. As the backfilling of the mine would substantially decrease the aeration rate of the mine, sulfide oxidization to sulfate and the acidification process would also be reduced significantly.

According to the original schedule the backfilling of the *Mátraszentimre* mine was supposed to start in 2011 but is still pending.

3.4 AMD treatment

A large amount of water accumulated in the mine field during operation because of the geological conditions was pumped to the surface. The volume of mine water sometimes fluctuated depending on the inrush of water from one of the veins (2000–6000 m³/day in running period). The volume of mine water decreased slightly after mining was ended but increased to 4500 m³/day in wet seasons.

Mine water had to be treated because of its iron and toxic metal content where the main pollutants were zinc, cadmium, arsenic, iron, and manganese.

A simple water treatment process was used before 2006: mine water was neutralized using lime milk passing through aerators where the precipitated sludge, mainly $Fe(OH)_2$, was oxidized. The treated water was fed into two open field basins for sedimentation, the sludge from the basins was pumped into a sludge repository (lagoon), and the cleaned water was discharged into the nearby *Toka Creek*. During more than 25 years of operation a large amount of sludge (solid content: 13%) was accumulated in the Bence Valley pond. The deposited sludge was hauled into a hazardous waste repository and the pond remediated. Figure 7.13 shows the partly remediated sedimentation pond.

A new mine water treatment technology was introduced in 2006 which eliminates on-site deposition of the sludge. The sludge is dewatered by a centrifuge resulting in 25–30% solid content and is transported to the final repository for toxic wastes. It is expected that mine water treatment has to be continued in the future. The composition of AMD is only expected to change after completing the *Mátraszentimre* mine backfilling.

3.5 ARD from waste rock piles

The long-lasting mining activity in the region produced a number of WRPs with various metal and pyrite contents. Records from the 1980s refer to 39 piles. The total amount of the waste piles was 1.150 million tons, of which 1.1 million tons was produced in the period 1949–1986. In 2004 only 23 piles were identified, plus the waste from the main transporting adit which was reused as foundation for industrial buildings, in particular for

Figure 7.13 Bence Valley: emptying the sedimentation pond and the valley after remediation.

the water treatment station, for the dams of the sludge repository and the industrial water reservoir along the *Toka Creek*. Therefore, only a small part of the original WRPs can currently be identified. At present most of the piles are hardly recognizable to the eye but are identifiable through chemical analysis. A tiered site risk assessment of the site included a multipoint XRF assessment (using a handheld mobile apparatus) and laboratory analysis of pre-levated samples (Gruiz, 2005, 2014; Gruiz *et al.*, 2000; Sarkadi, 2009). Some characteristic data are presented in Table 7.7 (pile names are Hungarian). The main toxic component is arsenic. In addition to the metals, all samples were analyzed for total inorganic carbon (TIC) and total sulfur to calculate the net acid-generating potential (NAP). Most of the wastes – even after 20–50 years – exhibited significant net acid generating potential.

The WRPs were grouped into three categories according to a qualitative score-based risk assessment (Gruiz, 2004), taking into account the quality and quantity of the waste rock, the potential transport and exposition pathways and the receptors:

- No need for any remediation (small piles);
- Piles to be remediated *in situ*;
- Piles to be removed and relocated to the top of the flotation tailings pond and to be covered and recultivated together with the tailings pond. This includes 43,025 m^3 of wastes.

Table 7.7 Toxic metal content of waste rock piles (Gruiz, 2005).

Waste rock pile, adit	As	Cd	Cu	Pb	Zn
	mg/kg				
Ezüst Bányabérci – 1	70.4	0.16	36.7	28.1	60.7
Hidegkúti táró – 1	27.9	<0.049	11.4	39.0	53.8
Katalin táró	910	10.2	14.1	299	109
József	332	3.84	20.7	367	221
Kistölgyesbérc – 1	468	12.9	135	4563	1920
Kistölgyesbérc – 2	583	4.20	93.1	2921	503
Lajos – 1	41.7	<0.049	18.1	161	133
Péter-Pál – 1	171	0.91	6.92	529	14.8
Péter-Pál shaft – 1	142	2.27	45.3	2223	392
Száka – 1	343	1.36	19.8	9.9	67.4
Új Károlytáró – 1	239	1.29	39.8	118	230
Új Károlytáró – 2	167	0.86	18.6	195	130
Vereskő – 1	16.0	<0.049	7.47	9.22	16.6
Nagytölgyesbérc – 1	293	2.44	42.4	173	361
Kistölgyesbérc – 1	96.9	0.06	4.9	217	240
Kistölgyesbérc – 2	29.4	<0.049	31.5	1.8	103
Vizeslyuk – 1	529	1.11	19.3	177	303
Pelyhes – 1/1	1919	6.62	2.7	2052	141
Pelyhes – 2/1	145	0.40	16.9	69.7	117
Altáró – 1	169	14.6	163	1425	2760

Some WRPs are small and contain only a few hundred tons. The photo in Figure 7.14 shows Szákacsurgó WRP for which there are plans for on-site remediation. Katalin WRP (Figure 7.15) and Új Károlytáró (Figure 7.16) are planned to be relocated to the flotation tailings pond for containment. Figure 7.17 shows Bányabérc WRP before remediation (1964) and the site after its removal (2011). The relocated mass will be remediated by *in situ* treatment and revegetation on the top of the tailings pond.

Figure 7.14 Szákacsurgó WRP from 1967 (to be remediated *in situ*).

Figure 7.15 Katalin adit WRP from 1957 to 1958 (to be relocated to the flotation tailing dump).

Figure 7.16 Új Károlytáró waste pile from 1955 (to be relocated to the flotation tailing dump for containment).

Figure 7.17 The Bányabérc WRP in 1964 and in 2011 after removal.

The risks of the questionable piles were assessed quantitatively as point sources and their contribution to the diffuse contamination of the watershed was quantified (Gruiz, 2004).

3.6   Ore processing and flotation tailings

Ore processing recovered lead and zinc from the ore and parts of the pyrite in the form of heavy concentrates. The concentrates also contained gold, silver, cadmium, and copper.

3.6.1   Ore processing

Figure 7.18 shows the principal technological process used by the company. Galena, sphalerite, and, for a short period, pyrite concentrate, was separated from the ROM by flotation (Kun *et al.*, 1988). Ore was crushed and classified by sieving. The 4- to 1-mm fraction was upgraded into heavy suspension using ferrosilicon slurry (r = 2.63 kg/L). This fraction was washed prior to upgrading. The slimes from this operation were pumped to the hydrocyclone's underflow, which was fed into the collecting bins. The overflow was added to the flotation tailings.

The washed fraction was upgraded in a heavy medium. Gangue minerals of lower density were separated and then handled as waste rock and used for road construction or as backfilling material.

The upgraded ore was directed for milling. The milled ore (92% below 0.20 mm) underwent flotation in three steps: galena was removed in the first step, sphalerite in the second, and finally pyrite. Water consumption of the process was 4 m³/t, water having been taken from the *Toka Creek*. The following reagents were used for flotation: 91 t NaCN and 1200 t ferrosilicon. The end product of the flotation was the concentrate and total production by 1986 was this:

Galena: 56,900 t (grade: Pb = 53.8%);
Sphalerite: 150,700 t (grade: Zn = 51.3%);
Pyrite: 68,000 t (grade: FeS$_2$ ≈ 90–95%).

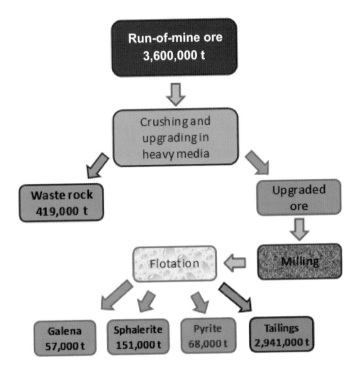

Figure 7.18 Material flow sheet of ore processing.

Table 7.8 Metal content and net acid-generation potential of the flotation tailings.

Content	Pb	Zn	Fe	Mn	As	Cu	Cd	Ni	S	Total inorganic carbon	Net acid-generation potential
	%	%	%	%	mg/kg	mg/kg	mg/kg	mg/kg	%	%	kg $CaCO_3$/t
ROM	0.99	2.85	5.1	0.1	370	1,500	150	~2	5.23		~80
Tailings	0.11	0.26	4.2	0.1	309	426	17	2	4.25	1.04	45

3.6.2 *Tailings pond*

Flotation "produced" 2.95 million tons of flotation tailings. The tailings were placed into tailings ponds of a total area of 22 ha.

Many authors have studied the composition of the tailings and suggested several alternatives for the remediation of the tailings ponds (GEOPARD Ltd, 2002; Gruiz, 2005; Csővári, 2007). Table 7.8 shows the average data from six drilling core samples together with the original ore composition (the study was carried out by MECSEKÉRC Co. within the TAIL-SAFE EU project, 2002).

The average net acid generating potential of the flotation tailings was 45 kg/t at a high sulfide content (4.25%). Arsenic content was practically the same as in the processed ore: 300 g/t on average.

At >80 mg/kg As content, Hg concentration was only a few mg/kg. A good correlation has been found between As and Hg concentrations in the tailings.

The southern dam of the tailings pond – as shown in Figure 7.19 – was strengthened with rocks during rehabilitation in 2005–2011. Figure 7.20 shows the construction of the dam and the surface water collection system around the tailings pond.

The stability of the tailings pond and the dam was checked several times in the past because its planned role as a central repository for relocated wastes of different origin such as WRPs, dredged sediments, and sludges required a comprehensive geotechnical stability study. Geotechnical stability proved to be lower than required, meaning that the pond in its present condition cannot accept all the waste. Thus the first step of rehabilitation was the strengthening of the southern and eastern part of the dam. Remediation started with dewatering the pond using a deep drainage system.

Figure 7.19 Flotation tailings pond (Gyöngyösoroszi in 2010).

Figure 7.20 Strengthening the tailings dam and constructing the water management system.

The drilling cores provided a visual picture of the exposure of tailings: the oxidation zone was at a depth of 1.2–1.4 m based on the color of the tailings core (see the brownish zone in Figure 7.21).

Loading of the tailings pond with wastes from WRPs and reservoirs was carried out under stringent control of the pore water pressure inside the pond.

After relocating the wastes and reinforcing the dam of the tailings pond, a multilayer cover system was applied to stabilize the tailings over the long term. The cover system also includes a 0.3-m thick clay component to minimize infiltration. The water content of the tailings varies between 35 and 40 wt%, as shown in Figure 7.22.

Figure 7.21 Sampling the tailings pond dam.

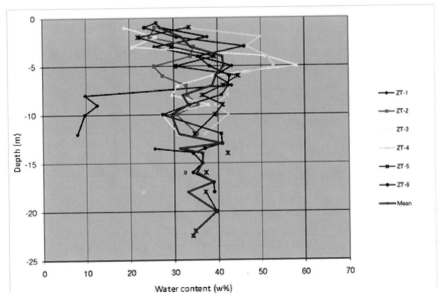

Figure 7.22 Water content of the tailings (ZT-1, ZT-2, etc.: the drilling number).

3.7   Surface water reservoirs and creeks

There are three water reservoirs along the *Toka Creek* which are connected with the ore mining area: (i) industrial water reservoir for the flotation facility, (ii) agricultural water reservoir, and (iii) Gyöngyösoroszi-Nagyréde reservoir. The remediation of the site also includes the cleanup of these three reservoirs.

Industrial water reservoir for the flotation facility

The most contaminated sediments were found in the industrial water reservoir, which has been fed partly with the discharge (treated water) from the mine water treatment station. It was determined that it included approximately 80,000 m³ sludge, part of which was hazardous waste due to its high As concentration (As >1000 mg/kg). In the course of the clean-up the sludge will be treated with lime for stabilization and the hazardous waste and the contaminated (but not hazardous) waste will be separated subsequently. The hazardous waste has to be conveyed into a hazardous waste repository. Its contaminated portion is planned to be relocated onto the tailings pond. The remediated reservoir will be used in the future as an overflow pond in the case of high water level.

Agricultural reservoir

This reservoir contains approximately 60,000 m³ of sediment contaminated with metals and pathogenic contaminants from the nearby settlement. The contaminated sediment will be mixed to biological waste for composting and treated with CaO for stabilization. The stabilized compost will be used as growing medium for the vegetation on the tailings pond.

Gyöngyösoroszi-Nagyréde reservoir

This water reservoir accommodates approximately 54,000 m³ of contaminated sediment with low solid content. Therefore it has to be removed by applying dry dredging and hydromechanization methods. The removed sludge will be treated on site (aeration, biodegradation for BOD reduction, dewatering, stabilization) and the stabilized sludge will be hauled to the tailings pond.

Toka Creek and Száraz Brook

The *Toka Creek* has been contaminated with untreated mine water and with minerals and sludge of different origins (Fügedi *et al.*, 2000; Gruiz *et al.*, 2009c). Breakdowns in the operation of the tailings ponds and the flotation facility have led to occasional contamination of the creek every 2–3 years, sometimes even of the shore and the nearby field, orchards etc. The contaminated length of the creek is approximately 10 km with dense vegetation on the shore.

The *Száraz Brook* was also contaminated occasionally during tailings pond breakdowns. The cleanup of the creek and brook will focus on removing the contaminated soil from orchards and shore as well as the legacies from earlier dredging. The soil contaminated with heavy metals will be placed on the tailings pond. The removed contaminated material will be replaced by sterile soil.

The remediation of the *Toka Creek* will restore the quality of the water and soil.

3.8   Summary of the Gyöngyösoroszi case study

1   The former metal mining site is under restoration. Cleanup of the creeks has started; relocation of the WRPs is underway. The tailings dam has been strengthened, the water management system is ready and in operation, and plans for the remediation have been made and accepted by the competent authorities.

2   Treatment of AMD is being continued in a new water treatment station, the water treatment sludge accumulated on the site for more than 25 years was disposed of to a hazardous waste repository. The sludge produced has also been conveyed to the hazardous waste repository.

3   *In situ* studies at the *Mátraszentimre* mine have concluded that fly ash can be used for backfilling the mine. Preparatory work for backfilling has been finished. Backfilling will be carried out step by step upon evaluation of the results of previous work.

REFERENCES

Aelion, M.C., Davis, H.T., Flora, J.R.V., Kirtland, B.C. & Amidon, M.B. (2009) Application of encapsulation (pH – Sensitive polymer and phosphate buffer macrocapsules), A novel approach to remediation of acidic ground water. *Environmental Pollution*, 157, 186–193.

Akcil, A. & Koldas, S. (2006) Acid Mine Drainage (AMD): Causes, treatment and case studies. *Journal of Cleaner Production*, 14, 1139–1145.

Ali, M.S. (2011) Remediation of acid mine waters. In: Rüde, T.R., Freund, A. & Wolkersdorfer, C. (eds.) *Mine Water – Managing the Challenges. 11th International Mine Water Association Congress*. IMWA, Aachen, Germany. pp. 253–258.

Aube, B.C. & Payant, S. (1997) The Geco process: A new high density sludge treatment for acid mine drainage. *Proceedings of the Fourth International Conference on Acid Rock Drainage, May 30– June 6, 1997*, Vancouver, BC, Canada. Vol. I, pp. 165–180.

Bagdy, I. & Sulyok, P. (2000) *Case Example on Acid-Resistant Bulkhead in Hungary*. Lisboa 90. Mine Water and the Environment. Available from: www.IMWA.info. [Accessed 3rd July 2018].

Barmettler, F., Castelberg, C., Fabbri, C. & Brandl, H. (2016) Microbial mobilization of rare earth elements (REE) from mineral solids—A mini review. *AIMS Microbiology*, 190–204. doi: 10.3934/microbiol.2016.2.190.

Barnes, H.L. & Romberger, S.B. (1968) The chemical aspect of acid mine drainage. *Journal of Water Pollution*, 40, 371–384.

Bayat, B. & Sari, B. (2010) Comparative evaluation of microbial and chemical leaching processes for heavymetal removal from dewatered metal plating sludge. *Journal of Hazardous Materials*, 174, 763–769.

Bebié, J. & Schoonen, M. (2000) Pyrite surface interaction with selected organic aqueous species under anoxic conditions. *Geochemical Transactions*, 1, 47–53. doi:10.1039/b005581f.

Benner, S.G., Blowes, D.W. & Ptacek, C.J. (1997) A full-scale porous reactive wall for prevention of acid mine drainage. *Groundwater Monitoring and Remediation*, 17(4), 99–107.

Bhattacharya, J., Ji, S.W., Lee, H.S., Cheong, Y.W., Yim, G.J., Min, J.S. & Choi, Y.S. (2008) Treatment of acidic coal mine drainage, design and operational challenges of successive alkalinity producing systems. *Mine Water and the Environment*, 27, 12–19.

Blowes, D.W., Ptacek, C.J., Jambor, J.L. & Weisener, C.J. (2003) The geochemistry of acid mine drainage. *Treatise on Geochemistry*, 9, 149–204.

Blowes, D.W., Ptacek, C.J., Jambor, J.L., Weisener, C.G., Paktunc, D., Gould, W.D. & Johnson, D.B. (2014) The geochemistry of Acid Mine Drainage. In: Holland, H.D. & Turekian, K.K. (eds.) *Treatise on Geochemistry*, 11. 2nd ed. Elsevier Science, Oxford, UK. pp. 131–190. doi:10.1016/B978-0-08-095975-7.00905-0.

Bosecker, K. (1997) Bioleaching, metal solubilisation by microorganisms. *FEMS Microbial Review*, 20, 591–604.

Brandl, H., Lehmann, S., Faramarzi, M.A. & Martinelli, D. (2008) Biomobilization of silver, gold, and platinum from solid waste materials by HCN-forming microorganisms. *Hydrometallurgy*, 94(1–4), 14–17.

Brierley, C.L. & Brierley, J.A. (2013) Progress in bioleaching, fundamentals and mechanisms of bacterial metal sulfide oxidation – Part A. *Applied Microbiology and Biotechnology*, 97(17), 7543–7552.

Brodie, G.A. (1993) Staged, aerobic constructed wetlands to treat acid drainage: Case history of Fabius Impoundment 1 and overview of the TVA's Program. In: Moshiri, G.A. (ed) *Constructed Wetlands for Water Quality Improvement*. Lewis Publishers, Boca Raton, FL, USA. pp. 157–166.

Brodie, G.A., Britt, C.R., Taylor, H.N., Tomaszewski, T.M. & Turner, D. (1990) Passive anoxic limestone drains to increase effectiveness of wetlands acid drainage treatment systems. *Proceedings of the 12th Annual Conference of the National Association of Abandoned Mine Land Programs*, Breckenridge, CO, USA.

Buzzi, D.C., Viegas, L.S., Rodrigues, S., Bernardes, A.M. & Tenório, J.A.S. (2013) Water recovery from acid mine drainage by electrodialysis. *Mining Engineering*, 4, 82–89.

Chen, X., Guo, C., Ma, H., Li, J., Zhou, T., Cao, L. & Kang, D. (2018) Organic reductants based leaching: A sustainable process for the recovery of valuable metals from spent lithium ion batteries. *Waste Management*, 75, 459–468. doi:10.1016/j.wasman.2018.01.021.

Clark, D.A. & Norris, P.R. (1996) Oxidation of mineral sulphides by thermophilic microorganisms. *Minerals Engineering*, 9, 1119–1125.

Coulton, R., Bullen, C. & Hallet, C. (2003) The design and optimization of active mine water treatment plants. *Land Contamination & Reclamation*, 11, 273–279.

Cravotta, C.A. & Trahan, M.K. (1999) Limestone drains to increase pH and remove dissolved metals from acidic mine drainage. *Applied Geochemistry*, 14, 581–606.

Cravotta, C.A. & Watzlaf, G.R. (2002) Design and performance of limestone drains to increase pH and remove metals from Acidic Mine Drainage. In: *Handbook of Groundwater Remediation Using Permeable Reactive Barriers*. Elsevier Science, Amsterdam, Netherlands, pp. 19–66.

Crundwell, F.K. (1988) The influence of the electronic structure of solids on the anodic dissolution and leaching of semiconducting sulphide minerals. *Hydrometallurgy*, 21, 155–190.

Csővári, M. (2007) *Environmental Consequences of Metal Mining and the Possibility of Their Mitigation* (In Hungarian). ELGOSCAR-2000, Budapest, Hungary. pp. 1–96.

Demchak, J., Morrow, T. & Skousen, J. (2001) Treatment of acid mine drainage by four vertical flow wetlands in Pennsylvania. *Geochemistry: Exploration, Environment, Analysis*, 1(1), 71–80.

Dempsey, B.A., Roscoe, H.C., Ames, R., Hedin, R. & Jeon, B.-H. (2001) Ferrous oxidation chemistry in passive abiotic systems for treatment of mine drainage. *Geochemistry: Exploration, Environment and Analysis*, 1(1), 81–88.

De Vegt, A.L., Exton, P.A., Bayer, H.G. & Buisman, C.J. (1998) Biological sulphate removal and metal recovery from mine waters. *Mining Engineering*, 50(11), 67–70.

Dold, B. (2010) Basic concepts in environmental geochemistry of sulphide mine waste management. In: Kumar, E.S. (ed) *Waste Management*. In Tech, Rijeka, Croatia. pp. 173–198.

Dugan, P.R. (1986) Prevention of formation of acid mine drainage from high-sulfur coal refuse by inhibition of iron – And sulphur oxidizing microorganisms. I. Preliminary experiments in controlled shake flasks. *Biotechnology and Bioengineering*, 29, 41–48.

Edelbro, R., Sandström, Å. & Paul, J. (2003) Full potential calculations on the bandstructures of sphalerite, pyrite and chalcopyrite, *Applied Surface Science*, 206, 300–313.

Egiebor, N.O. & Oni, B. (2007) Acid rock drainage formation and treatment: A review. *Asia-Pacific Journal of Chemical Engineering*, 2, 47–62. doi:10.1002/ apj.57.

Evangelou, V.P. (1995) *Pyrite Oxidation and Its Control*. CRC Press, New York, NY, USA. 275 p.

Evangelou, V.P. & Zhang, Y.L. (1995) A review, Pyrite oxidation mechanisms and acid mine drainage prevention. *Critical Reviews in Environmental Science and Technology*, 25, 141–199.

Fügedi, U. (2004) Geochemical background contamination on the Gyöngyösoroszi site. (In Hungarian). *Földtani Közlöny*, 134(2), 291–301.

Fügedi, U. (2006) *Geochemical Study of the Pollution of the Environment in GYÖNGYÖSOROSZI.* (In Hungarian). Available from: www.mafi.hu/static/microsites/geokem/ubul/pdf/Ubul_phd.pdf. [Accessed 3rd July 2018].

Fügedi, U., Horváth, I. & Ódor, L. (2000) Changes in the mineralogical composition of the Gyöngyösoroszi mine waste spread out on the floodplain of the *Toka Creek. Acta Mineralogica-Petrographica.* XLI. (B) *Abstracts of the Symposia on Environmental Mineralogy, 18–19 May 2000*, Budapest, Hungary. p. 29.

Fügedi, U., Horváth, I. & Ódor, L., (2001) *Pollution of the Environment in Gyöngyösoroszi.* (In Hungarian). Available from: www.mafi.hu/microsites/geokem/oroszi/SZINT1.htm. [Accessed 3rd July 2018].

Gaikwad, R.W. & Gupta, D.V. (2008) Review on removal of heavy metals from acid mine drainage. *Applied Ecology and Environmental Research*, 6(3), 81–98.

Garland, R. (2011) Acid mine drainage – Can it affect human health? *Quest*, 7(4), 46–47.

Garrett, W.E., Bartolucci, A.A., Pitt, R.R. & Vermace, M.E. (2002) Recirculating – Reducing and Alkalinity Producing System (RERAPS) for the Treatment of Acidic Coal Pile Runoff. *Proceedings of the 2002 National Meeting of the American Society of Mining and Reclamation*, Lexington, KY, USA. pp. 539–557.

Gehrke, T., Telegdi, J., Thierry, D. & Sand, W. (1998) Importance of extracellular polymeric substances from *Thiobacillus ferrooxidans* for Bioleaching. *Applied Environmental Microbiology*, 64, 2743–2747.

GEOPARD Ltd (2002): *Landscaping of the Tailings Pile in Gyöngyösoroszi "Száraz Valley".* (Manuscript in Hungarian). Pécs, December 19, 2002.

Gerhardt, A., de Bisthoven, L.J. & Soares, A.M.V.M. (2004) Macroinvertebrate response to acid mine drainage: Community metrics and on-line behavioural toxicity bioassay. *Environmental Pollution*, 130(2), 263–274.

Gibert, O., Rötting, T., Cortina, J.L., de Pablo, J., Ayora, C., Carrera, J. & Bolzicco, J. (2011) In – Situ remediation of acid mine drainage using a permeable reactive barrier in Aznalcóllar (Sw Spain). *Journal of Hazardous Materials*, 191, 287–295.

Gilbert, O., de Pablo, J., Cortina, J.L. & Ayora, C. (2002) Treatment of acid mine drainage by sulphate – Reducing bacteria using permeable reactive barriers. A review from laboratory to full – Scale experiments. *Reviews in Environmental Science and Biotechnology*, 1, 327–333.

Glombitza, F. & Reichel, S. (2014) Metal-containing residues from industry and in the environment: Geobiotechnological urban mining. *Advances in Biochemical Engineering/Biotechnology*, 141, 49–107.

González, A., Bellenberg, S., Mamani, S., Ruiz, L., Echeverría, A., Soulère, L., Doutheau, A., Demergasso, C., Sand, W., Queneau, Y., Vera, M. & Guiliani, N. (2012) AHL signaling molecules with a large acyl chain enhance biofilm formation on sulfur and metal sulfides by the bioleaching bacterium *Acidithiobacillus ferrooxidans. Applied Microbiology and Biotechnology*, 97, 3729–3737.

Gray, N.F. (1996) Environmental impact and remediation of acid mine drainage, a management problem. *Environmental Geology*, 30, 62–71.

Gruiz, K. (2004) *Risk Assessment of the Contaminated Areas in Gyöngyösoroszi.* Expert Report. (Manuscript in Hungarian). Available from: MECSEK-ÖKO Co. Archives.

Gruiz, K. (2005) *Data on Pollution in Gyöngyösoroszi (1990–2004). Wastes, Waters, Sediments, Soils and Vegetation. Summary of the Publications.* (Manuscript in Hungarian). Available from: MECSEK-ÖKO Co. Archives.

Gruiz, K. (2014) Abandoned and contaminated land. In: Gruiz, K., Meggyes, T. & Fenyvesi, É. (eds.) *Environmental Deterioration and Contamination – Problems and Their Management.* CRC Press, Boca Raton, FL, USA. pp. 81–83.

Gruiz, K., Horváth, B., Molnár, M. & Sipter, E. (2000) When the chemical bomb explodes – Chronic risk of toxic metals at a former mining site. *Proceedings of ConSoil 2000.* Thomas Telford, Leipzig. pp. 662–670.

Gruiz, K., Vaszita, E. & Siki, Z. (2009a) Environmental risk management of diffuse pollution of mining origin. In: Sarsby, R.W. & Meggyes, T. (eds.) *Construction for a Sustainable Environment.* CRC Press/Balkema, Leiden, The Netherlands. pp. 219–228.

Gruiz, K., Vaszita, E., Siki, Z. & Feigl, V. (2007) Environmental risk management of an abandoned mining site in Hungary. *Advanced Materials Research*, 20–21. Trans Tech Publications, Switzerland. pp. 221–225.

Gruiz, K., Vaszita, E., Siki, Z. & Feigl, V. (2009b) Environmental risk management of an abandoned mining site in Hungary. In: Schippers, A., Sand, W., Glombitza, F. & Willscher, S. (eds.) *Biohydrometallurgy: From the Single Cell to the Environment (IBS2007), Frankfurt am Main, Germany, 2–5 September 2007.* Transtech Publications, Stafa-Zurich, Switzerland. pp. 221–225.

Gruiz, K., Vaszita, E., Siki, Z., Feigl, V. & Fekete, F. (2009c) Complex environmental risk management of a former mining site. *Land Contamination Reclamation*, 17(3–4), 357–372.

Gruiz, K. & Vodicska, M. (1993) Assessing heavy-metal contamination in soil applying a bacterial biotest and x-ray fluorescent spectroscopy. In: Arendt, F., Annokkée, G.J., Bosman, R. & van den Brink, W.J. (eds.) *Contaminated Soil '93.* Kluwer Academic Publishers, Dordrecht, The Netherlands. pp. 931–932.

Gusek, J.J. (1995) Passive treatment of acid rock drainage: What is the bottom line? *Mining Engineering*, 47(3), 250–253.

Gusek, J.J. (1998) *Three Case Histories of Passive Treatment of Metal Mine Drainage.* South African Mining Delegation Seminar. Denver, CO. August 24–September 4, 1998.

Hansen, J.A., Welsh, P.G., Lipton, J. & Cacela, D. (2002) Effects of copper exposure on growth and survival of juvenile bull trout. *Transactions of the American Fisheries Society*, 131(4), 690–697.

Harneit, K., Göksel, A., Kock, D., Klock, J.-H., Gehrke, T. & Sand, W. (2006) Adhesion to metal sulfide surfaces by cells of *Acidithiobacillus ferrooxidans, Acidithiobacillus thiooxidans* and *Leptospirillum ferrooxidans. Hydrometallurgy*, 83, 245–254.

Harrison, A.P. (1978) Microbial succession and mineral leaching in an artificial coal spoil. *Applied Environmental Microbiology*, 36, 861–869.

Harrison, A.P. (1984) The acidophilic *thiobacilli* and other acidophilic bacteria that share their habitat. *Annual Review of Microbiology*, 38, 265–292.

Hedin, R.S. & Nairn, R.W. (1990) Sizing and performance of constructed wetlands: Case studies. *Proceedings of 1990 Mining and Reclamation Conference*, Morgantown, WV, USA. pp. 385–392.

Hedin, R.S., Nairn, R.W. & Kleinmann, R.L.P. (1994a) *Passive Treatment of Coal Mine Drainage.* USBM IC 9389, Pittsburgh, PA, USA. 35 pp.

Hedin, R.S., Watzlaf, G.R. & Nairn, R.W. (1994b) Passive treatment of acid mine drainage with limestone. *Journal of Environmental Quality*, 23, 1338–1345.

Hedin R.S. (2003) Recovery of marketable iron oxide from mine drainage in the USA. *Land Contamination and Reclamation*, 11, 93–97.

Hennebel, T., Boon, B., Maes, S. & Lenz, M. (2015) Biotechnologies for critical raw material recovery from primary and secondary sources: R&D priorities and future perspectives. *New Biotechnology*, 32, 121–127.

Henrot, J. & Wieder, R.K. (19909 Processes of iron and manganese retention in laboratory peat microcosms subjected to acid mine drainage. *Journal of Environmental Quality*, 19, 312–320.

Huang, Q., Yu, Z., Pang, Y., Wang, Y. & Cai, Z. (2015) Coupling bioleaching and electrokinetics to remediate heavy metal contaminated soils. *Bulletin of Environmental Contamination and Toxicology*, 94(4), 519–524. doi:10.1007/s00128-015-1500-1.

Hyman, D.M. & Watzlaf, G.R. (1995) Mine drainage characterization for the successful design and evaluation of passive treatment systems. *Proceedings of 17th Conference of National Association of Abandoned Mine Lands*, FrenchLick, IN, USA.

Jiwan, S. & Kalamdhad, A.S. (2011) Effects of heavy metals on soil, plants, human health and aquatic life. *International Journal of Research in Chemistry and Environment*, 1(2), 15–21.

Johnson, D.B. (2000) Biological removal of sulfurous compounds from inorganic wastewaters. In: Lens, P. & Hulshoff, P.L. (eds.) *Environ Mental Technologies to Treat Sulfur Pollution: Principles and Engineering*. International Association on Water Quality, London, UK. pp. 175–206.

Johnson, D.B. (2003) Chemical and microbiological characteristics of mineral spoils and drainage waters at abandoned coal and metal mines. *Water, Air, & Soil Pollution: Focus*, 3, 47–66. doi:10.1023/A:1023977617473.

Johnson, D.B. & Hallberg, K.B. (2003) The microbiology of acidic mine waters. *Research in Microbiology*, 154, 466–473.

Johnson, D.B. & Hallberg, K.B. (2005) Acid mine drainage remediation options, a review. *Science of the Total Environment*, 338, 3–14.

Johnson, D.B., Yajie, L. & Okibe, N. (2008) 'Bioshrouding' – A novel approach for securing reactive mineral tailings. *Biotechnology Letters*, 30, 445–449.

Jones, S.N. & Cetin, B. (2017) Evaluation of waste materials for acid mine drainage remediation. *Fuel*, 188, 294–309. doi:10.1016/j.fuel.2016.10.018.

Kalin, M., Fyson, A. & Wheeler, W.N. (2006) The chemistry of conventional and alternative treatment systems for the neutralization of acid mine drainage. *Science of the Total Environment*, 366, 395–408.

Kelly, D. & Wood, A. (2000) Reclassification of some species of *Thiobacillus* to the newly designated genera *Acidithiobacillus* gen. nov., *Halothiobacillus* gen. nov. and *Thermithiobacillus* gen. nov. *International Journal of Systematic & Evolutionary Microbiology*, 50, 511–516.

Kepler, D.A. & McCleary, E.C. (1994) Successive alkalinity-producing systems (SAPS) for the treatment of acidic mine drainage. *International Land Reclamation and Mine Drainage Conference, U.S. Bureau of Mines SP 06A-94*, Pittsburgh, PA, USA. pp. 195–204.

Kirby, C.S., Thomas, H.M., Southman, G. & Donald, R. (1999) Relative contributions of abiotic and biological factors in fe(II) oxidation in mine drainage. *Applied Geochemistry*, 14, 511–530.

Kleinmann, R.L.P., Hedin, R.S. & Nairn, R.W. (1998) Treatment of mine drainage by anoxic limestone drains and constructed wetlands. In: Geller, W., Klapper, H. & Salomons, W. (eds.) *Acidic Mining Lakes: Acid Mine Drainage, Limnology and Reclamation*. Springer, Berlin, Germany. pp. 303–319.

Kock, D., Graupner, T., Rammlmair, D. & Schippers, A. (2007) Quantification of microorganisms involved in cemented layer formation in sulfidic mine waste tailings (Freiberg, Saxony, Germany). *Advanced Materials Research*, 20–21, 481–484.

Kock, D. & Schippers, A. (2006) Geomicrobiological investigation of two different mine waste tailings generating acid mine drainage. *Hydrometallurgy*, 83, 167–175.

Kock, D. & Schippers, A. (2008) Quantitative microbial community analysis of three different sulfidic mine tailings dumps generating acid mine drainage. *Applied Environmental Microbiology*, 74, 5211–5219.

Koschorreck, M. (2008) Microbial sulphate reduction at a low pH. *FEMS Microbiology Ecology*, 64, 329–342.

Kourai, H., Yabuhara, T., Shirai, A., Maeda, T. & Nagamune, H. (2006) Syntheses and antimicrobial activities of a series of new bis – Quaternary ammonium compounds. *European Journal of Medicinal Chemistry*, 41, 437–444.

Krebs, W., Brombacher, C., Bosshard, P.P., Bachofen, R. & Brandl, H. (1997) Microbial recovery of metals from solids. *FEMS Microbiology Reviews*, 20(3–4), 605–617.

Kun, B. (1966) 25 years activity of Company Országos Érc- és Ásványbányák. In: Pantó, E. (ed) *Mining of Multimetal Ore at Gyöngyösoroszi and Manganese Ore in Bakony*. Országos Érc- és Ásványbányák, Budapest, Hungary.

Kun, B. (1980) Gyöngyösoroszi and the Central Mátra Mountains. In: Kun, B. (ed) *25 Years of the Activity of Company Országos Érc- és Ásványbányák*. OMBKE, Budapest, Hungary.

Kun, B. (1998) Metal ore mining on Gyöngyösoroszi site and its vicinity (in Hungarian). *Földtani Kutatás*, 35(4), 22–27.

Kun, B., Szigeti, K., Lovász, A. & Germus, B. (1988) *Overview of Metal Ore Mining on Gyöngyösoroszi Site and Its Vicinity, 1–2–3*. (Manuscript in Hungarian). MECSEK-ÖKO Co. Archive.

Kuyucak, N. (2002) Acid mine drainage prevention and control options. *Canadian Institute of Mining, Metallurgy Bulletin*, 95(1060), 96–102.

Leathen, W.W., Braley, S.A. & McIntyre, L.D. (1953) The role of bacteria in the formation of acid from certain sulfuritic constituents associated with bituminous coal, I. *Thiobacillus Thiooxidans. Applied Microbiology*, 1, 61–64.

Liu, C.Q., Plumb, J. & Hendry, P. (2006) Rapid specific detection and quantification of bacteria and archaea involved in mineral sulfide bioleaching using real – Time PCR. *Biotechnology and Bioengineering*, 94, 330–336.

Luptakova, A., Balintova, M., Jencarova, J., Macingova, E. & Prascakova, M. (2010) Metals recovery from acid mine drainage. *Nova Biotechnologica et Chimica*, 10(1), 23–32.

Macingova, E. & Luptakova, A. (2012) Recovery of metals from acid mine drainage. *Chemical Engineering Transactions*, 28, 109–114.

Matthies, R., Aplin, A.C. & Jarvis, A.P. (2010) Performance of a passive treatment system for net – Acidic coal mine drainage over five years of operation. *Science of the Total Environment*, 408, 4877–4885.

Meggyes, T., Csővári, M., Roehl, K.E. & Simon, F.-G. (2009) Enhancing the efficacy of permeable reactive barriers. *Land Contamination and Reclamation*, 17(3–4), 635–650.

Meggyes, T., Simon, F.-G. & Debreczeni, E. (2001) New developments in reactive barrier technology. In: Sarsby, R.W. & Meggyes, T. (eds.) *The Exploitation of Natural Resources and the Consequences. Proceedings of the 3rd International Symposium on Geotechnics Related to the European Environment, 21–23 June 2000*. Thomas Telford, London. pp. 474–483.

Menezes, C.T.B., Barros, E.C., Rufino, R.D., Luna, J.M. & Sarubbo, L.A. (2011) Replacing synthetic with microbial surfactants as collectors in the treatment of aqueous effluent produced by Acid Mine Drainage, using the dissolved air flotation technique. *Applied Biochemistry & Biotechnology*, 163(4), 540–546.

Mulopo, J. (2015) Continuous pilot scale assessment of the alkaline barium calcium desalination process for acid mine drainage treatment. *Journal of Environmental Chemical Engineering*, 3, 1295–1302. doi:10.1016/j.jece.2014.12.001.

Nancucheo, I. & Johnson, B. (2009) Production of glycolic acid by chemolithotrophic iron – And sulfur – Oxidizing bacteria and its role in delineating and sustaining acidophilic sulfide mineral – Oxidizing consortia. *Applied Environmental Microbiology*, 76, 461–467.

Nett, J.E., Guite, K.M., Ringeisen, A., Holoyoda, K.A. & Andes, D.R. (2008) Reduced biocide susceptibiliy in *Candida albicans* biofilms. *Antimicrobial Agents and Chemotherapie*, 52, 3411–3413.

Nicomrat, D., Dick, W.A. & Tuovinen, O.H. (2006) Assessment of the microbial community in a constructed wetland that receives acid coal mine drainage. *Microbial Ecology*, 51, 83–89.

Nixdorf, B., Uhlmann, W. & Lessmann, D. (2010) Potential for remediation of acidic mining lakes evaluated by hydrogeochemical modeling, case study Grünewalder Lauch (Plessa 117, Lusatia/Germany). *Limnologica*, 40, 167–174.

Nordstrom, D.K. (2011) Mine waters: Acidic to circumneutral. *Elements*, 7, 393–398.

Norris, R. (1997) Thermophiles and Bioleaching. In: Rawlings, D.E. (ed) *Biomining, Theory, Microbes and Industrial Processes*. Springer Verlag, Berlin, Germany.

Olsen, G.J., Brierley, J.A. & Brierley, C.L. (2003) Bioleaching review Part B, Progress in bioleaching, applications of microbial processes by the minerals industries. *Applied Microbiology and Biotechnology*, 63, 249–257.

Onysoko, S.J., Kleinmann, R.L.P. & Erickson, P.M. (1984) Ferrous iron oxidation by *Thiobacillus ferrooxidans*, inhibition with benzoic acid, sorbic acid and sodium lauryl sulphate. *Applied Environmental Microbiology*, 48, 229–231.

Ötvös, K., Juhász, Z., Wittinger, K., Földessy, M., Földing, G., Fekete, F., Szulimán *et al*. (2004) *Abandonment of the Gyöngyösoroszi Ore Mining Site*. (Manuscript in Hungarian) Ötvös és Társa Kft., Pécs. p. 259 + supplements.

Petrik, L., White, R., Klink, M., Burgers, C., Somerset, V., Key, D., Iwuoha, E., Burgers, C. & Fey, M.V. (2005) *Utilization of Fly Ash for Acid Mine Drainage Remediation*. WRC Report No.1242/1/05. Water Research Commission, Pretoria.

PIRAMID (2003) *Engineering Guidelines for the Passive Remediation of Acidic and/or Metalliferous Mine Drainage and Similar Wastewaters*. Report European Commission 5th Framework RTD Project No. EVK1-CT-1999-000021. University of Newcastle Upon Tyne, Newcastle Upon Tyne, UK. 166 pp.

Porsch, K., Meier, J., Kleinsteuber, S. & Wend-Potthoff, K. (2009) Importance of different physiological groups of iron reducing microorganisms in an acidic mining lake remediation experiment. *Microbial Ecology*, 57, 701–717.

Rawlings, D.E. (1997) Mesophilic, autotrophic bioleaching bacteria, description, physiology and role. In: Rawlings, D.E. (ed) *Biomining, Theory, Microbes and Industrial Processes*. Springer Verlag, Berlin, Germany.

Rawlings, D.E. (2002) Heavy metal mining using microbes. *Annual Review of Microbiology*, 56, 65–91.

Rawlings, D.E., Dew, D. & du Plessis, C. (2003) Biomineralization of metal – Containing ores and concentrates. *Trends in Biotechnology*, 21, 38–44.

Recio-Vazquez, L., Garcia-Guinea, J., Carral, P., Alvarez, A.M. & Garrido, F. (2010) Arsenic mining waste in the catchment area of the Madrid detrital aquifer (Spain). *Water Air Soil Pollution*, 214, 307–320.

Rice, K.C. & Bayles, W. (2008) Molecular control of bacterial death and lysis. *Microbiology and Molecular Biology Reviews*, 72, 85–109.

Roehl, K.E., Meggyes, T., Simon, F.-G. & Stewart, D.I. (eds.) (2005) *Long-term Performance of Permeable Reactive Barriers*. Trace Metals and Other Contaminants in the Environment. Volume 7 (Series editor: Nriagu, J.O.). Elsevier, Amsterdam, The Netherlands, Boston, MA, USA, Heidelberg, Germany & London, UK. p. 326. ISBN: 0-444-52536-4.

Rohwerder, T., Gehrke, T., Kinzler, K. & Sand, W. (2003) Bioleaching review part A: Progress in bioleaching, fundamentals and mechanisms of bacterial metal sulfide oxidation. *Applied Microbiology and Biotechnology*, 63, 239–248.

Rohwerder, T., Jozsa, P.-G., Gehrke, T. & Sand, W. (2002) Bioleaching. In: Bitton, G. (ed) *Encyclopedia of Environmental Microbiology*. Volume 2. John Wiley & Sons, New York, NY, USA. pp. 632–641.

Rötting, T.S., Thomas, R.C., Ayora, C. & Carrera, J. (2008) Passive treatment of acid mine drainage with high metal concentrations using dispersed alkaline substrate. *Journal of Environmental Quality*, 37, 1741–1751.

RoyChowdhury, A., Sarkar, D. & Datta, R. (2015) Remediation of acid mine drainage-impacted water. *Current Pollution Reports, Water Pollution*, 1, 131–141. doi:10.1007/s40726-015-0011-3.

Rust, C.M., Aelion, C.M. & Flora, J.R.V. (2002) Laboratory sand column study of encapsulated buffer release for potential *in situ* pH control. *Journal of Contaminant Hydrology*, 54, 81–98.

Sand, W. (1985) Influence of four detergents on the substrate oxidation by *Thiobacillus ferrooxidans*. *Environmental Technology Letters*, 6, 439–444.

Sand, W. & Gehrke, T. (2006) Extracellular polymeric substances mediate bioleaching/biocorrosion via interfacial processes involving iron(III) ions and acidophilic bacteria. *Research in Microbiology*, 157, 49–56.

Sand, W., Gehrke, T., Jozsa, P. & Schippers, A. (2001) (Bio)chemistry of bacterial leaching – Direct vs. indirect bioleaching. *Hydrometallurgy*, 59, 159–175.

Sand, W., Jozsa, P.G., Kovacs, Z.M., Sasaran, N. & Schippers, A. (2007) Long – Term evaluation of acid rock drainage mitigation measures in large lysimeters. *Journal of Geochemical Exploration*, 92, 205–211.

Sand, W., Rohde, K., Sobotke, B. & Zenneck, K. (1992) Evaluation of *Leptospirillum ferrooxidans* for leaching. *Applied Environmental Microbiology*, 58, 85–92.

Santelli, C.M., Pfister, D.H., Lazarus, D., Sun, L., Burgos, W.D. & Hansel, C.M. (2010) Promotion of Mn(II) oxidation and remediation of coal mine drainage in passive treatment systems by diverse fungal and bacterial communities. *Applied Environmental Microbiology*, 76, 4871–4875.

Santomartino, S. & Webb, J.A. (2007) Estimating the longevity of limestone drains in treating acid mine drainage containing high concentrations of iron. *Applied Geochemistry*, 22, 2344–2361.

Sarkadi, A., Vaszita, E., Tolner, M. & Gruiz, K. (2009) *In situ* site assessment: Short overview and description of the field portable XRF and its application. *Land Contamination and Reclamation*, 17(3–4), 431–442.

Schippers, A. (2007) Microorganisms involved in bioleaching and nucleic acid – Based molecular methods for their identification and quantification. In: Donati, E.R. & Sand, W. (eds.) *Microbial Processing of Metal Sulfides*. Springer Verlag, Dordrecht, The Netherlands.

Schippers, A., Hallmann, R., Wentzien, S. & Sand, W. (1995) Microbial diversity in uranium mine waste heaps. *Applied Environmental Microbiology*, 61, 2930–2935.

Schippers, A., Jozsa, P.G. & Sand, W. (1996) Sulfur chemistry in bacterial leaching of pyrite. *Applied Environmental Microbiology*, 62, 3424–3431.

Schippers, A., Kock, D., Schwartz, M., Böttcher, M.E., Vogel, H. & Hagger, M. (2007) Geomicrobiological and geochemical investigation of a pyrrhotite – Containing mine waste tailings dam near Selebi – Phikwe in Botswana. *J Geochem Explor Journal of Geochemical Exploration*, 92, 151–158.

Schippers, A. & Sand, W. (1999) Bacterial leaching of metal sulfides proceeds by two indirect mechanisms via thiosulfate or via polysulfides and sulfur. *Applied Microbiology and Biotechnology*, 65, 319–321.

Schmidt, T.S., Soucek, D.J. & Cherry, D.S. (2002) Modification of an ecotoxicological rating to bioassess small acid mine drainage-impacted watersheds exclusive of benthic macroinvertebrate analysis. *Environmental Toxicology and Chemistry*, 21(5), 1091–1097.

Schroeter, A.W. & Sand, W. (1993) Estimations on the degradability of ores and bacterial leaching activity using short – Time microcalorimetric tests. *FEMS Microbiological Review*, 11, 79–86.

Seidel, H., Ondruschka, J., Morgenstern, P., Wennrich, R. & Hoffmann, P. (2000) Bioleaching of heavy metal – Contaminated sediments by indigenous *Thiobacillus* spp, metal solubilization and sulfur oxidation in the presence of surfactants. *Applied Microbiology and Biotechnology*, 54, 854–857.

Sheoran, A.S. & Sheoran, V. (2006) Heavy metal removal mechanism of acid mine drainage in wetlands: A critical review. *Minerals Engineering*, 19, 105–116. doi:10.1016/j.mineng.2005.08.006.

Siebert, H.M., Marmulla, R. & Stahmann, K.P. (2011) Effect of SDS on planctonic *Acidithiobacillus thiooxidans* and bioleaching of sand samples. *Minerals Engineering*, 24(11), 1128–1131.

Siebert, H.M., Rohwerder, T., Sand, W., Strzodka, M. & Stahmann, K.P. (2009) Evidence for iron – And sulfur – Oxidizing bacteria and archaea in a currently active lignite mining area of Lusatia (Eastern Germany). *Advanced Material Research*, 71–73, 97–100.

Sikkema, J., de Bont, J.A.M. & Poolmann, B. (1995) Mechanisms of membrane toxicity of hydrocarbons. *Microbiological Review*, 59, 201–222.

Simate, G.S. & Ndlovu, S. (2014) Acid mine drainage: Challenges and opportunities. *Journal of Environmental Chemical Engineering*, 2, 1785–1803.

Simmons, J., Ziemkiewicz, P. & Black, D.C. (2002) Use of steel slag leach beds for the treatment of acid mine drainage. *Mine Water and the Environment*, 21(2), 91–99.

Simon, F.-G. & Meggyes, T. (2000) Removal of organic and inorganic pollutants from groundwater using permeable reactive barriers. Part 1. Treatment processes for pollutants. *Land Contamination & Reclamation*, 8(2), 103–116.

Simon, F.-G., Meggyes, T. & McDonald, C. (eds.) (2002) *Advanced Groundwater Remediation: Active and Passive Technologies*. Thomas Telford, London, UK. p. 356. Available from: www.ttbooks.co.uk/advanced-groundwater-remediation. [Accessed 3rd July 2018].

Simon, F.-G., Meggyes, T., Tünnermeier, T., Czurda, K. & Roehl, K.E. (2001) Long-term behaviour of permeable reactive barriers used for the remediation of contaminated groundwater. *Proceedings 8th*

International Conference on Radioactive Waste Management and Environmental Remediation, 30 September–4 October 2001. ASME, Tech. Inst. Royal Flemish Soc. of Engineers, Belgian Nuclear Society, Bruges, Belgium.

Singer, P.C. & Stumm, W. (1970) Acidic mine drainage: The rate-determining step. *Science*, 167, 1121–1123. doi:10.1126/sci- ence.167.3921.1121.17829406.

Skousen, J. (1995) Acid mine drainage. *Green Lands*, 25(2), 52–55.

Skousen, J. (1996) Overview of passive systems for treating acid mine drainage. *Green Lands*, 27(4), 34–43. Available from: http://anr.ext.wvu.edu/resources/295/1256049359.pdf.

Skousen, J. (2000) Overview of passive systems for treating acid mine drainage. Reclamation of drastically disturbed lands. *Agronomy*, 41, 1–214. Available from: www.wvu.edu/~agexten/landrec/passtrt/passtrt.htm. [Accessed 13th July 2018].

Skousen, J.G., Rose, A., Geidel, G., Foreman, J., Evans, R. & Hellier, W. (1998) *A Handbook of Technologies for Avoidance and Remediation of Acid Mine Drainage*. Acid Drainage Technology Initiative, National Mine Land Reclamation Ctr, WVU, Morgantown, WV, USA. 131 pp.

Skousen, J.G., Sexstone, A. & Ziemkiewicz, P.F. (2000) Acid mine drainage control and treatment. *Agronomy*, 41, 131–168.

Skousen, J.G., Simmons, J., McDonald, L.M. & Ziemkiewicz, P. (2002) Acid-base accounting to predict post-mining drainage quality on surface mines. *Journal of Environmental Quality*, 31(6), 2034–2044.

Smyth, D.J.A., Blowes, D.W., Ptacek, C.J. & Bain, J.G. (2003) Removal of dissolved metals from groundwater using permeable reactive barriers: Applications. *Paper presented at the 6th ICARD, July 12–18, Cairns, Australia*.

Soós, I. (1966) Ore mining in Mátra up to 1850 (In Hungarian). In: Pantó, E. (ed) *Mining of Multimetal Ore at Gyöngyösoroszi and Manganese ore in Bakony*. Országos Érc- és Ásványbányák, Budapest, Hungary.

Sumitomo, T., Maeda, T., Nagamune, H. & Kourai, H. (2004) Bacterioclastic action of a bis-quaternary ammonium compound against *Escherichia coli*. *Biocontrol Science*, 8, 145–149.

Soucek, D.J., Cherry, D.S., Currie, R.J., Latimer, H.A. & Trent, G.C. (2000) Laboratory and field validation in an integrative assessment of an acid mine drainage-impacted watershed. *Environmental Toxicology and Chemistry*, 19(4), 1036–1043.

TAILSAFE (2002) *Sustainable Improvement in Safety of Tailings Facilities*. TAILSAFE EVG1-CT-2002-00066. Available from: www.tailsafe.com/ [Accessed 3rd July 2018].

Taylor, J., Pape, S. & Murphy, N. (2005) A Summary of Passive and Active Treatment Technologies for Acid and Metalliferous Drainage (AMD). *Australian Centre for Minerals Extension and Research (ACMER), Fifth Australian Workshop on Acid Drainage, 29–31 August 2005, Fremantle, Australia*. Available from: www.earthsystems.com.au/wp-content/uploads/2012/02/AMD_Treatment_Technologies_06.pdf. [Accessed 16th July 2018].

Temple, K. & Colmer, A. (1951) The autotrophic oxidation of iron by a new bacterium, *Thiobacillus ferrooxidans*. *Journal of Bacteriology*, 62, 605–611.

Tributsch, H. & Bennett, J.C. (1981) Semiconductor- Electrochemical aspect of bacterial leaching. I. Oxidation of metal sulphides with large energy gaps. *Journal of Chemical Technology*, 31, 565–577.

Turner, D. & McCoy, D. (1990) Anoxic alkaline drain treatment system, a low cost Acid Mine Drainage treatment alternative. In: Graves, D.H. & De Vore, R.W. (eds.) *Proceedings of the 1990 National Symposium on Mining*, Lexington, KY, USA. pp. 73–75.

Tuttle, J.H., Dugan, P.R., MacMillan, C.B. & Randles, C.I. (1969) Microbial dissimilatory sulfur cycle in acid minewater. *Journal of Bacteriology*, 97, 594–602.

Urbieta, M.S., González Toril, E., Aguilera, A., Giaveno, M.A. & Donati, E. (2012) First prokaryotic biodiversity assessment using molecular techniques of an acidic river in Neuquén, Argentina. *Microbiological Ecology*, 64, 91–104.

US EPA (2014) *Reference Guide to Treatment Technologies for Mining-Influenced Water*. U.S. Environmental Protection Agency, Office of Superfund Remediation and Technology Innovation.

Available from: https://cluin.org/download/issues/mining/Reference_Guide_to_Treatment_Technologies_for_MIW.pdf. [Accessed 16th July 2018].

Vadapalli, V.R.K., Klink, M.J., Etchebers, O., Petrik, L.F., Gitari, W., White, R.A., Key, D. & Iwuoha, E. (2008) Neutralization of acid mine drainage using fly ash, and strength development on the resulting solid residues. *South African Journal of Science*, 104, 317–322.

Vaszita, E., Siki, Z. & Gruiz, K. (2009) GIS-based Quantitative Hazard and Risk Assessment of an abandoned mining site. *Land Contamination & Reclamation*, 17(3–4), 515–534.

Vera, M., Schippers, A. & Sand, W. (2013) Progress in bioleaching, fundamentals and mechanisms of bacterial metal sulfide oxidation – Part A. *Applied Microbiology and Biotechnology*, 97(17), 7529–7541.

Vidacs, A. (1966a) History of ore mining in Mátra in period 1850–1945. In: Pantó, E. (ed) *Mining of Multi-metal ore at Gyöngyösoroszi and Manganese ore in Bakony*. Országos Érc- és Ásványbányák, Budapest, Hungary.

Vidacs, A. (1966b) General geological knowledge on the formation of the ore deposits (In Hungarian). In: Jantsky, B. (ed) *Minerals' Geology. Raw Material Deposits in Hungary*. Műszaki Könyvkiadó, Budapest, Hungary. pp. 179–206.

Waksman, S.A. (1922) Microorganisms concerned in the oxidation of sulfur in the soil: IV. A solid medium for the isolation and cultivation of *Thiobacillus thiooxidans*. *Journal of Bacteriology*, 7, 605–660.

Watten, B.J., Lee, P.C., Sibrell, P.L. & Timmons, M.B. (2007) Effect of temperature, hydraulic residence time and elevated PCO_2 on acid neutralization within a pulsed limestone bed reactor. *Water Research*, 41, 1207–1214.

Watten, B.J., Sibrell, P.L. & Schwartz, M.F. (2005) Acid neutralization within limestone sand reactors receiving coal mine drainage. *Environmental Pollution*, 137, 295–304.

Watzlaf, G.R. & Hedin, R. (1993) *A Method for Predicting the Alkalinity Generated by Anoxic Limestone Drains*. Available from: www.researchgate.net/publication/255619191. [Accessed 3rd July 2018].

Watzlaf, G.R., Schroeder, K.T. & Kairie, C.L. (2000) *Long-Term Performance of Anoxic Limestone Drains*. U.S. Department of Energy, National Energy Technology Laboratory. Available from: www.imwa.info/bibliographie/19_2_098-110.pdf. [Accessed 3rd July 2018].

Watzlaf, G.R., Schroeder, K.T., Kleinmann, R.L.P., Kairies, C.L. & Nairn, R.W. (2002) *The Passive Treatment of Coal Mine Drainage*. DOE/NETL-2004/1202.

Watzlaf, G.R., Schroeder, K.T., Kleinmann, R.L.P., Kairies, C.L. & Nairn, R.W. (2004) *The Passive Treatment of Coal Mine Drainage*. DOE/NETL-2004/1202. Available from: www.netl.doe.gov/File%20Library/Research/Coal/ewr/water/Passive-Treatment.pdf. [Accessed 3rd July 2018].

Wieder, R.K. (1993) Ion input/output budgets for wetlands constructed for acid coal mine drainage treatment. *Water, Air, and Soil Pollution*, 71, 231–270.

Wieder, R.K. & Lang, G.E. (1982) Modification of Acid Mine Drainage in a Freshwater Wetland. *Proceedings of the Symposium on Wetlands of the Unglaciated Appalachian Region*. West Virginia University, Morgantown. pp. 43–53.

Willscher, S., Hertwig, T., Frenzel, M., Felix, M. & Starke, S. (2010) Results of remediation of hard coal overburden and tailings dumps after a few decades, insights and conclusions. *Hydrometallurgy*, 104, 506–517.

Yadav, S.K. (2010) Heavy metals toxicity in plants: An overview on the role of glutathione and phytochelatins in heavy metal stress tolerance of plants. *South African Journal of Botany*, 76, 167–179. doi:10.1016/j.sajb.2009.10.007.

Younger, P.L. (2000) The adoption and adaptation of passive treatment technologies for mine waters in the United Kingdom. *Mine Water and the Environment*, 19, 84–97.

Younger, P.L., Banwart, S.A. & Hedin, R.S. (2002) *Mine Water: Hydrology, Pollution, Remediation*. Kluwer Academic Publishers, Dordrecht, The Netherlands. 442 pp.

Younger, P.L., Jayaweera, A., Elliot, A., Wood, R., Amos, P., Daugherty, A.J., Martin, A., Bowden, L., Aplin, A.C. & Johnson, D.B. (2003) Passive treatment of acidic mine waters in subsurface flow systems: Exploring RAPS and permeable reactive barriers. *Land Contamination & Reclamation*, 11, 127–135.

Zhu, W., Xia, J., Yang, Y., Nie, Z., Zheng, L., Ma, C., Zhang, R., Peng, A., Tang, L. & Qiu, G. (2011) Sulfur oxidation activities of pure and mixed thermophiles and sulfur speciation in bioleaching of chalcopyrite. *Bioresource Technology*, 102, 3877–3882.

Zhuang, W-Q., Fitts, J.P., Ajo-Franklin, C.M., Maes, S., Alvarez-Cohen, L. & Hennebel, T. (2015) Recovery of critical metals using biometallurgy. *Current Opinion in Biotechnology*, 33, 327–335. doi:10.1016/j.copbio.2015.03.019.

Ziemkiewicz, P.F., Skousen, J.G. & Lovett, R. (1994) Open limestone channels for treating acid mine drainage: A new look at an old idea. *Green Lands*, 24(4), 36–41.

Ziemkiewicz, P.F., Skousen, J.G. & Simmons, J. (2003) Long-term performance of passive acid mine drainage treatment systems. *Mine Water and the Environment*, 22, 118–129.

Chapter 8

Remediation technologies for metal-contaminated soil and sediment – an overview and a case study of combined chemical and phytostabilization

V. Feigl

Department of Applied Biotechnology and Food Science, Budapest University of Technology and Economics, Budapest, Hungary

ABSTRACT

Hundreds of abandoned mines and mine waste disposal sites all over the world are waiting for restoration and soil remediation. The economic and social benefit of overexploited mining was enjoyed over the last 60–70 years without investing in proper management of environmental risks posed to the area during mining operations and its reduction afterwards. The footprints of former metal ore mining activities are the worst: lack of vegetation, drastically altered landscape, wastes piled up and left, damage to water and soil from acidic mine water and waste leachates, diffuse contaminated environment and deteriorated, destroyed ecosystem. Contaminants are derived originally from point sources which are dispersed during the years of careless abandonment, resulting in countless diffuse secondary sources polluting entire watersheds. This is a typical consequence of long-abandoned mine sites without restoration.

To reduce the environmental impact of mining, collaborative efforts are needed from a variety of stakeholders. Environmental engineering plays a key role in risk reduction, both before and after mining activities. Risk assessment and risk reduction of abandoned metal mines involve similar tasks all over the world and the innovative solutions could be largely utilized.

The problem with abandoned mines and its management was discussed in the first volume of this book series (Vaszita, 2014), the acid mine drainage formation and treatment in Chapter 7 of this volume, while risk reduction by remediation of soil contaminated from point and diffuse sources of metal-containing waste is discussed here. After a general overview of the existing technological solutions, a case study is presented about the former Hungarian zinc, lead mine in Gyöngyösoroszi, abandoned over 30 years ago.

1 REMEDIATION TECHNOLOGIES FOR METAL-CONTAMINATED SOIL AND SEDIMENT – AN OVERVIEW

1.1 Introduction

A huge number of technologies have been developed for the remediation of metal-contaminated soil and water. These technologies are generally classified as isolation, immobilization/stabilization, toxicity reduction, physical separation, and extraction (Evanko & Dzombak, 1997). More coherent classification is introduced in Chapter 1, where all remediation technologies are classified as mobilization- and immobilization-based methods including physicochemical, biological, ecological, and combined methods. The most frequently used technologies for metal-contaminated environments are (i) physicochemical mobilization with extraction and (ii) physicochemical stabilization with stabilizing agents. The same solutions work more efficiently with biological mediation (utilizing biological products or biological processes such as the reduction or increase of the redox potential). Usually, a combination of several approaches and technologies increases the cost-effectiveness of the remediation of contaminated sites. In this chapter, two main categories are introduced in detail: technologies based on mobilization and immobilization with the focus on physical and chemical processes.

Isolation/capping and containment are also mentioned in spite of the fact that the soil itself will not be healed (remediated), but the endangered environment can be protected from certain risks.

The remediation can be performed *ex situ* by removing the contaminated soil from its original location and treatment on site or off site, or implemented *in situ* when the contaminated soil is not excavated but treated at its original location. The decision on the *in-situ* or *ex-situ* solution depends on the amount, the surrounding environment, and spatial planning considerations (see more in Lens *et al.*, 2005).

1.2 Physical-chemical technologies

Traditional technologies for the treatment of contaminated soil include methods originating from waste management and civil engineering. In the past, excavation of contaminated soil and its safe deposition (dig and dump) was a common procedure. The excavated soil was then replaced with uncontaminated soils. This methodology was costly and not sustainable. It transferred the risk to another location and the excavated soil lost its value. The availability of the uncontaminated soil is limited and the impact of its transportation may be high. In addition, replacement can only be applied for point sources.

Excavation of the contaminated soil is acceptable if the risk can be significantly reduced by the removal of the contaminated proportion and the treatment technology of the excavated soil – possibly in the vicinity – can produce useful products. The more soil utilized means less of the remaining contaminated material needs to be further treated or disposed. Managerial arguments can also support contaminated soil removal, e.g. if the foundation of a building is planned to be where contaminated soil is to be excavated, or if sufficient replacement soil is available due to construction works nearby.

The counterpart of soil removal by excavation is the dredging of contaminated sediment from rivers and lakes waters. The reason for dredging may be the need for removal of the contaminated proportion, but, in most cases, its shipping or port functions that are served by dredging, which may, and typically does, result sediment of inappropriate quality. As sediments are removed in slurry form, they are treated by any reactor-based technology designed for slurries, such as washing, fractionation, extraction, degradation.

1.2.1 Isolation/containment

The aim of isolation is to prevent the transport of contaminants from fixed into mobile environmental compartments (atmosphere, waters) and physical phases (runoff, soil moisture, groundwater). In the case of metals it is mainly surface waters and groundwater that is exposed. Isolation by applying a cover is usually applied to high volumes of contaminated soil, waste rock, and other solid waste, typically of mining origin. *In situ*/on-site isolation/ containment creates waste disposal sites requiring long-term care. Isolation may also be applied temporarily in order to limit transport during site assessment and site remediation.

1.2.1.1 Cover systems

Cover systems are used to provide an impermeable barrier to water infiltration into contaminated soil so preventing further release of contaminants into the surrounding surface water or groundwater. Secondary objectives include controlling wind erosion, gas and odor emissions, improving aesthetics, and providing a stable surface over a contaminated site (Evanko & Dzombak, 1997). A variety of natural and synthetic materials can be used in capping systems. These may range from a single layer to multi layer cover systems. Covering methods vary depending upon site conditions including contaminant chemistry, soil type, climate, land use, site location, and budget.

1.2.1.2 Subsurface barriers

Subsurface barriers isolate contaminated soil and water by controlling the movement of groundwater in a contaminated site. These barriers are designed to reduce groundwater inflow and/or outflow from the site, i.e. the interaction of any water with the contaminated solid phase (see more in Rumer & Ryan, 1995). Vertical and horizontal barriers, slurry walls, grout curtains, and sheet piles are constructed, usually as barriers. Permeable barriers follow another concept; they influence water flow at a minimal scale while selectively retaining and eliminating the hazardous components potentially transported by the water flow.

1.2.1.3 Physical separation of the solid phase fractions

Physical separation may be based on particle size, density, surface characteristics, or magnetic properties of the soil constituents. It is generally an *ex situ* process that attempts to separate the contaminated material from the rest of the soil matrix by exploiting the actual characteristics of the soil and the contaminating metals as well as their interactions with each other. Such interactions are typically the dominant binding of metals to the colloidal size, inorganic soil fractions, or the influence of pH and redox potential on certain metal species and their solubility. Several techniques are available including screening, classification, gravity concentration, magnetic separation, and froth flotation. It is often used as pretreatment in order to reduce the amount of material requiring subsequent treatment. Most parts of the separation process are carried out in slurry form, consequently this approach is primarily used for the remediation of dredged sediment and for soils/wastes which need washing (with water).

1.2.2 Mobilization of metals

Mobilization techniques aim at extracting the contaminant and the contaminated physical phases or solid fractions from the rest of the soil/sediment. It can be achieved by contacting the contaminated solid medium with a liquid phase solution containing extracting agents (solvent, surfactant, pH, or redox-setting agent) in a soil washing or *in situ* soil flushing technology. Thermal and electrokinetic processes can also be applied for metal mobilization in soil. The contaminated fraction of soil and/or process water is separated from the remaining soil and disposed of or further treated.

1.2.2.1 Soil washing

Soil washing is used to remove metals from the soil by chemical or physical treatment methods in aqueous suspension (see scheme on Figure 8.1 and Figure 1.8 in Chapter 1). The process entails excavation of the contaminated soil, mechanical screening to remove various oversize materials, separation processes to generate coarse- and fine-grained fractions, treatment of those solid fractions in order to transfer metals from solid to the liquid/water phase by washing, extraction by other solvents and additives, chemical modification, etc. The final step of the multi-stage integrated technology covers the management of the treated residuals (both of the remediated and the ready-to-reuse fractions) and of those which should be further treated or contained. The contaminated solid phase can be drastically reduced; sustainability depends mainly on the feasibility of the treatment and the amount of the wash-water.

Surficial contaminants are removed through abrasive scouring and scrubbing actions in one step using wash-water that is sometimes doped with surfactants and other agent, such as acids, alkali, or chelating agents. The soil is then separated from the spent wash fluid and separated into two main fractions: a clean, coarse fraction and a contaminated fine fraction. The fine fraction concentrates the contaminant so that is why most of the contaminant can be removed by removing the fine solid (read more: US EPA, 1993).

The contaminated fractions need further treatment and the clean ones reused/recycled. Water can be reused in the washing process. The coarse fraction can be utilized as building material. For the treatment of the fine fraction, highly reduced in volume, more expensive methods such as acid, alkali or other chemical extraction may be cost efficient, mainly if the metal is valuable and/or the fine material can be recycled in ceramics production. The main advantage of soil washing is that it produces smaller volumes but more concentrated

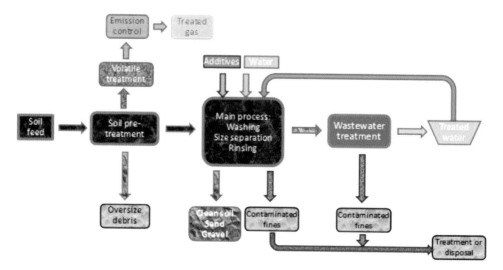

Figure 8.1 Schematic of aqueous soil washing.

contaminants. The other managerial advantage can be the establishment of a network for soil/sediment washing plants similar to wastewater or solid waste treatment facilities. Soil washing can be coupled to various wastewater and solid waste treatment technologies in order to compile the best possible arrangement for the actual contaminated solid or slurry. Where a large amount of contaminated solid is constantly produced – e.g. in harbors, or continuously dredged surface waters such a washing plant can be installed on site. A third version is the mobile-version of a soil washing plant, which can be placed in a temporary excavated/dredged location (US EPA, 1993).

1.2.2.2 In situ soil flushing

In situ soil flushing involves flooding a contaminated soil zone with an appropriate solution to facilitate the distribution/partitioning of the metals from as solid to liquid phase. After metal transfer to the wash-water, the metal containing liquid phase is extracted/removed. Repeated flooding and extraction may reduce metal content of the soil to the acceptable environmental level. Water or a liquid solution is injected or left to infiltrate into the area of contamination. The contaminants are mobilized by solubilization or by a chemical reaction (producing more mobile metal species) with the flushing solution. After passing through the contaminated soil zone, the contaminant-bearing fluid is collected and brought to the surface for disposal, recirculation, or on-site treatment and reinjection. Flushing solutions may be water, acidic aqueous solutions, alkaline solutions, chelating or complexing agents, reducing or oxidizing agents, cosolvents, or surfactants (US EPA, 2006). The applicability of *in situ* soil flushing technologies on contaminated sites will depend largely on site-specific properties, such as hydraulic conductivity that influence the ability of the extractant to make contact with contaminants and to effectively recover the flushing solution using collection wells (NRC, 1994). Intensively flushed/washed soils can release other, essential metals and ions, e.g. plant nutrients and biologically active substances. A site-specific complex assessment is needed to compare the benefits, the costs and the risks of applying such soil treatment technologies. Figure 8.2 provides an illustration of one type of *in situ* soil flushing process.

1.2.2.3 Pyrometallurgical separation/thermal desorption

Pyrometallurgical technologies use high temperature furnaces (200–700°C) to volatilize metals in contaminated solid matrices. Metals are then recovered (US EPA, 1996, 2018). This method is mainly applicable for mercury (Mulligan *et al.*, 2001).

1.2.2.4 Electrokinetic treatment

Electrokinetic remediation applies low density current to contaminated soil in order to mobilize contaminants in the form of charged metal species. The current is applied by inserting electrodes into the subsurface and relying on the natural conductivity of the soil (due to water and salts) to influence movement of water, ions, and particulates through the soil (Figure 9.1 in Chapter 9). The electric potential induces the transport of the contaminants by the following mechanisms: electroosmosis, electromigration, electrophoresis, and diffusion. Electromigration is the dominant process as positively charged metal ions migrate to the cathode while metal anions migrate to the anode (Smith *et al.*, 1995; Reddy *et al.*, 1997; Reddy & Cameselle, 2009; Reddy, 2013 and Chapter 9).

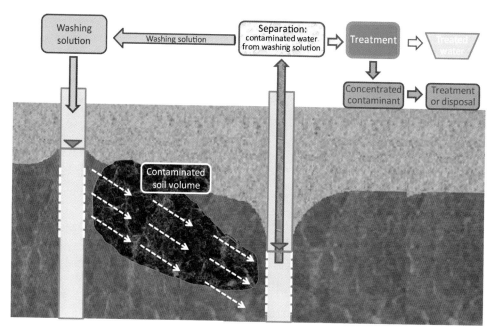

Figure 8.2 In situ soil flushing using vertical wells.

1.2.3 Immobilization

Immobilization technologies are designed to reduce the mobility of contaminants by changing the physicochemical form of the contaminant, the binding capacity of the soil and/or the interaction between the contaminant and the solid matrix, e.g. leaching characteristics, and partitioning the contaminants between solid and liquid by promoting strong binding and by limiting transport into the liquid phases. Mobility is usually decreased by physically restricting contact between the contaminant and the surrounding groundwater, or by chemically altering the contaminant to make it more stable with respect to dissolution in groundwater (Evanko & Dzombak, 1997). These changes can be implemented by solidification/stabilization (S/S) technologies. The metal content of the groundwater can be typically reduced by physical processes such as sorption or by chemical reactions, which result in immobile, strongly bound forms, e.g. precipitates. Physical and chemical immobilization/stabilization can be performed *ex situ* or *in situ*, groundwater can be treated in permeable reactive barriers, treatment walls or larger treatment volumes, *in situ* filled reactors. *in situ* immobilization of toxic metals in soil need permanent monitoring to prove that the immobile physical or chemical form is stabile in the long term.

1.2.3.1 Solidification

Solidified contaminants are physically bound or enclosed within a stabilized mass (FRTR, 2002). The product of solidification may be a monolithic block, a clay-based brick or ceramic-like material, a granular particulate, or any other physical form commonly considered "solid."

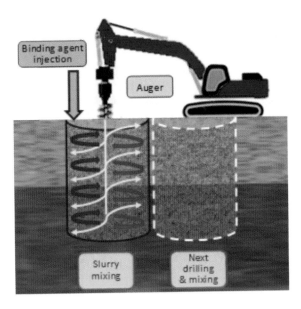

Figure 8.3 Typical solidification process: mixing of biding agents using augers.

The most common inorganic binders are Portland cement, pozzolans (siliceous or aluminous materials that can react with calcium hydroxide to form compounds with cementitious properties), and cement/pozzolan mixtures. Three basic approaches are used for mixing the binder with the matrix: vertical auger mixing, shallow in-place mixing, and injection grouting (US EPA, 2006, 2012) (Figure 8.3).

1.2.3.2   Vitrification

Vitrification uses an electric current to melt siliceous soil components at elevated temperatures. Contaminating metals get inserted into the glass/ceramic-like melted matrix. Upon cooling, the vitrification product is a chemically stable, usually leach-resistant, crystalline material similar to obsidian or basalt rock. The extremely high temperature (800–1200°C) destroys or volatilizes organic materials, and volatile metals, while radionuclides and heavy metals are retained within the vitrified product. It can be performed *in situ* or *ex situ* (US EPA, 1992). If the leaching test does not confirm complete stabilization, the residual mobile metal species could be washed out by acid extraction, before using the vitrified product. Figure 8.4 shows the vitrified soil residue from the Savannah River demonstration site (Infohouse, 1999).

1.2.3.3   Chemical stabilization

Chemical stabilization induces chemical reactions between the stabilizing agent and the contaminant to reduce its mobility (FRTR, 2002). The stabilization of contaminants can be achieved by enhanced sorption, by complexation or precipitation reactions. The processes

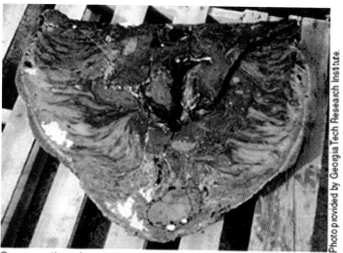

Cross section of a vitrified monolith from the Savannah River
Site demonstration.

Figure 8.4 Cross-section of a vitrified monolith (infohouse.p2ric.org).

are dependent on both the metal types and soil properties (pH, redox potential, type of soil constituents, cation exchange capacity etc.) as well as on the used amendment or catalyzer (Adriano *et al.*, 2004). *In situ* stabilization techniques use rather inexpensive natural and/or industrial by-products (Kumpiene *et al.*, 2008; Chang *et al.*, 2013):

-   Alkaline compounds with pozzolan activity, e.g. cyclonic ash, coal fly ash (Adriano *et al.*, 2004; Vangronsveld *et al.*, 1996, 1999; Feigl *et al.*, 2008, 2009a,b,c, 2010a,b,c);
-   Cement kiln dust (Rahman *et al.*, 2011);
-   Iron and manganese oxides, such as lime sludge from water softening (Baker *et al.*, 2005);
-   Sugar factory lime (Wasner *et al.*, 2005);
-   Elemental iron, e.g. steel shot (Boisson *et al.*, 1999) or steel slag (Zhuo *et al.*, 2012);
-   Red mud (Feigl *et al.*, 2012; Ujaczki *et al.*, 2016a,b);
-   Phosphate compounds (Xenidis, 2010);
-   Biosolids and composts (Christie *et al.*, 2001; McBride *et al.*, 2013) and charcoal originating from different agricultural wastes (Molnár *et al.*, 2016);
-   The technology is often combined with phytostabilization (Vangronsveld *et al.*, 1996, 1999; Feigl *et al.*, 2007, 2008, 2009b, c, 2010a,b,c).

1.2.3.4 Chemical treatments for mobility and toxicity reduction

Chemical treatment by reductive and oxidative agents may be used to decrease the mobility and/or toxicity of metal contaminants. The three types of reactions that can be used for this purpose are oxidation, reduction and neutralization reactions. Changing the oxidation state of metals by oxidation or reduction can precipitate or solubilize the metals and this change may result in reduced toxicity or complete detoxification. Chemical neutralization is used to adjust the pH balance of extremely acidic or alkaline soils and/or groundwater (Evanko & Dzombak, 1997). This procedure can be used to precipitate insoluble metal salts from contaminated water, groundwater or soil moisture *in situ* or *ex situ*.

1.2.4   Permeable reactive barriers

Permeable reactive barriers, treatment walls, or treatment zones all are flow-through filled reactors placed in the way of the contaminated water. While the contaminated water flows through the permeable reactor wall (subsurface trenches, walls, barriers) the filling removes contaminants from groundwater by sorbing, precipitating, or transforming them. The reactor is filled by reactive material and the water flow can be driven by gravity or by artificially created pressure difference (see also Vidic & Pohland, 1996).

1.3   Bioremediation of metal-contaminated soil

Bioremediation is the reduction of risk of contaminated soil, groundwater and surface water utilizing biological/biochemical processes (typically the biodegrading, bioaccumulation, or stabilizing potential of living cells or organisms), or their products (for example enzymes). The remedial technology ensures optimal conditions for the desired biological processes and the beneficial organisms. The technological parameters of the remediation – such as optimal temperature, pH, redox potential, nutrients, stimulants – can be set by physical, chemical, or biological means, using heat, additives, amendments, etc. (enfo. hu). Similarly to physicochemical methods, we can differentiate between biomobilization and biostabilization methods. Bioremediation can take place upon the action of biological entities (biotransformation, bioaccumulation, biodegradation) or under biologically mediated physicochemical conditions (bioleaching, biologically reduced redox potential, biosurfactants, etc.).

In this chapter bioremediation technologies that apply to microorganisms (or their products) and use them for the treatment of metal and metalloid contaminated soils are presented (for reviews, see Gadd, 2000, 2004; Lovely & Coates, 1997; Stephen & Macnaughton, 1999; Mulligan *et al.*, 2001; Tabak *et al.*, 2005) (see more at BioMineWiki, 2018). Usually bacteria and archea are applied in bioremediation, but fungi and algae can also be used for water and soil decontamination. Bioremediation can be performed with the application of single-cell animals (protozoa), macro-animals (e.g. vermiremediation using earthworms), and phytoremediation by plants.

The microbiologically mediated mobilization and immobilization of metals is based on several physical and chemical conditions of the soil/water and microbial surface as well as biochemical reaction such as redox transformation, degradation, and synthesis or complexation (Table 8.1).

1.3.1   Biological mobilization

Microorganisms can mobilize metals through autotrophic and heterotrophic leaching, chelation by microbial metabolites, and siderophores. Such processes can lead to dissolution of insoluble metal compounds and minerals, and desorption of metal species from exchange sites on clay minerals or organic matter in the soil (Gadd, 2004). Methylation can result in volatilization of Hg, Sn, Cd, Tl, and Sb.

Table 8.1 Summary of the microbially mediated transformation of metals (Me) in soil.

Mobilization
Volatilization of metals by methylation or dealkylation
Bioleaching: mobilization/solubilization by microorganisms and transport by water flows
Complexation with degraded organics and producing small Me-organic complexes
Immobilization
Biocoagulation is based on clustering microorganisms and metals together
Biosorption of metals on the large specific surface of the microorganisms
Bioprecipitation by inorganic or large organic microbial products
Bioaccumulation: uptake and storage of metals in the microbial cell

1.3.1.1 Bioleaching

Bioleaching can be accomplished by autotrophic or heterotrophic bacteria and fungi. They increase metals' mobility by changing their redox form, e.g. $MeS \rightarrow Me^{2+}$ by sulfur oxidation and sulfuric acid production or by organic acids produced by the microbial cell. The chemolitotrophic sulfur-oxidizing bacteria use reduced sulfur and iron containing substances as energy source, and as a result they produce sulfuric acid and Fe(III) ions. Metals are mobilized in the form of metal sulfates. The most common bacteria which can mobilize metals are *Acidithiobacillus* species (e.g. *Acidithiobacillus thiooxidans*, *Acidithiobacillus ferrooxidans*), which are also used for biomining at an industrial scale. Sulfate-reducing bacteria can be applied to precipitate metals from the bioleaching solution (Gadd, 2000, 2004; Tabak *et al.*, 2005; Schippers & Sand, 1999). In heterotrophic bioleaching, microorganisms produce organic acids which function as proton donors and as complexing agents (Gadd, 2000). For example, acids (citric acid, gluconic acid and oxalic acid) produced by *Aspergillus niger* can form water-soluble complexes with metals, such as copper. The technology requires cheap carbon sources (Mulligan *et al.*, 2001). The mechanism of bioleaching of metal sulfides is discussed in Chapter 7 and shown in Figures 7.1 and 7.2).

1.3.1.2 Siderophores

Siderophores are highly specific low molecular weight Fe(III) ligands that can bind other metals (Gadd, 2004). The method published by Diels *et al.* (1999) for the treatment of metal-contaminated sandy soil relies on siderophore-mediated metal solubilization by *Alcaligenes eutrophus*. Solubilized metals are adsorbed on the biomass and/or precipitated with biomass separated from soil slurry by flocculation.

1.3.1.3 Biomethylation

Methylation of Hg, As, Se, Sn, Te, and Pb can be mediated by a range of bacteria and fungi under aerobic and anaerobic conditions. Methyl groups are enzymatically transferred to the metal and a given species may transform a number of different metal(-loid)s. Methylated metal compounds formed by these processes differ in their solubility, volatility, and toxicity, usually the methylated form is more mobile and volatile. From Se: $(CH_3)_2Se$ or $(CH_3)_2Se_2$, from As: $(CH_3)_2HAs$ or $(CH_3)_3As$ are formed (Gadd, 2004).

1.3.1.4   Mobilization by redox transformation

Microorganisms can mobilize metals by redox transformations. Metal-reducing bacteria reduce metals as terminal electron acceptors during their anaerobic respiration. For example, by the reduction of Fe(III) and Mn(IV) their mobility increases. As Fe(III)- and Mn(IV)-oxides strongly adsorb metals, their reduction results in metal desorption/release.

1.3.2   Immobilization

Immobilization reduces the amount of mobile metal species. Biosorption and biocoagulation are biologically passive processes while in bioprecipitation and bioaccumulation play the active living cell a role by producing specific metal binding/mobility reducing molecules.

1.3.2.1   Biosorption and biocoalugation

During biosorption metals are sorbed on the surface of the living or dead microbial biomass. Biosorption is based on physicochemical binding and it is a metabolically passive process. Biocoagulation is a passive process too; it is based on clustering microorganisms and metals together due to their different zeta potentials.

1.3.2.2   Bioaccumulation

When metals are taken up by the cell and accumulated within the cell we speak about bioaccumulation. In the cell the metals can precipitate or attach to inner structures and organelles. Bioaccumulation of metals is a widespread biological phenomenon, mainly due to the low specificity of the transport processes used for microbial nutrient uptake. To utilize bioaccumulation for metal removal from water, both suspended or fixed bacteria, cyanobacteria, algal or fungal biomass can be used. Biofilm fixed on inert carriers may be applied in permeable reactive barriers too (Gadd, 2004; Lovely & Coates, 1997).

1.3.2.3   Metal binding by specific and non-specific biomolecules

Microorganisms produce specific and non-specific metal binding molecules. These can be simple organic acids, alcohols or macromolecules. They may be extracellular polymeric substances that are able to sorb and trap metals and granular substances, e.g. metal sulfides and oxides. Specific metal binding biomass can be produced (e.g. from *E. coli*) by expressing small molecular weight-specific metal binding proteins, called metallothioneins that are produced by microbes, plants, and animals in response to toxic substances (Gadd, 2004).

1.3.2.4   Bioprecipitation

Metals are transformed into a less mobile, oxidized, or reduced form by bioprecipitation. This kind of immobilization is mediated by biological products which reduce metals' mobility: e.g. inorganic species, such as OH^-, HS^-, HCO_3^-, or $H_2PO_4^-$, or large molecular organic complexing agents, e.g. extracellular polymers, polysaccharides, or proteins. Several organisms use metals as terminal electron acceptors during anaerobic respiration. Sulfate-reducing bacteria, which use Fe^{3+} ions and elemental sulfur as electron acceptors, precipitate metals in the form of metal sulfides. One of the most studied processes is the microbial reduction of Cr^{6+} to Cr^{3+}. Some iron-reducing bacteria are able to reduce U^{6+} to U^{4+}. These bacteria can utilize aromatic hydrocarbons as electron donors so they can be used for the remediation of soil polluted by organics, metals, and radionuclides. Se^{6+} can be reduced to Se^0 and microorganisms can oxidize As^{3+} to the less toxic and less mobile As^{5+} form (Gadd, 2004; Lovely & Coates, 1997).

2   COMBINED CHEMICAL AND PHYTOSTABILIZATION OF METAL-CONTAMINATED MINE WASTE AND SOIL – A DEMONSTRATION CASE

2.1   Introduction

In the Gyöngyösoroszi, Hungary, mining area (see Chapter 7) the remediation strategy consists of the removal of the point sources, covering of the collected solid waste and combined chemical and phytostabilization (CCP) of the diffuse and residual pollution. To find the best stabilizer and stabilizer-plant combinations scaled-up experiments were performed with agents and plants selected for this purpose in former experiments. The scale-up had three steps: laboratory microcosms in pots, open-air lysimeter studies, and field experiments of various sized field plots (Feigl *et al.*, 2008, 2010a).

2.2   Demonstration site and technological conditions

In this section of the site, contaminants, stabilizing agents, and plants are introduced. The agricultural area in Gyöngyösoroszi in the Toka Valley was chosen for the scaled-up experiments and as a demonstration site for the innovative CCP technology. The technology is introduced from the planning phase, through technology implementation and monitoring to the evaluation of the results and verification of the technology demonstration.

2.2.1   Mine wastes and soils at the site

For the technological experiments, three typical subareas were chosen: (i) a highly weathered mine waste from the *Bányabérc* pile, beyond the *Toka Creek* watershed; (ii) the Gyöngyösoroszi mining site in the Toka Valley: a relatively intact waste rock heap; and (iii) an agricultural site exposed to regular floods downstream the mine and the village. The soil of the village hobby gardens was heavily contaminated with metals due to highly contaminated deposited sediment deriving from the abandoned mine area (Figure 8.5).

The catchment area of the *Toka Creek* and the neighboring other watersheds have mine wastes of different ages, even from the 1800s. Most of the waste originates from the years 1950–1960. These uncovered waste deposits are exposed to intensive weathering, resulting in oxidation, acidification, and leaching. Two waste heaps were chosen for remediation experiments, both were ranked as highly risky during the risk assessment of the waste heaps in the Toka Valley (Vaszita & Gruiz, 2010).

1   A large mine waste heap near the Bányabérc shaft (BB) consists of heavily weathered mine waste that has a pH of 4.6. The highly acidic, metal containing leachate water from the BB area has affected the water quality of the nearby creek and the neighboring

Figure 8.5 Demonstration sites of the CCP technology – from north to south: (1) weathered mine waste; (2) non-weathered waste rock; (3) regularly flooded agricultural soil.

watershed (Tamás, 2007). After the removal and containment of the bulk majority of the highly risky BB heap, the residual soil (with lower level contamination) is to be treated by combined chemical and phytostabilization.

2   The huge waste rock heap under the main entrance of the mine (ME), the disposal site of the non-weathered waste rock resulting from excavation of the main mine adit. The material has a pH of 7.2 and highly a variable metal content. As it is usual in mine estab-lishment, mine buildings and the mine-water treatment plant were built on the top of the heap of waste rock. Therefore its removal was not an option.

The waste materials contain As, Cd, Pb, and Zn at high concentration however, As and Pb are dominantly in immobile forms. In the ME waste material 11.2% of the total Cd (22.7 mg/kg) and 4.6% of the total Zn (4631 mg/kg) is acetate-extractable. The BB waste has less metal content (4.9 mg/kg Cd and 1176 mg/kg Zn) than the ME, but both metals occur in mobile forms: 21.8% of Cd is extractable by acetate and 6.9% by water, and 16.8% Zn is extractable by acetate and 4.8% by water (Feigl *et al.*, 2009a).

3   Regular flood events result in contaminated sediment disposal on the agricultural soil near the creek. Therefore, the concentration of metals in these soils shows a gradient perpendicular to the flow direction: the highest total metal concentration was mea-sured close to the creek which decreases with increasing distance from the creek (Horváth & Gruiz, 1996; Tolner *et al.*, 2010) (Figure 8.6). A special characteristic of the flooded area is that the mobile metal forms do not follow this pattern due to the

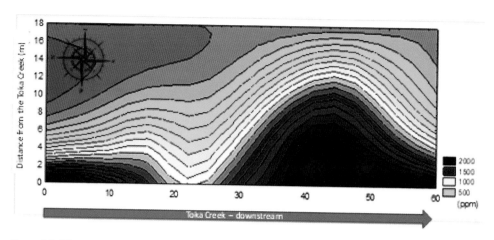

Figure 8.6 3D contour plot: concentration distribution of Zn in a hobby garden.

time-dependent mobilization from the disposed sediment. The risk assessment of the hobby gardens of village people showed that some vegetables, especially sorrel and rhubarb accumulate metals in large quantities, but most of the vegetables had higher metal contents than acceptable. Therefore the cultivation of non-accumulating berry-type plants was suggested to the owners of the gardens at flood risk (Horváth & Gruiz, 1996; Sipter *et al.*, 2008).

The agricultural soil involved in the technological experiments was typical polluted soil of the area with a pH of 5.4–7.6 and with a wide range metal content: total metal concentrations measured in aqua regia extracts were 57–330 mg/kg As, 4.1–11.1 mg/kg Cd, 227–1589 mg/kg Pb, and 871–1863 mg/kg Zn. However, it can be concluded from the acetate and water-extractable metal contents that Zn and Cd are the most mobile contaminants (acetate-extractable Cd and Zn are 11.3–34.4% and 8.9–24.7%, water-extractable Cd and Zn are a maximum 15.5% and 10.8%, respectively), therefore their immobilization has priority in agricultural soils (Feigl *et al.*, 2009a). Cadmium is characterized by high acute and chronic toxicity, carcinogenicity, and developmental toxicity, while Zn is ecotoxic. Chemical stabilization combined with phytostabilization significantly reduces metal mobility from soil and the risk of metal release into the aquatic habitat.

2.2.2   Fly ash as chemical stabilizer

The chemical stabilizers selected for the remediation of the Gyöngyösoroszi agricultural soils and mine wastes were evaluated in the first step of the technological experiments in laboratory microcosms (Feigl *et al.*, 2007, 2008). The action mechanism of fly ash is complex; alkalizing effect (provided that the fly ash has alkaline properties), and pozzolan activity are the most important ones, having an impact in the short term; clay mineral formation and metal binding into the molecular lattice in the long term.

Several successful applications proved the efficiency of fly ash as amendment for soil stabilization (reviews: Adriano *et al.*, 1980; Haering & Daniels, 1991; Asokan *et al.*, 2005).

Table 8.2 Characteristics of the fly ash applied in the studies.

Fly ash	pH	Total metal content (mg/kg)	Water-extractable metal content (mg/kg)
OA	12.6	As: 18.8, Cd: 0.87, Pb: 18.3, Zn: 223	As: <DL, Cd: <DL, Pb: 0.09, Zn: 0.43
OB	9.7	As: 20.4, Cd: 1.13, Pb: 19.1, Zn: 102	As: 0.61, Cd: <DL, Pb: 0.04, Zn: <DL
T	6.8	As: 16.7, Cd: 0.48, Cu: 53.2, Pb: 9.39, Zn: 56.4	As: <DL, Cd: <DL, Cu: <DL; Pb: <DL
V	6.4	As: 33.6, Cd: 1.02, Cu: 32.7, Pb: 35.5, Zn: 303	As: 0.107, Cd, Cu and Pb: <DL, Zn: 0.235

One of the first and most successful results for long-term metal immobilization by fly ash was obtained by Vangronsveld *et al.* (1995a,b) in the remediation of the metal-polluted area of a former zinc smelter in Belgium by beringite. Beringite is a modified aluminosilicate originating from fluidized bed combustion of coal refuse (low-quality coal mixed with rock). The soil near the abandoned zinc smelter contained up to 13,250 mg/kg Zn, making the soil barren. After treatment with beringite a healthy and sustainable vegetation cover was established (Vangronsveld *et al.*, 1999; Bouwman *et al.*, 2001). Twelve years after treatment, the Nematode fauna was still improving (Bouwman & Vangronsveld, 2004), metal availability remained low, and no increase in phytotoxicity was observed.

The two crucial factors determining the metal immobilizing potential of ash are the origin and the composition (Vangronsveld *et al.*, 1996). Besides sorption and crystal growth, the alkalizing effect also reportedly contributes to the immobilizing effect of fly ash. In the Gyöngyösoroszi experiments both composition and alkalinity of the tested alkaline and non-alkaline fly ash were studied.

Two alkaline fly ash (OA and OB) originating from lignite combustion in the Oroszlány, Hungary, power plant were applied for immobilizing metals in the Gyöngyösoroszi soils and wastes. The studied non-alkaline fly ash (T and V) originate from the burning of the Tatabánya lignite (T) and the other from burning the lignite of Visonta (V). Characteristics of the fly ash are summarized in Table 8.2.

2.2.3 Plants for phytostabilization

The plant types applied in the technological experiments were chosen based on successful preliminary experiments and pilot applications of the authors or based on published case studies. A grass mixture, industrial plants (two *Sorghum* species), and an agricultural cultivar (maize) were selected for phytostabilization coupled to chemical stabilization. Plants beneficial for phytostabilization should be tolerant to site conditions (e.g. low pH, high metal concentration, salinity), grow quickly and produce sufficient biomass to provide complete coverage of the soil. Plants should have a relatively dense root system for stabilization and should not accumulate metals in above-ground plant tissues that could be consumed by humans or animals. Plant roots may provide some additional surface area for metal precipitation or adsorption to occur (Berti *et al.*, 1998); it plays a most important role in marshlands and other aquatic habitats.

The grass mixture (pasture:meadow = 1:2, consisting of 20.0% *Festuca arundinacea*, 20.0% *Lolium perenne*, 18.3% *Bromus inermis*, 15.0% *Dactylis glomerata*, 13.3% *Festuca rubra*, 5.0% *Phalaris/Baldingera arundinacea*, 3.3% *Agropyron pectinatum*, 3.3% *Festuca pratensis*, 1.6% *Phleum pratensis*) was previously used successfully for covering red mud reservoirs in Hungary (Murányi, 2008). The mixture contained species with various needs of light, temperature, and water, therefore there is a higher chance in successful vegetation. Grasses create a closed vegetation cover, which hinders wind and water erosion – a main goal of phytostabilization. Furthermore, they do not accumulate metals in the parts above soil.

Broom corn (*Sorghum vulgare* Pers. var. Technikum) and sudan grass (*Sorghum sudanense*, strain: zöldözön) are fast growing plants, the latter with a possible applicability as an energy crop (Liu & Lin, 2009). Their roots can thoroughly net the soil close to the surface, they are drought tolerant and live together with arbuscular mycorrhiza fungi therefore they are frequently used for phytoremediation.

Maize was chosen to investigate the risks of metal uptake from the stabilized soil with a highly accumulative plant to predict maximal risk from measured data. Anton and Máthé-Gáspár (2005) and Máthé-Gáspár and Anton (2005) rated maize a metal accumulating and metal-tolerant plant therefore it was selected to represent a worse case situation in the Gyöngyösoroszi agricultural area.

2.3 Scaled-up studies

To test and evaluate the effect of combined chemical stabilizers and plants in scaling-up, the technology was carried out in three steps (Figure 8.7). In the microcosm study (step 1) the effect of chemical stabilizers was tested. The efficiency of metal immobilization and retention was characterized by measuring the leachability of toxic metals. Open-air lysimeters (step 2) were used to follow the fate of the water in the soil profile, to measure the effect of stabilizing agents mixed to the soil or installed as a reactive barrier, and to observe the

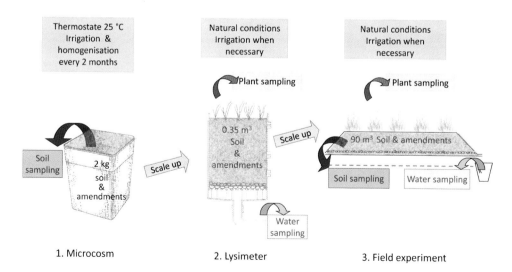

Figure 8.7 Scaled-up technological experiments for combined chemical and phytostabilization.

vegetation planted or grown spontaneously on the soil surface. Field experiments (step 3) were carried out on typical soils and wastes of the area: mine waste rock, highly weathered mine waste, and agricultural soil polluted by contaminated sediment through floods.

2.3.1 Lab-scale microcosm studies

For the evaluation of chemical stabilizers, the simplest and cheapest methodology is the application of laboratory microcosms (Gruiz *et al.*, 2015). Before selecting the best stabilizer for the Gyöngyösoroszi soils and mine wastes, several amendments were tested, such as alkaline and non-alkaline fly ash from different origins, lime, various waste materials such as (i) red mud; (ii) Fe-Mn-hydroxide from water treatment residuals; and (iii) raw phosphate, alginate, lignite, and a mixture of some additives. The microcosms composed of 1–2 kg soil mixed with different stabilizers at 1–10% in weight and were incubated at 25°C for 2–3 years. To monitor the microcosms, an integrated methodology was developed that included physical-chemical, biological and ecotoxicological methods for the assessment of metal mobility and availability in soil after treatment (Feigl *et al.*, 2007, 2008, 2009a,b,c, 2012). Advantages of these small-scale microcosms are the almost non-limited number of tests and the applicability of vegetation.

2.3.2 Field lysimeter studies

Lysimeters were used to examine the stabilization processes under natural conditions. Various processes can be studied in lysimeters such as the formation of acid mine drainage (Sand *et al.*, 2007), phytoextraction (Wenzel *et al.*, 2003), phytodegradation (Saison *et al.*, 2004), stabilization (Lidelöw *et al.*, 2007), as well as combined chemical and phytostabilization (Pathan *et al.*, 2003; Mench *et al.*, 2003; Ruttens *et al.*, 2006a; Kumpiene *et al.*, 2007). Lysimeter studies compared to field experiments have several advantages: (i) the number of parallels and combinations is not much restricted by space and costs; (ii) the construction allows pore water collection and (iii) soil profile dependent studies; and (iv) while plant growth and vegetation quality can also be tested. In the Gyöngyösoroszi case, ten 0.34-m³ lysimeters were constructed from concrete cylinders on the surface to study the stabilizing effect of fly ash on mine wastes and agricultural soil (Figures 8.8 and 8.9). Fly ash was applied

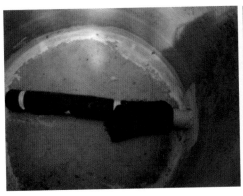

Figure 8.8 Leachate collecting drainage: collector pipe and coarse material on its top. The waste/soil profile was established above the coarse layer.

Figure 8.9 Field lysimeters in operation.

also as a separate layer in the lysimeter under the contaminated soil simulating a reactive barrier (Feigl *et al.*, 2008, 2010a). The experiments were followed for 3 years. The monitoring focused on the changes of the metal content and transport, covering leachate quantity and quality, soil quality, and plant growth and metal uptake.

2.3.3   Field demonstration

The suitability of combined chemical and phytostabilization was demonstrated in field experiments on three different sites in the Gyöngyösoroszi mining area. The field demonstrations were the last step of the scaled-up technological experiments before routine application.

From the heavily weathered mine waste of *Bányabérc* (BB) three field plots of 6×15 m were constructed, each containing 54 m³ of waste material. The plots were isolated from the underlying ground by a plastic foil, overlain by a 5-cm thick andesite gravel drainage layer (Figure 8.10). The water leaking through the soil profile was drained by the gravel layer and collected into subsurface vessels. This arrangement allowed the collection of an average water sample leaking through the highly heterogeneous waste material and made possible the assessment of risks connected to the transport of toxic metals by the infiltrated precipitation.

The BB waste was treated with the mixture of fly ash T and V (2.5% in weight each) and lime (1%) (1st plot), with fly ash T alone (5%) (2nd plot), and the 3rd plot was left untreated as a control. A mixture of *Sorghum sudanense* and *Sorghum vulgare* grass seed was sown on the plots to contribute to the stabilization process by erosion and runoff control, water infiltration reduction, and metals translocation (Feigl *et al.*, 2009b, 2010b). To reduce the

Figure 8.10 Construction of the field plots from weathered mine waste. Left: construction of the drainage system. Right: untreated control plot in front, plots treated with fly ash (darker) in the background.

Figure 8.11 Chemical stabilizers applied to field plots. Left: fly ash. Right: steel shots.

elevated concentration of arsenic in the drainage water from the plots treated with fly ash and lime, steel shots were added to the soil based on published cases (Bleeker *et al.*, 2002; Kumpiene *et al.*, 2006; Ruttens *et al.*, 2006b); 5 kg/m² of steel shot was evenly spread on the surface of the plots and mixed into the upper 20 cm of the plot material in the second year. Figure 8.11 shows the soil amendments of fly ash and steel shot.

The ME waste under the mine entrance was excavated straight to the experiment and 1.5×3-m plots constructed from the rock-like waste material. One part of the plots was treated with V fly ash and lime, the other with OA fly ash and lime.

Two other 3.3×3.3-m plots were established on top of the waste heap where the aim was to stabilize the top layer of ME. The top layer showed advanced weathering and soil-like appearance. Both of these plots were treated with T fly ash and lime. One plot was located in a sunny place, the other one under trees in the shadow to observe plant growth under different conditions. All ME plots were treated with steel shot in the second year. Grass mixture was sown on all plots (Klebercz, 2009).

In the agricultural area a 20×60-m field plot was constructed in a regularly flooded hobby garden. Half of the area was treated with 5% T fly ash, the other half was used as an untreated control. Three plant species were sown: Zea mays, *Sorghum vulgare*, *Sorghum sudanense*, and natural vegetation (invasive weed) was applied as plant control (Feigl *et al.*, 2009c).

All the field demonstrations with BB and ME waste and agricultural soil were monitored for 2 or 3 years. During this time period, the amount of precipitation, irrigation water and drainage water (in the case of BB plots) were recorded on site on a daily basis during the vegetation period. Samples from the drainage water, the solid waste material and the plants were regularly taken and analyzed.

2.3.4 Monitoring using integrated methodology

For monitoring the stabilization experiments, an integrated methodology (see also Soil Testing Triad in Gruiz, 2016) was developed and applied during all three steps of the scaled-up experiments. The integrated methodology means the combination of physicochemical analysis with biological and ecotoxicity tests (Figure 8.12). The integrated evaluation of the

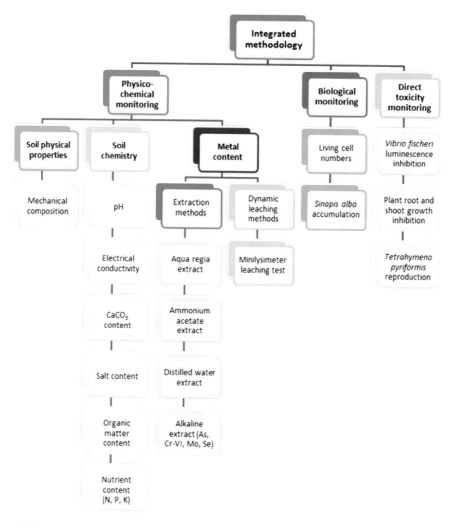

Figure 8.12 Integrated methodology for the monitoring of chemical stabilization in soil microcosms.

monitoring results provides a realistic picture on the risk, and its nature (Gruiz, 2005; Gruiz *et al.*, 2009; Gruiz, 2016).

Monitoring by chemical analysis included the application of soil extractants with growing acidity: distilled water, water solution of ammonium-acetate (pH = 4.5), and aqua regia. The metal content of the different extracts was determined by IPC-AES: Inductive Plasma Coupled Atomic Emission Spectrometry. Their aim was to characterize the extractability or the "chemical availability" of the metals (Feigl *et al.*, 2009a). Besides the metal content and mobility, other soil characteristics such as pH, EC, texture (according to Arany, 2010), organic matter content, ammonium-acetate extractable P and K were also determined. Metal content of the drainage water and plants was also measured by ICP-AES (Feigl *et al.*, 2010b).

The actual risk of the soil contaminated with a dynamically changing mixture of various metal species can be better characterized by direct toxicity assessment (DTA). DTA methodology uses soil-living test organisms and observes their response both in time and in comparison with non-contaminated soil (Gruiz, 2005, 2016). Generic soil activity was characterized by soil-living microbial cell numbers (e.g. aerobic heterotrophic cells). The soil and leachate toxicity was measured by test organisms at three trophic levels: bacteria (*Aliivibrio fischeri* luminescence inhibition test), plant (*Sinapis alba* root and shoot growth inhibition test) and animal (*Tetrahymena pyriformis* reproduction inhibition test). A rapid (5-day) plant accumulation test using S. alba was developed by the author to obtain direct information on the stabilizer from the point of view of the plants, used for the phytostabilization process (Feigl *et al.*, 2010b) to avoid additional loading of the food chain.

The results of chemical analysis and biological and toxicity assessment were evaluated together and a score system which was developed to characterize the soil stabilization options (Feigl *et al.*, 2010b). Complex verification was applied for the combined chemical and phytostabilization including (i) material balance; (ii) quantitative characterization of the environmental risk before, during, and after the remediation; (iii) cost efficiency assessment; and (iv) SWOT analysis (Gruiz *et al.*, 2008).

2.4   Results of the combined chemical and phytostabilization

The results of the microcosms and the lysimeters made it possible to select the best-fitting additives (fly ash, lime, and steel shot) to the mine wastes and soils to remediate them. The results of the field plots provide the foundation of the first field application of the combined chemical and phytostabilization technology in the Toka Valley for those areas which cannot be cleaned up by removing the dispersed waste or which remained after point source removal. The technology verification based on the field applications demonstrated and proved its future applicability, its technological, environmental, and socioeconomic efficiency.

2.4.1   Results of the pre-experiments

The alkaline OA fly ash showed the best immobilizing effect in microcosms over the long term (2 years). One single treatment with 5% OA fly ash reduced the acetate-extractable metal content of the soil by 45–49% and the water-soluble part by more than 99%. It also reduced soil toxicity for both bacteria and plants and decreased the bioavailable metal proportion by 70%. The pH-neutral fly ash types were not able to shift the pH of the acidic waste materials to the alkaline domain, therefore lime was mixed in. The addition of lime enhanced the immobilization capacity of the neutral fly ash types and their mixture was as effective as the alkaline fly ash (Feigl et al., 2009a).

In field lysimeters the stabilization processes were examined under natural conditions. The short-term results (2 months) from drain water analysis showed that both the OA alkaline fly ash and the non-alkaline V fly ash caused a 99% decrease in the concentration of Zn and Cd compared to the untreated mine waste, and the toxicity also decreased due to treatment (Feigl et al., 2010a). This efficiency remained the same over the 3 years (Feigl, unpublished results).

2.4.2   Results for highly acidic, sulfuric, weathered mine waste

The weathered acidic mine waste (BB) was treated with fly ash in three versions: (i) T fly ash (FA); (ii) Mixture of T and V fly ash and lime (FAL); (iii) Steel shot (SS) in the second year. The experiments were monitored for 3 years.

In the drain water from the untreated waste the concentration of Cd, Cu, Pb, and Zn was much above the site specific Maximum Effect Based Quality Criteria ($EBQC_{max}$) set by Gruiz et al. (2006) for local surface waters (Table 8.3). During the 3 years of monitoring, the leachable quantity of metals decreased to the third of the untreated plot, but it was still more than 100 times higher than the $EBQC_{max}$ in the case of Cd and Zn. The FA treatment decreased the mobile metal proportion in the waste material, but not sufficiently, and the mobile Pb content has increased due to the effect of the treatment. However, additional lime in FAL treatments reduced the metal contents sufficiently, below the $EBQC_{max}$ by the second year, except As. The concentration of As in the drain water increased with increasing pH. Steel shots managed to compensate for this increase under pH 5, but not above 7. The As concentration remained three times higher than $EBQC_{max}$ (Feigl et al., 2010c), contrary to the laboratory microcosm results, which showed that a 48–78% decrease in leachable As and Pb was sufficient to meet the criteria. The probable reason for this difference and the lower efficiency in the field may be that the steel shot granules have not been mixed as homogeneously into soil as in the microcosms (Bertalan, 2009).

Table 8.3 Metal content of the drain water from the BB mine waste field plots (Feigl et al., 2010c).

Treatment	Year	Metal content					pH
		As µg/L	Cd µg/L	Cu µg/L	Pb µg/L	Zn µg/L	
Untreated	2007	<1.8	441	1510	17.0	89,079	2.91
Untreated	2008	<1.8	180	714	16.2	37,286	2.88
Untreated	2009	11.2	157	433	12.5	24,126	3.25
FA	2007	<1.8	138	**88.7**	131	30,380	4.12
FA+SS	2008	<1.8	124	**77.4**	192	26,009	4.13
FA+SS	2009	**4.23**	111	**85.2**	184	17,111	4.39
FAL	2007	20.7	2.30	**14.1**	**1.96**	226	7.23
FAL+SS	2008	20.9	**0.42**	**11.7**	**<1.50**	**48.8**	7.77
FAL+SS	2009	33.3	**0.120**	**9.85**	**<1.50**	**29.3**	7.85
$EBQC_{max}$		10.0	1.0	200	10.0	100	

Values below the $EBQC_{max}$ are marked in bold face font.

As expected from the drain water results, the water- and acetate-extractable metal content of the solid waste has significantly decreased. In the FA-treated plot the amount of water-extractable Zn, Cd, and Pb decreased by 81–95%, in the FAL-treated plot by more than 99% (Feigl *et al.*, 2010c).

Toxicity results confirmed the effectiveness of the stabilizing treatments: drain water samples changed from toxic to non-toxic as measured both by *Aliivibrio fischeri* bacterial test and *Sinapis alba* plant growth test. The luminescence inhibition of *A. fischeri* decreased to half in the first year and showed no inhibition by the third year. The roots of the *S. alba* grown in treated waste were five times longer than in the untreated one. The inhibition of the reproduction of *T. pyriformis* (a single-cell animal) decreased in the first two years of treatment from 100% (inhibition of the untreated waste), to 40–15% (inhibition is quantified compared to a good quality natural soil without an inhibitory effect). The microbial cell numbers and activities in the mine waste increased ten times in the FA-treated plot and 100 times in the FAL-treated plot compared to the non-treated one (Feigl *et al.*, 2010c).

The untreated field plots remained bare, plants could not grow at all, justifying the importance of the application of chemical stabilizers before planting (Figures 8.13 and 8.14).

Figure 8.13 Field plots in August 2007, in the first year of the field study. Left: untreated control plot with plants unable to grow. Right: fly ash and lime treated plot with rich grass cover.

Figure 8.14 *Sorghum sudanense* and *Sorghum vulgare* grown on the fly ash and lime treated plot at the end of August 2007. Naturally grown vegetation appears at the edge of the plot.

Three plant species – grass, maize, and sorghum – did not show significant differences in the rate of accumulation, all of them fulfilled the criteria not to deliver metals into the food chain. Plants grown in the FAL field plots all met the Hungarian Quality Criteria for fodder, which is 2 mg/kg for As, 1 mg/kg for Cd, 100 mg/kg for Cu, 10 mg/kg for Pb, and 100 mg/kg for Zn (in dry weight). Plants from the FA-treated plot had some higher metal contents compared to plants from the FAL plots, but still below the criteria (Feigl *et al.*, 2010c).

2.4.3   Results for non-weathered waste rock

The freshly excavated (subsurface) waste rock (ME) and the soil-like surface material formed from the same waste rock (SME) were studied in small-size field plots. Stabilization with a mixture of fly ash (V, T, and O) and lime as well as steel shots was achieved in the second year.

Due to the treatments the acetate- and water-extractable Cd and Zn contents decreased by 99%, except for the soil-like surface (SME), which had originally higher total and mobile metal concentration: here the acetate-extractable metal proportion decreased "only" by 89–93% (Table 8.4). The Pb content of the acetate extract decreased by 99% and the distilled water extract by 83–89%. The As concentration decreased in the alkaline extract by 60% in the ME waste, but it increased by 20% in SME. The pH increased in both cases from 6.0 to 7.8–8.8. The steel shot addition did not show significant effects in this case.

The toxicity/inhibitory effect of the wastes decreased significantly: to 30% in the bioluminescence bioassay, to 30–15% based on *S. alba* root elongation, and to 50% according to the shoot growth tests.

The metal content of the grasses planted on the small plots shows differences due to the scale of weathering of the mine waste: as predicted, it was higher for SME than for ME.

Plants grown on treated ME waste accumulated Cd, Pb, and As at an acceptable rate, the metal contents were below or close to the Hungarian Quality Criteria (HQC) for fodder, except for As and Pb in one plot, which were higher in the first year, but decreased in the

Table 8.4 Metal contents of the stabilized ME mine waste in the second year – extracted by acetate and water.

	Non-treated ME waste	Fly ash V + lime treated ME waste	Fly ash O + lime treated ME waste	Non-treated, weathered SME	Fly ash T + lime treated weathered SME
	mg/kg	mg/kg	mg/kg	mg/kg	mg/kg
Total Cd[1]	7.81	7.81	7.81	27.6	27.6
Acetate-extractable Cd	1.21	0.158	0.097	5.71	63.0
Water-extractable Cd	0.237	<0.004	<0.004	0.311	<0.004
Total Zn[1]	1295	1295	1295	4814	4814
Acetate-extractable Zn	161	0.009	0.015	849	0.612
Water-extractable Zn	25.7	0.126	0.265	47.5	0.302

[1]Total metal content measured after aqua regia digestion.

second. The Zn concentration decreased from the original 100 times higher values to 2–3.5 times higher values than the HQC.

Plants grown on treated SME had three to five times higher Cd, Zn, and As concentrations than the HQC, and the Pb content of plants was ten times the HQC in the first year. However, by the second year all metal concentrations reached similar levels as in the plants grown in ME except for Pb which remained two to four times higher than the HQC.

2.4.4 Results for the CCP-treated agricultural soil

The agricultural soil at the Gyöngyösoroszi village contaminated with river sediment containing metals was treated with the non-alkaline T fly ash. One half of the area was left untreated and used as a control for the field study.

The acetate-extractable Zn, Cd and Pb content of the soil decreased by 80–82% due to the treatment, while the water-extractable metal content decreased by 92% near the creek (0–5 m), an area exposed to frequent floods (Table 8.5) (Feigl et al., 2010a). At 5–10 m distance from the creek the metal content decreased by 50% in the soil treated by fly ash, due to the lower initial metal content (Feigl et al., 2009c).

The biological activity of the soil shows that the soil microbiota is active despite the high toxicant concentrations. In spite of the fact that microorganisms in flooded areas have already been adapted to long-term toxic metal contamination, the fly ash treatment significantly increased cell numbers and activities. Bioluminescence and plant toxicity tests showed a 20–30% improvement, but many samples were still in the moderately toxic category (the test organisms used are rather sensitive to metals, cf. field-grown plants). (Feigl et al., 2010a).

Metal content of field-grown plants (maize and two sorghum species) has been reduced by the fly ash treatment to below the Hungarian Quality Criteria both for food and fodder (Table 8.6, Figures 8.15 and 8.16). In the two Sorghum species the Zn and Cd concentration decreased by 70–90%; in the Zea mays roots and leaves, the reduction reached only 35–80% compared to the plants grown in untreated soil. The results proved that the non-alkaline T fly ash was also able to reduce metal mobility in a pH-neutral soil (Feigl et al., 2009c).

Table 8.5 Decrease of metal content in acetate and water extracts of the stabilized agricultural soil.

Extraction method	Non-treated mg/kg	Fly ash-treated mg/kg	Decrease[1] %
Total Cd[2]	5.23		
Acetate-extractable Cd	1.54	0.275	82
Water-extractable Cd	0.051	<0.004	92
Total Zn[2]	1102		
Acetate-extractable Zn	237	47.7	80
Water-extractable Zn	4.11	0.315	92

[1]Decrease by treatment compared to non-treated control.
[2]Total metal content measured after aqua regia digestion.

Table 8.6 Accumulated metal quantities in plants grown in the agricultural study area.

Plant	Treatment	As	Cd	Pb	Zn
Metal content of plant leaves (mg/kg)					
Sorghum sudanense	Non-treated	0.78	3.00	3.32	348
	Fly ash-treated	0.45	0.90	2.20	104
Sorghum vulgare	Non-treated	0.51	6.63	6.25	503
	Fly ash-treated	0.53	0.72	1.86	108
Zea mays	Non-treated	4.66	5.29	25.42	665
	Fly ash-treated	1.00	1.59	5.62	301
HQC for animal fodder		2.0	1.0	10.0	
HQC for fresh vegetable		2.0	0.5	3.0	100
Decrease compared to non-treated plants (%)					
Sorghum sudanense		42	70	34	70
Sorghum vulgare		−4	89	70	79
Zea mays		79	70	78	55

[1]HQC= Hungarian Quality Criteria for animal fodder and fresh vegetables, calculated to dry biomass based on the FVM Decree 44/2003. (IV.26.) and the EüM Decree No. 17/1999. (VI. 16.).

Left: untreated plot, right: fly ash-treated plot shortly after emergence.

Figure 8.15 Sorghum sudanense grown on the agricultural field plots in 2007.

Figure 8.16 Difference at the border of the untreated and treated maize plots.

Table 8.7 Efficacy of treatments evaluated by a score system.

Treatment	Total score of 210 points	Percentage of actual score/2.1
Untreated	58.5	28
FA	109.3	52
FAL	151.0	72
FA+SS	128.5	61
FAL+SS	160.5	76

50%: points for chemical-analytical tests divided by their maximum (105 points),
30%: points for biological tests divided by their maximum (80 points),
20%: points for ecotoxicological tests divided by their maximum (25 points).

Table 8.8 Comparison of technological alternatives based on risk score and costs.

	No treatment	Dig & dump	Dig & dump on site	Soil washing	Combined chemical and phytostabilization
Risk score	1291	192	110	149	44
Specific cost (euro/ton, 2006)	3.4	91.7	12.1	52.1	2.4

2.4.5 Integrated technology evaluation and verification

The efficiency of the chemical stabilizers applied in the combined chemical and phytosta-bilization technology was evaluated based on a score system (Table 8.7). The score system included all the measured and available chemical, biological, and toxicological results. As an example, the results for the heavily weathered, acidic mine waste (BB) is shown here. FAL treatment proved to be better, the single FA treatment and the addition of SS improved the efficacy of both treatments by 5–10%. The best technology (FAL + SS) reached 75% of the maximum score, which means a 49% improvement compared to the initial (untreated) acidic mine waste (Feigl *et al.*, 2010b).

The complex verification system applied to the combined chemical and phytostabiliza-tion technology showed that CCP is the least risky option for the treatment of mine wastes and metal-contaminated agricultural soils in the Gyöngyösoroszi region compared to other alternatives. In terms of costs CCP is even better than the "no treatment" option, considering the monitoring required by law (Gruiz *et al.*, 2008) (Table 8.8).

2.5 Conclusions

The field demonstration of combined chemical and phytostabilization was implemented in the abandoned Gyöngyösoroszi mining area to demonstrate and verify its applicability for the toxic metal-polluted soil, sediment, and mine waste dispersed diffusely at the site. The subsites for study purposes were selected to represent the most widespread types of contaminated waste rocks, soils and sediment-loaded agricultural soils. The waste/soil contaminated with As, Cd, Cu, Pb, and Zn was stabilized by various fly ash types and lime, and the stabilizing plants were maize, sorghum, and grass. The integrated monitoring included physicochemical, biological, and toxicological methods and provided data for the integrated evaluation of the results of both technology application and residual risk. The results confirmed that fly ash as a chemical stabilizer can reduce metal mobility and toxicity of the wastes and contaminated soils. As a consequence, a healthy vegetation can develop without posing any risk to humans and the ecosystem or the food chains.

In summary, the demonstration of the innovative CCP technology justified the results of the small-scale preliminary tests and proved to be technologically efficient. In addition to the strict technological aspects, the complex verification, including environmental and socioeconomic criteria, proved the efficiency and applicability of combined chemical and phytostabilization for soils contaminated with toxic metals and waste rocks of mining origin.

REFERENCES

Adriano, D.C., Page, A.L., Elseewi, A.A., Chang, A.C. & Straughan, I. (1980) Utilization and disposal of fly ash and other coal residues in terrestrial ecosystems: A review. *Journal of Environmental Quality*, 9, 333–344.

Adriano, D.C., Wenzel, W.W., Vangronsveld, J. & Bolan, N.S. (2004) Role of assisted natural remediation in environmental cleanup. *Geoderma*, 122, 121–142.

Anton, A. & Máthé-Gáspár, G. (2005) Factors affecting heavy metal uptake in plant selection for phytoremediation. *Zeitschrift für Naturforschung*, C 60, 244–246.

Arany (2010) *Recommended Methods for Soil Investigation.* Hungarian Ministry of Environmnet and Water. Available from: www.kvvm.hu/szakmai/karmentes/kiadvanyok/karmkezikk2/2-09.htm. [Accessed 16th July 2018].

Asokan, P., Saxena, M. & Asolekar, S.R. (2005) Coal combustion residues – Environmental implications and recycling potentials. *Resources, Conservation and Recycling*, 43, 239–262.

Baker, R.J., Van Leeuwen, J.H. & White, D.J. (2005) *Applications for Reuse of Lime Sludge From Water Softening.* Final Report for TR-535. Iowa Department of Transportation Highway Division and the Iowa Highway Research Board.

Bertalan, Z. (2009) *Stabilisation of Toxic Metal Contaminated Mine Waste With Fly Ash and Lime.* Master Thesis at BME ABET, Budapest, Hungary.

Berti, W.R., Cunningham, S.C. & Cooper, E.M. (1998) Case studies in the field – In-place inactivation and phytorestoration of Pb-contained sites. In: Vangronsveld, J. & Cunningham, S.C. (eds.) *Metal Contaminated Soils: In Situ Inactivation and Phytorestoration.* Springer, Berlin, Germany & Heidelberg, Germany. pp. 235–248.

BioMineWiki (2018) *Biohydrometallurgy Knowledge Base.* Available from: http://wiki.biomine. skelleftea.se. [Accessed 16th July 2018].

Bleeker, P.M., Assunçao, A.G.L., Teiga, P.M., de Koe, T. & Verkleij, J.A.C. (2002) Revegetation of the acidic, As contaminated Jales mine spoil tips using a combination of spoil amendments and tolerant grasses. *Science of the Total Environment*, 300, 1–13.

Boisson, J., Ruttens, A., Mench, A.M. & Vangronsveld, J. (1999) Evaluation of hydroxyapatite as a metal immobilizing soil additive for the remediation of polluted soils. Part 1. Influence of hydroxyapatite on metal exchangeability in soil, plant growth and plant metal accumulation. *Environmental Pollution*, 104, 225–233.

Bouwman, L., Bloem, J., Römkens, P., Boon, G. & Vangronsveld, J. (2001) Beneficial effects of the growth of metal tolerant grass on biological and chemical parameters in copper and zinc contaminated sandy soils. *Minerva Biotecnologica*, 13, 19–26.

Bouwman, L. & Vangronsveld, J. (2004) Rehabilitation of the nematode fauna in phytostabilised heavily zinc-contaminated, sandy soil. *Journal of Soils and Sediments*, 4, 17–23.

Chang, Y.T., Hsi, H.C., Hseu, Z.Y. & Jheng, S.-L. (2013) Chemical stabilization of cadmium in acidic soil using alkaline agronomic and industrial by-products. *Journal of Environmental Science and Health. Part A, Toxic/Hazardous Substances & Environmental Engineering*, 48(13), 1748–1756. doi:10.1080/10934529.2013.815571.

Christie, P., Easson, D.L., Picton, J.R. & Love, S.C.P. (2001) Agronomic value of alkaline-stabilized sewage biosolids for spring barley. *Journal of Agronomy*, 93, 144–151.

Diels, L., De Smet, M., Hooyberghs, L. & Corbisier, P. (1999) Heavy metals bioremediation of soil. *Molecular Biotechnology*, 12, 149–158.

Evanko, C.R. & Dzombak, D.A. (1997) Remediation of metals-contaminated soils and groundwater. *Technology Evaluation Report, TE-97-01*. Ground-Water Remediation Technologies Analysis Center, Pittsburgh, PA, USA.

Feigl, V., Anton, A., Fekete, F. & Gruiz, K. (2008) Combined chemical and phytostabilisation of metal polluted soil: From microcosms to field experiments. *Conference Proceedings CD of Consoil 2008, 3–6 July 2008*. Theme E, Milan, Italy. pp. 823–830.

Feigl, V., Anton, A. & Gruiz, K. (2009b) Combined chemical and phytostabilisation: Field application. *Land Contamination & Reclamation*, 17(3–4), 577–584.

Feigl, V., Anton, A. & Gruiz, K. (2010a) An innovative technology for metal polluted soil – Combined chemical and phytostabilisation. In: Sarsby, R.W. & Meggyes, T. (eds.) *Construction for a Sustainable Environment Proceedings of the International Conference of Construction for a Sustainable Environment, Vilnius, Lithuania, 1–4 July, 2008*. Taylor & Francis Group, London, UK. pp. 187–195.

Feigl, V., Anton, A., Uzinger, N. & Gruiz, K. (2012) Red mud as a chemical stabilizer for soil contaminated with toxic metals. *Water, Air & Soil Pollution*, 223(3), 1237–1247.

Feigl, V., Atkári, Á., Anton, A. & Gruiz, K. (2007) Chemical stabilisation combined with phytostabilisation applied to mine waste contaminated soils in Hungary. *Advanced Materials Research*, 20–21, 315–318.

Feigl, V., Gruiz, K. & Anton, A. (2010b) Remediation of metal ore mine waste using combined chemical- and phytostabilisation. *Periodica Polytechnica Chemical Engineering*, 54(2), 71–80.

Feigl, V., Gruiz, K. & Anton, A. (2010c) Combined chemical and phytostabilisation of an acidic mine waste: Long-term field experiment. *Conference Proceedings CD of Consoil 2010, 22–24 September 2010*. Consoil 2010 Posters A3–24, Salzburg, Austria.

Feigl, V., Uzinger, N. & Gruiz, K. (2009a) Chemical stabilisation of toxic metals in soil microcosms. *Land Contamination & Reclamation*, 17(3–4), 483–494.

Feigl, V., Uzinger, N., Gruiz, K. & Anton, A. (2009c) Reduction of abiotic stress in a metal polluted agricultural area by combined chemical and phytostabilisation. *Cereal Research Communications*, 37(Suppl), 465–468.

FRTR (2002) *Remediation Technologies Screening Matrix and Reference Guide*. Version 4.0. USA. Available from: www.frtr.gov/matrix2. [Accessed 18th July 2018].

Gadd, G.M. (2000) Bioremedial potential of microbial mechanisms of metal mobilization and immobilization. *Current Opinion in Biotechnology*, 11, 271–279.

Gadd, G.M. (2004) Microbial influence on metal mobility and application for bioremediation. *Geoderma*, 122, 109–119.

Gruiz, K. (2005) Biological tools for soil ecotoxicity evaluation: Soil testing triad and the interactive ecotoxicity tests for contaminated soil. In: Fava, F. & Canepa, P. (eds.) *Soil Remediation Series No 6*, INCA, Venice, Italy. pp. 45–70.

Gruiz, K. (2016) Integrated and efficient characterization of contaminated sites. In: Gruiz, K., Meggyes, T. & Fenyvesi, É. (eds.) *Engineering Tools for Environmental Risk Management. Volume 3. Site Assessment and Monitoring Tools*. CRC Press, Boca Raton, FL, USA. pp. 1–98.

Gruiz, K., Molnár, M. & Feigl, V. (2009) Measuring adverse effects of contaminated soil using interactive and dynamic test methods. *Land Contamination & Reclamation*, 17(3–4), 445–462.

Gruiz, K., Molnár, M., Feigl, V., Vaszita, E. & Klebercz, O. (2015) Microcosm models and technological experiments. In: Gruiz, K., Meggyes, T. & Fenyvesi, É. (eds.) *Engineering Tools for Environmental Risk Management. Volume 2. Environmental Toxicology.* CRC Press, Boca Raton, FL, USA. pp. 401–444.

Gruiz, K., Molnár, M. & Fenyvesi, É. (2008) Evaluation and verification of soil remediation. In: Kurladze, G.V. (ed) *Environmental Microbiology Research Trends.* Nova Science Publishers, Inc., New York, NY, USA. pp. 1–57.

Gruiz, K., Vaszita, E. & Siki, Z. (2006) Quantitative Risk Assessment as part of the GIS based Environmental Risk Management of diffuse pollution of mining origin. *Conference Proceedings of Difpolmine Conference, 12–14 December 2006,* Montpellier, France.

Haering, K.C. & Daniels, W.L. (1991) Fly ash: Characteristics and use in mine land reclamation – A literature review. *Virginia Coal. Energy Journal,* 3, 33–46.

Horváth, B. & Gruiz, K. (1996) Impact of metalliferous ore mining activity on the environment in Gyöngyösoroszi, Hungary. *Science of the Total Environment,* 184, 215–227.

Infohouse (1999) *New Angles of in situ Vitrification.* Initiatives online 6. Available from: http://infohouse.p2ric.org/ref/14/0_initiatives/init/fall99/newangl.htm. [Accessed 18th July 2018].

Klebercz, O. (2009) *Stabilisation of Toxic Metal Contaminated Soils.* Master Thesis at BME ABET, Budapest, Hungary.

Kumpiene, J., Lagerkvist, A. & Maurice, C. (2007) Stabilization of Pb- and Cu-contaminated soil using coal fly ash and peat. *Environmental Pollution,* 145, 365–373.

Kumpiene, J., Lagerkvist, A. & Maurice, C. (2008) Stabilization of As, Cr, Cu, Pb, and Zn in soil using amendments – A review. *Waste Management,* 28, 215–225.

Kumpiene, J., Ore, S., Renella, G., Mench, M., Lagerkvist, A. & Maurice, C. (2006) Assessment of zerovalent iron for stabilization of chromium, copper, and arsenic in soil. *Environmental Pollution,* 144, 62–69.

Lens, P., Grotenhuis, T., Malina, G. & Tabak, H. (2005) *Soil and Sediment Remediation. Mechanisms, Technologies and Applications.* IWA Publishing, London, UK & Seattle, WA, USA.

Lidelöw, S., Ragnvaldsson, D., Leffler, P., Tesfalidet, S. & Maurice, C. (2007) Field trials to assess the use of iron-bearing industrial by-products for stabilisation of chromate copper arsenate-contaminated soil. *Science of the Total Environment,* 387, 68–78.

Liu, S.Y. & Lin, C.Y. (2009) Development and perspective of promising energy plants for bioethanol production in Taiwan. *Renewable Energy,* 34(8), 1902–1907. doi:10.1016/j.renene.2008.12.018.

Lovely, D.R. & Coates, J.D. (1997) Bioremediation of metal contamination. *Current Opinion in Biotechnology,* 8, 285–589.

Máthé-Gáspár, G. & Anton, A. (2005) Phytoremediation study: Factors influencing heavy metal uptake of plants. *Acta Biologica Szeged,* 49, 69–70.

McBride, M.B., Simon, T., Tam, G. & Wharton, S. (2013) Lead and arsenic uptake by leafy vegetables grown on contaminated soils: Effects of mineral and organic amendments. *Water, Air, & Soil Pollution,* 2224(1), 1378. doi:10.1007/s11270-012-1378-z.

Mench, M., Bussiere, S., Boisson, J., Castaing, E., Vangronsveld, J., Ruttens, A., De Koe, T., Bleeker, P., Assuncao, A. & Manceau, A. (2003) Progress in remediation and revegetation of the barren Jales gold mine spoil after *in situ* treatments. *Plant and Soil,* 249, 187–202.

Molnár, M., Vaszita, E., Farkas, É., Ujaczki, É., Fekete-Kertész, I., Tolner, M., Klebercz, O., Kirchkeszner, C., Gruiz, K., Uzinger, N. & Feigl, V. (2016) Acidic sandy soil improvement with biochar – A microcosm study. *The Science of the Total Environment,* 563–564, 855–865. doi:10.1016/j.scitotenv.2016.01.091.

Mulligan, C.R., Yong, R.N. & Gibbs, B.F. (2001) Remediation technologies for metal contaminated soils and groundwater: An evaluation. *Engineering Geology,* 60, 193–207.

Murányi, A. (2008) Innovative decision tools for risk based environmental management. MOKKA Project report on the 3rd work phase by RISSAC, Budapest, Hungary.

NRC (1994) *Alternatives for Ground Water Cleanup.* National Academy Press, Washington, DC, USA.

Pathan, S.M., Aylmore, L.A.G. & Colmer, T.D. (2003) Soil properties and turf growth on a sandy soil amended with fly ash. *Plant and Soil*, 256, 103–114.

Rahman, M.K., Rehman, S. & Al-Amoudi, O.S.B. (2011) Literature review on cement kiln dust usage in soil and waste stabilization and experimental investigation. *International Journal of Recent Research and Applied Studies (IJRRAS)*, 7(1). Available from: www.arpapress.com/Volumes/Vol7Issue1/IJRRAS_7_1_12.pdf. [Accessed 18th July 2018].

Reddy, K.R. (2013) Electrokinetic remediation of soils at complex contaminated sites: Technology status, challenges, and opportunities. In: Manassero *et al.* (eds.) *Coupled Phenomena in Environmental Geotechnics*. Taylor & Francis Group, London, UK.

Reddy, K.R. & Cameselle, C. (2009) (eds.) *Electrochemical Remediation Technologies for Polluted Soils, Sediments and Groundwater*. Available from: www.scopus.com/inward/record.url?eid=2-s2.0-84889389978&partnerID=40&md5=44064164eb24877831fec0ba842c3202. [Accessed 18th July 2018].

Reddy, K.R., Parupudi, U.S., Devulapalli, S.N. & Xu, C.Y. (1997) Effects of soil composition on the removal of chromium by electrokinetics. *Journal of Hazardous Materials*, 55(1–3), 135–158. doi:10.1016/S0304-3894(97)00020-4.

Rumer, R.R. & Ryan, M.E. (1995) *Barrier Containment Technologies for Environmental Remediation Applications*. John Wiley & Sons, New York, NY, USA.

Ruttens, A., Boisson, J., Jonca, G. & Pottecher, G. (2006b) Rehabilitation of the La Combe du Saut site – Phytostabilisation: field experiments. *Proceedings of Difpolmine Conference, 12–14 December 2006, Montpellier, France*. Available from: www.ademe.fr/difpolmine/Difpolmine_RapportFinal/communication/14_posters/Difpolmine_Poster06_Field.pdf. [Accessed 18th July 2018].

Ruttens, A., Mench, M., Colpaert, J., Boisson, J., Carleer, R. & Vangronsveld, J. (2006a) Phytostabilization of a metal contaminated sandy soil. I: Influence of compost and/or inorganic metal immobilizing soil additives on phytotoxicity and plant availability of metals. *Environmental Pollution*, 144(2), 533–541.

Saison, C., Perrin-Ganier, C., Schiavon, M. & Morel, J.L. (2004) Efect of cropping and tillage on the dissipation of PAH contamination in soil. *Environmental Pollution*, 130, 275–285.

Sand, W., Józsa, P.G., Kovács, Z.M., Sasaran, N. & Schippers, A. (2007) Long-term evaluation of acid rock drainage mitigation measures in large lysimeters. *Journal of Geochemical Exploration*, 92(2–3), 205–211.

Schippers, A. & Sand, W. (1999) Bacterial leaching of metal sulfides proceeds by two indirect mechanisms via thiosulfate or via polysulfides and sulfur. *Applied Environmental Microbiology*, 65, 319–321.

Sipter, E., Rózsa, E., Gruiz, K., Tátrai, E. & Morvai, V. (2008) Site-specific risk assessment in contaminated vegetable gardens. *Chemosphere*, 71, 1301–1307.

Smith, L.A., Means, J.L., Chen, A., Alleman, B., Chapman, C.C., Tixier, J.S. Jr., Brauning, S.E., Gavaskar, A.R. & Royer, M.D. (1995) *Remedial Options for Metals-Contaminated Sites*. Lewis Publishers, Boca Raton, FL, USA.

Stephen, J.R. & Macnaughton, S.J. (1999) Developments in terrestrial bacterial remediation of metals. *Current Opinion in Biotechnology*, 10, 230–233.

Tabak, H.H., Lens, P., van Hullebusch, E.D. & Dejonghe, W. (2005) Developments in bioremediation of soils and sediments polluted with metals and radionuclides – 1. Microbial processes and mechanisms affecting bioremediation of metal contamination and influencing metal toxicity and transport. *Reviews in Environmental Science and Bio/Technology*, 4, 115–156.

Tamás, P. (2007) Summary on the research work done in the BÁNYAREM Project, session: Field experiments, evaluation of results. BÁNYAREM Project Study by Mecsek-Öko, Pécs.

Tolner, M., Nagy, G., Vaszita, E. & Gruiz, K. (2010) *in situ* delineation of point sources and high resolution mapping of polluted sites by X-ray Fluorescence measuring device. In: Sarsby, R.W. & Meggyes, T. (eds.) *Construction for a Sustainable Environment*. CRC Press/Balkema, Leiden, The Netherlands. pp. 237–244.

Ujaczki, É., Feigl, V., Farkas, É., Vaszita, E., Gruiz, K. & Molnár, M. (2016a) Red mud as acidic sandy soil ameliorant: A microcosm incubation study. *Journal of Chemical Technology and Biotechnology*, 91(6), 1596–1606. doi:10.1002/jctb.4898.

Ujaczki, É., Feigl, V., Molnár, M., Vaszita, E., Uzinger, N., Erdélyi, A. & Gruiz, K. (2016b) The potential application of red mud and soil mixture as additive to the surface layer of a landfill cover system. *Journal of Environmental Sciences*, 44, 189–196. doi:10.1016/j.jes.2015.12.014.

US EPA (1992) Handbook: Vitrification technologies for treatment of hazardous and radioactive waste. EPA 625/R-92-002.

US EPA (1993) Innovative site remediation Technology: Soil washing/soil flushing. American Academy of Environmental Engineers.

US EPA (1996) A citizen's guide to thermal desorption. Technology Fact Sheet.

US EPA (2006) In situ treatment technologies for metal contaminated soils. Engineering forum issue paper.

US EPA (2012) A citizen's guide to solidification and stabilization.

US EPA (2018) *Thermal Treatment: In situ – Overview*. Available from: www.clu-in.org/techfocus/default.focus/sec/Thermal_Treatment%3A_In_Situ/cat/Overview/ [Accessed 18th July 2018].

Vangronsveld, J., Colpaert, J.V. & Tichelen, K.K. (1996) Reclamation of a bare industrial area contaminated by non-ferrous metals: Physico-chemical and biological evaluation of the durability of soil treatment and revegetation. *Environmental Pollution*, 94, 131–140.

Vangronsveld, J., Ruttens, A. & Clijsters, H. (1999) The use of cyclonic ashes of fluidized bed burning of coal mine refuse for long-term immobilization of metals in soils. In: Sajwan, K.S., Alva, A.K. & Keefer, R.F. (eds.) *Biogeochemistry of Trace Elements in Coal and Coal Combustion Byproducts*. Springer, Boston, MA, USA. pp. 223–233.

Vangronsveld, J., Sterckx, J., Van Assche, F. & Clijsters, H. (1995b) Rehabilitation studies on an old non-ferrous waste dumping ground: Effects of revegetation and metal immobilization by beringite. *Journal of Geochemical Exploration*, 52, 221–229.

Vangronsveld, J., Van Assche, F. & Clijsters, H. (1995a) Reclamation of a bare industrial area contaminated by non-ferrous metals: *In situ* metal immobilization and revegetation. *Environmental Pollution*, 87, 51–59.

Vaszita, E. & Gruiz, K. (2010) Scoring based Risk Assessment in an abandoned base metal sulphide mining area. *Conference Proceedings CD of Consoil 2010, 22–24 September 2010*. Consoil 2010 Posters C3–23, Salzburg, Austria.

Vaszita, E. (2014) Environmental risk of mining. In: Gruiz, K., Meggyes, T. & Fenyvesi, É. (eds.) *Engineering Tools for Environmental Risk Management. Volume 1. Environmental Detereioration and Contamination*. CRC Press, Boca Raton, FL, USA. pp. 113–134.

Vidic, R.D. & Pohland, F.G. (1996) Treatment walls. Technology evaluation. Report TE-96-01. Ground-Water Remediation Technologies Analysis Center, Pittsburgh, PA, USA.

Wasner, J., Liebhard, P. & Eigner, H. (2001) Application of carbonation lime on high pH soils in the Pannonian region of Austria. *Zuckerindustrie*, 126(3), 194–201.

Wenzel, W.W., Unterbrunner, R., Sommer, P. & Sacco, P. (2003) Chelate-assisted phytoextraction using canola (Brassica napus L.) in outdoors pot and lysimeter experiments. *Plant and Soil*, 249, 83–96.

Xenidis, A., Stouraiti, C. & Papassiopi, N. (2010) Stabilization of Pb and As in soils by applying combined treatment with phosphates and ferrous iron. *Journal of Hazardous Materials*, 177, 929–937.

Zhuo, L., Li, H., Cheng, F., Shi, Y., Zhang, Q. & Shi, W. (2012) Co-remediation of cadmium-polluted soil using stainless steel slag and ammonium humate. *Environmental Science and Pollution Research*, 19(7), 2842–2848. doi:10.1007/s11356-012-0790-7.

Chapter 9

Electrochemical remediation for contaminated soils, sediments and groundwater

C. Cameselle[1] & K.R. Reddy[2]

[1]Department of Chemical Engineering, University of Vigo, University Campus, Building Fundicion, Vigo, Spain
[2]Department of Civil and Materials Engineering, University of Illinois at Chicago, Chicago, Illinois, USA

ABSTRACT

Electrokinetic remediation is an environmental technology specially designed for the removal or degradation of contaminants in soils, sediments, sludges, and even solid wastes. This technology relies upon the application of a low intensity electric field directly to the porous material to be treated. Under the effect of an electric field, the contaminants are mobilized and transported towards the electrodes anode and cathode. The two main transport mechanisms are electromigration and electroosmosis. Electromigration is the transport of ions in solution in the interstitial fluid towards the electrode of opposite charge. Electroosmosis is the net flux of water induced by the electric field that, under natural conditions, flows from anode to cathode transporting the contaminants in solution out of the solid material. Electrokinetic remediation has shown several successes in the removal of inorganic and organic contaminants from soil. Furthermore, this technology is able to remediate fine grained and low permeability soils where other technologies have failed. In electrokinetics, the removal of contaminants requires their mobilization and solubilization in the interstitial fluid. The solubilization of contaminants can be enhanced with the addition of facilitating agents (acids or bases, complexing agents, surfactants etc.) and the pH adjustment in the solid material. The facilitating agents can be transported into the solid material by electromigration or electroosmosis, and the pH in the solid material can be controlled by the electrochemical split of water combined with the addition of an acid or base. Field applications of electrokinetics on contaminated sites have proved that this technology can be used on a large scale.

1 INTRODUCTION

Contamination of soils, sediments and groundwater raises increasing public awareness due to the serious impact of contaminants on the environment and public health. The treatment and remediation of contaminated sites is a priority in the agendas of politicians, but remediation efficacy depends on innovative technologies developed by scientists, engineers, and technicians. Improper waste disposal practices and accidental spills have been the prime sources of contamination. Contaminants include a wide range of toxic pollutants such as heavy metals, radionuclides, and recalcitrant organic compounds. Contaminants often are in different mixtures which makes the remediation process more difficult.

Several technologies based on physicochemical, thermal, and biological processes have been developed during the last 20 years to remediate soils, sediments, and groundwater (Sharma & Reddy, 2004). However, they are often costly, energy intensive, ineffective, and may themselves create adverse environmental impacts when dealing with difficult subsurface and contaminant conditions. Remediation has been inefficient at numerous polluted sites due to low permeability and heterogeneity of the soil and/or contaminant mixtures (multiple contaminants or combinations of different contaminant types such as co-existing heavy metals and organic pollutants).

Electrochemical and electrokinetic remediation technologies have been studied and tested for the remediation of contaminated soils, sludge and sediments. Laboratory and field results have confirmed the capability of this technology to address difficult contaminated site conditions (low permeability soils, mixed contaminants, soil heterogeneities, etc.) where other technologies failed.

The objective of this chapter is to describe the basics and technology status of electrokinetic remediation for the restoration of polluted sites and outline future perspectives of this technology.

2   ELECTROCHEMICAL TECHNOLOGIES FOR REMEDIATION OF POLLUTED SOILS

Electrokinetic remediation is a powerful technology that can be used for the remediation of contaminated soils, sludge, sediments, wastes and other porous matrices. It is also called *electrokinetics, electroremediation,* or *electroreclamation* in literature. Electrokinetic technology is based on a low intensity electric field applied to the contaminated soil (Acar, 1993). The electric field mobilizes the contaminant chemical species that migrate under the effect of the electric field towards one of the electrodes. The key transport processes are *electromigration* (the movement of ions towards the opposite electrode), *electroosmosis* (the net flux of water in the interstitial fluid induced by the electric field), *and electrophoresis* (the movement of charged particles towards the opposite electrode). *Diffusion* (the movement of chemical species due to a concentration gradient) is usually insignificant compared to other transport mechanisms.

The electrokinetic process is designed as *in situ* soil decontamination, although it can also be applied *ex situ* in specific equipment. In field applications wells are first drilled in the contaminated site, then a series of electrodes (anodes and cathodes) are implemented. Figure 9.1

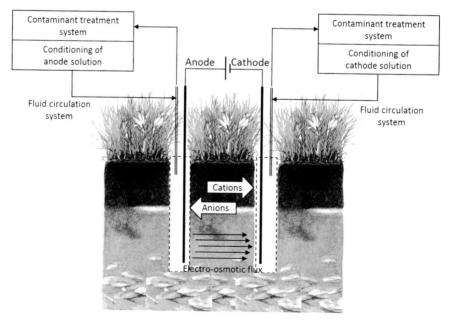

Figure 9.1   In situ application of electrokinetic treatment.

shows the arrangement especially designed for the circulation of the processing fluid. The contaminant species is dissolved in the processing fluid, migrates due to the effect of the electric field and is collected in the electrode wells. The processing fluid is pumped out to a treatment system engineered for each specific application. The electrokinetic technology is complex due to transfer and transformation processes, the buffer capacity, mineralogy, organic matter content, geochemistry, soil-contaminant interactions, and heterogeneity of soil. All these factors affect the efficacy of the process and the success of the remediation process (Reddy & Cameselle, 2009).

3 ELECTROCHEMICAL TRANSPORT

3.1 Electromigration

Electromigration is the movement of the dissolved ionic species that are present in the pore fluid toward the opposite electrode. Anions move toward the anode and cations migrate toward the cathode. Electromigration depends on the mobility of ionic species and the strength of the electric field. The electromigration flow of an ionic species can be expressed by Equations 1 and 2:

$$J_i = u_i^* c_i \nabla(-E) \tag{1}$$

$$u_i^* = u_i \tau n = \frac{D_i \tau n z_i F}{RT}, \tag{2}$$

where
- J_i = electromigration flow (mol/m²s)
- u_i = the ionic mobility (m/Vs)
- u_i^* = the effective ionic mobility, which includes the effect of
- τ = tortuosity (dimensionless) and
- n = porosity of the soil (dimensionless)
- c_i = the concentration (mol/m³)
- E = the electric field strength (V/m)
- D_i = the molecular diffusion coefficient (m²/s)
- z_i = the ionic valence, the charge of the chemical species (dimensionless)
- F = Faraday's constant (96487 C/mol)
- R = the universal gas constant (8.314 J/K•mol)
- T = the absolute temperature (K).

The extent of electromigration of a given ion depends on the conductivity and porosity of the soil, pH gradient, applied electric potential, initial concentration of the specific ion, and the presence of competitive ions. Electromigration is the major transport process for ionic metals, polar organic molecules, ionic micelles, and colloidal electrolytes.

3.2 Electroosmosis

Electroosmosis is the net flux of pore fluid in the soil induced by the electric field. Electroosmosis is able to remove all dissolved ionic and non-ionic species in the pore fluid. Generally, soil particle surfaces are charged, and counter-ions (positive ions or cations) concentrate within a diffuse double layer region adjacent to the particle surface (Figure 9.2). The electric field compels ions in local excess migrate in along a plane parallel to the particle surface towards the electrode charged oppositely. As they migrate, they transfer momentum to the surrounding fluid molecules via viscous forces producing an electroosmotic flow (Figure 9.3). Since soil particle surfaces are generally negatively charged, the counter-ions that neutralize the surface charge are cations. Under the effect of the electric field, those cations move toward the cathode, inducing an electroosmotic flow toward the cathode.

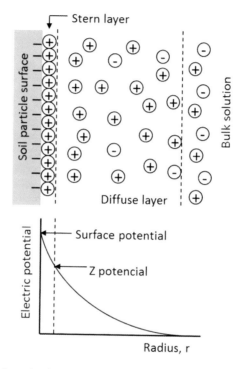

Figure 9.2 Double electric layer developed on a charged particle surface. Definition of the zeta potential.

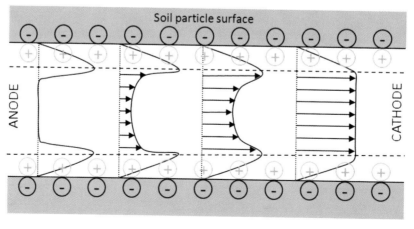

Figure 9.3 Development of the electroosmotic flow under the effect of an electric field.

The Helmholtz-Smoluchowski equation describes the electroosmotic flow (q_{eo}):

$$q_{eo} = nA \frac{D\zeta}{\eta} E_z \qquad (3)$$

where
-   q_{eo} = electroosmotic flow (m³/s)
-   n = porosity
-   A = cross-sectional area of the soil (m²)
-   D = dielectric constant of the fluid (C²/Nm²)
-   E_z = applied voltage gradient (V/m)
-   ζ = zeta potential (V)
-   η = fluid dynamic viscosity (kg/ms).

The electroosmotic flow depends on the dielectric constant and viscosity of the pore fluid as well as the surface charge of the solid matrix represented by the zeta potential (the electric potential at the interface between the fixed and mobile parts in the double layer). The zeta potential is a function of parameters including the type of clay minerals and ionic species that are present as well as the pH, ionic strength, and temperature. If cations and anions are evenly distributed, equal and opposite flow occurs, causing a zero net flow. However, when the momentum transferred to the fluid in one direction exceeds the momentum of the fluid traveling in the other direction, an electroosmotic flow results.

The pH changes induced in the soil by electrolysis reactions affect the zeta potential of soil particles, thereby affecting the electroosmotic flow. The low pH near the anode may be less than the PZC (point of zero charge) of the soil and the soil surfaces are positively charged, while high pH near the cathode may be higher than PZC of the soil, making the soil more negative. Electroosmotic flow may be reduced and even stopped as the soil is acidified near the anode. If a large part of the soil is acidified, the electroosmotic flow direction may even be reversed, from typical "anode-to-cathode," to "cathode-to-anode." This phenomenon is known as *electro-endosmosis*. Understanding of such electroosmotic flow variations is critical when remediating soils polluted by organic substances.

Electroosmosis is considered the dominant transport process for both organic and inorganic contaminants that are in dissolved, suspended, emulsified, or similar forms. Besides, electroosmotic flow through low-permeability regions can be significantly greater than a flow generated by an ordinary hydraulic gradient, thus electroosmotic flow is much more efficient than hydraulic gradient.

3.3   Electrophoresis

Electrophoresis (also known as cataphoresis) is the transport of charged particles of colloidal size and bound contaminants due to a low direct current or voltage gradient relative to the stationary pore fluid. Compared to ionic migration and electroosmosis, mass transport by electrophoresis is negligible in low permeability soil systems. However, it may become significant in soil suspension systems and it is the key mechanism for the transportation of biocolloids (e.g. those of bacterial origin) and micelles.

3.4   Diffusion

Diffusion is the motion of ionic and molecular contaminants from areas of higher to areas of lower concentration due to the concentration gradient or chemical kinetic activity. Estimates of ionic mobilities from the diffusion coefficients using the Nernst-Einstein-Townsend relation indicate that ionic mobility of a charged species is much higher (about 40 times) than the diffusion coefficient. Therefore, diffusive transport is often neglected.

4   ELECTROCHEMICAL REACTION AND TRANSFORMATION PROCESSES

The application of an electric field to a contaminated soil induces the movement of ions and water (the pore fluid) towards the electrodes. An electric field also triggers electrochemical reactions at the electrodes and in the soil.

The main electrochemical/electrokinetic reaction is the decomposition of water at the electrodes. Oxidation at the anode generates oxygen gas and hydrogen ions (H^+) and reduction at the cathode produces hydrogen gas and hydroxyl (OH^-) ions as shown by Equations 4 and 5.

Oxidation at the anode:

$$2H_2O \rightarrow O_{2(gas)} + 4H^+_{(aq)} + 4e^- \qquad E^0 = -1.229 \text{ V} \tag{4}$$

Reduction at the cathode:

$$4H_2O + 4e^- \rightarrow 2H_{2\,(gas)} + 4OH^-_{(aq)} \quad E^0 = -0.828 \text{ V} \tag{5}$$

Essentially, an acid is produced at the anode and base (alkali) at the cathode, therefore, pH decreases at the anode and increases at the cathode. The migration of H^+ from the anode and OH^- from the cathode in the soil leads to dynamic changes in soil pH. H^+ is about twice as mobile as OH^-, so protons dominate the system and an acid front moves across the soil until it meets the hydroxyl front in a zone near the cathode where the ions may recombine to generate water. Thus, the soil is divided in two zones with a sharp pH jump in between: a high-pH zone close to the cathode, and a low-pH zone near the anode (Ricart *et al.*, 1999). The actual soil pH values will depend on the extent of transport of H^+ and OH^- ions and the geochemical characteristics of the soil. The buffering capacity of the soil will exert a decisive influence on the transport of H^+ and OH^- and the pH in the soil.

The implications of these electrolysis reactions are enormous in the electrokinetic treatment since they affect the contaminant migration, the development of electroosmotic flow and the transformation and degradation of contaminants in the soil (Cameselle & Reddy, 2012). The most important geochemical reactions that must be considered include the following:

– Sorption – desorption reactions.
– Precipitation – dissolution reactions.
– Oxidation – reduction reactions.

Contaminants are often present in the soil sorbed to solid particle surfaces and/or attached to the organic matter in the soil. The transportation and removal of contaminants from the soil by electroosmosis and electromigration imply the desorption and solubilization of the contaminant in the pore fluid. The equilibrium between the solid phase and the liquid phase is dependent on the chemical nature of the contaminant and the composition of the soil and the pore fluid. The pH value has a major impact on sorption equilibrium. For instance, metals can be removed at low pH because metal ions can be exchanged form the solid surface when the H^+ concentration is increased. The pH-dependent sorption-desorption behavior is generally determined by performing batch experiments on the soil and contaminant of interest.

Precipitation and dissolution of the contaminant species during the electrokinetic process can significantly influence the removal efficiency of the process. The solubilization of precipitates is affected by the hydrogen ions generated at the anode migrating across the contaminated soil, which encourages the acidification of soil and the dissolution of metal hydroxides and carbonates. However, in some types of soils, the migration of hydrogen ions is hindered by the relatively high buffering capacity of the soil. In a high-pH environment, heavy metals will precipitate, and the movement of the contaminants will be impeded. During the electrokinetic treatment, metal ions migrate towards the cathode until they reach the high-pH zone where they accumulate and eventually precipitate, clogging soil pores and hindering the remediation process. It is essential to prevent precipitation and have the contaminants in dissolved form during the electrokinetic process to achieve efficient contaminant removal (Ricart et al., 2004).

Oxidation and reduction reactions are important when dealing with contaminants that can be transformed under the conditions developed at the electrodes or in the soil during electrokinetic treatment. Thus, organic contaminants can be degraded in the soil or at the electrodes by oxidation or reductive degradation. The degradation of organic contaminants results in simpler organic compounds, sometimes less toxic than the former contaminants, improving the effectiveness of the electrokinetic treatment.

Metallic contaminants such as chromium and arsenic are also affected by redox reactions. Chromium most commonly exists in two valence states: trivalent chromium [Cr(III)] and hexavalent chromium [Cr(VI)]. Cr(III) occurs in the form of cationic hydroxides such as $Cr(OH)^{2-}$ and it will migrate towards the cathode during electrokinetic remediation. However, Cr(VI) exists in the form of oxyanions such as chromate (CrO_4^{2-}) or dichromate ($Cr_2O_7^{2-}$), which migrate towards the anode. The valence state depends on the soil composition, especially the presence of reducing agents such as organic matter and Fe(II) and/or oxidizing agents such as Mn(IV). The chemical speciation of the contaminants and their movement across the soil depends on the valence state of metals and their possible redox chemistry (Reddy et al., 1997).

Arsenic is difficult to remove from the soil due to its complex chemistry. It forms a variety of anionic and cationic compounds that migrate toward the cathode and anode, respectively. The electric field and pH value induce redox reactions on arsenic compounds, changing the direction of migration making them difficult to be fully removed from the soil (Baek et al., 2009).

5 REMOVAL OF METALS AND INORGANIC CONTAMINANTS

5.1 Toxic metals

The electrokinetic treatment was initially used for the removal of heavy metals from contaminated soils. In the early 1990s, several papers were published dealing with the use of electric fields for the removal of heavy metals in model soils and real soils sampled in landfills, and industrial and mining areas. Kaolinite was commonly used to represent low-permeability soils, and electrokinetic treatment was carried out at the bench scale on kaolinite spiked with a cationic metal (e.g. Pb or Cd) in a specified concentration. Electric potential was applied to the soil sample in an electrokinetic setup. The cationic metals were dissolved in the pore fluid, which enabled their electrokinetic transportation. Furthermore, the acid front electrogenerated at the cathode decreases the pH in the soil, encouraging the desorption of the cationic metals and their transportation towards the cathode. However, the alkaline front electrogenerated at the cathode provokes the premature precipitation of the metal in the soil. The overall metal removal is then very limited due to the formation of the alkaline environment at the cathode (Ricart *et al.*, 1999).

Enhanced electrokinetic methods were developed to improve heavy metal removal in model soils. Mineral acids applied at the cathode can suppress the alkaline front by depolarization. Under these conditions, the acid front can acidify the soil and improve the desorption of cationic heavy metals that can electromigrate and accumulate in the cathode solution. The metals can then be recovered in their native form due to the reduction at the cathode surface. When selecting the mineral acid, it is important to consider its potential reactions with the contaminant metal and other compounds in the soil. The acid must not form insoluble salts with the contaminant metals to avoid precipitation. Organic acids are also often used to depolarize water. Advantages of the organic acids are that they are environmentally benign and can act as complexing agents for the contaminant metals, which enhances their solubility and stability in solution (Ottosen *et al.*, 2009).

Enhancing the metal removal by acidification of the soil is possible in model soils consisting of kaolinite due to their low buffering capacity. However, acidification is not possible in alkaline soils with high buffering capacity. In this case, metals can be desorbed using complexing agents.

The formation of metal complexes with organic acids and other organic compounds also improve the results of the electrochemical treatment of soils polluted with metals. Complexing agents encourage the extraction and solubility of metals in the pore fluid and they are stable in a wide range of pH, so they are not affected by the alkaline front. Complexes are often negative, so they can be recovered at the anode. The use of complexing agents can significantly enhance metal recovery in soils with high buffering capacity. In this kind of soil, it is not possible to extract the metals by acidification with the acid front generated at the anode; however, the metals can be mobilized using complexing agents, even at high pH. The best results can be achieved by combining the use of complexing agents with pH control at the cathode or anode depending on the chemistry of the contaminant metal (Cameselle & Pena, 2016; Figueroa *et al.*, 2016).

Oxidation and reduction are of particular importance for metallic chromium. The trivalent chromium Cr(III) exists in form of a cationic hydroxide, e.g. $Cr(OH)_3$, and migrates toward the anode. The hexavalent chromium Cr(VI) is present in oxyanions, e.g. CrO_4^{2-}, and migrates toward the anode. The valence state of chromium depends on soil properties, reductants, and/or oxidants, and determines the outcome of electrokinetic remediation (Reddy *et al.*, 1997).

5.2   Inorganic anionic pollutants

Inorganic compounds such as fluoride, chloride, nitrate, sulfate, phosphate, and other inorganic anions can be found in soil and may exert a negative effect on its properties and uses. The origin of anionic inorganic substances in soil is associated with the weathering of rocks and minerals and other natural processes. However, human activities have a major contribution to the accumulation of this kind of contaminants in soils. Land cultivation and the intensive use of chemical fertilizers may lead to nitrate contamination in soils and groundwater. The use of manure and other organic amendments on agricultural fields and greenhouses may result in increasing concentrations of salts (sulfate, chloride, phosphate etc.). The salinity of soils is a major problem mainly in paddy soils and greenhouses (Cho et al., 2010). Fluoride contamination in soil is usually associated with mineral processing and poor management of wastes such as red mud and spent pot linen from aluminum synthesis (Zhu et al., 2017). Although the toxicity of the inorganic anions is not as critical as that of cationic metals, it may create serious problems in groundwater and land use (Baek & Yang, 2009).

Electrokinetic remediation has been tested for the removal of inorganic anions such as nitrate, sulfate, fluorine, and others from contaminated soil. The two main transportation mechanisms in electrokinetics are electromigration and electroosmosis. Anionic compounds are removed from the soil mainly by electromigration towards the anode, in the opposite direction of the cationic metals. Under natural conditions, the pore water in soil flows by electroosmosis from anode to cathode, retarding the electromigration of anions towards the anode. Therefore, the electroosmotic flow should be limited in the design of an electrokinetic application for the removal of anions. Soil pH is another key parameter to be considered. Under natural conditions, soil particle surfaces show a negative charge. Inorganic anions are less attracted to negatively charged surfaces. Thus, anionic contaminants will be more easily desorbed in alkaline conditions than in acidic environments, unlike the cationic metals.

Electrokinetic remediation has also been tested for the removal of various inorganic anions. Literature results demonstrate that electrokinetics is very effective in the restoration of saline soil from a greenhouse (Cho et al., 2009). The electrokinetic treatment used a constant voltage gradient (1 V/cm) and lasted for only 6–48 hours. In these conditions, a significant amount of ions was removed from the soil as confirmed by a reduction in the electric conductivity after the test. The removal of nitrate was over 80% in just 2 days. The removal of cations by an EK process was negligible; in fact, the exchangeable concentration of potassium after EK treatment actually increased, compared to the initial value. Chlorine, sulfate, and phosphate can be removed from saline soil with efficiencies over 90% in 2–3 days (Cho et al., 2010). Changes in the soil pH near the electrodes after the electrokinetic treatment are the disadvantages of the electrokinetic restoration of saline soil. This problem can be fixed by neutralization, by inverting polarity. Fluorine can be removed by electromigration from soils and wastes such as red mud and spent pot linen. Zhu et al. (2017) reported that over 60% of F- in the red mud was removed in 7 days by electromigration towards the anode. The alkaline conditioning of anolyte favors the desorption and removal of fluorine, but, at the same time, it enhances the electroosmotic flow from anode to cathode, counteracting the electromigration of F-. The overall removal efficiency proved that the transportation of F- by electromigration is faster and more effective than the electroosmotic transportation.

6   REMOVAL OF ORGANIC POLLUTANTS

The electrokinetic removal of soluble organic compounds from soils is very efficient. The organic compound will be removed by electromigration or electroosmosis depending on the compound's chemical nature. pH conditions at the anode or cathode can be adjusted to induce desorption, solubilization, and ionization of the organic compounds. Flushing solution at the cathode or anode can be used in order to improve the desorption of soluble organic compounds from the soil particle surface. In fact, soluble organics can be naturally removed from the soil, so they are not a common environmental problem although they can contaminate groundwater (Ricart *et al.*, 2008).

Hydrophobic organic compounds pose a high risk to the soil because they are persistent, usually very toxic, and difficult to degrade (Kessler *et al.*, 2009; Lu & Yuan, 2009; Yang & Lee, 2009). Electrokinetic remediation of soil contaminated with organic contaminants has proved efficient for the following contaminants:

–   Polycyclic aromatic hydrocarbons (PAHs): while non-ionic surfactants (Tergitol 15-S-12, APG and Igepal CA-720), n-butylamine, HPCD (2-hydroxypropyl-beta-cyclodextrin), pH adjustment, and periodic voltage application have been found to have enhancing effects.
–   Chlorinated aliphatic hydrocarbons, chlorophenols, and chlorobenzenes: removal efficiency stems from electromigration and electroosmosis, while desorption efficiency may be increased by surfactants, cosolvents, and cyclodextrins.
–   Chlorinated pesticides are strongly bound to the soil, therefore surfactants and cosolvents are needed for solubilization and desorption to achieve high efficiency; herbicides exhibited the following removals: atrazine: 30–50 %, molinate: 90%, and betazone: 92%.
–   Nitroaromatic and other energetic compounds: cyclodextrins can significantly enhance removal, whereas nanotechnology and bioremediation may yield further improvements.

7   REMOVAL OF ORGANIC/INORGANIC CONTAMINANT MIXTURES

The problem of soils contaminated with mixed heavy metals and organic compounds is even more complex than cases of single contaminants because of the different chemistry of heavy metals and organic compounds (Elektorowicz, 2009). Some studies have shown synergistic effects that retard contaminant transport and removal, but a few other studies show the behavior of metals and organic compounds similar to those observed with either metals or organic compounds alone.

Metals can be removed predominantly by the electromigration process, while organic contaminants can be removed by electroosmosis. The presence of metals causes the zeta potential of the soil to be less negative and even results in a positive value, affecting electro-osmotic flow and sorption of the contaminants.

As it was explained before, metals are affected by various geochemical processes due to changes in the soil pH. The removal of cationic metals is hindered by their sorption and precipitation near the cathode; therefore, soil pH near the cathode must be lowered using weak organic acids. This can be done by forming soluble complexes at high pH using chelating agents or preventing high pH generation by the use of electrode membranes. In the case of anionic metals, pH near the anode should be increased using alkaline solutions such as NaOH to reduce sorption of these metals to the soil. When the metals exist in soluble form, they are transported and removed predominantly by the electromigration process.

For the simultaneous removal of heavy metals and organic contaminants, the organic compounds are solubilized using different solubilizing agents such as surfactants, cosolvents, or cyclodextrins. Once they are in solution, they can be transported and removed mainly by electroosmosis. Heavy metals can be solubilized as it was explained before, lowering soil pH or with the use of complexing agents. The transportation and removal of heavy metals is mainly by electromigration although electroosmosis may also help in the removal of heavy metals. It is important to keep all contaminants in soluble form and maintain electroosmotic flow for the removal of both heavy metals and organic compounds.

Sequential approaches are developed where (i) anionic metals are removed first and then cationic metals when mixed metal contaminants are present; and (ii) organic compounds are removed first followed by the removal of metals when co-existing organic contaminants and metals and are treated.

Efficient removal of multiple metals or co-existing metals and organic compounds (e.g. PAHs) is demonstrated in spiked soil conditions, but the performance was found inadequate in field soils from real contaminated sites.

Key factors of the feasibility of electrochemical remediation of soil contaminated with mixtures of metals and organics are: treatment duration, handling of secondary liquid waste and costs. These should be considered in decision making and planning.

8 ELECTROKINETIC/ELECTROCHEMICAL BARRIERS FOR POLLUTANT CONTAINMENT

The purpose of electrokinetic barriers is to prevent migration of contaminants from their current location. These barriers are similar to traditional passive containment barriers (such as vertical slurry walls) for soil and groundwater pollution containment and active containment barriers (such as pumping systems and drainage systems) for groundwater pollution containment (Sharma & Reddy, 2004). Such barriers are often used as an interim measure prior to implementing a permanent treatment system.

Electrokinetic barriers consist of a row of electrodes bordering a high-concentration area or polluted groundwater plume. The row of electrodes is set perpendicularly to the groundwater flow direction, while the depth of electrodes coincides with the lowest point where pollutants can be found. Electric current is induced into the ground by alternating anodes and cathodes. Anodes and cathodes are connected to separate closed loop pump systems and are used to circulate electrolytes. Depending on the contaminant, pH can be controlled by conditioning the electrolytes. Periodically, contaminants from the electrolytes are removed using different techniques such as sorption and ion exchange.

9 COUPLED TECHNOLOGIES

Conventional electrochemical remediation refers to the removal of contaminants from the contaminated media (soil, sediment, and groundwater), also known as electrokinetic extraction or electrokinetically enhanced flushing. Removal of contaminants (including contaminant mixtures) from spiked soils is possible using adaptive enhancement strategies. Despite efficient removal, practical implementation of such removal strategy is limited due to the following: (i) regulatory constraints on injecting the selected enhancement solutions into the subsurface, (ii) high cost, and (iii) long treatment times. Recently, electrochemical remediation is combined with other remediation technologies in order to overcome these issues as well as known deficiencies of conventional remediation methods. Such integrated or coupled technologies investigated include the following:

9.1 Electrochemical oxidation/reduction and electrokinetics

It is possible to remove a wide range of organic contaminants from soils using solubilizing agents such as surfactants, biosurfactants, cosolvents, and cyclodextrins. However, there may be regulatory objections for injecting these solubilizing agents into the subsurface. In addition, post-treatment of the extracted solutions at the electrodes makes this treatment costly. An alternative approach is to degrade the organic contaminants within the soil by injecting oxidants (e.g. hydrogen peroxide, permanganate, or persulfate) or reductants (zero-valent iron in the form of nanoscale iron particles). Chemical oxidation and reduction processes have been used for wastewater treatment for decades. However, it is challenging to introduce the oxidants/reductants into low-permeability clay soils. Combining electrokinetics with chemical oxidation/reduction facilitates delivery of the oxidants and reductants and increases contaminant availability in low-permeability soils. Oxidants such as Fenton's reagents (H_2O_2 and Fe^{2+}) produce hydroxyl radicals which break C–H bonds of organics into environmentally benign end products. Electrokinetics will also allow control of the soil pH and potential increase in temperature to create optimal conditions to achieve maximum oxidation.

It should be noted that the electrochemical-chemical reduction principles are the same as electrochemical-PRB using iron filing as reactive media, but one has to wait for the contaminated water to pass through the PRB for the remediation to occur. However, in electrochemical-chemical reduction approach, nanoscale iron particles are introduced into the contaminant source zone itself, thereby reducing the treatment time.

Generally, oxidants are stable only for a short period of time; therefore, electro-synthesis methods have been developed to produce oxidants *on site*. Ultrasonic methods involving high-intensity ultrasound are used to induce sonolysis of water to produce hydrogen peroxide and hydroxyl radicals. It should be noted that a simple combination of electrokinetics and ultrasonic waves has also been investigated by some researchers, but the purpose has been to enhance contaminant removal, not degradation (Yang, 2009).

Another technology known as electrochemical geo-oxidation (ECGO) has been proposed which involves application of low voltage and amperage to induce reduction-oxidation reactions at the micro-scale. This technology is based on the premise that soil particles act as micro-capacitors that charge and discharge in a cyclic fashion. Even though low voltage and amperage are used, the energy burst on discharge at the micro-scale is intense, resulting in destruction of organic contaminants, theoretically to carbon dioxide and water. ECGO itself generates the reductants (H as ion or radical) and oxygen (O elemental, OH and its radicals, HO_2 and its radicals) and, when combining the generation of H_2O_2 with the corrosion products of steel anodes, Fenton's reagent is produced. This technology is particularly attractive because no external oxidants need to be added.

9.2 PRB – electrokinetics

Permeable reactive barriers (PRBs) have been extensively used for the remediation of inorganic and organic pollutants in groundwater. PRBs are built by digging a trench in the path of flowing groundwater and then filling it with a selected permeable reactive material. As the contaminated groundwater passes through the PRB, organic contaminants may be degraded or sequestered, inorganic contaminants are sequestered, and clean groundwater exits the PRB. Reactive materials commonly considered include iron filings, limestone, hydroxyapatite, activated carbon, and zeolite. Monitoring data from several field PRB projects showed that high-concentration dissolved inorganic species flowing through the PRB tend to precipitate and clog the reactive material. In addition, the reactivity of the material used in the PRB may decrease. Coupling electrokinetics with PRBs has been found to eliminate clogging of the PRB system caused by mineral precipitation and improve the long-term performance of PRBs. More research is needed towards developing combined electrokinetic-PRB systems to induce favorable geochemical conditions within the PRB as needed during the course of the remediation process (Weng, 2009).

9.3   Bioremediation – electrokinetics

Bioremediation, which involves the degradation of organic compounds using microbes, has received great attention because it is environmentally friendly and inexpensive and requires low energy. However, it is a slow remediation processes and its effectiveness depends on the availability of nutrients, bioavailability of contaminants and physical conditions such as temperature, redox potential, and moisture. Coupling electrokinetics with bioremediation (also known as electro-bioremediation or electro-bioreclamation) can include injection of nutrients, electron acceptors, or microbes (if needed) and increase the bioavailability of contaminants, especially in low-permeability soils where hydraulic delivery techniques are ineffective. Interestingly, electrolysis reactions at the electrodes may be used to provide electron acceptors and donors. Hydrogen produced at the cathode may be used as electron donor for reductive degradation processes, while oxygen produced at the anode may be used for oxidative biodegradation. Although biodegradation has often been applied to organic pollutants, few studies have reported on biological immobilization of heavy metals enhanced by electrokinetics. Advances are being made through additional research to optimize electrokinetic effects on the microbial activity to achieve efficient contaminant degradation (Gill *et al.*, 2014).

9.4   Phytoremediation – electrokinetics

Phytoremediation involves the use of living plants and their associated root-microorganisms to remove, degrade, or sequester inorganic and organic pollutants from soil, sediment and groundwater. The main advantages of this method are low cost and ecologically friendliness. However, this method is limited to shallow depths (limited by the root depth). Moreover, slow plant growth and the solubility and bioavailability of the pollutants limits the remediation rate and capacity. Coupling electrokinetics with phytoremediation is aimed to increase the availability of the contaminants and also facilitate their transport towards the root zone. The effects of the electric field on soil pH, availability of nutrients, etc. may also help plant growth. Electrodes are placed strategically and a low direct current or electric gradient is applied and the contaminants are transported by electromigration and/or electroosmosis processes toward the plant root zone. Electrode solutions of low toxicity towards plants can be used to enhance solubilization of the contaminants. Small-scale experiments showed that plants grow faster when exposed to electric fields, enhancing overall contaminant removal efficiency. More research is needed to address organic contaminants and contaminant mixtures and possible effects on soil quality and biology of the coupled technology electro-phytoremediation (Cameselle *et al.*, 2013; Chirakkara *et al.*, 2014).

9.5 Thermal desorption and electrokinetics

Thermal effects during electrochemical remediation have been studied. *In situ* thermal methods such as electrical resistance heating (ERH) have been used for site remediation, but electromotive forces that may occur during ERH have not been investigated. Heating resulting from electrochemical treatment involves the resistance to the passage of electrical current through the soil moisture. It is this resistance to electrical flow that increases temperature. Heat transport from this joule heating takes place mainly through conduction and convection. The increase in temperature decreases viscosity of pore fluid, decreases sorption of contaminants (increasing bioavailability), and increases the volatilization of organic contaminants. Heating reduces the oxidation reduction potential in water, thereby affecting redox reactions. Elevated temperatures within the tolerable limits of microorganisms are also conducive to higher metabolic activity and enhanced biodegradation of contaminants. Proper control of increased temperature during electrochemical remediation may be exploited to enhance the overall remedial efficiency (Smith, 2009).

10 ECONOMIC ASPECTS

The cost of electrokinetic remediation of inorganic contaminants is estimated to be in the range of 115–400 US$/m³ (90–275 US$/m³ for organic contaminants) with an average of 200 US$/m³, i.e. 90 US$/ton, which is less than excavation and disposal as hazardous waste. Due to the limited number of full-scale electroremediation projects performed, cost estimates are usually based on pilot tests. As the technology matures, confidence is expected to rise and costs should drop (Cameselle, 2014).

11 FIELD APPLICATIONS

Several field applications of electrokinetics have been reported in the US, Europe, Korea, and Japan (Oonnittan *et al.*, 2009). The first field-scale electrokinetic application was carried out by Geokinetics in 1987 (Lageman, 1993) in a soil contaminated with heavy metals form a former paint factory. The subsoil was contaminated with Cu (> 5000 mg/kg) and lead (500–1000 mg/kg). The treated area was 70 m long, 3 m wide, and 1 m deep. They used a horizontal cathode installed close to the surface and a series of vertical anodes 2 m apart. After 43 days of operation the Cu content was reduced by 80% and the Pb content by 70%. Another remediation project carried out by Geokinetics was a big leap into the practical field-scale implementation of electrokinetics. In the US (USEPA, 1995), electrokinetic remediation was applied to various superfund sites for the removal of heavy metals. Some examples include Naval Air Weapons Station in Point Mugu (Cr and Cd); Honolulu, Hawaii (Pb); and Albuquerque, New Mexico (Cr). Monsanto and GE Corporate (Ho *et al.*, 1997) developed the so-called lasagna process, a novel technology based on directing contaminants by the electrokinetic process into certain soil layers for treatment/decontamination. The lasagna process would cut the costs compared to digging, storing, and/or burning contaminated soil. The first application was successfully completed in May 1995 at a DOE site in Paducah, Kentucky, in a clayey soil contaminated with trichloroethylene. In Korea, Kim *et al.* (2010) have developed an electrokinetic application for the removal of radionuclides (^{137}Cs and ^{60}Co) reaching efficiencies higher than 90% with the appropriate conditioning of the electrode solution using nitric and/or acetic acid.

Overall, field applications demonstrate that the electrokinetic treatment is an adequate tool for the remediation of sites polluted with toxic metals. However, the effectiveness of the process largely depends on the geochemical characteristics of and the interaction of the pollutants with the soil. Therefore, a detailed study of each case is necessary to specify the most adequate operating conditions, which include current intensity or voltage drop, electrode disposition, and chemical conditioning of electrode solutions. Laboratory studies can be carried out before the field operation in order to determine the effect of those operating conditions that influence solubilization and removal of heavy metals, but it must be considered that the results at field scale may differ from those obtained in the laboratory experiments due to the change in scale.

12   FUTURE PERSPECTIVES

This chapter has demonstrated significant advances towards establishing electrochemical remediation as a practical technology. However, field application lags far behind laboratory tests and other research studies. The following issues should be addressed in order to develop this technology towards commercial field applications for soil, sediment and groundwater remediation:

- Restrict small-scale laboratory experiments and encourage large-scale pilot tests in the laboratory or in the field.
- Reduce the single contaminant experiments in model soils spiked at laboratory and promote tests on actual soils contaminated with aged and multiple contaminants.
- Encourage the development of combined technologies.
- Extend the electrochemical technology to other contaminated porous matrixes such as industrial wastes in order to achieve their valorization.
- Promote the analysis of experimental results based on the physicochemical properties of contaminants, geochemistry of soil, reaction kinetics, equilibrium constants, and transport parameters, rather than a phenomenological analysis of the contaminant removal.
- Advance fundamental research and developing predictive modeling tools. In particular, geochemical reactions in various soil and contaminant conditions should be properly characterized.
- Establish quality control and quality assurance protocols for laboratory and field studies.
- Assess impacts of electrochemical remediation on soil quality and ecology.
- Investigate short-term and long-term effects on electrochemical processes.
- Develop new, innovative, and low-cost approaches to electrochemical remediation for practical use.
- Develop electrochemical remediation systems that are ecologically safe based on sustainability considerations.
- Perform and disseminate results of well-monitored pilot-scale field demonstrations.
- Establish communication among researchers and identifying areas that require future research.
- Develop guidance documents for the design, installation, and operation of typical electrochemical remediation systems.

In summary, the importance of full-scale field demonstration projects cannot be overemphasized. The lessons learned from these projects are invaluable for identifying the advantages and limitations of this technology and develop effective and economical adaptive field systems based on site-specific conditions.

REFERENCES

Acar, Y. (1993) Principles of electrokinetic remediation. *Environment Science Technology*, 27(13), 2638–2647.

Baek, K., Kim, D.-H., Park, S.-W., Ryu, B.-G., Bajargal, T. & Yang, J.-S. (2009) Electrolyte conditioning-enhanced electrokinetic remediation of arsenic-contaminated mine tailing. *Journal of Hazardous Materials*, 161(1), 457–462. doi:10.1016/j.jhazmat.2008.03.127.

Baek, K. & Yang, J.-S. (2009) Electrokinetic removal of Nitrate and Fluoride. In: Reddy, K.R. & Cameselle, C. (eds.) *Electrochemical Remediation Technologies for Polluted Soils, Sediments and Groundwater*. John Wiley & Sons, Inc., Hoboken, NJ, USA. pp. 141–148. doi:10.1002/9780470523650.ch6.

Cameselle, C. (2014) Electrokinetic remediation, cost estimation. In: Kreysa, G., Ota, K. & Savinell, R.F. (eds.) *Encyclopedia of Applied Electrochemistry*. Springer, New York, NY, USA. pp. 723–725. Retrieved from http://link.springer.com/10.1007/978-1-4419-6996-5_91.

Cameselle, C., Chirakkara, R.A. & Reddy, K.R. (2013) Electrokinetic-enhanced phytoremediation of soils: Status and opportunities. *Chemosphere*, 93(4), 626–636. doi:10.1016/j.chemosphere.2013.06.029.

Cameselle, C. & Pena, A. (2016) Enhanced electromigration and electro-osmosis for the remediation of an agricultural soil contaminated with multiple heavy metals. *Process Safety and Environmental Protection*, 104, 209–217. doi:10.1016/j.psep.2016.09.002.

Cameselle, C. & Reddy, K.R. (2012) Development and enhancement of electro-osmotic flow for the removal of contaminants from soils. *Electrochimica Acta*, 86, 10–22.

Chirakkara, R.A., Reddy, K.R. & Cameselle, C. (2015) Electrokinetic amendment in phytoremediation of mixed contaminated soil. *Electrochimica Acta* 181, 179–191. doi:10.1016/j.electacta.2015.01.025.

Cho, J.M., Kim, K.J., Chung, K.Y., Hyun, S. & Baek, K. (2009) Restoration of saline soil in cultivated land using electrokinetic process. *Separation Science and Technology*, 44, 2371–2384.

Cho, J.-M., Park, S.-Y. & Baek, K. (2010) Electrokinetic restoration of saline agricultural lands. *Journal of Applied Electrochemistry*, 40(6), 1085–1093.

Elektorowicz, M. (2009) Electrokinetic remediation of mixed metals and organic contaminants. In: Reddy, K.R. & Cameselle, C. (eds.) *Electrochemical Remediation Technologies for Polluted Soils, Sediments and Groundwater*. John Wiley & Sons, Inc., Hoboken, NJ, USA. pp. 315–331. doi:10.1002/9780470523650.ch15.

Figueroa, A., Cameselle, C., Gouveia, S. & Hansen, H.K. (2016) Electrokinetic treatment of an agricultural soil contaminated with heavy metals. *Journal of Environmental Science and Health – Part A Toxic/Hazardous Substances and Environmental Engineering*, 51(9), 691–700. doi:10.1080/10934529.2016.1170425.

Gill, R.T., Harbottle, M.J., Smith, J.W.N. & Thornton, S.F. (2014) Electrokinetic-enhanced bioremediation of organic contaminants: A review of processes and environmental applications. *Chemosphere*, 107, 31–42. doi:10.1016/j.chemosphere.2014.03.019.

Ho, S.V., Athmer, C.J., Sheridan, P.W. & Shapiro, A.P. (1997) Scale-up aspects of the lasagna® process for in situ soil decontamination. *Journal of Hazardous Materials*, 55(1–3), 39–60. doi:10.1016/S0304-3894(97)00016-2.

Kessler, D.A., Marsh, C.P. & Morefield, S. (2009) Electrokinetic removal of energetic compounds. In: Reddy, K.R. & Cameselle, C. (eds.) *Electrochemical Remediation Technologies for Polluted Soils, Sediments and Groundwater*. John Wiley & Sons, Inc., Hoboken, NJ, USA. pp. 265–284. doi:10.1002/9780470523650.ch13.

Kim, G., Lee, S., Shon, D., Lee, K. & Chung, U. (2010) Development of pilot-scale electrokinetic remediation technology to remove 60Co and 137Cs from soil. *Journal of Industrial and Engineering Chemistry*, 16(6), 986–991. doi:10.1016/j.jiec.2010.05.014.

Lageman, R. (1993) Electroreclamation: Applications in the Netherlands. *Environmental Science and Technology*, 27(13), 2648–2650. doi:10.1021/es00049a003.

Lu, X. & Yuan, S. (2009) Electrokinetic removal of chlorinated organic compounds. *Electrochemical Remediation Technologies for Polluted Soils, Sediments and Groundwater*, Generic. pp. 219–234.

Oonnittan, A., Sillanpaa, M., Cameselle, C. & Reddy, K.R. (2009) Field applications of electrokinetic remediation of soils contaminated with heavy metals. In: *Electrochemical Remediation Technologies for Polluted Soils, Sediments and Groundwater*. pp. 607–624. Retrieved from www.scopus. com/inward/record.url?eid=2-s2.0-79955995060&partnerID=40&md5=bd1a1842175b88caf894d a55d7951918.

Ottosen, L.M., Hansen, H.K. & Jensen, P.E. (2009) Electrokinetic removal of heavy metals. In: *Electrochemical Remediation Technologies for Polluted Soils, Sediments and Groundwater*. Wiley Online Library, John Wiley & Sons, Inc. pp. 95–126. doi:10.1002/9780470523650.ch4.

Reddy, K.R. & Cameselle, C. (eds.) (2009) *Electrochemical Remediation Technologies for Polluted Soils, Sediments and Groundwater*. Retrieved from www.scopus.com/inward/record.url?eid=2-s2. 0-84889389978&partnerID=40&md5=44064164eb24877831fec0ba842c3202.

Reddy, K.R., Parupudi, U.S., Devulapalli, S.N. & Xu, C.Y. (1997) Effects of soil composition on the removal of chromium by electrokinetics. *Journal of Hazardous Materials*, 55(1–3), 135–158. doi:10.1016/S0304-3894(97)00020-4.

Ricart, M.T., Cameselle, C., Lucas, T. & Lema, J.M. (1999) Manganese removal from spiked kaolinitic soil and sludge by electromigration. *Separation Science and Technology*, 34(16), 3227–3241. doi:10.1081/SS-100100832.

Ricart, M.T., Hansen, H.K., Cameselle, C. & Lema, J.M. (2004) Electrochemical treatment of a polluted sludge: Different methods and conditions for manganese removal. *Separation Science and Technology*, 39(15), 3679–3689. doi:10.1081/SS-200040079.

Ricart, M.T., Pazos, M., Gouveia, S., Cameselle, C. & Sanromán, M.A. (2008) Removal of organic pollutants and heavy metals in soils by electrokinetic remediation. *Journal of Environmental Science and Health – Part A Toxic/Hazardous Substances and Environmental Engineering*, 43(8), 871–875. doi:10.1080/10934520801974376.

Sharma, H.D. & Reddy, K.R. (2004) Geoenvironmental engineering: Site remediation, waste containment, and emerging waste management technologies. *Geoenvironmental Engineering: Site Remediation, Waste Containment, and Emerging Waste Management Technologies*. Retrieved from www.cabdirect.org/cabdirect/abstract/20053079233.

Smith, G.J. (2009) Coupled electrokinetic – Thermal desorption. In: Reddy, K.R. & Camesell, C. (eds.) *Electrochemical Remediation Technologies for Polluted Soils, Sediments and Groundwater*. John Wiley & Sons, Inc., Hoboken, NJ, USA. pp. 505–535. doi:10.1002/9780470523650.ch24.

USEPA (1995) In situ remediation technology: Electrokinetics. Report number: EPA542-K-94-007.

Weng, C.-H. (2009) Coupled electrokinetic – Permeable reactive barriers. In: Reddy, K.R. & Cameselle, C. (eds.) *Electrochemical Remediation Technologies for Polluted Soils, Sediments and Groundwater*. John Wiley & Sons, Inc., Hoboken, NJ, USA. pp. 483–503. doi:10.1002/9780470523650.ch23.

Yang, G.C.C. (2009) Electrokinetic – Chemical Oxidation/Reduction. In: Reddy, K.R. & Cameselle, C. (eds.) *Electrochemical Remediation Technologies for Polluted Soils, Sediments and Groundwater*. John Wiley & Sons, Inc., Hoboken, NJ, USA. pp. 439–471. doi:10.1002/9780470523650.ch21.

Yang, J.-W. & Lee, Y.-J. (2009) Electrokinetic Removal of PAHs. In: Reddy, K.R. & Cameselle, C. (eds.) *Electrochemical Remediation Technologies for Polluted Soils, Sediments and Groundwater*. John Wiley & Sons, Inc., Hoboken, NJ, USA. pp. 195–217. doi:10.1002/9780470523650.ch9.

Zhu, S., Zhu, D. & Wang, X. (2017) Removal of fluorine from red mud (bauxite residue) by electrokinetics. *Electrochimica Acta*, 242, 300–306.

Chapter 10

Elemental iron and other nanotechnologies for soil remediation

C. Cameselle[1] & K. R. Reddy[2]

[1]Department of Chemical Engineering, University of Vigo, Vigo, Spain
[2]Department of Civil and Materials Engineering, University of Illinois at Chicago, Chicago, IL, USA

ABSTRACT

Nanotechnology is an emerging engineering field with multiple applications in industry, medicine and in environmental engineering. The use of nanomaterials may help in the remediation of contaminated sites, soil and groundwater, alone or in combination with other remediation technologies. Nanomaterials may stabilize metals or inorganic anions and eliminate hydrophobic and recalcitrant organics that are difficult to remove with the available technologies. At the nanoscale, materials show new and unexpected properties, different than those at the macroscale. Thus, for instance, iron nanoparticles show a great reactivity and reduction capacity that can be used in the dehalogenation of recalcitrant organic contaminants. In this chapter, the properties of nanoparticles and the possibilities of nanotechnology in the remediation of contaminated sites are reviewed and discussed. Special attention is paid to nanoscale iron particles due to their wide range of application in environmental engineering, but the applications of other types of nanoparticles is also discussed. Finally, the application of nanotechnology in remediation at the field scale is reviewed, analyzing the cost and the possible risks associated with the release of nanoparticles to the environment.

1 NANOSCALE PARTICLES

The term "nanoparticle" refers to materials with a size in the range of 10–100 nm in any dimension. To have an idea of the size of a nanoparticle, the size of the glucose molecule is below 1 nm, the DNA is around 10 nm, a typical virus size is 10^2 nm, and the size of a eukaryotic cell is about 10^4–10^5 nm (Figure 10.1). Many materials can form nanoparticles such as carbon, graphite, silicon, metals, metal oxides, and others. Some of the different types of engineered nanoparticles include (i) carbon-based materials (e.g. carbon nanotubes), (ii) metal-based materials (quantum dot, nanoiron, nanogold, nanosilver, titanium dioxide nanoparticles and other nanoscale metal oxides), (iii) dendrimers (polymer branched units), and (iv) composites (nanoparticles made as a combination of different materials) (US EPA, 2007).

Although nanoparticles seem to be the result of the latest science and engineering research, materials at nanoscale can be found and produced unintentionally. Examples are airborne combustion by-products, exhaust particles in diesel engines, volcanic ash, viruses, etc. A curious case of unnoticed production of nanoparticles is the Damascus steel. In the Middle Ages, the city of Damascus (in the present Syria) was famous for the manufacturing of blades made of a special quality of steel that was very hard but, at the same time, superplastic; and the edge of the blade could be extraordinary sharpened. Recent studies reported that such properties for the Damascus steel are related to the presence of nanoparticles, steel

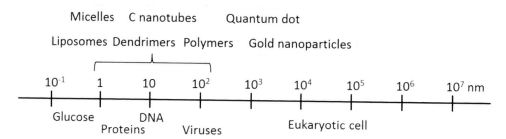

Figure 10.1 Comparative size scale of nanoparticles

wires, and carbon nanotubes (Reibold *et al.*, 2006), that were unnoticeable and formed during the special forge procedure, not known at present.

Recently, nanoparticle synthesis and application have received a lot of attention due to their special properties that largely differ from the original materials they are made of. This is mainly due to the large specific surface area compared with the original material. Any material at a macroscopic scale shows constant properties independent of its size, but at nanoscale, the number of atoms on the surface of the particle is very significant compared to the total number of atoms in the particle. It provokes the appearance of interesting and sometimes unexpected properties in the nanoparticles. These special properties are enhanced reactivity, specific catalyst activity, optical properties, superparamagnetism, high diffusion, and other unexpected properties as the possibility of forming suspensions despite the high difference in density between the nanoparticles and the solvent. Such special and specific properties have resulted in the use of nanoparticles in many scientific fields and technical applications. Thus, nanoparticles are used or are being tested for their application in medicine, chemical industry, electronics, fiber production, clothes, batteries, and also in environmental engineering. In fact, environmental applications of nanoparticles show very interesting perspectives, specially dealing with the remediation of soil and groundwater. This emerging application requires more research, especially at field scale in order to demonstrate the benefits of the application of this technology and evaluate the possible negative effects of releasing large amounts of nanoparticles to the environment.

2 ENVIRONMENTAL APPLICATION OF NANOPARTICLES

There are many sites around the globe contaminated with persistent pollutants such as PAHs, BTEX, PCBs, other organochlorides, pesticides, and nitroaromatic compounds, many of them listed as priority pollutants by the US EPA. Despite the efforts in the development of innovative technologies, the satisfactory treatment of contaminated soils and groundwater is still unresolved, mainly due to the difficulty to deal with persistent contaminants in heterogeneous soil matrixes and low permeability soils. The efficiency of remediation can be increased using aggressive technologies operating in harsh conditions. Unfortunately, the damage to the soil is so extensive that the resulting material cannot recover the former properties and characteristics of the original soil.

Permeable reactive barriers (PRB) is a technology that permits the treatment of contaminated groundwater avoiding the spread of the contamination far from its source zone.

Basically, PRB is a trench in the soil installed through the flow of groundwater and filled with a specific reactive material to adsorb, precipitate or degrade the contaminants in the groundwater. The effective use of a PRB depends on its location, dimensions, and the filled material. The hydrodynamic groundwater flow in the subsurface has to be studied to select the right location of the PRB. Its dimensions have to assure that the whole flow of groundwater passes through the PRB avoiding any bypass, and what it is more important, the width of the PRB and the permeability of the filled-in material needs to be designed for sufficient residence time of the groundwater in the PRB to assure complete transformation of the contaminants. The selection of the filled-in material is critical for the success of this technology. In general, specific reactive materials for the target contaminants have to be selected, but at the same time, the possible impact in the environment need to be considered. Elemental iron or zero-valent iron has been used due to its good reactivity and very low impact on the soil ecosystem.

Zero-valent iron has been proved to be an effective reductant in the treatment of halogenated organic compounds. The standard reductive potential of zero-valent iron is 447 mV. Elemental iron is converted to ferrous ions (Fe^{2+}) in the presence of an oxidizing agent and therefore releases electrons and becomes available to reduce other compounds. The possible reaction mechanisms for halogenated organic compounds (RX) are shown in Table 10.1. The predominant reactions in the degradation of the dissolved halogenated compounds RX take place on the surface of the metal rather than with hydrogen and ferrous iron in solution, although the corrosion reactions (aerobic or anaerobic) are the major source of reducing species for subsequent reduction of the halogenated compounds. The possible reaction mechanisms for the treatment of organic compounds are dehalogenation, β-elimination, dehydrogenation, hydrolysis, transformation, and hydrogenation depending on the type of contaminant.

Zero-valent iron fillings in PRB have proven its remedial capacity for several contaminants, e.g. chlorinated solvent plumes in groundwater (Muchitsch *et al.*, 2011). However, the zero-valent iron is liable to form an oxide surface film that, consequently reduce the reactivity. In fact, even in proper storage conditions, the reactivity of iron toward the target contaminant is reduced along the time. Reactivity of zero-valent iron also varies depending on the origin or manufacturer. These factors limit the applicability of zero-valent iron fillings in PRB for long-term in-situ treatment.

Table 10.1 Reactions of zero-valent iron (Fe^0) with a halogenated organic compound (RX) in water.

Reaction	Mechanism
Anaerobic corrosion of iron	$Fe + 2H_2O \rightarrow Fe^{2+} + H_2 + 2OH^-$
Aerobic corrosion of iron	$2Fe + O_2 + 2H_2O \rightarrow 2Fe^{2+} + 4OH^-$
Possible reductive reactions of iron with RX	
Reaction of RX with ferrous iron in the aqueous phase	$RX + 2Fe^{2+} + H^+ \rightarrow 2Fe^{3+} + RH + X^-$
Reaction of RX at the surface of the metal (electron transfer reaction)	$RX + Fe + H^+ \rightarrow RH + Fe^{2+} + X^-$
Adsorption of RX to the metal surface and the subsequent surface reaction of the organic radical R*	$Fe \rightarrow Fe^{2+} + 2e^-$ $RX + e^- \rightarrow R* + X^-$ $R* + H^+ + e^- \rightarrow RH$

The development of nanotechnology opens the possibility of using nanoparticles with specific properties for the removal of selected contaminants. Based on the success of permeable reactive barriers with ZVI with macroscale or micro-scale materials, nanoscale zero-valent iron particles (NIP) have been used for *in situ* contaminant reduction with a significant success (Bigg & Judd, 2000; Taghavy *et al.*, 2010). NIP can be used as filling in PRBs instead of micro-scale iron. So far, the studies with NIP as a remediating agent in soil proved to be more effective than micro-scale iron in PRBs. The active surface of NIP is 1100 times greater than micro-scale zero-valent iron; it results in faster and more effective dechlorination of TCE in soil with NIP (95%) compared to micro-scale iron (86%) (Cao *et al.*, 2005a,b). NIP are superior both in terms of initial rate of reduction and the amount of moles of contaminant reduced by mole of iron.

NIP can be injected directly in the contaminated source zone for a rapid and effective *in-situ* decontamination of soil. The effectiveness of direct injection of NIP in soil clearly depends on the uniform delivery and distribution to the entire contaminated zone. Unfortunately, the tendency of NIP to aggregate and settling, limits its transportation in the subsurface soil. Several investigators have developed surface-modified NIP to overcome the tendency to the aggregation of NIP, quantifying the decreasing in reactivity with target contaminants.

Nanoscale iron particles (NIP) are considered best suited for environmental remediation of contaminated sites due to their environmentally benign characteristics, favorable chemistry, relatively low cost, and easy use. Moreover, the reactivity of NIP with target contaminants is shown to be relatively fast, resulting less toxic or non-toxic transformation products.

3 NANOSCALE IRON PARTICLES (NIP): SYNTHESIS, PROPERTIES, AND REACTIVITY

Zero-valent iron has a reductive capacity that can be used in the degradation of several chlorinated organics, and in the reduction of heavy metals and nitrate. Nanoscale iron particles show the same reductive capacity than macroscale iron, but due to the large specific surface area of iron nanoparticles, their reactivity is much higher and, therefore, the reaction rate. NIPs have been tested in environmental applications. Thus, NIPs are reported to be able to reductively dechlorinate chlorinated hydrocarbons, replacing chloride atoms in the molecule by hydrogen atoms. The toxicity of the organic compound largely decreases and so its impact in the environment. The application of NIPs *in situ* also results in reaction with water releasing hydrogen gas. It is well known that molecular hydrogen is consumed by anaerobic microorganisms during the biotic reductive dechlorination. In very reducing environments, NIPs may contribute to the formation of ferrous iron from ferric iron present in the soil in minerals. It was also demonstrated that ferrous iron participates in the abiotic degradation of some chlorinated compounds such as chloroethane and chloroethene. Therefore, the environmental applications of NIPs are not limited to the reductive capacity of the iron but also the synergic combination of chemical and biochemical processes. NIPs can also remove heavy metals from groundwater: hexavalent chromium, mercury, lead, and arsenic can by removed by reduction, adsorption on nanoparticle surface and precipitation. Inorganic anions as chlorate and nitrate can be reduced to chloride ions and nitrogen gas by NIPs respectively (Table 10.2).

The environmental applications of NIPs imply the introduction of the nanoparticles in the contaminated area and the uniform distribution to achieve an effective remediation in

Table 10.2 Contaminants that can be treated by nanoscale iron particles (NIPs).

Chlorinated hydrocarbons	Carbon tetrachloride
	Chloroform
	Tetrachloroethene
	Trichloroethene
	Dichloroethene
Chlorinated benzenes	Hexachlorobenzene,
	Pentachlorobenzene
	Pentachlorophenol
Pesticides	DDT, DDD, DDE, lindane
Polychlorinated organics	PCBs, dioxins,
Inorganic anions	ClO_4^-, NO_3^-, As
Heavy metals	Cr, Hg, Pb, Se, Ni
Nitroaromatic compounds	TNT, RDX

the whole site. Obviously, NIP remediation will be affected by the type and concentration of contaminants but the specific conditions of the contaminated site will also affect the performance of the remediation process. These factors include geologic conditions: soil permeability, heterogeneity, and stratification of the soil; geochemical conditions: composition of the soil matrix, composition, and ionic strength of the groundwater, pH, dissolved oxygen (DO), ORP (redox potential), and the presence and concentration of some chemical species that can react with NIPs as nitrate, nitrite, etc. These factors affect both the reactivity of the NIP and the introduction, transportation, and distribution of the nanoparticles into the soil and groundwater.

The transportation and distribution of nanoparticles in contaminated site is greatly affected by the sedimentation and the agglomeration of the particles. The density of the nanoparticles is much higher that the solvent (typically water) and it makes the nanoparticles tend to settle in the bottom of the flask or in the well during the injection. Furthermore, the surface characteristics of NIP made the nanoparticles to attract to each other or to attach to the soil particles' surface. The agglomeration of the particles favors the sedimentation and impedes the distribution in the soil and groundwater. Agglomeration also increases the effective radius of the nanoparticles decreasing the surface/volume ratio, i.e. it decreases the exposed active surface. It will decrease the reactivity and the efficiency of the treatment. Agglomeration of NIP is affected by the pH of the subsurface environment and groundwater. The presence of some cationic species and the ionic strength of groundwater exerts a decisive impact on agglomeration and mobility of the particles.

In order to improve the mobility and stability of nanoparticles in the subsurface environment, many studies used coated NIP. Polyelectrolytes and polymers were used to modify the surface characteristics of the nanoparticles. Emulsions of nanoparticles with vegetable oils were also tested. NIPs were encased in very small drops of oil. It prevents the contact with the aqueous phase and the adsorption to the soil particles, and enhances the contact with hydrophobic contaminants. However, the modification of the surface characteristics of NIP may improve mobility and stability, and prevent the aggregation, but it is usually associated to loss in reactivity.

The high specific surface of the nanoparticles and its high reactivity make them susceptible of reaction in open air, creating a surface of iron oxide around the nanoparticle. It clearly decreases the chemical reactivity of iron nanoparticles and its utility for the effective remediation of the environment. This loss of activity is called passivation or deactivation. The original properties of the nanoparticles must be preserved during transportation, utilization, and injection in the contaminated site. Nanoparticles are usually commercialized as suspensions in water alone or with chemical additives to preserve stability and reactivity. As a general rule, during the manipulation and injection of the nanoparticles in the subsurface, NIPs must be prevented from having contact with air. In some cases, deoxygenated water can be used for the injection. Thus, passivation of the nanoparticles can be avoided until they reach the contaminated area in the subsurface. Once the nanoparticles have been injected in the contaminated site, their distribution and reactivity must be monitored to evaluate the efficiency of the remediation process. Iron concentration is easy to measure and the profile of iron concentration may help to know the distribution and transportation of NIP. However, iron concentration does not inform about the reduction activity of NIP. ORP may help to define the area where the iron is acting as a reducing agent. Finally, sampling and analyzing the soil and groundwater for contaminant concentration will establish the efficiency of the treatment.

3.1   Synthesis of NIP

NIP can be synthesized by different physical and chemical methods (Li *et al.*, 2006). The physical methods consist of inert gas condensation, severe plastic deformation, high-energy ball milling, and ultrasound shot peening. The chemical methods include the reverse micelle method (microemulsion), controlled chemical coprecipitation, chemical vapor condensation, pulse electrodeposition, liquid flame spray, liquid-phase reduction, and gas-phase reduction. Liquid-phase reduction and gas-phase reduction methods are the most widely used methods for synthesis of NIP that are used for site remediation purposes.

Wang and Zhang (1997) used the liquid-phase reduction method to produce NIP in size between 1 nm and 100 nm and a BET surface area of the particles of 33.5 m^2/g. The synthesis consists of ferrous iron reduction by sodium borohydride ($NaBH_4$), a strong reductant, to produce NIP as shown in Equation 1.

$$Fe(H_2O)_3^{3+} + 3BH_4^- + 3H_2O \rightarrow Fe^0 + 3B(OH)_3 + 10.5H_2 \qquad (10.1)$$

Toda America Inc. (Battle Creek, MI, US) uses the gas-phase reduction method to synthesize NIP from $FeSO_4$ solution (Okinaka *et al.*, 2005a). The synthesis process starts with the precipitation of acicular goethite [FeO(OH)] from the oxygenated ferrous sulfate solution. Then, the acicular goethite is reduced to alpha-Fe grains in a heated hydrogen gas atmosphere at a high temperature (350–600°C). Finally, the α-Fe grains are wet-milled to convert the surface to magnetite as indicated in Equations 2 and 3.

$$2\,FeO(OH) + H_2 \rightarrow 2\,Fe^0 + 2\,H_2O \qquad (10.2)$$
$$3\,Fe^0 + 4\,H_2O \rightarrow Fe_3O_4 + 4\,H_2 \qquad (10.3)$$

The NIP produced by this method shows a characteristic structure as shown in Figure 10.2A. It is formed by an elemental Fe core (alpha-Fe) and a magnetite shell (Fe_3O_4).

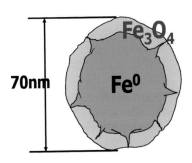

Figure 10.2 Structure (A) and SEM image (B) of NIP synthesized by TODA

Table 10.3 Properties of NIP synthesized by TODA.

Properties	Value
Coercive force (Hc)	408 Oe
Mass magnetization (σ_s)	149.6 emu/g
σ_p/σ_s (ratio of ferromagnetism to antiferromagnetism)	0.152
pH	10.7
Surface area (BET)	37.1 m²/g
Electrical conductivity	2.29×10^2 µS/cm
Particle size	50–300 nm
Aqueous suspension	20–30 wt%
Density of aqueous slurry	1.2–1.3 g/mL

The average particle size determined by a scanning electron microscope (SEM) is 70 nm (Figure 10.2B). Typical properties of NIP are summarized in Table 10.3. It is interesting to note that the particles possess electromagnetic properties.

3.2   Properties of NIP

Nanoparticles can be characterized using several methods and techniques to obtain their chemical composition, structure, size, shape and specific surface. These techniques involve X-ray diffractometry (XRD), scanning electron microscopy (SEM), transmission electron microscopy (TEM), and Fourier transform infrared spectroscopy (FTIR).

Liu *et al.* (2005a) compared Fe particles synthesized from sodium borohydride reduction of ferrous iron (Fe/B) with NIP produced by Toda America (RNIP). The results of this study are reported in Table 10.4. As can be seen, the borohydride reduction method results in smaller particles with a significantly greater specific surface. Similarly, Wang and Zhang

Table 10.4 Properties of iron nanoparticles produced by different methods.

Properties	Nanoiron	
	Fe/B	RNIP
Average primary size (nm)	30–40	40–60
Shape	Spherical	Spherical
Boron (wt%)	5	$<10^{-4}$
Specific surface area (m²/g)	36.5	23
Initial Fe^0 content (wt%)	97±8	26.9±0.3

(1997) used TEM to analyze the morphology of the particles and found that the BET surface area of NIP and micro-scale iron particles was 33.5 m²/g and 0.9 m²/g, respectively. NANO IRON, s.r.o. company, located in the Czech Republic, produces and markets iron nanoparticles with an average size of 50 nm, an average surface area of 20–25 m²/g, a narrow particle size distribution of 20–100 nm and a high content of iron in the range of 80–90%.

Schrick *et al.* (2002) used X-ray diffraction and transmission electron microscope to characterize NIP particles. The results showed values in the range of previous studies. NIP sizes vary from 3 to 30 nm in diameter and they appear like chains of beads. The BET surface area of NIP was 18 m²/g. Nurmi *et al.* (2005) compare the properties of NIP particles produced by different methods using TEM, XRD and XPS. The NIP particles from Toda America Inc. consisted of elemental iron (alpha-Fe0) and Fe_2O_4. The α-Fe^0 is about 30 nm and the oxide is ~60 nm in size. The surface consists mainly of Fe, O and small amounts of S, Na, and Ca. The presence of S is due to the use of $FeSO_4$ and that the sulfur (S) seems to play a role in its reactivity. The ratio of Fe/O ranges from 0.72 to 1.15. The size of borohydride reduced NIP is <1.5 nm and the particle aggregate diameter is between 20 and 100 nm. They contain less iron and sulfur, but more boron and the Fe/O ratio varies from 0.4 to 0.55. It has a thin oxide layer and an outer layer of mainly sodium borate.

3.3   Reactivity of NIP

The reactivity of nanoscale iron nanoparticles upon organic contaminants in soils and groundwater is affected by the concentration of NIP, reaction time and the concentration and chemical nature of the organic contaminant. Variables that affect the reaction rate are pH, ORP, temperature, oxygen concentration: i.e. aerobic or anaerobic conditions (Tratnyek & Johnson, 2006) and ionic strength, although its effect is only important at low concentration of NIP (Okinaka *et al.*, 2005b). When NIP is injected into a contaminated soil, the following steps are considered to occur in the degradation of halogenated organic contaminants: (i) the organic halogenated compound migrates towards the surface of NIP; (ii) where it is absorbed; (iii) a dehalogenation reaction takes place upon the nanoparticle surface; (iv) the reaction products are desorbed, and finally (v) the reaction products migrate to the bulk solution (Shih *et al.*, 2009).

Liu *et al.* (2005a) investigated the reaction mechanisms of NIP produced by Toda America Inc. (RNIP) and nanoparticles produced by borohydride reduction using TCE as a contaminant. About 92% of the iron was available for the reaction using borohydride reduction

NIP, whereas only 55% of the iron was available when RNIP was used. Despite the difference, the amount of TCE reduced by unit mass of nanoparticles was essentially the same. However, the resulting reaction products were very different: RNIP transformed TCE in unsaturated hydrocarbons, mainly acetylene and ethane, whereas the borohydride reduction NIP produced ethane (80%) and C3–C6 coupling products (20%). Liu et al. (2005b) investigated the effect of the structure (crystalline or amorphous) on reactivity and the possible addition of external H_2 to improve the reduction of TCE. The crystalline nanoparticles did not produce hydrogen nor could use external hydrogen and most of the reaction products were unsaturated hydrocarbons. However, the nano-crystalline nanoparticles were able to produce H_2 and to use external H_2, which help in the total hydrodechlorination of TCE. This fact was related to the weaker atomic interactions in the non-crystalline nanoparticles. On the other hand, the non-crystalline nanoparticles were easier to be oxidized compared to the crystalline particles.

The properties of NIP change with time, a process known as aging. In general, contact with oxygen produces the oxidation of the nanoparticles forming a layer of oxide, changing the surface properties and therefore its reactivity towards the target contaminant. Anyway, even under proper storage conditions, without the exposition to atmospheric oxygen, the reactivity of iron nanoparticles decrease along time. For instance, the company NANO IRON, s.r.o. (Czech Republic) reported a significant loss of activity as a function of time of NIP's being in contact with water. The amount of iron in the NIP suspension was stable for 20 days but it started to decrease rapidly at room temperature (20–22°C) reaching a 75% of loss in 70 days, whereas in a cooled sample (2–4°C) the loss of iron was minor, less than 10%. The loss of iron is associated with the dissolved oxygen in water, but also with the reaction with water. Temperature is critical to maintain activity for long time. Due to the importance of aging of NIP, as it affects the properties of the particle, Liu and Lowry (2006) investigated the effect of particle age of RNIP on reactivity of TCE. The study showed that the initial iron content tends to decrease with time over a period of two years despite having been stored in anaerobic conditions.

Tratnyek and Johnson (2006) reported that NIP possesses high reactivity due to its high surface area, greater density of reactive sites and higher intrinsic reactivity on reactive sites. They tested the treatment of three isomers of chlorophenol (2-CP, 3-CP, 4-CP). The reaction rate was higher with 2-CP confirming that the Cl position in the aromatic ring affects reactivity, probably due to a major electronic acceptance in the 2 position than in the other two positions tested. Also temperature affected the rate of removal and reaction pathway as dechlorination was predominant at higher temperature (300°C) whereas adsorption was the leading process at low temperature of 100°C (Cheng et al., 2007; Cheng et al., 2008).

Choe et al. (2001) used NIP for the degradation of several halogenated compounds in anaerobic conditions: trichloroethylene, chloroform, nitrotoluene, nitrobenzene, dinitrobenzene, and dinitrotoluene. They predicted degradation of organic contaminants by the following reaction pathways (Equations 4–8):

$$Fe^0 \rightarrow Fe^{2+} + 2e \tag{10.4}$$
$$R\text{-}CL + H^+ + 2e^- \rightarrow R\text{-}H + Cl^- \tag{10.5}$$
$$ArNO_2 + 6H^+ + 6e^- \rightarrow ArNH_2 + 2H_2O \tag{10.6}$$
$$2H_2O + 2e^- \rightarrow 2OH^- + H_2 \tag{10.7}$$
$$O_2\,(g) + 2H_2O + 4e^- \rightarrow 4OH^- \tag{10.8}$$

The study showed very interesting results for the degradation of such contaminants. Chloroform was completely dehalogenated within 5 minutes and 80% of the products was methane. Nitro compounds were also transformed in 5 minutes and the reaction products were different than the amine group: 90% of TCE was transformed in 30 minutes and 70% of the reaction products was ethane. After 30 minutes of reaction, 100% aniline was produced from nitrobenzene, 85% of toluidine from nitrotoluene, 80% of benzenediamine from dinitrobenzene, and 70% of diaminotoluene was produced from dinitrotoluene. NIP completely removed nitroaromatic compounds within 30 minutes of reaction. The reaction kinetics was pseudo-first order and showed an average rate of 0.216 L/min (Welch & Riefler, 2008; Choe *et al.*, 2001). Shih *et al.* (2009) were able to dechlorinate 60% of hexachlorobenzene (HCB) with NIP after 24 hours reaction. High energy is required for the degradation of multiple-bond compounds and more functional group compounds result in lower reaction rate with the nanoscale iron particles (Choe *et al.*, 2001).

4   MODIFIED NANOSCALE IRON PARTICLES (NIP)

NIPs are very reactive due to the intrinsic properties of the zero valent iron combined with the high specific surface of the particles at nanoscale. However, reactivity and transformation yield also depend on the target contaminants. In order to enhance the reactivity of specific contaminants, several studies have modified the structure or composition of NIP particles. On the other hand, NIPs are not stable due to agglomeration and sedimentation; therefore, modifications are proposed to achieve stable NIP that can be available for reaction with the target contaminants.

4.1   Modification to increase reactivity

Several studies have reported methods to increase the reactivity of NIP. One approach is to combine NIP with a noble metal to form composite nanoparticles. The particles are called bimetallic nanoparticles and examples are Ag/Fe, Ni/Fe, Pd/Fe, and Cu/Fe. The noble metal acts as a catalyst and does not allow NIP to be easily oxidized by protecting the iron core from rapid oxidation (Li *et al.*, 2006). Bimetallic particles have the potential to degrade most organic contaminants in aqueous solutions. Thus, for instance, nano palladized-iron (Pd/Fe) nanoparticles have been used in the degradation of TCE, reaching a complete transformation about seven times faster than NIP (Wang & Zhang, 1997). Nano-bimetallic particles have been used to degrade different contaminants such as dichloromethane, chloroform, carbon tetrachloride, TCE, VOCs, trichlorobenzene; and they were also used in the treatment of metal and non-metal contaminants like Pb(II), Cr(VI), and arsenate. The combination of iron with other metals (usually noble metals) promotes catalytic properties and therefore exhibits rapid initial rate of reaction (Wang *et al.*, 2009; Schrick *et al.*, 2002). Unfortunately, the initial fast reaction rate cannot be sustained and the cost of treatment using such bimetallic nanoparticles increases, making them less attractive for use. Another downside of the bimetallic nanoparticles is the toxicity of the metals which may create new metal contamination problem to deal with.

4.2   Modification to increase stability

Stability of NIP is another major concern. NIP exhibit certain properties that result in aggregation and premature precipitation, therefore, uniform delivery and distribution into the subsurface soil is very difficult to achieve. Those properties are magnetic attraction, electrostatic or van der Waals attraction, and high density compared to the fluid (usually water). Researchers have been trying to modify the surface of the NIP to make them more stable in suspension and ensure adequate reactivity as well as mobility when delivering them into the contaminated subsurface zones. To achieve this, several different modifiers have been investigated, including polyacrylic acid (PAA), guar gum, potato starch, triblock-polymer-modified, and commercially available polymer-modified nanoiron (Saleh *et al.*, 2007; Schrick *et al.*, 2004; Yang *et al.*, 2007; Tiraferri *et al.*, 2008).

Modifiers can change the properties of NIP both positively and negatively. Modified NIP are more dispersed and homogeneous, stable, and exhibit a lower sedimentation and aggregation rate (Yang *et al.*, 2007; Kanel *et al.*, 2008; Schrick *et al.*, 2004; Saleh *et al.*, 2007). However, the hydrodynamic diameter of the modified particles tends to change. For example, the addition of guar gum to 231 mg/L RNIP reduces the hydrodynamic radius but there was no effect when added to 1 g/L RNIP. Increase in ionic strength (NaCl) of bare RNIP tends to increase the hydrodynamic radius but it has no effect on RNIP modified with guar gum. However, addition of $CaCl_2$ increased the hydrodynamic radius of both bare and guar modified RNIP (Tiraferri *et al.*, 2008). The surface charge of NIP changes when they are modified depending on the modifier. Reactivity reduces by about a factor of 2–10 depending on the type of modifier (Saleh *et al.*, 2007). The charge on the particle became negative when modified with anionic polymer (Kanel *et al.*, 2008). Since the particles are surface mediated, a thin film is formed on the surface which can influence the reactivity negatively.

5 APPLICATIONS OF NIP FOR SOIL AND GROUNDWATER REMEDIATION

Many studies were conducted to investigate the reactivity of NIP with various target contaminants under different conditions (aerobic/anaerobic, pH, ORP, temperature, etc.). Those studies are very valuable to evaluate the possibilities of NIP in real contaminated sites. However, limited studies are available that deal with contaminants in subsurface soils and groundwater. Not only does the effectiveness of NIP at the field scale depend on the intrinsic reactivity of NIP but also on the proper delivery and uniform distribution in the subsurface.

5.1 Reactivity of NIP and modified NIP in soils

The small size of NIP particles make them suitable for direct injection into subsurface soil and groundwater to achieve degradation of contaminants *in situ*. It is very important to maintain high reactivity towards the target contaminants in order to achieve fast degradation and high contaminant transformation. Unfortunately, NIP reactivity in soils can be lower than in aqueous solutions due to rate-limited desorption or solubilization/dissolution of contaminants. For example, 38% of PCB was only degraded in soils compared to 90% in aqueous solutions and the difference in reaction rate is due to the PCB's difficulty in diffusing from the surface of the soil particles to the NIP surface for effective reaction (Varanasi *et al.*, 2007; Wang & Zhang, 1997). The loss in reactivity can be compensated with an increase of NIP concentration and reaction time (Chang *et al.*, 2005; Varanasi *et al.*, 2007; Chang *et al.*, 2007). Anyway, NIP effectiveness is always higher than micro-scale iron, which is also affected by a loss of reactivity in soils compared to aqueous solutions. Chang *et al.* (2007) reported that 60% of pyrene was removed by NIP from soils in Taiwan, while micro-scale iron was only able to achieve 11% removal from soil.

The loss of NIP reactivity in soils can be compensated with modified NIP particles. Thus, Elliott and Zhang (2001) compared the reactivity of NIP and modified nanoscale particles (bimetallic Fe/Pd) with TCE at the lab scale using groundwater and sediments from an actual contaminated site. The results showed that TCE was fully eliminated by 0.25 g Fe/Pd in 12 hours, whereas 0.1 g Fe/Pd took 2 days to completely remove TCE. The improvement in reactivity by the bimetallic nanoparticles may justify the use of more expensive nanoparticles.

The tendency of NIP to aggregate and settle can be avoided by the use of dispersants. Thus, for instance, Reddy and Karri (2008) and Darko-Kagya (2010) investigated the reactivity of NIP and lactate-modified NIP (LM-NIP) with pentachlorophenol and 2,4-dinitrotoluene in kaolin and field sand as representative low- and high-permeability soils, respectively. Lactate was used as a dispersant to improve the transportation and uniform distribution into the soil (Cameselle *et al.*, 2008, 2013). The use of higher NIP and LM-NIP improve the degradation of both contaminants in the two types of soil. However, the use of lactate as modifier of the NIP particles resulted in a loss of reactivity in all the cases tested. The degradation of both contaminants was significantly higher during initial stages, but tended to be similar in longer treatment times. The loss of reactivity was compensated with increasing treatment time.

5.2 Transport of NIP and modified NIP in soils

The successful use of NIP particles to remediate soils and groundwater depends on the intrinsic reactivity of the nanoparticles, but mainly on the injection and uniform distribution into the subsurface soil. In PRB the iron fillings act over the contaminants transported through the barrier by the groundwater, but the nanoparticles can be injected into the soil through wells or using other technologies such as soil fracturing. Thus, the remediation can be performed simultaneously in the whole contaminated area. This is possible due to the small size of the nanoparticles, which are smaller than the soil pores that permit the distribution of the nanoparticles to the entire soil in both the saturated and unsaturated zones.

Transportation of NIP can be affected or even impeded by the hydrodynamic characteristics of the soil, and the physicochemical interactions between the soil particles and NIP. Thus, the heterogeneity of the subsurface soil may lead to different NIP distributions in the soil, impeding the penetration of NIP in the low-permeability layers and reaching high NIP concentrations in high-permeability soil layers. Hydrophobic and hydrophilic interactions also lead to uneven NIP distributions. The aggregation of NIP particles and settling may result in clogging of soil pores impeding the transport of the nanoparticles, remaining concentrated around the injection point. The charge of the particle is important during transport and therefore zeta potential meter was used to investigate the charge of NIP. The zeta potential of the NIP has been found to decrease with increase in pH (Yang & Lee, 2005). The dispersants used to avoid NIP aggregation also change the zeta potential (Cameselle *et al.*, 2013).

The transport of bare NIP is very limited due to the agglomeration. Bare NIP was only transported for 2 cm in a sand-packed column, but there was no transport when higher ionic strength of 1 mM $NaHCO_3$ solution was used. Particle – particle interactions were more prominent during transport than particle – sand interactions (Saleh *et al.*, 2007). Two types of aggregation can occur during agglomeration: (i) iron particles can gather to form secluded aggregates of microsized particles and this tends to occur under 30 minutes in solution, and

(ii) the aggregates link to each other to form chains and this starts after 30 minutes. Particles tend to settle when they form aggregates since they become heavy. In loamy soil packed in a vertical column (Yang *et al.*, 2007) the estimated travel for bare NIP was even less than in sand, about 0.25 cm. The transport can be improved by increasing the flow rate when soil flushing is used in the delivery of the nanoparticles. Thus, NIPs are transported by advection and dispersion in the subsoil.

5.3   Enhanced transport of NIP and modified NIP in soils

NIP particles require modification to avoid premature aggregation and settling because these phenomena are mainly responsible for the impossible transportation of bare NIP into the subsurface soil. Modification of the NIP particles can be used to increase reactivity, but in this section, the modification to enhance transportation will be discussed. Anyway, any modification of NIPs aimed to improve transport has also to consider altering reactivity with respect to the target contaminants under subsurface conditions. This has been a major concern in the development of environmental processes using nanoparticles. Soil flushing, hydraulic delivery, and pressurized or pulsed pressurized systems were used to improve transport of NIP into the soil. Electrokinetic delivery of nanoparticles was also proposed and tested for bare and modified NIP to overcome the inefficacy of the mentioned techniques in low-permeability soils.

5.3.1   Enhanced transport of modified NIP

Various investigations used surfactants and polyelectrolytes to enhance transportation of NIP in soil while maintaining appropriate reactivity towards the target contaminants. Saleh *et al.* (2007) studied ways to enhance the NIP for effective transport using three different modifiers (triblock polymer modified, surfactant modifier and a commercially available polymer-modified nanoiron). Transport of bare RNIP through a sand column of low ionic strength showed that very low amount of iron eluted out about 1.4±3%. Bare RNIP got trapped within the first 1–2 cm of the column. When RNIP was modified, more particles were eluted (98% of the commercial modified RNIP and 95% of triblock polymer). The study showed that transport depends on particle concentration (180 mg/L RNIP eluted about 50% iron) and ionic strength. Increase in ionic strength decreases the transport abilities of both the bare RNIP and modified RNIP. In the case of bare RNIP, there was no transport when the ionic strength was increased more than 1 mM $NaHCO_3$. In the case of surface-modified RNIP there was no pore plugging and sticking which is a very interesting result for real soil applications. However, the concentration of additives or modifiers has to be optimized for each application. Polyacrylic acid was used as a modifier of NIP and better transportation was observed up to 6 g/L, but further increase hindered the NIP transport (Hydutsky *et al.*, 2007). Increase in flow rate also enhances delivery of NIP (Kanel & Choi, 2007).

Cameselle *et al.* (2013) investigated eight different dispersants at different concentrations to determine their ability to modify the surface characteristics and increase the stability of the NIP suspension, thereby minimizing agglomeration and sedimentation of the NIPs. The studied dispersants and their tested concentration ranges [as % of NIP suspension, weight-to-weight ratio (w/w)] were as follows: aluminum lactate (Al lactate; 2–15%), sodium lactate (Na lactate; 6–12%), ethyl lactate (EL; 6–12%), aspartic acid (ASP; 2–8%), polyacrylic acid (PAA; 2–8%), 2-hydroxypropyl-beta-cyclodextrin (1–4%), beta-cyclodextrin (BCD;

1–4%), and methyl-beta-cyclodextrin (MeBCD; 1–4%). Zeta potential measurements and column experiments were performed on NIP-dispersant suspensions. Results showed that the zeta potential of bare NIPs was 41.7±2.3 mV. The influence of the dispersants was found to vary significantly depending on the chemical nature of the dispersant and the electrical charge of the ions in solution. Aluminum lactate released Al^{3+} into the solution, resulting in a reduction of the NIP zeta potential from 37.7±1.8 mV at 2% concentration to 9.5±0.7 mV at 15% concentration. After their characterization, the relative effectiveness of each dispersant on the transport of NIP was investigated in column experiments. The porous media used was a clean natural fine to coarse sand. The study showed that 10% aluminum lactate exhibited the highest (93%) elution of the modified iron from the soil media (Figure 10.3). Aluminum lactate was selected as suitable dispersant for NIP to enhance transport in soils, based on the column experiments and zeta potential measurement results. Lactate is considered as a green (environmentally friendly) and cheap compound. Moreover, it also enhances bioremediation of contaminants in soils, making it a best choice if long-term residual treatment is relied on biodegradation. As previously explained, lactate modification also assures stability and adequate reactivity of NIP.

Reddy *et al.* (2008) investigated the transport and reactivity of bare and lactate-modified NIP by conducting horizontal column experiments using field sand contaminated with PCP. Bare NIP and modified NIP with 10% aluminum lactate were investigated at two different slurry concentrations of 1 g/L and 4 g/L. Lactate was found to prevent or slow agglomeration and settlement of NIP. Transport of NIP in experiments with bare NIP was not uniform and most of the PCP degradation occurred near the inlet where NIP could be transported during the initial stages of testing. Lactate enhanced the transport of NIP, but the reactivity

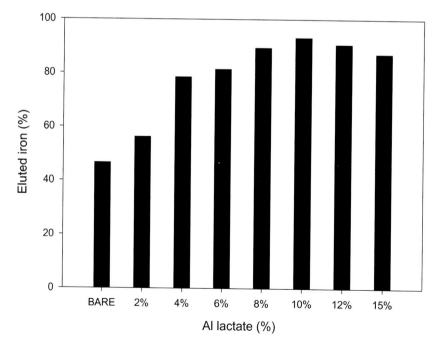

Figure 10.3 Elution of aluminum lactate modified NIP concentration.

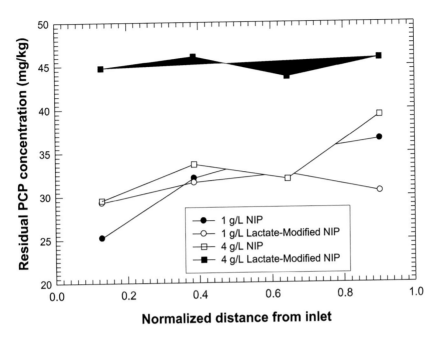

Figure 10.4 PCP distribution in the soil at the end of testing (Reddy *et al.*, 2008).

of NIP with PCP decreased as compared to the bare NIP experiments (Figure 10.4). The distribution of NIP was uniform in the 4 g/L modified NIP experiment compared to all other experiments. Hydraulic conductivity of the soil was measured during the course of each experiment and it remained approximately the same in all experiments except that it reduced in the experiment with bare NIP at 4 g/L concentration.

Darko-Kagya (2010) investigated the transport and reactivity of bare NIP and lactate-modified NIP (LM-NIP) in field sand spiked with 740 mg/kg 2,4-dinitrotoluene (DNT), a representative nitroaromatic organic compound found at munitions waste sites. Column tests were conducted using NIP and lactate-modified NIP (LM-NIP) at two different con-centrations (1 and 4 g/L) and two different flow velocities (0.75 and 1.2 cm/min). Changes in the magnetic susceptibility values across the soil column were measured during testing, and the effluent was collected and its volume was measured at different times. After the end of test, the soil was extruded and sectioned, and Fe and DNT concentrations were measured in each soil section and effluent sample. Measured iron concentrations in the soil and corresponding magnetic susceptibility were found to have a linear correlation. Therefore, the MS measurements made during the tests were used to assess the extent of the transport of NIP and LM-NIP at different times during the testing. The results showed that 4 g/L LM-NIP was transported more successfully than the 4 g/L bare NIP, but there was no significant difference in the transport of NIP and LM-NIP at the low concentration of 1 g/L. As a result of the enhanced transport, 4 g/L LM-NIP caused greater degradation of the DNT. The higher flow rate increased the transport of NIP and LM-NIP, thus increas-ing DNT degradation.

5.3.2 Pressurized/pulsed hydraulic delivery

Bare NIP cannot be transported in soils, not even in high-permeability sands. In order to overcome that limitation, the modification of NIP with polymers is a proposed solution even at commercial stage. Thus, Toda America Inc. developed three types of polymer-coated NIPs. Reddy *et al.* (2014) assessed the transport of the bare and the three modified NIPs in a sandy soil by performing a series of column experiments. The NIP suspension was added at the top of a 20-cm high sand column, and then simulated groundwater was used to flush the column. The eluting effluent was recovered at the bottom of the column and analyzed. The results showed that only one type of the modified NIP was effectively transported through subsurface soils under pressurized conditions. Darko-Kagya (2010) also demonstrated through two-dimensional bench-scale experiments that pressurized system is effective for the transport of lactate-modified NIP through different types of sand.

5.3.3 Electrokinetic delivery of NIP in soils

Electrokinetics is a technique for the transportation and removal of contaminants in soils. The transportation is carried out by electromigration and electroosmosis (Reddy & Cameselle, 2009) and these two transport mechanisms are very appropriate for the transportation of contaminants in low-permeability soils where the hydraulic gradients or pressurized flushing is completely ineffective. Similarly, the transportation of NIP in low permeability subsurface soil can be carried out by electrokinetics. Bare NIP marketed by Toda America Inc. shows a zeta potential of about 40 mV (Cameselle *et al.*, 2013), therefore NIP shows a positive electrical charge. The transportation of NIP into the soil matrix can be done by electromigration, or by electroosmosis if NIP is kept in suspension in the interstitial fluid.

Reddy *et al.* (2007) investigated the potential electrokinetic delivery of NIP in low-permeability soils contaminated with PCP and DNT. Kaolin soil was used as a model low-permeability soil and it was artificially spiked with PCP (1000 mg/kg of dry soil). During the experiment, substantial electroosmotic flow was observed and it was not hindered by the presence of NIP. The total iron in the soil increased from the anode to the cathode, indicating that NIP may have been transported towards the cathode. PCP was mainly degraded in the cathode by reductive dechlorination. Limited transport of PCP was observed due to its low solubility. It appears that NIP may not have contributed to PCP degradation.

Darko-Kagya (2010) investigated the transport and reactivity of NIP and lactate modified NIP (LM-NIP) in low permeability clayey soils contaminated with 1000 mg/L dinitrotoluene (DNT) under applied electric potential. NIP or lactate-modified NIP (LM-NIP) at a concentration of 4 g/L was injected at a distance of 3 cm from the anode to avoid the degradation of NIP due to the oxidizing conditions in the anode. The highest DNT degradation was achieved using LM-NIP. The total degradation of DNT was attributed to both NIP and electrochemical process. Overall it was found that the electrokinetic system can enhance the delivery of LM-NIP in low-permeability soils for the degradation of energetic organic contaminants such as DNT.

Yang *et al.* (2007) reported that electric field can enhance the transportation of NIP in loamy sand. Without the electric field the estimated travel of NIP was 0.25 m, whereas electric field increased the advance of NIP to 2.5 m, i.e. 10 times higher. The sticking coefficient of NIP to the loamy sand was reduced from 0.0061 to 0.0034 due to the application of the electric field.

Pamukcu *et al.* (2008) investigated the transport of modified NIP using electrokinetics. The NIP was coated with polyvinyl alcohol-co-vinyl acetate-co-itaconic acid. The particles were positively charged and moved from the anode to the cathode under an electric potential. A thin bed of clay (60% moisture content) was used as soil matrix. It was shown that electric field assisted the transport of surface-charged NIP from the anode to the cathode. However, diffusion did not help the migration of NIP as it was still stacked in the grove. Corrosion occurs at low pH and high ORP, whereas passivity occurs at the cathode end of the clay with high ORP.

Gomes *et al.* (2013) reported that limited transport, especially in low-permeability soils, is the biggest problem for the use of NIP in subsurface soil. These authors studied the enhanced transport of polymer-coated NIP at constant voltage application (5 DCV) in glass beads and kaolin clay. Experimental results indicate that the use of direct current can enhance the transport of the polymer-modified nanoparticles when compared with natural diffusion in low-permeability or surface-neutral porous medium. The applied electric field appeared to enhance the oxidation – reduction potential, creating a synergistic effect of NIP usage with electrokinetics. Aggregation of the nanoparticles, observed near the injection point, remained unsolved.

6   BIMETALLIC NANOPARTICLES

Bimetallic nanoparticles are a new emerging class of nanomaterials made of two different metals. Unlike zero valent iron nanoparticles (NIPs), bimetallic nanoparticles exhibit new functions because of the synergistic rather than merely additive effects of the metals.

Compared to NIPs, bimetallic nanoparticles show markedly enhanced physical and chemical properties, including magnetism and reducing ability. The most commonly cited and used bimetallic nanoparticles are made of iron and other metals such as Pd, Ru, Ni, La, Ti, Au, Pt, Cu, Ag, Co, and Sn.

The synthesis methods involve the preparation of a mixture of two salts that contain the two metals. Some bimetallic nanoparticles made of Fe with Pd, Pt or Ni can be prepared from a mixture of monometallic nanoparticles (Abdelsayed *et al.*, 2008). The typical and most commonly used synthesis procedures are physical or chemical process. For the physical process, mechanical alloying and radiolysis by microwave, ultrasonic, and pulsed laser are the most widely used methods. For the chemical process, thermal decomposition of metal complexes, chemical reduction, and electrochemical deposition are three of the most important methods. Depending on the synthesis procedure and the metals involved, the bimetallic nanoparticles show different structures: segregated core-shells, heterostructures, or alloyed structures (Figure 10.5).

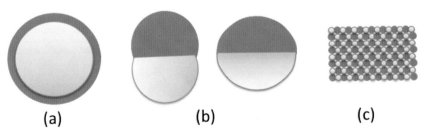

Figure 10.5 Structures of bimetallic nanoparticles: (A) core-shell segregated structure, (B) heterostructure, (C) alloyed structure.

Bimetallic nanoparticles have been widely tested to eliminate heavy metals as pollutants using their high redox activity. The removal mechanism implies redox reactions between metal contaminants and iron. The efficiency depends on the standard redox potentials of the metal contaminants and Fe. Considering that Fe has a low redox potential ($E° = -0.44$ V), it usually acts as a reductant. A typical example of heavy metal reduction with bimetallic nanoparticles is the case of Cr(IV): in the presence of zero valent iron, Cr(IV) is reduced to Cr(III) that is then easily precipitated with the appropriate anion forming an insoluble salt, or just increasing the pH, forming the corresponding hydroxide (Ponder *et al.*, 2000). Other metals or non-metals, such as As, Ni, Pb, Hg, Cd, and U can also be reduced to their zero valent form in the presence of iron nanoparticles, being finally immobilized *in situ* (Yan *et al.*, 2010a; Crane *et al.*, 2011). These metals can also be incorporated into the Fe nanoparticles to form *in situ* bimetallic nanoparticles in the reduction process with Fe^0, enhancing the removal efficiency. For example, Lien *et al.* (2007) have demonstrated that the co-existence of some metal ions [Pb^{2+}, Cu^{2+}, and Cr(VI)] in wastewater can significantly improve the dechlorination of carbon tetrachloride by NIP due to the in situ formed Fe-Pb, Fe-Cu, and Fe-Cr bimetallic nanoparticles in the reduction process.

Oxyanions nitrate, nitrite, bromate, chlorate, and perchlorate are also a concern as contaminants in soils and groundwater. They can be removed by reduction with bimetallic nanoparticles (Fe-Metal). Iron acts as an electron donor to produce hydrogen gas from H_2O, whereas the other metal activates the hydrogen gas and facilitates the formation of H_2O from the hydrodeoxygenation of oxyanions (Hurley & Shapley, 2007). The proposed pathway for the reduction of nitrate is described in Equations 9–13:

Electron transfer:

$$Fe \rightarrow Fe^{2+} + 2 e \qquad (10.9)$$
$$2 H_2O + 2 e^- \rightarrow H_2 + 2 OH^- \qquad (10.10)$$

Activate process:

$$2 Cu + H_2 \rightarrow 2 Cu\text{-}H \qquad (10.11)$$

Hydrogenation:

$$2 Cu\text{-}H + NO_3^- \rightarrow NO_2^- + H_2O + 2 Cu \qquad (10.12)$$
$$4 Cu\text{-}H + NO_2^- \rightarrow NH_4^+ + 2 H_2O + 4 Cu \qquad (10.13)$$

As can be seen, the final product of nitrate reduction is ammonium, which is considered a contaminant in water. The assimilation of ammonium is much faster than that of nitrate by soil microorganisms, so the combination of bimetallic nanoparticles with enhanced bio-remediation is a possible solution for nitrate contamination. The final product of the nitrite and nitrate reduction by bimetallic nanoparticles has been reported to be nitrogen gas or ammonium (Su & Puls, 2004). The selectivity of the treatment depends on the pH, ionic strength, initial nitrate concentration, and the chemical nature of the bimetallic nanoparticles. Selectivity towards ammonium production is favored by increasing the pH (Prüsse & Vorlop, 2001). The presence of HCO^{3-} also favors the selectivity of ammonium (Pintar *et al.*, 1998). The main effect for the formation of ammonium or N_2 is due to the intrinsic catalytic activity of the nanoparticles. Liou *et al.* (2009) confirmed that the Fe deposited Pd and Cu can promote the formation of N_2 in nitrate reduction because the adsorbed atomic hydrogen on the Cu and Pd surface can enhance the abstraction of oxygen from NO_x and the formation of N_2. Chaplin *et al.* (2012) demonstrated that the presence of Cu or Pd in the bimetallic catalyst favors the formation of nitrogen whereas the presence of Ni favors the formation of ammonia.

For the reduction of halogen oxyanions (BrO_3^-, ClO_3^-, ClO_4^-) with bimetallic nanoparticles, the final products are often halogen anions such as chloride or bromide due to the high reactivity of the intermediate products hypochlorite or chlorine (Srinivasan & Sorial, 2009; Ye *et al.*, 2012).

Similar to zero valent iron nanoparticles (NIPs), bimetallic nanoparticles can be used in the remediation of recalcitrant organic contaminants in soils and groundwater. Typical target organic contaminants are halogenated hydrocarbons, phenol and its derivatives, polychlorinated biphenyls, and other polyhalogenated aromatics (Kim *et al.*, 2008; Parshetti & Doong, 2012). As it was reported before, iron in the nanoparticles induces hydrodehalogenation reactions, but the use of bimetallic nanoparticles has many advantages compared to iron nanoparticles alone. First, the second metal in bimetallic nanoparticles, which has a lower hydrogen over-potential (e.g. Ni, Cu, and Pd), can also act as a catalyst that effectively enhances hydrodehalogenation reactions. Second, the metal can also act as an electron transfer medium that can overcome the self-inhibition of electron transfer in the reduction reaction. Third and last, bimetallic nanoparticles usually have a higher density of reductive surface sites (Liu *et al.*, 2014). Many factors affect the hydrodehalogenation reaction of bimetallic nanoparticles, among which pH is the most important. The system pH can affect the performance of Fe accelerating the corrosion of Fe at low pH, and passivating it by the formation of Fe hydroxides at high pH. Both processes clearly affect the reactivity of iron because hydrodehalogenation reactions occur on the surface of bimetallic nanoparticles. Apart from pH, other factors such as composition of bimetallic nanoparticles can also influence the hydrodehalogenation of organic contaminants. The molar ratio of two metals in the bimetallic nanoparticles plays an important role in determining the catalytic performance of bimetallic NPs because the addition of another metal with a different content can change the property of iron. Furthermore, the composition is closely related to the structure and surface chemistry, and changing the molar ratio of the two metals usually leads to a change in their structures, especially the atomic distribution of the two metals on the surface. In addition, the composition can also affect the morphology (crystal facet) and size of bimetallic nanoparticles, which is also important for their catalytic performance (Figure 10.5).

Nitroaromatic and azo compounds are two groups of important organic contaminants widely detected in the environment. Reduction through hydrogenation is a commonly used method to eliminate these pollutants. The first step is hydrogenation of the N–O bond to form nitroso compounds, which are further hydrogenated to hydroxylamine compounds, and final transformation into amine compounds. For azo compounds, the first step is hydrogenation of the $N{\equiv}N$ group to generate HN–NH, and then the N–N bond breaks to form amines. The amines can be further reduced to ammonium or nitrogen gas. Koutsospyros *et al.* (2012) used Fe–Cu and Fe–Ni bimetallic nanoparticles to degrade nitro compounds from munitions. Results indicate that bimetal NPs are highly effective for degrading these nitro compounds. pH and the dosage of nanoparticles are the main variables that affect the rate and yield of the reaction.

In recent years, significant progress has been made to apply bimetallic nanoparticles in environmental technologies. However, there are still some problems that hinder their application at a field scale. Some of these problems are common to the NIP particles because iron is the main component of the bimetallic nanoparticles. One of the most challenging problems is the inhibition and fouling of bimetallic NPs in the catalytic process by natural water constituents, such as humic acid, sulfur compounds and other inorganic solutes. It seems that one of the possible solutions is the design of fouling resistant nanoparticle structures. Considering that up to date most of the synthesis procedures of bimetallic nanoparticles are empiric or semiempiric, it will be very important basic research in the synthesis procedures to understand what are the key variables that govern the structure formation and their influence in the final nanoparticle reactivity. Another challenging problem is the leaching of metal species from bimetallic nanoparticles during remediation, which may generate secondary pollution in the remediated site. The cost of bimetallic nanoparticles and the associated synthesis procedures are still high, especially compared to NIP. Cost is a very limiting factor for the application of bimetallic nanoparticles at large scale in remediation projects.

7 OTHER NANOTECHNOLOGIES WITH ENVIRONMENTAL APPLICATIONS

7.1 Titanium dioxide

Titanium dioxide (TiO_2) has been used for the degradation of organic contaminants in water. Basically, TiO_2 is a solid catalyst of oxidation reactions activated by UV radiation. Titanium oxide can be used as a fine solid particle suspension in the water to be remediated or special catalysts with an inert support and TiO_2 as an active compound can be prepared and used. For instance, TiO_2 can be dispersed over porous silica with satisfactory results (Luo *et al.*, 2009). Recently, it has been found that titanium dioxide, when spiked with nitrogen ions or doped with metal or metal oxides also shows photocatalytic activity under visible light (Fan *et al.*, 2013).

Nanoparticles made of TiO_2 increase the specific area of the catalyst exposed to the medium, therefore, increasing both activity and reactivity towards the organic contaminants (Seitz *et al.*, 2012). Several types of nanoparticles have been synthesized and tested, including TiO_2 nanotubes. Bench-scale research has shown TiO_2 nanotubes are particularly effective at high temperatures and are capable of reducing a wide variety of organic contaminants (Huang *et al.*, 2010). The modification with nitrogen, metals or metal oxides also increases the catalytic activity of TiO_2 nanoparticles, and the transformation of the contaminants can be performed with visible light (Yang *et al.*, 2013). Anyway, the requirement of visible or UV light in the degradation of organic contaminants makes the use of TiO_2 nanoparticles an ex-situ technology because a good *in situ* illumination is not possible in the remediation site. Effective illumination usually implies the treatment to occur in a reactor that is designed for this purpose (Tratnyek & Johnson, 2006).

7.2 Self-assembled monolayers on mesoporous supports

Some materials can be specifically produced with surface functional groups to serve as adsorbents to scavenge specific contaminants from waste streams. Pacific Northwest National Laboratory (Richland, WA, US) has developed the SAMMS™ technology (self-assembled monolayers on mesoporous supports). SAMMS materials consist of a nanoporous ceramic substrate coated with a monolayer of functional groups tailored to specifically or preferentially bind to the target contaminant. The functional molecules are covalently bonded to the silica surface, leaving the other end of the functional group available to bind to a variety of contaminants. Although SAMMS materials are larger than the nanoscale range, they are considered a nanotechnology because of their nanoscale pores.

Contaminants successfully retained or sorbed to SAMMS materials include radionuclides, mercury, chromate, arsenate, and selenite (Tratnyek & Johnson, 2006). According to the information available from the Pacific Northwest National Laboratory, SAMMS materials have produced positive results in pilot-scale tests for remediation of mercury in water, using thiol-SAMMS that was specifically developed for the removal of mercury from (both aqueous and non-aqueous) liquid media. Since SAMMS materials adsorb the contaminants, these materials are used primarily as an *ex situ* treatment technology (Tratnyek & Johnson, 2006).

7.3   Dendrimers

Dendrimers are hyper-branched polymer molecules. The structure of dendrimers can be divided in three parts: core, branches, and end groups. In the outer surface of dendrimers, several functional groups can interact with the medium. Those functional groups can be designed or modified to show a specific chemical activity. Barakat *et al.* (2013) report on the effective immobilization of polyamidoamine dendrimers onto titania (TiO_2), and this composite was used for the effective remediation of Cu(II), Ni(II), and Cr(III), commonly found in industrial electroplating wastewater. Fe^0/FeS nanocomposites, synthesized using dendrimers as templates, could be used to construct permeable reactive barriers for remediation of contaminated groundwater. Dendrimers can be used *in situ* or *ex situ* (Xu & Zhao, 2006).

7.4   Swellable organic modified silica

Osorb is the commercial name for swellable, organically modified silica (SOMS) or glass capable of absorbing volatile organic compounds and other contaminants from water. The material has been nano-engineered to capture contaminants such as oil, pesticides, and pharmaceutical products. SOMS may be able to capture important amounts of organic contaminants present as vapors, liquids, or dissolved in water. Moreover, SOMS is hydrophobic and do not absorb water.

Some modification of SOMS may improve or widen its application field. For instance, the combination of SOMS with iron nanoparticles (NIPs) results in a material capable of dechlorinate captured organics. Furthermore, the silica matrix concentrates the organics to be degraded or dehalogenated by the embedded NIPs, avoiding any contact of NIPs with dissolved cations. Thus, the deactivation commonly observed in NIPs is avoided. Due to the stability of the NIPs embedded in SOMS, the material can be reused after the treatment, with a simple thermal desorption of the residual contaminants in the composite. The composite of palladium and SOMS is a similar metal nanoparticles–glass hybrid material used in *ex situ* treatment systems for remediation of chlorinated volatile organic compounds. Palladium nanoparticles are responsible for the degradation of halogenated compounds. The system requires the addition of H_2 to complete the reductive dehalogenation on Pd nanoparticles (Edmiston & Underwood, 2009).

8   FIELD APPLICATIONS

The application of nanotechnologies to the remediation of contaminated sites has to consider several factors: geology of the contaminated site (physicochemical properties and hydrodynamic behavior) and type and distribution of contaminants. Thus, the application of nanotechnologies is considered site specific. Once the site has been characterized, the design of the treatment has to identify the injection points, the dose, and the monitoring of the remediation.

Injection of nanoscale iron is typically done via direct injection through gravity feed or under pressure. Direct injection can be carried out through permanent or temporary wells, or by direct injection into the subsoil. Recirculation is another option that involves injecting nanoscale materials and extracting groundwater and reinjecting it into the treatment zone, possibly adding more nanoscale particles. This procedure assures a better distribution of nanoparticles in the subsoil while the spread of nanoparticles out of the contaminated area is significantly reduced. This method of *in situ* application keeps the groundwater in the aquifer in contact with the nanoparticles and also prevents larger agglomerated nanoparticles from premature settling, promoting continuous contact with the contaminant.

Additional methods for the delivery of nanoparticles in the subsoil for *in situ* treatment include pressure pulse technology, liquid atomization injection, pneumatic fracturing, and hydraulic fracturing. Pressure pulse technology uses large-amplitude pulses of pressure to insert the nanoparticle slurry into the subsoil below the water table to assure the transportation of nanoparticles by the groundwater. Liquid atomization injection is based on the direct injection of nanoparticle slurry in the subsurface using a carrier gas. The atomization of the liquid provoked by the gas assures a good distribution of nanoparticles in the subsoil. Pneumatic fracturing and hydraulic fracturing consist of the injection of liquid (water) or gas (air) to fracture the rock in the subsurface, creating a network of channels that favor the quick transportation of nanoparticles in soil layers of low permeability.

8.1   Field demonstrations and case studies

Nanotechnology is being considered as a promising technology for the remediation of contaminated sites, soil and groundwater, with heavy metals and other inorganics, hydrocarbons, organic solvents, chlorinated organics, and other hydrophobic and persistent organic contaminants. The main benefits of nanotechnology compared to other classical technologies (e.g. thermal desorption, soil flushing, chemical oxidation, etc.) is the promising results at lab and field scales and the relatively low cost. The most common nanoparticles used at a field scale are NIP, although other kind of particles were also tested: carbon nanotubes and fibers, bimetallic nanoparticles, nanoscale zeolites, and metal oxides (Patil *et al.*, 2016). Nanoscale ZVI is relatively easy to produce at a low cost. Thus, NIP is relatively cheap and readily available. Furthermore, their high surface area and reactivity make NIP an excellent reagent for soil and groundwater remediation at field scale. Other interesting properties that favor the selection of NIP are (Tosco *et al.*, 2014) the low standard reduction potential, favorable quantum size properties, and the possible transportation in the subsurface soil. NIP was satisfactorily tested at field scale in the transformation or degradation of arsenic, nitrate, heavy metals (e.g. Cr), organochlorides (solvents, PCBs, pesticides), and hydrocarbons, in both soil and groundwater (Karn *et al.*, 2009; Ghasemzadeh *et al.*, 2014). Bimetallic nanoparticles made of Ni, Ag, or Cu with Fe were tested in the immobilization or degradation of heavy metals and organochlorides, and energetic compounds (Nie *et al.*, 2013; Yan *et al.*, 2010b). Carbon nanotubes have been tested in the removal of organic contaminants from water due to the unique adsorption capacity of the nanotubes (Liang *et al.*, 2004; Savage & Diallo, 2005). Yu *et al.* (2014) reported the capacity of nanotubes in the adsorption of dyes, pesticides, pharmaceuticals, and other organics in water. Moreover, carbon nanotubes can be used in the absorption of arsenic and heavy metals.

Elliot and Zhang (2001) have reported an early study for *in situ* degradation of chlorinated organics in groundwater using nanoparticles of iron. These authors reported the tendency of the NIP to aggregate and settle in the injection well. Those nanoparticles that remained in suspension travel through the subsurface soil up to few feet. These results made that many studies focused on the stabilization of nanoparticle suspension to enhance the transportation in the subsurface soil and the remediation results.

Varadhi *et al.* (2005) applied modified NIP to remediate contaminated groundwater in Hamilton Township, NJ *in situ*. The site area was contaminated with 1,1-dichloroethane, 1,1-dichloroethene, 1,1,1-trichloroethane, 1,2-dichloroethane and 1,1,2-trichloroethane (TCE). Nanoiron concentration of 30 g/L was injected on a 20-foot grid pattern. The NanoFe Plus™ used was nanoiron particles modified to include a catalyst and support additive. Treatment was done in two phases and 3000 pounds and 1500 pounds were injected, respectively. Geologically the site consisted of 6 feet fill, underlain by 20 feet interbedded silt, clay, and lenses. The water table was at 21 feet below ground surface with perched water also at 2–8 feet below ground surface. The perched water zone contains high concentrations of dissolved hydrocarbons and metals, while the water table contains lower concentrations. The pH was less than 4, and this was solubilizing heavy metals present in the groundwater. The addition of nanoiron increased the pH to between 6 and 7, which helped to reduce the heavy metal concentrations. GeoProbe™ and truck-mounted drilling equipment was used for the injection and *on-site* mixing. Injection was performed at a pressure of 20psi, and nine monitoring wells were used for monitoring. Injection in the central portion of the site was easy due to the presence of mainly sandy soil as compared to the western and eastern portions where some difficulties were experienced due to the presence of silt and clay. The study reported 90% degradation of the dissolved contaminants, in particular in the central part, after two weeks of injection and this was confirmed by an increase in methane and ethane concentrations. The pH increased and ORP reduced.

Quinn *et al.* (2005) performed field application of the dehalogenation of TCE using emulsified nanoscale zero valent iron (EZVI) *in situ*. EZVI was injected into the saturated zone over five days. The team chose emulsified ZVI for the modification since it forms a membrane of oil-liquid droplets around the particles' biodegradable part, which helps degradation over the long term by donating electrons. The hydrophobic droplets prevent the reaction with unwanted targets. The RNIP was modified on site by the emulsifier. The lithology was characterized and slug tests were performed before injection. Soil and groundwater samples were analyzed before and after injection for TCE concentration. The study showed that more than 80% of TCE was degraded from the soil and about 68% was removed from the groundwater. Degradation was confirmed by observing the increase in concentration of cis-dichloroethene, vinylchloride, and ethane. However, the by-product of laboratory analyses was ethane. The difference was attributed to the fact that the degradation of TCE was not only due to dehalogenation but also biodegradation. The hydraulic conductivity before and after were 43 ft/day and 38.2 ft/day, respectively. The pressure pulse technology used for injection did not help the delivery of EZVI and therefore the pneumatic and direct push method was recommended.

Logan and Pastor (2007) performed a field pilot study experiment for the decontamination of VOCs using nanoscale zero valent iron particles. The site was a Nease chemical superfund site in Ohio. The facility operated between 1961 and 1973 whereby they produced household cleaning products, fire retardants, and pesticides. Mirex was the common chemical used. The soils had a mirex concentration of 2080 mg/kg and the VOCs concentration in the groundwater was 100 mg/L and consisted mainly of chlorinated ethane and ethene plus other chemicals such as benzene and toluene. The site geology consisted of fractured sedimentary bedrock that is overlain by glacial till. The area is hydraulically connected and the groundwater is about 1–9 feet below the ground surface. The NIP containing 20% organic dispersant and 1% palladium was injected in batches. A total of 100 kg NIP was injected at a rate of 0.15–1.54 gpm (0.5–5L/min) within 22 days. The volume of injection was about 2665 gallons (10 m³). Pressure injection systems were used and mixing was performed onsite. Water levels and geochemical parameters such as conductivity, pH, ORP, DO, temperature, and potentiometric head were monitored in four wells close by the injection area. The study included a bench-scale test prior to the field pilot tests. The bench-scale test was aimed to provide engineers the concentration to use and also to inform them about the by-products of dechlorination. After two weeks of the bench-scale test, almost all the contaminants were degraded with the exception of benzene. The pilot study showed 33–88% PCE degradation and 30–70% TCE degradation after 4 weeks of injection. Methane, ethane, and ethene were the major products but cis-DCE was also produced as a partial dechlorination of the compound. After 8–12 weeks of monitoring, the concentration was about the same or it had increased. The increase in concentration was attributed to the fact that the higher gradient was not treated and was polluting the low-gradient area. Monitoring is on-going as benzene still has not been degraded.

He *et al.* (2010) tested Fe-Pt nanoparticles stabilized with carboxymethyl cellulose in a contaminated aquifer in Alabama. These authors injected the nanoparticles in the aquifer and three monitoring wells permitted the monitoring of the remediation (one well was upstream and the other two downstream) of the groundwater contaminated with PCBs, TCE, and PCE. The suspension of Fe-Pt nanoparticles was prepared *in situ* to preserve the maximum activity of the nanoparticles. Two injections of nanoparticles were done in a period of 1 month. Then, the monitoring of groundwater revealed that Fe-Pt nanoparticles facilitated the rapid abiotic degradation of TCE in the first 2 weeks, and, most importantly, the nanoparticles favored the biotic dechlorination and degradation for the long term (about 2 years). The metal corrosion process act as a hydrogen source whereas the carboxymethyl cellulose served as a carbon source for the aerobic microorganisms. Overall, 88% of TCE was eliminated from the aquifer. Similar to this study, Bennett *et al.* (2010) found that NIP stabilized with carboxymethyl cellulose was able to rapidly (in about 2 hours) dechlorinate chlorinated ethane, proving the rapid abiotic degradation.

Velimirovic *et al.* (2014) stabilized NIP with guar gum to be used in the remediation of a Belgian contaminated site with chlorinated aliphatic hydrocarbons. The NIP was directly injected in the site through a well at depth between 8.5 and 10.5 m. The effect of the NIP in the chlorinated hydrocarbons was only detected around the injection well due to the limited diffusion/transportation of nanoparticles far from the injection well. The delivery of the nanoparticles in the subsurface can be improved by hydraulic fracturing and the complex electrical conductivity imaging can be used to monitor the transportation of the nanoparticles in the subsurface (Flores-Orozco *et al.*, 2015). Luna *et al.* (2015) was able to reach a better transportation of the NIP stabilized with guar gum because of the higher hydraulic permeability of the soil in the contaminated site. Thus, the NIP were transported about 1 m far from the injection point, and in this area almost complete removal of chlorinated solvents was observed in just 1 day after nanoparticle injection, due to the dechlorination reaction with the elemental iron.

Overall, the remediation studies at a field scale proved the promising capability of nanoparticles in the remediation of contaminated sites, resulting in rapid removal/stabilization of contaminants due to the direct reaction of the contaminants on the nanoparticle surface, then enhancing the biotic degradation at long term with the nanoparticles act as an electron acceptor/donor in the biotic respiration reactions. Variables such as temperature, organic matter in soil, pH ionic strength, type and nature of contaminants, and the possible effect of other additives (Henderson & Demond, 2007; Zhao *et al.*, 2016) are the main variables to control for a successful operation in the remediation of contaminated sites.

8.2   Cost

The cost of nanotechnology in environmental application depends on several factors. The first factor is the cost of the nanoparticles themselves. The most commonly used nanoparticles are made of zero valent iron, which is cheap and readily available. However, the cost of nanoparticles also depends on the synthesis procedure. The cost of NIP has decreased due to a better understanding of the synthesis procedures and the larger number of manufacturers and vendors. Thus, the NIP cost was nearly US$ 500/kg in 2001 and decreased to about US$ 50 to US$ 100/kg by 2006. Moreover, the quality of NIP increased. Decreasing NIP costs and improved quality control procedures that provide better homogeneity make this technology more viable in today's remediation market. In addition to the costs associated with the nanoscale material, other factors contribute to the cost of site remediation such as site type, contaminant type, contaminant concentration, extent of the contaminant plume, and any challenges that may occur during remediation.

9   ENVIRONMENTAL BENEFITS AND RISKS OF NIP (AND OTHER NANOPARTICLES)

Limited knowledge is available about the fate and transport of nanoscale materials in the environment. Nanoscale materials made of iron are the most widely used nanoscale materials in site remediation but there is a limited knowledge of the final fate of NIP after remediation. Moreover, limited research is available regarding the potential toxicological effects of nanoscale materials.

During and after remediation, nanoparticles may be transported by the aquifers and are expected to be oxidized forming precipitates or remaining soluble in the groundwater. Under standard environmental conditions, i.e. aerated water and pH = 5–9, Fe^{2+} will readily and spontaneously oxidize to Fe^{3+} and precipitate out of the groundwater as insoluble iron oxides and oxyhydroxides. However, NIPs have been modified to increase their reactivity and stability. They reduce agglomeration of nanoscale materials and maximize subsurface mobility, so the exposure of living microorganism to NIP is increased. Furthermore, substances considered non-toxic at the macroscale may have negative impacts on human health when nanoscale particles are inhaled, absorbed through the skin, or ingested. Because of the small size of nanoscale materials, the particles have the potential to migrate to, or accumulate in, places that larger particles cannot, such as the alveoli in the lungs thereby potentially increasing toxicity.

Moore (2006) points out the evidence for the harmful effects of nanoscale combustion-derived particulates (ultrafines), which when inhaled can cause a number of pulmonary pathologies in mammals and humans. However, release of manufactured nanoparticles into the aquatic environment is largely unknown. Moore (2006) suggests the possible nanoparticle association with naturally occurring colloids and particles and how this could affect their bioavailability and uptake into cells and organisms. Klaine *et al*. (2008) suggest the possible negative effects of nanoparticles in the environment, but the evaluation of the risk requires more research and standardized testing protocols.

U.S. Environmental Protection Agency's (EPA) Office of Research and Development (ORD) published a Nanomaterial Research Strategy (NRS) in June 2009 to act as a guide for nanotechnology research within ORD. The NRS centers around progressing EPA's understanding of nanomaterials for decision support and focuses on four research themes (US EPA, 2009):

- Identifying sources, fate, transport, and exposure.
- Understanding human health and ecological research to inform risk assessment and test methods.
- Developing risk assessment approaches.
- Preventing and mitigating risks.

Hjorth *et al*. (2015) have studied the ecotoxicological effects of various nanoparticles in organisms and ecosystems. These authors reported the technical limitation of ecotoxicity testing of nanoparticles in the environment and even questioned if the environmental risk assessment needed for engineered nanoparticles is feasible. Those limitations are mainly related with aggregation, agglomeration, sedimentation, and other physicochemical effects that the nanoparticles undergo in the environment and that may affect the measurement in the toxicity tests. Overall, Hjorth *et al*. (2015) found that nanoparticles made of iron oxides, iron zeolites, and iron-carbon composite showed low toxicity, below the regulatory limits of the UE. The milled iron nanoparticles showed significant ecotoxicity, detected in bacteria, earthworms, and plants. It can be concluded that the ecotoxicity of the nanoparticles in related to their compositions and the mode of fabrication. Furthermore, the nanoparticle stability and dispersion in the environment will be decisive for the assessment of their toxic effect at the mid- and long term.

REFERENCES

Abdelsayed, V., Glaspell, G., Nguyen, M., Howe, J.M. & Samy El-Shall, M. (2008) Laser synthesis of bimetallic nanoalloys in the vapor and liquid phases and the magnetic properties of PdM and PtM nanoparticles (M = Fe, Co and Ni). *Faraday Discussions*, 138, 163–180.
Barakat, M.A., Ramadan, M.H., Alghamdi, M.A., Algarny, S.S., Woodcock, H.L. & Kuhn, J.N. (2013) Remediation of Cu(II), Nni(II), and Ccr(III) ions from simulated wastewater by dendrimer/titania composites. *Journal of Environmental Management*, 117, 50–57.
Bennett, P., He, F., Zhao, D.Y., Aiken, B. & Feldman, L. (2010) *In situ* testing of metallic iron nanoparticle mobility and reactivity in a shallow granular aquifer. *Journal of Contaminant Hydrology*, 116(1–4), 35–46.

Bigg, T. & Judd, S.J. (2000) Zero-valent iron for water treatment. *Environmental Technology*, 21(6), 661–670.

Cameselle, C., Darko-Kagya, K., Khodadoust, A. & Reddy, K.R. (2008) Influence of type and concentration of dispersants on the zeta potential of reactive nanoiron particles. *Proceedings of International Environmental Nanotechnology Conference*. USEPA, Chicago, IL, USA.

Cameselle, C., Reddy, K.R., Darko-Kagya, K. & Khodadoust, A. (2013) Effect of dispersant on transport of nanoscale iron particles in soils: Zeta potential measurements and column experiments. *Journal of Environmental Engineering*, 139(1), 23–33.

Cao, J., Clasen, P. & Zhang, W. (2005a) Nanoporous zero-valent iron. *Journal of Materials Research*, 20(12), 3238–3243.

Cao, J., Elliott, D. & Zhang, W. (2005b) Perchlorate reduction by nanoscale iron particles. *Journal of Nanoparticle Research*, 7, 499–506.

Chang, M., Shu, H., Hsieh, W. & Wang, M. (2005) Using nanoscale zero-valent iron for the remediation of polycyclic aromatic hydrocarbons contaminated soil. *Journal of the Air and Waste Management Association*, 55(8), 1200–1207.

Chang, M., Shu, H., Hsieh, W. & Wang, M. (2007) Remediation of soil contaminated with pyrene using ground nanoscale zero-valent iron. *Journal of the Air and Waste Management Association*, 57(2), 221–227.

Chaplin, B.P., Reinhard, M., Schneider, W.F., Schüth, C., Shapley, J.R., Strathmann, T.J. & Werth, C.J. (2012) Critical review of Pd-based catalytic treatment of priority contaminants in water. *Environmental Science and Technology*, 46(7), 3655–3670.

Cheng, R., Wang, J. & Zhang, W. (2007) Comparison of reductive dechlorination of p-chlorophenol using Fe^0 and nanosized Fe^0. *Journal of Hazardous Materials*, 144(1–2), 334–339.

Cheng, R., Wang, J. & Zhang, W. (2008) Degradation of chlorinated phenols by nanoscale zero-valent iron. *Frontiers of Environmental Science and Engineering in China*, 2(1), 103–108.

Choe, S., Lee, S., Chang, Y., Hwang, K. & Khim, J. (2001) Rapid reductive destruction of hazardous organic compounds by nanoscale Fe^0. *Chemosphere*, 42, 367–372.

Crane, R.A., Dickinson, M., Popescu, I.C. & Scott, T.B. (2011) Magnetite and zero-valent iron nanoparticles for the remediation of uranium contaminated environmental water. *Water Research*, 45, 2931–2942.

Darko-Kagya, K. (2010) *Stability, Transport and Reactivity of Nanoscale Iron Particles for in-situ Remediation of Organic Pollutants in Subsurface*. PhD Thesis, University of Illinois at Chicago, IL, USA.

Edmiston, P.L. & Underwood, L.A. (2009) Remediation of dissolved organic pollutants in water using organosilica-based materials that rapidly and reversibly swell. *Materials Research Society Symposium Proceedings*, 1169, 35–41.

Elliott, D.W. & Zhang, W.X. (2001) Field assessment of nanoscale bimetallic particles for groundwater treatment. *Environmental Science and Technology*, 35(24), 4922–4926.

Fan, W., Bai, H., Zhang, G., Yan, Y., Liu, C. & Shi, W. (2013) Titanium dioxide macroporous materials doped with iron: Synthesis and photo-catalytic properties. *CrystEngComm*, 16(1), 116–122.

Flores-Orozco, A., Velimirovic, M., Tosco, T., Kemna, A., Sapion, H., Klaas, N., Sethi, R. & Bastiaens, L. (2015) Monitoring the injection of microscale zerovalent iron particles for groundwater remediation by means of complex electrical conductivity imaging. *Environmental Science and Technology*, 49(9), 5593–5600.

Ghasemzadeh, G., Momenpour, M., Omidi, F., Hosseini, M.R., Ahani, M. & Barzegari, A. (2014) Applications of nanomaterials in water treatment and environmental remediation. *Frontiers of Environmental Science & Engineering*, 8(4), 471–482.

Gomes, H.I., Dias-Ferreira, C., Ribeiro, A.B. & Pamukcu, S. (2013) Enhanced transport and transformation of zerovalent nanoiron in clay using direct electric current. *Water, Air and Soil Pollution*, 224(12), 1–12.

He, F., Zhao, D. & Paul, C. (2010) Field assessment of carboxymethyl cellulose stabilized iron nanoparticles for in situ destruction of chlorinated solvents in source zones. *Water Research*, 44(7), 2360–2370.

Henderson, A.D. & Demond, A.H. (2007) Long-term performance of zero-valent iron permeable reactive barriers: A critical review. *Environmental Engineering Science*, 24(4), 401–423.

Hjorth, R., Coutris, C., Nguyen, N., Sevcu, A., Baun, A. & Joner, E. (2015) Ecotoxicity testing of nanoparticles for remediation of contaminated soil and groundwater. *13th International UFZ-Deltares Conference on Sustainable Use and Management of Soil, Sediment and Water Resources, Copenhagen, Denmark*.

Huang, H., Zhou, J., Liu, H., Zhou, Y. & Feng, Y. (2010) Selective photoreduction of nitrobenzene to aniline on TiO_2 nanoparticles modified with amino acid. *Journal of Hazardous Materials*, 178(1–3), 994–998.

Hurley, K.D. & Shapley, J.R. (2007) Efficient heterogeneous catalytic reduction of perchlorate in water, *Environmental Science and Technology*, 41, 2044–2049.

Hydutsky, B.W., Mack, E.J., Beckerman, B.B., Skluzacek, J.M. & Mallouk, T.E. (2007) Optimization of nano and microiron transport through sand columns using polyelectrolyte mixtures. *Environmental Science and Technology*, 41, 6418–6424.

Kanel, S.R. & Choi, H. (2007) Transport characteristics of surface-modified nanoscale zero-valent Iron in porous media. *Water Science & Technology*, 55(1), 157–162.

Kanel, S.R., Goswami, R.R., Clement, T.P., Barnett, M.O. & Zhao, D. (2008) Two dimensional transport characteristics of surface stabilized zero-valent iron nanoparticles in porous media. *Environmental Science and Technology*, 42, 896–900.

Karn, B., Kuiken, T. & Otto, M. (2009) Nanotechnology and in situ remediation: A review of the benefits and potential risks. *Environmental Health Perspectives*, 117, 1823–1831.

Kim, J.-H., Tratnyek, P.G. & Chang, Y.-S. (2008) Rapid dechlorination of polychlorinated dibenzo-p-dioxins by bimetallic and nanosized zerovalent iron. *Environmental Science and Technology*, 42, 4106–4112.

Klaine, S.J., Alvarez, P.J.J., Batley, G.E., Fernandes, T.F., Handy, R.D., Lyon, D.Y., Mahendra, S., McLaughlin, M.J. & Lead, J.R. (2008) Nanomaterials in the environment: Behavior, fate, bioavailability, and effects. *Environmental Toxicology and Chemistry*, 27, 1825–1851.

Koutsospyros, A., Pavlov, J., Fawcett, J., Strickland, D., Smolinski, B. & Braida, W. (2012) Degradation of high energetic and insensitive munitions compounds by Fe/Cu bimetal reduction. *Journal of Hazardous Materials*, 219–220, 75–81.

Li, L., Fan, M., Brown, R.C. & Leeuwen, J.V. (2006) Synthesis, properties, and environmental applications of nanoscale iron-based materials: A review. *Environmental Science and Technology*, 36, 405–431.

Liang, P., Liu, Y., Guo, L., Zeng, J. & Pei, H.L. (2004) Multiwalled carbon nanotubes as solid-phase extraction adsorbent for the pre concentration of trace metal ions and their determination by inductively coupled plasma atomic emission spectrometry. *Journal of Analytical Atomic Spectrometry*, 19, 489–1492.

Lien, H., Jhuo, Y. & Chen, L. (2007) Effect of heavy metals on dechlorination of carbon tetrachloride by iron nanoparticles. *Environmental Engineering Science*, 24(1), 21–30.

Liou, Y.H., Lin, C.J., Weng, S.C., Ou, H.H. & Lo, S.L. (2009) Selective decomposition of aqueous nitrate into nitrogen using iron deposited bimetals. *Environmental Science and Technology*, 43(7), 2482–2488.

Liu, Y., Choi, H., Dionysiou, D. & Lowry, G.V. (2005a) Trichloroethene hydrodechlorination in water by highly disordered monometallic nanoiron. *Chemistry of Materials*, 17, 5315–5322.

Liu, Y. & Lowry, G.V. (2006) Effect of particle age (Fe^0 content) and solution pH on nZVI reactivity: H_2 evolution and TCE dechlorination. *Environmental Science and Technology*, 40(19), 6085–6090.

Liu, Y., Majetich, S.A., Tilton, R.D., Sholl, D.S. & Lowry, G.V. (2005b) TCE dechlorination rates, pathways, and efficiency of nanoscale iron particles with different properties. *Environmental Science and Technology*, 39, 1338–1345.

Liu, W., Qian, T. & Jiang, H. (2014) Bimetallic Fe nanoparticles: Recent advances in synthesis and application in catalytic elimination of environmental pollutants. *Chemical Engineering Journal*, 236, 448–463.

Logan, M. & Pastor, S. (2007) *Technology Update #2: Nanotechnology*. Environmental protection agency document. Nease Chemical Site. Columbiana County, OH, USA.

Luna, M., Gastone, F., Tosco, T., Sethi, R., Velimirovic, M., Gemoets, J., Muyshond, R., Sapion, H., Klaas, N. & Bastiaens, L. (2015) Pressure-controlled injection of guar gum stabilized microscale zerovalent iron for groundwater remediation. *Journal of Contaminant Hydrology*, 181, 46–58.

Luo, M., Bowden, D. & Brimblecombe, P. (2009) Removal of dyes from water using a TiO_2 photocatalyst supported on black sand. *Water, Air, and Soil Pollution*, 198(1–4), 233–241.

Moore, M.N. (2006) Environmental risk management – The state of the art. *Environment International*, 32(8), 967–976.

Muchitsch, N., Van Nooten, T., Bastiaens, L. & Kjeldsen, P. (2011) Integrated evaluation of the performance of a more than seven year old permeable reactive barrier at a site contaminated with chlorinated aliphatic hydrocarbons (CAHs). *Journal of Contaminant Hydrology*, 126(3–4), 258–270.

Nie, X., Liu, J., Zeng, X. & Yue, D. (2013) Rapid degradation of hexachlorobenzene by micron Ag/Fe bimetal particles. *Journal of Environmental Science*, 25, 473–478.

Nurmi, J.T., Trantnyek, P.G., Sarathy, V., Baer, D.R., Amonette, J.E., Pecher, K., Wandg, C., Linehan, J.C., Matson, D.W., Leepenn, R. & Driessen, M.D. (2005) Characterization and properties of metallic iron nanoparticles: Spectroscopy, electrochemistry, and kinetics. *Environmental Science & Technology*, 39, 1221–1230.

Okinaka, K., Jazdanian, A.D., Dahmani, A.M., Nakano, J., Okita, T. & Kakuya, K. (2005b). Degradation of trichloroethene with reactive nanoscale iron particles in simulated ground water. *Paper Presented at the ACS, Division of Environmental Chemistry – Preprints of Extended Abstracts*, 45(2), 662–666.

Okinaka, K., Jazdanian, A.D., Nakano, J., Kakuya, K., Uegami, M. & Okita, T. (2005a) Removal of arsenic, chromium and lead from simulated groundwater with reactive nanoscale iron particles. In: *Proceedings of Environmental. Nanotechnology. Symposia Papers Presented Before the Division of Environmental Chemistry*. American Chemical Society Washington, DC, USA. 6 pp.

Pamukcu, S., Hannum, L. & Wittle, J.K. (2008) Delivery and activation of nano-iron by DC electric field. *Journal of Environmental Science and Health*, 43, 934–944.

Parshetti, G.K. & Doong, R.A. (2012) Dechlorination of chlorinated hydrocarbons by bimetallic Ni/Fe immobilized on polyethylene glycol-grafted microfiltration membranes under anoxic conditions. *Chemosphere*, 86, 392–399.

Patil, S.S., Shedbalkar, U.U., Truskewycz, A., Chopade, B.A. & Ball, A.S. (2016) Nanoparticles for environmental clean-up: A review of potential risks and emerging solutions. *Environmental Technology & Innovation*, 5, 10–21.

Pintar, A., Vetinc, M. & Levec, J. (1998) Hardness and salt effects on catalytic hydrogenation of aqueous nitrate solutions. *Journal of Catalysis*, 174, 72–87.

Ponder, S.M., Darab, J.G. & Mallouk, T.E. (2000) Remediation of Cr(VI) and Pb(II) aqueous solutions using supported, nanoscale zero-valent iron. *Environmental Science and Technology*, 34, 2564–2569.

Prüsse, U. & Vorlop, K.D. (2001) Supported bimetallic palladium catalysts for water phase nitrate reduction. *Journal of Molecular Catalysis A*, 173, 313–328.

Quinn, J., Geiger, C., Clausen, C., Brooks, K., Coon, C., O'hara, S., Krug, T., Major, D., Yoon, W., Gavaskar, A. & Holdsworth, T. (2005) Field demonstration of DNAPL dehalogenation using emulsified zero-valent iron. *Environmental Science and Technology*, 39, 1309–1318.

Reddy, K.R. & Cameselle, C. (2009) *Electrochemical Remediation Technologies for Polluted Soils, Sediments and Groundwater*. Wiley, New York, NY, USA.

Reddy, K.R. & Karri, M.R. (2008) Removal and degradation of pentachlorophenol in clayey soil using nanoscale iron particles. *Proceedings of Geotechnics of Waste Management and Remediation*, Reston, Virginia, ASCE Press, *Geotechnical Special Publication*, 177, 463–469.

Reddy, K.R., Khodadoust, A.P. & Darko-Kagya, K. (2008) Transport and reactivity of lactate-modified nanoscale iron particles in PCP-Contaminated field sand. *Proceeding of the. International Environmental Nanotechnology Conference*. USEPA, Chicago, IL, USA. pp. 261–267.

Reddy, K.R., Khodadoust, A.P. & Darko-Kagya, K. (2014) Transport and reactivity of lactate-modified nanoscale iron particles for remediation of DNT in subsurface soils. *Journal of Environmental Engineering* (United States), 140(12). doi:10.1061/(ASCE)EE.1943-7870.0000870.

Reddy, K.R., Khodadoust, A.P. & Karri, M.R. (2007) Electrokinetic delivery of nanoscale iron particles for remediation of pentachlorophenol in clay soil. *Proceedings of the 6th Symposium on Electrokinetic Remediation*. Vigo, Spain. pp. 7–8.

Reibold, M., Paufler, P., Levin, A.A., Kochmann, W., Pätzke, N. & Meyer, D.C. (2006) Materials: Carbon nanotubes in an ancient Damascus sabre. *Nature*, 444(7117), 286.

Saleh, N., Sirk, K., Liu, Y., Phenrat, T., Dufour, B., Matyjaszewski, K., Tilton, R.D. & Lowry, G.V. (2007) Surface modifications enhance nanoiron transport and NAPL targeting in saturated porous media. *Environmental Engineering Science*, 24(1), 45–57.

Savage, N. & Diallo, M.S. (2005) Nanomaterials and water purification: Opportunities and challenges. *Journal of Nanoparticle Research*, 7, 331–342.

Schrick, B., Blough, J.L., Jones, A.D. & Mallouk, T.E. (2002) Hydrodechlorination of trichloroethylene to hydrocarbons using bimetallic nickel-iron nanoparticles. *Chemistry of Materials*, 14, 5140–5147.

Schrick, B., Hydutsky, B.W., Blough, J.L. & Mallouk, T.E. (2004) Delivery vehicles for zerovalent metal nanoparticles in soil and groundwater. *Chemistry of Materials*, 16, 2187–2193.

Seitz, F., Bundschuh, M., Dabrunz, A., Bandow, N., Schaumann, G.E. & Schulz, R. (2012) Titanium dioxide nanoparticles detoxify pirimicarb under UV irradiation at ambient intensities. *Environmental Toxicology and Chemistry*, 31(3), 518–523.

Shih, Y., Chen, Y., Chen, M., Tai, Y. & Tso, C. (2009) Dechlorination of hexachlorobenzene by using nanoscale fe and nanoscale Pd/Fe bimetallic particles. *Colloids and Surfaces A: Physicochemical Engineering Aspects*, 332, 84–98.

Srinivasan, R. & Sorial, G.A. (2009) Treatment of perchlorate in drinking water: A critical review. *Separation and Purification Technology*, 69, 7–21.

Su, C. & Puls, R.W. (2004) Nitrate reduction by zerovalent iron: Effects of formate, oxalate, citrate, chloride, sulfate, borate, and phosphate. *Environmental Science and Technology*, 38, 2715–2720.

Taghavy, A., Costanza, J., Pennell, K.D. & Abriola, L.M. (2010) Effectiveness of nanoscale zero-valent iron for treatment of a PCE-DNAPL source zone. *Journal of Contaminant Hydrology*, 118(3–4), 128–142.

Tiraferri, A., Chen, K.L., Sethi, R. & Elimelech, M. (2008) Reduced aggregation and sedimentation of zero-valent iron nanoparticles in the presence of guar gum. *Journal of Colloid and Interface Science*, 324, 71–79.

Tosco, T., Papini, M.P., Viggi, C.C. & Sethi, R. (2014) Nanoscale zero valent iron particles for groundwater remediation: A review. *Journal of Cleaner Production*, 77, 10–21.

Tratnyek, P.G. & Johnson, R.L. (2006) Nanotechnology for environmental cleanup. *Nanotoday*, 1, 2.

US EPA (2007) *Nanotechnology White Paper*. Report no. EPA 100/B-07/001. Available from: https://nepis.epa.gov/Exe/tiff2png.cgi/60000EHU.PNG?-r+75+-g+7+D%3A%5CZYFILES%5CINDEX%20DATA%5C06THRU10%5CTIFF%5C00000064%5C60000EHU.TIF. [Accessed 20th September 2017].

US EPA (2009) *Nanomaterial Research Strategy*. Report no. EPA 620/K-09/011. Available from: https://nepis.epa.gov/Exe/tiff2png.cgi/P10051V1.PNG?-r+75+-g+7+D%3A%5CZYFILES%5CINDEX%20DATA%5C06THRU10%5CTIFF%5C00000518%5CP10051V1.TIF. [Accessed 20th September 2017].

Varadhi, S.N., Gill, H., Liao, K., Blackman, R.A. & Wittman, W.K. (2005) Full-scale nanoiron injection for treatment of groundwater contaminated with chlorinated hydrocarbons. *Proceedings of Natural Gas Technologies, Orlando, FL*. 10 pp.

Varanasi, P., Fullana, A. & Sidhu, S. (2007) Remediation of PCB contaminated soils using iron nanoparticles. *Chemosphere*, 66, 1031–1038.

Velimirovic, M., Tosco, T., Uytteboek, M., Luna, M., Gastone, F., De Boer, C., Klaas, N., Spaion, H., Eisenmann, H., Lasson, P.O., Braun, J., Sethi, R. & Bastiaens, L. (2014) Field assessment of guar gum stabilized microscale zerovalent iron particles for in situ remediation of 1,1,1-trichloroethane. *Journal of Contaminant Hydrology*, 164, 88–99.

Wang, C. & Zhang, W. (1997) Synthesizing nanoscale iron particles for rapid and complete dechlorination of TCE and PCBs. *Environmental Science and Technology*, 31(7), 2154–2156.

Wang, X., Chen, C., Chang, Y. & Liu, H. (2009) Dechlorination of chlorinated methanes by Pd/Fe bimetallic nanoparticles. *Journal of Hazardous Materials*, 161, 815–823.

Welch, R. & Riefler, R.G. (2008) Estimating treatment capacity of nanoscale zero-valent iron reducing 2,4,6-trinitrotoluene. *Environmental Engineering Science*, 25(9), 1255–1262.

Xu, Y. & Zhao, D. (2006) Removal of lead from contaminated soils using poly (amidoamine) dendrimers. *Industrial and Engineering Chemistry Research*, 45(5), 1758–1765.

Yan, W., Herzing, A.A., Kiely, C.J. & Zhang, W.X. (2010a) Nanoscale zero-valent iron (nZVI): Aspects of the core – Shell structure and reactions with inorganic species in water. *Journal of Contaminant Hydrology*, 118, 96–104.

Yan, W., Herzing, A.A., Li, X.Q., Kiely, C.J. & Zhang, W.X. (2010b) Structural evolution of Pd-doped nanoscale zero-valent iron (nZVI) in aqueous media and implications for particle ageing and reactivity. *Environmental Science and Technology*, 44, 4288–4294.

Yang, G.C.C. & Lee, H. (2005) Chemical reduction of nitrate by nanosized iron: Kinetics and pathways. *Water Research*, 39(5), 884–894. doi:10.1016/j.watres.2004.11.030.

Yang, G.C.C., Tu, H. & Hung, C. (2007) Stability of nanoiron slurries and their transport in the subsurface environment. *Separation and Purification Technology*, 58(1), 166–172.

Yang, X.H., Fu, H.T., Wong, K., Jiang, X.C. & Yu, A B. (2013) Hybrid $Ag@TiO_2$ core-shell nanostructures with highly enhanced photocatalytic performance. *Nanotechnology*, 24(41), 415601.

Ye, L., You, H., Yao, J. & Su, H. (2012) Water treatment technologies for perchlorate: A review. *Desalination*, 298, 1–12.

Yu, J.G., Zhao, X.H., Yang, H., Chen, X.H., Yang, Q., Yu, L.Y., Jiang, J.H. & Chen, X.Q. (2014) Aqueous adsorption and removal of organic contaminants by carbon nanotubes. *Science of the Total Environment*, 482–483, 241–251.

Zhao, X., Liu, W., Cai, Z., Han, B., Qian, T. & Zhao, D. (2016) An overview of preparation and applications of stabilized zero-valent iron nanoparticles for soil and groundwater remediation. *Water Research*, 100, 245–266. doi:10.1016/j.watres.2016.05.019.

Chapter 11

Planning, monitoring, verification, and sustainability of soil remediation

K. Gruiz[1], M. Molnár[1] & É. Fenyvesi[2]

[1]Department of Applied Biotechnology and Food Science,
 Budapest University of Technology and Economics, Budapest, Hungary
[2]CycloLab Cyclodextrin Research & Development Laboratory Ltd, Budapest, Hungary

ABSTRACT

Sustainability as a decision-making criterion is increasingly gaining significance in contaminated site management. The state of the art of the development is reflected by regulations, management systems, standards, and protocols. The uniform assessment of environmental technology performance is the main component of sustainability analysis. A holistic approach is applied that evaluates the efficiencies not only from a strict technological and cost analysis perspective but also from socioeconomic and wider environmental/ecological aspects characterized by the sustainability footprint.

This chapter deals with sustainability from the perspective of environmental remediation, it gives definitions and describes the state of the art of sustainability management, the regulatory environmental technology verification (ETV) and sustainability assessment used in project implementation. The initiatives developed by experts in technology, economy, environment/ecology, and society are introduced. It shows that irrespective of the starting point many existing evaluation methods can be upgraded by sustainability indicators. "Green remediation" is very close to sustainable remediation, thus socioeconomic sustainability indicators can easily be added, similar to soil quality and function evaluation. Fully quantitative assessment methods such as cost–benefit assessment or mass balance-based evaluations can also include sustainability aspects if they are based on their cost and benefits or mass balance. Multi-criteria Analysis (MCA) can aggregate quantitative and qualitative information of various metrics.

In situ remediation attracts the greatest attention because the technology is in intimate contact with soil and groundwater which blurs the border between the negative impact of operation, emissions and disruptions to the local environment and the overall beneficial impact of the intervention on the wider environment. Another reason why *in situ* remediation at the center of interest is that its action is almost invisible (nothing happens on the surface), it is almost unknown and little trusted. Consequently it is not included in the operators' repertoire regardless of its capability of saving energy and cost and at the same time, fulfilling a number of sustainability requirements.

1 INTRODUCTION

Sustainable remediation and inexpensive, environmentally friendly, "green" technologies are in great demand both for the remediation of the still growing number of contaminated sites and for long-term quality management of soil and water. With the increasing number of innovative remediation technologies, information is needed for decision makers, owners, and other stakeholders regarding the characteristics and applicability of the best available technology. This complex information, which can properly control the decision-making process must be uniform, easily interpreted, and capable of supporting the entire management and decision-making process including decisions in the initial phases, corrective actions during technology application, and verification of the technology at the end of remediation and in long-term aftercare.

The main fields to be evaluated in support of the site management are technology, the local and wider environment – ecosystems and humans – and socioeconomic conditions together with the consequences. This framework of technology evaluation involves the concept, scope, and tools of the all-round assessment. The more complex the evaluation, the more versatile sets of assessment tools are required. The main parts of the evaluation tool are those which evaluate the performance of the technology in terms of its ability to accomplish remediation goals and at what cost. The "cost" is not only interpreted in economic sense but also including the burden and loss in ecosystem services and any other negative impacts on social and human health issues. The same can be said about the evaluation of the benefits and the comparison of all costs and benefits due to ecological, social, and economic changes. Sustainability means that the project within which the technology is applied can harmonize seemingly contradictory requirements and fulfill the goals of human development while ecological sustainability (ecosystem quality and services) is maintained during implementing remediation. Verification of the whole management process justifies all decisions and operations that have served the holistic improvement in order to minimize negative impacts caused by technology application and maximize benefits from the remediation. The scope of evaluation involves not only the three areas of environment, society, and economy, but covers spatial and temporal dimensions and extents and several interactions between the compartments.

The holistic approach in sustainability assessment should encourage soft technologies, "green" innovations, *in situ* and other knowledge- and experience-based solutions. *In situ* technologies based on natural biological processes are especially mistrusted and suffer from lack of confidence due to missing information, objective evaluation, and transparent verification, although they are the most promising tools in future environmental management.

2   SUSTAINABILITY, SUSTAINABLE REMEDIATION

Verified innovative site assessment and remediation techniques can achieve higher environmental quality at the same or lower costs compared to conventional solutions (Gruiz *et al.*, 2008). A practical and uniform evaluation tool can shorten the time between development and application and increase confidence and promote the commercialization of new remediation technologies.

Verification should not only include the demonstration of a good technological performance and check if it is "green," but also approve sustainability in the widest sense including socioeconomic and cultural elements (see also Gruiz, 2014). The definition of sustainability derived from the Brundtland Report (1987) says:

1   "Sustainable development meets the need of the present without compromising the need of future generations, while minimizing overall burdens to society."
2   "Sustainable development is the organizing principle for meeting human development goals while at the same time satisfying the ability of natural systems to provide the natural resources and ecosystem services upon which the economy and society depend."

Sustainability means much more today, not just the usual economy-based approach of sustainable (economic) development, but rather the aspects to protect human health and the environment as well as to serve socioeconomic and cultural developments and ethics – as the definition in the UK Sustainable Remediation Forum (SuRF UK) states: to demonstrate in terms of environmental, economic and social indicators that the benefit of undertaking remediation is greater than its impact (Smith *et al.*, 2010; Bardos *et al.*, 2011).

Sustainability and sustainable remediation need holistic thinking involving positive and negative social, economic and environmental impacts of a remediation project. The study of these disciplines improves project quality as the sustainable remediation white paper (Ellis & Hadley, 2009) summarizes: "It makes it possible to identify opportunities to improve the net benefit of the project and highlight specific negative project impacts that can be mitigated to limit their adverse socioeconomic and environmental impacts."

Green and sustainable remediation can easily be combined as the requirement of being green is the practice of considering all environmental effects of remedy implementation and incorporating options to maximize net environmental benefit of cleanup actions. Green remediation reduces the demand placed on the environment (footprint) during cleanup actions and avoids the potential for collateral environmental damage (US EPA, 2008).

The focus of green remediation is on reducing impacts on air pollution, water cycles, soil deterioration, ecological diversity and densities, and emission of greenhouse gases. Sustainability adds to the green side of the balance, socioeconomic elements to the other side and tries to find the optimum between these, often competing elements by considering the social and economic limitations. Environmental protection does not preclude economic development, and economic development is ecologically viable today and in the long run (US EPA, 2008, 2009). This means that economic development does not use more resources than can be replenished and does not produce more wastes that can be reabsorbed by the environment.

The optimum in a remediation project can be achieved using best management practices (BMPs) which are to be applied throughout the project from the first step of decision making and design, during construction, operation, monitoring, and the site-specific adaptations including evaluation. The dynamic character of BMP enables management to adapt to site-specific characteristics and challenges. BMP strives to achieve maximum environmental and social benefit at minimum socioeconomic costs involving current and long-term sustainable land uses. BMP helps decision makers, communities and other stakeholders to be aware and find the points where technological, environmental, social, and cost efficiencies can be improved. Innovative techniques or new strategies in terms of sustainability can also be involved.

2.1 State of the art

The technology verification programs all over the world try to establish a framework and databases for verified technologies. Most of them include environmental monitoring methods, water treatment technologies, and material and waste treatment technologies.

The EPA's Environmental Technology Verification (EPA ETV, 2018) program was established in 1994 and concluded in 2014. Twelve technology areas were covered by the program such as advanced monitoring systems for air, water, and soil. From 2014, EPA ETV changed and the Environmental and Sustainable Technology Evaluation (ESTE, 2018) program started with the aim of providing credible performance information on technologies that addressed high-risk environmental problems. Some soil remediation technologies have already been included in ESTE, e.g. ISCO (Dindal, 2016).

CLUE-IN Remediation (2018) has close cooperation with the ETV program concerning superfund sites and other properties where industrial or commercial activities have left a legacy of hazardous contaminants that limit future development. CLUE-IN is linked to Green Remediation (2018) through its website especially in terms of the basic principles and objectives of green remediation and best practices for reducing the environmental footprint of cleanup projects. Green remediation (2008) guides professionals how to incorporate sustainable environmental practices into the remediation of contaminated sites (see Section 3.3).

In a memorandum (GR Memo, 2016), the US EPA summarized the need to mitigate potential environmental impacts when implementing clean-up actions under the Comprehensive Environmental Response, Compensation and Liability Act (CERCLA, 1980). This document also emphasizes that greener cleanup activities are required throughout all phases of clean-up processes, including the following:

– Site characterization;
– Prospective evaluation of the technology (technological performance and costs);
– Feasibility study;
– Remedy selection;
– Technology implementation (design and construction of both technology and monitoring);
– Operation and maintenance.

A standard guide (ASTM E2893–16e1, 2016) provides the process for identifying, prioritizing, selecting, implementing, documenting, and reporting activities to reduce the environmental footprint of a clean-up activity by minimizing

– air pollutants and greenhouse gas emissions;
– water use and impacts to water resources;
– materials and waste;
– total energy use

while maximizing

– renewable energy;
– protecting land and ecosystems.

The guide also deals with the uncertainties in "greener cleanups."

Many other countries all over the world have started their own environmental technology verification and sustainability assessment programs such as ETV Canada (2018); ETV Japan (2018); the Korean NETV (2018) or ETV Philippines (2018). In Australia CRC CARE established a program in 2009 to develop a sustainable remediation framework applicable to their existing policies. SuRF-Australia has close links to the SuRF UK framework (CL:AIRE, 2018).

In Europe in addition to the intention of the European Union, Denmark, and the UK devoted significant efforts to develop uniform indicators and methods to properly assess remedy efficiency and sustainability (State of green, 2018).

The Danish Center for Verification of Climate and Environmental Technologies verification body (DANETV, 2018) accredited under the EU ETV Pilot Programme is responsible for testing, verifying, and documenting the performance of an environmental technology and offers independent testing of environmental and climate impacts within the areas of:

- Materials, waste, and resources;
- Environmental technologies for water treatment and monitoring;
- Energy efficiency and production;
- Environmental technologies for air cleaning and monitoring;
- Environmental technologies for soil and groundwater remediation and monitoring;
- Environmental technologies for agriculture;
- Cleaner production and processes.

The SuRF UK (2018) framework (Bardos, 2014, 2016a,b, 2018; CL:AIRE, 2010a,b) include the application of sustainable development principles for contaminated land management both in general terms and risk-based site remediation (Defra, 2004). The first comprehensive publications about SuRF UK (CL:AIRE, 2009, 2010a, 2011; 2014a,b; Bardos, 2010; Bardos et al., 2011) emphasized that a remedy improves the quality of a site which results in a less risky situation and better protects ecological and human receptors. However, this does not suffice to classify an environmental technology as sustainable, even if it supports sustainability to a certain scale. "Sustainable remediation" should by definition cover all required fields of sustainability such as the ecosystem and the socioeconomic and cultural items. But the process itself and methodologies identifying the optimum management strategy that maximizes the benefits while limiting the impacts of a remedy should be developed. SuRF UK, which is similar to the US EPA "green remediation" concept, identifies land contamination management criteria and strategies to find and implement the best alternative for a specific site (CL:AIRE, 2010a).

At the European level several initiatives and developments preceded the existing ETV (Environmental Technology Verification) and the Eco-innovation Action Plan (ECOAP, 2018) such as EURODEMO (2004–2008) and EURODEMO+ focusing on innovative remediation and the evaluation of the technology demonstrations. TESTNET was a project "Towards a European verification system for Environmentally sound Technology" with the aim of designing, developing and testing an ETV system for Europe. NICOLE (2018), the "Network of Industrially Co-ordinated Sustainable Land Management in Europe" (formerly Network for Industrially Contaminated Land), shifted its objective toward enabling European industry to identify, assess and manage industrially contaminated land efficiently, cost-effectively, and within a framework of sustainability. In order to achieve this objective, NICOLE provides a European forum for the dissemination and exchange of good practice, practical and scientific knowledge, and ideas to manage contaminated land in a sustainable way. NICOLE focuses on the issue of how to measure sustainability during a remediation project, how to use socioeconomic indicators for decision making, and sharing experience and information in the form of case studies. Another European platform is the Common Forum (2018), a network of contaminated land policy makers, regulators and technical advisors. Its mission is to exchange knowledge and experience, initiating international projects and function as a discussion platform on policy, research, technical, and managerial concepts of contaminated land.

ETAP (2018), Environmental Technologies Action Plan has established a mechanism to objectively validate the performance of environmental technologies and to increase purchasers' confidence in new ones. The European Commission has set up the pilot program of *Environmental Technologies Verification* (EU-ETV, 2018) in order to improve the competitiveness of European technologies on the Global market. It aims to develop pre-normative protocols for credible environmental technology verification that will be based on a performance verification concept.

The EU ETV Pilot Programme operates under the Eco-Innovation Action Plan of the European Commission (ECOAP, 2018) and works at a European level tool promoting eco-innovation. The program is targeted at environmental technologies whose value cannot be proved through existing standards or certification schemes and whose claims could benefit from a credible verification procedure as a guarantee to investors (What is ETV, 2018).

The International Working Group on ETV (IWG-ETV, 2018) is an international level cooperation on environmental technology verification. IWG ETV includes the Canadian, Korean, Philippine, European, and US ETV programs. The goal is to develop an international approach to verification that will allow mutual recognition: "Verify once, accept everywhere." The ISO ETV working group initiated an international standard on ETV and ISO, the International Standardization Organization (ISO) drafted a new standard on Environmental Technology Verification (ETV) and Performance Evaluation in 2013. The new standard was edited by the ISO Technical Committee 207 (Environmental Management) in 2016 (ISO 14034, 2016).

2.2   Technology verification and management sustainability

Decreasing adverse impacts/risks and improving environmental quality/benefits can characterize and prove performance quality of an environmental technology. Risk reduction and quality improvement may refer to the contaminated site itself, its surroundings, humans and ecosystems within the site or the wider environment. All ecosystem services related to adverse impacts over the short and long term must be involved in this analysis. In addition to direct environmental impacts, socioeconomic and cultural consequences should also be assessed. The performance of an environmental technology can be best characterized by the following three scenarios:

1   Assessing the contaminated site before remediation in its actual state: potential short- and long-term impacts and risks to environmental (ecosystem) and human health (both workers and residents) and the socioeconomic status (economic and social costs and benefits).
2   Impacts during remediation due to the operations: additional temporal costs, emissions and the relevant adverse impacts, typically energy, water and material use, emissions, and impacts on workers.
3   Assessing the site after remediation when the planned new land use has been established: long-term risks/costs and benefits in environmental and socioeconomic terms.

The generic evaluation of a technology – without knowing the actual site of application – cannot consider the site-specific characteristics, risks and benefits, only items closely related to the operations and machinery, i.e. the (energy and material consumption, labor demand, etc.) costs of and emissions from the technology (transport, machines operations, etc.). The disturbances to the surface and subsurface and degradation of the ecosystem caused by the technology can be characterized "in general," defined as:

(i) An average of former applications (if such data exists); or
(ii) As an application to a generic environment.

The latter one results in a relative outcome useful for comparison, for prospective evaluation and decision making.

A remedial technology can be

– sustainable (in general) based only on generic characteristics;
– having low energy and material consumption;
– no or low hazardous substance emission and other sustainability indicators.

An assessment method, e.g. life cycle assessment (LCA), can provide proper tools for such kind of generic evaluation, but it is not suitable for good decision making at the case level as the site-specific characteristics and quantitative values can modify the generic view. For example, natural attenuation (see Chapter 3) or other soft remediation can be the best option for a biodegradable toxicant from a sustainability point of view, but the site-specific environmental or social situation may require an urgent solution even at higher financial and environmental costs (temporarily increased risk). Green/soft remediations – generally classified as "sustainable" – may be accompanied by significant uncertainties due to the long-time requirement and unpredictable changes in the environmental conditions.

Sustainability assessment and technology verification are an integral part of environmental and quality management. It is useful to have an overview of the existing solutions, but it is even more important to see the details of the problem and the site from environmental and socioeconomic perspectives and to be able to predict the consequences of the technology and the project, including future land uses. This is why most guidelines recommend a *multi-stakeholder approach*; the involvement of different professionals, various interests and attitudes may enlarge the scope of sustainability and the valid representation of socioeconomic and ecological expectations.

2.3 Management process steps in terms of sustainability assessment

Sustainability management of an environmental technology should involve the following:

– A preliminary prospective evaluation of the available options to make the best decision on the best available remediation;
– Design of the selected technology;
– Monitoring planning to achieve information on operation and emissions and to compile the best, if possible, quantitative data acquisition methods;
– Continuous evaluation of monitoring data during operation, their comparison and adaptation to an optimum;
– Retrospective evaluation of the technological, environmental, and socioeconomic performance and efficiencies, i.e. case-specific verification of the technology and the project. Retrospective verification can validate the initial predictions and prove the good

performance of the technology and good quality of the site in terms of sustainability (technological and environmental performance, socioeconomic impact, etc.). Predicted and achieved results are linked by monitoring.

The prospective evaluation for decision making is often based on generic experience and on evaluation of former similar technology applications and typically uses database information. Even if the technology is not innovative and the application is "routine," every single case differs from the next. Thus, the prospective generic sustainability assessment can be based on the technology's intrinsic hazards and generic energy, water and material consumption, manpower need and potential environmental impacts (emissions, effect on ecosystem services, etc.). These results can be used for the evaluation and comparison of technological options, but they may be accompanied by large uncertainties for an actual case, especially for those that differ greatly from the average.

Continuous monitoring represents an iterative relationship between the measured parameters and the results derived from them and the predicted outcome (Figure 11.1). The difference is the driving force making changes in technological and monitoring parameters. The initial monitoring plan – with the selected parameters to be measured – reflects the knowledge and expectation of the professionals on the technology's effects and performance. This is based on generic information such as on a validated theoretical concept (e.g. volatile contaminants can be mobilized by heat), database information (e.g. LCA, or other generic databases), or on the results of other applications at sites and problems different from the actual one. Thus, the expected variability and uncertainties of the monitoring parameters should be checked in advance, and concentrate on those which may be responsible for the site-specific differences between planned and realized results. A dynamic monitoring plan integrating iterative loops makes modifications and changes during monitoring and can result in the best possible procurement of information. This is essential in view of the outcome of technology verification as it is strongly affected by the type and proportion of site-specifically measured and default/generic/read across data.

Verification using retrospective evaluation can strongly support the market entry of an innovative technology and enhance trust in less-popular, unknown technologies.

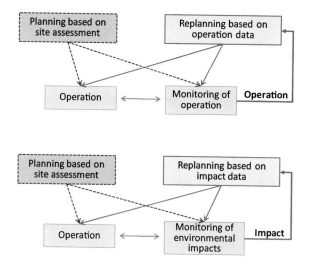

Figure 11.1 Adaptation of remediation to optimum operation and minimum environmental impact.

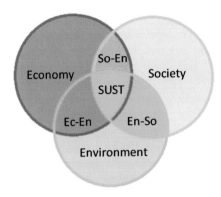

Figure 11.2 The three main components of sustainability.

Environmental technology verification systems all over the world pursue the collection and publication of controlled information on technology applications and innovative technology demonstrations. In relation to this goal, standardized assessment tools, accreditation or authorization systems, trademarks, and databases have been created in the last few years for environmental technologies.

The three main components of sustainability are the environment, society, and economy (Figure 11.2). A perfect balance is represented by the center where the three overlap. In this area the aggregated socioeconomic and ecological risks and costs are balanced by their aggregated benefits. If the benefit exceeds costs, an improved quality evolves and this should be kept in balance. The improved quality of the environment results in a better social/quality life. As the interaction between humanity and the environment/ecosystem is increasingly intensive, the two can only develop together with greater and greater overlap (Figure 11.2). Increasing knowledge and holistic care is needed to manage this development and keep the improved balance.

2.4 History of sustainability assessment and verification tools

The evaluation of a technology, especially of an innovative one, went through significant development toward a more holistic approach in the last 20 years. The basic requirement of

1   Technological performance – to fulfill the target quantity or quality – has constantly been expanded and integrated with the next point.
2   Economically based characteristics such as *cost effectiveness*, and later on cost–benefit assessment and the BATNEEC (Best Available Technology Not Entailing Excessive Costs) concept.
3   From the 1990s *the risk-based concept* came in addition to the technological performance and considers the impact on the environment in the form of risk and sets site-specific remediation goals. This approach made environmental impacts more manageable, even in a quantitative manner, but its shortcoming was that it has not included external social and economic impacts beyond identified environmental impacts. Nevertheless it may provide a strong basis for the shift toward sustainability.

4   From the 2000s the sustainable remediation concept has included the requirement of being beneficial for the society (health, education, employment, life quality, etc.), in addition to technological performance and economic and environmental efficiencies.

The versatility and universality of the holistic sustainability approach is nicely demonstrated by the fact that any of the former evaluation types can be put together under its umbrella, and what is even more convincing is that any of the concepts can be expanded to include all three domains. The economic approach can involve, and even can monetize, environmental values and services and social and health items and manage all the three together using cost–benefit thinking and tools. The risk-based concept can also involve wider ecological, and socioeconomic risks and manage all three risks together with risk management tools. The "green" concept puts the ecosystem in first place and links it to the socioeconomic domain. Finally, the engineering tools based on balances (mass balance, energy balance, material flows, and balances) can also be utilized for the holistic approach and extended toward cost–benefit, social, and ecological balances.

The inclusion of multiple-scale environmental and socioeconomic aspects – not just the site and its surroundings but the watershed and global footprints plus short and long terms – raised a large number of questions and debates. The main questions arise in scoping (temporal and spatial scales and evaluation scoping), and in selecting proper indicators (fitting to the case) which correctly characterize the technological, economic, environmental, and social efficiencies. The example in Figure 11.3 demonstrates that the integral of both positive and negative environmental and socioeconomic impacts can compensate each other and may differ over the short and the long terms.

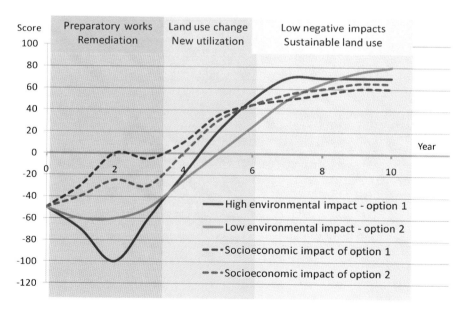

Figure 11.3 Environmental and socioeconomic impacts of two remedy options. The integrals of the + and – impacts (the area bounded by the graph under and above the "0" axis) give the balance of the (i) environmental impacts of the two options (green and red-brown lines) and (ii) the socioeconomic impacts of the same options (dotted lines). Future land uses and the time interval largely influence the integral impact value.

The requirement of sustainable remediation initiated intensive research and resulted in many application case studies. Several recommendations have been made on sustainability assessment methodology such as these:

– The inclusion of routine methodologies from other fields such as economy, sociology, life cycle assessment, quality management;
– The selection of indicators, to gather information for sustainability and verification;
– To define what kind of information would be appropriate for each indicator: default/ generic information, read-across or measured data;
– Tools to evaluate the collected information such as comparative evaluation of the options, qualitative and quantitative assessments, scoring, weighting, and aggregation of the information.

Developments in sustainability assessment from the 2000s endeavor to consider the three fields – economy, environment, and society – together as shown by Figure 11.2. There is a consensus in the involvement of these three elements, but there are different views and solutions, recommendations, and methods about the execution in terms of protocols, guidelines, software, and so on. Many of the developed tools still have not defined their scope in the complex project management process, i.e. it has not been specified whether they apply to the initial phase of decision making, to planning, to monitoring or to the retrospective evaluation of the technology or to the entire project. The spatial and temporal scale also influences the selection of the proper evaluation tool. Seemingly beneficial processes such as a reduction in contaminant concentration may cause accumulation over the long term, or a local economic benefit may cause adverse impacts many times higher over a wider range. Information on the scope and validity of the evaluation tools together with a realistic conceptual model of the project can ensure the proper and efficient compilation and application of the evaluation/verification tool box. Some typical scopes and goals of sustainability assessments related to contaminated site management are the following:

– Sustainability of the site before remediation. Assessment involves the fate and transport of the contaminant, the impacts at various scales from local to global and the impacted ecological and human receptors from an ecological and socioeconomic point of view.
– Sustainability of the remediation project can be evaluated throughout the management tasks, including decision, planning, implementation, and monitoring of the technology and its environmental and socioeconomic consequences.
– Optimizing feedback: how to reach minimized negative impacts/temporary imbalances and maximized benefits by the modification and control of the process or system by its own results or effects. A comprehensive and dynamic monitoring is needed to use this tool.
– Final retrospective evaluation, i.e. the verification of the outcome compared to the planned performance and comparing the site's sustainability before and after remediation.
– Long-term sustainability of the site considering the new and future land uses. The predictions should be validated from time to time to demonstrate continuous development in sustainability.

Several indicators were collected to measure sustainability in general and on special fields – e.g. forestry or water extraction – several hundred initiatives were created and published. A few publications introduced new approaches and methods or metrics (e.g. WBCSD, 2011; Hopton et al., 2010) or tried to collect sustainability initiatives and indicator types (Singh et al., 2009; IISD, 2018a,b). Tables 11.1 and 11.2 show well-known and often-used sustainability indicators.

Table 11.1 Some typical indicators used for sustainability assessment.

Reduced environmental impact indicators	Increased societal value indicators
– Energy intensity	– Human health and safety
– Water intensity	– Poverty alleviation
– Material intensity, waste reduction	– Human dignity
– Greenhouse gas emission	– Resource conservation
– Toxic emissions with persistent toxicants in focus	– Asset recovery
– Ecological impacts, ecosystem services	– Biodiversity and ecological resilience
– Land uses	– Prosperity and economic resilience

Table 11.2 Indicators of sustainable development used by the United Nations and the World Bank (UN, 2007).

Social indicators:	Environmental indicators:
Poverty:	Atmospheric impacts:
– Poverty index, population below poverty line	– Greenhouse gas emissions
– Unemployment rate	– Sulfur and nitrogen oxides emissions
– Demographics (population stability, density)	– Ozone depleting emissions
Human health:	Generation of hazardous waste:
– Average life expectancy	– Municipal waste
– Access to safe drinking water	– Hazardous and radioactive waste
– Access to basic sanitation	– Land occupied by waste
– Infant mortality rate	– Transportation
Living conditions and accessibility:	Consumption of ecosystem products:
– Urban population growth rate	– Forest area change
– Floor area per capita	– Annual energy consumption
– Housing cost	– Mineral and fossil fuel reserves
– Telephone lines per capita	– Groundwater reserves
– Information access	– Material intensity
Economic indicators:	Ecosystem indicators:
– Economic growth	– Ecosystem stability
– GNP	– Threatened species
– National debt/GNP	– Annual rainfall
– Average income	– Coastal protection (algae index)
– Capital imports and foreign investment	– Agriculture (pesticide/fertilizer/irrigation)

The overviews in the next few sections demonstrate how the different approaches and methodologies try to involve sustainability indicators and how the attitude of these existing methodologies converge. This is a promising process and development of the disciplines of life cycle assessment (LCA), cost efficiency and cost–benefit assessment (CBA), risk-benefit assessment (RBA), green remediation (GSR), and the mass-balance type engineering tools will help to find the best possible type of evaluation for each management phase and for individual contaminated sites.

3   INTEGRATION OF SUSTAINABILITY ASSESSMENT INTO EXISTING REMEDIATION EVALUATION METHODS

Some practical solutions are introduced in this section that were developed from existing technology evaluation for decision making and verification of remedies at generic and site-specific levels. Experts of the relevant areas have integrated technology performance, economic feasibility, and environmental (or eco-) and social effectiveness of site clean-up technologies and the entire projects. The ecology focused on "green remediation" has included more generic ecological and socioeconomic indicators, the economy-based cost–benefit assessment has been expanded to socioeconomic costs and benefits by monetizing if possible environmental and social "costs" and "benefits." Life cycle assessment originally focused on products has been extended to involve more risk-related and locally occurring items such as human health risk. Material balance being the tool of engineering meets both main approaches: it is both quantitative and balanced. And MCA has also been involved to solve the problem of aggregation of quantitative and qualitative information of various metrics. Going deeper in each relevant specialization, more and more aspects of sustainability came up and are waiting for the proper solution such as scoping, scaling, and quantifying the indicators. The overview in the followings introduces the promising development in the field of sustainability evaluation of remediation and management of remedial projects.

3.1   LCA for the evaluation of remediation options

When looking for the best practice in environmental remediation evaluation, some approaches tend to use LCA tools in spite of the differences in aims and scopes, just because LCA has developed quantitative metrics for measuring environmental impacts at a generic level (ISO 14040, 2006 and ISO 14044, 2006). LCA provides a simplified tool which helps

– to compile the inventory of energy and material inputs and environmental releases;
– to evaluate the potential impacts of the material used, energy and also of releases/ emissions;
– to make a proper decision between the available options.

LCA has also been developed and refined during the years and current LCA versions include human toxicity data and consider not only the amount but also the type of energy (renewable or not) or the Eco-LCA which includes the assessment of direct and indirect impacts of human activities on ecological resources and surrounding ecosystems.

The life cycle management (LCM) approach became popular for the characterization of remediation options using LCA and supports the decision by selecting the one with the minimum impact on the ecosystem and human health.

LCA provides a generic and comparative estimate on the impact of remediation but without evaluating the technology performance and socioeconomic consequences, thus it is not suitable for decision making. LCA is neither capable of evaluating future land uses, or the consequences of reintroducing the remediated site. Stakeholder involvement into the decision-making process is not possible when using only LCA. Another shortcoming of LCA is the difference between LCA databases: even a comparative evaluation can only be performed if the same database is used for the evaluation of the technologies to be compared. Unpredictable differences between the characteristics of generic and case-specific databases cause further uncertainty.

LCA – similar to any other assessment tools – cannot answer all questions about the site and the remedial technology, but it is most beneficial for site "fingerprinting" and can protect natural resources during remedial technology application and site development. The result of LCA is best integrated into a complex assessment tool system and use for the selection of the best technology option from a LCA point of view.

The first applications of LCA to the evaluation of remediation options was a combination with site-specific risk assessment in 1997–1998 (Beinat et al., 1997; Bender et al., 1998; Volkwein et al., 1999). The REC (risk reduction, environmental merit, and costs) decision support tool and the software tool have been developed by joining a streamlined life cycle assessment (LCA) and risk assessment of the contaminated site.

Cappuyns and Kessen (2012) collected case studies which used REC and ReCiPe, two LCA-based evaluation methods in the management of contaminated site remediation and decision making. Both REC and ReCiPe yielded a single score for the environmental impact of the soil remediation process and allowed the same conclusion to be drawn (when parallel applied) on the soil remediation alternatives assessed. The ReCiPe method takes into account more impact categories but is also more complex and needs more data input. The author's opinion is that within the preliminary evaluation phase of soil remediation alternatives, the use of the REC method will be sufficient in most cases compared to the much more labor-intensive ReCiPe (Cappuyns, 2013). If we use various tools for technology or site evaluation, the result may become better, but only easy-to-use tools are feasible in such cases.

Sparrevik et al. (2011) demonstrated the importance of the assessment scope. The authors applied LCA to investigate the environmental footprint of different active and passive thin-layer covers compared to natural recovery for the Grenland fjord, Norway, contaminated with polychlorinated dibenzo-p-dioxins and -furans (PCDD/F). Covers turned out to be better than natural recovery when the assessment included site contamination alone. However, considering cover construction, its impacts and resource and energy use during implementation increased the environmental footprint by an order of magnitude. The evaluation of several technological options helped identify the version having a footprint as low as that of natural recovery.

Lemming et al. (2010) carried out a survey on the use of LCA for remediation and they found that only very few LCA assessments were used for in situ remediation where the impact on groundwater is in focus. They identified the lack of impact category covering human toxicity via groundwater, the associated characterization models and normalization procedures.

Suèr et al. (2010) discussed the problem of the scale of impacts, e.g. cases where remediation provides a better quality remediated site but had negative environmental impacts on the local, regional, and global scales. The authors reviewed nine case studies and concluded that the limitation of the LCA methodologies for space, time, and secondary processes are different, and this strongly influenced the results. The choice of impact categories and land uses strongly affected the results. In general, the negative impact of site remediation was due to energy consumption. For excavation combined with *ex situ* treatment, the transport of contaminated soil to the treatment facility or landfill required the most energy. Pumping consumed the most energy for *in situ* treatment of soil and groundwater. Bioremediation was the best from an energy consumption perspective.

If one would like to use LCA for the prospective evaluation of environmental remediation, the most important things are to identify and harmonize the scope, the temporal and spatial scales, the impact categories, and other key characteristics both of the remedial project and the LCA methodology.

3.2   Green remediation

In a survey, the US EPA found huge energy and water consumption for environmental remediation using pump & treat, thermal desorption, multiphase extraction, *in situ* thermal treatment, air sparging, and/or soil vapor extraction technologies at Superfund sites. The estimated yearly value of energy consumption is 631,000 MWh and the carbon footprint based on used fuel was 435,357 metric tons of carbon dioxide equivalent.

Ecologically efficient remediation (called green) is an important step toward sustainability but is not sufficient in itself, given that no socioeconomic and direct human health requirement and indicators are included into the evaluation. Another useful concept in green remediation is the requirement of eco-efficiency. The US EPA (2008) emphasized that eco-efficiency should dominate the entire project from site investigation through the design of remediation and technology monitoring, construction work up to maintenance and post-monitoring. The green remediation concept can easily be supplemented with the cost efficiency requirement and extended to site reuse planning and to overall project management.

"Green remediation" in US EPA terminology means taking into account the "costs and benefits" of the environment (in addition to economic costs and benefits) and lowering environmental impact and consumption of natural resources such as water and energy to reduce the pressure on the environment and ecosystems. Remediation in itself reduces environmental risk and improves the quality of the environment, but as it covers various technology applications with highly diverse negative impacts on the environment, this should also be taken into consideration. Discrepancy between the targeted quality improvement and reduced risk as a primary aim of remediation, and the environmental impacts due to the energy and water consumption plus emissions into air, water, and soil, still causes confusion. This problem may be solved by displaying the clear concept of the project and its management and the fitting evaluation tools for each step. This confusion is even more critical for *in situ* soft remediation, e.g. natural attenuation, where the same process can be risky and costly or beneficial.

Green remediation by definition is "the practice of considering all environmental effects of remedy implementation and incorporating options to minimize the environmental footprints of cleanup" (US EPA, 2011b). Green remediation concentrates on soft technologies which can conserve the ecological value of soil and water and can significantly reduce the impact of the technology to the site and its surroundings by the following:

–   Reducing the total energy use (energy-efficient equipment and optimized use) and increasing the percentage of energy from renewable resources and passive energy technologies;

–   Reducing air pollutants and greenhouse gas emissions by excluding heavy equipment with high fuel consumption – especially diesel – and reducing release of toxicants (O_3, CO, NO, SO_2, toxic metals), PM10 and contaminated dust;

–   Reducing water use and negative impacts on water resources by minimizing fresh water consumption and maximizing water reuse during operations, reusing technological waters, using vegetation not needing irrigation, preventing nutrient loading of water bodies;

–   Protecting ecosystem services during clean-up by minimizing impacts on land and ecosystem through minimally invasive *in situ* technologies, passive energy technologies such as bioremediation and phytoremediation, minimizing soil and habitat disturbance and preventing ecosystem from contaminants through contaminant source and plume controls and reducing noise and lighting disturbance;

–   Improving materials and waste management by minimal waste generation, materials reuse, recycling, minimizing natural resource extraction and disposal, and by using passive sampling devices.

3.3   Green and sustainable remediation

Green remediation is based on eco-efficiency and is an essential part of sustainability. Best management practices (BMPs) of green remediation help balance key elements of sustainability such as resource conservation and material and energy efficiency. Thus green remediation objectives fulfill a significant proportion of sustainability in remediation (US EPA, 2008) in as much they do the following:

–   Achieve remedial action goals;
–   Support use and reuse of remediated sites;
–   Increase operational efficiencies;
–   Reduce total pollutant and waste burdens on the environment;
–   Minimize degradation or enhance ecology of the site and other affected areas;
–   Reduce air emissions and greenhouse gas production;
–   Minimize impacts to water quality and water cycles;
–   Conserve natural resources;
–   Achieve greater long-term financial return from investments;
–   Increase sustainability of site clean-ups.

Green and sustainable remediation (GSR) is the site-specific employment of products, processes, technologies, and procedures that mitigate contaminant risk to receptors while making decisions that are cognizant of balancing community goals, economic impacts and environmental effects (US EPA, 2011a)

GSR is an approach integrating environmental, social, and economic considerations into site management decisions addressing environmental, social, and economic impacts to be improved in the project, while still meeting regulatory objectives. In addition to the original "green" requirements (conservation of ecological values and reduction of the impact of the technology), GSR added the requirement of "long-term stewardship actions" to better fulfill the socioeconomic requirements of sustainability:

–   Reduce emissions contributing to climate change;
–   Integrate an adaptive management approach into long-term controls for a site;

–   Install renewable energy systems to power long-term clean-up and future activities on redeveloped land;
–   Use passive sampling devices for long-term monitoring where feasible;
–   Solicit community involvement to increase public acceptance and awareness of long-term activities and restrictions.

GSR utilizes some well-known US EPA developed management tools to serve sustainability:

–   The TRIAD approach in site investigation (see Gruiz, 2016), which reduces field mobilizations, results in fewer samples being shipped for laboratory analysis and minimizes resampling (ITRC, 2003).
–   Direct-push wells both in site investigation and in remediation implementation result in cost saving due the speed and ease of installation. Faster installation also means decreased contaminant exposure and waste production (ITRC, 2006b).
–   Remediation process optimization for identifying opportunities for enhanced and more efficient site remediation (ITRC, 2004).
–   Performance-based environmental management is a strategic, goal-oriented uncertainty management methodology that is implemented through effective planning (ITRC, 2007).
–   Project risk management for site remediation cuts across the entire project addressing and interrelating cost, schedule, and performance/operational risks. Performance risk is a consideration in the planning, design, and execution of remediation activities (ITRC, 2011b).
–   Life cycle cost analysis to compare the net present value of different remediation alternatives, evaluate the cost-effectiveness of a remediation system, and perform a cost–benefit analysis of remediation alternatives (ITRC, 2006a).

3.3.1   Planning the GSR process

The GSR planning process comprises four general steps that are meant to provide a conceptual structure for guiding the integration of GSR into each phase of the clean-up project. It is intended to be flexible and scalable to site circumstances (ITRC, 2011a).

Such a flowchart (e.g. in Figure 11.4) can be created for each phase of remedy implementation, the site assessment phase, planning, implementation/monitoring of the technology

Figure 11.4 Flowchart for planning green and sustainable remediation.

application, and the whole management process with scoping, evaluation, stakeholder involvement, future land use planning, etc.

3.3.2 Phases and aspects of the GSR process

GSR should be applied during all phases of a remediation project:

- Investigation;
- Remedy evaluation and selection;
- Remedy design;
- Remedy construction;
- Operations, maintenance, and monitoring;
- Remedy optimization;
- Closeout.

The GSR aspects should be collected and listed in each of the four phases shown in Figure 11.4:

1 Evaluation/updating the conceptual site model of ecological significance.

This step means several mainly managerial supplements can be used to add sustainability information to the site's original conceptual model, for example:

- Available material and energy sources, renewable energy sources;
- Feasible water treatment and reuse options, e.g. water reuse for irrigation;
- Waste treatment, e.g. nearby soil washing or recycling plant;
- Soil requirement for geotechnical purposes or other maximum benefit reuses;
- On-site establishments for renewable energy production and waste treatment.

2 Establishment of GSR goals.

GSR goals are the general sustainability targets such as low energy, water, and material consumption; reduced manpower; minimum duration and low emission and waste; and maximal ecological and socioeconomic benefits. These goals are formulated in the initial phase of the project where all stakeholders' interests should be reviewed, listed and prioritized, in addition to the primary "technology performance" goals of the remediation to improve site quality and satisfy future land uses. In addition, specific qualitative goals should be formulated in terms of energy consumption, duration, etc.

3 Sustainability evaluation of the technologies and the project.

This task covers the sustainability evaluation of

- Site assessment;
- Remediation options in the initial phase;
- Implementation of the technology;
- The remediation;
- The technological and environmental monitoring during remediation.

Based on the figures of monitored indicators, sustainability evaluation enables changes and adaptations in technology application to optimize the remediation for most priority requirements.

Some typical issues where sustainability is the key point in management and in selecting the option which can best reduce the project's sustainability footprint are:

- Off-site, on-site, and *in situ* solutions for assessment and remediation;
- Waste produced by site investigation and remediation;
- Emissions and transport from site investigation and remediation;
- Portable *in situ* assessment and monitoring tools;
- Minimum disturbance and negative impacts on the ecosystem;
- Ecosystem habitat considerations;
- Reducing travel and using field portable tools result in lower costs and economic benefits;
- Communication and stakeholder involvement.

4   Metrics, levels, boundaries.

GSR metrics may be quantitative or qualitative, they can be part of a mass balance or monetized costs. All GSR goals are characterized by sustainability indicators. These indicators should be defined by their use, boundaries, and units of measurement:

- Quantitative indicators: unit of measurement;
- Qualitative indicators: definition should be given for their classification and scoring;
- Identification of the relevant step of the management process or the level of evaluation where the indicator is used;
- Boundaries where the indicator is valid: temporal and spatial scope.

5   All GSR steps should be documented and demonstrated how GSR goals have been satisfied in the course of the whole project.

In summary, the green and sustainable remediation concept was created by incorporating socioeconomic sustainability goals into the existing green remediation goals. The goals are translated into indicators which can be measured or otherwise observed/monitored to control the process and finally verify project sustainability. The holistic approach to remediation that considers ancillary environmental impacts and aims to optimize net effects to the environment is concisely and thoughtfully characterized by Reddy and Adams (2015). It addresses a broad range of environmental, social, and economic impacts during all remediation phases, and achieves remedial goals through more efficient, sustainable strategies that conserve resources and protect air, water, and soil quality through reduced emissions and other waste burdens. GSR also simultaneously encourages the reuse of remediated land and enhanced long-term financial returns for investments.

The same authors also mention that though the potential benefits are enormous, many environmental professionals and project stakeholders do not utilize green and sustainable technologies because they are unaware of methods for selection and implementation.

3.4   System-based indicators

The US EPA has developed an integrated approach in the last years relying on system-based indicators, which provide a more holistic and flexible system for sustainability evaluation of environmental technologies and management processes (Fiksel *et al.*, 2012). The goal of this framework is to support the application of sustainability indicators in decision making and program evaluation by the US EPA, but the same approach can also be useful for contaminated site management. To fulfill this goal, the indicators are selected, organized, and interpreted from different points of view which have been specified by US EPA as follows:

– Get real information on the current state of the environment and the observed trends (in the US);
– Incorporate sustainability considerations into all (EPA level) decisions;
– Involve sustainability indicators into research planning;
– Evaluate programs and projects and measure and analyze their productivity and effectiveness.

System-based indicators capture not only the characteristic features of ecological, societal, and industrial/economic systems but can manage dynamic interactions among these domains (Fiksel, 2009). The dynamic interactions are symbolized by the overlapping parts of the three sustainability compartments (Figure 11.2). Four major categories of indicators are recommended by Fiksel *et al.* (2012) that are applicable to evaluation of any project:

– Adverse outcome: indicates destruction of values due to impacts on individuals, communities, business enterprises, or the natural environment.
– Resource flow: indicates pressures associated with the rate of resource consumption including materials, energy, water, land, or biota.
– System condition: indicates the state of the relevant systems, i.e., individuals, communities, business enterprises, or the natural environment.
– Value creation: indicates creation of value (both economic and well-being) through enrichment of individuals, communities, business enterprises, or the natural environment.

Table 11.3 shows some system-based indicators within the four major categories.

3.5   Multi-criteria analysis for sustainability assessment by scoring and weighting

Whatever method we use for sustainability assessment, there are typically too many data to handle and aggregate. The holistic approach addresses a broad range of environmental, social, and economic impacts (both negative and positive ones) during all remediation phases from the first step of site assessment and decision making to the final evaluation. The MCA tool enables aggregation of qualitative, semi-qualitative, and quantitative data by using a scoring and weighting system as objective as possible.

3.5.1   SNOWMAN – multi-criteria analysis (MCA)

SNOWMAN-MCA has been developed for the evaluation of remediation alternatives to access their overall impacts and cost–benefit ratio with a focus on soil function (ecosystem services and goods) and sustainability (SNOWMAN-MCA Project, 2009–2013; Rosèn *et al.*, 2015).

Table 11.3 System-based indicators (according to Fiksel *et al.*, 2012).

Major category	Indicator	Example
Adverse outcome	Exposure	Health impact of air pollution
	Impact	Public safety
	Risk	Footprint of energy use
	Incidence	Footprint of transport
	Loss	Species extinction
	Impairment health	Ecosystem/human health damage
Resource flow	Volume	Greenhouse gas emission
	Intensity	Material flow volume
	Recovery	Resource depletion rate
	Impact	Water treatment efficacy
	Quality	Recycling rate, land use
System condition	Health	Air quality
	Wealth	Water quality
	Satisfaction	Employment
	Growth	Household income
	Dignity	Housing density
	Capacity	Infrastructure durability
	Quality of life rate	Community educational equity
Value creation	Profitability	Cost (reduction)
	Economic output	Fuel efficiency (gain)
	Income	Energy efficiency (gain)
	Capital investment	Vehicle use (miles per capita)
	Human development	Education level, health conditions

SNOWMAN-MCA approaches ecological effects by using the concept of soil function (SF), e.g. the soil's ability to provide services and goods. Specific goals of the project were these:

– Incorporate effects on soil system functions to evaluate ecological sustainability, and
– Prepare a prototype for MCA to identify sustainable remediation alternatives.

Key results of the SNOWMAN-MCA project include the following (Volchko *et al.*, 2014):

– A suggested hierarchy between soil functions, soil processes, soil services, and ecosystem services, resulting in a set of soil function-related ecological, socio-cultural, and economic criteria and sub-criteria;
– A suggested minimum data set (MDS) of soil quality indicators for soil function evaluation and a software tool (SF Box) for calculating changes in soil quality based on the proposed MDS;
– A suggested structured and transparent approach for incorporating soil function and soil use aspects into sustainability appraisal of remediation alternatives.

An MCA software tool has also been developed, it is the SCORE model: Sustainable Choice Of REmediation method (Rosèn *et al.*, 2013 and 2015). It is based on a MCA prototype by Rosèn *et al.* (2009).

The effects of remediation alternatives on soil functions is evaluated in this system using *quality indicators* (SQIs), i.e. the measurable physical, chemical and biological properties of soil. SQIs are used to evaluate the degree to which the soil quality matches the soil functions determined by the intended end use of the soil.

There is no universally accepted definition of soil quality, none of the soil quality indicators are uniformly selected for the evaluation of soil and the impact of soil remediation. On the other hand soil quality has been the subject of soil science and investigated extensively and exhaustively for a long time.

Physical, chemical, and biological SQIs are usually suggested to describe the capacity of soil to function within managed or natural boundaries. The minimum dataset of soil quality indicators (SQI) has been suggested by SNOWMAN-MCA as follows:

- Soil texture;
- Content of coarse material;
- Available water capacity;
- Organic matter content;
- Potentially mineralizable nitrogen;
- pH;
- Available phosphorus.

Soil classification based on soil quality indicators (SQIs) forms the basis for evaluating the effects of remediation alternatives on ecological soil functions (read more in Volchko *et al.*, 2014).

Another possibility to evaluate the effects of remediation options is based on soil *service indicators* (SSIs), i.e. value measurements that indicate to what degree a management action contributes to human well-being by preserving, restoring, and/or enhancing a soil ecosystem service. These value-related metrics can be expressed in terms of the following (SAB, 2009; Volchko *et al.*, 2014):

1   Community-based values (attitudes, preferences, and intentions associated with a soil ecosystem service).
2   Economic values revealed by market data (if any) about a soil service (price of groundwater, agricultural land, building site, agricultural products, etc.), or the willingness to pay (WTP) for the service provided by the end use of the soil.

3.6   Sustainability in combination with cost–benefit assessment

A guidance document outlining the use of cost-benefit assessment (CBA) together with sustainability assessment (SA) in selecting remedial options was issued by the Australian CRC CARE in 2015 (Barnes *et al.*, 2015). The aim was to consider sustainability aspects and incorporate them into the selection of remediation options and evaluation of soil remediation projects (CBA and SA). The objectives laid down in the guideline were intended to do the following:

- Develop the principles and the process of CBA and SA within the context of contaminated site remediation;

– Establish requirements for the economic and sustainability evaluation of contaminated site remediation options;
– Encourage consideration of every reasonable option in the evaluation process as early as possible;
– Provide guidance on identifying and assessing the full range of costs and benefits associated with these contaminated site remediation options;
– Assist in identifying and choosing the preferred site remediation option.

The guideline supports the user step by step through the process of CBA and SA, from the creation of indicators and evaluating land uses through valuing/monetizing and comparing the economic, social, and environmental benefits and costs to the final decisions on implementing a project or to undertake a particular investment. CBA is a quantitative analytical tool to aid decision makers in the efficient allocation of resources. Sustainability indicators cannot be fully quantified or monetized, thus CBA should be conducted on the indicators that can be monetized, and an MCA should be conducted on the indicators that can be quantified or qualified. At the end, CBA is integrated into the MCA. MCA enables a structured and robust approach to assessing the likely economic, environmental, and social impacts of projects and is therefore well suited to CBA and SA.

The more accurate the picture is about the sustainability of remediation, and remediation projects, the more questions and uncertainties arise about sustainability management. Some of these are listed here:

– The time frame of the assessment: 1, 5, 50, or 100 years?
– The spatial extent of the assessment: only the remediated site, within 1 km, 5 km, 20 km vicinity, in the watershed concerned, for the country or globally?
– What are the best indicators to characterize sustainability?
– How to weight indicators?
– How to gather good quality data, information on environmental and socioeconomic impacts?
– What to measure, and what to accept from other resources?
– How to score qualitative and quantitative data?
– How to manage uncertainties?

There are no uniform answers to these questions, the different evaluation methods have to find their own solution case by case.

3.7   SuRF UK – a framework for evaluating sustainable remediation options

The SuRF framework provides a systematic, process-based, holistic approach for the consideration, application, and documentation of sustainability parameters during the remediation process in a way that complements and builds on existing sustainable remediation guidance documents (Bardos, 2014; CL:AIRE, 2014a,b). The goals of the SuRF framework are to accomplish the following:

– Be accessible and helpful to all stakeholders involved in remediation projects;
– Be applicable to different phases of a remediation project;
– Be applicable to different regulatory programs in the United States.

By using the framework, site-specific parameters, stakeholder concerns, and preferred end and future use(s) can be evaluated throughout the remediation life cycle and balanced with sustainability parameters. Fifteen categories of indicators spread over five environmental, five social, and five economic factors that can be used for sustainability assessment in support of remediation decision making is presented by Bardos *et al.* (2011). A new paper about SuRF UK (Bardos *et al.*, 2018; CL:AIRE, 2018) gives an overview on the indicators used and the rationale behind their structure, how the indicator categories might be further refined in the future and on the use of the SuRF UK framework in the practice in the UK. It is widely recognized as the most appropriate mechanism to support sustainability-based decision making in contaminated land management.

The SuRF Framework describes an approach for the following:

– Performing a tiered sustainability evaluation;
– Updating the conceptual site model (CSM) based on the results of the sustainability evaluation;
– Identifying and implementing sustainability impact measures;
– Balancing sustainability and other considerations during the remediation decision-making process.

The SuRF UK Framework highlights the importance of considering remediation sustainability issues from the outset of a project and identifies opportunities for considering sustainability at a number of key points in a site's (re)development or risk management process.

4 VERIFICATION OF *IN SITU* BIOREMEDIATION

4.1 Introduction

Environmental remediation in the last 40 years became a must and accordingly hundreds of innovations, technology transfers, and application cases have been publicized. Scientists, researchers, and engineers developed technologies for a more sustainable future. With the increasing number of environmental remedies, decision makers, owners, and other stakeholders became less well advised due to the lack of uniform technology assessment and evaluation tools. It is still a problem today, in spite of the well-established environmental technology verification (ETV) systems in the United States, Canada, Japan (see Volume 1: Gruiz *et al.*, 2014, and some other countries and the sustainability assessment tools (this chapter). Several protocols and guidelines and decision support tools are available, but prospective and retrospective technology assessment remained a highly case-specific practice requiring case-specific indicator sets for their evaluation. The proper indicator set can be selected from a uniform list but, at the end of the day, only professionals who know the contaminated site, its history and future, its hydrogeochemical features, land uses, and land users can select the relevant ones for the site in question.

As the remediation technology assessment methodology has not yet been clarified, the different approaches and practical solutions can be important information for decision makers and developers. A combined chemical and a biological remediation from the authors' experience is introduced in detail – the cyclodextrin-enhanced *in situ* bioremediation (Schwartz & Bar, 1995; Gruiz *et al.*, 1996; Molnár *et al.*, 2003 and 2005; Molnár, 2007). The evaluation of CDT is used to demonstrate the prospective and retrospective evaluation of an innovative technology for a certain site. The assessment and verification tool evaluates

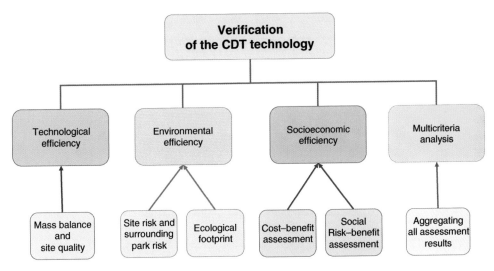

Figure 11.5 The tool box used for the verification of the cyclodextrin enhanced *in situ* bioremediation.

the technological and environmental performance of the technology focusing on the site, and its ecological and socioeconomic impact in a wider context.

The holistic approach was used for the verification system developed for soil bioremediation and especially for *in situ* soil and groundwater treatment. Some elements of the system were used for prospective evaluation in support of the decision made on the selection of the best technology option. For this innovation very little information was available, evaluation had to rely on technological experiments and pilot studies combined with read-across from comparable technologies. The "control matrix" of verification is the conceptual model of the projects and the risk model of the site, which is used in the same way as in the course of the site management from site investigation and decision making, through the case-specific design of the remediation and its monitoring to the end of the clean-up that is followed by verification. The monitoring is designed based on the site conceptual model and is aimed at satisfying the data requirement of verification. The conceptual model of the case introduced in this chapter is shown in Figure 11.7 as an example.

4.1.1   Assessment of the efficiency of in situ *bioremediation*

The verification tool box introduced here (see Figure 11.5) includes the assessment of the following:

1   *Technological efficiency* characterized by material balance: the mass flow balance of the soil phases and the pollutant amount. Mass balance results are compared to the targeted ones, typically represented by environmental quality criteria. This part of the tool is fully quantitative and is limited to the area under treatment.

2   *Environmental efficiency* consists of two elements: (i) the quantitative characterization of the site risk before and after remediation and comparing residual risk to the acceptable risk level related to future land uses. This part of the results characterizes the

treated site and its neighborhood which is directly impacted by emissions or transport. (ii) The environmental footprint of the remedy includes energy, water, and material uses and atmospheric emissions (typically greenhouse gases and ozone depleting substances). Evaluation of the footprint involves biodiversity and ecosystem services, and covers wider contexts, even at global scales.

3   *Cost efficiency* and *cost–benefit* assessment are priority tools in many cases. Cost efficiency assessment enables the comparative evaluation of technological and management options and is prospective, while cost–benefit assessment can be prospective or retrospective and quantitative to a certain extent.

4   *Social risk and benefit* evaluation is based mainly on qualitative indicators and their aggregation just by a SWOT or by MCA. Various indicators can be selected for evaluation depending on stakeholder participation.

The preliminary evaluation in support of decision making involves large uncertainties and several qualitative indicators. That is why the iterative approach has to be used in the management process (in technology planning and application as is usual), so allowing decision makers and other stakeholders to modify the management scheme and schedule. The retrospective evaluation is mostly quantitative and is mainly based on monitoring data.

The constituents on the technology evaluation system can be characterized by indicators. The selection of the best fitting indicator set is typically an iterative procedure, similar to project monitoring and control. The initial indicator set can be selected from recommendations and sustainability indicator lists in knowledge of remediation and management goals, technology type, and the wider context (e.g. legal background) but it would need modification and specialization for the assessed case. In the view of the requirement to establish, as far as possible, a quantitative indicator set, one has to strive to maximize quantitative sustainability indicators.

The technology evaluation process steps are as follows (Figure 11.6):

–   Learning the case, the site, stakeholders and the context;
–   Overview of existing and available technologies, which can be an option for the case;
–   Selection of the characteristic indicators for technological performance, eco-efficiency, socioeconomic efficiency;
–   Planning technology- and footprint-monitoring: data need of indicator creation and the method for acquiring these data (database, measurement and testing);
–   Evaluation of monitoring data and derive information from them; balancing the indicators of the three sustainability constituents (environment, economy, society);
–   Determine the unit of measurement and the qualitative or quantitative character of the indicator and providing them with scores and weights;
–   Dividing the indicators into positive and negative impacts groups;
–   Aggregating positive and negative indicator values;
–   Calculating the balance of the two;
–   Apply the integrated evaluation just to soil remediation or to the entire project and its management: stakeholders involvement, communication, link to spatial planning, long-term sustainability.

Most of these steps involve an iterative loop that makes the tool box dynamic and ensures the enhancement of the tool box reliability and quality during the project in parallel with the increasing amount of information being derived from monitoring.

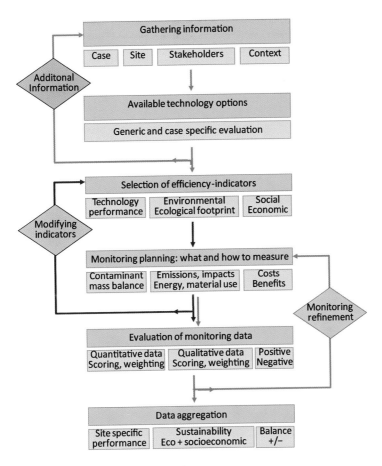

Figure 11.6 Technology evaluation process step-by-step.

The verification system developed and applied for the characterization and evaluation of innovative *in situ* soil remediation technologies is introduced on the cyclodextrin technology (CDT) which is an innovative *in situ* bioremediation developed and used for soil contaminated with hydrocarbons of low bioavailability.

The randomly methylated beta-cyclodextrin (RAMEB) was found to significantly enhance the bioremediation and detoxification of the (non-PCB) transformer oil-contaminated soil, increasing the bioavailability of the pollutants and the activity of indigenous microorganisms. The field demonstration was the basis of the complex verification of the CDT. An integrated technology monitoring was established to get as much data as possible for technology verification. The application of the verification method proved both the suitability of the verification tool box and the efficiency and the competitiveness of the CDT in terms of technological, economic, ecological, and social efficiencies.

The usability and feasibility of the newly developed remediation technology verification tool was demonstrated on several conventional and innovative *in situ* remediation sites including those introduced here. One site was polluted with mixed organics of petroleum origin, other sites with transformer oil, creosote oil, and mazout, and one mining site with

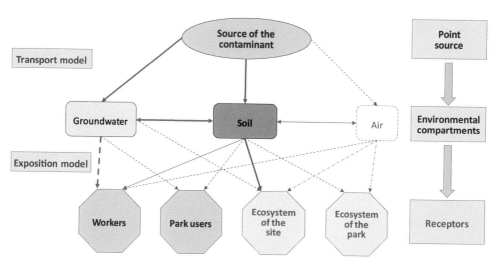

Figure 11.7 Conceptual model of the transformer oil-contaminated site in Budapest.

mine waste containing toxic metals (see also Chapter 8). The tool box was also applied for the innovative reclamation developed for degraded (sandy acid) agricultural soil by using coal char and bioaugmentation. Several specific sustainability indicators were involved to check benefits for ecosystem and agroecosystem services characteristic for agricultural soils. The verification tool was confirmed as flexible, controllable, and modifiable based on the conceptual site model (Figure 11.7) and in this way was generally applicable as an efficient decision support and quality management tool.

It can be considered also as an educational or knowledge transfer tool as the whole management process is better understood. Mainly the starting and closing points are highlighted and the awareness is raised in general. As a beneficial consequence it supports increase in trust in remediation; mainly *in situ* bioremediation based on enhanced natural soil processes. One of the cases, the transformer-oil contaminated site remediation, is introduced here.

The term "environmental technology verification" (ETV) is more often used for the process of generic characterization of a technology using a uniform standardized methodology to support decision makers and raise trust in owners and managers. These verification systems all over the world (US, Canada, Japan, and nowadays in Europe) (see also Chapter 1) are based on typical (successful) applications and on generic footprinting and sustainability assessment (see US EPA ETV, 2018; EU ECOAP, 2018; PROMOTE, 2018; TESTNET, 2018; ETV Canada, 2018; ETV Japan, 2018). Generic database information such as average energy and material use of certain operations, generic emission values from activities, global carbon footprint values, environmental services and their values, etc. are found in databases; much of this data being held by LCA databases.

4.1.2   Evaluation tools for in situ *bioremediation*

The evaluation of the performance and efficiencies of the planned, ongoing, or closed remediations is covered by the term "sustainability assessment" which follows the holistic approach of the harmonic balance between environment, economy, and society. The compartments of the MOKKA technology evaluation tool box are the following (MOKKA, 2004–2008):

1   *Technological performance* tool: traditional chemical-engineering tool, material balance, is used for the quantitative characterization of the efficiency of the remediation process and fulfillment of clean-up goals. It relies dominantly on the transport model included by the conceptual risk model of the site. To draw a true picture of contaminant transport, transformation, degradation, removal, etc. and additives usage and fate in the environment, physicochemical measurement data are acquired by technology monitoring and used for calculation material balance. In addition to technological performance, this measured data may serve for environmental efficiency (introducing hazardous substances into the soil and removing them by the technology) and cost–benefit assessment.

2   *Environmental efficiency* tool includes three further sub-tools:

    –   Risk characterization tools: the quantitative assessment of the environmental risk of the site before and after remediation: the scale of reduction of the risk is a quantitative measure: the difference of risk characterization ratio valid for the initial and the final quality of the site. As the risk characterization ratio (RCR) shows the ratio of the actual risk to the acceptable risk, the RCR can truly characterize the site in itself and can be based on the measured values of physicochemical, biological, ecological or toxicological indicators.
    –   The assessment of the site-specific environmental impact of the technology itself due to its emission and direct disturbances in the near environment during and after the technology application is a partly quantitative tool.
    –   Footprinting the technology: the assessment of the more generic environmental impact due to energy and water consumption, greenhouse effect, ozone depletion, acid rain, the use of non-renewable and renewable resources, etc. is a partly quantitative tool.

3   *Cost efficiency* and *cost–benefit* assessment tools: cost efficiency is mainly used as a decision support tool to help the selection of the technology that best fulfills cost efficiency requirements. The initially calculated costs is validated within the technology verification. Cost-benefit analysis can be used in both phases to support decisions and to verify the project. Cost efficiency assessment is a quantitative tool, with absolute or relative monetary results. Cost-benefit analysis is a fully quantitative tool with monetary values in spite of the difficulty in monetizing all socio-environmental costs and benefits.

4   *Social risks and benefits* are based mainly on qualitative indicators such as stakeholder involvement, risk communication, community impacts (including job opportunity, healthy environment, better life quality, access to the area), environmental justice, future land-use potential, and cultural resources. These indicators can be characterized even by a simple SWOT or by MCA after scoring and weighting.

The prospective and retrospective evaluation set can be flexibly compiled from the previous four assessment types of the MOKKA evaluation system.

The MOKKA verification tool is applicable for any remediation technology. It is part of the management concept and can fit the management process of the project and can be applied to the site assessment and technology selection. It also harmonizes with the design of the monitoring tool and technological interventions, modifications, and overall. It may have special importance in the case of *in situ* remediation when material uses, emissions, and current and future land uses are closely linked or combined with the technology implementation and natural process cannot be distinguished from those controlled by the technology. The reactor approach (see Chapter 2) may clarify the cooperation of natural and technological processes, but still there are several inputs which are beneficial for the technology making it more efficient and faster but pose a negative impact on the environment. In such cases it is necessary to investigate the size, life span and the spatial scope of these impacts and execute the assessment for different temporal and spatial scales. The verification of the remediation also contains site-specific elements. The same technology can reach a very good rating at one site but may show poor results on another. This means, in practice, that the selection of the suitable technology for the site is the first step of the verification. Technology-related information used for technology selection and decision making should be gathered during site assessment/monitoring (Gruiz *et al.*, 2008). Site assessment, site risk assessment, and technology-related assessment at contaminated sites should be applied in an integrated way.

Technology monitoring should ensure not only the follow up of the technology application by measuring the technological parameters but also the evaluation of all the environmental and social (emissions, load on renewable resources, load on the society, etc.) as well as the economic impacts (used energy and material, manpower, and the related costs and benefits) both the positive and negative ones. Monitoring data are crucial for technology verification and sustainable project management. The overall monitoring has the following main goals:

1   Monitoring of the technological parameters such as temperature, pH, redox potential, aeration, water flow, and flow of additives and products to control the efficient functioning of the technology. It can be evaluated in comparison with the planned values and the expected final outcomes. Based on the differences, the state of the art can be evaluated and the necessary modifications implemented.
2   Controlling possible emissions from the technology to minimize local and regional environmental risks and global footprints. The two main goals are to identify deviation from the planned (acceptable/allowable) emission values, and to observe extreme emissions due to operational failures.
3   Measuring energy, water, and material consumption to characterize eco-efficiency of the technology.
4   Observing ecological consequences.
5   Accounting costs or calculating from measured values.
6   Assessing social aspects.
7   Gathering positive and negative indicators separately in each category.
8   Aggregating the impact categories.
9   Characterizing the technology and the project from the point of view of pollution, the site and its uses, the wider environment, watershed scale or global footprinting, the social environment and social consequences, the economic aspects and all of these in an integrated way by aggregating all acquired information.

The demonstration of the use of the verification tool is introduced using the combined biological and chemical soil treatment technology, the cyclodextrin-enhanced bioremediation (CDT): the soil and groundwater polluted with transformer oil were remediated by using the combination of *in situ* bioventing, temporary *in situ* soil flushing accelerated with the solubilizing agent cyclodextrin and *ex situ* physicochemical groundwater treatment (Molnár *et al.*, 2003; Fenyvesi *et al.*, 2003; Molnár *et al.*, 2005; Molnár, 2007).

4.1.3 Monitoring of in situ soil remediation

Technology verification and some of the sustainability assessment are based on the monitoring data from the CDT demonstration project. The quality of the evaluation depends on the quality of the monitoring data. Applying an *in situ* remediation, one is faced with two basic problems when sampling and analyzing whole soil. Core samples from boreholes give a false picture of the site due to the heterogeneity of both the soil and the contaminant. These heterogeneities may exceed the scale of changes in time, therefore to get an "average" sample or to take samples in different times from the same place is not possible. *In situ* measurements and sensors may help to a certain extent. In the case of sensitive hydrogeology or a treatment applying water and soil gas circulation or directed flow, a borehole may cause disturbance and increase the risk of pollutant transport. To avoid these problems during the monitoring of *in situ* remediation it is recommended that the mobile soil phases, i.e. soil gas and groundwater, are sampled and analyzed. It is advantageous also because these phases represent the quasi-average of the total treated volume. As they are circulated for technological reasons, sampling does not disturb the inner soil. From the data measured on soil gas and soil water it is possible to extrapolate to the whole soil, knowing the partition of the contaminant between soil phases, effects of soil pH and redox potential on the chemical form and effects of the contaminant, mechanism of biodegradation, etc. Averaging the sample from different depths can jeopardize the clear picture but can be compensated by innovative water sampling for depth-specific sample taking.

To follow bioremediation, the rate of the microbial conversion has to be quantified. The biological processes in the soil, like biodegradation, bioleaching, biological stabilization, bioaccumulation, can be described by the following general equation, where S is the substrate and P is the product of the microbial metabolic activity:

$$S + \text{additive} \xrightarrow{\text{soil microbiota}} P + \text{side-product}$$

The contaminant serves as a substrate for the soil microbes, going through the microbial metabolic pathway producing harmless end products (P) and side products. Any variables of this equation can be used for monitoring to follow and control the biological process: the decrease of the substrate or the additive, the increase of the product and the side products, the quantitative and qualitative characteristics of the soil and the soil microbiota, like nutrient supply, pH, redox potential of the soil and cell numbers, cell activities, enzyme activities, indicator species, etc. These variables can be measured in a direct or in an indirect way: e.g. instead of the concentration of the sorbed contaminant, the dissolved value can be measured and the sorbed value calculated by the partition coefficient (K_p). Another option is measuring the pH instead of the product concentration if the product is an acid or measuring the CO_2 as a product of aerobic biodegradation instead of measuring the substrate decrease.

Monitoring parameters can be grouped as relevant to soil, soil microbiota, substrate, and to the product. The relatively homogeneous mobile soil phases, i.e. soil gas and the groundwater enjoy priority for *in situ* technology monitoring and verification. Nevertheless the same matrix and measurement data can be used for the control and regulation of bioremediation to ensure an optimal environment for the desired soil microbial activity. The best technology indicators are the volatile and water-soluble small molecules, which move from the biofilm smoothly into the easy-to-sample mobile soil phases by simple partition. Substrates or products bound or strongly sorbed into the solid phase can be analyzed only from the solid phase, which causes heterogeneity problems and high standard deviation in analytical results.

Besides technology monitoring, environmental monitoring of *in situ* remediation plays an important role in controlling emission and other risky processes to prevent nature and humans from adverse impacts. The environmental monitoring data are used in the complex technology verification, project sustainability assessment both on local and regional levels as well as on the generic level considering footprinting. Environmental monitoring plays an extremely important role in the verification of *in situ* soil and groundwater remediation due to the formerly detailed characteristic of the direct contact and interaction between the technological volume and content and the natural or constructed environment (Gruiz *et al.*, 2015; Gruiz, 2016).

Integrated technology monitoring means that the bioremediation process is characterized by a suitable tool box of selected methods to measure both the physicochemical characteristics of the soil and the biological state of the soil microbiota. For good-quality environmental monitoring, an integrated methodology is required: the physicochemical analyses should be combined with environmental toxicity testing, and ecological and biological assessments. This is extremely important when the polluted site is an old, inherited one with an incompletely identified mixture of contaminants. In such cases the assessment program is not able to investigate all possible contaminants and intermediates, also the interactions between contaminants, contaminant and soil matrix, and contaminant and soil microflora may result in endless risky effects, so the integration of toxicity and other effect-measuring methods is essential (Gruiz, 2005). The physicochemical analyses serve the characterization of the contaminant, the soil and groundwater, the biological activity by measuring respiration rate and metabolites. Ecotoxicity testing serves safety and gives information on bioavailability of the contaminants. In the case of *in situ* remediation, ecotoxicity testing plays a role in the monitoring of emissions from the technology and the risk characterization of the end product.

Aftercare of treated contaminated sites is an important issue because one can never be sure that the pollutant has been removed completely. More sensitive new land uses increase the risk and require higher precaution. The aftercare of *in situ* treatments assumes monitoring not only of the treated soil, but of the whole site.

4.2 Evaluation of the cyclodextrin technology (CDT) by the MOKKA verification tool

CDT is an innovative bioremediation method which applies cyclodextrin (CD) to further intensify a combined bioventing and bioflushing type remediation technology based on aerobic biodegradation. It eliminates the bottleneck caused by low mobility and availability of the contaminants in soil. Low bioavailability of hydrophobic pollutants may limit their biodegradation by the "cell factory" in soils. Cyclodextrins can solubilize and mobilize high K_{ow} (octanol–water partition coefficient: the partition between soil water and solid phase is proportional to K_{ow}) contaminants and make them more available to the degrading soil microbiota. The use of CD may increase the bioavailability of poorly degradable pollutants and in turn intensify soil bioremediation (Gruiz *et al.*, 1996; Fava *et al.*, 1998; Wang *et al.*, 1998; Bardi *et al.*, 2003). It has been demonstrated that randomly methylated beta-cyclodextrin

(RAMEB) can decrease the toxic effects of contaminants on soil microbes, plants, and animals (Gruiz et al., 1996). RAMEB was used successfully as a bioavailability-enhancing additive in soil bioremediation for diesel oil in unsaturated soil by Molnár et al. (2003), for transformer oil both in saturated and unsaturated soil zones by Leitgib et al. (2003) and Molnár et al. (2005). It was found that RAMEB, with 1 year half-life time, is slowly biodegradable in the three-phase soil (Fenyvesi et al., 2005). The calculated time requirement of the CDT from the results is considerably less, only 0.5–1.5 years compared to the 2.5 years needed by the next best technology alternative.

4.2.1 Short description of the site

The selected site for technology demonstration was an inherited contaminated site with long-term pollution derived from a leaking spare transformer of the Népliget (City Park) Transformer Station in Budapest, Hungary. The contaminated site for CDT demonstration was relatively small, 30 m³ (50 t) of contaminated soil was treated which was only a small part of a much larger transformer station contaminated from subsurface transformer oil pipelines and awaiting clean-up.

The contaminant – a non-PCB transformer oil (TO 40A) – had reached the groundwater. The chemical analyses of environmental samples showed high contaminant concentration (20,000–30,000 mg/kg extractable petroleum hydrocarbon (EPH) content and 20,000–44,000 mg/kg extractable organic material (EOM) content in the soil at the upper 1–2 m only under the spare transformer. The oil content of the groundwater was 1 mg/L. The concentration of the volatile fraction was under the detectable limit. Both the saturated and the unsaturated zones had to be decontaminated.

The selected subsite that the innovative CDT technology was going to be applied to was physically unreachable from the surface. It was an ideal subsite to prove that both bioventing and in situ flushing can be applied for such barely accessible soil volumes.

The specialty of this transformer station is that it is a closed area within the largest city park, a popular recreational area in Budapest. It was not just a simple transformer station but it belonged to the transformer producing industry and the oil for the transformers produced was stored here and transported by a subsurface pipeline system to the production sites.

4.2.2 Management and decision making

In the course of the preparatory phase of the contaminated site remediation several conventional remedies were selected as realistic options. In between, CDT has been developed and offered as an innovative technology option for the site management team and included as an additional option to the list and involved in the prospective comparative evaluation. The evaluated options and some characteristics are presented in the following:

A "Zero" option: no remediation, let natural processes proceed by themselves with a 15-year long monitoring (Monitored Natural Attenuation, MNA). More information is needed about the natural attenuation process and the spread of the contaminants. The risk is that after spending money for the long-term monitoring, the contaminant will still be present and spread outside the fence.

B "Dig & dump": excavation and transport to the disposal site or a soil treatment plant and soil replacement on the site. After soil removal groundwater is treated for 1.5 years by "pump & treat" using activated carbon absorption. The excavation and transport are calculated for 0.2 year. In Table 11.4 the two activities' time requirement is given as 0.2/1.5.

C   *Ex situ*, on-site biological soil treatment of the excavated soil and backfilling with the treated soil on the site. As long as the *ex situ* soil bioremediation (in prisms) is going on, the groundwater (accumulating in the pit) will be treated (aeration and recirculation) in a free surface pond.

D   "Pump & treat": *in situ* soil washing combined with *ex situ* water treatment. The soil is continuously washed by infiltrating the treated and recycled water between the well screening and the trenches established for water inlet. Depending on the partition of the contaminant between solid phase and groundwater, it needs several decades (see mass balance of CDT: while 1000 kg is biodegraded only about 1.6 kg of transformer oil was removed by the water within about one year). The main disadvantage of this technology, i.e. very low water solubility and slow removal of the contaminant, can be avoided by the addition of mobilizing agents (surfactants or cyclodextrin), which can lower the time requirement by a half.

E   *In situ* bioventing of the unsaturated zone combined with the *ex situ* physicochemical treatment of the groundwater and the temporary flushing of the unsaturated soil (without additives).

F   CDT – *in situ* complex bioremediation using cyclodextrin: *In situ* bioventing of the unsaturated zone combined with the *ex situ* physicochemical treatment of the groundwater and the temporary flushing of the unsaturated soil and the capillary fringe with the recycled water containing cyclodextrin.

In Table 11.4, the sustainability indicators chosen for the evaluation of the technology options are shown as negative scores. Scores were created from both the qualitative and quantitative indicators enabling their aggregation. The main items in the table are these:

–   Emissions: including emission of volatile compounds; transport of water-soluble compounds via groundwater and leachate; transport of sorbed contaminants via erosion; dust emission; noise emission and emissions from transportation.
–   Regional/global items: including CO_2 production (kg/ton) (by biodegradation, due to excavation, transport and activated carbon regeneration/burning); water consumption (liter/ton), energy consumption (pumps, ventilators, machines, transport, waste (soil, activated carbon).
–   Material consumption: new soil, nutrients, like N, P, K and additives, like CD.
–   Risks; mainly health risk to workers and community (toxicants, transport, accidents).
–   Ecological indicators: impact on the ecosystem and ecosystem services.
–   Socioeconomic indicators involving land uses before, during and after remediation works, in the property value and use value perspective, access of the community to the space, future land use potential, cultural losses.

Dig and dump represents the highest risk in spite of the short time period of transportation activities, emission score is close to C and D, but the total risk score is twice as much as in the case of C and D due to energy and material consumption. The risk score of the *ex situ* on-site treatment (C) and "pump and treat" (D) is half of dig and dump, but four times that of the scores of the *in situ* technologies E and F. The "0" option (A) has a value of 265 in our

Table 11.4 Comparative sustainability assessment of selected technology options.

Score	"0" MNA	Dig & dump GW: ex situ	Dig & on-site treatment GW: pond	Pump & treat GW: ex situ	In situ bioventing GW: ex situ	CDT GW: ex situ
	A	B	C	D	E	F
Treatment duration (years)	15	0.2/1.5*	2.5	10/5**	2.5	1.5
Emission x years	37	50	35	40/20	10	6
– Air pollution, dust, waste, noise, contaminant migration						
Total CO₂ +W+E score x year	70	247	122	102/61	30	28
– CO₂ score	20	42	42	80	22	22
– Water consumption score	0	50	25	0	0	0
– Energy consumption score (non-renewing)	50	155	55	22	8	5.6
Total material score x year	4	100	4	2/4	2	4
– Non-reuse of materials, use of additives						
Total risks score	4	100	60	4	3	3
– Workers risks (health, accidents)						
– Community risk (transport, accidents)						
Ecosystem – total impact score	45	90	60	40/20	12	7
– Ecosystem damage						
– Habitat disturbances						
Total land value/use loss score	100	15	40	55/35	16	12
– Lack of land reuse score x years	45	5	30	30/15	8	5
– No access to the space score	15	3	3	10/5	3	2
– Property value loss score	15	2	2	10	3	3
– Future land-use potential loss score (residual pollution)	25	5	5	5	2	2
Cultural resources loss score	5	5	5	2	2	2
Total impact score	265	607	326	245/146	75	62

*0.2/1.5: 0.2 year of soil excavation, and 1.5-year water treatment **10/5: 10 years without surfactant, 5: with

CO_2

transformer-oil pollution case. The lowest impact scores (75 and 62) were calculated for *in situ* bioventing and cyclodextrin-enhanced *in situ* bioventing. The difference between the two is explained with the lower time requirement of the CDT technology.

4.2.3 Cost efficiency of the CDT

Cost efficiency assessment was used for the comparison of technological options ensuring the same target value and identical land uses. Specific cost indicators were created such as cost per volume unit. These indicators were then used for ranking the alternatives (see Table 11.4).

The transformer station is an isolated site in the Budapest City Park. The access to ecosystem and humans is limited, but not completely restricted, as the ecosystem "uses" the site and is in connection with the neighboring park through plant and animals. The land use of the site in the near future will be unchanged, so no benefit from this element can be calculated (it is expected that it becomes part of the park in the long run, but the date is uncertain). The model site is a little, separate volume, ideal for demonstration but too small for cost estimation. The application of the same technology is more realistic for the whole transformer station; therefore the cost calculation was done for 50, 1000, and 5000 tons. The costs out of the mass range are not proportional to the quantity. The results are given in Table 11.5.

The operational cost is proportional to the time and includes monitoring and administrative costs. Investment is generally a one-off expenditure but it can be given on yearly basis, considering the turnover of the capital. Costs estimation was prepared on the basis of bids or on the basis of references. Future costs (aftercare, monitoring, etc.) were also included. About one-third of the costs are spent in site assessment, planning, and aftercare, which are necessary expenditures even if remediation is not implemented (Gruiz et al., 2008). According to the cost efficiency analysis the costs, mainly the specific costs (cost/ton), are strongly dependent on the treated soil amount. Specific cost is 3–10 times more in the 50 tons case than in the 5000 tons case. Time requirement is also a dominant factor; its reduction may decrease the costs significantly.

The zero, "0" option (A) has relatively high costs (long-term monitoring and site supervision), similar to any of high-cost options and the site cannot be utilized or can be used only with restrictions for minimum 15 years. The selection of the "0" version is acceptable if 15 years waiting is scheduled by spatial planning and land use management, and the risk is

Table 11.5 Costs of the remediation alternatives for the transformer station.

Total and specific costs for 50 t, 1000 t and 5000 t treated soil	"0" MNA	Dig & dump GW: ex situ	Dig & treat on site GW: pond	Pump & treat* in situ soil washing GW: ex situ	In situ bioventing GW: ex situ	CDT GW: ex situ
	A	B	C	D*	E	F
Treatment duration (years)	15	0.2 for soil 1.5 for water	2.5	10/5	2.5	1.5
Total costs for 50 t (EUR)	10,900	13,400	9750	27,700 / 19,200	11,300	11,000
Specific costs 50 t case (EURO/t)	218	268	195	554 / 384	226	220
Total cost for 1000 ton (EURO)	74,400	135,500	76,900	228,000 / 144,000	78,000	82,800
Specific cost: 1000 t case (EURO/t)	**74.4**	**135.5**	**76.9**	**228 / 144**	**78.0**	**82.8**
Total cost for 5000 ton (EURO)	110,000	485,000	257,000	770,000 / 401,000	200,000	234,000
Specific cost: 5000 t case (EURO/t)	**22.0**	**97.0**	**51.4**	**154 / 80.2**	**40.0**	**46.8**

*Without/with surfactant

restricted to the contaminated area. An additional requirement to this case is the biodegradability of the contaminant and the definite existence of a risk-reducing natural attenuation process. Otherwise the "0" version has no right to come into existence because after spending the money for 15 years monitoring, one will still own contaminated land.

The dig and dump option (B) means only 1.2–2.5 times higher expenditure than the more environmentally efficient *in situ* or *ex situ* biodegradation-based technologies (not accounting for environmental and social "costs" and sustainability). Excavation and transportation of 50 tons of contaminated soil can be a realistic version if the pollution has high risk and the land use brings high benefit. (Compared to European prices, the cost of new soil and the disposal of the polluted soil is lower in Hungary.) Nevertheless, from the point of view of environmental and eco-efficiency, this solution should not have a priority, even if the costs would allow this. Decision makers tend to choose this option if they lack holistic thinking, especially if small sites hinder larger investments.

The *ex situ* biological treatment in prisms (option C) combined with water treatment in the pit, is a financially acceptable solution, but the surface area of the site cannot be used and it restricts land use and lowers benefits from land use during the treatment. The other shortcoming is the risk posed by the open pond containing contaminated water.

In situ soil washing and groundwater treatment by a pump and treat type technology (D) shows extremely high costs due to the time requirement of the inefficient removal of the insoluble contaminant by water. If time requirement can be lowered, e.g. by surfactant application, the cost is reduced to a value which is only twice the *in situ* bioventing (E and F).

The application of RAMEB in addition to bioventing and *ex situ* water treatment can be advantageous in those cases when the saved time compensates for higher costs. The enhancement with cyclodextrin can be economically beneficial if some arguments and benefits originate from saving time, for example, the company can save the rental costs of buildings or storage costs, or the change in land-use ensures a high benefit.

The duration of remediation has significant influence on the costs. The CDT was calculated with a 1.5-year duration, compared to the 2.5 years of the similar, non-enhanced *in situ* technology. In the demonstration case, 24 weeks were enough for remediation, but the continuation for another 23 weeks ensured safety: 0.5 years instead of 2.5 is crucial, the time saving may lower the costs of CDT for the level of 3 and 5. The cost estimates are shown by Table 11.5.

When selecting and implementing CDT or any of the alternatives after decision making by the environmental managers, an integrated evaluation is necessary, considering cost evaluation, technological and environmental efficiency, and SWOT analysis.

Based on the conclusion of the comparative prospective sustainability and cost assessment, decision makers tended to choose the CDT technology, but given that it is an innovative technology, a subsite-scale demonstration was necessary to verify the technology and the CDT will be implemented at full scale after a successful demonstration.

4.2.4 Short description of CDT and its application

The CDT was tested and demonstrated in several laboratory tests, microcosms and pilots, and – as the top of the increasing size of the experiments – on an inherited contaminated site with long-term pollution derived from a leaking spare transformer in a transformer station in Budapest. A complex methodology was applied for the remediation of the soil and groundwater. The *in situ* bioventing of the unsaturated zone was combined with the *ex situ* physicochemical treatment of the groundwater and the temporary flushing of the unsaturated soil. Randomly methylated beta-cyclodextrin (RAMEB) was applied to improve desorption, solubilization and biological availability of hydrophobic contaminants. The field experiment has confirmed the laboratory findings. The calculated time requirement of the CDT from the results is considerably less, only 0.5–1.5 years compared to the 2.5 years needed by the next best technology alternative.

The scale-up from the laboratory to the field included four steps during the development of the innovative CDT technology. Contaminants of different biodegradability and cyclodextrins of different mode of actions were tested in various combinations in small-scale biodegradation experiments. The influence of the soil type, cyclodextrin type, cyclodextrin concentration, aeration, pH, and moisture content on the biodegradation, the influence of other additives, nutrients and microbes on the technology was tested in soil microcosms simulating real situations. Some parameters, like cyclodextrin type and concentration, necessary level of aeration and nutrients have been optimized for larger size application. To be able to follow the remediation in the long term, a larger amount of soil is needed. This was the main reason for the next scale-up step: pilot-scale laboratory experiments (see Figure 5.2 and Chapter 5). The methodology for the integrated technology monitoring was also worked out during these experiments.

The results and experience gained from laboratory experiments with gradually increasing size and a detailed contaminated site assessment were used to design the field experiment.

Biological characterization of the contaminated soil originated from the transformer station showed that the microbiota has been adapted and able to biodegrade the contaminant, so the technology could be based upon the activity of the indigenous microbiota. On a site which has been contaminated for a long time (3–4 years), a biotechnology able to utilize the activity of the cell factory is the most promising. The technological parameters found to be optimal in the laboratory experiments were assumed to provide optimal conditions for the transformer oil-degrading microbes. Based on the results of the site-assessment and the previous lab-scale technological experiments, a site-specific combination of three technologies was selected and implemented:

- Treatment of the unsaturated zone by bioventing and intensification of the biodegradation with additives such as nutrients and cyclodextrin (RAMEB);
- *Ex situ* treatment of the groundwater after pumping it to the surface;
- Continuous moisture supply by slow infiltration of the treated water;
- Temporary flushing the unsaturated and the saturated zones with part of the water treated on the surface.

Figure 11.8 shows the complex technology scheme with the passive wells used for atmospheric air introduction and the active wells for water pumping and air exhaust (one well for the air and one combined well for air and water exhaust) and with the equipment on the surface for the water treatment and injection of the necessary additives.

Oxygen was supplied by bioventing for the biodegradation of hydrocarbons. Maximal air demand was calculated on the basis of the McCarty's equation (1988) of biodegradation of an average petroleum-hydrocarbon:

$$C_7H_{12} + 5O_2 + NH_3 = C_5H_7O_2N + 2CO_2 + 4H_2O \tag{1}$$

Molecular weights: 96   160   17   113 (biomass) 88   72

Fresh air
injection

Cyclodextrin
Nutrients

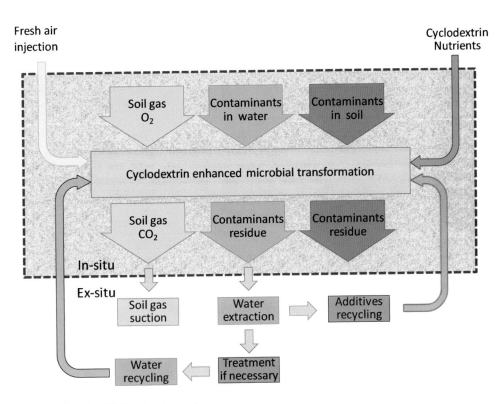

Figure 11.8 In situ CDT technology scheme.

A slow airflow was produced by a Siemens ELMO 2BH7–3G ventilator of low performance through aeration wells equipped with perforated casing. An air flow of 20–30 m³/h was ensured.

A Grundfos SP5 type plunger pump placed into the combined well was used for groundwater extraction. The treatment of the water occurred in three steps: phase separation in a sand filter followed by adsorption on activated carbon. The treated water was reserved in a closed tank and used in the technology for (i) moistening of the treated soil volume by slow infiltration of a part of the treated groundwater (15–16 m³/day) through small ditches on the surface, (ii) injection of nutrients, and (iii) temporary flushing of the contaminated soil volume by the RAMEB solution.

Nutrients (40 kg chemical garden-fertilizer, containing 15% P_2O_5, 15% N, and 15% K_2O) were added three times, i.e. in the 9th, 13th, and 21st week, during the 47-week experiment.

Randomly methylated beta-cyclodextrin (RAMEB) was applied to improve desorption of the contaminants strongly adsorbed on the surface of the soil particles and increase the contaminant concentration in the aqueous biofilms where the microbes work: 10 kg aqueous RAMEB solution containing 50% RAMEB (Cawasol W7 MTL, Wacker Chemie, Munich) was added three times together or separately with the nutrients dissolved in 1 or 2 m³ water. Nutrients and additives have been supplied through the perforated casing of both the active and passive wells.

The additives were injected after reaching the steady-state of the soil system, beside permanent water pumping and air exhaust, a constant O_2 and CO_2 value was measured in the exhaust gas and more-or-less constant oil content in the groundwater. The changes in the soil gas and groundwater on the effect of nutrient supply and CD addition were measured and evaluated.

4.2.5   Mass balance of the CDT technology

The material balances indicate the efficiency of the technology; it tells us what the mass flux is in the different soil phases during remediation. With full knowledge of the numerical information about the mass (or element) fluxes, the technologists design, perform, monitor, and intervene in the applied technology to ensure optimal operation.

Our technology monitoring was mainly based on the analyses of the carbon content in the mobile soil phases: carbon dioxide in soil gas and total petroleum hydrocarbons in groundwater. On the basis of the results of soil gas and groundwater analyses, the majority of the contaminating hydrocarbon has been biodegraded by microbes.

Calculation of the substrate's mass (S)

> Treated soil mass: 50 t; volume of contaminated water: 1200 m³
> Initial oil content of the soil: average concentration in the soil at start is 25,000 mg/kg; 50 tons of soil contain 1250 kg of hydrocarbons.
> Initial transformer oil in water: ~1 mg/L → 1000 g (1 kg) of hydrocarbon in the groundwater.
> Total amount of hydrocarbons to be removed is **1251 kg**.

Calculation of products' mass (P)

> Contaminant removed by water:
> Water extracted and treated: 0.7 m³/h, 250 days of treatment, 4200 m³ of water treated
> Final concentration of hydrocarbons in the groundwater: ~0.3 g hydrocarbon/m³
> The amount of hydrocarbons removed by water treatment: 2240 g

Biodegraded transformer oil mass calculated on the basis of the CO_2 production

> Volume of the surplus CO_2 produced (compared to uncontaminated soil) was 534 m³, which is equivalent to 1049 kg CO_2; 1049 kg CO_2 has been produced from **1144 kg** of hydrocarbon according to Equation 1. It suggests that from the initial **1251 kg** of transformer oil (in soil and water) ~**1145 kg** of hydrocarbon was removed by the complex CDT during 47 weeks of treatment. At the end of remediation the residual hydrocarbon content measured was about 12 kg (the average of the EPH content: 240 mg/kg in the treated soil). The mass balance accounts for **93%** of the contaminants removed.

Conclusion: contaminant removal with degradation and flushing resulted in the elimination of 93% of the contaminant of which 99.8% was biodegraded and 0.2% extracted and treated *in situ*. This ratio shows the overall difference in efficiency of biodegradation and pump and treat technology for biodegradable contaminants.

4.2.6   Environmental and eco-efficiencies: site risks and footprints

Soil compartments – soil gas, groundwater, and solid soil – exert various impacts on the environment, the users of the site itself and the vicinity.

The soil gas does not represent a direct risk for users (as it has no volatiles content) but may increase the carbon footprint of the contaminated site. The author tends to consider this increased CO_2 emission from biodegradation not as the result of remediation because this value remains the same even on an abandoned site or when the soil is exposed to natural attenuation or another biodegradation process. It is "natural" that a biodegradable contaminant (hydrocarbon) is recycled into the global biogeochemical cycles in the form of CO_2. The extra load on the footprint was created by hydrocarbon mining. On this site, this contaminant does not contain volatile components: gas chromatography of the soil extract showed that the contaminant is in the carbon number range of C15–C36, and the analysis of the soil gas has confirmed that no volatiles were present. Thus it was not necessary to treat the soil gas in the field application.

The groundwater contains a high concentration of petroleum hydrocarbons with additional supply from the soil. The technology applied (continuous water extraction) ensures stringent control and helps to avoid the risk of spreading the contaminant through the natural movements of groundwater. So the site was under depression except the few cases where the vadose zone was temporarily flooded by the cyclodextrin-containing wash-water used for flushing. The CD-supplemented wash-water did not leave the site with the groundwater, as the monitoring wells in the outskirts justified. The extracted wash-water was reused until its CD content was measurable. Otherwise the extracted water went through *ex situ* treatment and was subsequently recycled into the soil. Pollutant concentration was controlled; the treatment facility constantly produced permitted concentrations in the released water.

The soil itself was investigated on the core samples prepared before and after the implementation of the CDT technology. Contaminant content was determined by chemical analyses and the potential toxic impact of the soil on soil organisms was measured using direct toxicity assessment (DTA) (see Volumes 2 and 3 of this series). The measurements have provided essential information for the biodegradation-based technology. DTA measures the toxicity of the sample in its actual state for biological activities and in direct contact with the test organism for toxicity. Direct contact simulates the real conditions and interactions with the real environment. The results of ecotoxicity tests performed on test organisms of three trophic levels (microbes, plants, animals) show the toxic impacts. The results enable the risk to be determined directly from the toxicity of the actual sample and its dilution series. The latter one can identify the no-effect dilution, from which the risk characterization ratio (effect/no effect concentrations) can easily be determined.

4.2.6.1 Initial and final risk of the contaminated soil and groundwater: measured data and calculated risks

The risk-based approach is utilized in the verification tool box to quantify the potential impacts in the beginning and at the end of the project. The initial and residual risk is calculated by the comparison of the Predicted Environmental Concentration (PEC) of the site at the beginning and at the end of remediation to the Predicted No Effect Concentration (PNEC). This ratio is called the risk characterization ratio (RCR), which can be the ratio of two concentrations (the measured to the allowable) or two effect values: the measured to the no effect. The term "predicted" refers to the uncertainties, that must be taken into consideration. PEC is calculated from the values measured at the transformer station and the same concentration is thought to be outside the station fence of the City Park's territory. The applied PNEC value is a site-specific target value – representing a "no risk" situation for groundwater, soil, or other compartments also considering land uses. The risk-based quality criteria for groundwater (w) is this: $RBQC_{iw}$ = 0.5 mg/L in the territory of the transformer station (an industrial site [i] with restricted access), but only $RBQC_{rw}$ = 0.2 mg/L in the area of the surrounding City Park, a recreational (r) area.

"No risk" value for industrial soil uses i.e. for the transformer station soil is $RBQC_{is}$ = 300 mg/kg, the target value for multifunctional soil use is $RBQC_{rs}$ = 100 mg/kg. This may not be a realistic scenario currently because the transformer station is completely isolated from the surrounding City Park but can be expected to be included in the park through a future land use change.

Local risk posed by the groundwater within the transformer station:

RCR_{iw} initial = 1.0 mg/L / 0.5 mg/L = 2.0 > 1 medium risk;
RCR_{iw} final = 0.3 mg/L / 0.5 mg/L = 0.6 < 1 acceptable risk.

Local risk of groundwater out of the transformer station within the City Park: based on conservative estimates, not considering dilution, sorption and biodegradation during transport by flow from the transformer station toward City Park:

RCR_{rw} initial = 1.0 mg/L / 0.2 mg/L = 5.0 > high risk – pessimistic estimate;
RCR_{rw} final = 0.3 mg/L / 0.2 mg/L = 1.5 >1 small risk – pessimistic.

Local risk of groundwater out of the transformer station within the City Park: based on a model considering dilution, sorption and biodegradation:

RCR_{rw} initial = 1.0 mg/L / 0.2 mg/L = 5.0 > high risk – realistic estimate;
RCR_{rw} final = 0.1 mg/L / 0.2 mg/L = 0.5 <1 acceptable risk – realistic estimate.

Local risk of the transformer station soil:

RCR_{is} initial = 25,000 mg/kg / 300 mg/kg = 83 > very high risk;
RCR_{is} final = 240 mg/kg / 300 mg/kg = 0.8 < 1 acceptable risk.

To conclude the results of site risk it is clear that the CDT technology efficiently lowered the risk to an acceptable value. If the transformer station is to be terminated and land use changed in the future, the remediated area is ready to join the park as it has reached the quality that allows it to become part of the recreational area. Of course the surface should be covered by vegetation which can further reduce environmental risk.

Further decrease in soils' contaminant content can be prognosed as the degrading microbiota is healthy and active, and the degrading capacity is untouched.

The soil was tested by environmental toxicity tests to control the real effect (*in situ*, real time) of the treated soil on soil-living organisms: bacteria, plant and animal. The soil samples proved to be very toxic at the beginning of the treatment and became non-toxic at the end (Table 11.6; Molnár *et al.*,2005). After 47 weeks of treatment, all soil samples were non-toxic in all of the applied tests.

Soil toxicity assessment at the end was part of safety management: the non-toxic qualification of soil proves the efficiency of the remediation and justifies the technology selection.

Table 11.6 Environmental toxicity of the contaminated soil before, during, and after CDT application.

Test organisms and end points of the DTA	Depth of contaminated soil					
	10–30 cm			80–90 cm		
	Date of testing					
	Start	week 24	week 47	Start	week 24	week 47
Vibrio fischeri luminescence	toxic	slightly toxic	non-toxic	very toxic	non-toxic	non-toxic
Sinapis alba root/shoot elongation	toxic	non-toxic	non-toxic	toxic	non-toxic	non-toxic
Folsomia candida mortality	toxic	non-toxic	non-toxic	toxic	non-toxic	non-toxic
Aerobic biodegradation activity scale	++	+	+	++	+	+

Figure 11.9 Transformer station after remediation.

The CDT technology can reduce the risk of the soil to an acceptable level, it is proven both by the calculated RCR value and the "non-toxic" soil qualification by DTA. The RCR of the groundwater decreased to 0.6 from 2.0 (it is 1.5 when compared to multifunctionality quality criteria). The results are influenced by the untreated surroundings 50 times larger and with the same initial pollution at the treated subsite.

After the CDT treatment of the subsite, the whole area has been remedied, and although the transformer station function has remained, it fits now better into the park's ecosystem as a clean, organized, grassed area (Figure 11.9).

4.2.6.2 Environmental and socioeconomic impacts of in situ CDT – some sustainability indicators

CDT is an innovative technology: the combination of bioventing with the application of the available increasing solubilizing agent, cyclodextrin. The rating of some sustainability indicators and the relevant comments are summarized in Table 11.7.

Treatment duration could be decreased by CD application, thus the biodegradation-based remediation could be terminated within 23 weeks, somewhat more than a half year (about one-third of the predicted duration). As many of the impacts are proportional to time, this acceleration can be beneficial both from cost and socioeconomic efficiency perspectives.

Air pollution by contaminants has not appeared, minimal dust, waste, and noise emission impacted the narrow surrounding site for the period of technology installation (1 week). The surrounding park is not inhabited, people use it just for recreational purposes. The emitted CO_2 was equivalent to the biodegraded contaminant.

The contaminants have been mobilized as part of the technology, resulting in higher transformer oil concentration in the groundwater than formerly when the dominant fraction of the contaminants was strongly bound to soil solid. This increased water contamination is temporary and only lasts for a few days or weeks because the mobilized hydrocarbons are intensively degraded by soil microorganisms.

An important characteristic of the technology is that the site is kept under depression, functioning as a hydraulic barrier. The continuous groundwater extraction is a technological component of the CDT and of other implementations on the site. Thanks to that, contaminant migration is restricted to the treated area, inside the fence. Another consequence is the lower water table increasing the depth of the three-phase soil to support aerobic microorganisms. The measures ensured and the monitoring proved that no volatile components and no contaminated water were emitted from the demonstration site. Water was completely recycled; the only waste is the activated carbon filling the absorber used for water treatment. For the treatment of the small-scale demonstration site, less than 10% of the capacity of the absorber was used.

Water consumption of CDT tends to zero, and the treated water is recycled after treatment for supplying soil moisture, injecting nutrients and flooding the vadose zone and capillary fringe to flush it with CD solution. The total extracted water from the site is reused.

Material requirement consists of the nutrients and the cyclodextrin (CD). Nutrients stimulate bacterial growth: these are consumed shortly after injection by soil microorganisms, so the transport of nutrients out of the treated site has no possibility, and as no drinking water exists in the park, this kind of impact of nutrient injection is negligible. After flooding the soil, the CD solution is extracted, collected and reinjected from time-to-time. The randomly methylated beta-CD (RAMEB) used is a natural molecule, a starch-like organic substance, which is not toxic and has no other adverse effects. RAMEB injection into the soil could theoretically have some risk, but in reality the transport of RAMEB within the soil is highly limited and the hydraulic barrier also hinders its transport. RAMEB is biodegradable, its half-life is about 1 year, an ideal life-time for application in soil remediation.

Risk to workers can be kept to a minimum, the additives are non-hazardous, and the contaminated water goes to the treatment plant in a closed system. Community risk can be excluded, as the site is closed.

The closed industrial site is not a proper habitat for the ecosystem. The concrete pavement allows only a small surface free of vegetation. But the transformer station is surrounded

Table 11.7 Evaluation of selected sustainability indicators of CDT.

Indicator	Evaluation (+) impact (–) impact	Comment
Time requirement	+++	CDT is less time consuming than other options, time requirement can be lowered to 30–70%. Shortened treatment duration results in earlier utilization of the site and benefits from future land use. Most of the biodegradation-based technologies can be accelerated by various CDs
Development, innovation	++	CDT is an innovative technology, increasing the selection of technology options. The use of CD in ecotechnologies is recommended especially when time available for treatment is limited. Its spread is hindered by too few references and low awareness. The price of CD is high, but shorter time requirement can compensate for the excess
Water consumption	+++	No additional water is used
Energy consumption	–	Conventional energy use: periodic soil venting and water extraction
Material use, additives	++	Additives (RAMEB and nutrients) can be controlled throughout the process. They are natural compounds, non-toxic and beneficial for the soil. RAMEB is moderately biodegradable, its biodegradation half-life is 0.5–1.5 years, which is comparable to the duration of the technology. After fulfilling its role, it disappears from the soil. CD increases the bioavailability and is able to accelerate biodegradation of poorly bioavailable, high K_{ow} contaminants
Reuse of materials, additives	+++	Extracted CD is reinjected into the soil
Workers risks (health, accidents)	+	The technology works in a closed system
Community risk (transport, accidents)	+	Soil or contaminants are not transported, treatment is *in situ*. The site is closed to transportation and construction works at the beginning and end of the project
Ecosystem damage	+++	CDT utilizes natural microbiological activity in soil. It has no impact on surrounding ecosystems in this case applied to an industrial site
Habitat disturbances	++	CDT is an environmentally friendly technology, soil quality is positively influenced. Does not disturb surrounding habitats, deteriorated soil can be reclaimed to become healthy again
Land reuse	++	The site can be used as a "green industrial site" after remediation or as part of the surrounding park, when industrial use is terminated
Public access to the space	++	After remediation unlimited public access is possible
Property value	++	Increase in property value due to remediation
Future land-use potential (residual pollution)	+++	Residual pollution is under quality criteria, fulfills multifunction use criteria
Cultural resources	++	The City Park damage can be remediated. Knowledge on soil microbiology and biodegradation is increased by CDT application. The popular City Park improves stakeholders' (e.g. the people) involvement.

by natural vegetation of the 130-ha park. It means that the park animals may be in contact with the soil and rare vegetation of the site. The CD technology may have a positive impact on the site as habitat and for the development of a healthy ecosystem, and at the same time a recreational area. The site was in industrial use for 75 years, but after remediation it can become a valuable part of the park or used for other recreational or cultural purposes.

4.2.7 Summary and conclusion on the CDT technology and the verification tool

A short summary of the key findings of the four-stage evaluation of CDT is in this section.

The prospective comparative evaluation has drawn attention to the cyclodextrin-enhanced bioventing showing lowest time requirements and good results in all time-dependent indicators. As it is a "green" technology, not using toxic substances and not emitting toxicants or other harmful gases, it performed well in the comparison. Its estimated price was 10% higher than the second best option, but the time requirement was largely overestimated, in reality half of the time was enough to reach the target, meaning that the price was also comparable to the other best techniques.

CDT could reduce the risk significantly at the site and resulted in acceptable groundwater and soil quality for any land uses, e.g. for recreational use.

The risk due to contaminant- and additive-emission is low: there was no measurable emission at all. Emission of dust is not relevant because the surface is vegetated and isolated by plants or paved by concrete. Noise emission is not considerable, the pump and the blower are low performance machines and there is no residential area close to the station. Background noise is stronger during the day. Installation and dismantling of the technology only took a few days and did not emit dust or noise.

The complex verification tool could be efficiently applied for the CDT technology, and the other (not introduced here) *in situ* technologies. The results of the prospective assessment can be utilized for decision making in environmental management.

– The MOKKA verification system is applicable for an *in situ* bioremediation based on biodegradation.
– Mass balance can be calculated from the technology monitoring data.
– The risk of the site before and after remediation can be calculated from the initial and final contaminant concentrations and toxicity assessment data.
– The risk due to emissions from the source and the technology can be easily monitored and reduced during the technology. In the case of *in situ* technology application, better control of contaminant transport can be reached than in the case of *ex situ* technologies.
– The sustainability indicators and the ecological footprint components are partly quantitative (CO_2 production, water and energy consumption), partly qualitative (i.e. based on subjective judgment), and can be evaluated in comparison with other technology options.
– The majority of the cost elements can be quantified, cost efficiency is a good tool for the comparison of remediation options for technology selection. Cost–benefit assessment upgraded with eco-efficiency requirements can be a good concept, but ecosystem costs and benefits can only be partly monetized, so the evaluation remains subjective and qualitative.

The CDT has proven its applicability and has been verified by the new verification tool. CDT can be characterized as suitable for transformer oil and other moderately bioavailable contaminants, environment- and cost-effective and sustainable technology.

The introduced technology verification methodology supports decision making and the selection of the best possible technology or in the case of verified innovative technologies, a better technology than the former best ones. This procedure can be applied generally and its uniform application is also possible to make progress in the acceptance of innovative *in situ* technologies to increase trust towards them and to support their market entry.

REFERENCES

ASTM E2893–16e1 (2016) *Standard Guide for Greener Cleanups*. ASTM International. Available from: www.astm.org; www.astm.org/Standards/E2893.htm; www.astm.org/cgi-bin/resolver. cgi?E2893-16e1. [Accessed 22th July 2018].

Bardi, L., Ricci, R. & Marzona, M. (2003) *In situ* bioremediation of a hydrocarbon-polluted site with cyclodextrin as a coadjuvant to increase bioavailability. *Water, Air, & Soil Pollution Focus*, 3(3), 15–23.

Bardos, P.R. (2014) Progress in sustainable remediation. *Remediation Journal*, 25, 23–32. doi:10.1002/rem.21412.

Bardos, P.R., Bone, B., Boyle, R., Ellis, D., Evans, F., Harries, N.D. & Smith, J.W.N. (2011) Applying sustainable development principles to contaminated land management using the SuRF-UK Framework. *Remediation*, 21(2), 77–100. doi:10.1002/rem.20283.

Bardos, P.R., Thomas, H.F., Smith, J.W.N., Harries, N.D., Evans, F., Boyle, R., Howard, T., Lewis, R., Thomas, A.O. & Haslam, A. (2018) The development and use of sustainability criteria in SuRF-UK's sustainable remediation framework. *Sustainability*, 10(6), 1781. doi:10.3390/su10061781.

Bardos, R.P. (2010) SuRF-UK framework for evaluating sustainable remediation options. *Archive of July 12, 2010 Seminar US and EU Perspectives on Green and Sustainable Remediation*. Available from: www.cluin.org/consoil. [Accessed 22th July 2018].

Bardos, R.P., Bone, B.D., Boyl, R., Evans, F., Harries, N.D., Howard, T. & Smith, J.W.N. (2016b) The rationale for simple approaches for sustainability assessment and management in contaminated land practice. *Science of the Total Environment*, 563–564(1), 755–768. doi:10.1016/j.scitotenv.2015.12.001.

Bardos, R.P., Cundy, A.B., Smith, J.W.N. & Harries, N. (2016a) Sustainable remediation. *Journal of Environmental Management*, 184, 1–3. Available from: www.sciencedirect.com/science/journal/03014797/184/part/P1. [Accessed 22th July 2018].

Barnes, S., Somek, D., Pitzler, D. et al. (2015) *Guideline for Performing Cost-Benefit and Sustainability Analysis of Remedial Alternatives*. CRC for Contamination Assessment and Remediation of the Environment (CRC CARE). Available from: www.crccare.com/files/dmfile/482847_Cost-BenefitandSustainabilityAnalysis_FinalDraftGuideline_Rev2.pdf. [Accessed 22th July 2018].

Beinat, E., Drunen, van M.A., Janssen, R., Njiboer, M.H., Kohlenbrander, J.G.M. & Okx, J.P. (1997) *The REC Decision Support System for Comparing Soil Remediation Options. A Methodology Based on Risk Reduction, Environmental Merit and Costs*. CUR/NOBIS, Gouda, The Netherlands.

Bender, A., Volkwein, S., Battermann, G., Klöpffer, W., Hurtig, H.W. & Kohler, W. (1998) Life cycle assessment for remedial action techniques: Methodology and application. In: *ConSoil '98, Sixth International FZK/TNO Conference on Contaminated soil*. Thomas Telford, London, UK. pp. 367–376.

Brundtland Report (1987) *Our Common Future*. Report of the World Commission on Environment and Development. Available from: www.un-documents.net/wced-ocf.htm. [Accessed 22th July 2018].

Cappuyns, V. (2013) LCA based evaluation of site remediation – Opportunities and limitations. *Chimica Oggi – Chemistry Today*, 31(2), 18–21. Available from: www.teknoscienze.com/tks_article/lca-based-evaluation-of-site-remediation-opportunities-and-limitations/. [Accessed 22th July 2018].

Cappuyns, V. & Kessen, B. (2012) Evaluation of the environmental impact of brownfield remediation options: Comparison of two life cycle assessment-based evaluation tools. *Environmental Technology*, 33(19–21), 2447–2459. doi:10.1080/09593330.2012.671854.

CERCLA (1980) Comprehensive Environmental Response, Compensation and Liability Act of 1980, as amended.

CL:AIRE (2009) *A Review of Published Sustainability Indicator Sets: How Applicable Are They to Contaminated Land Remediation Indicator-set Development?* CL:AIRE, London, UK. ISBN: 978-1-905046-18-8. Available from: www.claire.co.uk/surfuk. [Accessed 22th July 2018].

CL:AIRE (2010a) *A Framework for Assessing the Sustainability of Soil and Groundwater Remediation.* Available from: www.claire.co.uk/projects-and-initiatives/surf-uk/20-framework-and-guidance. [Accessed 22th July 2018].

CL:AIRE (2010b) *A Framework for Assessing the Sustainability of Soil and Groundwater Remediation.* Public Consultation, March 2010, CL:AIRE, London, UK. ISBN 978-1-905046-19-5. Available from: www.claire.co.uk/surfuk. [Accessed 22th July 2018].

CL:AIRE (2011) *The SuRF-UK Indicator Set for Sustainable Remediation Assessment.* Available from: www.claire.co.uk/component/phocadownload/category/8-initiatives?download=262:annex-1-the-surf-uk-indicator-set-for-sustainable-remediation-assessment. [Accessed 22th July 2018].

CL:AIRE (2014a) *Sustainable Management Practices for Management of Land Contamination.* Available from: www.claire.co.uk/projects-and-initiatives/surf-uk/21-executing-sustainable-remediation. [Accessed 22th July 2018].

CL:AIRE (2014b) *SuRF UK SMP Supporting Spreadsheet.* Available from: www.claire.co.uk/component/phocadownload/file/403-surf-uk-smps. [Accessed 22th July 2018].

CL:AIRE (2018) *Sustainable-Remediation.* Available from: www.claire.co.uk/projects-and-initiatives/surf-uk/77-sustainable-remediation. [Accessed 22th July 2018].

Clue-in Remediation (2018) *Contaminated Site Clean-up Information.* Available from: https://clu-in.org/remediation/. [Accessed 22th July 2018].

Common Forum (2018) *Common Forum on Contaminated Land in Europe.* Available from: www.commonforum.eu/. [Accessed 22th July 2018].

DANETV (2018) *Danish Center for Verification of Climate and Environmental Technologies.* Available from: www.etv-denmark.com. [Accessed 22th July 2018].

DEFRA (2004) *Model Procedures for the Management of Land Contamination.* Contaminated Land Report 11. DEFRA and Environmental Agency, Bristol, Rio House, UK.

Dindal, A. (2016) *Environmental Technology Verification Program Materials Management and Remediation Center Generic Protocol for Verification of in situ Chemical Oxidation.* U.S. Environmental Protection Agency, Washington, DC, EPA/600/R-14/415. Available from: https://cfpub.epa.gov/si/si_public_record_report.cfm?dirEntryId=311439. [Accessed 22th July 2018].

ECOAP (2018) *Eco-Innovation Action Plan: Europe.* https://ec.europa.eu/environment/ecoap/etv/documents/165_en. [Accessed 22th July 2018].

Ellis, D.E. & Hadley, P.W. (2009) *Sustainable Remediation White Paper: Integrating Sustainable Principles, Practices, and Metrics into Remediation Projects.* U.S. Sustainable Remediation Forum. Available from: https://doi.org/10.1002/rem.20210. [Accessed 22th July 2018].

EPA ETV (2018) *Environmental Technology Verification (ETV) Program.* Available from: https://archive.epa.gov/nrmrl/archive-etv/web/html/este.html. [Accessed 22th July 2018].

ESTE (2018) *Environmental and Sustainable Technology Evaluations.* Available from: https://archive.epa.gov/nrmrl/archive-etv/web/html/este.html. [Accessed 22th July 2018].

ETAP (2018) *Environmental Technologies Action Plan: Europe* (see ECOAP, 2018). Available from: http://ec.europa.eu/environment/archives/ecoinnovation2007/2nd_forum/pdf/etv_booklet.pdf. [Accessed 22th July 2018].

ETV Canada (2018) *Environmental Technology Verification: Canada*. Available from: http://etvcanada.ca. [Accessed 22th July 2018].

ETV Japan (2018) *Environmental Technology Verification: Japan*. Available from: www.env.go.jp/policy/etv. [Accessed 22th July 2018].

ETV Philippines (2018) *Environmental Technology Verification: Philippines*. Available from: http://cptech.dost.gov.ph/ETV.php. [Accessed 22th July 2018].

EU ECOAP (2018) *Eco-Innovation Action Plan: Europe* (former Environmental Technologies Action Plan). Available from: https://ec.europa.eu/environment/ecoap/etv_en and http://ec.europa.eu/environment/archives/etv/. [Accessed 22th July 2018].

EU-ETV (2018) *European Environmental Technology Verification*. Available from: http://ec.europa.eu/environment/archives/etv/. [Accessed 22th July 2018].

EURODEMO (2004–2008) *European Platform for Demonstration of Efficient Soil and Groundwater Remediation*. Available from: www.eurodemo.info/. [Accessed 22th July 2018].

Fava, F., Di Gioia, D. & Marchetti, L. (1998) Cyclodextrin effects on the *ex-situ* bioremediation of a chronically polychlorobiphenyl-contaminated soil. *Biotechnology and Bioengineering*, 58, 345–355.

Fenyvesi, É., Csabai, K., Molnár, M., Gruiz, K., Murányi, A. & Szejtli, J. (2003) Quantitative and qualitative analysis of RAMEB in Soil. *Journal of Inclusion Phenomena and Macrocyclic Chemistry*, 44, 413–416.

Fenyvesi, É., Gruiz, K., Verstichel, S., De Wilde, B., Leitgib, L., Csabai, K. & Szaniszló, N. (2005) Biodegradation of cyclodextrins in soil. *Chemosphere*, 60, 1001–1008.

Fiksel, J. (2009) *Design for the Environment*. A Guide to Sustainable Product Development. 2nd ed. McGraw-Hill, New York, NY, USA. ISBN: 978-0-07-160556-4.

Fiksel, J., Eason, T. & Frederickson, H. (2012) A framework for sustainability indicators at EPA. Available from: National Risk Management Research Laboratory. EPA/600/R/12/687.

Green Remediation (2008): *Incorporating Sustainable Environmental Practices into Remediation of Contaminated Sites*. Available from: https://clu-in.org/greenremediation/docs/Green-Remediation-Primer.pdf.

Green Remediation (2018) *Green Remediation Focus*. Available from: https://clu-in.org/greenremediation/. [Accessed 22th July 2018].

GR Memo (2016) *Green Remediation Memorandum*. US EPA. Available from: https://semspub.epa.gov/work/HQ/100000160.pdf.

Gruiz, K. (2005) Biological tools for the soil ecotoxicity evaluation: Soil testing triad and the interactive ecotoxicity tests for contaminated soil. In: Fava, F. & Canepa, P. (eds.) *Innovative Approaches to the Bioremediation of Contaminated Sites*. Soil Remediation Series NO 6, INCA, Venice, Italy. pp. 45–70.

Gruiz, K. (2014) Environmental problems – An overview. In: Gruiz, K., Meggyes, T. & Fenyvesi, É. (eds.) *Environmental Contamination and Deterioration*. CRC Press, Boca Raton, FL, USA. pp. 1–40.

Gruiz, K. (2016) Integrated and efficient characterization of contaminated sites. In: Gruiz, K., Meggyes, T. & Fenyvesi, É. (eds.) *Site Assessment and Monitoring Tools*. CRC Press, Boca Raton, FL, USA. pp. 1–98.Gruiz, K., Fenyvesi, É., Kriston, É., Molnár, M. & Horváth, B. (1996) Potential use of cyclodextrins in soil bioremediation. *Journal of Inclusion Phenomena and Macrocyclic Chemistry*, 25, 233–236.

Gruiz, K., Meggyes, T. & Fenyvesi, É. (eds.) (2015) *Environmental Toxicology*. CRC Press, Boca Raton, FL, USA.

Gruiz, K., Molnár, M. & Fenyvesi, É. (2008) Evaluation and Verification of Soil Remediation. In: Kurladze, G.V. (ed) *Environmental Microbiology Research Trends*. NOVA Science Publishers, Inc., New York, NY, USA. pp. 1–57.

Gruiz, K., Sára, B. & Vaszita, E. (2014) Risk management of contaminated land – from planning to verification. In: In: Gruiz, K., Meggyes, T. & Fenyvesi, É. (eds.) Environmental Contamination and Deterioration. CRC Press, Boca Raton, FL, USA. pp. 227–312.

Hopton, M.E., Cabezas, H., Campbell, D.E., Eason, T., Garmestani, A.S., Heberling, M.T., Karunanithi, A.T., Templeton, J.J., White, D. & Zanowick, M. (2010) Development of a multidisciplinary approach to assess regional sustainability. *International Journal of Sustainable Development & World Ecology*, 17(1), 48–56. doi:10.1080/13504500903488297.

IISD (2018a) *SDG Indicator Portal*. The International Institute for Sustainable Development. Available from: www.iisd.org; https://sustainable-development-goals.iisd.org/country-data. [Accessed 22th July 2018].

IISD (2018b) *International Institute for Sustainable Development*. Available from: www.iisd.org/

ISO 14034 (2016) *Environmental Management/ Environmental Technology Verification (ETV)*. Available from: www.iso.org/standard/43256.html. [Accessed 22th July 2018].

ISO 14040 (2006) *Environmental Management – Life Cycle Assessment – Principles and Framework*. International Organisation for Standardisation (ISO), Geneve. Available from: www.iso.org/standard/37456.html. [Accessed 22th July 2018].

ISO 14044 (2006) *Environmental Management – Life Cycle Assessment – Requirements and Guidelines*. International Organisation for Standardisation (ISO), Geneve. Available from: www.iso.org/standard/38498.html. [Accessed 22th July 2018].

ITRC (2003) *Technical and Regulatory Guidance for the Triad Approach: A New Paradigm for Environmental Project Management*. SCM-1. Interstate Technology & Regulatory Council. Available from: www.itrcweb.org. [Accessed 22th July 2018].

ITRC (2004) *Remediation Process Optimization: Identifying Opportunities for Enhanced and More Efficient Site Remediation*. RPO-1. Interstate Technology & Regulatory Council. Available from: www.itrcweb.org. [Accessed 22th July 2018].

ITRC (2006a) *Life Cycle Cost Analysis*. RPO-2. Interstate Technology & Regulatory Council. Available from: www.itrcweb.org. [Accessed 22th July 2018].

ITRC (2006b) *The Use of Direct Push Well Technology for Long-Term Environmental Monitoring in Groundwater*. SCM-2. Interstate Technology & Regulatory Council. Available from: www.itrcweb.org. [Accessed 22th July 2018].

ITRC (2007) *Improving Environmental Site Remediation through Performance-based Environmental Management*. RPO-7. Interstate Technology & Regulatory Council. Available from: www.itrcweb.org. [Accessed 22th July 2018].

ITRC (2011a) *Green and Sustainable Remediation: State of the Science and Practice*. GSR2009–1. Interstate Technology & Regulatory Council, Green and Sustainable Remediation Team, Washington, DC, USA. Available from: www.itrcweb.org. [Accessed 22th July 2018].

ITRC (2011b) *Project Risk Management for Site Remediation*. RRM-1. Interstate Technology & Regulatory Council. Available from: www.itrcweb.org. [Accessed 22th July 2018].

IWG-ETV (2018) *International Working Group on ETV*. Available from: https://ec.europa.eu/environment/ecoap/etv/international-activities_en. [Accessed 22th July 2018].

Korean NETV (2018) *New Excellent Technology and Verification*. Available from: www.koetv.or.kr/eng/. [Accessed 22th July 2018].

Leitgib, L., Gruiz, K., Molnár, M. & Fenyvesi, É. (2003) Bioremediation of Transformer Oil Contaminated Soil. In: Annokkée, G.J., Arendt, F. & Uhlmann, O. (eds.) *Wissenschaftliche Berichte*. FZKA Publisher, Karlsruhe, Germany. pp. 2762–2771.

Lemming, G., Hauschild, M.Z. & Bjerg, P.J. (2010) Life cycle assessment of soil and groundwater remediation technologies: Literature review. *International Journal of Life Cycle Assessment*, 15, 115–127. doi:10.1007/s11367-009-0129-x.

McCarty, P.L. (1988) Bioengineering issues related to *in situ* remediation of contaminated soils and groundwater. *Basic Life Sciences*, 45, 143–162.

MOKKA (2004–2008) *Innovative Decision Support Tools for Risk-based Environmental Management in Hungary*. Hungarian Research and Development Grant NKFP-3-0020/2005. Available from: www.mokkka.hu. [Accessed 22th July 2018].

Molnár, M. (2007) *Intensified Bioremediation of Contaminated Soils with Cyclodextrin: From the Laboratory to the Field*. PhD Thesis, Budapest University of Technology and Economics, Hungary.

Molnár, M., Fenyvesi, É., Gruiz, K., Leitgib, L., Balogh, G., Murányi, A. & Szejtli, J. (2003) Effects of RAMEB on bioremediation of different soils contaminated with hydrocarbons. *Journal of Inclusion Phenomena and Macrocyclic Chemistry*, 44, 447–452.

Molnár, M., Leitgib, L., Gruiz, K., Fenyvesi, É., Szaniszló, N., Szejtli, J. & Fava, F. (2005) Enhanced biodegradation of transformer oil in soils with cyclodextrin: From the laboratory to the field. *Biodegradation*, 16, 159–168.

NICOLE (2018) *Network of Industrially Co-ordinated Sustainable Land Management in Europe*. Available from: www.nicole.org/pagina/2/Organisation.html. [Accessed 22th July 2018].

PROMOTE (2018) *Environmental Technology Verification*. Available from: www.promote-etv.org/. [Accessed 22th July 2018].

Reddy, K. & Adams, J. (2015) *Sustainable Remediation of Contaminated Sites*. Momentum Press, New York, NY, USA.

Rosèn, L., Back, P.-E., Söderqvist, T., Norrman, J., Brinkhoff, P., Norberg, T., Volchko, Y., Norin, M., Bergknut, M. & Döberl, G. (2015) SCORE: Multi-criteria decision analysis for assessing the sustainability of remediation at contaminated sites. *Science of the Total Environment*, 511, 621–638.

Rosèn, L., Back, P.-E., Norrman, J., Söderqvist, T., Norberg, T., Volchko, Y., Brinkhoff, P., Norin, M., Bergknut, M. & Döberl, G. (2013) SCORE: Multi-Criteria Analysis (MCA) for sustainability appraisal of remedial alternatives. *Proceedings of the Second International Symposium on Bioremediation and Sustainable Environmental Technologies, 10–13 June 2013*. Batelle, Jacksonville, FL, USA.

Rosèn, L., Söderqvist, T., Back, P.E., Soutukorva, Å., Brodd, P. & Grahn, L. (2009) Multi-criteria analysis (MCA) for sustainable remediation at contaminated sites. Method development and examples. Report 5891. Swedish Environmental Protection Agency, Stockholm, Sweden.

SAB (2009) Valuing the Protection of Ecological Systems and Services: A report of the EPA Science Advisory Board. Washington, USA. EPA-SAB-09-012.

Schwartz, A. & Bar, R. (1995) Cyclodextrin-enhanced degradation of toluene and p-toluic acid by *Pseudomonas putida*. *Applied Environmental Microbiology*, 61, 2727–2731.

Singh, R., Murty, H., Gupta, S. & Dikshit, A. (2009) An overview of sustainability assessment methodologies. *Ecological Indicators*, 9(2), 189–212.

Smith, J., Bardos, P., Bone, B., Boyle, R., Ellis, D., Evans, F. & Harries, N. (2010) *SuRF-UK: A Framework for Evaluating Sustainable Remediation Options, and Its Use in a European Regulatory Context*. Available from: www.researchgate.net/publication/228481485_SuRF-UK_A_framework_for_evaluating_sustainable_remediation_options_and_its_use_in_a_European_regulatory_context. [Accessed 22th July 2018].

SNOWMAN-MCA Project (2009–2013) SNOWMAN-MCA Multi-criteria analysis (MCA) of remediation alternatives to access their overall impact and cost/benefit, with focus on soil function (ecosystem services and goods) and sustainability. *EUGRIS Database*. Available from: www.eugris.info/DisplayProject.asp?p=4713. [Accessed 22th July 2018].

Sparrevik, M., Saloranta, T., Cornelissen, G., Eek, E., Magerholm Fet, A., Breedveld, G.D. & Linkov, I. (2011) Use of life cycle assessments to evaluate the environmental footprint of contaminated sediment remediation. *Environmental Science and Technology*, 45(10), 4235–4241. doi:10.1021/es103925u.

State of Green (2018) *Environmental Technology Verification, Denmark*. Available from: https://stateofgreen.com/en/partners/danetv/solutions/environmental-technology-verification. [Accessed 22th July 2018].

Suèr, P., Nilsson-Påledal, S. & Norrman, J. (2010) LCA for site remediation: A literature review. *Journal Soil and Sediment Contamination: An International Journal*, 13(4), 415–425. doi:10.1080/10588330490471304.

SuRF UK (2018) *Sustainable Remediation Forum: UK LinkedIn Page*. Available from: www.claire.co.uk/home/news/56-surf-uk-linkedin-page. [Accessed 22th July 2018].

TESTNET (2018) *Towards a European Verification System for Environmentally Sound Technology.* Available from: www.est-testnet.net. [Accessed 22th July 2018].

UN (2007) *Indicators of Sustainable Development: Guidelines and Methodologies.* Third edition. Methodology Sheets. The World Bank. Available from: www.un.org/esa/sustdev/natlinfo/indicators/methodology_sheets.pdf. [Accessed 22th July 2018].

US EPA (2008) *Green Remediation: Incorporating Sustainable Environmental Practices into Remediation of Contaminated Sites.* EPA 542-R-08-002. Available from: www.cluin.org/greenremediation. [Accessed 22th July 2018].

US EPA (2009) *Principles for Greener Cleanups.* Available from: www.epa.gov/oswer/green cleanups/principles.html. [Accessed 22th July 2018].

US EPA (2011a) *Green and Sustainable Remediation: A Practical Framework.* Technical/Regulatory Guidance. ITRC. Available from: www.itrcweb.org/guidancedocuments/gsr-2.pdf. [Accessed 22th July 2018].

US EPA (2011b) *Green Remediation Focus: Superfund Remediation and Technology Innovation.* Available from: www.epa.gov/climatechange/wycd/waste/calculators/Warm_home.html. www.clu-in.org/greenremediation. [Accessed 22th July 2018].

US EPA ETV (2018) *Environmental Technology Verification Program.* Available from: https://archive.epa.gov/nrmrl/archive-etv/web/html/. [Accessed 22th July 2018].

Volchko, Y., Rosèn, L., Norrman, J., Bergknut, M., Döberl, G., Anderson, R., Tysklind, M. & Müller-Grabherr, D. (2014) *Multi-Criteria Analysis of Remediation Alternatives to Access their Overall Impact and Cost/Benefit, with Focus on Soil Function (Ecosystem Services and Goods) and Sustainability.* Report 2014:6. Final Report of the SNOWMAN-MCA project carried out as part of the SNOWMAN Network coordinated call II. CHALMERS University of Technology, Göteborg. Available from: http://publications.lib.chalmers.se/records/fulltext/204158/local_204158.pdf. [Accessed 22th July 2018].

Volkwein, S., Hurtig, H.W. & Klöpffer, W. (1999) Life cycle assessment of contaminated sites remediation. *The International Journal of Life Cycle Assessment*, 4, 263. doi:10.1007/BF02979178.

Wang, J.-M., Marlowe, E.M., Miller-Maier, R.M. & Brusseau, M.L. (1998) Cyclodextrin-enhanced biodegradation of phenanthrene. *Environmental Science and Technology*, 32, 1907–1912.

WBCSD (2011) *Guide to Corporate Ecosystem Valuation: A Framework for Improving Corporate Decision-making.* Available from: http://wbcsdpublications.org/project/guide-to-corporate-ecosystem-valuation/. [Accessed 22th July 2018].

What is ETV (2018) *About ETV.* Available from: https://ec.europa.eu/environment/ecoap/etv/about-etv_en. [Accessed 22th July 2018].

Subject index

Engineering Tools for Environmental Risk Management

Editors: Katalin Gruiz, Tamás Meggyes & Éva Fenyvesi

Engineering Tools for Environmental Risk Management: 1
Environmental Deterioration and Contamination – Problems and their Management
©2014
Editors: Katalin Gruiz, Tamás Meggyes & Éva Fenyvesi
ISBN: 9781138001541 (Hardback)
e-book ISBN: 9781315778785
Cat# K22815

Engineering Tools for Environmental Risk Management: 2
Environmental Toxicology
©2015
Editors: Katalin Gruiz, Tamás Meggyes & Éva Fenyvesi
ISBN: 9781138001558 (Hardback)
e-book ISBN: 9781315778778
Cat# K22816

Engineering Tools for Environmental Risk Management: 3
Site Assessment and Monitoring Tools
Editors: Katalin Gruiz, Tamás Meggyes & Éva Fenyvesi
ISBN: 9781138001565 (Hardback)
e-book ISBN: 9781315778761
Cat# K22817

Engineering Tools for Environmental Risk Management: 4
Risk Reduction Technologies and Case Studies
Editors: Katalin Gruiz, Tamás Meggyes & Éva Fenyvesi
ISBN: 9781138001572 (Hardback)
e-book ISBN: 9781315778754
Cat# K22818